Regenerative Engineering and Developmental Biology

CRC PRESS SERIES IN
REGENERATIVE ENGINEERING

SERIES EDITOR

Cato T. Laurencin

University of Connecticut Health Center, Farmington, USA

PUBLISHED TITLES

Regenerative Engineering and Developmental Biology:
Principles and Applications
David M. Gardiner

Regenerative Engineering and Developmental Biology

Principles and Applications

Edited by

David M. Gardiner

Department of Developmental and Cell Biology
University of California Irvine
Irvine, California

CRC Press
Taylor & Francis Group
Boca Raton London New York

CRC Press is an imprint of the
Taylor & Francis Group, an **informa** business

CRC Press
Taylor & Francis Group
6000 Broken Sound Parkway NW, Suite 300
Boca Raton, FL 33487-2742

First issued in paperback 2019

© 2018 by Taylor & Francis Group, LLC
CRC Press is an imprint of Taylor & Francis Group, an Informa business

No claim to original U.S. Government works

ISBN-13: 978-1-4987-2331-2 (hbk)
ISBN-13: 978-0-367-87309-7 (pbk)

Visit the Taylor & Francis Web site at
http://www.taylorandfrancis.com

and the CRC Press Web site at
http://www.crcpress.com

Contents

SECTION II   HOW CELLS COMMUNICATE TO REMAKE THE PATTERN AND RESTORE FUNCTION

SECTION III INTEGRATION OF NEW
STRUCTURES WITH THE OLD

SECTION IV   PRINCIPLES OF ORGAN DEVELOPMENT AND REGENERATION

Preface

This book represents one of the next steps in the development of the field of regenerative engineering. Beginning with the view that we can achieve human regeneration through the convergence of developmental biology and engineering, this new field has moved forward as evidenced in publications, birth of the new journal *Regenerative Engineering and Translational Medicine*, and the establishment of the new Regenerative Engineering Society as part of the American Institute of Chemical Engineers. Collectively, these activities embrace the view of the importance of integrating developmental biology and tissue engineering to be successful moving beyond repairing damage to body parts in order to regenerate tissues and organs.

This new book is envisioned as the next step beyond the seminal textbook *Regenerative Engineering*[*] that was written for an advanced undergraduate audience of biomedical engineering students. Unlike this book that laid out the broad range of topics and scientific disciplines to be integrated into regenerative engineering, *Regenerative Engineering and Developmental Biology: Principles and Applications* focuses on one of the core scientific disciplines, developmental biology. This is part of a series of books in the *Regenerative Engineering Series* with Dr. Cato Laurencin as the series editor. Collectively, these books will focus on one or multiples of the five legs of the field of regenerative engineering: materials, developmental biology, stem cells, physics, and clinical translation. The chapters within this book were written by leading developmental biologists who were challenged to think about how the processes that they study could be controlled by using the tools available through tissue engineering/biomaterials. Rather than discussing a list of genes and developmental pathways, the authors approached their topics more broadly, and provided insights and speculations about how to take the next step toward proregenerative therapies. For each chapter, there is enough detail to support these conclusions and thoughts about what to do next, along with references to guide the reader as to where to go for more details.

---

[*] Laurencin, C. T. and Khan, Y. (2013). *Regenerative Engineering.* CRC Press, Taylor & Francis Group. Boca Raton, FL.

This book is intended to achieve two goals. First, for the developmental biology audience, I encourage everyone to begin thinking about how inducing regeneration is an engineering problem. Historically, developmental biology has been about the discovery of how things work, and it is time to take that knowledge and use it to make something work. This is what engineers do; they build things that work. For the engineering audience, I encourage you to recognize that developmental biologists have discovered a lot about how embryos develop and animals regenerate, and have identified a number of key properties and processes that need to be engineered in order to regenerate body parts. At the end of this book, I have attempted to synthesize the major themes that were addressed repeatedly by the authors. I see this book as an experiment in which leading researchers in developmental biology were asked to write about what is most important from their perspective of how regeneration works. I assume that the important points are those that were written about the most.

Beyond the complexity and diversity of the biology of the many different model organisms discussed, three unifying principles emerged. The first is the recognition that the underlying cellular and molecular mechanisms of embryonic and regenerative development are highly conserved. When thinking about regeneration in a broad sense, it is evident the mechanisms for regeneration (mitosis, migration, and differentiation) evolved very early in the history of life on earth. Thus, it follows that all animals can regenerate to some degree, and that some can do it very well. Historically, developmental genetics has allowed us to discover the signals and pathways that regulate these cellular behaviors. Today, with advances in genomics and tools for genetic manipulation, we can study these mechanisms in an unlimited variety of organisms beyond the traditional, genetically amenable model organisms. We thus are able to revisit the classic models of regeneration in order to tease apart the sequence of events that begin with injury and lead eventually to regeneration of damaged or lost structures.

Equally important is the second principle that the pathways for making an organ in the embryo can be reused for remaking it (regeneration) in the adult. Although there might be differences in the details of controlling these mechanisms in embryos or adults, the developmental pathways worked at least once (during embryonic development), even in animals that do not regenerate well as adults. Therefore, accessing and controlling these developmental pathways are essential, and viable, for engineering regeneration. The conservation of pathways is the remarkable discovery of modern developmental genetics, and the key to regeneration is discovering how to regulate the pathways after injury.

Finally, and most importantly for what actually needs to be engineered, is the principle that we need both the building blocks (the cells) and the instruction manual for how to put the blocks together in order to rebuild a lost or damaged structure. A focus of several of the chapters is on what cells can do (proliferate, migrate, and differentiate/dedifferentiate) leading to the production of a population of regeneration-competent cells. But we know that you need more than just the cells; you also need the information that tells the cells where to go and what to make. This is the regeneration blueprint that we now refer to as the Information Grid. In the end, regeneration is about the replacement of the lost function and not just the structure. Function is an emergent property that is created when different cells become appropriately organized into a functional structure, and it is this information for organizing the cells that is encoded in the Information Grid.

Regenerating the Information Grid by necessity is the goal of regenerative engineering. Advances in how to make regeneration-competent cells (e.g., stem cells, the topic of one of the books in the *Regenerative Engineering Series*) will almost certainly take care of the challenge of getting the building blocks for regeneration in the near future. However, without the information required to orchestrate the behaviors of these cells and their progeny, we will not be able to regenerate structures that restore the lost function. The challenge lies in learning how to provide the necessary information, and I argue that this information has to come from the regeneration engineers. There may come a day in the distant future when it will be possible to induce regeneration of the Information Grid endogenous, but we need to remember the urgency for patients today in need of effective regeneration therapies. The challenge today is twofold; we need to understand what the information is and we need to use the engineering toolbox to provide and to control that information.

Acknowledgments

Many people have put in tremendous effort to think and write about the problem of how to engineer regeneration. They have shared their views about how regeneration works in the context of their particular areas of expertise. They also have shared their best and most creative ideas. In the end, this volume has captured a view of where the field of regeneration biology is in 2017, and we will be able to look back and see how the field has moved forward and built upon these ideas. Most importantly, the goal of this book is to catalyze interactions between developmental biologists and regenerative engineers. On behalf of the many authors, I hope that we have been successful in communicating the views from the side of embryonic and regenerative developmental biology. I therefore thank all the authors for their efforts and their patience. Your contributions are very much appreciated.

I also thank Michael Slaughter, our production assistant at CRC Press, for his patience and support from the beginning to the end, and Dr. Cato Laurence for his vision and efforts leading to the inception and completion of this project. Finally, I express my special thanks to Dr. Susan Bryant for her never-ending support, encouragement, and patience as I worked through the process of bringing this book to completion.

Editor

David M. Gardiner is a professor in the department of developmental and cell biology at the University of California Irvine (UCI), Irvine, California. He received his BA degree in biology from Occidental College, Los Angeles, California; his PhD degree from the Scripps Institution of Oceanography at University of California, San Diego, San Diego, California; and his postdoctoral training at The University of California, Davis, Davis, California. His research has been focused on discovering the mechanisms regulating limb regeneration in salamanders. He pioneered the use of the axolotl (*Ambystoma mexicanum*) as a model system for studies of vertebrate regeneration and developed the accessory limb model as an assay for bioactive compounds that induce dedifferentiation, blastema formation, and limb regeneration. This novel assay is the basis for ongoing studies to identify molecular pathways that regulate regeneration in humans. Professor Gardiner is a fellow of the American Association for the Advancement of Science, a recipient of the Marcus Singer Medal for Excellence in Regeneration Research, and a recipient of the Frontiers in Stem Cell and Regeneration Biology Pioneer Award. He is an author of more than 100 articles and has served on numerous peer review committees, journal editorial boards, and scientific advisory boards. He serves as the associate dean for research and academic affairs in the Francisco J. Ayala School of Biological Sciences at the UCI.

Contributors

Alaa Abdelhamid
Tissue Engineering and
    Biomaterials Research Unit
College of Dentistry
Qassim University
Al-Mulida, Kingdom of
    Saudi Arabia

Cristian Aguilar
Department of Biology and
    Chemistry
Azusa Pacific University
Azusa, California

Md. Ferdous Anower-E-Khuda
Department of Cellular and
    Molecular Medicine
University of California
La Jolla, California

Joelle Baddour
Department of Chemical and
    Biomolecular Engineering
Rice University
Houston, Texas

Caroline Beck
Department of Zoology
University of Otago
Dunedin, New Zealand

Susan V. Bryant
Department of Developmental
    and Cell Biology
University of California Irvine
Irvine, California

Diane Carlisle
Department of Neurological
    Surgery
and
Department of Medicine
Presbyterian Hospital
University of Pittsburg
Pittsburgh, Pennsylvania

David Chambers
Wolfson Centre for Age-related
    Diseases
King's College London
London, United Kingdom

Cheng-Ming Chuong
Department of Pathology
Keck School of Medicine
University of Southern California
and
Department of Pathology
Herman Ostrow School of
    Dentistry
University of Southern California
Los Angeles, California

and

Research Center for
    Developmental Biology and
    Regenerative Medicine
Graduate School of Clinical
    Dentistry
National Taiwan University
Taipei, Taiwan

and

Integrative Stem Cell Center
China Medical University
China Medical University
    Hospital
Taichung, Taiwan

Peter Currie
Australian Regenerative Medicine
    Institute
Monash University
Clayton, Victoria, Australia

Carlos Díaz-Castillo
Department of Developmental
    and Cell Biology
and
Department of Evolution and
    Ecology
University of California Irvine
Irvine, California

Karen Echeverri
Department of Genetics, Cell
    Biology and Development
University of Minnesota
Minneapolis, Minnesota

Ophelia Ehrlich
Australian Regenerative Medicine
    Institute
Monash University
Clayton, Victoria, Australia

José E. García-Arrarás
Department of Biology
University of Puerto Rico
San Juan, Puerto Rico

David M. Gardiner
Department of Developmental
    and Cell Biology
University of California Irvine
Irvine, California

James Godwin
The Jackson Laboratory
Bar Harbor, Maine

Yona Goldshnmit
Australian Regenerative Medicine
    Institute
Monash University
Clayton, Victoria, Australia

Hannah Grover
Bioengineering Program
Univeristy of Maine
Orono, Maine

Justin Guay
Department of Biology
Allen Discovery Center
Tufts University
Medford, Massachusetts

Celia Herrera-Rincon
Department of Biology
Allen Discovery Center
Tufts University
Medford, Massachusetts

Michael S. Hu
Division of Plastic and
    Reconstructive Surgery
Hagey Laboratory for Pediatric
    Regenerative Medicine
Department of Surgery
Stanford University School
    of Medicine
and
Institute for Stem Cell Biology
    and Regenerative Medicine
Stanford University
Stanford, California

and

Department of Surgery
John A. Burns School of
    Medicine
University of Hawai'i
Honolulu, Hawai'i

Mayumi Ito
The Ronald O. Perelman
    Department of Dermatology
and
Department of Cell Biology
School of Medicine
New York University
New York, New York

Henrik Lauridsen
Department of Clinical Medicine
Comparative Medicine
    Laboratory
Aarhus University
Aarhus N, Denmark

and

Department of Biomedical
    Engineering
Cardiovascular Developmental
    Bioengineering Laboratory
Cornell University
Ithaca, New York

Michael Levin
Department of Biology
Allen Discovery Center
Tufts University
Medford, Massachusetts

Michael T. Longaker
Division of Plastic and
    Reconstructive Surgery
Hagey Laboratory for Pediatric
    Regenerative Medicine
Department of Surgery
Stanford University School of
    Medicine
and
Institute for Stem Cell
    Biology and Regenerative
    Medicine
Stanford University
Stanford, California

Malcolm Maden
Department of Biology
UF Genetics Institute
University of Florida
Gainesville, Florida

Catherine D. McCusker
Department of Biology
University of Massachusetts
Boston, Massachusetts

James Monaghan
Department of Biology
Northeastern University
Boston, Massachusetts

Jennifer R. Morgan
Marine Biological Laboratory
Eugene Bell Center for
    Regenerative Biology and
    Tissue Engineering
Woods Hole, Massachusetts

Leonardo Morsut
Department of Stem Cell Biology
    and Regenerative Medicine
Broad CIRM Center
Keck School of Medicine
University of Southern California
Los Angeles, California

and

Department of Cellular and
    Molecular Pharmacology
University of California San
    Francisco
San Francisco, California

Ken Muneoka
Department of Veterinary
    Physiology and Pharmacology
College of Veterinary Medicine &
    Biomedical Sciences
Texas A&M Univerity
College Station, Texas

Masamitsu Oshima
Department of Oral
    Rehabilitation and
    Regenerative Medicine
Graduate School of Medicine,
    Dentistry and Pharmaceutical
    Sciences
Okayama University
Okayama, Japan

and

RIKEN Center for Developmental
    Biology
Kobe, Japan

H. Peter Lorenz
Division of Plastic and
    Reconstructive Surgery
Hagey Laboratory for Pediatric
    Regenerative Medicine
Department of Surgery
Stanford University School of
    Medicine
Stanford University
Stanford, California

Anne Q. Phan
Department of Cellular and
    Molecular Medicine
University of California
La Jolla, California

Sandra Rieger
MDI Biological Laboratory
Kathryn W. Davis Center for
    Regenerative Biology and
    Medicine
Salisbury Cove, Maine

Keith Sabin
Department of Genetics, Cell
    Biology and Development
University of Minnesota
Minneapolis, Minnesota

Florenci Serras
Department of Genetics
Faculty of Biology
Institute of Biomedicine of the
    University of Barcelona
Barcelona, Spain

Konstantinos Sousounis
Department of Neuroscience
Baylor College of Medicine
Houston, Texas

Makoto Takeo
Laboratory for Organ
    Regeneration
RIKEN Center for Developmental
    Biology
Kobe, Japan

and

The Ronald O. Perelman
    Department of Dermatology
and
Department of Cell Biology
School of Medicine
New York University
New York, New York

Stephanie Tsai
Department of Pathology
Keck School of Medicine
University of Southern California
and
Department of Pathology
Herman Ostrow School of
    Dentistry
University of Southern California
Los Angeles, California

and

Research Center for
    Developmental Biology and
    Regenerative Medicine
Graduate School of Clinical
    Dentistry
National Taiwan University
Taipei, Taiwan

Panagiotis A. Tsonis
Department of Biology
Center for Tissue Regeneration
    and Engineering
University of Dayton
Dayton, Ohio

Takashi Tsuji
Laboratory for Organ
    Regeneration
RIKEN Center for Developmental
    Biology
Kobe, Japan

and

Organ Technologies Inc.
Tokyo, Japan

Warren A. Vieira
Department of Biology
University of Massachusetts
Boston, Massachusetts

Randal B. Widelitz
Department of Pathology
Keck School of Medicine
University of Southern California
and
Department of Pathology
Herman Ostrow School of
    Dentistry
University of Southern California
Los Angeles, California

Chapter 1 Introduction to regenerative engineering and developmental biology

David M. Gardiner

Contents

Key concepts

   a. Convergence of developmental biology and engineering can use the vast knowledge about developmental biology to build new body parts, make them work, and make them work better over time.

   b. The conservation of developmental mechanisms between organisms and between embryonic development and regenerative development will enable an engineering approach to regeneration.

   c. Regeneration is a stepwise process, and thus, ultimately, each step can be identified and regulated sequentially (engineered) to achieve successful regeneration.

   d. Although new cells are needed to replace the missing parts, the information about how to put the right cells in the right place in order to restore lost function is also required.

   e. Achieving endogenous regeneration will necessitate learning how to talk to the cells.

1.1 Introduction

We are at a time of convergence of developmental biology and engineering. From the early days in experimental biology when it was first recognized that body parts could regrow and function again, scientists have been investigating and discovering the processes and steps involved in regeneration. Meanwhile, for millennia, engineers have been building complex structures and devices that have transformed how we live our lives. The focus of this book is on the next grand challenge for engineering, which is to build the ultimate machine, the human body. We argue that although we will continue to learn more about the details of development, we are at a time when we know enough to be able to engineer functional body parts from scratch. Each of the authors has been tasked with identifying the most important concepts and principles of development that can be used by engineers to design strategies for manipulating developmental processes for regeneration. Our premise is that this is the right time to undertake this challenge.

The challenge in terms of bringing these two different approaches together is that they start at different places and move in opposite directions. Embryonic development and regenerative development are progressive, and complex structures emerge from more simple progenitors in a bottom-up way. To date, the engineering approach to regeneration has been a top-down approach that starts at the more complex final stage in which the form is already established (e.g., decellularized organs) and then attempts to coax cells to populate the engineered structure. Consequently, tissue engineering has focused on fabrication and material science.

Each approach has led to remarkable successes independent of each other, for example, stem cell therapies and artificial joints, both of which can lead to restoration of lost structure and function. Nevertheless, the vision of convergence is to bring these disparate fields together, with the vision that we achieve more than the sum total of each. Although they are very different scientific disciplines, they both are moving toward a common place. By analogy, this is like the building of the transcontinental railroad across the United States that was created by the coming together of the Central Pacific and Union Pacific railroads in 1869. With this connection, engineering and developmental biology can move forward together to discover how to control the intrinsic regenerative properties of the human body in order to rebuild and replace damaged body parts. The goal of this book is to identify the key principles and processes of embryonic development and regenerative development that could be engineered in order to enhance and induce endogenous regeneration in humans.

1.2 The view from developmental biology

From the earliest days of experimental biology, scientists have been discovering the principles and processes of development. This all began with Lazzaro Spallanzani in the eighteenth century. He demonstrated that embryonic development was initiated by fertilization of the ovum by sperm and discovered regenerative development of an amputated limb of a newt (Morgan 1901; Dinsmore 1992). Over the next two centuries, many of the phenomena of development and regeneration were discovered through experimental studies involving detailed descriptions of developing embryos and the response to surgical manipulations. Central to these studies was the recognition that cells are the structural and functional units of biology, as formalized in the Cell Theory in 1838 by Matthias Schleiden and Theodor Schwann (Tavassoli 1980). With the subsequent advances in instrumentation and technology, the sciences of cell biology, genetics, and now molecular biology have led to a detailed and comprehensive understanding of the phenomenology of both embryonic development and regenerative development.

Although the rate of discovery of the details of development has accelerated with time, the basic principles and concepts were conceptualized centuries ago. Questions about how a single cell (the egg) can give rise to the complex, multicellular animal and how new body parts can arise from old parts are among the earliest questions asked by experimental biologists. The fundamental questions and the experiments they inspired led to the use of terms such as pluripotency, determination, specification, differentiation, dedifferentiation, and pattern formation (Gilbert 2013; Wolpert et al. 2015). With subsequent advances in cell biology, it was possible to understand the basic cellular processes underlying these phenomena, such as proliferation, migration, and morphogenesis. Eventually, these early studies have led to our current understanding of the genetic and molecular regulation of cellular behaviors that result in embryogenesis and regeneration. In this book, we explore these principles and concepts at these multiple levels of complexity in order to begin the dialog between developmental biologist and engineers as to how to engineer regeneration.

The explosive growth of information from genetics and genomics of the past several years has identified the key signaling networks controlling development, and also how these networks can be regulated through both genetic and epigenetic mechanisms. There is no question that with more time, even more details about these processes will be discovered. Developmental biology has been driven by the quest for more data about complexity, and therefore, the strategy for how to make something happen based on the knowledge we have today may not come from the developmental biologists. It is reasonable to presume that at some point, we will know enough to begin applying this information to devise new therapies to induce regeneration. The question then becomes how to do this rather than when to try. If we can succeed, then history will show that the time was now.

1.3 The view from engineering

As a non-engineer, it appears to me that engineers are fascinated by building things and making them work. As discussed earlier, experimental biology traces its history back a couple of centuries, but engineering goes back to the days when early humans used to make tools and then work on them to improve their function. One example of the success of engineering from modern times is the remarkable advances of human flight, starting with the flight of Wright brothers in 1903 to Apollo 11 taking humans to the moon, which occurred less than seven decades later. Once an engineer can figure out how to build something and make it work, the advances leading to it working better can be rapid.

In modern times, engineers have begun to approach biological problems through tissue engineering involving approaches to manipulating the interactions between cells and engineered materials. The growth of this discipline was dramatically advanced with the establishment of the Whitaker Foundation in 1975 that funded the establishment of biomedical engineering programs throughout the world. Driven by the challenge of engineering devices and synthetic tissues, advances in material science, biomaterials, and bioactive scaffoldings have led to the engineering of functional replacement tissue for clinical use. The engineering approach focuses attention on the fact that, ultimately, the goal of regeneration is to restore lost function. Replacing the lost structure with a copy of the original would presumably restore the same function. However, regenerating a different structure could also restore function and might work better. For example, engineered prostheses can function better for specialized functions, leading to the ironic situation of athletes with amputated limbs being banned from competition because their engineered prostheses give them an unfair advantage.

For developmental biologists, an engineering approach will focus attention on what is more important and what is less important. As noted earlier, modern biology has become increasingly focused on collecting massive amounts of data and then using computational techniques to yield an unbiased analysis of the dynamics of the process being studied. In contrast, the engineering approach is biased toward sifting through the data to find pieces of the puzzle that will allow for improving and optimizing the functional outcome. An engineering approach is thus a biased approach that would allow biologists to focus on what is most important for improving the outcome of regeneration. In addition, engineering would enable biologists to know when there is something missing in their body of knowledge that they did not know was missing. This later point is underappreciated in that there is an assumption that if you keep discovering new facts, you eventually will know everything, and nothing will be undiscovered. By taking an engineering approach, when you set out to build something (e.g., a regenerated limb), there is a presumption that you know the essential components, and if what you build does not work, you then realize that something is missing from your body of knowledge. With this realization, you then need to go back and discover what is missing.

For engineers, the next challenge is to control the developmental process directly to orchestrate redevelopment of biological structures in time and space or, in other words, to engineer regeneration itself. To date, the focus of tissue engineering has largely been to create the final structure and then recruit the cells to add function to the structure. The goal of regenerative engineering is to discover how to regulate the endogenous regenerative potential of cells and tissues to have them progressively rebuild the missing structure, as it is done during embryonic development and regenerative development. The core rationale for this approach is that regeneration is a fundamental biological property, with conserved mechanisms for processes that can be controlled through the application of principles of engineering. The goal of this book is to identify these mechanisms and explore ways to control them.

1.4 Principles of development

The authors explore a broad range of topics in order to identify the mechanisms of development that could be engineered in order to induce regeneration. The goal for each author is to share thoughts and perspectives pertaining to his or her particular areas of expertise and experimental model system. The authors provide guidance as to how to explore these topics in more depth, and thus, this book provides a breadth and depth of understanding of both embryonic and regenerative development biology. By way of introduction, the goal of this chapter is to lay a broad framework that is based on six principles of development that I consider fundamental to approaching the challenge of regenerative engineering.

1.4.1 Development begins with fertilization and ends with death

Development of an organism is a continuous process throughout life. In recent decades, studies of developmental biology have focused primarily on the processes involved in embryogenesis; however, historically, developmental biology encompassed both embryogenesis and regeneration as functionally related processes. Before the late twentieth century, these two areas of research were considered related areas of biological research, and many of the founders of modern genetics (e.g., Thomas Hunt Morgan, who established *Drosophila* as a genetic model organism) began their careers studying regeneration. With the spectacular successes of mutagenesis screens over the past couple of decades, first in *Drosophila* and subsequently in other model organisms for developmental genetics (yeast, worms, zebrafish, and mice), the field of developmental biology morphed into developmental genetics. Since none of the classic models for regeneration was selected for genetic studies, the field of regeneration biology and developmental biology became largely non-overlapping. With the tools of genomics and genetic manipulation now widely available for all model systems, there is a renaissance of experimental work with the classic models for regeneration (e.g., salamander, planarians, and crickets). As discussed later, this reunified view of developmental biology leads to the premise that regeneration involves reactivating the developmental mechanisms used during embryogenesis to make a tissue or organ in order to remake it in the adult.

Although embryonic and regenerative developmental mechanisms are related, the outcome of the latter is highly variable compared with the former. By definition, embryonic development always leads to the formation of a complete organism, typically with complex structures. In contrast, the outcome of regenerative development is highly variable among different organisms. At the level of tissues (e.g., epithelia), regeneration is robust and is required for maintaining tissue homeostasis throughout the life of the post-embryonic organism. It is at the level of more complex structures that regeneration is continuously variable, with some animals having the ability to regenerate an entire body (e.g., planarians) or body parts (e.g., salamanders) and others such as humans having very limited regenerative abilities (e.g., finger tips). Nevertheless, renewal and replacement of cells in the adult are essential and continuous processes shared among all organisms, and the mechanisms are the same as those utilized in organ regeneration in animals that have a high regenerative capacity. Most of the underlying mechanisms are functional, and the failure of humans to regenerate complex structures is likely a consequence of failure to control these mechanisms appropriately in time and space. Thus, the challenge for the regenerative engineer is not how to invent new ways to do something but rather how to control the already existing processes in order to achieve a different outcome.

The view of embryonic and post-embryonic development (regeneration) as a continuous process leads to the realization that both cancer (dysregulated regeneration) and aging share these core developmental mechanisms. Most cancers arise in the tissues that are highly regenerative in all animals (e.g., carcinomas arising from epithelia) and are characterized by a loss of differentiation (comparable to the process of dedifferentiation associated with regeneration) and continuous proliferation. This phenotype lacks the function associated with normal development, leading eventually to failure of the affected organ(s) and death of the animal. In the case of aging, there is a balance throughout life between processes that tear the structure down (injury and disease) and the ability to counteract this degeneration through regeneration. Since regenerative abilities characteristically decline with age, this balance shifts toward the progressive accumulation of defects over time, ultimately resulting in organ failure and death. Thus, advances in discovering how to engineer regeneration will lead to increased longevity and eventually to rejuvenation.

1.4.2 Mechanisms of development are conserved between organisms

The most exciting discovery of modern developmental biology is the remarkable degree of conservation of developmental mechanisms among different species. Previously, there was a presumption that the complexity of the organism was a consequence of a correspondingly complex genetic code. Thus, simple organisms would have small genomes with few genes, and increasingly more complex organisms would have increasingly larger numbers of genes. There was also a presumption that phenotypic differences are a reflection of genetic differences, and thus, the genes that control wing development in a fly would be different than those controlling leg development in tetrapod vertebrates. Both of these presumptions proved to be wrong (Gilbert 2013; Wolpert et al. 2015).

Before the discovery of developmental regulatory genes and sequencing of genomes that led to an understanding of the genetic mechanisms controlling development, the basic principles of how regeneration occurs were formalized in the Polar Coordinate Model (PCM) (French et al. 1976[*]; Bryant et al. 1981[†]). Based on pre-molecular biology studies of appendage regeneration in three evolutionarily divergent model

---

[*] This is the original Polar Coordinate Model paper that made the case for conservation of mechanism for regeneration of salamander limbs, *Drosophila* imaginal discs, and cockroach legs.

[†] This is the second of the Polar Coordinate papers that established the *rules* that cells follow during regeneration.

organisms (salamanders, fruit flies, and cockroaches), it was evident that the fundamental principles of regeneration were common to all three, thus predicting that the underlying mechanisms would be conserved. The PCM predicted that regeneration is regulated by interactions between cells with different cell surface identities, and that these interactions occurring on the outside of the cells resulted in mitosis in the cells and a change in the state of differentiation of the daughter cells. The conserved rules of the PCM predated the discovery of conservation of developmental genetic pathways and the mechanisms of receptor–ligand interactions and signal transduction.

The paradigm shift in realizing that the genetic pathways controlling development are conserved was triggered largely by the discovery of the homeobox genes. Based on early studies of homeotic transformations in which one body part develops in the place of another (e.g., the antennapedia mutation in *Drosophila*), the sequencing of mRNA identified a highly conserved DNA-binding sequence, the homeodomain (Duboule and Dollé 1989; De Robertis et al. 1990; Gehring 1993). With the sequencing of more genes, it quickly became evident that the homeodomain is highly conserved among a family of related genes (the Hox complex), and that highly conserved homologous genes are present in all animals. What now is a fundamental concept of developmental biology was revolutionary at that time and led quickly to the discovery that gene function and sequence are conserved. Thus, homologous mammalian genes could substitute for the function of *Drosophila* genes. Today, it is recognized that gene regulatory networks and signaling pathways are conserved between species, and are reutilized multiple times at different developmental stages to make an embryo (Gilbert 2013).

Decades later, with advances in whole genome sequencing, it was surprising to discover that the number of genes was remarkably similar between organisms, regardless of the complexity of their body plans. Thus, the early estimates of there being approximately 100,000 genes in the human genome have been revised downward such that there are about the same number of genes in human genome as in the nematode worm *Caenorhabditis elegans* (20,000) (Gilbert 2013; Wolpert et al. 2015). These data have led to the realization that the development of complexity is a function of how a relatively small number of conserved genes are differentially regulated. The importance of this principle of development is that when thinking about how to engineer development, the strategy should be to change the way in which genes are regulated rather than to change the protein-coding regions of the genome. Finally, the conservation of mechanism is good news for engineering, since what is discovered from studies of one system will be applicable to all. Once the basic rules that regulate the function are discovered, it will be possible to fine-tune the details specific to different systems and outcomes.

1.4.3 Mechanisms are conserved between embryonic development and regenerative development

The fact that the mechanisms controlling embryonic development are conserved between species leads to the question of whether these same mechanisms can also remake body parts in the adult (regeneration). Since embryonic development and regenerative development start at different stages in the life history of an animal (zygote vs. adult) and are induced by different signals (sperm–egg fusion vs. injury), it would be expected that the early pathways are different. However, since the same structure is formed eventually, it is likely that at some point there is a convergence to a common stage of development, at which point the mechanisms are conserved (Bryant et al. 2002). One early experimental study to address this question was performed in the axolotl, in which reciprocal grafting between limb buds and blastemas indicated that these are developmental comparable structures, leading to the presumption of shared developmental mechanisms controlling formation of a functional limb (Muneoka and Bryant 1982[*]).

More recently, studies of the mechanisms of regenerative development have identified many pathways that are conserved in embryonic development. In the case of limb regeneration, formation of the regeneration blastema involves the re-expression of embryonic genes and the genesis of a population of undifferentiated cells, with the same developmental potential as that in the limb bud of the embryo (Han et al. 2005; Muneoka et al. 2008; McCusker et al. 2015[†]). This process, referred to historically as *dedifferentiation*, involves changes in the state of differentiation of some cell lineages (e.g., connective tissues) and the activation of adult stem cells (e.g., satellite cells of the skeletal muscle). The signaling factors that control blastema formation are the same as those in embryogenesis (e.g., growth factors such as fibroblast growth factor [FGF], bone morphogenetic protein [BMP], and Wnts) and appear to function through the same pathways involving conserved growth factor receptors and signal transduction pathways. The conservation of these mechanisms has been demonstrated directly by the use of human growth factors and extracellular matrix (ECM) proteins to induce blastema formation and pattern formation in salamander wounds (Makanae et al. 2014, 2016; Satoh et al. 2015). Therefore, the factors that are necessary for regeneration maybe used in special ways, but they are not special regeneration factors, and the lack of regeneration in mammalian wounds may not be a consequence of a lack of pro-regenerative signaling factors.

---

[*] This presents the most direct evidence that the mechanisms of limb development are reutilized during limb regeneration.

[†] This report demonstrates that positional information is plastic and is reprogrammed during formation of the early blastema.

Central to this unified view is the assumption that regeneration involves reactivating of the developmental mechanisms used during embryogenesis to make a tissue or an organ in order to remake it in the adult (Bryant et al. 2002). By this view, the basic structures such as appendages (e.g., arms and wings) evolved once, and the underlying genetics controlling how to make an appendage is conserved, whether in the embryo or in the adult. As a consequence, we can draw upon the vast body of data on the genetics involved in embryonic development to make strong inferences about what is happening during regeneration. Although there may be subtle differences, it will likely not be necessary to go back and repeat all the studies from embryonic development in order to verify that these conserved pathways are functioning in the same way during regenerative development. We should move forward with confidence that we already have a good understanding of how these mechanisms work, which means we have the genetic toolbox, and now we need to discover the instructions on how to use the tools.

Finally, there is a possibility that it may not be necessary to engineer regeneration from start to finish. Since there appears to be a point of convergence between embryonic development and regenerative development (e.g., the limb bud and blastema), we may only need to engineer formation of the blastema, at which point it will make a new limb, just as it occurred with the limb bud in the embryo (Bryant et al. 2002; McCusker et al. 2015). In fact, evidence for this comes from experiments in which blastemas at different stages of development are grafted to an ectopic location. Under appropriate conditions, the blastema functions as a self-organized unit and can form an entire limb, independent of signaling from the host tissues (Stocum 1968; McCusker and Gardiner 2013). Regardless, the strategy is to study regeneration in organisms in which it happens and then draw on the knowledge and insights from developmental genetics in embryos in order to devise strategies to engineer regeneration in organisms in which it does not happen.

1.4.4 Complexity is an emergent property of the behavior of the cells

The importance of the cell as the unit of biology was formalized in the Cell Theory and was accredited to Theodor Schwann and Matthias Schleiden in the mid-nineteenth century (Tavassoli 1980). Thus began the modern era of cell biology that led to our present-day understanding of the complexity of cell structure and function. The focus of this book is on how cells behave in order to build complex structures during embryogenesis and to regenerate these structures in the adult. In the case of embryogenesis, this process begins with a single cell, the zygote, which gives rise to all structures of the embryo and adult. With regeneration, the process starts with the recruitment of a relatively small population of progenitor cells that give rise to the many cells needed to redevelop the missing structures. Thus, both development and regeneration begin with one or a few cells that give rise to complex, multicellular structures over time.

Thus, the goal of regenerative engineering is to coordinate the behaviors of these cells over time to achieve the desired outcome in terms of rebuilding complex structures.

Although the molecular pathways that control cell behaviors are complex, as demonstrated by the increasingly complex gene regulatory pathways being published, the repertoire of cell behaviors is simple. Cells can (1) proliferate, (2) change shape (leading to migration or epithelial morphogenesis), (3) differentiate, or (4) die. To further simplify this view of cellular behavior, it is obvious that cells can engage in only one of these behaviors at a time. Thus, cells stop migrating in order to divide, and dividing cells do not simultaneously migrate. Finally, each of these behaviors (except for cell death) is dynamic and therefore can be regulated in order to achieve different outcomes. For example, undifferentiated, induced pluripotent stem cells (iPSCs) can be derived from differentiated cells and can migrate and proliferate. The challenge of regenerative engineering is to discover how to orchestrate these fundamental behaviors of cells to progressively build complex structures. To do this, there must be a source of cells that undergo cell division to replace the lost cells (proliferation). These cells need to end up in the right place relative to each other in order to re-establish the spatial pattern (migration). Finally, they need to differentiate into the appropriate functional cell types (e.g., muscle and bone).

1.4.5 Successful regeneration requires both the cells and the blueprint (the information grid)

Much like when building a house, regenerating a missing body part requires both new cells with the appropriate behaviors and a blueprint that instructs the cells where to go relative to the other cells. In the case of building a house, all the components (e.g., plumbing, wiring, windows, and doors) need to be put in the right place in order to end up with different parts of the house having different functions (e.g., kitchen, bathroom, bedroom, and living room). Similarly, the different cellular components of the missing body part (e.g., a limb) need to be put in the right place in order to form a functional structure (e.g., the origin and insertion of a muscle must span a joint such that when the muscle contracts, the limb bends in the right direction). This regeneration blueprint is classically referred to as positional information and is a topic that is discussed repeatedly throughout this book. Based on our present-day understanding of how positional information is encoded by the cells, the blueprint can be conceptualized as a grid of intercommunicating cells within the loose connective tissue. Thus, the cells of the positional information grid provide the blueprint to guide the other cells to the appropriate location.

Many laboratories are investigating how to create regeneration competent cells, such as stem cells, and advances in this area of research are rapid and exciting. The biology of stem cells and the challenge of recruiting these cells in order to engineer regeneration are the

11

focus of another book in this series and are discussed only to a limited extent in this book. In contrast, relatively few laboratories are working to understand the nature of positional information and the information grid. In the end, learning how to engineer the information grid will be essential to achieve regeneration. Thus, understanding the molecular nature of the grid, how it forms, and how it regenerates will likely prove to be a major challenge for successful regenerative engineering.

1.4.6 We need to learn how to talk to the cells

Although it may prove to be the case that there are special regeneration genes that encode for special regeneration factors, the remarkable degree of conservation of developmental mechanisms between species and between embryonic development and regenerative development (see previously) clearly suggests that this is not the case or at least need not be the case. Thus, we have the genes to redevelop a limb as an adult since we have the genes to make a limb, which is what we already did as an embryo. The challenges for engineering regeneration are to learn how to recruit regeneration-competent cells and how to control their behavior in response to the information grid. This process occurred during embryonic development, which indicates that the cells in our body have the necessary behaviors and can respond to the information grid to make functional body parts. Not only can the cells respond to the positional information, but they also make this information, and thus, the cells are talking to themselves and to each other during development. In order to make this happen again during regenerative development, we need to learn the language that the cells use.

We already know the alphabets that underlie this language, since we have now sequenced multiple genomes. Classical grafting experiments and more recent studies of developmental genetics have led to insights into many of the words that the cells use. This vocabulary is not highly complex, and the same signals (e.g., FGF, BMP, Wnts, transforming growth factor beta [TGF-β], and sonic hedgehog [SHH]) are used repeatedly in development and in regeneration (Gilbert 2013; Wolpert et al. 2015). Therefore, the complexity of the conversation between cells appears to lie in how the words are arranged into sentences (also known as the syntax). Thus, we need to appreciate that cell–communication is a dynamic process in space and time. In hindsight, this is not surprising, since a fundamental concept in regeneration and developmental biology is that there are temporal windows during which the outcome depends on the signals produced and received. As a consequence, the same signal will result in very different outcomes, depending on the window of time during which the signal is present. This is a recurring theme in many of the chapters and is very important when thinking about how to explore and engineer a complex conversation with the cells in order to talk to them through the process of regeneration.

Finally, although not a topic in this book, this conversation between cells lies at the heart of cancer. During both embryonic development and regenerative development, undifferentiated and proliferative cells end up making functional body parts, and when they are done, they stop dividing and become functional. Thus, the language of development tightly regulates and coordinates the behavior of the cells. It is increasingly evident that the behavior of cancers arising from highly regenerative tissues (e.g., carcinomas arising from epithelia) is dysregulated, as if the cells no longer communicate with each other and with their environment to maintain functional tissues. Presumably, this failure to communicate is a consequence of either the cells not responding to the information grid or an error in the information encoded in the grid.

1.5 Organization of this book

Since regeneration has a strong temporal aspect beginning with injury, the first three sections of this book are organized to capture the sequence of processes involved in replacing the missing body parts. The signals for the initiation of regeneration are associated with injury. In response, undifferentiated cells with the ability to participate in regeneration (regeneration-competent cells) are recruited from the tissues that remain after injury. Over a period of time, these cells communicate with each other and the surrounding environment to rebuild the complex pattern of tissues of the lost structure. In order for this newly formed body part to function, it needs to integrate structurally with the rest of the body that was not lost to injury.

The fourth section of this book focuses on specific examples of organ systems that are being, or potentially could be, engineered based on the principles of embryonic and regenerative engineering. In some cases, for example, *ex vivo* engineering and implantation of teeth, the proof of principle has been well established. In others, such as regeneration of connective tissues without scar formation, the way forward remains challenging, despite decades of intensive research efforts.

The authors of each chapter have focused on their view of the most important principles to understand in order to begin strategizing how to engineer each of these steps along the way to successful regeneration. Most have not cataloged lists of genes, but they do provide references that will guide a reader in that direction. Most importantly, they have identified the key concepts from their perspective based on their expertise in this area of developmental biology. Some have also annotated the bibliography for their chapter to guide the readers who want to dig deeper into a particular topic. Collectively, we have covered a wide range of topics that capture the important principles of embryonic and regenerative development from diverse perspectives.

13

1.5.1 Signals associated with injury that initiate regeneration

What cells do and when they do it depend on changes in the environment. There are extracellular signals that control the behavior of cells, and these signals can be, and often are, produced by other cells. Since regeneration is, by definition, a response to injury, the signals in the environment that are essential for regeneration are produced as a consequence of injury. The issues of what these signals are, how they are regulated, and how they regulate the behavior of cells are the topics covered in the first section of this book.

The regulation of these early signals is critical in terms of the final outcome. Regeneration occurs as the outcome of a stepwise sequence of events, and hence, the failure to progress beyond any one step will result in regenerative failure. We do not know how many steps are required for a successful regenerative outcome, but it is intuitively obvious that at least one of those steps does not occur in humans. The good news is that regenerative failure need not be a consequence of failures at multiple steps, and therefore, it maybe, and likely is, the case that many or even most of the steps required for successful regeneration can occur in wounds in humans if the appropriate regulatory signals are provided at the right time. The engineering strategy thus is to start at the beginning of the regeneration cascade, identify steps that are barriers because they do not occur, and then provide the signals required to progress beyond that barrier (Muller et al. 1999). This is an iterative process in that having overcome one barrier, regeneration will progress until the next barrier is encountered and overcome. At the point when there are no further barriers, regeneration will progress endogenously to completion.

A number of signals and changes in signals are covered in the chapters in this section of this book. Among the earliest are changes in the wound microenvironment, resulting in changes in the presence of reactive oxygen species (ROS), to which cells respond (Chapter 2). As cells of the immune system are recruited to the wound, they release a large number of cytokines and chemokines that are required for successful regeneration (Chapter 3). Injury disrupts the integrity of the epithelial barrier of the body, resulting in changes in the flux of ions into and out of the tissues that generate the flow of electrical currents. There is a long and rich history of the role of bioelectrical currents in controlling wound healing and regeneration, and these issues are revisited from the modern perspective of molecular mechanisms (Chapter 4). Similarly, the role of nerves and nerve signaling has been an important and consistent theme in the regeneration literature. This topic has been addressed previously in the context of limb regeneration but has not been explored on a broader basis—the role of nerves in the mammalian injury response or the role of nerves in the development of organs other than limbs. In recent years, the gain-of-function model for limb regeneration in the axolotl (the Accessory

Limb Model) (Endo et al. 2004*; Satoh et al. 2007) has led to the identification of a cocktail of growth factors that can substitute for the function of nerves in terms of recruiting regeneration-competent cells for blastema formation (Makanae et al. 2014, 2016; Satoh et al. 2015). These topics addressing the function of nerves in regeneration are explored in Chapter 5. Finally, the availability of oxygen is regulated dynamically, starting with injury and persisting through the process of blastema formation, and it regulates the behavior of cells at the site of injury (Chapter 6).

1.5.2 How cells communicate to remake the pattern and restore the function

As with both embryonic development and regenerative development, the functional unit for regenerative engineering is the cell. There are countless books and reviews on the complex details of cell biology, and therefore, it is not necessary to cover these topics yet again. Rather, the chapters in this section focus on how the behavior of cells is orchestrated in order to achieve the desired outcome (regeneration). As discussed earlier, the behavioral repertoire of cells is limited (proliferate, migrate, and differentiate/dedifferentiate), yet it is sufficient to form all of our body parts. The vast numbers of cells to build body parts are a result of cell proliferation; the movement and sorting out of cells to end up in the correct location occur via migration; and the changes in cell shape and ultimately cell function occur via differentiation. Ultimately, the different tissues are assembled in association with each other to yield organ function via the control of pattern formation.

It is obvious that you need the building blocks (regeneration-competent cells) in order to regenerate a structure. However, regeneration is about the replacement of the lost function and not just the structure. Function is an emergent property that is created when different cells of different tissues become appropriately organized into a functional structure; for example, muscles have origins and insertions that span joints in the skeleton, and the motor neurons innervate the correct muscles. We know from the literature on amphibian regeneration that in addition to the cells, you need the information that tells the cells where to go and what to make (see previously). Some of these cells are already lineage-restricted (e.g., myoprogenitor cells) and only need to be instructed as to where to migrate and then differentiate. Others, such as the multiple cell types of the connective tissues, appear to be derived from a common progenitor cells, and thus, this information is also instructive in terms of determining a cell's fate. For limb regeneration in salamander, we now refer to this information as the information grid, and the grid is required for remaking the pattern and restoring the function during regeneration.

---

* This is the first gain-of-function model for vertebrate limb regeneration that is based on the induction of accessory limbs (also referred to as the Accessory Limb Model [ALM]).

The question of where the cells come from, which has been investigated since the early days of regeneration research, has been revisited in recent years by applying the tools of molecular biology (e.g., grafting of cells from transgenic animals expressing green fluorescent protein [GFP]) to lineage analysis and is discussed later in the third section (Chapter 19). The more complex challenge of understanding how pattern formation is controlled is the focus of the chapters in this section. Over the years, a number of viewpoints have emerged as to how cellular behaviors can be regulated in order to control pattern formation, including the role of retinoic acid (Chapter 7) and microRNA signaling (Chapter 8). These signals appear to occur over long ranges (Chapter 9) and by short-range cell–cell interactions (Chapters 10 and 11) and cell–ECM interactions (Chapter 12). The result of these interactions is the regulation of growth and pattern formation (Chapters 13 and 14). As discussed in the Introduction, positional information for limb regeneration in salamander is encoded in an information grid, and regeneration of the grid is at the core of limb regeneration (Chapter 14). Finally, it is important to recognize that models of how biological processes, such as regeneration, work imply that there is a regulatory network that more or less precisely controls the outcome (e.g., see complex gene regulatory network diagrams). The aim of Chapter 15 is to address the importance of recognizing that there is stochastic variation (genetic noise) associated with biological processes, and it makes a case for how this variation could affect the response of a biological system to a major perturbation such as amputation.

1.5.3 Integration of new structures with the old ones

During embryogenesis, all tissues are at comparable stages of development, and their development is coordinately regulated in time and space. With regeneration (and thus with regenerative engineering), it is necessary to replace a missing or damaged part with something new and different than the pre-existing body part. This new part ultimately needs to become integrated into the host, which is comparable to the challenge faced with engineered prosthetic devices. There have been impressive advances in making these replacement body parts that can function as well as or even better than a normally developed body part, but discovering how to have these devices integrate seamlessly and permanently into the human body remains a challenge.

Although the issue of integration during endogenous regeneration has not been studied extensively, it is evident that regeneration of the new part need not lead to its integration into the host. For example, when ectopic limbs are made in the Accessory Limb Model, they can be remarkably normal and complete in terms of pattern, but typically, they do not connect up to the host skeleton (Endo et al. 2004; Satoh et al. 2007). These ectopic limbs arise from injury to the skin, without damage to the underlying soft tissues and skeleton; however, if the host skeletal tissues are also injured at the site of the wound, the ectopic limb skeleton integrates seamlessly into the skeleton of the host arm (Satoh et al. 2010). Presumably, injury to the host skeleton induces those cells to respond to pro-regenerative signaling associated with formation of the ectopic limb. In this section of this book, we explore the hypothesis that the injured host cells are induced to dedifferentiate, and thus, integration is dependent on dedifferentiation. If this is the case, then it will be necessary to discover how to engineer the processes controlling the state of differentiation.

Like other cellular behaviors, differentiation is dynamically regulated. During embryonic development, the early embryonic cells are undifferentiated and become progressively more differentiated as the embryo develops. Once differentiated, either the cells can reverse this process endogenously, as occurs in animals that can regenerate, or this reversal can be induced to occur via cell reprogramming techniques. The topics pertaining to the biology of undifferentiated embryonic stem cells, adult stem cells, and induced pluripotent stem cells (iPSC) are covered in another book in this series on regenerative engineering. Regardless, it is important to present an overview of how the differentiated state of cells can be regulated (Chapter 16) and how dedifferentiation occurs endogenously during regeneration (Chapter 17). It is becoming increasingly evident that both differentiation and dedifferentiation are regulated by epigenetic mechanisms (Chapter 18). Thus, the epigenetic regulation of differentiation allows for the developmental plasticity required to make new tissues from pre-existing tissues and for the new tissues to integrate with the old ones (Chapter 19).

1.5.4 Principles of organ development and regeneration

Efforts to induce tissue repair and regeneration have been ongoing since the early discoveries that regeneration can occur naturally in some animals. In recent years, remarkable progress has been made in understanding how to go about regenerating body parts. In part, these advances have resulted from a paradigm shift in thinking about inducing human regeneration. Although the ability of animals such as worms, fish, and salamanders was taken for granted, there was a presumption that it would never occur in humans. Over the past few decades, that attitude has changed, and it is time to consider regeneration as fundamental biological property that is shared to varying degrees by all animals. This change in thinking about human regeneration came about as a result of key discoveries such as the cloning of Dolly, the derivation of human embryonic stem cells, and the discovery that differentiated somatic cells could be reprogrammed to a pluripotent state by iPSC techniques. In the past, the question was whether or not human regeneration was possible; whereas, today, the question is how long will it take.

The final chapters of this book focus on areas of research where the application of principles of development are leading to advances that will likely lead to therapies in the not-too-distant future. One approach is to reverse engineer a developing organ. If we can take it apart, put it back together, and make it work, then we are well on the way to engineering it from scratch. This approach has been successful for a number of organs (e.g., teeth) and is the focus of Chapter 20. The complex structures that regenerate well in mammals are the digit tip and the associated nail bed (Chapter 21). Regeneration in the nervous system, once thought to be impossible in humans, has become the focus of major research efforts over the past two decades and has a great potential for success with advances in neurobiology, in general, and stem cell biology, in particular (Chapter 22). Only recently have researchers recognized the potential for regeneration of heart tissues in humans, despite decades of research on non-mammalian vertebrates (Chapter 23). Similarly, one of the classic models for endogenous regeneration is the eye, which in recent years has become promising for application of stem cell therapies to repair and regenerate the tissues of the eye (Chapter 24). Perhaps, the greatest challenge is to induce human regeneration in the skin (Chapter 25). The issues of scar and scar-free healing come up repeatedly throughout this book, and being able to control wound healing and fibrosis will be about learning how to talk to fibroblasts. In particular, the cells of the connective tissue in salamander make the position information grid, but in mammals, they make scars. The challenges are to understand the biology of these cells and to learn how to regulate their behavior in order to heal wounds by regenerating the skin rather than by fibrosis (excess connective tissue deposition) or by failing to heal the wound at all (ulceration). This section of this book ends with a discussion of the relationship between regenerative engineering and synthetic biology (Chapter 26), which in reality are the same, and together will enable us to achieve the long desired goal of induced human regeneration.

The authors of each chapter have identified the most important concepts from their point of view. During the early stages of organizing this book, it became evident that there would be an overlap between these concepts. Rather than trying to avoid this during the editorial process, it seemed better to view this overlap as indicative of the importance of those concepts. Chapter 27 is a summary of the most commonly identified concepts of embryonic development and regenerative development that are relevant to the challenges of regenerative engineering.

References

Bryant, S. V., T. Endo, and D. M. Gardiner. 2002. Vertebrate limb regeneration and the origin of limb stem cells. *The International Journal of Developmental Biology* 46 (7): 887–896.

Bryant, S. V., V. French, and P. J. Bryant. 1981. Distal regeneration and symmetry. *Science* 212: 993–1002.

De Robertis, E. M., G. Oliver, and C. V. Wright. 1990. Homeobox genes and the vertebrate body plan. *Scientific American* 263 (1): 46–52.

Dinsmore, C. E. 1992. The foundations of contemporary regeneration research: Historical perspectives. *Monographs in Developmental Biology* 23: 1–27.

Duboule, D., and P. Dollé. 1989. The structural and functional organization of the murine HOX gene family resembles that of Drosophila homeotic genes. *EMBO J* 8 (5): 1497–1505.

Endo, T., S. V. Bryant, and D. M. Gardiner. 2004. A stepwise model system for limb regeneration. *Developmental Biology* 270 (1): 135–145.

French, V., P. J. Bryant, and S. V. Bryant. 1976. Pattern regulation in epimorphic fields. *Science* 193: 969–981.

Gehring, W. J. 1993. Exploring the homeobox. *Gene* 135 (1–2): 215–221.

Gilbert, S. F. 2013. *Developmental Biology*. 10th ed. Sunderland, MA: Sinauer Associates.

Han, M., X. Yang, G. Taylor, C. A. Burdsal, R. A. Anderson, and K. Muneoka. 2005. Limb regeneration in higher vertebrates: Developing a roadmap. *Anatomical Record. Part B, New Anatomist* 287 (1): 14–24. doi:10.1002/ar.b.20082.

Makanae, A., K. Mitogawa, and A. Satoh. 2014. Co-operative Bmp- and Fgf-signaling inputs convert skin wound healing to limb formation in urodele amphibians. *Developmental Biology* 396 (1): 57–66. doi:10.1016/j.ydbio.2014.09.021.

Makanae, A., K. Mitogawa, and A. Satoh. 2016. Cooperative inputs of Bmp and Fgf signaling induce tail regeneration in urodele amphibians. *Developmental Biology* 410 (1): 45–55. doi:10.1016/j.ydbio.2015.12.012.

McCusker, C. D., and D. M. Gardiner. 2013. Positional information is reprogrammed in blastema cells of the regenerating limb of the axolotl (Ambystoma mexicanum). *Plos One* 8 (9): e77064. doi:10.1371/journal.pone.0077064.

McCusker, C. D., D. M. Gardiner, and S. V. Bryant. 2015. The axolotl limb blastema: Cellular and molecular mechanisms driving blastema formation and limb regeneration in tetrapods. *Regeneration* 2: 54–71.

Morgan, T. H. 1901. *Regeneration*. New York: The Macmillan Company.

Muller, T. L., V. Ngo-Muller, A. Reginelli, G. Taylor, R. Anderson, and K. Muneoka. 1999. Regeneration in higher vertebrates: Limb buds and digit tips. *Seminars in Cell & Developmental Biology* 10 (4): 405–413. doi:10.1006/scdb.1999.0327.

Muneoka, K., and S. V. Bryant. 1982. Evidence that patterning mechanisms in developing and regenerating limbs are the same. *Nature* 298: 369–371.

Muneoka, K., M. Han, and D. M. Gardiner. 2008. Regrowing human limbs. *Scientific American* 298 (4): 56–63.

Satoh, A., G. M. Cummings, S. V. Bryant, and D. M. Gardiner. 2010. Regulation of proximal-distal intercalation during limb regeneration in the axolotl (Ambystoma mexicanum). *Development, Growth & Differentiation* 52 (9): 785–798. doi:10.1111/j.1440-169X.2010.01214.x.

Satoh, A., D. M. Gardiner, S. V. Bryant, and T. Endo. 2007. Nerve-induced ectopic limb blastemas in the axolotl are equivalent to amputation-induced blastemas. *Developmental Biology* 312 (1): 231–244.

Satoh, A., K. Mitogawa, and A. Makanae. 2015. Regeneration inducers in limb regeneration. *Development, Growth & Differentiation* 57 (6): 421–429. doi:10.1111/dgd.12230.

Stocum, D. L. 1968. The urodele limb regeneration blastema: A self-organizing system. I. Differentiation in vitro. *Developmental Biology* 18 (5): 441–456.

Tavassoli, M. 1980. The cell theory: A foundation to the edifice of biology. *The American Journal of Pathology* 98 (1): 44.

Wolpert, L., C. Tickle, and A. M. Arias. 2015. *Principles of Development*. Oxford: Oxford University Press.

Section I
Signals associated with injury that inititate regeneration

As regeneration has a strong temporal aspect beginning with injury, the Chapters 2 through 6 of this book focus on these early responses and the signals that control them. By definition, the signals for the initiation of regeneration are associated with the injury. It all begins with a wound, either an accident or a surgical intervention (e.g., to implant an engineered construct). At that moment, a cascade of events is initiated that determines the eventual outcome depending on the subsequent downstream events. Thus regeneration is a stepwise process, and what the cells do and when they do it are determined by their interactions with other cells and with their environment. These extracellular signals control the behavior of cells; and the issue of what the early signals are, how they are regulated, and how they regulate the behavior of cells are the topics covered in this first section of this book.

The regulation of these early signals is critical in terms of the final outcome such that the failure to progress beyond any one step will result in regenerative failure. We do not know how many steps are required for a successful regenerative outcome, but it is intuitively obvious that at least one of those steps does not occur in humans. Since we do not know all the signaling events that are required to progress from one step to the next, the engineering strategy must be to start at the beginning of the regeneration cascade and identify the steps that are barriers because they do not occur, and then provide the signals required to progress beyond that barrier. This is an iterative process in that having overcome one barrier, regeneration will progress until the next barrier is encountered and overcome. At this point, when there are no further barriers, regeneration will progress endogenously to completion.

Chapter 2 Reactive oxygen species and neuroepithelial interactions during wound healing

Hannah Grover and
Sandra Rieger

Contents

Key Concepts

a. Hydrogen peroxide is generated by keratinocytes of the wound epidermis. It promotes repair processes, such as cell migration, angiogenesis, and somatosensory axon regeneration.

b. Hydrogen peroxide at low concentrations acts as a second messenger molecule to activate signaling pathways involved in wound repair.

c. The role of hydrogen peroxide in wound repair and appendage regeneration is conserved.

d. Somatosensory axons secrete neuroinflammatory peptides after skin injury, which is essential to promote re-epithelialization of the wound bed.

2.1 Introduction

The skin is a barrier tissue that protects animals from the harsh outside environment. This barrier must be rapidly repaired on injury to prevent internal fluid loss and pathogen invasion. Nonhealing wounds after trauma, which are common in the elderly population, or denervation of the skin due to diabetes and spinal cord injury bear the risk of mutated stem cells migrating to the skin surface, where they can trigger tumor formation when predisposed to cancer (Wong and Reiter 2011). Impaired wound healing also presents a particular risk factor for limb amputations due to the development of infections. For instance, diabetic foot ulcers and infections precede 85% of lower limb amputations in diabetic patients (Apelqvist 2012). Despite advances in the wound healing field and the identification of molecular mechanisms that are involved in wound repair, chronic nonhealing wounds still represent a major clinical challenge. In contrast, advances in the tissue-engineering field have vastly improved trauma-related wound repair. These treatments primarily include skin grafting techniques, which have been shown to significantly enhance the healing process (Mcheik et al. 2014). Nevertheless, these techniques are associated with undesirable outcomes, such as aberrant cosmetic appearances, discomfort, and lasting complications such as infections, shearing of the grafts from the wound bed, and persistent pain at the donor graft site. Grafted tissues further suffer from the lack of sensory nerve reinnervation, which represents another challenge that typically leaves patients with numbness and unpleasant feelings. Approaches in which cell suspensions of keratinocytes are directly placed onto burn wounds have shown efficacy, but they must be derived from a highly proliferative donor site, which may prove difficult if the skin is largely destroyed after extensive trauma (Mcheik et al. 2014).

To overcome these challenges, strategies that are more effective will be necessary and will need to take into consideration the cellular behaviors within grafted tissues and the intercellular interactions between the graft and the host environment. Ideally, future approaches should largely focus on the stimulation of natural repair processes from tissue-resident cells, leading to the reactivation of signaling pathways that promote cell division and the migration of epidermal and dermal cells into the wound bed. However, this task is difficult, as manipulations of these processes may also promote cancer. To tackle this problem, it will be essential to elucidate wound repair–signaling mechanisms in different species, as this may help identify general facilitators of this process, which have been shown to be efficacious in nature.

In this chapter, we do not attempt to cover all known aspects of wound healing, but we particularly emphasize on the important roles of the reactive oxygen species (ROS), hydrogen peroxide (H_2O_2), and sensory nerve endings in this process. Hydrogen peroxide has emerged as an important factor that is essential for the stimulation of various paracrine interactions in the wound. Recent studies have demonstrated that this molecule is critical for tissue regeneration in species with high regenerative capacity. We will therefore also briefly summarize these newly emerging functions. In addition, we will focus on the regeneration and functions of cutaneous sensory nerve endings during wound repair. Regeneration of nerve endings is stimulated by H_2O_2, and it has been shown that successful wound repair depends on pro-inflammatory neuropeptides that are secreted by nerve endings into the wound environment. The information presented here may facilitate the design of future therapeutic strategies with which to stimulate repair mechanisms, either through current or emerging tissue-engineering approaches or on stimulation of natural repair and regeneration processes.

2.2 Hydrogen peroxide–dependent skin wound repair

The skin, though only a few millimeters thick, makes up about 16% of the body mass and weighs 8–10 pounds in an average adult person (Smalls et al. 2006). The skin is important for the survival of an organism, as it forms a protective barrier from ultraviolet (UV) radiation, extreme temperatures, toxins, and invading pathogens. Hydrogen peroxide was shown to be critical for various wound repair processes after acute tissue injury, and it has also emerged as an important regulator of appendage regeneration (Table 2.1). It has often been portrayed as a harmful molecule that causes widespread cellular death as a byproduct of mitochondrial respiration. Although under pathological conditions, concentrations of H_2O_2 can be chronically elevated and damaging, its signaling functions during wound healing and appendage regeneration, in contrast, are tightly regulated. Hydrogen peroxide belongs to the class of ROS, which are radicals that are formed during the sequential reduction of molecular oxygen. Three major products are known to

Table 2.1 Species Comparison of Functions of Reactive Oxygen Species during Tissue Repair and Regeneration

Species	Time Until ROS Formation	Persistence of ROS after Injury	Source of ROS	ROS-Dependent Processes	ROS-Dependent Molecular Factors	References
Humans	20 min	Until wound closure	Cytoplasmic NADPH oxidases	Keratinocyte migration, proliferation	EGFR, HSPLAI, ERCC6, HMOX1, MMP9, MMP13, NUDT1, IL6ST, CYP24A1, FOSL2, ATF3	Nam et al. 2010; Lisse and Rieger 2017
Mice	Unknown	Unknown	Exogenous H_2O_2 addition	Angiogenesis, accelerated wound closure	Unknown	Loo et al. 2012
Zebrafish embryo	20 min	Up to 6 hours	DUOX1, cyba	Fin regeneration, leukocyte and keratinocyte migration, neutrophil reverse migration, sensory axon regeneration	Lyn, Fyn	Niethammer et al. 2009; Rieger and Sagasti 2011; Yoo et al. 2011; Yoo et al. 2012; Tauzin et al. 2014
Adult zebrafish	Few min	Up to 18 hours	NADPH oxidases	Wound repair, blastema formation, apoptosis, proliferation	Shh, ERK, klf4, Dio3, FGF20	Gauron et al. 2013
Xenopus tadpoles	<1 h	Until completion of tail regeneration	NADPH oxidases	Tail length regrowth, proliferation	Wnt/β-cat signaling, FGF20 expression	Love et al. 2013
Drosophila	1 min	Minimum 3 min	DUOX1 in wound epithelium	Wound closure, hemocyte migration	Src42A, Draper-I, Shark	Moreira et al. 2010; Evans et al. 2015
C. elegans	100 sec	Several minutes	Mitochondria	Wound closure	Oxidative inhibition of Rho-GTPase	Xu and Chisholm 2014
Planaria	<30 min post-amputation	Unclear	Anterior and posterior amputation wound	Neuroblast differentiation, brain regeneration, cephalic ganglia formation	Smed-gpas, prohormone convertase 2 (pc2), a neurotransmitter convertor, nou-darake (ndk), an FGFR-related gene	Pirotte et al. 2015
Arabidopsis	Peak 10–12 min	90 min	Presumably epidermis	Defense against pathogenic fungus	Unclear	Beneloujaephajri et al. 2013

result from this process, superoxide anion (O_2^*), peroxide (H_2O_2), and hydroxyl radicals. Many of these radicals are generated continually as part of normal cellular processes, such as during respiration along the electron transport chain in the mitochondria. Most free radicals are rapidly scavenged under normal conditions. Hydrogen peroxide is an exception, as it has been shown to act as an important secondary messenger molecule within cells. Research in the past decade has revealed that H_2O_2 has critical signaling functions during wound repair. Injury-induced H_2O_2 signaling is remarkably conserved and has been observed in plants, planarians, fruit flies, nematodes, zebrafish, frogs, mice, and humans (Orozco-Cárdenas et al. 2001; Young et al. 2008; Niethammer et al. 2009; Moreira et al. 2010; Rieger and Sagasti 2011; Yoo et al. 2011; Loo et al. 2012; Love et al. 2013; Xu and Chisholm 2014; Pirotte et al. 2015), underscoring its importance. In tomato plants, for instance, H_2O_2 is produced in the wound epidermis and is required for the induction of defense genes (Orozco-Cárdenas et al. 2001). In post-hatching zebrafish larvae, H_2O_2 serves as a chemoattractant for leukocytes that are migrating toward the wound (Niethammer et al. 2009), and it has also been found to stimulate cutaneous sensory axon regeneration (Rieger and Sagasti 2011). Hydrogen peroxide is produced in wound keratinocytes within 20 minutes after injury, a process that is also conserved in humans (Lisse et al. 2016). It has also proven to be beneficial when topically applied to the skin at low concentrations. For instance, addition of 0.01% H_2O_2 to the media of larval zebrafish stimulates cutaneous axon growth in the absence of injury. Moreover, mammals benefit from topical H_2O_2 application, since treatment of human keratinocytes in culture with low H_2O_2 concentrations enhances scratch wound repair *in vitro* (Loo et al. 2011). Studies in mice further demonstrated that 0.5% H_2O_2 applied to excisional wounds on the back of the animal promotes angiogenesis and wound closure, whereas higher concentrations at 3%, which are typically used for wound disinfection, delay this process. This delay has been attributed to increased expression of matrix metalloproteinase 8, leading to decreased connective tissue formation and lack of inflammatory resolution.

2.3 NOX-dependent H_2O_2 formation after injury

Injury-induced H_2O_2 production depends on NADPH oxidases (NOX/DUOX). These enzymes transfer electrons across biological membranes and are consequently found at the plasma membrane. NOX/DUOX enzymes produce H_2O_2 by the transfer of an electron from NADPH to molecular oxygen, leading to the formation of superoxide. Since superoxide is highly reactive, it is rapidly converted to H_2O_2. However, most NADPH oxidases catalyze only the first part of the reaction and form the superoxide, and thus, additional enzymes are required for the formation of H_2O_2. These have been identified as superoxide dismutases (SODs), which are located in the extracellular matrix, cytoplasm, and mitochondria. NOX4 and DUOX1, in contrast, catalyze H_2O_2 without the need of SODs through their NADPH oxidase domain (Donkó et al. 2005). The NOX/DUOX family consists of seven members, NOX1–5, DUOX1, and DUOX2. These appear to be expressed in a cell-type and context-dependent manner (Bedard and Krause 2007). The first discovered member was NOX2 (gp91phox), which was shown to regulate H_2O_2 production during oxidative bursts in phagocytes (Bedard and Krause 2007). The subsequently identified additional members of this family have been shown to be expressed in a variety of other cell types and tissues, including the epithelium, endothelium, ovary, pancreas, smooth and skeletal muscles, cardiomyocytes, and keratinocytes (Bedard and Krause 2007). NOX enzymes are multi-subunit complexes, which need to be activated to be functional. NOX1–4 require the common subunit p22phox for activation, which is encoded by the gene cytochrome b-245 (CYBA). In contrast, NOX5, DUOX1, and DUOX2 are activated by calcium, evident by the presence of two calcium-binding EF-hand motifs, which are absent from the other NOX enzymes (Donkó et al. 2005). These EF-hand motifs have been shown to be important in wound-induced H_2O_2 formation. Deletion of both motifs in *Drosophila* DUOX eliminates H_2O_2 and impairs immune cell migration after laser-induced wounding, which depends on instantaneous calcium flashes produced in the wound (Razzell et al. 2013). DUOX1 is also required for H_2O_2-dependent leukocyte migration and cutaneous axon regeneration in post-embryonic zebrafish (Niethammer et al. 2009), since knockdown of DUOX1 impairs both processes (Rieger and Sagasti 2011). The injury-specific induction of DUOX1 is further supported by the fact that DUOX1 knockdown has no effects on development of the skin or sensory axons. The function of DUOX in wound repair and regeneration is not conserved. Tail regeneration in *Xenopus* tadpoles, for instance, is regulated by cyba and not by DUOX (Love et al. 2013).

2.4 Hydrogen peroxide–dependent signaling processes in the wound

Until recently, little was known about the precise molecular mechanisms by which H_2O_2 signaling occurred in the wound. One key criterion is that relatively low concentrations are generated, as these stimulate specific signaling events (Lisse and Rieger 2017). These low concentrations can be regulated by a number of factors, including (1) the limited availability of cytoplasmic NADPH oxidases within cells, (2) the presence of antioxidant proteins that scavenge H_2O_2, and (3) the controlled diffusion of H_2O_2 into cells via aquaglyceroporin channels. Activation of NADPH oxidase after an injury likely depends on transcriptional or post-transcriptional regulation. Since the time of H_2O_2 production is rather fast (~20 minutes in vertebrates), post-transcriptional events may initially play a role, while sustained H_2O_2 production may be regulated on a transcriptional level.

The initial post-transcriptional regulation is supported by the rapid calcium-dependent activation of DUOX in *Drosophila*. Evidence also suggests that the subcellular localization of NADPH oxidases is critical in the activation of specific downstream signaling pathways. For instance, NOX2 and NOX4 are localized to different regions in endothelial cells, and they activate distinct pathways in an H_2O_2-dependent manner, leading to survival and proliferation, respectively (Peshavariya et al. 2009). Thus, NADPH oxidase-dependent H_2O_2 production and signaling within cells are highly localized. This could explain how increased H_2O_2 levels within cells, such as under pathological conditions, might lead to widespread diffusion and non-specific oxidation, culminating in cellular death. Localized H_2O_2 domains are likely also controlled by antioxidant scavenger proteins, such as catalase and glutathione peroxidase.

Furthermore, it is likely that intra- and extracellular H_2O_2 concentrations are regulated by H_2O_2 diffusion into cells. Owing to its physical properties that are similar to water, it was long speculated that H_2O_2 could freely cross plasma membranes. However, increasing evidence in recent years indicates that free diffusion is negligible and that specialized membrane channels, aquaporins 3 and 8 (AQP3/8), regulate its transport. These channels are also known to be permeated by other small molecules, such as glycerin and water. Studies using human epithelial cell lines have shown that overexpression and knockdown of both aquaporins increase and decrease H_2O_2 diffusion, respectively, and modulate downstream signaling pathways (Miller et al. 2010). In addition to these cell culture studies, AQP3 has also been shown to play a role in H_2O_2-dependent T-cell migration toward chemokines within the skin of mice. Hydrogen peroxide uptake into T cells via AQP3 activates the Rho family GTPase Cdc42, leading to modulation of actin dynamics and stimulation of migration (Hara-Chikuma et al. 2012). Additionally, signaling proteins have been implicated in the regulation of H_2O_2 concentrations, such as the forkhead transcription factor FOXO1. FOXO1 was shown to negatively regulate H_2O_2 levels within the wound environment and protect keratinocytes against oxidative stress. Small interfering RNA (siRNA)-mediated knockdown of FOXO1 in human primary keratinocytes increases endogenous H_2O_2 levels, which impairs scratch wound repair (Ponugoti et al. 2013).

The second messenger functions of H_2O_2 are largely achieved via oxidation of specific cysteine thiols in signaling proteins, which are often found in catalytic domains of signaling enzymes (Claiborne et al. 1999). Oxidation of cysteine thiols stimulates the formation of sulfenic acid (sulfenylation), a highly unstable metabolite that can rapidly convert to other metabolites, such as sulfinic acid and sulfonic acid, or nitrosothiol (Leonard et al. 2009). However, a more common metabolic process in which sulfenic acid participates is to promote disulfide bond formation. It was found that this can lead to a conformational change in signaling proteins, which modulates their enzymatic activity, either activating or inactivating the enzyme (Stone and Yang 2006; Leonard and Carroll 2011; Truong and Carroll 2012). This has been shown to lead to the modulation of phosphorylation cascades within cells, often by activating kinases and deactivating phosphatases (Claiborne et al. 1999; Gough and Cotter 2011). In zebrafish, oxidative activation of the kinase, Lyn, a member of the Src-family kinase (SFK), recruits neutrophils toward the wound site (Niethammer et al. 2009; Yoo et al. 2011). Hydrogen peroxide oxidizes a specific cysteine residue (C466) in the kinase domain, which promotes autophosphorylation and Lyn activation (Yoo et al. 2011). A homolog of Lyn in *Drosophila* is Src42A, which also harbors this oxidation-sensitive cysteine residue 466, indicating the conservation of this signaling mechanism in fruit flies. It was demonstrated that Src42A similarly stimulates immune cell migration through interactions with two downstream targets, Draper/ CED1 and Shark, which have been implicated in the vertebrate adaptive immune response (Underhill and Goodridge 2007). Oxidative regulation of wound repair is also conserved in the nematode *Caenorhabditis elegans*. However, mitochondrial superoxide rather than H_2O_2 mediates this process. The skin in nematodes, the hypodermis, is an epithelium that consists of large syncytial cells. Puncture or laser wounding of the hypodermis leads to a calcium-dependent repair mechanism by which an actin purse string contracts around the wound to seal the injury (Xu and Chisholm 2014). Injuries rapidly trigger the formation of calcium flashes emanating from the wound site. The calcium enters the mitochondria via the mitochondrial calcium uniporter (MCU-1). Once in the mitochondria, calcium stimulates the production of superoxide (mtROS) and subsequent local inhibition of Rho-1 GTPase activity via a redox-sensitive motif and oxidation of Cys16, leading to actin reorganization and closure.

Although low H_2O_2 levels are critical for specific oxidation of signaling proteins, they do not explain how only some signaling proteins are oxidized but others are not. A computational study predicted that susceptible cysteine residues must be in the vicinity of another cysteine sulfur atom, accessible to solvents, and possess a pKa less than 9.05 (Sanchez et al. 2008). Furthermore, oxidation-prone cysteine residues must be present in an active site or regulatory motif in order to change the enzymatic function of a signaling protein after reversible disulfide bond formation. Recent data show that cysteine oxidation is not a linear process, which might explain the high specificity of signaling protein activation. It was found that peroxiredoxins 1 and 2 (Prx1 and Prx2), which are known to be efficient H_2O_2 scavengers, participate in the oxidative modification of oxidation-sensitive cysteine residues. Peroxiredoxins have been shown to transiently form intermolecular disulfide bonds with signaling proteins, which in a second step result in intramolecular disulfide bond formation within signaling proteins (Sobotta et al. 2015).

2.5 Hydrogen peroxide signaling functions during appendage regeneration

Given the important signaling functions of H_2O_2 during wound repair and the discovery that H_2O_2 is produced after fin amputation in zebrafish, a few studies have investigated whether it also has specific functions in appendage regeneration. A few vertebrates such as *Xenopus* tadpoles, salamanders, and zebrafish have a remarkable capacity to regenerate their appendages and many of their organs. The limb regeneration process has been well characterized and proceeds roughly in four major steps: (1) wound repair, during which wound epidermal keratinocytes migrate over the stump, (2) formation of the apical ectodermal cap (AEC) via migration and proliferation of keratinocytes, (3) formation of a blastema, a mound of undifferentiated cells that aggregates below the AEC, and (4) outgrowth of newly formed tissue from differentiating blastema cells (Endo et al. 2004; Iovine 2007; Yokoyama 2008). In frogs, regeneration occurs only during tadpole stages, whereas metamorphosed frogs have lost this regenerative potential. Tadpoles regenerate lost tails through reactivation of classical developmental transcription factors, such as Wnt, fibroblast growth factor (FGF), bone morphogenetic protein (BMP), Notch, and transforming growth factor beta (TGF-β). Hydrogen peroxide was shown to promote tail regeneration by inducing Wnt-dependent transcription of its target gene *fgf20* and blastema proliferation (Love et al. 2013). Interestingly, H_2O_2 production is sustained throughout the regeneration process, implicating it in distinct wound- and regeneration-specific mechanisms.

Hydrogen peroxide has also been found to be critical for fin regeneration in post-embryonic and adult zebrafish. Unlike in *Xenopus* tadpoles, where H_2O_2 is produced and required throughout the duration of the regeneration process, caudal fin regeneration in post-embryonic zebrafish depends on a short burst of H_2O_2 and calcium signaling during the first hour after amputation, leading to phosphorylation and activation of the Src-family kinase Fynb (Yoo et al. 2012). Thus, a transient time window of oxidation shortly after injury sets the stage for both wound repair and regeneration. Since post-embryonic zebrafish are still undergoing active growth, and the caudal fin at this age is rather anatomically simple, repair and regeneration processes may underlie the same signaling mechanisms. In adult zebrafish, ROS/H_2O_2 appears to have two functions: one function in wound repair and a later function in fin regeneration (Gauron et al. 2013). Incisional wounding stimulates a short burst of ROS/H_2O_2 at lower levels than amputation. Amputation-induced ROS/H_2O_2 formation increases more slowly and is sustained over longer time periods. The peak of ROS/H_2O_2 formation correlates with increased apoptosis and induction of JNK signaling, which was shown to be important for epidermal and blastema cell proliferation. Other key genes that are regulated by ROS/H_2O_2 during adult caudal fin regeneration are *fgf20*, *sdf1*, *klf4*, and *dio3*.

31

2.6 Wound-induced hydrogen peroxide stimulates cutaneous sensory axon regeneration

A hallmark of wound repair in vertebrates is the regeneration of cutaneous sensory axons into the wounded area. In rats, Dorsal root ganglion (DRG) axons were found to hyperinnervate the wound, and this hyperinnervation pattern persisted up to 12 weeks after injury (Reynolds and Fitzgerald 1995). Most of our current knowledge about the molecular mechanisms leading to somatosensory axon regeneration stems from rodent studies in which axotomies were performed. During an axotomy, either the brachial or sciatic nerve is exposed via an incision made into the limb, and axons are subsequently severed via nerve crush or transection. These injuries result in the disconnection of the distal axonal portion from the cell body, and the distal axon subsequently undergoes rapid degeneration. This process was originally identified by Augustus Volney Waller in 1850 and is defined by a stereotypic sequence of events leading to fragmentation, degradation, and clearance of axon debris. Nerve injury by axotomy triggers the release of two calcium waves at the injury site: a fast wave and a slow wave. These waves propagate along the proximal axon toward the soma. The fast wave leads to the release of internal calcium stores from the endoplasmic reticulum, which promotes nuclear export of histone deacetylase (HDAC5) in the soma (Cho et al. 2013). This primes the chromatin for transcription via acetylation (Ac) of histone 3 (H3). A second calcium wave stimulates local protein translation. One pathway involves importin-mediated retrograde transport of nuclear localization sequence (NLS)-containing transcription factors into the nucleus, which activate the expression of downstream genes. In addition, dual leucine zipper kinase (DLK-1) is locally translated, which leads to retrograde STAT3 transport into the soma. This process also involves Jun N-terminal kinase (JNK), leading to activation of C-JUN and activating transcription factor-3 (ATF3) (Rishal and Fainzilber 2014).

On the other hand, axon regeneration due to skin injury seems to proceed in a different way than axotomy-dependent regeneration. This has been shown in larval zebrafish by using *in vivo* time-lapse imaging. In this model, individual axon branches of a single sensory neuron innervating the epidermis were injured via laser axotomy, without damaging the surrounding keratinocytes (Rieger and Sagasti 2011). Epidermis-specific axotomy results in increased axon growth and retraction behavior, but regeneration is marginal. Although this could be related to intrinsic differences between mammalian and zebrafish sensory neurons, another possibility is that the epidermis prevents axons from growing, possibly due to the presence of repulsive cues. This is supported by observations in mice, where axotomy of the sciatic nerve promoted regrowth of sensory fibers into the dermis but nerve endings were largely prevented from re-entering the epidermis (Navarro et al. 1997). These repulsive cues might be abolished by injury, which is supported by the observation that sensory axons in zebrafish larvae fully regenerate when axons and epidermal keratinocytes are simultaneously injured. It was revealed that H_2O_2 is the responsible molecule, which is secreted by wound keratinocytes (Rieger and Sagasti 2011). Knockdown of DUOX1, which abolishes H_2O_2 production, impairs axon regeneration, but this process can be rescued on addition of exogenous H_2O_2 to the media of DUOX1 knockdown larvae. Furthermore, axon growth is stimulated when H_2O_2 is added to the larval media in the absence of injury. The axon growth-promoting functions of H_2O_2 have also been shown in adult zebrafish (Meda et al. 2016) and during DRG neurite outgrowth in culture and during development. In DRG neurons, H_2O_2 is produced by molecule interacting with CasL/MICAL at the growth cone tip. The MICAL-dependent H_2O_2 production is stimulated by semaphorin 3a binding to its receptor, resulting in oxidation of collapsin response mediator protein 2 (Crmp2). This leads to Crmp2 homodimer formation and transient complex formation with thioredoxin, which induces growth cone collapse and navigational turning (Morinaka et al. 2011). Moreover, bag cell neurons in the marine mollusk *Aplysia* utilize H_2O_2 for neurite outgrowth. This process depends on a functional NOX2/gp91phox/p40phox complex at cell adhesion sites in the growth cone, which stimulates F-actin flow and growth cone activity (Munnamalai et al. 2014). In addition to its functions in peripheral axons, H_2O_2 has proven beneficial effects during central nervous system regeneration. In the planarian *Schmidtea mediterranea*, H_2O_2 promotes brain regeneration (Pirotte et al. 2015). Inhibition of H_2O_2 production by Diphenyleneiodonium (DPI) treatment induces smaller brains and a reduction in cephalic ganglia. Brain regeneration is mediated by effects of H_2O_2 on neoblast differentiation toward neuronal lineages, without affecting their proliferation. Neoblasts are planarian stem cells that are abundant in the animal and that give rise to regenerating tissues.

2.7 The role of cutaneous nerve endings in wound repair

The ability of cutaneous sensory axons to regenerate after injury is critical, since their regeneration contributes to the restoration of somatosensation and tissue repair (Table 2.2). Classical experiments in the chick model demonstrate that denervated skin is impaired in healing. When a piece of neural tube that normally gives rise to DRG neurons innervating the wing bud skin was UV-irradiated to prevent sensory nerve innervation, wounds healed at a much slower rate compared with innervated skin (Harsum et al. 2001). Similar observations have been made in a rat injury model, where partial denervation of the skin with capsaicin treatment significantly delayed the re-epithelialization process after skin injury (Smith and Liu 2002). Using an artificial skin graft model constructed from human and mouse cells, it was shown that the lack of re-epithelialization is due to a defect in keratinocyte migration. In this model, type I and type III collagen sponges were co-cultured with

Table 2.2 Tissue Repair and Regeneration Processes Depending on Nerve Signals

Organism	Estimated Time of Axon Regeneration after Skin Injury	Nerve-Dependent Repair Processes	Nerve-Dependent Molecular Factors Involved in Repair	References
Rodents	<7 days post-wounding	Keratinocyte migration	Substance P	Reynolds and Fitzgerald 1995; Blais et al. 2013
Human	Few days	Keratinocyte migration	Substance P	Blais et al. 2013
Zebrafish	1 day	Fin regeneration, blastema proliferation	Schwann cell-derived sonic hedgehog	Rieger and Sagasti 2011; Meda et al. 2016
Urodeles	Unknown	Newt limb regeneration, blastema proliferation	Schwann cell-induced newt anterior gradient protein	Kumar et al. 2007

fibroblasts, keratinocytes, and endothelial cells, either in the absence or in the presence of sensory neurons. Skin grafts without innervation showed defects in keratinocyte migration. These defects were attributed to reduced levels of the neuropeptide Substance P in the wound (Blais et al. 2014), which is secreted by unmyelinated cutaneous nerve endings after tissue injury. Substance P promotes vasodilatation and local immune functions by binding to its receptor, tachykinin 1 (Holzer 1988; Antezana et al. 2002). The same mechanism seems to be responsible for impaired wound healing in human patients with diabetic peripheral neuropathy, where reduced Substance P levels have been found. Diabetic wounds also show increased expression levels of neutral endopeptidase, a Substance P–degrading enzyme, indicating that additional feedback mechanisms between nerve endings and skin exist to regulate Substance P levels (Antezana et al. 2002).

2.8 Future perspectives

Although the repair of large wounds is still a challenge and the regeneration of mammalian limbs is not yet realistic, it will be important to identify the molecular and physiological barriers that hamper these processes. This understanding will be key to natural repair and regeneration, which should be the goal of future tissue-engineering approaches. The mammalian response to tissue injury is characterized by healing that is accompanied by scar formation. Traumatic injuries with extensive tissue damage or loss heal slowly and bear the risk of infection. Moreover, missing tissues do not regenerate, leading to unwanted disfigurement of affected individuals. In contrast, species such as zebrafish, frog tadpoles, and salamanders have the capacity to rapidly heal their wounds without scar formation, and these animals regenerate missing portions of their body. Intriguingly, scar-free wound healing is also inherent to mammals during fetal development, and postnatal mice have been shown to naturally regenerate portions of their heart when dissected. Although it appears that most mammalian species lose this regenerative potential at some point during development, certain species have been identified that regenerate well even as adults. For instance, rabbits can regenerate large ear holes via blastema formation, and the African spiny mouse, *Acomys*, can heal large skin wounds that penetrate deep into the muscle. In addition, the Murphy Roths Large (MRL) mouse can regenerate ear holes and is considered a *super-healer* strain. These examples indicate that mammals possess the necessary molecular signatures to naturally heal and even regenerate portions of missing tissues. However, these signaling mechanisms lie dormant, thus raising the possibility that these processes may be reactivated on manipulation in humans. Carefully studying these mechanisms that lead to natural healing and regeneration will ultimately lead the way to the design of strategies with which to promote healing and regeneration in humans.

Current tissue-engineering approaches utilize cell culture techniques and skin grafts to replace damaged tissues. Although these approaches have led to substantial improvements in the wound healing field, they face many challenges that often lead to unwanted clinical outcomes and long-term complications. For instance, skin grafts are incompletely reinnervated by sensory nerve endings, and thus, patients experience numbness in the grafted area. Autologous skin grafts typically also damage the donor site, leading to significant pain that can last over prolonged time periods. Furthermore, healing of the wound edges is accompanied with infections, and hematomas can form beneath the graft. Designing strategies that promote natural skin wound healing based on the examples provided in this chapter may help improve or prevent these complications. Another avenue of the tissue-engineering field is to replace lost tissues by utilizing stem cells. In this approach, cadaver tissues are decellularized, and these scaffolds are re-populated with stem cells that replace the original organ. In addition, three-dimensional bioprinting of human organs has become a popular idea, and first strategies have been implemented with the goal to use these tissues and organs for transplantation (Arslan-Yildiz et al. 2016). Challenges of these approaches, as with any other transplantation method, are that foreign tissues are rejected and patients need to stay on immunosuppressant medications throughout the rest of their lives. To circumvent this, the patient's own cells are being induced into pluripotent stem cells. Despite the promising nature of these approaches, it will remain problematic to properly integrate engineered organs into the host to achieve full functionality. This chapter has summarized the current knowledge of the role of H_2O_2 and nerves in the paracrine interactions between various cell types under injury conditions and during appendage regeneration. Recognizing these interactions will be critical for the successful restoration and integration of tissues in the future.

References

Antezana, M., S. R. Sullivan, M. Usui, et al. 2002. Neutral endopeptidase activity is increased in the skin of subjects with diabetic ulcers. *J Invest Dermatol* 119 (6):1400–1404.

Apelqvist, J. 2012. Diagnostics and treatment of the diabetic foot. *Endocrine* 41 (3):384–397.

Arslan-Yildiz, A., R. E. Assal, P. Chen, S. Guven, F. Inci, and U. Demirci. 2016. Towards artificial tissue models: past, present, and future of 3D bioprinting. *Biofabrication* 8 (1):014103.

Bedard, K., and K. H. Krause. 2007. The NOX family of ROS-generating NADPH oxidases: Physiology and pathophysiology. *Physiol Rev* 87 (1):245–313.

Beneloujaephajri, E., A. Costa, F. L'Haridon, J. P. Métraux, and M. Binda. 2013. Production of reactive oxygen species and wound-induced resistance in Arabidopsis thaliana against Botrytis cinerea are preceded and depend on a burst of calcium. *BMC Plant Biol* 13: 160.

Blais, M., L. Mottier, M. A. Germain, S. Bellenfant, S. Cadau, and F. Berthod. 2014. Sensory neurons accelerate skin reepithelialization via substance P in an innervated tissue-engineered wound healing model. *Tissue Eng Part A* 20 (15–16):2180–2188.

Cho, Y., R. Sloutsky, K. M. Naegle, and V. Cavalli. 2013. Injury-induced HDAC5 nuclear export is essential for axon regeneration. *Cell* 155 (4):894–908.

Claiborne, A., J. I. Yeh, T. C. Mallett, et al. 1999. Protein-sulfenic acids: Diverse roles for an unlikely player in enzyme catalysis and redox regulation. *Biochemistry* 38 (47):15407–15416.

Donkó, A., Z. Péterfi, A. Sum, T. Leto, and M. Geiszt. 2005. Dual oxidases. *Philos Trans R Soc Lond B Biol Sci* 360 (1464):2301–2308.

Endo, T., S. V. Bryant, and D. M. Gardiner. 2004. A stepwise model system for limb regeneration. *Dev Biol* 270 (1):135–145.

Evans, I. R., F. S. Rodrigues, E. L. Armitage, and W. Wood. 2015. Draper/CED-1 mediates an ancient damage response to control inflammatory blood cell migration in vivo. *Curr Biol* 25 (12): 1606–1612.

Gauron, C., C. Rampon, M. Bouzaffour, et al. 2013. Sustained production of ROS triggers compensatory proliferation and is required for regeneration to proceed. *Sci Rep* 3:2084.

Gough, D. R., and T. G. Cotter. 2011. Hydrogen peroxide: A Jekyll and Hyde signalling molecule. *Cell Death Dis* 2:e213.

Hara-Chikuma, M., S. Chikuma, Y. Sugiyama, et al. 2012. Chemokine-dependent T cell migration requires aquaporin-3-mediated hydrogen peroxide uptake. *J Exp Med* 209 (10):1743–1752.

Harsum, S., J. D. Clarke, and P. Martin. 2001. A reciprocal relationship between cutaneous nerves and repairing skin wounds in the developing chick embryo. *Dev Biol* 238 (1):27–39.

Holzer, P. 1988. Local effector functions of capsaicin-sensitive sensory nerve endings: Involvement of tachykinins, calcitonin gene-related peptide and other neuropeptides. *Neuroscience* 24 (3):739–768.

Iovine, M. K. 2007. Conserved mechanisms regulate outgrowth in zebrafish fins. *Nat Chem Biol* 3 (10):613–618.

Kumar, A., J. W. Godwin, P. B. Gates, A. A. Garza-Garcia, and J. P. Brockes. 2007. Molecular basis for the nerve dependence of limb regeneration in an adult vertebrate. *Science* 318 (5851): 772–777.

Leonard, S. E., and K. S. Carroll. 2011. Chemical 'omics' approaches for understanding protein cysteine oxidation in biology. *Curr Opin Chem Biol* 15 (1):88–102.

Leonard, S. E., K. G. Reddie, and K. S. Carroll. 2009. Mining the thiol proteome for sulfenic acid modifications reveals new targets for oxidation in cells. *ACS Chem Biol* 4 (9):783–799.

Lisse, T. S., and S. Rieger. 2017. IKKα regulates human keratinocyte migration through surveillance of the redox environment. *J Cell Sci* 130 (5): 975–988.

Lisse, T. S., L. J. Middleton, A. D. Pellegrini, et al. 2016. Paclitaxel-induced epithelial damage and ectopic MMP-13 expression promotes neurotoxicity in zebrafish. *Proc Natl Acad Sci U S A* 113:E2189–E2198.

Loo, A. E., R. Ho, and B. Halliwell. 2011. Mechanism of hydrogen peroxide-induced keratinocyte migration in a scratch-wound model. *Free Radic Biol Med* 51 (4):884–892.

Loo, A. E., Y. T. Wong, R. Ho, et al. 2012. Effects of hydrogen peroxide on wound healing in mice in relation to oxidative damage. *PLoS One* 7 (11):e49215.

Love, N. R., Y. Chen, S. Ishibashi, et al. 2013. Amputation-induced reactive oxygen species are required for successful Xenopus tadpole tail regeneration. *Nat Cell Biol* 15 (2):222–228.

Mcheik, J. N., C. Barrault, G. Levard, F. Morel, F. X. Bernard, and J. C. Lecron. 2014. Epidermal healing in burns: Autologous keratinocyte transplantation as a standard procedure: Update and perspective. *Plast Reconstr Surg Glob Open* 2 (9):e218.

Meda, F., C. Gauron, C. Rampon, J. Teillon, M. Volovitch, and S. Vriz. 2016. Nerves control redox levels in mature tissues through Schwann cells and hedgehog signaling. *Antioxid Redox Signal* 24 (6):299–311.

Miller, E. W., B. C. Dickinson, and C. J. Chang. 2010. Aquaporin-3 mediates hydrogen peroxide uptake to regulate downstream intracellular signaling. *Proc Natl Acad Sci U S A* 107 (36):15681–15686.

Moreira, S., B. Stramer, I. Evans, W. Wood, and P. Martin. 2010. Prioritization of competing damage and developmental signals by migrating macrophages in the Drosophila embryo. *Curr Biol* 20 (5):464–470.

Morinaka, A., M. Yamada, R. Itofusa, et al. 2011. Thioredoxin mediates oxidation-dependent phosphorylation of CRMP2 and growth cone collapse. *Sci Signal* 4 (170):ra26.

Munnamalai, V., C. J. Weaver, C. E. Weisheit, et al. 2014. Bidirectional interactions between NOX2-type NADPH oxidase and the F-actin cytoskeleton in neuronal growth cones. *J Neurochem* 130 (4):526–540.

Nam, H. J., Y. Y., Park, G., Yoon, H., Cho, and J. H. Lee. 2010. Co-treatment with hepatocyte growth factor and TGF-beta1 enhances migration of HaCaT cells through NADPH oxidase-dependent ROS generation. *Exp Mol Med* 42 (4): 270–279.

Navarro, X., E. Verdu, G. Wendelschafer-Crabb, and W. R. Kennedy. 1997. Immunohistochemical study of skin reinnervation by regenerative axons. *J Comp Neurol* 380 (2):164–174.

Niethammer, P., C. Grabher, A. T. Look, and T. J. Mitchison. 2009. A tissue-scale gradient of hydrogen peroxide mediates rapid wound detection in zebrafish. *Nature* 459 (7249):996–999.

Orozco-Cárdenas, M. L., J. Narváez-Vásquez, and C. A. Ryan. 2001. Hydrogen peroxide acts as a second messenger for the induction of defense genes in tomato plants in response to wounding, systemin, and methyl jasmonate. *Plant Cell* 13 (1):179–191.

37

Peshavariya, H., G. J. Dusting, F. Jiang, et al. 2009. NADPH oxidase isoform selective regulation of endothelial cell proliferation and survival. *Naunyn Schmiedebergs Arch Pharmacol* 380 (2):193–204.

Pirotte, N., A. S. Stevens, S. Fraguas, et al. 2015. Reactive oxygen species in planarian regeneration: An upstream necessity for correct patterning and brain formation. *Oxid Med Cell Longev* 2015:392476.

Ponugoti, B., F. Xu, C. Zhang, C. Tian, S. Pacios, and D. T. Graves. 2013. FOXO1 promotes wound healing through the up-regulation of TGF-β1 and prevention of oxidative stress. *J Cell Biol* 203 (2):327–343.

Razzell, W., I. R. Evans, P. Martin, and W. Wood. 2013. Calcium flashes orchestrate the wound inflammatory response through DUOX activation and hydrogen peroxide release. *Curr Biol* 23 (5):424–429.

Reynolds, M. L., and M. Fitzgerald. 1995. Long-term sensory hyperinnervation following neonatal skin wounds. *J Comp Neurol* 358 (4):487–498.

Rieger, S., and A. Sagasti. 2011. Hydrogen peroxide promotes injury-induced peripheral sensory axon regeneration in the zebrafish skin. *PLoS Biol* 9 (5):e1000621.

Rishal, I., and M. Fainzilber. 2014. Axon-soma communication in neuronal injury. *Nat Rev Neurosci* 15 (1):32–42.

Sanchez, R., M. Riddle, J. Woo, and J. Momand. 2008. Prediction of reversibly oxidized protein cysteine thiols using protein structure properties. *Protein Sci* 17 (3):473–481.

Smalls, L. K., R. Randall Wickett, and M. O. Visscher. 2006. Effect of dermal thickness, tissue composition, and body site on skin biomechanical properties. *Skin Res Technol* 12 (1):43–49.

Smith, P. G., and M. Liu. 2002. Impaired cutaneous wound healing after sensory denervation in developing rats: Effects on cell proliferation and apoptosis. *Cell Tissue Res* 307 (3):281–291.

Sobotta, M. C., W. Liou, S. Stöcker, et al. 2015. Peroxiredoxin-2 and STAT3 form a redox relay for H_2O_2 signaling. *Nat Chem Biol* 11 (1):64–70.

Stone, J. R., and S. Yang. 2006. Hydrogen peroxide: A signaling messenger. *Antioxid Redox Signal* 8 (3–4):243–270.

Tauzin, S., T. W. Starnes, F. B. Becker, P. Y. Lam, and A. Huttenlocher. 2014. Redox and Src family kinase signaling control leukocyte wound attraction and neutrophil reverse migration. *J Cell Biol* 207 (5): 589–598.

Truong, T., and K. Carroll. 2012. Bioorthogonal chemical reporters for analyzing protein sulfenylation in cells. *Curr Protoc Chem Biol* 4:101–122.

Underhill, D. M., and H. S. Goodridge. 2007. The many faces of ITAMs. *Trends Immunol* 28 (2):66–73.

Wong, S. Y., and J. F. Reiter. 2011. Wounding mobilizes hair follicle stem cells to form tumors. *Proc Natl Acad Sci U S A* 108 (10):4093–4098.

Xu, S., and A. D. Chisholm. 2014. C. elegans epidermal wounding induces a mitochondrial ROS burst that promotes wound repair. *Dev Cell* 31 (1):48–60.

Yokoyama, H. 2008. Initiation of limb regeneration: The critical steps for regenerative capacity. *Dev Growth Differ* 50 (1):13–22.

Yoo, S. K., C. M. Freisinger, D. C. LeBert, and A. Huttenlocher. 2012. Early redox, Src family kinase, and calcium signaling integrate wound responses and tissue regeneration in zebrafish. *J Cell Biol* 199 (2):225–34.

Yoo, S. K., T. W. Starnes, Q. Deng, and A. Huttenlocher. 2011. Lyn is a redox sensor that mediates leukocyte wound attraction in vivo. *Nature* 480 (7375):109–112.

Young, C.N., J. I. Koepke, L. J. Terlecky, et al. 2008. Reactive oxygen species in tumor necrosis factor-alpha-activated primary human keratinocytes: Implications for psoriasis and inflammatory skin disease. *J Invest Dermatol* 128 (11):2606–2614.

Chapter 3 Controlling both the constructive power and the destructive power of inflammation to promote repair and regeneration

James Godwin

Contents

> **Key concepts**
>
> a. Inflammation is a robust response associated with the disruption of tissue homeostasis.
> b. Inflammation has been shown to be necessary for many contexts of repair and regeneration.
> c. Prolonged inflammation results in tissue destruction, fibrosis, and potentially tumorigenesis.
> d. The inflammatory network and associated immune response are a complex and coordinated sequence, dependent on both tissue context and environmental factors.
> e. Mechanical, chemical, oxidative, and metabolic stress can lead to inflammatory disease and tissue damage.
> f. Regenerative engineering must incorporate principles of biocompatibility and safety *in vivo*.
> g. Regulating the timing and profile of inflammatory responses associated with a regenerative therapy will be critical to achieving successful repair outcomes.

3.1 Introduction

Regenerative medicine aims to provide therapeutic strategies that restore normal tissue function after injury or destructive disease processes. At present, frontline approaches include the integration of biocompatible medical devices, implantation of functionalized scaffolds that allow cell attachment and repair, and delivering interventions that assist local cell populations to achieve functional repair. Each of these approaches will elicit acute or chronic inflammatory responses that will shape the success of the intervention. Regenerative engineering will require precise knowledge of how the host tissue and immune system will tolerate these interventions. An increasing body of knowledge has shed new light onto both the positive and negative roles of inflammation in different repair contexts. This chapter will discuss the implications for controlling the profile and timing of inflammatory signaling to drive a successful repair program.

3.2 Classical inflammation

The term *inflammation* broadly describes the complex microenvironment resulting from injury or infection that occurs due to the secretion of extracellular signaling molecules (typically cytokines, chemokines, or lipids) from both immune and non-immune cells. These signaling molecules shape the molecular response and phenotypic changes of cells, both local to and invading the injury/infection site.

At the level of the organism, inflammation is associated with redness, swelling, and pain. At the tissue level, these signals serve as a beacon for immune cell recruitment and repair. The phases of classical inflammation are well defined and are reviewed elsewhere (Gurtner et al. 2008; Reinke and Sorg 2012; Eming et al. 2014). Although inflammation is an important process for host defense, the duration of the acute response can recruit cells with priorities aimed at eliminating potential infection rather than preserving healthy tissue and functional repair. A new paradigm is emerging that subpopulations of immune cells have both positive and negative roles for repair, and the balance is both tissue-dependent and subject to environmental conditions, landscape of microorganisms, and severity of injury (Brancato and Albina 2011; Liddiard et al. 2011; Seledtsov and Seledtsova 2012; Wynn et al. 2013). These signals can cascade into events, both positive and negative for tissue homeostasis, stem cell differentiation, and regeneration.

The exact nature of the immune response can be determined by a number of factors. First, the type of injury or infection will dramatically shape the response of the cells that encounter the problem as part of their normal surveillance. A sterile injury will have a different response to an infection or injury and infection combined. However, whatever the insult, secreted signals are released locally, and these enter the systemic circulation to recruit *early responder cells* from both local and peripheral reservoirs. These *early responders* take their cues from what they have encountered and determine the amplification magnitude of the inflammatory signal. The next group of immune cells that are recruited to the wound/infection site (secondary responders) then determines the length and type of the inflammatory response either by secreting opposing signaling profiles or by promoting reverse migration of early responder immune cells. Secondary responder cells and their interaction with the damaged/infected tissue mainly determine the resolution of inflammatory signaling and are the major focus of many research efforts aimed at promoting regeneration (Brown et al. 2014; Forbes and Rosenthal 2014).

In the context of injury/surgery, the resolution of inflammation and return to homeostasis are critical for efficient repair (Eming et al. 2014). Inflammation must be controlled, as persistent (chronic) inflammation leads to a range of neurodegenerative and cardiovascular diseases, auto-immunity, and cancer, along with various metabolic disorders (Medzhitov 2008; Manabe 2011; De Lerma Barbaro et al. 2014; Matsushima 2014; Crupi et al. 2015). It is for these reasons that the deployment of implantable devices or treatments in regenerative medicine must be performed with caution. For the safety of the patient, there must be careful evaluation of the potential for the intervention to elicit low-level persistent inflammation.

3.3 Immune cells involved in sensing, amplifying, or resolving inflammatory signaling

The immune system is broadly divided into two complementary arms. The innate arm of the immune system represents a more evolutionary ancient group of cells with germline-encoded receptors. These cells are present in both invertebrates and vertebrates, maintaining tissue integrity and providing some protection from microorganisms. The adaptive arm of the immune system developed more recently in evolution and provided the additional functions of *immunological memory* with non-germline–encoded receptors. Adaptive immune cells produce antigen-specific responses that improve pathogen clearance rates with repeat exposure. The adult adaptive immune system in humans is particularly sophisticated and has been thought to pose restrictions on the ability of humans to reactivate regenerative cells that could carry more *embryonic* physical signatures that are detected as foreign. Adaptive immunity is strongly correlated with poor regenerative outcomes and is reviewed elsewhere (Mescher and Neff 2006; Godwin and Rosenthal 2014; Epelman et al. 2015).

Both adaptive and innate immune cells have evolved to provide complex crosstalk with each other and host tissues, creating complex dynamic relationships with multiple levels of redundancy. Despite the enormous complexity in immune cell interactions, these cells represent the most intensively studied group of cells in the body. Detailed profiling with surface markers has allowed the mapping of each branch and sub-branch of human/mouse immune cells with high fidelity (Chattopadhyay and Roederer 2015). This has allowed researchers to interrogate the functions of particular populations or even use specific antibodies to inactivate or promote cellular depletion (Proserpio and Mahata 2015). This has been complemented by transgenic labeling approaches that are targeted to specific immune cells, opening up the potential to deplete (loss of function) or alter the function (loss or gain of function) of particular immune cell players in contexts of repair or disease.

The whole immune system of a mouse can be destroyed by whole body irradiation and then replaced with a single healthy human donor hematopoietic stem cell, which will repopulate the host with a completely new donor-derived immune system (Notta et al. 2011). These tools provide opportunities to dissect the cellular processes important for repair and regeneration, allowing potential to reverse engineer aspects of regeneration. Recent gene-editing tools such as CRISPR/Cas9 have also removed much of the cost and difficulties of such approaches (Sander and Joung 2014).

The innate immune cells are a diverse group of cells that include granulocytes (neutrophils, eosinophils, basophils, and mast cells), monocytes, macrophages, and dendritic cells (DC), along with natural killer (NK) cells (Rosenbauer and Tenen 2007). Macrophages and DCs are the most important cells in maintaining tissue homeostasis, as they are equipped with the ability to both phagocytose debris and sample their environment through the uptake of tissue fluids (macropinocytosis) (Pinto et al. 2014). The innate immune system is an ancient mechanism of maintaining tissue integrity by recognizing a range of molecules that are not normally found in healthy tissue. These molecules are referred to as damage-associated molecular patterns (DAMPs) and pathogen-associated molecular patterns (PAMPs) (Tang et al. 2012). Exposure of resident innate immune cells to these molecules will trigger inflammatory responses that will recruit the appropriate help from cooperating immune cells to mount an antimicrobial or wound healing program.

Neutrophils are rapidly recruited by wound signals and are equipped with chemical weapons that can destroy microorganisms. However, their lack of specificity can result in extensive local tissue damage (Mantovani et al. 2011; Epstein and Weiss 1989). The persistence of pro-inflammatory neutrophils in the wound is regulated by the invading monocyte/macrophages, which actively signal neutrophil exit from the wound site. Macrophages resolve inflammation in several ways: by removing pro-inflammatory neutrophils, by secreting their own anti-inflammatory molecules, and by actively modulating local tissue inflammatory responses via fibroblasts or adaptive immune cells (Chazaud 2014) (Figure 3.1). These cells have been shown to facilitate successful repair and regeneration in a range of vertebrate and invertebrate systems (Stefater et al. 2011; Godwin et al. 2013; Aurora et al. 2014; Petrie et al. 2015).

3.4 Inflammation as a positive signal and an early requirement for tissue repair and regeneration

In examples of natural regeneration from various animal models and range of contexts, inflammatory signals are implicated as essential positive signals for regeneration. Several experiments addressing the role of inflammation in amphibian regeneration imply that inflammatory signaling is critical for regeneration. Other studies modulating inflammatory signaling by various agents such as glucocorticoids, non-steroidal anti-inflammatory agents, and immunosuppressant molecules have shown to improve regeneration success in both frog and salamander limb regeneration models (reviewed in Godwin and Rosenthal 2014). All these experiments indicate that an inflammatory response is required in amphibian regeneration, at least during the initial phase of induction.

In the zebrafish model of brain regeneration, immunosuppression with dexamethasone blocks the proliferation and activation of neurogenic progenitors and inhibits functional repair. Induction of acute inflammation by using the leukotriene-C4 (lipid mediator) rescues regeneration in this model and can induce a regeneration-specific molecular program (Kyritsis et al. 2012).

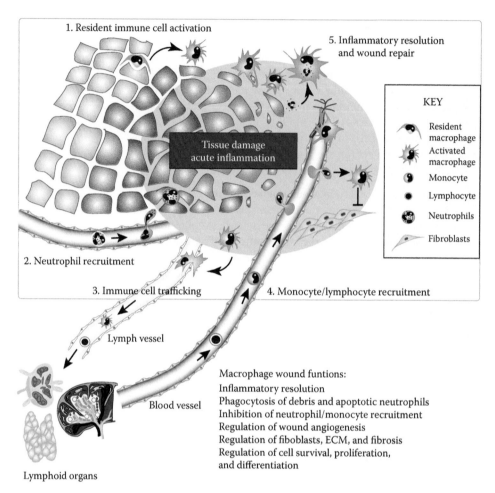

Figure 3.1 A simplified sequence of tissue damage–induced inflammation and the events leading to inflammatory resolution. (1) Tissue damage or device implantation will provide physical cues leading to the activation of tissue-resident immune cells (macrophages, Langerhans, dendritic, and T cells) along with endothelial cells and fibroblasts. (2) These cells provide extracellular signaling molecules (cytokines, lipids, and chemical signals) that recruit inflammatory polymorphonuclear (PMN) cells (neutrophils and granulocytes) to the wound. Clotting of blood also traps platelets and other circulating blood cells in the wound, amplifying the inflammatory response. (3) Immune cells that have encountered foreign or damage-associated antigens within the wound will drain from wound exudate via the lymphatic vessels to lymphoid organs (lymph nodes, thymus, Peyer's patches, or spleen) to present antigens and mount adaptive immune responses or monocyte recruitment. (4) Bone marrow–derived monocytes are recruited from the spleen and differentiate into macrophages that restore homeostasis to the wound in a number of ways. (5) Diverse macrophage populations coordinate inflammatory resolution and downstream repair. Macrophages facilitate the first step in resolving inflammation by engulfing apoptotic neutrophils, forcing the reverse migration of PMN cells from the wound and blocking further neutrophil recruitment. Macrophages can facilitate repair by regulating inflammation, angiogenesis, and the extracellular matrix, through both enzymatic release or direct actions on fibroblast phenotype. Macrophages may also regulate proliferation/differentiation of various progenitor cells in different tissues.

In mammals, inflammatory cytokines tumor necrosis factor alpha (TNFα) and interleukin-6 (IL6) and complement are required for liver regeneration (Webber 1994; Cressman et al. 1996; Mastellos et al. 2001). In models of optic nerve crush injury, inflammation enhances axonal regeneration (Lu and Richardson 1991; Yin et al. 2009). In an elegant study of hair regeneration, inflammatory signals from one damaged follicle can diffuse and promote regeneration in a neighboring follicle (Chen et al. 2015). Although most mice strains do not heal ear-hole biopsies, some strains exhibit advanced healing and closure closer to perfect regeneration. Mapping studies of quantitative trait loci (QTL) have identified genetic inflammatory modifiers controlling the improved regeneration. These studies indicate that acute inflammation loci clearly influence tissue regeneration in a positive manner (Canhamero et al. 2013).

Both pro-inflammatory and anti-inflammatory macrophages have been identified as critical regulators important for wound healing and regeneration in a range of contexts in various animal models. Extensive literature now exists on how macrophages shape the inflammatory response, and a detailed discussion is outside the scope of this chapter and is reviewed elsewhere (Liddiard et al. 2011; Novak and Koh 2013; Wynn et al. 2013; Forbes and Rosenthal 2014; Pinto et al. 2014; Braga et al. 2015).

Reactive oxygen species (ROS) molecules are produced by cells involved in host defense and can act as both signaling molecules and inflammatory signals important for regeneration (Love et al. 2013; Han et al. 2014; van der Vliet and Janssen-Heininger 2014). The previous chapter has discussed ROS signaling in regeneration. However, it should be noted that early signals such as the leaking of current on wounding and generation of ROS, reactive nitrogen species (NOS), and hydrogen peroxide all intersect and may even precede the induction of inflammatory signaling (Niethammer et al. 2009; Yoo and Huttenlocher 2009). Feedback loops between inflammation and the generation of these chemical mediators have also been established (Blaser et al. 2016). Immune cells regulate ROS production and cell survival. This may represent an evolutionary conserved mechanism to sense and activate repair and development.

Inflammation is now appreciated to have both positive and negative, context-dependent effects on stem cell niches and is reviewed elsewhere (Schuettpelz and Link 2013; De Lerma Barbaro et al. 2014; Forbes and Rosenthal 2014; Hsu et al. 2014; Kizil et al. 2015). In examples of progenitor cell transplantation, efficiency of both nuclear reprogramming and survival has been observed to be dependent on inflammatory signals (Park et al. 2009; Mathieu et al. 2010). In fact, the activation of inflammatory pathways is required for efficient nuclear reprogramming pluripotency induction (Lee et al. 2012a). In some contexts, inflammation has been shown to be important for homing, proliferation, and survival

(Wolf et al. 2009). In other experiments, mesenchymal stem cell (MSC) differentiation is dependent on inflammatory signals (IL-1β) from macrophages (Lee et al. 2012b). Mobilization of MSCs is also dependent on inflammatory signals such as interferon gamma (Koning et al. 2013), and the anti-fibrotic effects of MSC delivery have been shown to be dependent on anti-inflammatory or immunosuppressive activity (reviewed in Low et al. 2015). Taken together, precise activation of inflammatory signaling appears to be essential for efficient repair and regeneration.

3.5 The failure of injury resolution is a major roadblock to regeneration

Although aspects of inflammation can be instructive and promote tissue repair, exaggerated or chronic inflammatory signaling can corrupt the repair process. In some contexts, inflammatory signals can drive the destruction of healthy tissue and promote pro-fibrotic events or even malignant transformation, leading to cancer (Mantovani et al. 2008; Candido and Hagemann 2012; Wynn and Ramalingam 2012; Wallach, Kang, and Kovalenko 2013). The resolution of inflammation is thus an essential step in minimizing these unwanted clinical outcomes (Figure 3.2).

Most contexts of mammalian tissue repair are dependent on mobilization of progenitor cells or the activation of proliferation in existing cells within the injured tissue. Overstimulation with secreted inflammatory mediators such as TNFα can play a destructive role in liver, muscle, and nervous system repair (Mohammed et al. 2004; Alexander and Popovich 2009; Ruffell et al. 2009). Poorly controlled inflammation can also have a devastating effect on progenitor cell viability and cellular phenotype. Excess inflammation in the pancreas induces potent apoptotic cellular death of vital insulin-producing beta cells (Eizirik et al. 2009) and negatively regulates hair follicle stem cell activity (Doles et al. 2012). Chronic inflammation within the intestinal crypt leads to cellular transformation of stem cells, leading to tumor development (De Lerma Barbaro et al. 2014; Shalapour and Karin 2015). In other tissues, chronic inflammation promotes a range of metabolic disorders and pathologies (Manabe 2011; Medzhitov 2008).

In both salamander and frog models of limb and tail regeneration, experiments inducing chronic inflammation with the light metal beryllium indicate that inflammation must be resolved for regeneration to proceed (Thornton 1951; King et al. 2012; Godwin and Rosenthal 2014). These experiments reinforce the importance of understanding both the instructive and destructive influences of inflammation during different phases of the regeneration process.

Ultimately, the success of wound healing is also determined by the balance between pro-regenerative extracellular matrix (ECM) formation and the

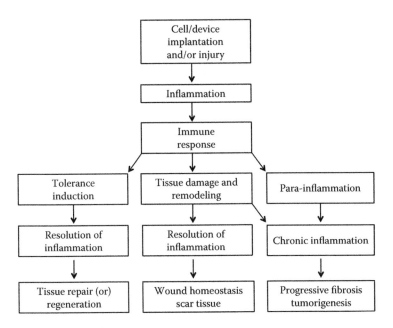

Figure 3.2 Inflammation plays a critical part in determining the outcome of regenerative therapies. Damaged or infected tissue will activate acute inflammatory reactions. After the acute phase, stressed or malfunctioning tissue can develop *para-inflammation* that can progress to chronic states of inflammation. The induction of inflammatory resolution pathways and tolerance are critical to the success of regeneration. A failure to activate these pathways can result in either poor healing outcomes and even progressive fibrosis or cancer.

formation of fibrotic scar tissue (Godwin et al. 2014). Fibroblasts are the critical cell type that shape the ECM environment during wounding and homeostasis and are highly responsive to inflammatory signaling. The profile of the inflammatory response is a major factor that influences the phenotype of fibroblasts and their role in fibrotic disease. The balance between the pro-fibrotic signals and the anti-fibrotic signals is, in turn, primarily shaped by the interactions between macrophages and T cells. If the microenvironment of the fibroblasts is pro-fibrotic, then these cells can differentiate into pro-fibrotic cells such as myofibroblasts, which secrete and participate in crosslinking ECM components into scar tissue. This fibrotic scar tissue is a major impediment to the migration and function of regeneration-competent cells and thus will thwart the regenerative process. The control of fibroblast function in wound repair, either targeting the fibroblast directly or by macrophage regulation and inflammation, has gained significant appreciation as an important point of intervention (Wynn 2008; Hinz et al. 2012; Wynn and Ramalingam 2012; Braga et al. 2015; Darby et al. 2015).

3.6 Practical engineering incorporating biocompatibility, efficacy, and safety *in vivo*

Current approaches in tissue engineering of organ or tissue regeneration can be expected to include a structural and molecular scaffold with or without seeded cells useful in supporting repair and regeneration. Depending on the context and design of the tissue-engineering approach, implanted biomaterials usually lead to inflammatory responses and fibrosis that can lead to the failure of the intervention (reviewed in Godwin et al. 2014; Crupi et al. 2015).

The use of systemic immunosuppression is no longer considered a viable treatment strategy for enhancing repair due to the positive role that inflammatory pathways and immune cells play in many regeneration contexts (Ben-Mordechai et al. 2013; Pinto et al. 2014). A new way of thinking has emerged that aims to shape the immune system to tolerate tissue-engineering interventions by exploiting positive aspects of inflammation and immune signaling (Boehler et al. 2011; Brown et al. 2014). One such strategy would be to use biomaterials that are both low in adjuvant activity and modified (functionalized) to promote the right type of inflammation to be favorable to the desired repair outcome. Functionalized bioactive materials can be used not only to influence the differentiation and engraftment of regenerative cells but also to provide local cues to dampen immune reactions that restrict regeneration or promote active pro-resolution pathways (Serhan et al. 2008; Franz et al. 2011).

Biomaterials suitable for regenerative medicine must also meet the structural requirements of the tissue to be repaired. In the case of implantable devices such as the bionic eye or cochlear implant, the electrode materials used must be optimised for both technical properties such as pixel size and conductance, but try to limit the expected fibrotic response that will eventually destroy function. New emerging biomaterials such as conductive plastics have shown promise in solving aspects of this problem, but the long-term suitability of these materials is yet to be determined (Balint et al. 2014).

Although cellular therapy is an attractive way of delivering a rare or important cell type to an injury or implanted device, immune-mediated clearance of transplanted cells or inflammatory stimulus deleterious to regeneration is a major practical consideration. Immune modulatory strategies to induce tolerance of transplanted cells have been in development for many years, and considerable progress has been made (Hlavaty et al. 2015). The use of iPS-based reprogramming of patient-derived cells for transplantation could overcome many of the immune rejection issues, and it is conceivable that protocols for manufacture of rare cells useful in regeneration and engraftment could be developed. *Off-the-shelf* cellular therapies have been developed for immune-privileged sites, such as the eye, using tissue-matched donors and have been approved for human use (Gamm et al. 2014; Chakradhar 2016). The use of non-patient–derived cells in other contexts still presents a significant immunological challenge.

3.7 Engineering regeneration through shaping the timing and profile of the inflammatory microenvironment

The role of inflammation is highly context-dependent. One major consideration when considering engineering strategies is the potential for *collateral damage*. Systemic modulation is not ideal, as care must be taken not to disturb other niches that maybe affected negatively by the treatment. Spatiotemporal dynamics in different compartments and niches is a major area yet to be fully explored. The harnessing of endogenous stem cells and progenitor cells is an emerging focus. Strategies for interrogating the niche *in vivo* still present several practical problems, where tools need to be developed.

Another challenge in regenerative engineering is to provide processes to dissect the individual steps at which particular inflammatory responses should be shaped and for how long. Optimal healing and regeneration will require changes in the signaling environment over time to facilitate different parts of the regeneration process. How to adjust the intervention over time and how to acquire this knowledge in a systematic way are important considerations going forward. Stem cell biologists have wrestled with ways to identify the combination of factors required to maintain pluripotency or to push cells toward particular cell fates. Identifying the timing and combination of factors required to produce many important cell types is still a major challenge. However, multidisciplinary approaches combining microfluidics engineered devices with live imaging and genetic-engineered cells have recently simplified the identification of cell fate *recipes in vitro* (reviewed in Titmarsh et al. 2014).

Acquiring specific knowledge on each cellular player participating in regeneration across various tissue contexts and animal models is a long and daunting task. However, studies in animals with extraordinary regenerative capabilities such as the salamander and fish have already yielded important insights. This knowledge has shaped our understanding of evolutionary conserved pathways relevant to regeneration in humans. This understanding will define the critical events and shape future attempts in designing therapeutic interventions.

3.8 Conclusions

Regenerative medicine is a broad term that describes a range of approaches to improve the repair and function in a range of human tissues. All of these approaches must consider immune-related signaling in order to provide the best outcome for the patient. There is an urgent need to understand how to shape the profile of the inflammatory milieu to minimize any negative effects and maximize pro-regenerative influences. Engineering principles could be harnessed to provide local-timed release of immunomodulatory agents and functionalized biomaterials capable of directing the immune response in the support of regeneration. Reverse engineering natural regeneration in order to create therapeutic treatments is an attractive goal in modern medicine. Although significant challenges still exist in interpreting the differential requirements of each phase of scar-free healing and regeneration, these could be addressed through convergent science. Coupling the disciplines of basic developmental biology, immunology, biomaterials, and engineering is likely to deepen our understanding and accelerate the path to translation.

References

Alexander, J. K., and P. G. Popovich. 2009. Neuroinflammation in spinal cord injury: Therapeutic targets for neuroprotection and regeneration. *Progress in Brain Research* 175:125–137. doi:10.1016/S0079-6123(09)17508-8.

Aurora, A. B., E. R. Porrello, W. Tan, A. I. Mahmoud, J. A. Hill, R. Bassel-Duby, H. A. Sadek, and E. N. Olson. 2014. Macrophages are required for neonatal heart regeneration. *Journal of Clinical Investigation* 124 (3): 1382–1392. doi:10.1172/JCI72181.

Balint, R., N. J. Cassidy, and S. H. Cartmell. 2014. Conductive polymers: Towards a smart biomaterial for tissue engineering. *Acta Biomaterialia* 10 (6): 2341–2353. doi:10.1016/j.actbio.2014.02.015.

Ben-Mordechai, T., R. Holbova, N. Landa-Rouben, T. Harel-Adar, M. S. Feinberg, I. A. Elrahman, G. Blum, et al. 2013. Macrophage subpopulations are essential for infarct repair with and without stem cell therapy. *Journal of the American College of Cardiology* 62 (20): 1890-1901. doi:10.1016/j.jacc.2013.07.057.

Blaser, H., C. Dostert, T. W. Mak, and D. Brenner. 2016. TNF and ROS crosstalk in inflammation. *Trends in Cell Biology* 26 (4): 249–261. doi:10.1016/j.tcb.2015.12.002.

Boehler, R. M., J. G. Graham, and L. D. Shea. 2011. Tissue engineering tools for modulation of the immune response. *BioTechniques* 51 (4): 239–240, 242, 244 passim. doi:10.2144/000113754.

Braga, T. T., J. S. H. Agudelo, and N. O. S. Camara. 2015. Macrophages during the fibrotic process: M2 as friend and foe. *Frontiers in Immunology* 6 (12): 953. doi:10.3389/fimmu.2015.00602.

Brancato, S. K., and J. E. Albina. 2011. Wound macrophages as key regulators of repair. *The American Journal of Pathology* 178 (1): 19–25. doi:10.1016/j.ajpath.2010.08.003.

Brown, B. N., B. M. Sicari, and S. F. Badylak. 2014. Rethinking regenerative medicine: A macrophage-centered approach. *Frontiers in Immunology* 5 (8): 349. doi:10.3389/fimmu.2014.00510.

Candido, J., and T. Hagemann. 2012. Cancer-related inflammation. *Journal of Clinical Immunology* 33 (S1): 79–84. doi:10.1007/s10875-012-9847-0.

Canhamero, T., M. A. Corrêa, and J. G. Fernandes. 2013. Frontiers | Acute inflammation loci influence tissue repair in mice. In *15th International Congress of Immunology (ICI)*, Milan, Italy, 22–27 August. doi:10.3389/conf.fimmu.2013.02.00535/event_abstract.

Chakradhar, S. 2016. An eye to the future: Researchers debate best path for stem cell–derived therapies. *Nature Medicine* 22 (2): 116–119. doi:10.1038/nm0216-116.

Chattopadhyay, P. K., and M. Roederer. 2015. A mine is a terrible thing to waste: High content, single cell technologies for comprehensive immune analysis. *American Journal of Transplantation* 15 (5): 1155–1161. doi:10.1111/ajt.13193.

Chazaud, B. 2014. Macrophages: Supportive cells for tissue repair and regeneration. *Immunobiology* 219 (3): 172–178. doi:10.1016/j.imbio.2013.09.001.

Chen, C.-C., L. Wang, M. V. Plikus, T. X. Jiang, P. J. Murray, R. Ramos, C. F. Guerrero-Juarez, et al. 2015. Organ-level quorum sensing directs regeneration in hair stem cell populations. *Cell* 161 (2): 277–290. doi:10.1016/j.cell.2015.02.016.

Cressman, D. E., L. E. Greenbaum, R. A. DeAngelis, G. Ciliberto, E. E. Furth, V. Poli, and R. Taub. 1996. Liver failure and defective hepatocyte regeneration in interleukin-6-deficient mice. *Science (New York, N.Y.)* 274 (5291): 1379–1383. doi:10.1126/science.274.5291.1379.

Crupi, A., A. Costa, A. Tarnok, S. Melzer, and L. Teodori. 2015. Inflammation in tissue engineering: The Janus between engraftment and rejection. *European Journal of Immunology* 45 (12): 3222–3236. doi:10.1002/eji.201545818.

Darby, I. A., N. Zakuan, F. Billet, and A. Desmoulière. 2015. The myofibroblast, a key cell in normal and pathological tissue repair. *Cellular and Molecular Life Sciences* 73 (6): 1145–1157. doi:10.1007/s00018-015-2110-0.

De Lerma Barbaro, A., G. Perletti, I. Bonapace, and E. Monti. 2014. Inflammatory cues acting on the adult intestinal stem cells and the early onset of cancer (Review). *International Journal of Oncology.* doi:10.3892/ijo.2014.2490.

Doles, J., M. Storer, L. Cozzuto, G. Roma, and W. M. Keyes. 2012. Age-associated inflammation inhibits epidermal stem cell function. *Genes & Development* 26 (19): 2144–2153. doi:10.1101/gad.192294.112.

Eizirik, D. L., M. L. Colli, and F. Ortis. 2009. The role of inflammation in insulitis and B-cell loss in type 1 diabetes. *Nature Reviews Endocrinology* 5 (4): 219–226. doi:10.1038/nrendo.2009.21.

Eming, S. A., P. Martin, and M. Tomic-Canic. 2014. Wound repair and regeneration: Mechanisms, signaling, and translation. *Science Translational Medicine* 6 (265): 265sr6–265sr6. doi:10.1126/scitranslmed.3009337.

Epelman, S., P. P. Liu, and D. L. Mann. 2015. Role of innate and adaptive immune mechanisms in cardiac injury and repair. *Nature Reviews Immunology* 15 (2): 117–129. doi:10.1038/nri3800.

Epstein, F. H., and S. J. Weiss. 1989. Tissue destruction by neutrophils. *New England Journal of Medicine* 320 (6): 365–376. doi:10.1056/NEJM198902093200606.

Forbes, S. J., and N. Rosenthal. 2014. Preparing the ground for tissue regeneration: From mechanism to therapy. *Nature Medicine* 20 (8): 857–869. doi:10.1038/nm.3653.

Franz, S., S. Rammelt, D. Scharnweber, and J. C. Simon. 2011. Immune responses to implants - A review of the implications for the design of immunomodulatory biomaterials. *Biomaterials* 32 (28): 6692–6709. doi:10.1016/j.biomaterials.2011.05.078.

Gamm, D. M., M. Joseph Phillips, and R. Singh. 2014. Modeling retinal degenerative diseases with human iPS-derived cells: Current status and future implications. *Expert Review of Ophthalmology* 8 (3): 213–216. doi:10.1586/eop.13.14.

Godwin, J., D. Kuraitis, and N. Rosenthal. 2014. Extracellular matrix considerations for scar-free repair and regeneration: Insights from regenerative diversity among vertebrates. *The International Journal of Biochemistry & Cell Biology* 56 (November): 47–55. doi:10.1016/j.biocel.2014.10.011.

Godwin, J. W., A. R. Pinto, and N. A. Rosenthal. 2013. Macrophages are required for adult salamander limb regeneration. *Proceedings of the National Academy of Sciences* 110 (23): 9415–9420. doi:10.1073/pnas.1300290110.

Godwin, J. W., and N. Rosenthal. 2014. Scar-free wound healing and regeneration in amphibians: Immunological influences on regenerative success. *Differentiation; Research in Biological Diversity* 87 (1–2): 66–75. doi:10.1016/j.diff.2014.02.002.

Gurtner, G. C., S. Werner, Y. Barrandon, and M. T. Longaker. 2008. Wound repair and regeneration. *Nature* 453 (7193): 314–321. doi:10.1038/nature07039.

Han, P., X.-H. Zhou, N. Chang, C.-L. Xiao, S. Yan, H. Ren, X.-Z. Yang, et al. 2014. Hydrogen peroxide primes heart regeneration with a derepression mechanism. *Cell Research* 24 (9): 1091–1107. doi:10.1038/cr.2014.108.

Hinz, B., S. H. Phan, V. J. Thannickal, M. Prunotto, A. Desmoulière, J. Varga, O. De Wever, M. Mareel, and G. Gabbiani. 2012. Recent developments in myofibroblast biology. *The American Journal of Pathology* 180 (4): 1340–1355. doi:10.1016/j.ajpath.2012.02.004.

Hlavaty, K. A., X. Luo, L. D. Shea, and S. D. Miller. 2015. Cellular and molecular targeting for nanotherapeutics in transplantation tolerance. *Clinical Immunology* 160 (1): 14–23. doi:10.1016/j.clim.2015.03.013.

Hsu, Y.-C., L. Li, and E. Fuchs. 2014. Emerging interactions between skin stem cells and their niches. *Nature Medicine* 20 (8): 847–856. doi:10.1038/nm.3643.

King, M. W., A. W. Neff, and A. L. Mescher. 2012. The developing XenopusLimb as a model for studies on the balance between inflammation and regeneration. *The Anatomical Record* 295 (10): 1552–1561. doi:10.1002/ar.22443.

Kizil, C., N. Kyritsis, and M. Brand. 2015. Effects of inflammation on stem cells: Together they strive? *EMBO Reports* 16 (4): 416–426. doi:10.15252/embr.201439702.

Koning, J. J., G. Kooij, H. E. de Vries, M. A. Nolte, and R. E. Mebius. 2013. Mesenchymal stem cells are mobilized from the bone marrow during inflammation. *Frontiers in Immunology* 4. doi:10.3389/fimmu.2013.00049.

Kyritsis, N., C. Kizil, S. Zocher, V. Kroehne, J. Kaslin, D. Freudenreich, A. Iltzsche, and M. Brand. 2012. Acute inflammation initiates the regenerative response in the adult zebrafish brain. *Science (New York, N.Y.)* 338 (6112): 1353–1356. doi:10.1126/science.1228773.

Lee, J., N. Sayed, A. Hunter, K. F. Au, W. H. Wong, E. S. Mocarski, R. R. Pera, E. Yakubov, and J. P. Cooke. 2012a. Activation of innate immunity is required for efficient nuclear reprogramming. *Cell* 151 (3): 547–558. doi:10.1016/j.cell.2012.09.034.

Lee, M. J., M. Y. Kim, S. C. Heo, Y. W. Kwon, Y. M. Kim, E. K. Do, J. H. Park, J. S. Lee, J. Han, and J. H. Kim. 2012b. Macrophages regulate smooth muscle differentiation of mesenchymal stem cells via a prostaglandin F2-mediated paracrine mechanism. *Arteriosclerosis, Thrombosis, and Vascular Biology* 32 (11): 2733–2740. doi:10.1161/ATVBAHA.112.300230.

Liddiard, K., M. Rosas, L. C. Davies, S. A. Jones, and P. R. Taylor. 2011. Macrophage heterogeneity and acute inflammation. *European Journal of Immunology* 41 (9): 2503–2508. doi:10.1002/eji.201141743.

Love, N. R., Y. Chen, S. Ishibashi, P. Kritsiligkou, R. Lea, Y. Koh, J. L. Gallop, K. Dorey, and E. Amaya. 2013. Amputation-induced reactive oxygen species are required for successful Xenopus tadpole tail regeneration. *Nature Cell Biology* 15 (2): 222–228. doi:10.1038/ncb2659.

Low, J. H., P. Ramdas, and A. K. Radhakrishnan. 2015. Modulatory effects of mesenchymal stem cells on leucocytes and leukemic cells: A double-edged sword? *Blood Cells, Molecules, and Diseases* 55 (4): 351–357. doi:10.1016/j.bcmd.2015.07.017.

Lu, X., and P. M. Richardson. 1991. Inflammation near the nerve cell body enhances axonal regeneration. *Journal of Neuroscience* 11 (4): 972–978.

Manabe, I. 2011. Chronic inflammation links cardiovascular, metabolic and renal diseases. *Circulation Journal* 75 (12): 2739–2748. doi:10.1253/circj.CJ-11-1184.

Mantovani, A., P. Allavena, A. Sica, and F. Balkwill. 2008. Cancer-related inflammation. *Nature* 454 (7203): 436–444. doi:10.1038/nature07205.

Mantovani, A., M. A. Cassatella, C. Costantini, and S. Jaillon. 2011. Neutrophils in the activation and regulation of innate and adaptive immunity. *Nature Reviews Immunology* 11 (8): 519–531. doi:10.1038/nri3024.

Mastellos, D., J. C. Papadimitriou, S. Franchini, P.A. Tsonis, and J. D. Lambris. 2001. A novel role of complement: Mice deficient in the fifth component of complement (C5) exhibit impaired liver regeneration. *The Journal of Immunology* 166 (4): 2479–2486. doi:10.4049/jimmunol.166.4.2479.

Mathieu, P., D. Battista, A. Depino, V. Roca, M. Graciarena, and F. Pitossi. 2010. The more you have, the less you get: The functional role of inflammation on neuronal differentiation of endogenous and transplanted neural stem cells in the adult brain. *Journal of Neurochemistry* 112 (6): 1368–1385. doi:10.1111/j.1471-4159.2009.06548.x.

Matsushima, K. 2014. Cellular and molecular mechanisms of chronic inflammation-associated organ fibrosis, April, 1–6. doi:10.3389/fimmu.2012.00071/abstract.

Medzhitov, R. 2008. Origin and physiological roles of inflammation. *Nature* 454 (7203): 428–435. doi:10.1038/nature07201.

Mescher, A. L., and A. W. Neff. 2006. Limb regeneration in amphibians: Immunological considerations. *The Scientific World Journal* 6: 1–11. doi:10.1100/tsw.2006.323.

Mohammed, F. F., D. S. Smookler, S. E. M. Taylor, B. Fingleton, Z. Kassiri, O. H. Sanchez, J. L. English, et al. 2004. Abnormal TNF activity in Timp3–/– mice leads to chronic hepatic inflammation and failure of liver regeneration. *Nature Genetics* 36 (9): 969–977. doi:10.1038/ng1413.

Niethammer, P., C. Grabher, A. T. Look, and T. J. Mitchison. 2009. A tissue-scale gradient of hydrogen peroxide mediates rapid wound detection in zebrafish. *Nature* 459 (7249): 996–999. doi:10.1038/nature08119.

Notta, F., S. Doulatov, E. Laurenti, A. Poeppl, I. Jurisica, and J. E. Dick. 2011. Isolation of single human hematopoietic stem cells capable of long-term multilineage engraftment. *Science (New York, N.Y.)* 333 (6039): 218–221. doi:10.1126/science.1201219.

Novak, M. L., and T. J. Koh. 2013. Macrophage phenotypes during tissue repair. *Journal of Leukocyte Biology* 93 (6): 875–881. doi:10.1189/jlb.1012512.

Park, D.-H., D. J. Eve, J. Musso III, S. K. Klasko, E. Cruz, C. V. Borlongan, and P. R. Sanberg. 2009. Inflammation and stem cell migration to the injured brain in higher organisms. *Stem Cells and Development* 18 (5): 693–702. doi:10.1089/scd.2009.0008.

Petrie, T. A., N. S. Strand, C. T. Yang, J. S. Rabinowitz, and R. T. Moon. 2015. Macrophages modulate adult zebrafish tail fin regeneration. *Development (Cambridge, England)* 142 (2): 406–406. doi:10.1242/dev.120642.

Pinto, A. R., J. W. Godwin, and N. A. Rosenthal. 2014. Macrophages in cardiac homeostasis, injury responses and progenitor cell mobilisation. *Stem Cell Research* 13 (3): 705–714. doi:10.1016/j.scr.2014.06.004.

Proserpio, V., and B. Mahata. 2015. Single-cell technologies to study the immune system. *Immunology* 147 (2): 133–140. doi:10.1111/imm.12553.

Reinke, J. M., and H. Sorg. 2012. Wound repair and regeneration. *European Surgical Research* 49 (1): 35–43. doi:10.1159/000339613.

Rosenbauer, F., and D. G. Tenen. 2007. Transcription factors in myeloid development: Balancing differentiation with transformation. *Nature Reviews Immunology* 7 (2): 105–117. doi:10.1038/nri2024.

Ruffell, D., F. Mourkioti, A. Gambardella, P. Kirstetter, R. G. Lopez, N. Rosenthal, and C. Nerlov. 2009. A CREB-C/EBP cascade induces M2 macrophage-specific gene expression and promotes muscle injury repair. *Proceedings of the National Academy of Sciences* 106 (41): 17475–17480. doi:10.1073/pnas.0908641106.

Sander, J. D., and J. K. Joung. 2014. CRISPR-cas systems for editing, regulating and targeting genomes. *Nature Biotechnology* 32 (4): 347–355. doi:10.1038/nbt.2842.

Schuettpelz, L. G., and D. C. Link. 2013. Regulation of hematopoietic stem cell activity by inflammation. *Frontiers in Immunology* 4. doi:10.3389/fimmu.2013.00204.

Seledtsov, V. I., and G. V. Seledtsova. 2012. A balance between tissue-destructive and tissue-protective immunities: A role of toll-like receptors in regulation of adaptive immunity. *Immunobiology* 217 (4): 430–435. doi:10.1016/j.imbio.2011.10.011.

Serhan, C. N., N. Chiang, and T. E. Van Dyke. 2008. Resolving inflammation: Dual anti-inflammatory and pro-resolution lipid mediators. *Nature Reviews Immunology* 8 (5): 349–361. doi:10.1038/nri2294.

Shalapour, S., and M. Karin. 2015. Immunity, inflammation, and cancer: An eternal fight between good and evil. *Journal of Clinical Investigation* 125 (9): 3347–3355. doi:10.1172/JCI80007.

Stefater, J. A., III, S. Ren, R. A. Lang, and J. S. Duffield. 2011. Metchnikoff's policemen: Macrophages in development, homeostasis and regeneration. *Trends in Molecular Medicine* 17 (12): 743–752. doi:10.1016/j.molmed.2011.07.009.

Tang, D., R. Kang, C. B. Coyne, H. J. Zeh, and M. T. Lotze. 2012. PAMPs and DAMPs: Signal 0s that spur autophagy and immunity. *Immunological Reviews* 249 (1): 158–175. doi:10.1111/j.1600-065X.2012.01146.x.

Thornton, C. S. 1951. Beryllium inhibition of regeneration. III. Histological effects of beryllium on the amputated fore limbs of amblystoma larvae. *The Journal of Experimental Zoology* 118 (3): 467–493. doi:10.1002/jez.1401180307.

Titmarsh, D. M., H. Chen, N. R. Glass, and J. J. Cooper-White. 2014. Concise review: Microfluidic technology platforms: Poised to accelerate development and translation of stem cell-derived therapies. *Stem Cells Translational Medicine* 3 (1): 81–90. doi:10.5966/sctm.2013-0118.

van der Vliet, A., and Y. M. W. Janssen-Heininger. 2014. Hydrogen peroxide as a damage signal in tissue injury and inflammation: Murderer, mediator, or messenger? *Journal of Cellular Biochemistry* 115 (3): 427–435. doi:10.1002/jcb.24683.

Wallach, D., T.-B. Kang, and A. Kovalenko. 2013. Concepts of tissue injury and cell death in inflammation: A historical perspective. *Nature Reviews Immunology* 14 (1): 51–59. doi:10.1038/nri3561.

Webber, E. 1994. In vivo response of hepatocytes to growth factors requires an initial priming stimulus. *Hepatology* 19 (2): 489–497. doi:10.1016/0270-9139(94)90029-9.

Wolf, S. A., B. Steiner, A. Wengner, M. Lipp, T. Kammertoens, and G. Kempermann. 2009. Adaptive peripheral immune response increases proliferation of neural precursor cells in the adult hippocampus. *The FASEB Journal* 23 (9): 3121–3128. doi:10.1096/fj.08-113944.

Wynn, T. A. 2008. Cellular and molecular mechanisms of fibrosis. *The Journal of Pathology* 214 (2): 199–210. doi:10.1002/path.2277.

Wynn, T. A., A. Chawla, and J. W. Pollard. 2013. Macrophage biology in development, homeostasis and disease. *Nature* 496 (7446): 445–455. doi:10.1038/nature12034.

Wynn, T. A., and T. R. Ramalingam. 2012. Mechanisms of fibrosis: Therapeutic translation for fibrotic disease. *Nature Medicine* 18 (7): 1028–1040. doi:10.1038/nm.2807.

Yin, Y., Q. Cui, H. Y. Gilbert, Y. Yang, Z. Yang, C. Berlinicke, Z. Li, et al. 2009. Oncomodulin links inflammation to optic nerve regeneration. *Proceedings of the National Academy of Sciences* 106 (46): 19587–19592. doi:10.1073/pnas.0907085106.

Yoo, S. K., and A. Huttenlocher. 2009. Innate immunity: Wounds burst H_2O_2 signals to leukocytes. *Current Biology* 19 (14): R553–R555. doi:10.1016/j.cub.2009.06.025.

Chapter 4 Bioelectrical coordination of cell activity toward anatomical target states
An engineering perspective on regeneration

Celia Herrera-Rincon,
Justin Guay, and Michael Levin

Contents

Key concepts

a. Regeneration is a kind of shape homeostasis—a closed-loop process that regulates cell behavior toward a large-scale anatomical goal state.

b. A complete understanding of regeneration will involve not only a bottom-up view of signaling molecules but also a top-down perspective on the information flow and computations performed by the system to effect repairs and stop when the target morphology is reached.

c. Coordination of cellular activity during regeneration is performed in part by bioelectrical signaling: the ability of all cells (not just excitable nerve and muscle) to produce and receive ionic signals via ion channels and electrical synapses known as gap junctions (GJs).

d. Endogenous bioelectric signaling, in the form of anatomical gradients of distinct resting potentials across cell membranes, regulates not only cell-level behaviors such as proliferation and differentiation but also large-scale properties such as organ size, axial patterning, and topological relationships among structures.

e. Brains implement memory, goal-directed activity, and integrated decision-making by virtue of computations within electrical networks; however, ion channels, neurotransmitters, and electrical synapses pre-date the appearance of brains and of multicellularity, being used for somatic information processing during development and physiology.

f. Given evolution's exploitation of unique properties of bioelectrical signaling for neural cognition and developmental pattern control, it is likely that the conceptual approaches of computational neuroscience will be critical to unraveling the ability of regenerating systems to detect damage, guide growth, and approach a specific morphogenetic outcome.

4.1 Introduction

4.1.1 The challenge of regeneration as dynamic morphostasis

Some animals, such as salamanders, are able to regenerate their entire limbs (Maden 2008). In many animals, this remarkable capability extends beyond traumatic injury (Figure 4.1) to directed remodeling. Other examples of dynamic pattern regulation include the ability of flatworms to

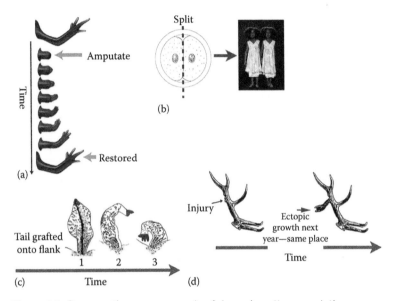

Figure 4.1 Regeneration as an example of dynamic pattern regulation. Large-scale patterning during regeneration and embryogenesis often exhibits flexible growth programs that work to achieve a specific target morphology. (a) Salamander limbs can regenerate perfectly after amputation, and the process stops when a correct limb is rebuilt.Image produced by Jeremy Guay, of Peregrine Creative. (From Pezzulo, G. and M. Levin, *Integr Biol (Camb)*, 7, 1487–1517, 2015. With permission.) (b) However, regenerative ability begins in embryogenesis. Embryos of many species can be split into half but result in two perfectly normal individuals—monozygotic twins (photo by Oudeschool via Wikimedia Commons). (c) In addition to regeneration, this morphological plasticity extends to remodeling intact structures. A tail grafted onto a flank of an amphibian slowly remodels into a limb—a structure more appropriate for its new anatomical position (Farinella-Ferruzza 1956); this includes re-specification of the distal-most tip into fingers, showing that the process is non-local (because the immediate environment of the tail tip is its expected *tail* context, and it should have no reason to change unless it received long-range signals). (d) Importantly, the target morphology can be edited. (From Pezzulo, G. and M. Levin, *Integr Biol (Camb)*, 7, 1487–1517, 2015. With permission.) In some species of deer, damage at a particular spot on the invariant branched structure will result in an ectopic tine appearing *in that same location next year* after the antlers are shed and regrow. (From Naselaris, T. et al., *Neuron*, 63, 902–915, 2009. With permission.)

continuously re-scale their bodies to available cell number (Oviedo et al. 2003), the remodeling of salamander tails transplanted to the flank into limbs—a structure more appropriate for their new location (Farinella-Ferruzza 1956), the rearrangement of craniofacial structures from experimentally imposed abnormal starting conditions to correct face morphologies during frog metamorphosis (Vandenberg et al. 2012), and even the reprogramming of transplanted cancer cells into normal embryonic components (Illmensee and Mintz 1976). All of these are the provinces of regenerative biology, in which we seek to understand pattern homeostasis that builds, remodels, and fully regenerates complex structures via the flexible control of cell behavior despite numerous perturbations.

What is an appropriate formalism within which to understand these phenomena and to translate that understanding into augmentation technologies for regenerative medicine (Levin 2012b)? The current paradigm views pattern and pattern-homeostatic capabilities as emergent from a set of cellular or molecular interactions (Doursat et al. 2013, Doursat and Sanchez 2014) iterated through time in parallel. On this view, the biologist's goal is then to understand signaling networks at the level of genes and proteins and thus to drive advances in complexity theory or computational biology to explain how shape homeostasis arises from local, low-level signals (Figure 4.2a). The advantage of this approach is that it adheres closely to the implementation details—the components without which the process does not work. Indeed, some of our most widely used assays are unbiased loss-of-function approaches designed to identify the needed list of components without which aspects of regeneration do not occur (Newmark 2005).

However, this approach also has inherent limitations (e.g., what kind of molecular model would explain antler trophic memory, Figure 4.1d). Models at the level of genes or protein pathways do not easily reveal the origin of specific shapes; no current biochemical or genetic dataset can be directly inspected to reveal whether it encodes a shape like that of a tree, an octopus, or a human hand. Such models generally do not constrain topology and thus make it difficult to explain any particular pattern or what signals would have to be provided in order to change the current pattern into a desired pattern. Thus, although swarm models have enjoyed great success for some phenomena (ant colony behavior), they generally run afoul of the inverse problem (Lobo et al. 2014c); even when you know all the rules of an iterative, many-agent system, it is overall extremely difficult to reverse the process to know what low-level properties should be tweaked to exert rational control over the resulting shape.

Importantly, significant recent work in developmental biology has begun to produce more generative models that make explicit the steps that are sufficient, and not merely necessary, for specific patterning to occur (Barkai and Ben-Zvi 2009, Lander et al. 2009, Ben-Zvi et al. 2013, Sheeba et al. 2014, Uzkudun et al. 2015). However, such models are still comparatively rare in the field of regeneration (Figure 4.1; Slack 1980). Thus, here we take an engineer's view of the problem (Lazebnik 2002,

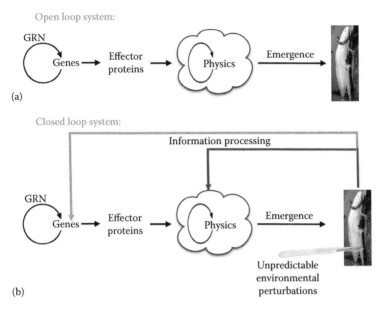

Figure 4.2 Regeneration as a closed-loop, cybernetic process. (a) Current approaches in this field largely treat morphogenesis as a feed-forward process, where gene-regulatory networks specify effector proteins, whose activity (physical interactions) results in complex pattern formation by emergence. (b) However, importantly, there is a feedback loop in many species that underlies pattern morphostasis. This allows unpredictable environmental events (such as amputation of specific body regions or deformation) to trigger remodeling events that correct back to the target morphology. An important goal of this field in the future is to understand what algorithms and mechanisms guide this error-correcting closed-loop process. (From Slack, J.M., *J Theor Biol*, 82, 105–140, 1980. With permission.)

Lobo et al. 2012). Engineering approaches have several advantages. First, engineers do not have an a priori commitment to a particular level of explanation (e.g., our field has chosen molecules, instead of tissues or subatomic field dynamics). Engineers use any level of description of any system (an appropriate coarse graining and modularization) that gives the optimal control and predictive power. We argue that regenerative medicine badly needs to question the assumption that genes or molecules are the only appropriate levels of explanation (Noble 2012).

A second advantage is that engineers are comfortable with *implementation independence*—as long as functional relationships are preserved, the underlying details are free to vary. The first question they ask of any system is: which are the details that can be ignored, and which are the dominant modes, functions, conserved properties, symmetries, or control loops that could provide rational control. This is handy when trying to take lessons learned in one model system and apply them to not only regenerative medicine but also the construction of artificial hybrid bioengineered construct (Doursat et al. 2012, 2013, Kamm and

Bashir 2014). Finally, engineers pay a lot of attention to modularity: which functions are encapsulated in such a way that the whole lot can be triggered with a simple *subroutine call* or trigger. This is a very powerful trick that evolution has certainly exploited (von Dassow and Munro 1999, Hansen 2003, Schlosser and Wagner 2004), and it provides the opportunity to defeat the complexity barrier that is sure to stymie attempts to micromanage patterning outcomes directly when it comes to complex structures such as limbs.

4.1.2 An engineering perspective on regeneration

One of the most salient (but less-studied) aspects of regeneration is that it stops at the right time. Regeneration is a closed-loop (feedback-based) process (Figure 4.2b), which ceases when the correct pattern has been restored. One of the implications of this is that regeneration is relevant not only for biomedical repair but also for the problem of cancer—fundamentally a disregulation of normal patterns of growth and form (Donaldson and Mason 1975, Wolsky 1978, Brockes 1998). Engineering has long dealt with systems that have such goal state–limiting properties, which can be viewed as a homeostatic cycle or a feedback loop (Ashby 1956, Apter 1966). We propose here that an essential level at which to model and control regenerative systems is a set of information-processing control loops of a cybernetic process guided by measurements of morphogenetic state.

An engineer views limb regeneration as a system that has the following main functions: (1) it detects damage (deviation from its target morphology); (2) it recalls the correct state as a kind of pattern memory; (3) it issues control commands to agents (cells) that bring the system closer to the correct state; and (4) it decides whether the process is finished; if not, it iterates from step 2. This is a classic Test-Operate-Test-Exit (TOTE) model (Rosenblueth et al. 1943, Miller 1960), first formalized in cybernetics and more recently exploited widely as control theory. Note that this description is in the form of information and computation—the basic units here are not molecules but messages, states, and comparisons. This model describes a homeostatic or goal-directed system. Given such a model, it is fairly clear what has to be done to derive appropriate outcomes from different starting conditions. What is needed is to understand how cells *in vivo* execute these steps. Therefore, we propose that a full explanation will inevitably not only be the reductive molecular details but also include the integration of those details into a story that explains computation and patterning homeostasis of large-scale anatomical properties (Pezzulo and Levin 2015). In other words, a true multi-scale understanding is needed.

How can concepts such as size, adjacency, topological arrangement, and organ identity be included in a realistic model of large-scale repair and remodeling when these do not exist at the level of proteins, genes, or cells? Fortunately, we have two mature examples of sciences in which these layers of explanation have been successfully integrated: computer science and computational neuroscience (Pezzulo et al. 2014, 2015*). In both of these areas, we have successful theories that enabled enormous scientific progress by taking higher levels of description seriously. Here, we will focus on one area, and we suggest the counter-paradigm proposal that regenerative biology can benefit from borrowing from the successful exploitation of top-down analysis in cognitive neuroscience. Work in this area over the last century has shown the power and necessity of operating not only at the machine code of the gene or molecule but also at the object-oriented, high-level language of memories, goals, perceptions, computations, and other concepts that do not exist at the level of molecular pathways or cells but are effectively implemented by cell networks.

It is important to note that hypothesizing memory or goal-directed activity in cells is not a call to pre-scientific magical thinking but a reminder that regenerative biology may trail behind other sciences in their successful, quantitative approaches to the integration of high-level properties with molecular-level details. We make no claims about teleology in evolution, consciousness, or anything else that does not have direct implications for functional interventions in regenerative repair. Our main theses are that computational neuroscience provides a roadmap for advances in our field by showing (1) how goal-directed control of cell activity by global pattern encoding in tissue can be explained by mechanisms operating in cellular networks, and (2) techniques and strategies for rational modification of large-scale properties, without micromanaging the molecular details (e.g., memory inception and behavior shaping).

4.1.3 Linking cognitive science and regeneration: An outline of this perspective

As an example, consider frog metamorphosis, where components of the tadpole face are rearranged into the very different frog face. When the initial tadpole's face is rearranged into abnormal positions, it is revealed that the movements of metamorphosis are not hardcoded (resulting in an abnormal frog, since each organ is starting from the wrong initial position). Rather, the body is a dynamic, flexible system that results in a normal frog face by driving cell rearrangements until the correct final pattern is achieved (different from the evolutionarily conserved normal movements). How can we understand not only the genes needed for this to occur but also the high-level (and thus transferrable beyond frogs)

---

* This paper presents an overview of a set of concepts that have not yet been integrated into regenerative biology but may have the potential to conceptually revolutionize this field. It covers the development of quantitative theory for understanding goal-directed activity in complex signaling systems (in this case, the brain, but extendable to non-neural cell networks during pattern regulation and regenerative repair).

aspects such as *detecting abnormal pattern*, *driving cells to reduce distance between current pattern and target pattern*, and *cease remodeling when correct pattern has been achieved* (TOTE)?

Having introduced top-down, cybernetic views of the regeneration/ remodeling process, we suggest that this is precisely the kind of cross-level problem that is faced by computational neuroscience (Figure 4.3), from whose strategies our field could benefit. How does the brain implement these kinds of computational, goal-directed functions in the control of behavior? They are implemented by electrochemical networks driven by the dynamics of ion channels and downstream signaling molecules in cell networks. Importantly, all cells, not just neurons, have the ability to drive these kinds of signaling dynamics. Thus, in the next section, we review bioelectricity in regeneration: the role of ion- and voltage-mediated control mechanisms in regeneration. Finally, having shown that regeneration and brain function share conserved mechanisms at the cellular and molecular levels, we suggest that not only the molecules but also the control algorithms might be conserved. It is very likely that brains did not invent their functions *de novo* but likely refined and speed-optimized information-processing mechanisms that cells used for metazoan development long before nervous systems appeared on the scene (Buznikov and Shmukler 1981, Keijzer et al. 2013[*]). To facilitate the development of such models in the field of regeneration, we then provide a brief overview of relevant neuroscience concepts, giving only enough detail to make it clear that goal-directed, representational control structures are routinely and uncontroversially implemented by cells in electrically coupled networks (in the brain). We conclude with some thoughts about the future of this exciting emerging field.

4.2 Bioelectricity and patterning: A new (old) control modality

4.2.1 Non-neural bioelectricity

A long history of work implicates bioelectric events in patterning (Lund 1947, Jaffe 1981, Nuccitelli 2003, McCaig et al. 2005). This field was somewhat eclipsed by the advent of molecular biology that focused on chemical gradients, as unlike bioelectric state, these can be visualized in fixed (dead) samples. However, recent advances in molecular physiology have revealed GJs, ion channels, and neurotransmitter pathway molecules

---

[*] This paper presents a radical theory, the skin brain thesis (SBT), on a goal-directed evolution of the primitive nervous systems: the creation of an *effective* effector, the muscle. According to the SBT, the synaptic communication and other typical elements of neuronal activity appeared to generate and control the dynamic patterns resulting from its own network activity (the whole body movement).

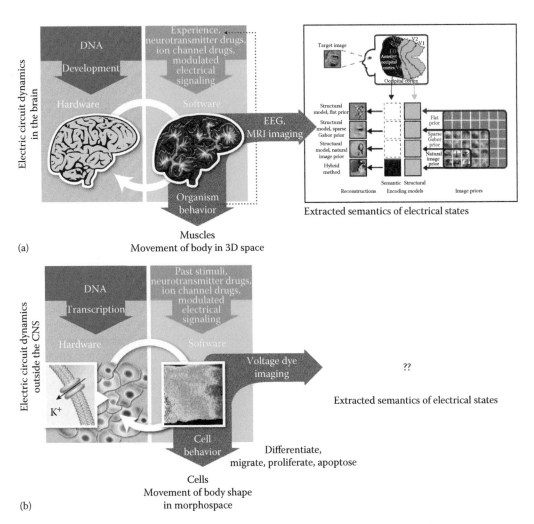

Figure 4.3 Parallelisms between brain function and regeneration control. There are fundamental parallels between brain function and regenerative bioelectricity. (a) The DNA-specified hardware of the brain supports electrical dynamics in the neural cells. These electrical dynamics are not themselves determined by DNA, being a function of intrinsic transition rules and experience. The output of this bioelectric network is muscle activity during behavior, which moves the body through 3D configuration space. Work is currently under way to develop computational pipelines to infer mental imagery content from brain readings. (b) Similarly, in non-neural somatic tissues, DNA specifies the ion channels, which enable bioelectric dynamics but do not determine the resulting voltage gradients directly, which develop according to transition rules and external signaling (experience from cells' somatic environment). The outputs of these bioelectric circuits are cell activities during remodeling (proliferation, differentiation, apoptosis, etc.), which move the body through a virtual space that describes its morphology (the morphospace). One of the key goals of the regenerative bioelectricity field is to learn to extract target morphology data from measurements of bioelectric gradients. Left side of panel A and panel B are courtesy of Jeremy Guay, of Peregrine Creative. (Right side of panel A is used by permission from Naselaris et al., *Neuron*, 63(6): 092–915.)

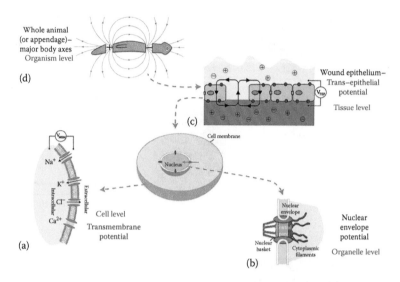

Figure 4.4 Bioelectric signals at multiple scales. Endogenous bioelectric signals comprise a set of biophysical properties, including voltage gradients, electric fields, and individual ion flows. *In vivo*, these originate at multiple levels of organization. (a) Organelle membranes generate voltage gradients, such as the nuclear envelope potential (largely unexplored) and the well-understood mitochondrial potentials. In recent years, the role of resting potential across the plasma membrane of the cell (b) has become to be known as an important determinant of cell fate; spatial gradients of such voltage values over cell fields are now known as regulators of pattern formation in embryogenesis and regeneration. Decades ago, it was recognized that the transepithelial electric field (resulting from the parallel activities of polarized cell layers) (c) was an important factor for guidance of migratory cell types during development and wound healing. Finally, at the level of entire appendages or even whole organisms (d), large-scale potential differences presage and control anatomical polarity and organ identity. (Courtesy of Maria Lobikin.)

ubiquitously expressed throughout the body, beginning before fertilization. Analogous to the brain, non-neural tissues continuously regulate resting potential (V_{mem}), local field potentials (LFPs), extracellular electrical fields, and the movement of neurotrasmitters among cells (Figure 4.4) (Pullar 2011, Bates 2015). Techniques have now been developed to detect these physiological parameters *in vivo*, using pharmacological and genetically encoded voltage reporters and microelectrodes, and to manipulate them functionally by using ion channel misexpression and drug blockers/activators (Adams 2008[*], Adams and Levin 2013). As in the brain, there

---

[*] This paper reviews techniques and approaches for investigating the cellular and tissue-level dynamics of bioelectrical gradients across plasma membranes (cell resting potential or V_{mem}). These strategies and reagents can now be used to address the biophysical signaling and global computational capabilities of neural and non-neural bioelectric networks that regulate growth and form.

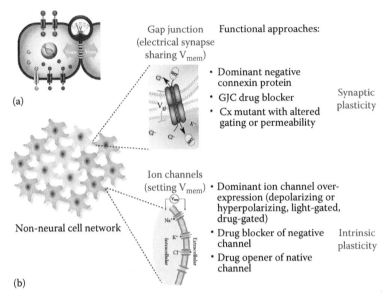

Figure 4.5 Bioelectric networks. (a) The two most important components of bioelectric signaling on a cellular level are the ion channels that set resting potential and the gap junctions (electric synapses) that share physiological state with neighboring cells. (b) Gap junctions are themselves voltage-sensitive, and hence, they enable complex dynamics in cellular networks. These dynamics can be functionally modulated in two basic ways: changes of resting potential of the cells (similar to intrinsic plasticity in the CNS) and changes of connectivity to other cells (similar to changes of network topology in synaptic plasticity of the CNS). Both of these can now be achieved by using pharmacological agents that target ion channels or gap junctions or the misexpression of dominant channels, pumps, and connexins with desired physiological properties. (Courtesy of Alexis Pietak and Jeremy Guay.)

are basically two functional approaches (Figure 4.5): modulating synaptic plasticity by altering the way in which cells communicate electrically (target GJ connectivity and neurotransmitter exchange), and modulating intrinsic plasticity by directly changing V_{mem} in cells (by altering the expression or function of ion channels and pumps).

4.2.2 Bioelectric networks process information for pattern regulation

Recent work has shown that signaling mediated by bioelectric events plays a crucial, instructive role in pattern formation (Funk 2013, Levin 2014a). Ion channel–mediated changes in V_{mem} not only affect individual cell behaviors such as proliferation, differentiation, apoptosis, and migration (Sundelacruz et al. 2009) but also determine large-scale parameters such as organ size, shape, and axial patterning of the entire body (Beane et al. 2011, Perathoner et al. 2014). In a range of model systems, we now know that V_{mem} regulates the formation of the brain, eye, wing, and face and controls patterning along the anterior–posterior and left–right axes during embryonic development (Levin et al. 2002, Dahal et al. 2012, Pai et al. 2015a*). Moreover, experimental control of bioelectric gradients has enabled induction of regenerative ability in non-regenerative contexts (Tseng et al. 2010, Leppik et al. 2015), induced reprogramming of gut tissue into complete eyes (Pai et al. 2012), and normalized tumors (Chernet and Levin 2013). Electrical synapses (GJs) and neurotransmitters such as serotonin are key components of several patterning systems, having been implicated in embryonic left–right asymmetry, bone patterning, tumor suppression, and brain size control (Levin and Mercola 1998, Iovine et al. 2005, Levin 2007, Chernet et al. 2015, Pai et al. 2015a). As in the brain, these elements often work together, such as the bioelectrically controlled movement of serotonin through GJs during left–right patterning and control of nerve growth (Levin et al. 2006, Blackiston et al. 2015). The molecular pieces are now being identified, but the idea of neurotransmitters being ancient *pre-nervous* developmental signaling molecules is an old one (Buznikov and Shmukler 1981).

In the invertebrate flatworm *Planaria*, a gradient of voltage membrane potential that regulates patterning during regeneration has been identified. In normal worms, amputated fragments faithfully regenerate the appropriate structures in the correct places (Salo et al. 2009, Lobo et al. 2012). However, if the natural V_{mem} gradient is pharmacologically duplicated during the initial stages of regeneration, the worms will regenerate two heads instead of a head and a tail. Disrupting H^+/K^+-ATPase, which regulates membrane voltage in planaria, drastically alters the size of regenerate organs, creating small heads respective to pharynxes (Beane et al. 2013). Thus, in planaria, bioelectric signaling is upstream of both size control and axial polarity decisions.

4.3 Bioelectricity and appendage regeneration

Interest in the role of bioelectricity in regeneration reaches back through the last century of science. At the turn of the twentieth century, studies were limited to the simple application of electrical current through a trough containing limb-amputated tadpoles. By these crude means, a marginal increase in regenerate tissue in treatment over control animals could be observed, which encouraged further research into this phenomenon (Frazee 1909). Studies by Rose in the 1940s by using exposure of limbs to a high-salt solution showed subtle but important differences in

---

* This excellent work demonstrates not only that appropriate V_{mem} is critical to eye development during development but also that altering V_{mem} in a subset of cells across an organism is sufficient to induce ectopic eyes composed of both manipulated and recruited, unaltered, cells. This demonstrates the higherorder signaling role that V_{mem} can play in generating complex structures.

treated animals versus controls. This effect was described as being due to *increased activity of the wound epidermis*, causing it to act like a younger, more regenerative epidermis (Rose 1945). Similar studies were repeated two decades later by Bodemer, who applied high-salt solutions as a form of trauma in combination with electrical stimulation and likewise saw a slight increase in the size and quality of regenerative tissue (Bodemer 1964). Coupling these findings with what is now known about ion flux in regeneration, as will be discussed later, more sophisticated hypotheses can be drawn regarding the effect of the ionic environment on regeneration.

4.3.1 Electrical fields

Early studies applying electrical stimulation showed that weak currents could induce more regenerative tissue and improved morphology in tadpole tails (Rose 1945). A recent study applied low current in a model of rat forelimb amputation. Although regeneration of complex structures was not achieved, treated rat stumps were more proliferative and showed more vascularization. Interestingly, the bone marrow remained open up to 7 days postamputation in the treated animals and grew significantly more new cartilage and bone at 28 days postamputation (Leppik et al. 2015). Similarly, Hechavarria et al. applied 6.4 µA to a mouse digit amputation in the context of a liquid-filled bioreactor surrounding the wound site and saw a similar improvement in complex tissue growth (Hechavarria et al. 2010).

As far back as the early 1960s, it was hypothesized that electrical signaling was involved in bone growth. This idea was derived from the observation that under mechanical strain, bone generates an electrical potential, which could be instructive for bone strengthening and healing (Bassett and Becker 1962). Later work showed significant improvements in bone healing under exposure to electric currents and electromagnetic fields (reviewed in Norton et al. 1984). In addition to individual tissue growth, regeneration appears to involve a number of rearrangements of motile cell types (Salo and Baguna 1985, Scott et al. 1995, Scott and Hansen 1997, Wang and Zhao 2010, Godwin et al. 2013, Munoz-Canoves and Serrano 2015).

Some of the most detailed molecular work in developmental bioelectricity was done to address galvanotaxis (McCaig et al. 2005, Pullar and Isseroff 2005, Pullar et al. 2006, Rajnicek et al. 2006, 2007)—the migration of cells in electric fields, with particular relevance to wound healing (Zhao 2009, Kucerova et al. 2011) and embryonic development (Yamashita 2013, Yamashita et al. 2013). Breaking the epidermis results in a wound electrical field in rodents and humans in the order of 100–200 mV/mm (Barker et al. 1982, Nuccitelli et al. 2011), which directs nerve regeneration and healing (Song et al. 2004). Human keratinocytes migrate directionally in electric fields, as do rat cornea cells (Nishimura et al. 1996, Lois et al. 2010, Kucerova et al. 2011, Saltukoglu et al. 2015). More recently, macrophages were observed to move in accordance with physiological electric fields of 150 mV/mm and to increase their phago-cytic activity, thereby connecting bioelectricity to inflammation, another key factor in regeneration (Hoare et al. 2015).

4.3.2 Ion flux and resting potentials

In a model of the axolotl spinal injury, Sabin et al. observed a rapid endog-enous membrane depolarization of resident neural stem cells around the wound site. When the return to resting potential was pharmacologically blocked, thereby keeping the cells in a depolarized state, inhibition of stem cell proliferation and regeneration was observed (Sabin et al. 2015). The axolotl is an excellent system for these studies, having been used to demonstrate electric gradients as a developmental coordinate system (Shi and Borgens 1995) and more recently to examine the electrophysiology of tail regeneration (Ozkucur et al. 2010). Interestingly, the transcriptional targets of bioelectric signaling in the axolotl spinal cord share many over-laps with genes downstream of depolarization in frog embryogenesis and human mesenchymal stem cell differentiation (Barghouth et al. 2015).

Xenopus tail regeneration depends on a strong proton efflux due to an upregulated plasma membrane V-ATPase pump that begins to function 6 hours after amputation (Adams et al. 2007). This hyperpolarizing effect is required to induce necessary proliferation and innervation in the nascent tail. Downstream of proton-mediated hyperpolarization lies a sodium-specific set of signals. At approximately 18 hours postamputation, an Na^+ influx and upregulation of the NaV1.2 channel were observed. Inhibition of the voltage-gated Na^+ channel NaV1.2 inhibited regenera-tion, via salt-inducible kinase (SIK) (Tseng et al. 2010).

Importantly, it was also shown that a variety of non-regenerative states (e.g., age-dependent decline) could be overcome by forced proton pump-ing. This could be achieved by using a yeast P-type proton pump, which has no sequence or structure homology to the endogenous V-ATPase, showing that it is the physiological state of the cells, and not the genetic identity of the ion pump, that was the necessary and sufficient step for regenera-tion. The same induction of tail regeneration was recently also induced by using light-triggered archaerhodopsin—a new application for optogenetics (Adams et al. 2013). The ability to swap out channels and pumps, as long as the appropriate V_{mem} gradient is set up, is a general theme in this field. Counter to typical genetic approaches, it is not correct to think of specific ion channel/pump genes (e.g., *NaV1.2*) as *regeneration genes*; rather, we must characterize the specific bioelectrical state that induces specific out-comes. This is a kind of functional *implementation independence* that is well familiar to engineers and is a major advantage for regenerative medicine. Any convenient reagent can be used to impose a regenerative state—it is not necessary to mimic expression patterns of endogenous ion translocator.

Two other aspects of inducing tail regeneration by V_{mem} control should be noted. These examples illustrate the ability of bioelectric signals to serve as *subroutine calls* or master triggers: a simple, low information-content signal (*pump protons*) kick-starts a complex cascade that builds a whole tail, of correct size and orientation, and stops growing when the tail is complete. Indeed, the sodium treatment (Tseng et al. 2010), which was done by an ionophore cocktail, avoiding gene therapy was required only for 1 hour: once it triggers downstream cascades, the bioelectric signal does not have to be continuously imposed. This offers great advantage to regenerative medicine because the many steps required to build a tail and its internal pattern did not have to be micromanaged—a single, brief stimulus was sufficient to trigger the tail-building module. However, a limitation of this field currently is that it is not yet known how to control the resulting pattern: these techniques have not yet shown how to grow something other than a tail at a tail blastema.

4.3.3 Downstream mechanisms

Bioelectrical signals become transduced into epigenetic and transcriptional responses by a number of mechanisms. The most obvious is calcium, as voltage-gated calcium channels are the main responders to electrical activity in the brain and regulate numerous downstream transcription sites via calmodulin and other calcium sensors (Barbado et al. 2009). In addition to Ca^{2+} signaling, Na^+ is coming to light as a regulator of development through the salt-inducible kinases (SIKs), which have been shown to regulate the Hippo pathway, a regulator of organ size in *Drosophila* (Wehr et al. 2013) and is required in *Xenopus* for tail regeneration (Tseng et al. 2010). Na^+ signaling through SIK is also tied to intracellular Ca^{2+} signaling, where increased levels of Na^+ are translated into increased Ca^{2+} by Na^+/Ca^{2+} exchange proteins (Bertorello and Zhu 2009).

Bioelectrical activity also regulates chromatin state. One mechanism is via voltage-dependent movement of butyrate and resulting changes of acetylation. The Na^+-coupled monocarboxylate transporter, SLC5A8/SMCT1, couples bioelectric state with regulation of cell activity via uptake of the histone deacetylase (HDAC) inhibitor butyrate (Kruh 1982, Miyauchi et al. 2004). Interestingly, HDAC inhibitors, when applied ectopically inhibit regeneration, and SMCT1 is highly expressed in tadpole tails in the refractory period of development (Taylor and Beck 2012, Tseng and Levin 2012). It can readily be hypothesized that Na^+ flux is able to act on an epigenetic scale to regulate regeneration (Tseng et al. 2010, Taylor and Beck 2012, Chernet and Levin 2014).

Sensing of membrane voltage, independent of specific ions, has been observed in many cases (Adams et al. 2007). A variety of voltage-sensitive proteins exist, including voltage-sensitive phosphatase and tensin homolog (PTEN) phosphatases (Murata et al. 2005, Okamura and Dixon 2011) and RAS clustering (Zhou et al. 2015). Another mechanism by which bioelectric signals are transduced through a voltage-sensitive protein is via regulation of the movement of serotonin. Serotonin is a well-known neurotransmitter and has been shown to be required for axon regeneration in *Caenorhabditis elegans*, innervation of transplanted organs in *Xenopus* (Blackiston et al. 2015), and regeneration of spinal motor neurons in zebrafish (Barreiro-Iglesias et al. 2015, Alam et al. 2016). Serotonin levels inside of cells are modulated by voltage-dependent electrophoresis (through gap junctional paths) or by regulation of the serotonin transporter (Fukumoto et al. 2005a, 2005b, Li et al. 2006, Blackiston et al. 2011, Lobikin et al. 2012).

Many of the common intracellular pathways have been implicated in bioelectric signaling. In the case of wound healing, PI3K has been identified as a mediator of electric field–mediated cell migration and healing in epithelial wound models (Sato et al. 2009, Arocena et al. 2010, Meng et al. 2011, Sun et al. 2013). In contrast, PTEN was identified as an agonist in the same context (Ozkucur et al. 2010). In an axolotl spine injury model, c-Fos and ERK were associated with V_{mem} flux after damage (Sabin et al. 2015). Interestingly, c-Fos is a common target of action potentials in the brain (Dragunow and Faull 1989, Dragunow et al. 1989), consistent with the theme of many molecular similarities between electrical activity in the central nervous system (CNS) and pattern formation during regeneration.

4.3.4 The nervous system: Is it related to bioelectrical control?

It has long been known that the nervous system is critical for regeneration in vertebrates. In salamander, severing the primary nerve before limb amputation inhibits regeneration (Singer 1952). Interestingly, the story is more complex: in limbs that have never been innervated, regeneration proceeds normally. However, nerve dependence in regeneration can be acquired in these aneural limbs by reintroduction of nerves (Filoni et al. 1995). Thus, the presence of nerves in the limb creates a dependence on innervation (Filoni et al. 1995, Kumar et al. 2011). Studies by Kumar et al. have identified a molecular mechanism through the AG protein as part of this acquired dependency (Kumar et al. 2011). Is there any role of bioelectricity in the relationship between nerve and regeneration (Borgens 1988a*, 1988b, Borgens et al. 1990, Jenkins et al. 1996)?

Several studies in the 1970s and 1980s found that neuron outgrowth and directionality could be altered by imposed electrical fields (Hinkle et al. 1981, Patel and Poo 1982). This phenomenon could be applied to disrupt normal nerve growth or to enhance it, with significant effects on regenerative outcomes, as in the case of *Xenopus* limb stimulation, where amputated limbs that typically possess poor regenerative ability showed increased innervation and improved regenerate morphology

---

* This paper, by one of the key figures in developmental bioelectricity, describes now classic data on the function of endogenous electric currents (transepithelial potentials) in limb ontogeny and regeneration. It also covers the use of applied electric fields to stimulate regenerative responses for the nervous system and for complex appendages.

when a properly oriented electrical field was applied (Borgens et al. 1977). Thus, in addition to direct effects on proliferation and patterning, endogenous bioelectric gradients may exert effects indirectly, by guiding innervation to the wound site (Borgens 1982).

However, the role of nerves may not be simply to provide trophic *permissive* factors. Neural connections could provide instructive information for regeneration and patterning. This was observed in the ability of the two different ends of a bisected nerve cord to induce heads versus tails when deviated to flank wounds of *Spirographis spallanzani* (Kiortsis and Moraitou 1965). This even extends to maintenance of adult tissues, as mature tongue papillae disorganize after the lingual nerve is cut (Takeda et al. 1996, Sollars et al. 2002).

Focal damage to the spine far anterior to the site of tail amputation in *Xenopus* tadpoles alters the shape of regenerate tail (Mondia et al. 2011). A simple hypothesis that can be made is that some information necessary for normal tail patterning travels from the site of spinal damage to other somatic tissues. It is interesting to note that the further anterior from the amputations site the spinal damage is, the more severe is the phenotype. However, this effect is compounded by multiple co-linear lesions to the spinal cord, confounding simple models in which the brain sends all of the necessary signals posteriorly along a linear path. It is possible that the spinal cord and related nerve tissues along the main body axis maybe participating in generating the needed information.

4.4 Learning from neuroscience: How bioelectric networks implement goals

One way to view regeneration is as a control loop over a homeostatic property: maintaining a target morphology. Whether via direct control by the CNS or by neural-like information processing occurring in bioelectric networks of non-neural tissues, we hypothesize that the study of regenerative biology may advance significantly by exploiting the understanding of how brains implement goal-directed activity. In the brain, electric, transcriptional, and neurotransmitter-based signaling across the whole body is regulated by cell networks that integrate information and make global decisions that flexibly guide the transition into appropriate states, despite environment-induced deviations. In this section, we discuss the basics of these remarkable systems, so as to facilitate application of these ideas to the dynamic regulation of shape during regeneration and the formulation of new molecular and conceptual models of regenerative repair.

There is no single *nucleus for reaching a defined goal* in the brain. Rather, multiple cognitive processes are integrated to create a goal-directed behavior (Figure 4.6) (Pezzulo et al. 2014): perceptual evidence, memory, planning, decision-making, motivation, and so on. Thus, the outcome will be a function of the interaction between several layers

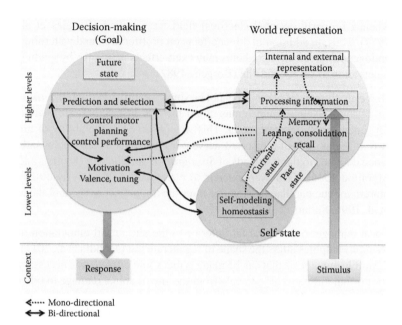

Figure 4.6 Representing models in neural networks. This figure schematizes goal-directed behavior, indicating the levels of control (vertical labels on the left), the main cognitive systems involved (circles: self-state, world representation, and decision-making), and their functions (rectangles). The final outcome (response for achieving a goal in a specific context) will be a function of the tight coupling (arrows indicate interaction) between the representations of the external and internal worlds in the past, current, and future states of the animal. Although the brain uses these systems to generate a behavioral motor task induced by a relevant stimulus, similar processing may occur during regeneration, controlling cells to maintain and reach the target morphology.

of control, each of which uses a specific language and circuit motifs (Verschure et al. 2014). We will discuss here the algorithms and the molecular mechanisms (Petersen and Sporns 2015) that underlie cognitive functions. Although the brain uses these systems to operate muscles and glands to move an animal through physical and physiological state space, the same processes may occur during regeneration to control the movement and differentiation of other cell types to shift the body to a better region of morphospace (Stone 1997).

4.4.1 Sensory processing and perception

Key parts of regeneration are to know when damage has occurred and to detect that correct shape has been restored; thus, measurement of specific properties (from molecular signals to overall shape or size) is a key component. From a cognitive perspective, sensory processing and perception imply that the system is capable of detecting and converting internal or external sensory information (mediated by some specific physical signal) into a perceptual quality-bearing salient information. The final perceptual representation will contain not only measurable data (such as timing, space, and quantity) but also the necessary information for initiating a circuit (in case the information is important enough to make a decision), which leads to an action for reaching or maintaining current goals (Mechelli et al. 2004, Kiebel et al. 2009, Hogendoorn 2015). For the system, the stimulus is more than the sum of its individual physical characteristics; a stimulus has a behavioral meaning in a relevant context (Kim and Biederman 2011, Maren et al. 2013). It should be noted that this aspect of context permeates the molecular biology of pattern regulation, where the same molecules (e.g., Wnt-5, sonic hedgehog, and calcium) are used in many different structures for many different outcomes, revealing the need for informational context in cells' interpretation of chemical or physical stimuli.

This complex multi-scale process can occur in the brain because each sensory stimulus is first encoded in a powerful language: electrical signals (signal reception and transduction). Electrical activity is able to move forward across increasingly complex anatomical and functional stations, where the information is further integrated (information processing), and eventually reaches a higher-order station, where the neural representations of the experienced percepts are generated (perception) (Hubel and Wiesel 1962) (see excellent reviews in Mountcastle 1998, Purves 2013). The different sensory systems (vision, audition, touch, pressure, temperature, olfaction, and gustation) transduce the different forms of energy (physical, mechanical, chemical) into a common one: electrical activity. Note that receptors for all of these exist at the cell and tissue levels, allowing regenerating systems to potentially measure anatomical states by sensing signals mediated by the same modalities that the CNS receives from sensory organs: chemical, physical, and so on. The energy of signals being detected is not only transduced but also continuously modified to be transformed into a *message* (Engel et al. 2012, Lobanov et al. 2014) that will constitute the perceptual representation in the higher stations (giving rise to the successful behavior) (Park and Friston 2013). This information processing leading to a message represents the neural code (Bialek et al. 1991; reviewed in Stanley 2013).

In the periphery, the energy of the stimulus (chemical, physical, and mechanical) acts on the membrane of the receptor cells, often directly involving ion channels and altering their permeability. These ion fluxes (Hille 2001) change the membrane resting potential (V_{mem}), constituting the receptor potential (RP) (analogous to the slow ionic currents or bioelectricity happening in non-excitable cells; Juusola et al. 1996). This RP, in turn, is modulated by both the electrical inputs arriving to the same cell (temporal and/or spatial summation) and the self-electrical state of the cell. An output or action potential (AP) will only be generated if the final potential resulting from the total integration is powerful enough and reaches a particular cellular region (axon hillock, with the highest density of voltage-dependent sodium channels; Hodgkin and Huxley 1952; reviewed in Bean 2007). The AP, an all-or-none process, is the way in which the system communicates across both short and long distances. The features of the stimulus are coded within properties of the trains of AP generated, such as frequency, amplitude, and duration (Clark and Hausser 2006). The communication (both horizontal and vertical) between different neurons occurs through the synapses, where the characteristics of the AP result in Ca^{2+}-mediated release of neurotransmitters (Nt) in the case of chemical synapses, and/or through the direct coupling of cells in the case of electrical synapses or gap junctions (Whalley 2011, Pereda 2014).

Perception is the integrated representation of the (external and internal) world, happening when sensory information reaches the higher station in the afferent processing pathway (or cerebral cortex, *higher levels*). The initial target of the input for any sensory modality is called the primary sensory cortex for that modality. From and to here, other *higher-order* cortical areas, or association cortices, extensively connected by feedforward and feedback circuitry, integrate the qualities of a given modality (e.g., color, brightness, and form in vision) with information from other sensory modalities and from brain regions carrying out other functions (e.g., attention and memory, reviewed in Gainotti 2011, Purves 2013).

4.4.2 Self-modeling

It is crucial that the brain is not only a physico-chemical system but also the one that encodes self-models—internal representations (Rosch 1975, Rosen 1985, Putnam 1988, Hinton 2007, Langston et al. 2010, Bonnici et al. 2011, Marstaller et al. 2013) of itself and of the outside world, to guide decision-making (Figure 4.6). The idea that somatic tissues during regeneration use explicit models of current state and target states has not been explored in the molecular age. A starting point in the regenerative process could be an internal model of the body (analogous to internal representations of somatic configuration in space [Blanke 2012]). By means of this representation (i.e., mimicking the input/output characteristics (Kawato 1999, Schilling and Cruse 2012), animals could *anticipate* the next-step consequences of muscle activity or cell differentiation (for behavior or morphogenesis, respectively).

A proof of concept for applicability of self-models in entirely unconscious systems is its implementation in artificial systems. In 2006, Bongard et al. (2006) engineered an artificial four-legged machine (*starfish*) that, by optimizing the parameters of its own resulting self-model (actuation-sensation), could infer its own structure *de novo* and use this representation to recover performance after a part of its leg was removed (creating compensatory behaviors).

Importantly, computational neuroscience increasingly reveals the mechanisms by which cellular networks achieve this capability. One of the most remarkable examples comes from the *place cells* in rodents (Kubie et al. 1990) and the way by which discharge patterns of single cells (hippocampal neurons) and their collective activity encode dynamical spatial patterns (Kubie et al. 1990, Leutgeb et al. 2005, Colgin et al. 2008). Place cells are neurons that are activated selectively when an animal occupies a particular location in its environment (the *place field* of the neuron; Kubie et al. 1990, Muller et al. 1996, Ferbinteanu and Shapiro 2003, Ainge et al. 2007, 2012). The hippocampal networks are also thought to retain sequential information. Past, present, and future events (or sequences) could be separated spatially by cell assemblies and temporally by network theta oscillations (Buzsaki 2002; reviewed in Leutgeb et al. 2005). The encoding of the spatial and temporal patterns allows animals to navigate from their current position to a goal position (Barca and Pezzulo 2015). The study of this example, in which electrical activity of cell networks implements spatial memory, could lead to new models with which to understand the guidance of trajectory of cell behavior toward specific spatial patterns in regeneration.

The brain uses different domains, or layers, to process the *internal needs* (related to self-homeostasis, managed by *lower levels*) or to create an integrated representation of both *internal and external* worlds (related to higher information processing, managed by *higher levels*). The self-state (monitoring and control of, among others, body heat, energy, body volume, sexual behavior or respiratory, digestive, and cardiovascular states) is usually encoded in *lower-level* regions (hypothalamus and brainstem, respectively). The communication between these two levels needs to be, most of the time, reciprocal. This means that it is possible to (1) first detect a stimulus from the external world (multisensory event, pain, etc.) and then, in order to generate a goal response, poll the low levels as to the state of the system in that moment, and (2) first detect an internal need (homeostasis broken, need for food, etc.) and then transmit this information to the *processing* high level in order to check how are the world and the system (self) with respect to the world (by self-modeling). It is likely that this example of cross-level interactions could help regenerative biologists understand and synthesize across information present in diverse levels such as molecular networks and anatomical structures.

In order to create a stable and coherent perception of the world, the final *construct* needs to be, first, multisensorial. The cortical networks (both at the level of layers in a single cortical column and through connections between different cortical areas) are laid out in such a way that the stimulation of a specific region influences the activity of other areas (supporting a different sensory modality, just as touch can modulate visual cortex activity) (Mountcastle 1998, Laurienti et al. 2003, Calvert et al. 2004). The outcome of this multisensory and cross-modal integration is the understanding of the environment to make a goal-directed decision and plan a successful behavior (David et al. 2005, Friston and Kiebel 2009). In this sense, the primary function of the neocortex could be seen as information mixing (Garey 1999). What information must generate a response (relevant or salience stimulus) depends on multiple goal-related cognitive facts, such as attention, memory, and limbic influence. It remains to be seen whether regeneration likewise makes use of centralized structures that integrate distinct types of signal (e.g., chemical gradients and bioelectric properties).

The brain has finite processing capacity and therefore needs mechanisms to selectively enhance the information that is most relevant to one's current behavior (Rowe et al. 2002, Brown et al. 2011, Buschman and Kastner 2015). Recent work showed that an artificial peripheral stimulus (coming from a sensory neuroprosthesis) can be fully processed and understood only if simultaneous stimulation happens in the primary cortical area (Herrera-Rincon and Panetsos 2014). These two inputs should come, respectively, from the peripheral stimulus per se and from a *lower* region (the nucleus basalis magnocellularis [NBM], mostly through its cholinergic neurons), traditionally linked to attention. The signal that would provoke the output from NBM (this nucleus does not have direct connection with periphery) could come of the region mostly implied in planning (prefrontal cortex). In this sense, perception could be promoted by Hebbian-like, activity-dependent, synapse-based learning rules (see later) that strengthen and refine the circuits if (and only if) there is some circuitry to route sensory signals to the brain and motor commands out of it (Murphy and Corbett 2009). The mechanisms by which the body directs attention (in the sense of increased priority of measured properties for subsequent cellular responses) are a key aspect of fully understanding and augmenting regeneration.

4.4.3 Memory: Creation, stabilization, persistence, and recall

A fundamental property of regeneration is a kind of morphogenetic memory; most systems regenerate to one specific pattern after experimental deviations, but some allow re-writing of the target morphology, such that subsequent rounds of regeneration will restore the new shape (reviewed in Durant et al. 2016). Of course, some systems do not regenerate at all. It is possible to view these cases as instances of long-term memory, re-writing pattern memory, and lack of recall capability, respectively. Thus, we next discuss the mechanisms that implement memory in the brain, with an eye toward conserved aspects that could be carried over into the regeneration field.

Memory is a central ingredient for the goal-satisfaction circuit (Pezzulo et al. 2014, Ambrosini et al. 2015). In sensory processing and perception, the system (brain) compares the qualities of a new event with those of past experience. This *recall* will influence the value given to the stimulus (meaning relevance with respect to internal needs and external opportunities) and, consequently, to the planning and decision-making process for the present behavior. Then, the outcome of the goal-satisfaction circuit, or the current state, will be evaluated as a function of the past state (comparison, expectation, and *new learning*). This means that a system with memory is able to infer the past state (or experiences) from the current state (or present behavior). The link between past events and present behavior is measurable changes (physical, biochemical, structural, and functional) in the system, leading to memory traces or engrams (Semon and Simon 1921). The engrams are thus the circuit motifs underlying the creation, stabilization, persistence, and recovery of the traces left in bioelectrical circuits by experience (Dudai 2004, 2012). These sequential processes indicate that the memory is not something unitary but rather a gradual and continuous process (learning, consolidation, storage, and retrieval).

For applications in regeneration, it is important to keep in mind that *memory* refers not only to full-blown (self-aware) perceptual experiences but also to the whole range of phenomena at all scales of biological organization, where past events influence how the system responds to future events. In the brain, there are two major distinctions (duration and quality) in the taxonomy of memory systems. In terms of *duration*, we can distinguish between the maintenance and manipulation of information for brief periods (working memory, especially important for a goal-satisfaction circuit) and the acquisition and recovery of information over longer periods (long-term memory, especially important for remembering a pattern). Nondeclarative memory is revealed through task performance, without the requirement of *conscious content*, and indeed, *positional information memory* is a concept that has been explored in the regeneration field (Carlson 1983, Chakravarthy and Ghosh 1997, Brugger et al. 2002, Chang et al. 2002, Kragl et al. 2009, Hecht et al. 2011). These memory systems operate in parallel (working cooperatively and/or competitively) to support and guide behavior (Poldrack and Packard 2003, McDonald and Hong 2013).

4.4.4 Learning: Forming and encoding of memory traces

To support the essential latency of any memory system, it must contain a physical substrate where induced inputs can be materialized or recorded (Cajal 1894). In the case of the brain, *neuronal assemblies* compose this substrate, where neurons encoding an event physically change the properties of their synaptic contacts (Kandel 2001) to define the final engram. In regeneration, biochemical, electrical, or tensile properties could define the target anatomical state—the tissue's engram. The challenge behind this hypothesis is to define the cellular mechanism that allows materializing or recording such inputs that shape the tissue engram. In regenerating tissues, electrical synapses known as gap junctions are a key candidate for mediating the required cellular plasticity (Goel and Mehta 2013, Palacios-Prado et al. 2013, Pereda et al. 2013); however, other aspects of V_{mem}-based bioelectric circuit dynamics (stable attractor states) could also implement this property in purely physiological circuits. The *neurons that fire together wire together* scheme proposed by Hebb (2002) is readily transferred to bioelectrical signaling among cells during regeneration. Harnessing and exploiting the synchrony (in terms of activity level) of different cell networks (or assemblies) are potential future directions to explore in the control of patterning and regenerating tissue. Moreover, the appreciation of *reciprocal* signaling between pre- and post-synaptic elements is consistent with other forms of intercellular communication in somatic tissues.

Bliss and Lomo (Bliss and Lomo 1973, Bliss and Collingridge 1993) showed that after a train of electrical stimuli, neurons in an area of the hippocampus exhibit an increased and sustained capacity to respond to later stimulation by the same route. This increase in synaptic transmission could last for hours *in vitro* and days *in vivo*. This phenomenon is called long-term potentiation (LTP) and is the basis for synaptic plasticity. This kind of associative conditioning is enabled by the properties of the key player in memory formation, the *N*-methyl-D-aspartate (NMDA) receptor (Tsien et al. 1996, Nicoll and Malenka 1999). The NMDA receptor responds to the neurotransmitter glutamate. Opening of the NMDA receptor depends on both binding of the neurotransmitter and the pre-existing membrane potential. Therefore, neurotransmitter binding is not enough to open the NMDA receptor; in addition, the membrane needs to be depolarized to some threshold level before glutamate can produce NMDA receptor opening. This illustrates how cells exploit bioelectricity to implement time-dependent plasticity.

Epigenetic marking (histone modification and DNA methylation) seems to play a key role in the formation of cellular memory (Miller et al. 2008, 2010) in the adult brain. Interestingly, enzymes responsible for acetylation/deacetylation have been implied both in behavioral memory (Swank and Sweatt 2001, Levenson et al. 2004, Ganai et al. 2016) and in regulation of regeneration (Tseng et al. 2011, Taylor and Beck 2012, Robb and Sanchez Alvarado 2014), as have been other molecules such as cAMP response element binding protein (CREB) (Chera et al. 2007, Benito and Barco 2010, Stewart et al. 2011). As with gap junction and ion channel proteins, cognition and patterning share molecules, signaling and communication pathways, intracellular and extracellular mechanisms, and, perhaps, a bioelectric code. Indeed, endogenous electrical fields in the cortex, recorded as LFPs (electrophysiological mass signals reflecting analog measurement of network-level bioelectric state) influence neuronal behavior in real time (Frohlich and McCormick 2010*). Local field potentials, and not only the spiking activity, could encode intended goals during the planning (Kuang et al. 2016). The information storage and information transfer in the brain via LFPs (not necessarily the functional integration of the spiking activity for a given brain area) (Belitski et al. 2008, Deco et al. 2008, Pinotsis and Friston 2011, Friston et al. 2015) are consistent with the role of voltage gradients in many patterning processes as not only instructive signals but also coding spatial, temporal, and semantic information.

4.4.5 Goal-directed decision-making and action planning

We define goal-directed decision-making process as the ability of a system to (1) infer the future state as a function of both current and past states (self and environment), and (2) regulate its behavior, in a flexible manner, for orienting it to the goal (Verschure et al. 2003, 2014, Pezzulo et al. 2014, 2015). In engineering terms, this is an optimization or minimization (least-action) process, which seeks to reduce some quantity (e.g., the difference between current, damaged, anatomical state and the target morphology) (Kaila and Annila 2008, Friston et al. 2012, Friston et al. 2015). The decision-making process integrates four different cognitive modules: perception and attention (selection of relevant information—self-state and external world and goals), memory (mostly recall), prediction and valuation (reward), and selection and monitoring (planning of active behavior, such as motor sequences, and performance monitoring, with subsequent error detection and adjustment).

The first layer of this process, the representation of motivational state (Pezzulo and Rigoli 2011, Botvinick and Braver 2015), starts when the system needs a reason (Aarts et al. 2008)—an internal or external homeostatic range (Maddox and Markman 2010, Simpson and Balsam 2015). In order to generate a goal response, the system polls the lower levels for current state information. An internal need (e.g., homeostatic range

---

* This paper provides the first experimental evidence that the weak electric fields in the neocortex, responsible of small changes in V_{mem} of neurons that generate them, influence network spatiotemporal dynamics. Using both *in vivo* and *in vitro* electrophysiological recordings and computational modeling techniques, the authors show that this closed loop could guide the neuronal behavior in real time.

exceeded) is transmitted as information to the *processing* high level in order to check external state (world) and internal system (self) state with respect to the world. It is important to note that definitions of self and world vary, depending on the entity in question. Somatic cells have other somatic cells as their environment, whereas the animal as a whole faces the external environment. Healthy organs, tumors, and whole organisms may have very different internal representations of their *self* and their relationship to the immediate external environment in which they exist.

Second, the system assigns a value to each possible alternative, in function of current (self-modeling), past (memory), and future (prediction) states. This second layer is the valuation process. A goal-directed valuation system could use functional architectures of different cognitive motifs (Rangel et al. 2008). In the Pavlovian systems, the valence to a stimulus depends on conditioning, based on reward or punishment. The habit systems are established by means of a trial-error process in specific internal and external contexts. The goal-centered valuation could have to overcome these weighted pathways, which could mean to move away from the stability of the system. The goal-focused decisions are flexible, and the system is able to update the valence of stimulus-response association as soon as the internal or external context changes. This top-down rearranging could be supported by two basic mechanisms: gain modulation and communication through coherence (CTC) (Kerr et al. 2014). The gain modulation implies the increase in the intensity of the neuronal responses when they are activated by both feed forward and feedback connections (Larkum et al. 2004). The CTC hypothesizes that communication can occur between separate neuronal groups when they are in the same excitability state at the same time. This is important because it implies that communication in the brain is not only about spiking activity but also involves rhythms and oscillations of electrical states to carry information (Fries 2005, Engel et al. 2012).

After the action selection, executive control takes over in order to evaluate and predict the performance and the outcome of the selected action. It is important in regeneration, as in the brain, to understand the relationship between distributed parallel control processes and centralized controls. For example, knowing what organ to regenerate at an amputation site has to be a single (centralized) decision. However, aspects of tissue- or cell-level growth are independent. A zebrafish fin amputated at different distances from the body regenerates at a specific speed in order to reach the correct target morphology (Monteiro et al. 2014, Perathoner et al. 2014). A system displays executive control when it is able to generate flexible, integrated behavioral responses to different stimuli, overcoming the more limited reflex behavior, in order to reach a goal (Stokes et al. 2013). This implies that the system has resources for predicting or simulating the actions and their consequences, inhibiting an automatic and enhanced behavior (reflex, *habit*, and so on) that is not appropriate in the current context, facilitating (or initiating) the response that leads to an effective behavior, and dynamic task switching by tracking the outcomes of actions and behavior, depending on the context. In a broad sense, the controllers code and support both the goal (motivation) and the effects of action, or potential outcomes, on the future state (reward). Its activity is detected before (prospective value of reward [Stott and Redish 2015]), during (motor control), and after (sensorimotor loop contributing to stop signal [Jahfari et al. 2012]) the action takes place. Lastly, feedback allows the system to analyze and encode the outcomes and its potential consequences to improve future actions (learning). These different motifs or layers (representation, valuation, action selection, outcome evaluation, and learning) are not static or imperatively sequential. It is interesting to consider how many of these functions may have tractable analogs in the regulatory loops that implement regeneration.

Malfunctioning of the executive control and valuation layers entails several pathologies and psychiatric disorders (*non-effective* behaviors) (Gradin et al. 2011), including for example addictions, depression and anxiety, apathy (initiation impairment). Some of these may have analogies in patterning and regeneration, such as acquired neural dependence in limb regeneration, known as *nerve addiction* (Kumar and Brockes 2012*). One of the most interesting examples of the nerve addiction process is the aneurogenic limb (ANL) regeneration. In this model, if a salamander's limb develops without innervation, such an ANL can regenerate without nerve in adulthood. If the ANL is exposed to nerve (by transplantation), it becomes *addicted* to nerves (in the sense of later regenerating only if nerves are present; Fekete and Brockes 1988, Tassava and Olsen-Winner 2003, Kumar et al. 2011, Kumar and Brockes 2012). Nerve dependence could be seen as a parallel process to addiction or, in other words, reinforcement of learned associations (Di Chiara 1999, Everitt et al. 2001). In cognitive processing, if dependence exists, it is because the system has developed an adaptive state (with long-term changes—first, molecular and cellular) to compensate the excessive exposure (overstimulation; excellent review in Chao and Nestler 2004) by a *reinforcer* (drug of abuse). These (irreversible) compensatory changes occur, mainly, in the reward-motivation circuits for decision-making, resulting in inflexible behaviors or behavioral sensitization (Brebner et al. 2005). Viewing nerve dependence from this perspective suggests testable hypotheses for intervention (see Implications later and Figure 4.7).

---

* This review discusses the role of innervation in regeneration, with specific attention to nerve addiction, and the molecular mechanisms of nervedependent regeneration, including AG protein expression.

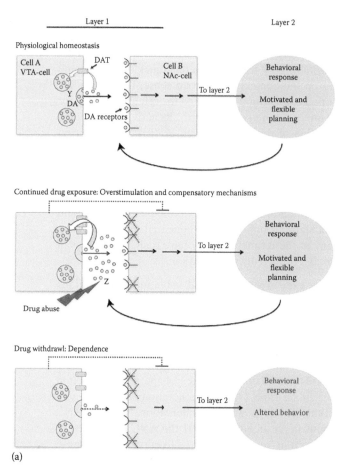

(a)

Figure 4.7 Addiction pathways and application to neural dependence in regeneration. Lessons from the brain applied to a specific model: parallelism between molecular/cellular substrate for drug addiction in brain and nerve addiction in aneurogenic limb (ANL) regeneration (see text and Table 4.1 for more details). (a) In the brain, the drugs of abuse (reinforcers or Z element) share a common mechanism: the overstimulation of synapsis between ventral tegmental area (VTA, cell A) and nucleus accumbens (NAc, cell B) neurons (or reward circuit, key for motivation and motor circuits responsible for execution of motivated behaviors). The overstimulation (red lines) occurs because the reinforcers (red ray) and produce a dysregulation of the synaptic homeostasis by means of excitation of presynaptic neuron or inhibition of dopamine (DA) reuptake (by dopamine transporter [DAT]), resulting in sustained release of DA. The elevated concentrations of synaptic DA induce cellular and molecular remodeling of the system, aiming to compensate for the excessive stimulation (compensatory mechanisms, dashed red lines). These changes are maintained for the long term and include, for example, desensitization and/or internalization of receptors, and synthesis of *inhibitory* molecules. Nevertheless, once the drug of abuse is no longer available, the previously mentioned compensatory mechanisms induce a dependence and withdrawal syndrome (anhedonic and/or craving behavior). In other words, to reestablish activation and motivation in the system, the drug intake is necessary. *(Continued)*

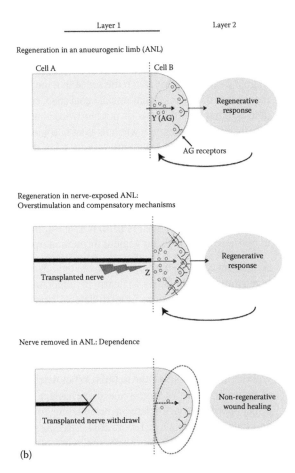

Figure 4.7 (Continued) (b) Representation of transition from nerve-independent to nerve-dependent regeneration in a salamander ANL after exposing it to a foreign nerve. The salamander ANL (developed without innervation) can regenerate in the absence of nerves (i.e., regeneration in ANL is nerve-independent; Yntema 1959). The levels of AG proteins and gland in the blastema are high enough to promote the proliferation and differentiation of cells to form the regenerated limb in the absence of nerves (Kumar et al. 2011). However, if the limb is exposed to nerves (by transplantation to a normal host), regeneration becomes nerve-dependent (Thornton and Thornton 1970), in the sense that the regeneration will now not be possible without nerve-regenerating supply. After ANL transplantation into a normal host, the nerve exposure leads to a dysregulation in homeostasis at the level of AG proteins and gland-derived signals. As consequence, the blastema acquires a permanent low tissue AG and hormonal state (dashed red lines). We hypothesize that this low-activated state is the consequence of nerve overstimulation–induced compensatory mechanisms, in line with the processes leading to addiction and dependence in the brain circuit. The ANL becomes addicted to the nerve, in the sense that it will not be able to regenerate after denervation. Future experiments on nerve dependence in regeneration should be directed to avoid the underlying causes to this compensatory state.

4.5 Summary: Where do we go from here?

Both normal regeneration and regulative embryogenesis stop precisely when the correct anatomical shape has been produced; this is a process akin to goal-directed behaviors, in the sense that the system can pursue multiple paths toward the same (anatomical) goal state, can accommodate unpredictable external perturbations (is not hardwired but flexible), and rests when it is satisfied (can recognize when its goal is achieved). All of these examples show the remarkable information processing that cells carry out, in order to create and maintain specific shapes (Levin 2012b). Likewise, non-neural cell networks process information about current and future anatomical shape. Although the brain operates muscles and glands in service of activity in ecological space, the computational processes of non-neural somatic networks control cell behaviors (differentiation, migration, and proliferation) to optimize the body's movement through morphospace.

A primary goal of developmental biology, synthetic bioengineering, and regenerative medicine is to learn to understand and control patterning networks, for applications in birth defects, organ regeneration, and cancer reprogramming (Ingber and Levin 2007, Doursat et al. 2013). In particular, it is crucial to tame the endogenous closed-loop pattern regulatory systems (flexible remodeling and regeneration pathways), as these offer the opportunity to exploit modularity to achieve needed changes in growth and form without micromanaging the details. What mechanisms underlie the ability of tissues to measure large-scale shape, detect deviations from the correct target morphology, implement remodeling toward repairing that shape, and know when to stop (Levin 2011)? Recent work has shown that, as in the brain, these control networks make use of ion channels, gap junctions (electrical synapses), and neurotransmitters (Levin 2012a, Tseng and Levin 2013). A parsimonious hypothesis is that this is no accident and that the brain learned its prodigious computational tricks from far more ancient developmental pathways, co-opting developmental bioelectricity and optimizing it for the speed needed for behavior. Although the spiking in the brain operates on millisecond scale, developmental bioelectricity involves steady, slow changes in ion fluxes, resting potentials, and electric fields.

4.5.1 New technologies

New technologies are required to continue pushing the study of bioelectricity in regeneration and development forward. To that end, a number of exciting new tools have been developed to report bioelectric signaling events and to manipulate voltage membrane potentials.

One of the most challenging aspects of studies in bioelectricity is to analyze endogenous and induced membrane voltages *in vivo*. A number of charged fluorescent dyes that localize specifically to hyperpolarized or depolarized regions of cells have been discovered. These include carbocyanine dyes, rhodamines, and octanols and have been used in several studies mentioned here. The application of these dyes in combination with one another or with other bioelectricity reporters is still being developed, and there are many effective ways to make use of them in bioelectricity studies (Adams and Levin 2012). Several such dyes and methods were applied in an elegant study of ion flux, including, Ca^{2+}, Na^+, K^+, H^+, and V_{mem} in the axolotl tail regeneration (Ozkucur et al. 2010).

A new and exciting alternative to dye labeling of tissues for voltage membrane visualization is the development of genetically encoded reporters (Lundby et al. 2008, Perron et al. 2012, Tsutsui et al. 2014, Empson et al. 2015, Knopfel et al. 2015). These transgenes can be incorporated into animal cells and will be expressed as fluorescent membrane voltage reporting proteins. One such promising reagent is the *Butterfly* reporter, which functions on the principle of fluorescence resonance energy transfer (FRET). The conformation of the Butterfly reporter proteins is such that the fluorescence donor end of the protein moves toward or away from the fluorescence receptor end of the protein in a membrane voltage–dependent manner, thus reporting the membrane voltage as a function of donor/receptor fluorescence (Empson et al. 2015). Another promising alternative to dye reporters is the use of highly sensitive biocompatible nanosensor ion reporters (Tyner et al. 2007). Sahari et al. and Dubach et al. reported on biocompatible nanospheres embedded with a pH-sensitive fluorophore and an ion-specific ionophore. Within the nanosensor, hydrogen ions are exchanged for ions attracted by the specific ionophore, thus changing the local pH inside the nanosensor and the fluorescence of the pH-sensitive fluorophore (Dubach et al. 2011, Sahari et al. 2015). This new panel of sensors also allows tracking of individual ions (Markova et al. 2008, Waseem et al. 2010, Ruckh et al. 2013), in addition to overall voltage. Of course, advances have also been made in vibrating probe technology for direct measurement of ion flux (Smith and Trimarchi 2001, Reid et al. 2007, Reid and Zhao 2011a, 2011b).

One of the more intriguing developments in the last decade for direct control of membrane voltage is the development of optogenetics (Boyden et al. 2005, Wyart et al. 2009, Arrenberg et al. 2010, Boyden 2011, Bernstein et al. 2012). The first applications made use of endogenous proteins derived from a variety of plants and lower organisms, which enabled light-triggered ion flow. The first application of this kind was channelrhodopsin-2, derived from algae, which when ectopically expressed can activate neurons by permitting cation flux in response to blue light (Boyden et al. 2005). Halorhodopsin from the archaebacterium *Natronomonas pharaonis* hyperpolarizes cells and has been used in neural suppression studies (Han and Boyden 2007). The light-sensitive H^+ pump archaerhodopsin has been exploited in amputated tails of tadpole to restore regeneration in non-regenerative developmental stages (Adams et al. 2007). Ongoing work is resulting in the development of magnetically (Long et al. 2015) and acoustically (Ibsen et al. 2015) triggered ion channels, which will help transition this technology to practical use in thick, optically opaque regenerating structures.

For addressing steps downstream of electrical signaling, a number of tools offer optogenetic control of neurotransmitter (Amatrudo et al. 2014) and second-messenger pathways such as cAMP (Weissenberger et al. 2011). For example, light-activatable serotonin molecules have been developed. These compounds are biologically inert until their *caged* side groups are released by specific wavelengths of light, thereby releasing the active serotonin (Rea et al. 2013). Such compounds will be useful in understanding the movement and signaling of serotonin through bioelectric networks such as those that regulate neural outgrowth (Blackiston et al. 2015). With the development of these high-resolution functional tools, and the increasing identification of transcriptional targets of bioelectrical activity (Pai et al. 2015b), testing successful strategies from computational neuroscience in regenerative contexts is becoming increasingly feasible.

4.5.2 Implications: Testable predictions and suggested approaches

The analogy between the brain and somatic pattern control makes several specific predictions. One prediction is that ion channels, GJs, and neurotransmitters should play a role in development; this has been amply demonstrated by the identification of a number of patterning channelopathies (Levin 2013), functional experiments in regenerative and developmental biology (Stewart et al. 2007), the teratogenic effects of numerous psychoactive drugs (Hernandez-Diaz and Levin 2014), and the pre-nervous roles of different neurotransmitters in embryonic patterning (Sullivan and Levin 2016). More specifically, our laboratory is now testing direct predictions of this model by exploring the ability of memory blockers, cognitive enhancers, and nootropic agents to alter regenerative morphology. Additional molecular targets that are suggested by this perspective for investigation as to roles in regeneration include CREB and c-Fos-hub transcriptional targets of electrical activity in the brain.

One layer above the molecular response lies the bioelectric dynamics itself. The ability of ion channel circuits to form stable memories in non-neural cells is only now beginning to be understood (Law and Levin 2015). In addition to the ongoing work on optogenetic targeting of resting potentials, approaches to be explored are weak applied currents: using new electrode technologies being developed for the CNS as part of the DARPA/GSK effort (Sinha 2013), it may be able to improve regenerative memory in the same way that transcranial direct current stimulation (tDCS) has shown to improve cognitive performance (Chi et al. 2010, Paulus 2011, Chi and Snyder 2012, Kuo and Nitsche 2012). It is possible that multitasking (parallel information processing) in regenerative tissues could be enhanced, as it occurs in the brain (Filmer et al. 2013) by altering the extracellular potential. Advances in understanding the message

Table 4.1 A Cognitive Perspective on Nerve Dependence for Regeneration

Label	Cognitive	Regeneration
Layer 1	Motivation station (nucleus accumbens [NAc])	Blastema
A cell	VTA neuron (DA-terminal from ventral tegmental area)	Blastema cell (glial cells)
B cell	NAc cell	Gland cells underlying Wound epithelium
Layer 2	High levels in goal circuit (Cx)	Wound epithelium
Y molecule	Dopamine (DA)	AG proteins (after [Kumar et al. 2007]) or gland-derived signals (after [Aberger et al. 1998, Kumar et al. 2010])
Z element (*reinforcer*)	Drug of abuse	Regenerating axons

Legend: Summary of possible analog elements participating in drug addiction and nerve dependence, respectively, showing their parallel structure.

encoding and passing in the brain oscillations could be applied to cracking and predictive control of the bioelectricity in regeneration, and vice versa.

The cognitive perspective suggests a new way to address nerve dependence in limb regeneration (Table 4.1 and Figure 4.7). In principle, strategies used to treat addiction could be tested for the ability to eliminate the innervation dependence on normal limbs. Addiction mechanisms can be described with the following simple circuit (Figure 4.7a): two cells in Layer 1, A and B, communicate with each other through molecule Y (dopamine [DA]). The output of this A–B interaction triggers a response in a higher-level Layer 2 (i.e., goal-centered decision-making). The Layer 2 response, in turn, influences Layer 1, closing the loop. Drugs of abuse such as cocaine and opiates (Figure 4.7, Z element) reinforce this A–B communication, overstimulating the circuit repeatedly. Addiction develops when the circuit responds by compensating for this overstimulation, either by inactivating Y receptors in the B cells or by synthesizing molecules to suppress the effects of Y (Figure 4.7, dashed red lines). Withdrawal of the drug of abuse, or Z element, produces a low-activity state in the circuit, since the compensatory mechanisms cannot be quickly reversed, giving rise to withdrawal symptoms, craving, anhedonia, and so on. In turn, the output to Layer 2 is affected, leading to the behavioral symptoms of addiction. We can translate this circuit to the regenerative context and, specifically, to the nerve-dependence event (Figure 4.7b). In this context, the goal of A–B communication in Layer 1 is to initiate regeneration, whereas Layer 2 responds by releasing holocrine signals for regeneration.

Following our hypothesis, the nerves (Z element or reinforcer) disrupt homeostasis at the level of AG proteins and gland-derived signals (Y molecules). Consequently, the system is encouraged to develop *compensatory* and non-reversible mechanisms to maintain homeostasis. Thus, the niche acquires a permanent low tissue AG and hormonal state, as has been experimentally observed (Kumar et al. 2011). Only when the system is again exposed to the reinforcer (as happens in regeneration with regenerating nerves in the blastema), activity levels are recovered, and the commands for initiating wound epithelium (WE) formation and, consequently, regeneration process are elaborated. Thus, we propose that similar mechanisms rule both development and regeneration and a possible role in other processes such as homeostasis and shape maintenance may have been underestimated. In fact, genetic manipulation (by plasmid electroporation), directed to increase the expression of newt anterior gradient (nAG) protein (ectopic Z signaling), is enough to rescue a denervated newt blastema and allow regeneration of the digit (Kumar et al. 2007). This approach can be seen as a kind of *replacement therapy* in drug addiction and indicates that the Layer 1 output, mediated by nAG and gland-derived signals, is necessary to trigger response in Layer 2 (WE). Our cognitive perspective applied to regenerative field is clearly visualized when the ANL model is considered (Figure 4.7b). The salamander aneurogenic limb (developed without innervation) can regenerate in the absence of nerves (i.e., regeneration in ANL is nerve-independent) (Yntema 1959). However, if the limb is exposed to nerves (by transplantation to a normal host), regeneration becomes nerve-dependent (Thornton and Thornton 1970), in the sense that the regeneration will now not be possible without a nerve supply. Thus, nerves could be acting like reinforcers, over-stimulating the system and leading to the subsequent establishment of sensitization and compensatory mechanisms (in line with the addiction circuit in the brain).

In the context of regeneration, a strategy to prevent nerve dependence could be to block the compensatory mechanisms triggered to maintain homeostasis, before dependence is firmly in place: (1) prevent AG downregulation by means of therapeutic intervention, (2) prevent changes in AG receptor activity (or Prod1 protein; da Silva et al. 2002, Kumar and Brockes 2012), for example, by transient inactivation to prevent AG-level downregulation, or (3) since nerve exposure results in a decreased number of glands (Kumar et al. 2011), possible strategies could include a hormonal replacement therapy. Another strategy, following the learning model for addiction, would be to prevent the re-wiring of the pathway in blastema-gland cells during development. We hypothesize that this re-wiring could be mediated by the co-activity between fibroblast growth factor (FGF) (probably a nerve-derived signal) and matrix metalloproteinase (MMP) (Satoh et al. 2011). One interesting approach will include the synchronous activation of alternative drivers for WE or blastema formation (e.g., by bioelectric or metabolic control, using Hebb-like rules) and the downregulation of FGF–MMP activity.

Additional approaches of this kind include training. Training paradigms are an excellent example of top-down control, because they offload the computational complexity of the task onto the natural plasticity of the cell networks. By rewarding and punishing appropriate outcomes, the operator specifies high-level target goals and lets the neural networks organize their topology and synapse strengths to result in the goal behavior, without having to micromanage this process. Similarly, could regeneration be augmented by a closed-loop *in vivo* bioreactor system that continuously rewarded cells for appropriate growth and patterning changes, with trophic factors or opioids? The guidance of regenerative patterning by rewarding for outcomes, instead of directing microstates (at the level of molecular pathways), represents a new frontier in this field, which our laboratory is currently exploring.

4.5.3 Bioelectric memories: Rewriting the default genome pattern in regeneration

Another key prediction of our view concerns the encoding of instructive information. In the brain, genetics establish the hardware—genes encode the available components and thus define the limits of cellular activity. However, the information content of the brain is not directly encoded by the genome, but rather, it arises dynamically through environmental stimuli (learning) and self-organizing dynamics of the electrochemical circuitry (plasticity). Is this the case in pattern formation as well? The current paradigm focuses entirely on DNA (sequence and chromatin structure) as the source of all patterning information. However, bioelectric signaling can modify pattern formation and edit the body plan by producing ectopic organs or revising axial organization (Tseng and Levin 2013). Indeed, a recent study underscored the dissociation between genomic and physiological control of shape by showing that simply altering the overall connectivity level among cells during flatworm regeneration (by inhibiting GJs) caused planaria to regenerate heads of other species of planaria despite a normal genomic sequence (Emmons-Bell et al. 2015).

However, more impressive would be the ability to rewrite pattern permanently. Could *long-term somatic memory* be edited, in the context of a wild-type genome, leading to a permanent change? A first example of this was shown in a different species of planaria (Nogi and Levin 2005), where targeting GJs for just 48 hours in a chunk of tissue caused planaria to regenerate two heads—one at the former anterior end (normal) and one at the posterior-facing end (which would normally grow a tail). Remarkably, these two-headed worms continue to regenerate as two-headed when cut in subsequent rounds of regeneration, in plain water, months after the GJ-blocking reagent is long gone from the tissue (Oviedo et al. 2010). The target morphology—the shape to which this animal regenerates on damage—has been permanently rewritten by temporarily editing the physiological network (Figure 4.8). This finding has clear similarity to plasticity (well known to be exhibited by electrical synapses) (Pereda et al. 2013): a brief induced change of GJ connectivity becomes stabilized to a long-term change (Levin 2014b). This interaction between bioelectric activity and voltage-gated GJs makes developmental bioelectrical networks especially suitable as a labile yet stable memory medium (Palacios-Prado and Bukauskas 2009). Another brain-like property exhibited in this effect is its holographic nature: in each

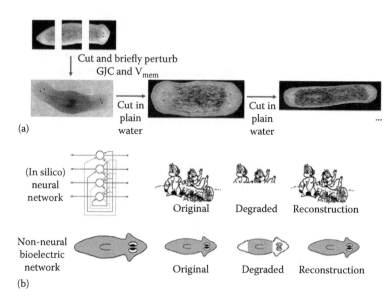

Figure 4.8 Re-writing regenerative memory. (a) Planaria regeneration has extremely high fidelity, normally regenerating exactly what's missing, no more no less, to give rise to perfectly normal worms. However, after inhibiting electrical synaptic connections among the network, a 2-headed form regenerates from middle fragments. Remarkably, this process continues in subsequent cuts in plain water, many weeks after all traces of the gap junctional blocker compound is gone from the tissues (Oviedo et al. 2010). (b) One way to understand the pattern memory and its revision is to visualize the bioelectric network of the worm's body as a large *neural network*, where many cell types couple to execute neural-like dynamics. These dynamics are known to store memories, able to supply missing components to inputs (for example supplying the missing portions of visual shapes that they have memorized previously when those shapes have been damaged by deletion). If planarian patterns represent the outcomes of bioelectric networks' attractor states, regeneration can be seen as recall of pattern memories from partial inputs.

round of cutting, the ectopic head (perhaps *epigenetically reprogrammed*) is removed and a middle fragment containing intestine and no head tissues still knows that it must make two heads if cut out. The patterning information is distributed non-locally throughout the network. The key open question concerns the encoding of head number within the GJ network's bioelectrical states, so that new patterns could be induced at will.

4.5.4 What does an answer look like: In search of a comprehensive model

This field is advancing rapidly in its mechanistic details at the cellular level: the genetics of endogenous ion channels causing the gradients, the transduction mechanisms that control transcription after V_{mem} change, and the gene expression changes downstream of bioelectrical signaling

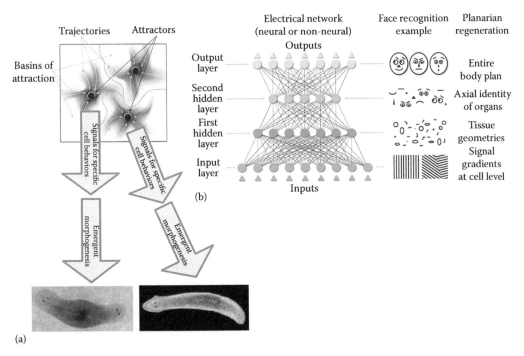

Figure 4.9 Hypothesis: Patterning outcomes as neural network attractors. (a) Memories are often analyzed in the neural network field as stable attractors in the dynamical state space of the network. We suggest a hypothesis where each attractor state issues differential instructions to the cells. Moreover, the network can be shifted into different attractors corresponding to stable, alternative configurations for the planarian body plan. (b) Another important feature of the neural net paradigm is that it provides a quantitative explanation of how *higher-order* (tissue, organ, and body plan) features are processed by cell networks making patterning decisions. Each deeper layer of neural nets processes increasingly complex features: in the case of face recognition, for example, the input layer sees pixels (analogous to protein levels), the next layer generalizes to extract edges (analogous to cell types), the next layer identifies larger structures (analogous to organs), and the next layer outputs final decisions about face identity (or anatomical body plans). Image courtesy of Jeremy Guay, of Peregrine Creative.

are all becoming clear (Yang and Brackenbury 2013, Pai et al. 2015b). What are largely missing are the models that explain large-scale features of pattern (the specific size and shape of individual organs during regeneration, such as planarian heads and amphibian hands). This gap is being addressed by the development of computational models that go beyond necessary molecular pathways and try to infer large-scale dynamics from functional data to explain *sufficient* dynamics that gives rise to specific shape and its regulation (Lobo et al. 2013a, 2013b, Lobo et al. 2014a, 2014b, Bessonov et al. 2015, Friston et al. 2015, Lobo and Levin 2015, Tosenberger et al. 2015). The continued development of these models will be essential to the inference of specific interventions for desired anatomical configuration changes. One source of quantitative models for this process is the field of artificial neural networks, which provides mechanistic explanations of capabilities such as memory and higher-order decision-making (Figure 4.9).

Modeling of self-organizing patterns and memory in non-neural circuits is starting to be developed (Cervera et al. 2014, 2015, Law and Levin 2015). In addition, techniques, such as optogenetics (Adams et al. 2013, 2014), are starting to come online for imposing desired voltage patterns onto tissue *in vivo*. As in the brain, where optogenetics is used to insert memories directly into brains (Ramirez et al. 2013, Liu et al. 2014), these techniques will be crucial to learn to rewrite pattern memories during regeneration or embryogenesis. However, as in neuroscience, there is more than one level at which progress needs to be made. A mature understanding of the brain requires synthesis of data from people working on the genetics and biochemistry of specific neurotransmitter receptors and their downstream molecular signaling, with the insights of workers at the level of circuits, behavior, cognitive science, and psychology.

Thus, neuroscience offers developmental bioelectricity more than just tools and molecular mechanisms: it offers a unique paradigm, otherwise unavailable to molecular and cell biologists, of the emergence of higher levels of organization that have both causal potency and experimental tractability. What developmental bioelectricity is missing most is not ever-finer resolution of molecular detail but new formalisms and conceptual tools for linking the dynamics of physiological circuits with downstream patterning outcomes. Developmental biology is currently focused entirely in a bottom-up mode, with molecules being the preferred level of explanation. Neuroscience teaches us that we must look upward and downward, for emergent levels with their own rules and advantages (Friston et al. 2015). For example, training an animal to a particular complex behavior is far more efficient than attempting to elicit the same behavior by manipulating individual molecules within the brain. Moreover, it is now known that changes at the genetic and chemical levels can be induced by cognitive therapies—top-down control of tissue structure and function by mental events and experiences. If patterning tissues are *primitive cognitive agents*, in the sense that they can be profitably understood as memory-bearing, information-processing, goal-directed cybernetic systems (Pezzulo and Levin 2015), then a whole new set of approaches becomes available for regenerative medicine.

If we understood the bioelectric code, we could interact with it at these higher levels of organization, taking advantage of endogenous modularity and perhaps rationally controlling anatomical outcomes, without having to micromanage molecular networks. Already a number of non-neural systems, such as pancreatic physiology and its disorders (Goel and Mehta 2013), cardiac memory (Chakravarthy and Ghosh 1997), and bone remodeling (Spencer and Genever 2003), are being explained in the context of learning models; a fundamental open direction in this field is to build a computational pipeline to extract goal patterns from bioelectric state data. This is quite parallel to efforts to extract image data from brain measurements (Nishimoto et al. 2011) and will greatly improve the control of endogenous and bioengineered pattern formation. Mature progress in this field will require smooth integration across levels: from molecular to physiological to anatomical (Figure 4.10). Excitingly, the

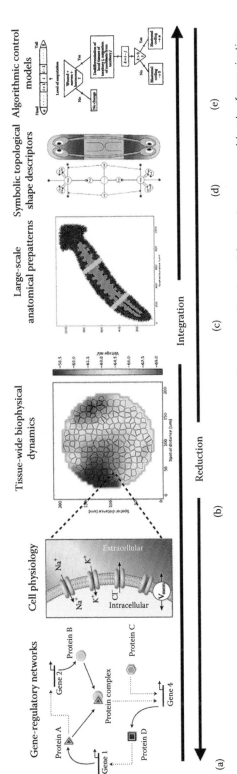

Figure 4.10 Integrating multiple scales of pattern regulations. Regeneration (and remodeling broadly) operates on several levels of organization, and effective intervention for anatomical repair will require knowledge of not only molecular mechanisms but also the integration of lower levels into higher ones. (a) In cells, protein and gene-regulatory networks determine the presence of specific ion channels and other physiological signaling molecules. (b) Cells group into bioelectric networks, which underlie circuits that establish stable spatio-temporal patterns of resting potential and electric fields at the level of tissues. (c) These bioelectric gradients integrate along the whole organism to establish regions that will turn into specific organs. (d) The memories are established at the levels of major anatomical structures, visualized here in an AI-tractable graph representation. (e) At the most global level, the process can be represented as an algorithmic loop that reveals the logic of pattern regulation. A mature science of regeneration must enable the quantitative integration of lower-level information into progressively higher levels, so that control can be exerted at the optimally efficient level. Images in panels A–C are courtesy of Alexis Pietak and Jeremy Guay of Peregrine Creative. Image in panel D is courtesy of Daniel Lobo. Image in panel E is taken with permission from Slack, J. M., (1980), JTB, 82(1): 105–40.

knowledge flow is likely not all in one direction: cracking the bioelectric code in patterning tissues is likely to, in turn, benefit fundamental neuroscience by showing, in perhaps a simpler context, how to extract semantic content from bioelectrical cell states in the brain.

4.6 Conclusion

It is clear that transformative advances in regeneration will require linking the higher levels of anatomy and its regulation to the molecular pathways to which we have direct access. We have outlined here one area of science in which a successful program of synthesis and reduction is linking molecular pathways to higher-order, goal-directed activity. Many important analogies remain to be explored computationally and experimentally, providing a fertile research program for the coming decades. Other areas promising similar kinds of insights, such as variational principles in physics and control theory, remain to be explored. This task requires that we recruit engineers, physicists, and computer scientists, to apply deep, fundamental concepts from these fields. This would be facilitated by creation of more reviews and introductory materials on regeneration that do not drown readers in molecular details but highlight functional, logical aspects of key problems in regeneration. We believe that the future of our field in its practical implementation lies in capitalizing on inherent modularity and computational properties of cellular networks. This will allow us to overcome inherent complexity of functional organs and develop therapies for the biomedical control of growth and form.

Acknowledgments

We gratefully acknowledge the support of the National Institutes of Health (NIH) (HD81401-01 and AR061988), W. M. KECK Foundation, the Paul G. Allen Family Foundation, the Templeton World Charity Foundation—TWCF0089/AB55, and the G. Harold and Leila Y. Mathers Charitable Foundation.

References

Aarts, H., R. Custers, and M. Veltkamp. 2008. Goal priming and the affective-motivational route to nonconscious goal pursuit. *Social Cognition* 26 (5):555–577. doi:10.1521/soco.2008.26.5.555.

Aberger, F., G. Weidinger, H. Grunz, and K. Richter. 1998. Anterior specification of embryonic ectoderm: The role of the Xenopus cement gland-specific gene XAG-2. *Mechanisms of Development* 72 (1–2):115–130.

Adams, D. S. 2008. A new tool for tissue engineers: Ions as regulators of morphogenesis during development and regeneration. *Tissue Engineering Part A* 14 (9):1461–1468.

Adams, D. S., J. M. Lemire, R. H. Kramer, and M. Levin. 2014. Optogenetics in Developmental Biology: Using light to control ion flux-dependent signals in Xenopus embryos. *The International Journal of Developmental Biology* 58:851–861. doi:10.1387/ijdb.140207ml.

Adams, D. S., and M. Levin. 2012. General principles for measuring resting membrane potential and ion concentration using fluorescent bioelectricity reporters. *Cold Spring Harbor protocols* 2012 (4):385–397. doi:10.1101/pdb.top067710.

Adams, D. S., and M. Levin. 2013. Endogenous voltage gradients as mediators of cell-cell communication: Strategies for investigating bioelectrical signals during pattern formation. *Cell and Tissue Research* 352 (1):95–122. doi:10.1007/s00441-012-1329-4.

Adams, D. S., A. Masi, and M. Levin. 2007. H+ pump-dependent changes in membrane voltage are an early mechanism necessary and sufficient to induce Xenopus tail regeneration. *Development* 134 (7):1323–1335.

Adams, D. S., A. S. Tseng, and M. Levin. 2013. Light-activation of the Archaerhodopsin H(+)-pump reverses age-dependent loss of vertebrate regeneration: Sparking system-level controls in vivo. *Biology Open* 2 (3):306–313. doi:10.1242/bio.20133665.

Ainge, J. A., M. Tamosiunaite, F. Woergoetter, and P. A. Dudchenko. 2007. Hippocampal CA1 place cells encode intended destination on a maze with multiple choice points. *Journal of Neuroscience* 27 (36):9769–9779. doi:10.1523/jneurosci.2011-07.2007.

Ainge, J. A., M. Tamosiunaite, F. Worgotter, and P. A. Dudchenko. 2012. Hippocampal place cells encode intended destination, and not a discriminative stimulus, in a conditional T-maze task. *Hippocampus* 22 (3):534–543. doi:10.1002/hipo.20919.

Alam, T., H. Maruyama, C. Li, S. I. Pastuhov, P. Nix, M. Bastiani, N. Hisamoto, and K. Matsumoto. 2016. Axotomy-induced HIF-serotonin signalling axis promotes axon regeneration in C. elegans. *Nature Communications* 7:10388. doi:10.1038/ncomms10388.

Amatrudo, J. M., J. P. Olson, G. Lur, C. Q. Chiu, M. J. Higley, and G. C. Ellis-Davies. 2014. Wavelength-selective one- and two-photon uncaging of GABA. *ACS Chemical Neuroscience* 5 (1):64–70. doi:10.1021/cn400185r.

Ambrosini, E., G. Pezzulo, and M. Costantini. 2015. The eye in hand: Predicting others' behavior by integrating multiple sources of information. *Journal of Neurophysiology* 113 (7):2271–2279. doi:10.1152/jn.00464.2014.

Apter, M. J. 1966. *Cybernetics and Development*. New York: Pergamon Press.

Arocena, M., M. Zhao, J. M. Collinson, and B. Song. 2010. A time-lapse and quantitative modelling analysis of neural stem cell motion in the absence of directional cues and in electric fields. *Journal of Neuroscience Research* 88 (15):3267–3274. doi:10.1002/jnr.22502.

Arrenberg, A. B., D. Y. Stainier, H. Baier, and J. Huisken. 2010. Optogenetic control of cardiac function. *Science* 330 (6006):971–974. doi:10.1126/science.1195929.

Ashby, W. R. 1956. *An Introduction to Cybernetics*. New York: J. Wiley.

Barbado, M., K. Fablet, M. Ronjat, and M. De Waard. 2009. Gene regulation by voltage-dependent calcium channels. *Biochimica et Biophysica Acta* 1793 (6):1096–1104. doi:10.1016/j.bbamcr.2009.02.004.

Barca, L., and G. Pezzulo. 2015. Tracking second thoughts: Continuous and discrete revision processes during visual lexical decision. *PLoS One* 10 (2):e0116193. doi:10.1371/journal.pone.0116193.

Barghouth, P. G., M. Thiruvalluvan, and N. J. Oviedo. 2015. Bioelectrical regulation of cell cycle and the planarian model system. *Biochimica et Biophysica Acta*. doi:10.1016/j.bbamem.2015.02.024.

Barkai, N., and D. Ben-Zvi. 2009. 'Big frog, small frog'—Maintaining proportions in embryonic development. *FEBS J* 276 (5):1196–1207.

Barker, A. T., L. F. Jaffe, and J. W. Vanable, Jr. 1982. The glabrous epidermis of cavies contains a powerful battery. *American Journal of Physiology* 242 (3):R358–R366.

Barreiro-Iglesias, A., K. S. Mysiak, A. L. Scott, M. M. Reimer, Y. Yang, C. G. Becker, and T. Becker. 2015. Serotonin promotes development and regeneration of spinal motor neurons in zebrafish. *Cell Reports* 13 (5):924–932. doi:10.1016/j.celrep.2015.09.050.

Bassett, C. A., and R. O. Becker. 1962. Generation of electric potentials by bone in response to mechanical stress. *Science* 137 (3535):1063–1064.

Bates, E. 2015. Ion channels in development and cancer. *Annual Review of Cell and Developmental Biology* 31:231–247. doi:10.1146/annurev-cellbio-100814-125338.

Bean, B. P. 2007. The action potential in mammalian central neurons. *Nature Review Neuroscience* 8 (6):451–465. doi:10.1038/nrn2148.

Beane, W. S., J. Morokuma, D. S. Adams, and M. Levin. 2011. A chemical genetics approach reveals H, K-ATPase-mediated membrane voltage is required for planarian head regeneration. *Chemistry & Biology* 18 (1):77–89.

Beane, W. S., J. Morokuma, J. M. Lemire, and M. Levin. 2013. Bioelectric signaling regulates head and organ size during planarian regeneration. *Development* 140 (2):313–322. doi:10.1242/dev.086900.

Belitski, A., A. Gretton, C. Magri, Y. Murayama, M. A. Montemurro, N. K. Logothetis, and S. Panzeri. 2008. Low-frequency local field potentials and spikes in primary visual cortex convey independent visual information. *Journal of Neuroscience* 28 (22):5696–5709. doi:10.1523/jneurosci.0009-08.2008.

Ben-Zvi, D., A. Fainsod, B. Z. Shilo, and N. Barkai. 2013. Scaling of dorsal-ventral patterning in the Xenopus laevis embryo. *BioEssays: News and Reviews in Molecular, Cellular and Developmental Biology*. doi:10.1002/bies.201300136.

Benito, E., and A. Barco. 2010. CREB's control of intrinsic and synaptic plasticity: Implications for CREB-dependent memory models. *Trends in Neurosciences* 33 (5):230–240. doi:10.1016/j.tins.2010.02.001.

Bernstein, J. G., P. A. Garrity, and E. S. Boyden. 2012. Optogenetics and thermogenetics: Technologies for controlling the activity of targeted cells within intact neural circuits. *Current Opinion in Neurobiology* 22 (1):61–71. doi:10.1016/j.conb.2011.10.023.

Bertorello, A. M., and J. K. Zhu. 2009. SIK1/SOS2 networks: Decoding sodium signals via calcium-responsive protein kinase pathways. *Pflugers Archiv* 458 (3):613–619. doi:10.1007/s00424-009-0646-2.

Bessonov, N., M. Levin, N. Morozova, N. Reinberg, A. Tosenberger, and V. Volpert. 2015. On a model of pattern regeneration based on cell memory. *PloS one* 10 (2):e0118091. doi:10.1371/journal.pone.0118091.

Bialek, W., F. Rieke, R. R. de Ruyter van Steveninck, and D. Warland. 1991. Reading a neural code. *Science* 252 (5014):1854–1857.

Blackiston, D., D. S. Adams, J. M. Lemire, M. Lobikin, and M. Levin. 2011. Transmembrane potential of GlyCl-expressing instructor cells induces a neoplastic-like conversion of melanocytes via a serotonergic pathway. *Disease Models & Mechanisms* 4 (1):67–85. doi:10.1242/dmm.005561.

Blackiston, D. J., G. M. Anderson, N. Rahman, C. Bieck, and M. Levin. 2015. A novel method for inducing nerve growth via modulation of host resting potential: Gap junction-mediated and serotonergic signaling mechanisms. *Neurotherapeutics* 12 (1):170–184. doi:10.1007/s13311-014-0317-7.

Blanke, O. 2012. Multisensory brain mechanisms of bodily self-consciousness. *Nature Reviews Neuroscience* 13 (8):556–571. doi:10.1038/nrn3292.

Bliss, T. V., and G. L. Collingridge. 1993. A synaptic model of memory: Long-term potentiation in the hippocampus. *Nature* 361 (6407):31–39.

Bliss, T. V., and T. Lomo. 1973. Long-lasting potentiation of synaptic transmission in the dentate area of the anaesthetized rabbit following stimulation of the perforant path. *Journal of Physiology* 232 (2):331–356.

Bodemer, C. W. 1964. Evocation of regrowth phenomena in anuran limbs by electrical stimulation of the nerve supply. *The Anatomical Record* 148:441–457.

Bongard, J., V. Zykov, and H. Lipson. 2006. Resilient machines through continuous self-modeling. *Science* 314 (5802):1118–1121. doi:10.1126/science.1133687.

Bonnici, H. M., D. Kumaran, M. J. Chadwick, N. Weiskopf, D. Hassabis, and E. A. Maguire. 2011. Decoding representations of scenes in the medial temporal lobes. *Hippocampus*. doi:10.1002/hipo.20960.

Borgens, R. B. 1982. What is the role of naturally produced electric current in vertebrate regeneration and healing. *International Review of Cytology* 76:245–298.

Borgens, R. B. 1988a. Stimulation of neuronal regeneration and development by steady electrical fields. *Advances in Neurology* 47:547–564.

Borgens, R. B. 1988b. Voltage gradients and ionic currents in injured and regenerating axons. *Advances in Neurology* 47:51–66.

Borgens, R. B., A. R. Blight, and M. E. McGinnis. 1990. Functional recovery after spinal cord hemisection in guinea pigs: The effects of applied electric fields. *Journal of Comparative Neurology* 296 (4):634–653.

Borgens, R. B., J. W. Vanable, Jr., and L. F. Jaffe. 1977. Bioelectricity and regeneration. I. Initiation of frog limb regeneration by minute currents. *Journal of Experimental Zoology* 200 (3):403–416.

Botvinick, M., and T. Braver. 2015. Motivation and cognitive control: From behavior to neural mechanism. *Annual Review of Psychology* 66:83–113. doi:10.1146/annurev-psych-010814-015044.

Boyden, E. S. 2011. A history of optogenetics: The development of tools for controlling brain circuits with light. *F1000 Biology Reports* 3:11. doi:10.3410/b3-11.

Boyden, E. S., F. Zhang, E. Bamberg, G. Nagel, and K. Deisseroth. 2005. Millisecond-timescale, genetically targeted optical control of neural activity. *Nature Neuroscience* 8 (9):1263–1268. doi:10.1038/nn1525.

Brebner, K., T. P. Wong, L. Liu, Y. Liu, P. Campsall, S. Gray, L. Phelps, A. G. Phillips, and Y. T. Wang. 2005. Nucleus accumbens long-term depression and the expression of behavioral sensitization. *Science* 310 (5752):1340–1343. doi:10.1126/science.1116894.

Brockes, J. P. 1998. Regeneration and cancer. *Biochimica et Biophysica Acta* 1377 (1):M1–11.

Brown, H., K. Friston, and S. Bestmann. 2011. Active inference, attention, and motor preparation. *Frontiers in Psychology* 2:218. doi:10.3389/fpsyg.2011.00218.

Brugger, P., E. Macas, and J. Ihlemann. 2002. Do sperm cells remember? *Behavioural Brain Research* 136 (1):325–328.

Bubenik, A. B., and R. Pavlansky. 1965. Trophic responses to trauma in growing antlers. *Journal of Experimental Zoology* 159 (3):289–302.

Buschman, T. J., and S. Kastner. 2015. From behavior to neural dynamics: An integrated theory of attention. *Neuron* 88 (1):127–144. doi:10.1016/j.neuron.2015.09.017.

Buznikov, G. A., and Y. B. Shmukler. 1981. Possible role of "prenervous" neurotransmitters in cellular interactions of early embryogenesis: A hypothesis. *Neurochemical Research* 6 (1):55–68.

Buzsaki, G. 2002. Theta oscillations in the hippocampus. *Neuron* 33 (3):325–340.

Cajal, S. R. Y. 1894. The Croonian Lecture: La Fine Structure des Centres Nerveux. *Proceedings of the Royal Society of London* 55 (331–335):444–468. doi:10.1098/rspl.1894.0063.

Calvert, G., C. Spence, and B. E. Stein. 2004. *The Handbook of Multisensory Processes.* Cambridge, MA: MIT Press.

Carlson, B. M. 1983. Positional memory in vertebrate limb development and regeneration. In *Limb Development and Regeneration*, Edited by J. F. Fallon and A. I. Caplan. pp. 433–443. New York: Alan R. Liss.

Cervera, J., A. Alcaraz, and S. Mafe. 2014. Membrane potential bistability in nonexcitable cells as described by inward and outward voltage-gated ion channels. *The Journal of Physical Chemistry B* 118 (43):12444–12450. doi:10.1021/jp508304h.

Cervera, J., J. A. Manzanares, and S. Mafe. 2015. Electrical coupling in ensembles of nonexcitable cells: Modeling the spatial map of single cell potentials. *The Journal of Physical Chemistry B* 119 (7):2968–2978. doi:10.1021/jp512900x.

Chakravarthy, S. V., and J. Ghosh. 1997. On Hebbian-like adaptation in heart muscle: A proposal for 'cardiac memory'. *Biological Cybernetics* 76 (3):207–215.

Chang, H. Y., J. T. Chi, S. Dudoit, C. Bondre, M. van de Rijn, D. Botstein, and P. O. Brown. 2002. Diversity, topographic differentiation, and positional memory in human fibroblasts. *Proceedings of the National Academy of Sciences of the United States of America* 99 (20):12877–12882. doi:10.1073/pnas.162488599.

Chao, J., and E. J. Nestler. 2004. Molecular neurobiology of drug addiction. *Annual Review of Medicine* 55:113–132. doi:10.1146/annurev.med.55.091902.103730.

Chera, S., K. Kaloulis, and B. Galliot. 2007. The cAMP response element binding protein (CREB) as an integrative HUB selector in metazoans: Clues from the hydra model system. *Biosystems* 87 (2–3):191–203.

Chernet, B. T., C. Fields, and M. Levin. 2015. Long-range gap junctional signaling controls oncogene-mediated tumorigenesis in Xenopus laevis embryos. *Frontiers in Physiology* 5:519. doi:10.3389/fphys.2014.00519.

Chernet, B. T., and M. Levin. 2013. Transmembrane voltage potential is an essential cellular parameter for the detection and control of tumor development in a Xenopus model. *Disease Models & Mechanisms* 6 (3):595–607. doi:10.1242/dmm.010835.

Chernet, B. T., and M. Levin. 2014. Transmembrane voltage potential of somatic cells controls oncogene-mediated tumorigenesis at long-range. *Oncotarget* 5 (10):3287–3306.

Chi, R. P., F. Fregni, and A. W. Snyder. 2010. Visual memory improved by non-invasive brain stimulation. *Brain Research* 1353:168–175. doi:10.1016/j.brainres.2010.07.062.

Chi, R. P., and A. W. Snyder. 2012. Brain stimulation enables the solution of an inherently difficult problem. *Neuroscience Letters* 515 (2):121–124. doi:10.1016/j.neulet.2012.03.012.

Clark, B., and M. Hausser. 2006. Neural coding: Hybrid analog and digital signalling in axons. *Current Biology* 16 (15):R585–R588. doi:10.1016/j.cub.2006.07.007.

Colgin, L. L., E. I. Moser, and M. B. Moser. 2008. Understanding memory through hippocampal remapping. *Trends in Neurosciences* 31 (9):469–477. doi:10.1016/j.tins.2008.06.008.

da Silva, S. M., P. B. Gates, and J. P. Brockes. 2002. The newt ortholog of CD59 is implicated in proximodistal identity during amphibian limb regeneration. *Developmental Cell* 3 (4):547–555.

Dahal, G. R., J. Rawson, B. Gassaway, B. Kwok, Y. Tong, L. J. Ptacek, and E. Bates. 2012. An inwardly rectifying K+ channel is required for patterning. *Development* 139 (19):3653–3664. doi:10.1242/dev.078592.

David, O., L. Harrison, and K. J. Friston. 2005. Modelling event-related responses in the brain. *Neuroimage* 25 (3):756–770. doi:10.1016/j.neuroimage.2004.12.030.

Deco, G., V. K. Jirsa, P. A. Robinson, M. Breakspear, and K. Friston. 2008. The dynamic brain: From spiking neurons to neural masses and cortical fields. *PLoS Computational Biology* 4 (8):e1000092. doi:10.1371/journal.pcbi.1000092.

Di Chiara, G. 1999. Drug addiction as dopamine-dependent associative learning disorder. *European Journal of Pharmacology* 375 (1–3):13–30.

Donaldson, D. J., and J. M. Mason. 1975. Cancer-related aspects of regeneration research: A review. *Growth* 39 (4):475–496.

Doursat, R., and C. Sanchez. 2014. Growing fine-grained multicellular robots. *Soft Robotics* 1 (2):110–121.

Doursat, R., H. Sayama, and O. Michel. 2012. Morphogenetic engineering: Reconciling self-organization and architecture. *Morphogenetic Engineering: Toward Programmable Complex Systems* 1–24. doi:10.1007/978-3-642-33902-8_1.

Doursat, R., H. Sayama, and O. Michel. 2013. A review of morphogenetic engineering. *Natural Computing* 12 (4):517–535. doi:10.1007/S11047-013-9398-1.

Dragunow, M., W. C. Abraham, M. Goulding, S. E. Mason, H. A. Robertson, and R. L. Faull. 1989. Long-term potentiation and the induction of c-Fos mRNA and proteins in the dentate gyrus of unanesthetized rats. *Neuroscience Letters* 101 (3):274–280.

Dragunow, M., and R. Faull. 1989. The use of c-Fos as a metabolic marker in neuronal pathway tracing. *Journal of Neuroscience Methods* 29 (3):261–265.

Dubach, J. M., E. Lim, N. Zhang, K. P. Francis, and H. Clark. 2011. In vivo sodium concentration continuously monitored with fluorescent sensors. *Integrative Biology (Camb)* 3 (2):142–148. doi:10.1039/c0ib00020e.

Dudai, Y. 2004. The neurobiology of consolidations, or, how stable is the engram? *Annual Review of Psychology* 55:51–86.

Dudai, Y. 2012. The restless engram: Consolidations never end. *Annual Review of Neurosciences* 35:227–247. doi:10.1146/annurev-neuro-062111-150500.

Durant, F., D. Lobo, J. Hammelman, and M. Levin. 2016. Physiological controls of large-scale patterning in planarian regeneration. *Regeneration* 3:78–102.

Emmons-Bell, M., F. Durant, J. Hammelman, N. Bessonov, V. Volpert, J. Morokuma, K. Pinet et al. 2015. Gap junctional blockade stochastically induces different species-specific head anatomies in genetically wild-type Girardia dorotocephala flatworms. *International Journal of Molecular Sciences* 16 (11):27865–27896. doi:10.3390/ijms161126065.

Empson, R. M., C. Goulton, D. Scholtz, Y. Gallero-Salas, H. Zeng, and T. Knopfel. 2015. Validation of optical voltage reporting by the genetically encoded voltage indicator VSFP-Butterfly from cortical layer 2/3 pyramidal neurons in mouse brain slices. *Physiological Reports* 3 (7):e12468. doi:10.14814/phy2.12468.

Engel, A. K., D. Senkowski, and T. R. Schneider. 2012. *Multisensory Integration through Neural Coherence the Neural Bases of Multisensory Processes.* Edited by M. M. Murray and M. T. Wallace. Boca Raton, FL: Llc.

Everitt, B. J., A. Dickinson, and T. W. Robbins. 2001. The neuropsychological basis of addictive behaviour. *Brain Research Reviews* 36 (2–3):129–138.

Farinella-Ferruzza, N. 1956. The transformation of a tail into a limb after xenoplastic transformation. *Experientia* 15:304–305.

Fekete, D. M., and J. P. Brockes. 1988. Evidence that the nerve controls molecular identity of progenitor cells for limb regeneration. *Development* 103 (3):567–573.

Ferbinteanu, J., and M. L. Shapiro. 2003. Prospective and retrospective memory coding in the hippocampus. *Neuron* 40 (6):1227–1239.

Filmer, H. L., J. B. Mattingley, and P. E. Dux. 2013. Improved multitasking following prefrontal tDCS. *Cortex* 49 (10):2845–2852. doi:10.1016/j.cortex.2013.08.015.

Filoni, S., C. P. Velloso, S. Bernardini, and S. M. Cannata. 1995. Acquisition of nerve dependence for the formation of a regeneration blastema in amputated hindlimbs of larval Xenopus laevis: The role of limb innervation and that of limb differentiation. *Journal of Experimental Zoology* 273 (4):327–341. doi:10.1002/jez.1402730407.

Frazee, O. E. 1909. The effect of electrical stimulation upon the rate of regeneration in Rana pipiens and Amblystoma jeffersonianum. *Journal of Experimental Zoology* 7:457–476.

Fries, P. 2005. A mechanism for cognitive dynamics: Neuronal communication through neuronal coherence. *Trends in Cognitive Sciences* 9 (10):474–480. doi:10.1016/j.tics.2005.08.011.

Friston, K., and S. Kiebel. 2009. Cortical circuits for perceptual inference. *Neural Networks* 22 (8):1093–1104. doi:10.1016/j.neunet.2009.07.023.

Friston, K., M. Levin, B. Sengupta, and G. Pezzulo. 2015. Knowing one's place: A free-energy approach to pattern regulation. *Journal of the Royal Society Interface* 12 (105). doi:10.1098/rsif.2014.1383.

Friston, K., S. Samothrakis, and R. Montague. 2012. Active inference and agency: Optimal control without cost functions. *Biological Cybernetics* 106 (8–9):523–541. doi:10.1007/s00422-012-0512-8.

Frohlich, F., and D. A. McCormick. 2010. Endogenous electric fields may guide neocortical network activity. *Neuron* 67 (1):129–143. doi:10.1016/j.neuron.2010.06.005.

Fukumoto, T., R. Blakely, and M. Levin. 2005a. Serotonin transporter function is an early step in left-right patterning in chick and frog embryos. *Developmental Neuroscience* 27 (6):349–363. doi:10.1159/000088451.

Fukumoto, T., I. P. Kema, and M. Levin. 2005b. Serotonin signaling is a very early step in patterning of the left-right axis in chick and frog embryos. *Current Biology* 15 (9):794–803. doi:10.1016/j.cub.2005.03.044.

Funk, R. 2013. Ion gradients in tissue and organ biology. *Biological Systems* 2:105. doi:10.4172/bso.1000105.

Gainotti, G. 2011. The organization and dissolution of semantic-conceptual knowledge: Is the 'amodal hub' the only plausible model? *Brain and Cognition* 75 (3):299–309. doi:10.1016/j.bandc.2010.12.001.

Ganai, S. A., M. Ramadoss, and V. Mahadevan. 2016. Histone deacetylase (HDAC) inhibitors - Emerging roles in neuronal memory, learning, synaptic plasticity and neural regeneration. *Current Neuropharmacology* 14 (1):55–71.

Garey, L. 1999. Cortex: Statistics and geometry of neuronal connectivity, 2nd ed. *Journal of Anatomy* 194 (Pt 1):153–157. doi:10.1046/j.1469-7580.1999.194101535.x.

Godwin, J. W., A. R. Pinto, and N. A. Rosenthal. 2013. Macrophages are required for adult salamander limb regeneration. *Proceedings of the National Academy of Sciences of the United States of America* 110 (23):9415–9420. doi:10.1073/pnas.1300290110.

Goel, P., and A. Mehta. 2013. Learning theories reveal loss of pancreatic electrical connectivity in diabetes as an adaptive response. *PLoS One* 8 (8):e70366. doi:10.1371/journal.pone.0070366.

Gradin, V. B., P. Kumar, G. Waiter, T. Ahearn, C. Stickle, M. Milders, I. Reid, J. Hall, and J. D. Steele. 2011. Expected value and prediction error abnormalities in depression and schizophrenia. *Brain* 134 (Pt 6):1751–1764. doi:10.1093/brain/awr059.

Han, X., and E. S. Boyden. 2007. Multiple-color optical activation, silencing, and desynchronization of neural activity, with single-spike temporal resolution. *PLoS One* 2 (3):e299. doi:10.1371/journal.pone.0000299.

Hansen, T. F. 2003. Is modularity necessary for evolvability? Remarks on the relationship between pleiotropy and evolvability. *Biosystems* 69 (2–3):83–94.

Hebb, D. O. 2002. *The Organization of Behavior: A Neuropsychological Theory.* Mahwah, NJ: L. Erlbaum Associates.

Hechavarria, D., A. Dewilde, S. Braunhut, M. Levin, and D. L. Kaplan. 2010. BioDome regenerative sleeve for biochemical and biophysical stimulation of tissue regeneration. *Medical Engineering & Physics* 32 (9):1065–1073. doi:10.1016/j.medengphy.2010.07.010.

Hecht, I., H. Levine, W. J. Rappel, and E. Ben-Jacob. 2011. "Self-assisted" amoeboid navigation in complex environments. *PLoS one* 6 (8):e21955. doi:10.1371/journal.pone.0021955.

Hernandez-Diaz, S., and M. Levin. 2014. Alteration of bioelectrically-controlled processes in the embryo: A teratogenic mechanism for anticonvulsants. *Reproductive Toxicology* 47:111–114. doi:10.1016/j.reprotox.2014.04.008.

Herrera-Rincon, C., and F. Panetsos. 2014. Substitution of natural sensory input by artificial neurostimulation of an amputated trigeminal nerve does not prevent the degeneration of basal forebrain cholinergic circuits projecting to the somatosensory cortex. *Frontiers in Cellular Neuroscience* 8:385. doi:10.3389/fncel.2014.00385.

Hille, B. 2001. *Ion Channels of Excitable Membranes*. 3rd ed. Sunderland, MA: Sinauer.

Hinkle, L., C. D. McCaig, and K. R. Robinson. 1981. The direction of growth of differentiating neurones and myoblasts from frog embryos in an applied electric field. *Journal of Physiology* 314:121–135.

Hinton, G. E. 2007. Learning multiple layers of representation. *Trends in Cognitive Sciences* 11 (10):428–434. doi:10.1016/J.Tics.2007.09.004.

Hoare, J. I., A. M. Rajnicek, C. D. McCaig, R. N. Barker, and H. M. Wilson. 2015. Electric fields are novel determinants of human macrophage functions. *Journal of Leukocyte Biology*. doi: 10.1189/jlb.3A0815-390R.

Hodgkin, A. L., and A. F. Huxley. 1952. A quantitative description of membrane current and its application to conduction and excitation in nerve. *The Journal of physiology* 117 (4):500–544.

Hogendoorn, H. 2015. From sensation to perception: Using multivariate classification of visual illusions to identify neural correlates of conscious awareness in space and time. *Perception* 44 (1):71–78.

Hubel, D. H., and T. N. Wiesel. 1962. Receptive fields, binocular interaction and functional architecture in the cat's visual cortex. *Journal of Physiology* 160:106–154.

Ibsen, S., A. Tong, C. Schutt, S. Esener, and S. H. Chalasani. 2015. Sonogenetics is a non-invasive approach to activating neurons in Caenorhabditis elegans. *Nature Communications* 6:8264. doi:10.1038/ncomms9264.

Illmensee, K., and B. Mintz. 1976. Totipotency and normal differentiation of single teratocarcinoma cells cloned by injection into blastocysts. *Proceedings of the National Academy of Sciences of the United States of America* 73 (2):549–553.

Ingber, D. E., and M. Levin. 2007. What lies at the interface of regenerative medicine and developmental biology? *Development* 134 (14):2541–2517.

Iovine, M. K., E. P. Higgins, A. Hindes, B. Coblitz, and S. L. Johnson. 2005. Mutations in connexin43 (GJA1) perturb bone growth in zebrafish fins. *Developmental Biology* 278 (1):208–219. doi:10.1016/j.ydbio.2004.11.005.

Jaffe, L. F. 1981. The role of ionic currents in establishing developmental pattern. *Philosophical Transactions of the Royal Society of London. Series B, Biological Sciences* 295 (1078):553–566.

Jahfari, S., F. Verbruggen, M. J. Frank, L. J. Waldorp, L. Colzato, K. R. Ridderinkhof, and B. U. Forstmann. 2012. How preparation changes the need for top-down control of the basal ganglia when inhibiting premature actions. *Journal of Neuroscience* 32 (32):10870–10878. doi:10.1523/jneurosci.0902-12.2012.

Jenkins, L. S., B. S. Duerstock, and R. B. Borgens. 1996. Reduction of the current of injury leaving the amputation inhibits limb regeneration in the red spotted newt. *Developmental Biology* 178 (2):251–262.

Juusola, M., A. S. French, R. O. Uusitalo, and M. Weckstrom. 1996. Information processing by graded-potential transmission through tonically active synapses. *Trends in Neurosciences* 19 (7):292–297. doi:10.1016/s0166-2236(96)10028-x.

Kaila, V. R. I., and A. Annila. 2008. Natural selection for least action. *Proceedings of the Royal Society A-Mathematical Physical and Engineering Sciences* 464 (2099):3055–3070. doi:10.1098/Rspa.2008.0178.

Kamm, R. D., and R. Bashir. 2014. Creating living cellular machines. *Annals of Biomedical Engineering* 42 (2):445–459. doi:10.1007/s10439-013-0902-7.

Kandel, E. R. 2001. The molecular biology of memory storage: A dialogue between genes and synapses. *Science* 294 (5544):1030–1038. doi:10.1126/science.1067020.

Kawato, M. 1999. Internal models for motor control and trajectory planning. *Current Opinion in Neurobiology* 9 (6):718–727.

Keijzer, F., M. van Duijn, and P. Lyon. 2013. What nervous systems do: Early evolution, input-output, and the skin brain thesis. *Adaptive Behavior* 21 (2):67–85. doi:10.1177/1059712312465330.

Kerr, R. R., D. B. Grayden, D. A. Thomas, M. Gilson, and A. N. Burkitt. 2014. Goal-directed control with cortical units that are gated by both top-down feedback and oscillatory coherence. *Frontiers in Neural Circuits* 8:94. doi:10.3389/fncir.2014.00094.

Kiebel, S. J., J. Daunizeau, and K. J. Friston. 2009. Perception and hierarchical dynamics. *Frontiers in Neuroinformatics* 3:20. doi:10.3389/neuro.11.020.2009.

Kim, J. G., and I. Biederman. 2011. Where do objects become scenes? *Cerebral Cortex* 21 (8):1738–1746. doi:10.1093/cercor/bhq240.

Kiortsis, V., and M. Moraitou. 1965. Factors of regeneration in Spirographis spallanzanii. In *Regeneration in Animals and Related Problems*, edited by V. Kiortsis and H. A. L. Trampusch, pp. 250–261. Amsterdam, the Netherlands: North-Holland.

Knopfel, T., Y. Gallero-Salas, and C. Song. 2015. Genetically encoded voltage indicators for large scale cortical imaging come of age. *Current Opinion in Chemical Biology* 27:75–83. doi:10.1016/j.cbpa.2015.06.006.

Kragl, M., D. Knapp, E. Nacu, S. Khattak, M. Maden, H. H. Epperlein, and E. M. Tanaka. 2009. Cells keep a memory of their tissue origin during axolotl limb regeneration. *Nature* 460 (7251):60–65. doi:10.1038/nature08152.

Kruh, J. 1982. Effects of sodium butyrate, a new pharmacological agent, on cells in culture. *Molecular and Cellular Biochemistry* 42 (2):65–82.

Kuang, S., P. Morel, and A. Gail. 2016. Planning movements in visual and physical space in monkey posterior parietal cortex. *Cerebral Cortex* 26 (2):731–747. doi:10.1093/cercor/bhu312.

Kubie, J. L., R. U. Muller, and E. Bostock. 1990. Spatial firing properties of hippocampal theta cells. *Journal of Neuroscience* 10 (4):1110–1123.

Kucerova, R., P. Walczysko, B. Reid, J. Ou, L. J. Leiper, A. M. Rajnicek, C. D. McCaig, M. Zhao, and J. M. Collinson. 2011. The role of electrical signals in murine corneal wound re-epithelialization. *Journal of Cellular Physiology* 226 (6):1544–1553. doi:10.1002/jcp.22488.

Kumar, A., and J. P. Brockes. 2012. Nerve dependence in tissue, organ, and appendage regeneration. *Trends in Neurosciences* 35 (11):691–699. doi:10.1016/j.tins.2012.08.003.

Kumar, A., J. P. Delgado, P. B. Gates, G. Neville, A. Forge, and J. P. Brockes. 2011. The aneurogenic limb identifies developmental cell interactions underlying vertebrate limb regeneration. *Proceedings of the National Academy of Sciences of the United States of America* 108 (33):13588–13593. doi:10.1073/pnas.1108472108.

Kumar, A., J. W. Godwin, P. B. Gates, A. A. Garza-Garcia, and J. P. Brockes. 2007. Molecular basis for the nerve dependence of limb regeneration in an adult vertebrate. *Science* 318 (5851):772–777.

Kumar, A., G. Nevill, J. P. Brockes, and A. Forge. 2010. A comparative study of gland cells implicated in the nerve dependence of salamander limb regeneration. *Journal of Anatomy* 217 (1):16–25. doi:10.1111/j.1469-7580.2010.01239.x.

Kuo, M. F., and M. A. Nitsche. 2012. Effects of transcranial electrical stimulation on cognition. *Clinical EEG and Neuroscience* 43 (3):192–199. doi:10.1177/1550059412444975.

Lander, A. D., K. K. Gokoffski, F. Y. Wan, Q. Nie, and A. L. Calof. 2009. Cell lineages and the logic of proliferative control. *PLoS Biology* 7 (1):e15. doi:10.1371/journal.pbio.1000015.

Langston, R. F., J. A. Ainge, J. J. Couey, C. B. Canto, T. L. Bjerknes, M. P. Witter, E. I. Moser, and M. B. Moser. 2010. Development of the spatial representation system in the rat. *Science* 328 (5985):1576–1580. doi:10.1126/science.1188210.

Larkum, M. E., W. Senn, and H. R. Luscher. 2004. Top-down dendritic input increases the gain of layer 5 pyramidal neurons. *Cerebral Cortex* 14 (10):1059–1070. doi:10.1093/cercor/bhh065.

Laurienti, P. J., M. T. Wallace, J. A. Maldjian, C. M. Susi, B. E. Stein, and J. H. Burdette. 2003. Cross-modal sensory processing in the anterior cingulate and medial prefrontal cortices. *Human Brain Mapping* 19 (4):213–223. doi:10.1002/hbm.10112.

Law, R., and M. Levin. 2015. Bioelectric memory: Modeling resting potential bistability in amphibian embryos and mammalian cells. *Theoretical Biology & Medical Modelling* 12 (1):22. doi:10.1186/s12976-015-0019-9.

Lazebnik, Y. 2002. Can a biologist fix a radio?—Or, what I learned while studying apoptosis. *Cancer Cell* 2 (3):179–182.

Leppik, L. P., D. Froemel, A. Slavici, Z. N. Ovadia, L. Hudak, D. Henrich, I. Marzi, and J. H. Barker. 2015. Effects of electrical stimulation on rat limb regeneration, a new look at an old model. *Scientific Reports* 5:18353. doi:10.1038/srep18353.

Leutgeb, S., J. K. Leutgeb, M. B. Moser, and E. I. Moser. 2005. Place cells, spatial maps and the population code for memory. *Current Opinion in Neurobiology* 15 (6):738–746. doi:10.1016/j.conb.2005.10.002.

Levenson, J. M., K. J. O'Riordan, K. D. Brown, M. A. Trinh, D. L. Molfese, and J. D. Sweatt. 2004. Regulation of histone acetylation during memory formation in the hippocampus. *Journal of Biological Chemistry* 279 (39):40545–40559. doi:10.1074/jbc.M402229200.

Levin, M. 2007. Gap junctional communication in morphogenesis. *Progress in Biophysics and Molecular Biology* 94 (1–2):186–206.

Levin, M. 2011. The wisdom of the body: Future techniques and approaches to morphogenetic fields in regenerative medicine, developmental biology and cancer. *Regenerative Medicine* 6 (6):667–673. doi:10.2217/rme.11.69.

Levin, M. 2012a. Molecular bioelectricity in developmental biology: New tools and recent discoveries: Control of cell behavior and pattern formation by transmembrane potential gradients. *Bioessays* 34 (3):205–217. doi:10.1002/bies.201100136.

Levin, M. 2012b. Morphogenetic fields in embryogenesis, regeneration, and cancer: Non-local control of complex patterning. *Bio Systems* 109 (3):243–261. doi:10.1016/j.biosystems.2012.04.005.

Levin, M. 2013. Reprogramming cells and tissue patterning via bioelectrical pathways: Molecular mechanisms and biomedical opportunities. *Wiley Interdisciplinary Reviews: Systems Biology and Medicine* 5 (6):657–676. doi:10.1002/wsbm.1236.

Levin, M. 2014a. Molecular bioelectricity: How endogenous voltage potentials control cell behavior and instruct pattern regulation in vivo. *Molecular Biology of the Cell* 25 (24):3835–3850. doi:10.1091/mbc.E13-12-0708.

Levin, M. 2014b. Endogenous bioelectrical networks store non-genetic patterning information during development and regeneration. *The Journal of Physiology* 592 (11):2295–2305. doi:10.1113/jphysiol.2014.271940.

Levin, M., G. A. Buznikov, and J. M. Lauder. 2006. Of minds and embryos: Left-right asymmetry and the serotonergic controls of pre-neural morphogenesis. *Developmental Neuroscience* 28 (3):171–185.

Levin, M., and M. Mercola. 1998. Gap junctions are involved in the early generation of left-right asymmetry. *Developmental Biology* 203 (1):90–105.

Levin, M., T. Thorlin, K. R. Robinson, T. Nogi, and M. Mercola. 2002. Asymmetries in H+/K+-ATPase and cell membrane potentials comprise a very early step in left-right patterning. *Cell* 111 (1):77–89.

Li, C., H. Zhong, Y. Wang, H. Wang, Z. Yang, Y. Zheng, K. Liu, and Y. Liu. 2006. Voltage and ionic regulation of human serotonin transporter in Xenopus oocytes. *Clinical and Experimental Pharmacology and Physiology* 33 (11):1088–1092. doi:10.1111/j.1440-1681.2006.04491.x.

Liu, X., S. Ramirez, and S. Tonegawa. 2014. Inception of a false memory by opto-genetic manipulation of a hippocampal memory engram. *Philosophical Transactions of the Royal Society of London B: Biological Sciences* 369 (1633):20130142. doi:10.1098/rstb.2013.0142.

Lobanov, O. V., F. Zeidan, J. G. McHaffie, R. A. Kraft, and R. C. Coghill. 2014. From cue to meaning: brain mechanisms supporting the construction of expectations of pain. *Pain* 155 (1):129–136. doi:10.1016/j.pain.2013.09.014.

Lobikin, M., B. Chernet, D. Lobo, and M. Levin. 2012. Resting potential, oncogene-induced tumorigenesis, and metastasis: The bioelectric basis of cancer in vivo. *Physical Biology* 9 (6):065002. doi:10.1088/1478-3975/9/6/065002.

Lobo, D., W. S. Beane, and M. Levin. 2012. Modeling planarian regeneration: A primer for reverse-engineering the worm. *PLoS Computational Biology* 8 (4):e1002481. doi:10.1371/journal.pcbi.1002481.

Lobo, D., E. B. Feldman, M. Shah, T. J. Malone, and M. Levin. 2014a. Limbform: A functional ontology-based database of limb regeneration experiments. *Bioinformatics* 30 (24):3598–3600. doi:10.1093/bioinformatics/btu582.

Lobo, D., E. B. Feldman, M. Shah, T. J. Malone, and M. Levin. 2014b. A bioinfor-matics expert system linking functional data to anatomical outcomes in limb regeneration. *Regeneration.* 1(2): 37–56. doi:10.1002/reg2.13.

Lobo, D., and M. Levin. 2015. Inferring regulatory networks from experimental mor-phological phenotypes: A computational method reverse-engineers planarian regeneration. *PLoS Computational Biology* 11 (6):e1004295. doi:10.1371/journal.pcbi.1004295.

Lobo, D., T. J. Malone, and M. Levin. 2013a. Planform: An application and data-base of graph-encoded planarian regenerative experiments. *Bioinformatics.* doi:10.1093/bioinformatics/btt088.

Lobo, D., T. J. Malone, and M. Levin. 2013b. Towards a bioinformatics of patterning: A computational approach to understanding regulative morphogenesis. *Biology Open* 2 (2):156–169. doi:10.1242/bio.20123400.

Lobo, D., M. Solano, G. A. Bubenik, and M. Levin. 2014c. A linear-encoding model explains the variability of the target morphology in regeneration. *Journal of the Royal Society, Interface/The Royal Society* 11 (92):20130918. doi:10.1098/rsif.2013.0918.

Lois, N., B. Reid, B. Song, M. Zhao, J. Forrester, and C. McCaig. 2010. Electric currents and lens regeneration in the rat. *Experimental Eye Research* 90 (2):316–323. doi:10.1016/j.exer.2009.11.007.

Long, X., J. Ye, D. Zhao, and S.-J. Zhang. 2015. Magnetogenetics: Remote non-invasive magnetic activation of neuronal activity with a magnetoreceptor. *Science Bulletin* 1–13. doi:10.1007/s11434-015-0902-0.

Lund, E. J. 1947. *Bioelectric Fields and Growth.* Austin, TX: University of Texas Press.

Lundby, A., H. Mutoh, D. Dimitrov, W. Akemann, and T. Knopfel. 2008. Engineering of a genetically encodable fluorescent voltage sensor exploiting fast Ci-VSP voltage-sensing movements. *PLoS One* 3 (6):e2514. doi:10.1371/journal.pone.0002514.

Maddox, W. T., and A. B. Markman. 2010. The motivation-cognition interface in learning and decision-making. *Current Directions in Psychological Science* 19 (2):106–110. doi:10.1177/0963721410364008.

Maden, M. 2008. Axolotl/newt. *Methods in Molecular Biology* 461:467–480. doi:10.1007/978-1-60327-483-8_32.

Maren, S., K. L. Phan, and I. Liberzon. 2013. The contextual brain: Implications for fear conditioning, extinction and psychopathology. *Nature Reviews Neuroscience* 14 (6):417–428. doi:10.1038/nrn3492.

Markova, O., M. Mukhtarov, E. Real, Y. Jacob, and P. Bregestovski. 2008. Genetically encoded chloride indicator with improved sensitivity. *Journal of Neuroscience Methods* 170 (1):67–76.

Marstaller, L., A. Hintze, and C. Adami. 2013. The evolution of representation in simple cognitive networks. *Neural Computation* 25 (8):2079–2107. doi:10.1162/NECO_a_00475.

McCaig, C. D., A. M. Rajnicek, B. Song, and M. Zhao. 2005. Controlling cell behavior electrically: Current views and future potential. *Physiological Reviews* 85 (3):943–978.

McDonald, R. J., and N. S. Hong. 2013. How does a specific learning and memory system in the mammalian brain gain control of behavior? *Hippocampus* 23 (11):1084–1102. doi:10.1002/hipo.22177.

Mechelli, A., C. J. Price, K. J. Friston, and A. Ishai. 2004. Where bottom-up meets top-down: Neuronal interactions during perception and imagery. *Cerebral Cortex* 14 (11):1256–1265. doi:10.1093/cercor/bhh087.

Meng, X., M. Arocena, J. Penninger, F. H. Gage, M. Zhao, and B. Song. 2011. PI3K mediated electrotaxis of embryonic and adult neural progenitor cells in the presence of growth factors. *Experimental Neurology* 227 (1):210–217. doi:10.1016/j.expneurol.2010.11.002.

Miller, C. A., S. L. Campbell, and J. D. Sweatt. 2008. DNA methylation and histone acetylation work in concert to regulate memory formation and synaptic plasticity. *Neurobiology of Learning and Memory* 89 (4):599–603. doi:10.1016/j.nlm.2007.07.016.

Miller, C. A., C. F. Gavin, J. A. White, R. R. Parrish, A. Honasoge, C. R. Yancey, I. M. Rivera, M. D. Rubio, G. Rumbaugh, and J. D. Sweatt. 2010. Cortical DNA methylation maintains remote memory. *Nature Neuroscience* 13 (6):664–666. doi:10.1038/nn.2560.

Miller, G. A. 1960. *Plans and the Structure of Behavior.* New York: Holt.

Miyauchi, S., E. Gopal, Y. J. Fei, and V. Ganapathy. 2004. Functional identification of SLC5A8, a tumor suppressor down-regulated in colon cancer, as a Na(+)-coupled transporter for short-chain fatty acids. *The Journal of Biological Chemistry* 279 (14):13293–13296. doi:10.1074/jbc.C400059200.

Mondia, J. P., M. Levin, F. G. Omenetto, R. D. Orendorff, M. R. Branch, and D. S. Adams. 2011. Long-distance signals are required for morphogenesis of the regenerating Xenopus tadpole tail, as shown by femtosecond-laser ablation. *PloS one* 6 (9):e24953. doi:10.1371/journal.pone.0024953.

Monteiro, J., R. Aires, J. D. Becker, A. Jacinto, A. C. Certal, and J. Rodriguez-Leon. 2014. V-ATPase proton pumping activity is required for adult zebrafish appendage regeneration. *PloS one* 9 (3):e92594. doi:10.1371/journal.pone.0092594.

Mountcastle, V. B. 1998. *Perceptual Neuroscience: The Cerebral Cortex.* Cambridge, MA: Harvard University Press.

Muller, R. U., M. Stead, and J. Pach. 1996. The hippocampus as a cognitive graph. *The Journal of General Physiology* 107 (6):663–694.

Munoz-Canoves, P., and A. L. Serrano. 2015. Macrophages decide between regeneration and fibrosis in muscle. *Trends in Endocrinology and Metabolism: TEM.* doi:10.1016/j.tem.2015.07.005.

Murata, Y., H. Iwasaki, M. Sasaki, K. Inaba, and Y. Okamura. 2005. Phosphoinositide phosphatase activity coupled to an intrinsic voltage sensor. *Nature* 435 (7046):1239–1243.

Murphy, T. H., and D. Corbett. 2009. Plasticity during stroke recovery: From synapse to behaviour. *Nature Reviews Neuroscience* 10 (12):861–872. doi:10.1038/nrn2735.

Naselaris, T., R. J. Prenger, K. N. Kay, M. Oliver, and J. L. Gallant. 2009. Bayesian reconstruction of natural images from human brain activity. *Neuron* 63 (6):902–915.

Newmark, P. A. 2005. Opening a new can of worms: A large-scale RNAi screen in planarians. *Developmental Cell* 8 (5):623–624.

105

Nicoll, R. A., and R. C. Malenka. 1999. Expression mechanisms underlying NMDA receptor-dependent long-term potentiation. *Annals of the New York Academy of Sciences* 868:515–525.

Nishimoto, S., A. T. Vu, T. Naselaris, Y. Benjamini, B. Yu, and J. L. Gallant. 2011. Reconstructing visual experiences from brain activity evoked by natural movies. *Current Biology: CB* 21 (19):1641–1646. doi:10.1016/j.cub.2011.08.031.

Nishimura, K. Y., R. R. Isseroff, and R. Nuccitelli. 1996. Human keratinocytes migrate to the negative pole in direct current electric fields comparable to those measured in mammalian wounds. *Journal of Cell Science* 109 (Pt 1):199–207.

Noble, D. 2012. A theory of biological relativity: No privileged level of causation. *Interface Focus* 2 (1):55–64. doi:10.1098/Rsfs.2011.0067.

Nogi, T., and M. Levin. 2005. Characterization of innexin gene expression and functional roles of gap-junctional communication in planarian regeneration. *Developmental Biology* 287 (2):314–335.

Norton, L. A., K. J. Hanley, and J. Turkewicz. 1984. Bioelectric perturbations of bone. Research directions and clinical applications. *The Angle Orthodontist* 54 (1):73–87. doi:10.1043/0003-3219(1984)054<0073:bpob>2.0.co;2.

Nuccitelli, R. 2003. Endogenous electric fields in embryos during development, regeneration and wound healing. *Radiation Protection Dosimetry* 106 (4):375–383.

Nuccitelli, R., P. Nuccitelli, C. Li, S. Narsing, D. M. Pariser, and K. Lui. 2011. The electric field near human skin wounds declines with age and provides a noninvasive indicator of wound healing. *Wound Repair and Regeneration* 19 (5):645–655. doi:10.1111/j.1524-475X.2011.00723.x.

Okamura, Y., and J. E. Dixon. 2011. Voltage-sensing phosphatase: Its molecular relationship with PTEN. *Physiology (Bethesda)* 26 (1):6–13. doi:10.1152/physiol.00035.2010.

Oviedo, N. J., J. Morokuma, P. Walentek, I. P. Kema, M. B. Gu, J. M. Ahn, J. S. Hwang, T. Gojobori, and M. Levin. 2010. Long-range neural and gap junction protein-mediated cues control polarity during planarian regeneration. *Developmental Biology* 339 (1):188–199. doi:10.1016/j.ydbio.2009.12.012.

Oviedo, N. J., P. A. Newmark, and A. Sanchez Alvarado. 2003. Allometric scaling and proportion regulation in the freshwater planarian *Schmidtea mediterranea*. *Developmental Dynamics* 226 (2):326–333.

Ozkucur, N., H. H. Epperlein, and R. H. Funk. 2010. Ion imaging during axolotl tail regeneration in vivo. *Developmental Dynamics* 239 (7):2048–2057. doi:10.1002/dvdy.22323.

Pai, V. P., S. Aw, T. Shomrat, J. M. Lemire, and M. Levin. 2012. Transmembrane voltage potential controls embryonic eye patterning in Xenopus laevis. *Development* 139 (2):313–323. doi:10.1242/dev.073759.

Pai, V. P., J. M. Lemire, J. F. Pare, G. Lin, Y. Chen, and M. Levin. 2015a. Endogenous gradients of resting potential instructively pattern embryonic neural tissue via notch signaling and regulation of proliferation. *The Journal of Neuroscience* 35 (10):4366–4385. doi:10.1523/JNEUROSCI.1877-14.2015.

Pai, V. P., C. J. Martyniuk, K. Echeverri, S. Sundelacruz, D. L. Kaplan, and M. Levin. 2015b. Genome-wide analysis reveals conserved transcriptional responses downstream of resting potential change in Xenopus embryos, axolotl regeneration, and human mesenchymal cell differentiation. *Regeneration.* 3 (1):3–25. doi:10.1002/reg2.48.

Palacios-Prado, N., and F. F. Bukauskas. 2009. Heterotypic gap junction channels as voltage-sensitive valves for intercellular signaling. *Proceedings of the National Academy of Sciences of the United States of America* 106 (35):14855–14860. doi:10.1073/pnas.0901923106.

Palacios-Prado, N., G. Hoge, A. Marandykina, L. Rimkute, S. Chapuis, N. Paulauskas, V. A. Skeberdis et al. 2013. Intracellular magnesium-dependent modulation of gap junction channels formed by neuronal connexin36. *The Journal of Neuroscience: The Official Journal of the Society for Neuroscience* 33 (11):4741–4753. doi:10.1523/JNEUROSCI.2825-12.2013.

Park, H. J., and K. Friston. 2013. Structural and functional brain networks: From connections to cognition. *Science* 342 (6158):1238411. doi:10.1126/science.1238411.

Patel, N., and M. M. Poo. 1982. Orientation of neurite growth by extracellular electric fields. *Journal of Neuroscience* 2 (4):483–496.

Paulus, W. 2011. Transcranial electrical stimulation (tES - tDCS; tRNS, tACS) methods. *Neuropsychological Rehabilitation* 21 (5):602–617. doi:10.1080/096020 11.2011.557292.

Perathoner, S., J. M. Daane, U. Henrion, G. Seebohm, C. W. Higdon, S. L. Johnson, C. Nusslein-Volhard, and M. P. Harris. 2014. Bioelectric signaling regulates size in zebrafish fins. *PLoS Genetics* 10 (1):e1004080. doi:10.1371/journal. pgen.1004080.

Pereda, A. E. 2014. Electrical synapses and their functional interactions with chemical synapses. *Nature Reviews Neuroscience* 15 (4):250–263. doi:10.1038/nrn3708.

Pereda, A. E., S. Curti, G. Hoge, R. Cachope, C. E. Flores, and J. E. Rash. 2013. Gap junction-mediated electrical transmission: Regulatory mechanisms and plasticity. *Biochimica et Biophysica Acta* 1828 (1):134–146. doi:10.1016/j. bbamem.2012.05.026.

Perron, A., W. Akemann, H. Mutoh, and T. Knopfel. 2012. Genetically encoded probes for optical imaging of brain electrical activity. *Progress in Brain Research* 196:63–77. doi:10.1016/b978-0-444-59426-6.00004-5.

Petersen, S. E., and O. Sporns. 2015. Brain networks and cognitive architectures. *Neuron* 88 (1):207–219. doi:10.1016/j.neuron.2015.09.027.

Pezzulo, G., and M. Levin. 2015. Re-membering the body: Applications of computational neuroscience to the top-down control of regeneration of limbs and other complex organs. *Integrative Biology (Camb)* 7 (12):1487–1517. doi:10.1039/ c5ib00221d.

Pezzulo, G., and F. Rigoli. 2011. The value of foresight: How prospection affects decision-making. *Frontiers in Neuroscience* 5:79. doi:10.3389/fnins.2011.00079.

Pezzulo, G., F. Rigoli, and K. Friston. 2015. Active Inference, homeostatic regulation and adaptive behavioural control. *Progress in Neurobiology.* doi:10.1016/j. pneurobio.2015.09.001.

Pezzulo, G., P. F. Verschure, C. Balkenius, and C. M. Pennartz. 2014. The principles of goal-directed decision-making: From neural mechanisms to computation and robotics. *Philosophical Transactions of the Royal Society of London. Series B, Biological Sciences* 369 (1655). doi:10.1098/rstb.2013.0470.

Pinotsis, D. A., and K. J. Friston. 2011. Neural fields, spectral responses and lateral connections. *Neuroimage* 55 (1):39–48. doi:10.1016/j.neuroimage.2010.11.081.

Poldrack, R. A., and M. G. Packard. 2003. Competition among multiple memory systems: Converging evidence from animal and human brain studies. *Neuropsychologia* 41 (3):245–251.

Pullar, C. E. 2011. *The Physiology of Bioelectricity in Development, Tissue Regeneration, and Cancer, Biological Effects of Electromagnetics Series.* Boca Raton, FL: CRC Press.

Pullar, C. E., B. S. Baier, Y. Kariya, A. J. Russell, B. A. Horst, M. P. Marinkovich, and R. R. Isseroff. 2006. beta4 integrin and epidermal growth factor coordinately regulate electric field-mediated directional migration via Rac1. *Molecular Biology of the Cell* 17 (11):4925–4935.

Pullar, C. E., and R. R. Isseroff. 2005. Cyclic AMP mediates keratinocyte directional migration in an electric field. *Journal of the Cell Science* 118 (Pt 9):2023–2034.

Purves, D. 2013. *Principles of Cognitive Neuroscience*. 2nd ed. Sunderland, MA: Sinauer Associates Inc. Publishers.

Putnam, H. 1988. *Representation and Reality*. Cambridge, MA: MIT Press.

Rajnicek, A. M., L. E. Foubister, and C. D. McCaig. 2006. Growth cone steering by a physiological electric field requires dynamic microtubules, microfilaments and Rac-mediated filopodial asymmetry. *Journal of the Cell Science* 119 (Pt 9):1736–1745.

Rajnicek, A. M., L. E. Foubister, and C. D. McCaig. 2007. Prioritising guidance cues: Directional migration induced by substratum contours and electrical gradients is controlled by a rho/cdc42 switch. *Developmental Biology* 312 (1):448–460.

Ramirez, S., X. Liu, P. A. Lin, J. Suh, M. Pignatelli, R. L. Redondo, T. J. Ryan, and S. Tonegawa. 2013. Creating a false memory in the hippocampus. *Science* 341 (6144):387–391. doi:10.1126/science.1239073.

Rangel, A., C. Camerer, and P. R. Montague. 2008. A framework for studying the neurobiology of value-based decision making. *Nature Reviews Neuroscience* 9 (7):545–556. doi:10.1038/nrn2357.

Rea, A. C., L. N. Vandenberg, R. E. Ball, A. A. Snouffer, A. G. Hudson, Y. Zhu, D. E. McLain et al. 2013. Light-activated serotonin for exploring its action in biological systems. *Chemistry & Biology* 20 (12):1536–1546. doi:10.1016/j.chembiol.2013.11.005.

Reid, B., R. Nuccitelli, and M. Zhao. 2007. Non-invasive measurement of bioelectric currents with a vibrating probe. *Nature Protocols* 2 (3):661–669.

Reid, B., and M. Zhao. 2011a. Ion-selective self-referencing probes for measuring specific ion flux. *Communicative & Integrative Biology* 4 (5):524–527. doi:10.4161/cib.4.5.16182.

Reid, B., and M. Zhao. 2011b. Measurement of bioelectric current with a vibrating probe. *Journal of Visualized Experiments* (47). doi:10.3791/2358.

Robb, S. M., and A. Sanchez Alvarado. 2014. Histone modifications and regeneration in the planarian Schmidtea mediterranea. *Current Topics in Developmental Biology* 108:71–93. doi:10.1016/B978-0-12-391498-9.00004-8.

Rosch, E. 1975. Cognitive representations of semantic categories. *Journal of Experimental Psychology-General* 104 (3):192–233. doi:10.1037//0096-3445.104.3.192.

Rose, S. M. 1945. The effect of NaCl in stimulating regeneration of limbs of frogs. *Journal of Morphology* 77 (2):119–139.

Rosen, R. 1985. *Anticipatory Systems: Philosophical, Mathematical, and Methodological Foundations*. 1st ed, IFSR International Series on Systems Science and Engineering; vol. 1. Oxford: Pergamon Press.

Rosenblueth, A., N. Wiener, and J. Bigelow. 1943. Behavior, purpose, and teleology. *Philosophy of Science* 10:18–24.

Rowe, J., K. Friston, R. Frackowiak, and R. Passingham. 2002. Attention to action: Specific modulation of corticocortical interactions in humans. *Neuroimage* 17 (2):988–998.

Ruckh, T. T., A. A. Mehta, J. M. Dubach, and H. A. Clark. 2013. Polymer-free optode nanosensors for dynamic, reversible, and ratiometric sodium imaging in the physiological range. *Scientific Reports* 3:3366. doi:10.1038/srep03366.

Sabin, K., T. Santos-Ferreira, J. Essig, S. Rudasill, and K. Echeverri. 2015. Dynamic membrane depolarization is an early regulator of ependymoglial cell response to spinal cord injury in axolotl. *Developmental Biology* 408 (1):14–25. doi:10.1016/j.ydbio.2015.10.012.

Sahari, A., T. T. Ruckh, R. Hutchings, and H. A. Clark. 2015. Development of an optical nanosensor incorporating a pH-sensitive quencher dye for potassium imaging. *Analytical Chemistry* 87 (21):10684–10687. doi:10.1021/acs.analchem.5b03080.

Salo, E., J. F. Abril, T. Adell, F. Cebria, K. Eckelt, E. Fernandez-Taboada, M. Handberg-Thorsager, M. Iglesias, M. D. Molina, and G. Rodriguez-Esteban. 2009. Planarian regeneration: achievements and future directions after 20 years of research. *The International Journal of Developmental Biology* 53 (8–10):1317–1327. doi:10.1387/ijdb.072414es.

Salo, E., and J. Baguna. 1985. Cell movement in intact and regenerating planarians. Quantitation using chromosomal, nuclear and cytoplasmic markers. *Journal of Embryology & Experimental Morphology* 89:57–70.

Saltukoglu, D., J. Grunewald, N. Strohmeyer, R. Bensch, M. H. Ulbrich, O. Ronneberger, and M. Simons. 2015. Spontaneous and electric field-controlled front-rear polarization of human keratinocytes. *Molecular Biology of the Cell* 26 (24):4373–4386. doi:10.1091/mbc.E14-12-1580.

Sato, M. J., H. Kuwayama, W. N. van Egmond, A. L. Takayama, H. Takagi, P. J. van Haastert, T. Yanagida, and M. Ueda. 2009. Switching direction in electric-signal-induced cell migration by cyclic guanosine monophosphate and phosphatidylinositol signaling. *Proceedings of the National Academy of Sciences of the United States of America* 106 (16):6667–6672.

Satoh, A., A. makanae, A. Hirata, and Y. Satou. 2011. Blastema induction in aneurogenic state and Prrx-1 regulation by MMPs and FGFs in Ambystoma mexicanum limb regeneration. *Developmental Biology* 355 (2):263–274. doi:10.1016/j.ydbio.2011.04.017.

Schilling, M., and H. Cruse. 2012. What's next: Recruitment of a grounded predictive body model for planning a robot's actions. *Frontiers in Psychology* 3:383. doi:10.3389/fpsyg.2012.00383.

Schlosser, G., and G. P. Wagner. 2004. *Modularity in Development and Evolution*. Chicago, IL: University of Chicago Press.

Scott, D. E., and S. L. Hansen. 1997. Post-traumatic regeneration, neurogenesis and neuronal migration in the adult mammalian brain. *Virginia Medical Quarterly* 124 (4):249–261.

Scott, D. E., W. Wu, J. Slusser, A. Depto, and S. Hansen. 1995. Neural regeneration and neuronal migration following injury. I. The endocrine hypothalamus and neurohypophyseal system. *Experimental Neurology* 131 (1):23–38.

Semon, R. W., and L. Simon. 1921. *The Mneme*. London: The Macmillan company.

Sheeba, C. J., R. P. Andrade, and I. Palmeirim. 2014. Limb patterning: From signaling gradients to molecular oscillations. *Journal of Molecular Biology* 426 (4):780–784. doi:10.1016/j.jmb.2013.11.022.

Shi, R., and R. B. Borgens. 1995. Three-dimensional gradients of voltage during development of the nervous system as invisible coordinates for the establishment of embryonic pattern. *Developmental Dynamics* 202 (2):101–114. doi:10.1002/aja.1002020202.

Simpson, E. H., and P. D. Balsam. 2015. The behavioral neuroscience of motivation: An overview of concepts, measures, and translational applications. *Current Topics in Behavioral Neurosciences*. doi:10.1007/7854_2015_402.

Singer, M. 1952. The influence of the nerve in regeneration of the amphibian extremity. *The Quarterly Review of the Biology* 27 (2):169–200.

Sinha, G. 2013. Charged by GSK investment, battery of electroceuticals advance. *Nature Medicine* 19 (6):654. doi:10.1038/nm0613-654.

Slack, J. M. 1980. A serial threshold theory of regeneration. *Journal of Theoretical Biology* 82 (1):105–140.

Smith, P. J., and J. Trimarchi. 2001. Noninvasive measurement of hydrogen and potassium ion flux from single cells and epithelial structures. *American Journal of Physiology - Cell Physiology* 280 (1):C1–11.

Sollars, S. I., P. C. Smith, and D. L. Hill. 2002. Time course of morphological alterations of fungiform papillae and taste buds following chorda tympani transection in neonatal rats. *Journal of Neurobiology* 51 (3):223–236.

Song, B., M. Zhao, J. Forrester, and C. McCaig. 2004. Nerve regeneration and wound healing are stimulated and directed by an endogenous electrical field in vivo. *Journal of Cell Science* 117 (Pt 20):4681–4690. doi:10.1242/jcs.01341.

Spencer, G. J., and P. G. Genever. 2003. Long-term potentiation in bone—A role for glutamate in strain-induced cellular memory? *BMC Cell Biology* 4:9. doi:10.1186/1471-2121-4-9.

Stanley, G. B. 2013. Reading and writing the neural code. *Nature Neuroscience* 16 (3):259–263. doi:10.1038/nn.3330.

Stewart, R., L. Flechner, M. Montminy, and R. Berdeaux. 2011. CREB is activated by muscle injury and promotes muscle regeneration. *PLoS One* 6 (9):e24714. doi:10.1371/journal.pone.0024714.

Stewart, S., A. Rojas-Munoz, and J. C. Izpisua Belmonte. 2007. Bioelectricity and epimorphic regeneration. *BioEssays* 29 (11):1133–1137. doi:10.1002/bies.20656.

Stokes, M. G., M. Kusunoki, N. Sigala, H. Nili, D. Gaffan, and J. Duncan. 2013. Dynamic coding for cognitive control in prefrontal cortex. *Neuron* 78 (2):364–375. doi:10.1016/j.neuron.2013.01.039.

Stone, J. R. 1997. The spirit of D'arcy Thompson dwells in empirical morphospace. *Mathematical Biosciences* 142 (1):13–30.

Stott, J. J., and A. D. Redish. 2015. Representations of value in the Brain: An embarrassment of riches? *PLoS Biology* 13 (6):e1002174. doi:10.1371/journal.pbio.1002174.

Sullivan, K. G., and M. Levin. 2016. Neurotransmitter signaling pathways required for normal development in Xenopus laevis embryos: A pharmacological survey screen. *Journal of Anatomy.* doi:10.1111/joa.12467.

Sun, Y., H. Do, J. Gao, R. Zhao, M. Zhao, and A. Mogilner. 2013. Keratocyte fragments and cells utilize competing pathways to move in opposite directions in an electric field. *Current Biology: CB* 23 (7):569–574. doi:10.1016/j.cub.2013.02.026.

Sundelacruz, S., M. Levin, and D. L. Kaplan. 2009. Role of membrane potential in the regulation of cell proliferation and differentiation. *Stem Cell Reviews and Reports* 5 (3):231–246.

Swank, M. W., and J. D. Sweatt. 2001. Increased histone acetyltransferase and lysine acetyltransferase activity and biphasic activation of the ERK/RSK cascade in insular cortex during novel taste learning. *Journal of Neuroscience* 21 (10):3383–3391.

Takeda, M., Y. Suzuki, N. Obara, and Y. Nagai. 1996. Apoptosis in mouse taste buds after denervation. *Cell and Tissue Research* 286 (1):55–62.

Tassava, R. A., and C. L. Olsen-Winner. 2003. Responses to amputation of denervated ambystoma limbs containing aneurogenic limb grafts. *Journal of Experimental Zoology Part A: Comparative Experimental Biology* 297 (1):64–79.

Taylor, A. J., and C. W. Beck. 2012. Histone deacetylases are required for amphibian tail and limb regeneration but not development. *Mechanisms of Development* 129 (9–12):208–218. doi:10.1016/j.mod.2012.08.001.

Thornton, C. S., and M. T. Thornton. 1970. Recuperation of regeneration in denervated limbs of Ambystoma larvae. *Journal of Experimental Zoology* 173 (3):293–301. doi:10.1002/jez.1401730308.

Tosenberger, A., N. Bessonov, M. Levin, N. Reinberg, V. Volpert, and N. Morozova. 2015. A conceptual model of morphogenesis and regeneration. *Acta Biotheoretica* 63 (3):283–294. doi:10.1007/s10441-015-9249-9.

Tseng, A. S., W. S. Beane, J. M. Lemire, A. Masi, and M. Levin. 2010. Induction of vertebrate regeneration by a transient sodium current. *Journal of Neuroscience* 30 (39):13192–13200. doi:10.1523/JNEUROSCI.3315-10.2010.

Tseng, A. S., K. Carneiro, J. M. Lemire, and M. Levin. 2011. HDAC activity is required during Xenopus tail regeneration. *PLoS one* 6 (10):e26382. doi:10.1371/journal.pone.0026382.

Tseng, A. S., and M. Levin. 2012. Transducing bioelectric signals into epigenetic pathways during tadpole tail regeneration. *Anatomical Record* 295 (10):1541–1551. doi:10.1002/ar.22495.

Tseng, A. S., and M. Levin. 2013. Cracking the bioelectric code: Probing endogenous ionic controls of pattern formation. *Communicative & Integrative Biology* 6 (1):1–8.

Tsien, J. Z., P. T. Huerta, and S. Tonegawa. 1996. The essential role of hippocampal CA1 NMDA receptor-dependent synaptic plasticity in spatial memory. *Cell* 87 (7):1327–1338.

Tsutsui, H., Y. Jinno, A. Tomita, and Y. Okamura. 2014. Rapid evaluation of a protein-based voltage probe using a field-induced membrane potential change. *Biochimica et Biophysica Acta*. doi:10.1016/j.bbamem.2014.03.002.

Tyner, K. M., R. Kopelman, and M. A. Philbert. 2007. "Nanosized voltmeter" enables cellular-wide electric field mapping. *Biophysical Journal* 93 (4):1163–1174.

Uzkudun, M., L. Marcon, and J. Sharpe. 2015. Data-driven modelling of a gene regulatory network for cell fate decisions in the growing limb bud. *Molecular Systems Biology* 11 (7):815. doi:10.15252/msb.20145882.

Vandenberg, L. N., D. S. Adams, and M. Levin. 2012. Normalized shape and location of perturbed craniofacial structures in the Xenopus tadpole reveal an innate ability to achieve correct morphology. *Developmental Dynamics* 241 (5):863–878. doi:10.1002/dvdy.23770.

Verschure, P., C. Pennartz, and G. Pezzulo. 2014. The why, what, where, when and how of goal directed choice: Neuronal and computational principles. *Philosophical Transaction of the Royal Society of London: Series B, Biological Sciences* 369:20130483.

Verschure, P. F., T. Voegtlin, and R. J. Douglas. 2003. Environmentally mediated synergy between perception and behaviour in mobile robots. *Nature* 425 (6958):620–624. doi:10.1038/nature02024.

von Dassow, G., and E. Munro. 1999. Modularity in animal development and evolution: Elements of a conceptual framework for EvoDevo. *The Journal of Experimental Zoology* 285 (4):307–325.

Wang, E. T., and M. Zhao. 2010. Regulation of tissue repair and regeneration by electric fields. *Chinese Journal of Traumatology* 13 (1):55–61.

Waseem, T., M. Mukhtarov, S. Buldakova, I. Medina, and P. Bregestovski. 2010. Genetically encoded Cl-Sensor as a tool for monitoring of Cl-dependent processes in small neuronal compartments. *Journal of Neuroscience Methods*. doi:10.1016/j.jneumeth.2010.08.002.

Wehr, M. C., M. V. Holder, I. Gailite, R. E. Saunders, T. M. Maile, E. Ciirdaeva, R. Instrell et al. 2013. Salt-inducible kinases regulate growth through the Hippo signalling pathway in Drosophila. *Nature Cell Biology* 15 (1):61–71. doi:10.1038/ncb2658.

Weissenberger, S., C. Schultheis, J. F. Liewald, K. Erbguth, G. Nagel, and A. Gottschalk. 2011. PACalpha—An optogenetic tool for in vivo manipulation of cellular cAMP levels, neurotransmitter release, and behavior in Caenorhabditis elegans. *Journal of Neurochemistry* 116 (4):616–625. doi:10.1111/j.1471-4159. 2010.07148.x.

Whalley, K. 2011. Synaptic plasticity: Tuning electrical synapses. *Nature Reviews Neuroscience* 12 (12):705. doi:10.1038/nrn3142.

Wolsky, A. 1978. Regeneration and cancer. *Growth* 42 (4):425–426.

Wyart, C., F. Del Bene, E. Warp, E. K. Scott, D. Trauner, H. Baier, and E. Y. Isacoff. 2009. Optogenetic dissection of a behavioural module in the vertebrate spinal cord. *Nature* 461 (7262):407–410.

Yamashita, M. 2013. Electric axon guidance in embryonic retina: Galvanotropism revisited. *Biochemical and Biophysical Research Communications* 431 (2):280–283. doi:10.1016/j.bbrc.2012.12.115.

Yamashita, T., A. Pala, L. Pedrido, Y. Kremer, E. Welker, and C. C. Petersen. 2013. Membrane potential dynamics of neocortical projection neurons driving target-specific signals. *Neuron* 80 (6):1477–1490. doi:10.1016/j.neuron.2013.10.059.

Yang, M., and W. J. Brackenbury. 2013. Membrane potential and cancer progression. *Frontiers in Physiology* 4:185. doi:10.3389/fphys.2013.00185.

Yntema, C. L. 1959. Regeneration in sparsely innervated and aneurogenic forelimbs of Amblystoma larvae. *Journal of Experimental Zoology* 140 (1):101–123. doi:10.1002/jez.1401400106.

Zhao, M. 2009. Electrical fields in wound healing-An overriding signal that directs cell migration. *Seminars in Cell and Developmental Biology* 20 (6):674–682.

Zhou, Y., C. O. Wong, K. J. Cho, D. van der Hoeven, H. Liang, D. P. Thakur, J. Luo et al. 2015. Signal transduction. Membrane potential modulates plasma membrane phospholipid dynamics and K-Ras signaling. *Science* 349 (6250):873–876. doi:10.1126/science.aaa5619.

Chapter 5 The role of nerves in the regulation of regeneration

David M. Gardiner

Contents

Key concepts

a. Interactions between nerves and target tissues, especially epithelial tissues, are critical to both embryonic development and regenerative development.

b. The factors that account for the pro-regenerative function of nerves are conserved in axolotls and humans, and the challenge is to discover how to control these signals in time and space.

c. Nerves function during regeneration to control cell migration, dedifferentiation, proliferation, and differentiation; engineering regeneration will necessitate the same.

d. Achieving successful regeneration will necessitate a better understanding of the dynamics of the interactions between nerves and target tissues.

5.1 Introduction

The influence of nerves on the process of regeneration in adults has long been recognized (Wallace 1981), and in recent years, it is also becoming evident that signaling from nerves influences both the development and regeneration of tissues in the embryo (Knox et al. 2010, 2013). This widespread and conserved function of nerve signaling emphasizes the importance of understanding the pathways involved that must be successfully engineered in order to achieve induced regeneration in humans.

Historically, the focus on the role of nerves in regeneration has been on how nerves control the regeneration of the non-neuron tissues (mostly in the amphibian limb). What has largely escaped the attention of the regeneration research community is the reality that the nerves themselves are also regenerating in coordination with the other tissue of the limb (Figure 5.1). The ability of the nerve to regenerate along with the tissues that it must eventually innervate is noteworthy in that peripheral nerves can regenerate so long as the pre-existing nerve sheath persists (e.g., after a nerve crush), but they do not do so in the absence of guidance cues provided by the sheath. Thus, the nerve regenerates in response to amputation by much the same way that it developed in the embryo, in that it grows and undergoes path finding without the guidance of the pre-existing nerve sheath.

Although studies of how the pattern of nerves is re-established during regeneration are somewhat limited, there is an entire field of biology (developmental neurobiology) focused on this problem during embryonic development. The early mutant screens in *Drosophila* identified a large number of genes involved in regulating neurite outgrowth, path finding, and synaptogenesis (Gilbert 2013; Wolpert et al. 2015). These studies, in turn, led to the characterization of large numbers of markers for many different types of nerves, which, combined with recent advance in real-time high-resolution imaging, are driving an explosive growth in our understanding of the dynamic regulation of nerve development. Assuming that many of these pathways and mechanisms are conversed and reused during limb regeneration, advances in learning how to engineer nerve regeneration de novo should progress rapidly.

Aside from the events associated with regeneration of the nerve itself, little is known about the mechanisms of the interaction between nerves and target cells during regeneration. This is likely a consequence of the early view that nerves are the source of factors (see later) that induced regeneration. Thus, the quest was, and continues to be, to identify and purify the neurotrophic factor(s) from an animal that can regenerate (e.g., salamander) and that would allow for the induction of regeneration in a non-regenerating animal (e.g., human). Although there maybe a regeneration elixir produced by nerves, I would make the argument that this is unlikely to be the case. It is more appropriate to think of the pro-regenerative activity of nerves to be an emergent property of the interaction of nerves and target cells. It is likely that the mechanisms of these interactions are

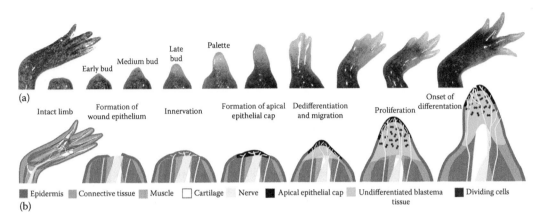

Figure 5.1 Stages of blastema formation and nerve regrowth during limb regeneration. (a) Photographs of progressive stages of blastema formation during regeneration of an amputated axolotl limb. The early-bud blastema forms several days after amputation as a consequence of the migration of connective tissue cells to the center of the wound. With proliferation and continued migration of cells, the blastema increases in size (medium and late bud) and then begins to flatten along the dorsal–ventral axis (palette). At later stages, structures of the limb begin to reform, as first evident by the condensation of cartilage to reform the digits. Before the palette stage, blastema development is dependent on signaling from nerves, such that if the limb is denervated, the blastema regresses. In contrast, at the palette and later stages, the blastema on a denervated limb stops growing and differentiates to form a small but normally patterned limb. (b) Cartoon of the early stages of regeneration illustrating the regeneration of the severed nerve (yellow) in concert with the formation and growth of the blastema. Keratinocytes of the peripheral epidermis (green) migrate centrally to form the wound epithelium (WE) within hours after amputation, during which time the severed nerves regress. As the WE increases in thickness over the next few days, the nerves begin to regrow distally and branch to hyperinnervate the apical epithelium. At about this time, the epithelium acquires its function of controlling recruitment and proliferation of blastema mesenchymal cells and is referred to as the apical epithelial cap (AEC, blue). The growth of the blastema involves the recruitment and proliferation of undifferentiated cells (turquoise) in response to signals arising from the interaction of the nerves with the AEC. (Courtesy of Catherine D. McCusker, Susan V. Bryant, and David M. Gardiner.)

highly conserved, and thus humans fail to regenerate because these critical interactions are not occurring in time and space, rather than because we lack a salamander-specific regeneration-inducing factor(s). Thus, we need to learn about the details of the interaction between nerves and their target cells. Advances in the availability of cell-type specific markers and high-resolution imaging in living tissues provide the opportunity to better understand these critical cell–cell interactions.

5.2 History of research on the nerve dependency of regeneration

The essential role of the nerve during salamander regeneration was described for the first time in 1823 by Tweedy John Todd (Wallace 1981). This initial discovery and most subsequent research involved surgically

115

denervating the limb, which inhibited regeneration (loss of function). The goal then was to figure out how to rescue regeneration of the denervated limb by an array of techniques. The problem with this approach is that nerves appear to be essential at multiple steps in the regeneration process and therefore might provide multiple signals. It is clear that nerves are required from the onset of regeneration through later blastema stages, a period of many days to a couple of weeks (Figure 5.1). Until recently (see later), it has not been possible to identify specific signaling pathways that both initiate and sustain limb regeneration in the absence of the nerve.

A large number of laboratories have worked on the role of nerves in regeneration, particularly for the salamander limb, and several basic principles that characterize this phenomenon have been identified. Many of these studies were the focus of work in the Marcus Singer laboratory in the mid-twentieth century. These principles include the fact that nerve dependency is a quantitative threshold phenomenon, such that regeneration occurs when there are enough nerves above the threshold and fails when there are not enough nerves (Singer 1974). This level of signaling is necessary for both initial blastema formation and subsequent growth and development of the blastema during the early and mid-stages of regeneration. The threshold phenomenon also was demonstrated in denervated limbs that become progressively re-innervated with time as a result of the proximal to distal progression of regeneration of the severed nerves (Salley and Tassava 1981; Salley-Guydon and Tassava 2006). Thus, there is a point during nerve regeneration at which the number of regenerated nerves exceeds the threshold required for regeneration, at which point a sufficient level of re-injury of the limb induces a nerve-dependent regenerative response, even though the full complement of innervation has not yet been restored (Salley and Tassava 1981; Salley-Guydon and Tassava 2006). It is at this threshold of partial re-innervation that electroporation of plasmids expressing the newt anterior gradient (nAG) signaling molecule has been demonstrated to induce regeneration of previously denervated newt limbs (Kumar et al. 2007).

The dependence of regeneration on nerves is quantitative (see above paragraph) but not qualitative, inasmuch as sensory, motor, and autonomic nerves can be substituted for one another, so long as the threshold is exceeded. The latter phenomenon is important, given that it is the interaction between nerves and the apical epithelium of the wound and blastema that is important for controlling regeneration. The specific identity of these nerves has not been characterized in detail, but many of them are sensory. In addition to the neural cells of the nerve, there have been suggestions about the possible role of non-neural cells (e.g., Schwann cells) in the phenomenon of nerve-dependent regeneration. In the end, once the signals that are provided by the nerves are identified and the developmental pathways are characterized, progress in engineering ways to control these pathways will be possible. Recent advances in identifying the involvement of fibroblast growth factor (FGF) and bone morphogenetic protein (BMP) signaling (see later) are encouraging in that specific details regarding the regulation of the events will soon be discovered.

Another phenomenon is the observation that although initiation and early stages of blastema formation are *dependent* on nerves (i.e., fail to initiate or progress along the regeneration pathway), at later stages, the blastema will progress in terms of differentiation and form a normally patterned limb after the limb is denervated (Figure 5.1; Wallace 1981). In the litera-ture, this phenomenon is referred to as a transition from an early nerve-*dependent* stage to the later nerve-*independent* stage. This terminology is misleading in that denervation at later stages actually does have a dramatic effect in terms of inhibiting the growth of the regenerate, resulting in a nor-mally patterned but small limb. Regulation of this later phase of growth to normal size (corresponding to the size of the unamputated portion of the limb) is yet to be understood, but it does appear to be nerve-dependent.

Until recently, there have been little experimental data that shed light on what might be occurring at this transitional stage of blastema forma-tion. We now realize that the blastema cells of early-stage blastemas are developmentally plastic and are maintained in this state by signaling from the nerve (McCusker and Gardiner 2013*, 2014; McCusker et al. 2015a). In contrast, at later stages, the cells in the basal region of the blastema are no longer under the influence of nerve signaling, which is distally restricted (Satoh et al. 2008a, 2012), and their positional iden-tity and developmental fate become stabilized (McCusker and Gardiner 2013). This change in nerve-dependent plasticity explains why early-bud blastemas and the apical region of late-bud blastemas integrate into the host site without forming ectopic structures; whereas, late-bud blastemas form supernumerary limbs when grafted to an ectopic host site (Figure 5.1; Stocum 1968; McCusker and Gardiner 2013). Thus, it appears that nerve signaling is required to maintain blastema cells in an undifferentiated and proliferative state until the final pattern and size of the missing limb are restored. When nerve signaling is terminated prematurely, the undifferentiated apical blastema cells differentiate and cease proliferation, resulting in a complete but miniaturized pattern.

There has been much effort over many years to identify the *neurotrophic* factor(s). Many candidates have been proposed and investigated, including neuropeptides, organic molecules, cyclic nucleotides, growth factors, and even bioelectric signals (Singer 1978; Wallace 1981; Mescher 1996; Bryant et al. 2002; Makanae et al. 2014; Farkas et al. 2016). One candidate that has attracted attention in recent years is nAG protein, which is a ligand for *Prod1* that is implicated in cellular identification along the proximal–distal limb axis. Expression of nAG is induced after amputation, is localized within Schwann cells of nerves at 5 days postamputation, and then appears within multicellular dermal glands underlying the apical epithelial cap (AEC) about

---

* This is the first demonstration that the positional identity of cells in the early blastema is plastic but is subsequently stabilized at later stages of regeneration.

a week later (Kumar et al. 2007). As noted earlier, induced expression of nAG can rescue regeneration in partially innervated newt limbs (Salley and Tassava 1981; Kumar et al. 2007), and thus, it is hypothesized to function as a nerve-derived neurotrophic factor. Overall, nAG appears to function via a newt-specific signaling pathway at relatively later time points in regeneration (starting at day 5), after the initial wound has already been induced by nerve signals (within 1–3 days post-injury) to progress along the blastema formation pathway (Endo et al. 2004*; Satoh et al. 2007). The recent demonstration that the axolotl wounds can be induced to form a blastema in response to human growth factors (Makanae et al. 2013†, 2014; Satoh et al. 2015) is consistent with the hypothesis that early growth factor signaling within the first 1–3 days post-injury is sufficient to trigger downstream events required for blastema formation and limb regeneration. The specific function of nAG in newt limb regeneration is unclear, since the presence of nAG from the glands underlying the AEC is thought to be essential. Axolotls lack these glands (Kumar et al. 2010; Seifert et al. 2012; unpublished observations), but they still regenerate. If the mechanisms of regeneration in newts and axolotls are conserved, then nAG does not appear to be essential for successful regeneration.

Before 2004, studies to identify the *neurotrophic* factor(s) were limited by the lack of an appropriate gain-of-function assay for regeneration (Mescher 1996). All early studies were based on either the limb amputation model or the treatment of blastema cells *in vitro*. As discussed later, the latter approach is problematic in that blastema cells lose their *in vivo* characteristic behaviors very soon after being dissociated and cultured *in vitro*. As for the limb amputation model, it is intrinsically limited, because it is a loss-of-function assay. It was effective early on in demonstrating that nerves are required for regeneration (denervated limbs do not regenerate) but cannot identify the signals that can induce a non-regenerating wound to regenerate (gain of function). In addition, amputation induces a multitude of changes, as evidenced by changes in expression of a thousand of genes (Monaghan et al. 2009, 2012; Voss et al. 2015); however, many of these are associated with injury (as in mammals) and cannot be identified as being specifically required for regeneration (since mammals do not regenerate, even though they express these genes). This is the *needle-in-a-haystack* challenge and cannot be solved without a gain-of-function assay.

The Accessory Limb Model (ALM) was developed to address both the *needle-in-a-haystack* and *gain-of-function* challenges (Figure 5.2; Endo et al. 2004; Satoh et al. 2007). In this assay, a skin wound is created surgically, without damage to the underlying tissues (thus eliminating the non-specific trauma associated with amputation). If untreated, this

---

* This paper establishes the Accessory Limb Model as a gain-of-function assay for signals that control limb regeneration.

† This is the first demonstration that a cocktail of mammalian growth factors can substitute for the function of the nerve in the induction of ectopic blastema and limb formation in the Accessory Limb Model.

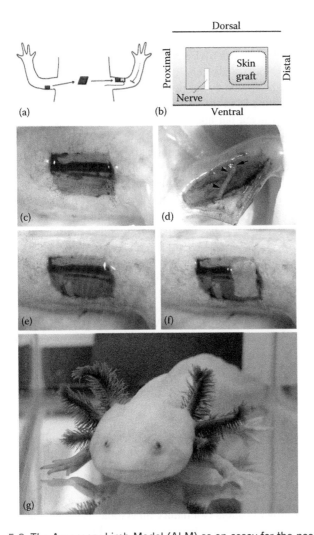

Figure 5.2 The Accessory Limb Model (ALM) as an assay for the necessary
and sufficient signals for limb regeneration. (a) Ectopic limbs with normal
pattern can be induced to form in response to a surgically created full-
thickness skin wound to which a nerve (blue) is surgically deviated, and a
piece of skin (red) from the opposite side of the limb (posterior) is grafted
into the host site (anterior). (b) Illustration of the final arrangement of
wound, deviated nerve, and skin graft. (c–f) Images of the surgical steps
in the ALM, beginning with making the full-thickness skin wound (c),
then isolating the brachial nerve (arrow heads) before being surgically
deviated (d), then deviating nerve into the wound bed (f), as illustrated
in (b), and finally grafting a piece of posterior skin in combination with
the deviated nerve (f), as illustrated in (b). An ectopic blastema then
forms and regenerates a normally patterned limb, as in (g), on an animal
with two left arms. Formation of ectopic limbs occurs over the same time
course and progresses through the same stages of regeneration as does
an amputated limb (illustrated in Figure 5.1). (Courtesy of Susan V. Bryant
and David M. Gardiner.)

wound will heal and the skin will be regenerated. If additional signals are provided, the wound can be induced to form a blastema and even an ectopic limb (gain-of-function) (Figure 5.2). As is evident from the discussion earlier, one critical source of signals is the nerve, and these signals are provided to the wound by surgically deviating the brachial nerve to the wound. Signaling from the nerve is sufficient to induce the formation of an ectopic blastema but not an ectopic limb. From discoveries that led to formalizing the Polar Coordinate Model (French et al. 1976*; Bryant et al. 1981†), regeneration also requires positional information (PI) from opposite sides of the limb. In the ALM, this information can be provided by a graft of skin from the side of the limb opposite the wound (e.g., posterior skin grafted to an anterior wound). This PI then induces the ectopic blastema to form a normally patterned ectopic limb. Use of the ALM as a gain-of-function assay has identified factors that can substitute for both the deviated nerve and the skin graft (Phan et al. 2015; Satoh et al. 2015).

Studies of embryonic development have repeatedly demonstrated a critical function for FGF and sonic hedgehog (SHH) signaling in making a limb in the first place (Laufer et al. 1994; Niswander et al. 1994; Li and Muneoka 1999‡; Mariani and Martin 2003). A number of experiments over many years have indicated that the function and mechanisms of FGF signaling are conserved in regenerative development. The FGFs are expressed in the apical blastema where blastema mesenchymal cells interact with the AEC (Mullen et al. 1996; Han et al. 2001; Christensen et al. 2002). Nerve-dependent expression of the transcription factor *Dlx3* is rescued in denervated axolotl limbs by implanting FGF2-soaked beads (Mullen et al. 1996). Keratinocyte growth factor (*FGF7*) is expressed in regenerating nerves and maintains the function of the AEC (Satoh et al. 2008b). These data led to recent efforts to screen for mammalian growth factors, FGFs in particular, that function to induce blastema formation in the ALM. No single factor has been identified; however, a number of combinations of growth factors can induce an ectopic blastema that can progress to make an ectopic limb when posterior PI is provided (posterior skin graft to an anterior wound). Among these combinations, a cocktail of FGF2, FGF8, and BMP2, all of which are expressed by regenerating axolotl nerves, can substitute for a deviated nerve and induce blastema formation in the ALM (Makanae et al. 2013, 2016).

As an alternative to grating posterior dermal cells to an anterior wound, host anterior cells can be reprogrammed to function as posterior cells and thus induce formation of an ectopic limb. This was discovered by treating axolotls systemically with retinoic acid (RA) (McCusker et al. 2014).

---

* This is the original Polar Coordinate Model that articulated the view that the mechanisms for pattern formation are conserved across species and involve short-range cell–cell signaling, leading to the growth and intercalation.

† This paper refines the original Polar Coordinate Model published in 1976.

‡ This report demonstrated that the primary function of FGF4 likely is to control migration of limb bud cells.

Multiple ectopic limbs were induced to form from anterior wounds, but not posterior wounds, as a consequence of the ability of RA to concert blastema cells to a posterior positional identity (Wanek et al. 1991; Bryant and Gardiner 1992*), which would be equivalent to grafting posterior cells. More recently, anterior cells that were engineered to express SHH also induced ectopic limb formation (Nacu et al. 2016). In this case, ectopic SHH expression presumably functions the same as when posterior, SHH-expressing cells are grafted to the anterior, as demonstrated for developing limbs. In the ALM, the pattern-inducing activity of SHH is dependent on the interaction with anterior cells expressing FGF (Nacu et al. 2016), an interaction that also occurs and has been well characterized in limb development in which maintenance of the zone of polarizing activity (ZPA) signaling by FGF4 is linked to the expression of SHH by ZPA cells (Laufer et al. 1994; Niswander et al. 1994; Li and Muneoka 1999).

Given that a number of growth factors, and especially FGFs, are involved in the interactions between nerves and target cells, the extracellular matrix (ECM) will, by necessity, also play a crucial role in regeneration. It is well established that the ECM has an obligatory function in the regulation of FGF signaling and appears to have an equivalent role in BMP and Wnt signaling as well (Rapraeger et al. 1991; Yayon et al. 1991; Sarrazin et al. 2011). The ability of FGFs to interact with their receptors on target cells is dependent on specific modifications of sulfated residues on glycosaminoglycans in the ECM. Recent studies based on the ALM have identified heparan sulfate proteoglycans (HSPGs) in both axolotl and mouse skin as important in both the initiation and inhibition of regeneration and the encoding of PI during regeneration (Phan et al. 2015). Although the glycobiology of the ECM is complex, advances in the field will soon allow us to better decipher these sulfation codes to reveal how growth factor signaling is regulated in time and space to induce regeneration.

The role of nerves in the regeneration of tissues other than limbs, and in other species (e.g., mouse digit, deer antlers, and invertebrates) in general, has not been as extensively studied as in the case of salamanders (and to a lesser extent, the anurans), and a review of that literature is beyond the scope of this chapter. In many of these other examples, it is not clear as to the specific effects of nerve signaling, even though denervation typically results in some sort of perturbation of the regenerative processes and/or outcomes. The complexity of nerve–target tissue interactions will ultimately become clear as the underlying mechanisms are discovered. At this time, it is evident that regeneration is dependent to variable degrees on nerves and that, in the end, the regenerated structure will have nerves. Thus, there must be crosstalk that leads to the coordinated regeneration of both nerves and their target tissues (Suzuki et al. 2005; Kumar and Brockes 2012).

---

* This essay established the model that retinoic acid (RA) induces supernumerary pattern formation by reprograming anterior cells rather than functioning as a diffusible morphogen.

Even in salamanders, where the overall regeneration process is nerve-dependent, it is evident that many of the steps need not be, or are not, nerve-dependent. For example, re-epithelialization and histolysis of the amputated stump tissues occur in denervated limbs (Wallace 1981). This makes sense since, initially, the most distal region of the amputated limb is denervated as a result of degeneration of severed nerves. Once these injured nerves heal and begin to regenerate, they innervate the wound epithelium (WE) and subsequently the AEC and are required for WE/AEC function (Satoh et al. 2008b; McCusker et al. 2015a). The nerve independence of at least some steps in regeneration is further confounded by the phenomenon of aneurogenic regeneration (Yntema 1959; Kumar and Brockes 2012; Pirotte et al. 2016). Limb buds in developing salamander embryos regenerate, even though there is a relatively low level of innervation, and if limbs are prevented from becoming innervated at later stages of larval development, they will regenerate when amputated, even though they are not innervated (Wallace 1981; Mescher 1996; Kumar and Brockes 2012). However, once the limb becomes innervated during normal development, it also becomes dependent on nerves for regeneration. Nerve-independent development of nAG in the limb bud epidermis and glands in larva has been hypothesized to account for this phenomenon (Kumar and Brockes 2012). The important point here is that regeneration is not absolutely dependent on nerve signaling; rather, there are regeneration pathways that need to be regulated, either positively or negatively. Although the nerve functions to accomplish this in some animals that regenerate exceptionally well (e.g., salamanders), the challenge is to discover the pathways in order to develop therapies to regulate them to achieve a pro-regenerative outcome.

5.3 Role of nerves in the regulation of embryonic development

In contrast to the extensive phenomenology with regard to the role of nerves in regeneration, relatively little is known about the function of nerves in controlling embryonic development. The behavior of nerves during embryogenesis has been well studied for both the central nervous system (CNS) and the peripheral nervous system (PNS) and is the basis for the expanding field of developmental neurobiology. Such studies have focused on how axons find their way to target cells and tissues and have led to a comprehensive understanding of axon guidance and path finding. Moreover, much focus is now on what happens when two nerves interact to form the complex synaptic network of the CNS. Similarly, the growth cone of the extending axon in the PNS reaches its target muscle cells, leading to the formation of the neuromuscular junction (NMJ). From these studies, it is evident that these interactions are mutual and reciprocal, such that neurons that are unable to find a target or targets that do not get innervated fail to survive (Gilbert 2013;

Wolpert et al. 2015). The signals regulating these interactions between developing nerves and muscle (e.g., neurotropic factors) have been well characterized.

There is a more limited but growing appreciation of the role of interactions between nerves and other cell types in the developing embryo. Evidence that the embryonic development of other target tissues is dependent on interactions with nerves includes an early report that taste buds of mouse undergo apoptosis if denervated (Takeda et al. 1996). This result is comparable to the NMJ studies referenced earlier, such that failure to become innervated results in failure to survive. More recently, the role of nerves has been demonstrated for the development of glandular tissues derived from epithelial progenitor cells, specifically salivary gland and prostate (Knox et al. 2010, 2013). In these studies, removal of the parasympathetic ganglion in explant organ culture from mouse decreased the number of epithelial progenitor cells and inhibited morphogenesis, resulting in disrupted development of the organ. As hypothesized for regenerating limbs (see later), the nerve appears to function to maintain cells in an undifferentiated state, which is required for organogenesis (Knox et al. 2010, 2013).

The mechanism of nerve-dependent regulation of embryonic development of these glands is conserved during their regeneration after damage by therapeutic irradiation. If the salivary gland is denervated by parasympathectomy, injured adult organs do not regenerate, despite the presence of stem and progenitor cells. This finding has led to the hypothesis that the surviving progenitors require innervation in order to participate in organ regeneration. In other words, salivary gland regeneration is nerve-dependent (Knox et al. 2013). In general, nerve dependency may be a phenomenon associated with epithelial tubulogenesis in other organs that are also innervated by the autonomic nervous system, including the lungs, prostate, and salivary glands, and other epithelial tissues that develop via the process of lumenization, such as the pancreas, kidney, and intestine (Nedvetsky et al. 2014).

5.4 Role of nerves in the regulation of regenerative development

5.4.1 Target(s) of nerve signals: The apical epithelial cap

Classically, all studies of nerve dependency during limb regeneration have been loss-of-function studies (denervated limbs do not regenerate). Although it is clear from such studies that you need the nerve, they tell you little about what the nerve does mechanistically and nothing about the mechanisms of regeneration, other than that it is nerve-dependent. Despite these limitations, it is noteworthy that the *lack-of-nerve* phenotype is the same as observed when the function of the WE/AEC is inhibited (i.e., regenerative failure), which has led to the hypothesis that the nerve functions to maintain the ability of the WE/AEC to support blastema formation and outgrowth (Wallace 1981; McCusker et al. 2015a). By extension, it has been hypothesized that a failure in the interaction between nerves and the WE/AEC could account for regenerative failure in mammals (Tassava and Olsen 1982). Regenerative failure is also observed in response to inhibition of the early inflammatory response mediated by macrophages (Godwin et al. 2013). By similar reasoning, this could reflect a functional interaction between nerve signaling and the early immune response, whereby the recruitment of macrophages is nerve-dependent. This interaction could be a direct effect of nerves and macrophages on each other or an indirect effect, whereby nerve–WE/AEC interactions recruit macrophages. Macrophages are a major source of enzymes and regulatory factors (e.g., cytokines and chemokines) that could function to regulate ECM remodeling, cell recruitment, proliferation, and dedifferentiation, all of which are important for blastema formation. The immune system has long been considered likely to be important in regeneration and has received renewed interest in recent years (Mescher 1996; Godwin et al. 2013; Mescher et al. 2013). Overall, nerve–epithelial interaction is a potential target for regenerative engineering, as the specific molecular mechanisms become understood (see later).

It appears that nerve signaling maintains the WE/AEC as a signaling center by controlling cell-cycle kinetics of the basal keratinocytes (Satoh et al. 2012). It has been known for a long time that the cells of the AEC exhibit the fascinating behavior of withdrawing from the cell cycle (Hay and Fischman 1961). This phenomenon has been reinvestigated, and cell-cycle withdrawal of the basal keratinocytes of the WE/AEC occurs in the axolotl limb blastema, as well as in the AER of chick and mouse limb buds (Satoh, Bryant, and Gardiner 2012). Cell-cycle withdrawal was associated with the formation of gap junctions between the WE/AEC basal keratinocytes and the expression of the embryonic AEC marker gene, *Sp9*. When the limb was denervated, the gap junction protein *Connexin 43* and *Sp9* were no longer expressed, basal keratinocytes re-entered the cell cycle, and regeneration was inhibited. In addition to the AER of mouse and chick, a number of other developmentally important signaling centers are characterized by cell-cycle withdrawal, or very long cell cycles (e.g., ZPA of chick limb buds; Bryant and Gardiner [2016]*). A model for the functional relationship between pattern formation and cell-cycle length has been proposed (Bryant and Gardiner 2016), and nerve signaling appears to play a role in regulating these interrelated processes.

---

* This essay presents a model for how regulation of cell-cycle kinetics can function upstream of pattern formation and that morphogens function to control the length of the cell cycle.

5.4.2 The outcome of neural–epithelial interaction

Whether the nerve functions directly or indirectly via interactions with the WE/AEC and/or immune system, there are a number of well-characterized nerve-dependent outcomes. Each is essential, and if any one fails, then regeneration fails; therefore, successful regenerative engineering will necessarily achieve these outcomes.

5.4.2.1 The outcome of neural–epithelial interaction: Migration and recruitment of blastema cells   One of the earliest events in regeneration, once the wound has healed by epithelial migration and formation of the WE (Satoh et al. 2008b; Tanner et al. 2009; Ferris et al. 2010), is the recruitment of the early blastema cells from the loose connective tissues of the wound bed and dermal tissues surrounding the wound margin (Gardiner et al. 1986; Muneoka et al. 1986; Endo et al. 2004). Although not demonstrated directly, the directed migration and accumulation of early blastema cells is considered a consequence of signals arising from the interaction of nerves with the overlying WE/AEC (Wallace 1981; Tassava and Olsen 1982; Endo et al. 2004; Satoh et al. 2008b). By this view, the WE/AEC functions like the AER of developing limb buds by producing a chemoattractant signal that regulates cell migration during blastema formation and outgrowth. The most likely candidate signals are FGF proteins produced by the WE/AEC and AER (Niswander et al. 1994; Li and Muneoka 1999; Christensen et al. 2002), specifically FGF4, which directs the migration of chick limb bud cells when delivered on microcarrier beads (Li and Muneoka 1999).

Surprisingly, little is known about the timing and regulation of early cell migration during regeneration (Gardiner et al. 1986). Migration is delayed for a few days after amputation, which is thought to be a consequence of the need to degrade the ECM by matrix-degrading enzymes that are produced rapidly in response to injury (Gardiner et al. 1986; Yang and Bryant 1994; Satoh et al. 2008b). Once the fibroblasts are liberated from the ECM, they begin to accumulate at the margin of the wound (Satoh et al. 2008b) and subsequently begin migrating toward the center of the wound. An increase in the labeling index of stump tissues is not observed until several days after amputation, which coincides with the time when the migrating early blastema cells begin to accumulate under the WE/AEC (Wallace 1981; Gardiner et al. 1986). Blastema cell migration precedes the onset of proliferation, which is consistent with the hypothesis that interaction between cells with different PI (migrating toward the center of the wound from different positions around the limb circumference) stimulates proliferation, as predicted by the Polar Coordinate Model (PCM) (French et al. 1976; Bryant et al. 1981). Thus the recruitment and migration of cells from the connective tissue are two of the earliest events in regeneration and precedes reentry into the cell cycle. By this view (discussed later), the function of the nerve in the stimulation of blastema cell proliferation is secondary and downstream of its function to control cell migration.

5.4.2.2 The outcome of neural–epithelial interaction: Dedifferentiation The role of nerves in controlling differentiation is challenging to understand, since so little is known about the process of *dedifferentiation*. This term has been used for most of the history of regeneration research to describe a variety of processes associated with regeneration. Presumably, it would not be appropriate to use this term to describe the activation of adult stem cells (e.g., satellite cells that can participate in the regeneration of skeletal muscle), since stem cells are already undifferentiated and possess regenerative potential. The term thus would be applied to differentiated cells that revert to a more embryonic-like state, as evidenced by the re-expression of genes associated with earlier stages of development. In so doing, these cells acquire increased developmental potential, so as to participate in regeneration (Han et al. 2005; McCusker and Gardiner 2013, 2014; McCusker et al. 2015a), much like what occurs with reprograming of induced pluripotent stem (iPS) cells. Thus, dedifferentiation during regeneration can be viewed as being equivalent to endogenous reprogramming, resulting in increased developmental plasticity (McCusker and Gardiner 2013, 2014). Although the concept of dedifferentiation is deeply embedded in the regeneration literature, little is known mechanistically about what it is and how it is regulated.

It is obvious that the steps leading to blastema formation are nerve-dependent and that once the blastema has formed, it is equivalent to an embryonic limb bud (Bryant et al. 2002; McCusker et al. 2015a). Given that dedifferentiation is a process that is associated with blastema formation, it does not happen when blastema formation is inhibited (e.g., by denervation). Although it is not possible at this point to know whether nerves directly or indirectly regulate dedifferentiation, there are a number of results that are consistent with the former. The ALM allows for making wounds that either do or do not form a blastema (deviated nerve vs. no deviated nerve). In terms of re-expression of embryonic genes, there are examples of those that are expressed in wounds that do not make a blastema (e.g., *Msx2* and *MMP9*), whereas others are expressed only in response to a deviated nerve (*Prrx-1, Sp-9, HoxA13*, and *Tbx5*) (Satoh et al. 2007).

Evidence for a direct neural regulation of the differentiated state comes from a study of regeneration of critical-size defects (CSD) in the radius of axolotl arms (Satoh et al. 2010). In these wounds, a piece of the radius is excised, such that beyond a crucial size, the bone does not regenerate, which is equivalent to what is observed in mammals. If dedifferentiated blastema cells are grafted into the CSD, regeneration occurs, but if the differentiated blastema progenitor cells (dermal fibroblasts) or differentiated chondrocytes are grafted into the wound, regeneration does not occur (fibrous tissue or fibrocartilage forms instead). However, if a nerve is surgically deviated to the CSD and the wound is left open to allow a WE to form (equivalent to what occurs when the limb is amputated), regeneration occurs. The regeneration-competent cells are recruited by the wound environment and by induced dedifferentiation of grafted dermal fibroblasts (Satoh et al. 2010). The ability to induce dedifferentiation in this injury model is dependent on signaling from both the nerve and the WE/AEC, as is the case during regeneration of an amputated limb or formation of an ectopic limb in the ALM.

Evidence that the nerve controls the state of blastema cell differentiation also comes from studies in which undifferentiated and developmentally plastic blastema cells are grafted to a new location (McCusker and Gardiner 2013, 2014). If the positional identity of the grafted blastema cells is stabilized, then ectopic limb structures form. In contrast, if undifferentiated, positionally plastic cells are grafted (e.g., early blastemas or the apical region of late blastemas), the cells adopt the positional identity of the host site and become integrated into the host structures. However, if the limb is denervated before grafting the blastemas, the cells lose positional plasticity and form ectopic structure when grafted to another location. Again, this maybe a consequence of an emergent signaling function of the interaction of nerves and the WE/AEC rather than a direct effect of nerve signaling. From these experiments, it appears that nerve signaling maintains cells in an undifferentiated state, whereas the CSD experiments indicate that nerve signals also induce dedifferentiation.

In addition to affecting the state of differentiation of blastema mesenchymal cells, nerves influence the behavior of the epithelial cells (WE/AEC). *Sp9* is a marker gene for the embryonic AEC and is induced in response to amputation and to a deviated nerve in the ALM. Expression of this gene is also induced in response to keratinocyte growth factor (KGF) (aka FGF7) that is synthesized in the injured nerves of the limb (Satoh et al. 2008b). Expression of the de novo DNA methyltransferase, *DNMT3a*, in the WE/AEC is regulated by nerve signaling. This enzyme leads to epigenetic modifications of DNA, such that treatment of skin wounds in the ALM with decitabine, a DNA methyltransferase inhibitor, induced changes in gene expression and cellular behaviors associated with a regenerative response, whereas untreated wounds inhibited a regenerative response (Aguilar and Gardiner 2015).

5.4.2.3 The outcome of neural–epithelial interaction: Proliferation When a regenerating limb is denervated, proliferation ceases, and thus, of all the possible functions of nerve signaling during regeneration, the regulation of blastema cell proliferation has received the most research attention. Historically, experiments have involved treating denervated blastemas with putative neurotrophic factors to rescue proliferation or treating cells *in vitro* with such factors to stimulate proliferation. The limitation of the first approach is the challenge of long-term delivery of exogenous factors, since it takes weeks for the blastema to form and develop. This is complicated by the fact that during this time, severed nerves grow back and the limb needs to be re-denervated every couple of weeks. Consequently, successful attempts to rescue

127

regeneration have occurred when limbs were denervated and treated at a late stage of blastema formation at around the time the blastema transitions from being nerve-dependent to nerve-independent (see Section 5.2); for example, FGF2 (Mullen et al. 1996) and neuregulin-1 (Farkas et al. 2016). Similarly, as nerves regenerate and reinnervate the limb, there is a threshold level of nerves at which regeneration can be induced, depending on the degree of re-injury to the limb stump (Salley and Tassava 1981; Salley-Guydon and Tassava 2006). It is at this threshold that over-expression of nAG protein can rescue regeneration (Kumar et al. 2007).

Understanding the mechanisms for regulating the proliferation of blastema cells will require *in vitro* studies. Although blastema tissue can be dissociated and blastema cells can be maintained in monolayer culture, the cells quickly lose properties associated with blastema cells *in vivo*. Their positional identity, as assayed by the induction of ectopic limb structure when grafted into host blastemas, decreases dramatically within the first day of culture and is lost entirely in about a week (Groell et al. 1993). Similarly, cultured blastema cells have highly variable and low rates of proliferation; for example, long-term labeling indices (growth fraction) ranging from 0.2% to 10%, as compared with *in vivo*, short-term labeling indices of 20% to 40%. It appears that the behavior of blastema cells is dependent on maintaining the *in vivo* cell– and cell–ECM interactions, and thus, it is unclear how relevant past *in vitro* studies are for understanding the mechanisms controlling blastema cell proliferation. This challenge of how to maintain *in vivo* cellular interactions while studying the cells *in vitro* has been overcome by neurobiologists by using organotypic slice cultures (OSCs). Use of the technique is promising for blastema studies such that *in vivo* proliferation rates are maintained over a period of several days in OSC (Lehrberg and Gardiner 2015). During that period of culture, the proliferation of blastema cells is stimulated by nerve signaling from co-cultured dorsal root ganglia (DRG) and by BMP2 (Lehrberg and Gardiner 2015).

The discussion to this point has implied that nerve signaling has multiple functions over the course of regeneration. There is an early function (1) involved in the recruitment of cells, which are then (2) induced to dedifferentiate, which is followed by (3) sustained proliferation of blastema cells. These different phases may reflect different underlying functions of nerve signals; however, there could be a common function for all of these phenomena. As discussed earlier, migration of pre-blastema cells into the wound precedes the onset of proliferation. Based on the PCM, proliferation is stimulated by the interaction between cells with different PI (from different positions around the wound margin) that are brought together as a result of wound closure and migration. By this view, the primary function of the nerve would be to control cell migration, and proliferation would be a secondary consequence. The FGFs are among the many signals that have been associated with nerve function, and they are known to play a role in controlling migration. Although referred to as growth factors (implying a function in controlling proliferation), the homologs and associated pathway genes in *Drosophila* function to control cell migration. Similarly, FGF4 expressed in the apical epithelium of the chick limb bud appears to function as an endogenous chemotactic signal that directs the migration of limb bud cells distally (Li and Muneoka 1999). Since cells cannot simultaneously migrate and proliferate, and differentiated cells cannot do either, these three processes (migration, proliferation, and differentiation) could be coordinately regulated by FGFs, provided either directly by the nerve and/or indirectly as a consequence of the interaction of the nerve with the WE/AEC (Figure 5.3).

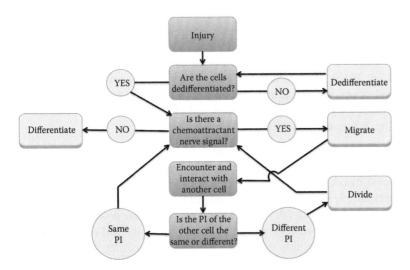

Figure 5.3 A flow diagram illustrating the coordination of differentiation, dedifferentiation, migration, and proliferation during regeneration. As formalized by the Polar Coordinate Model, when cells with different positional information (PI) interact, cell division is stimulated and the daughter cells adopt new PI that is intermediate between that of the two interacting cells. This process, referred to as intercalation, continues until all positional disparities are eliminated and the lost pattern has been restored. By the model illustrated here, these cell–cell interactions occur as a consequence of cells from different positions in the limb migrating to a localized source of a chemoattractant signal (e.g., FGF produced in the apical region of the blastema). Since migrating and/or proliferating cells are not simultaneously differentiated, an early response to injury is that cells dedifferentiate in association with the onset of migration in response to the chemoattractant. As cells move toward the source of the signal, they encounter and interact with other cells that have either the same PI or different PI. In the latter case, the cells are stimulated to divide. This iterative process of intercalation continues until regeneration is completed and the lost pattern is restored, at which point the chemoattractant signal is turned off and the cells differentiate since they are no longer induced to either migrate or divide.

By this view, FGFs function as a migration-inducing and -directing signal (i.e., chemoattractant), which is its well-established role in *Drosophila* development. In response to FGF signaling, cells migrate toward the center of the wound from around the limb circumference, where they encounter cells with different PI (Gardiner et al. 1986). These interactions between cells from different limb positions with different PI stimulate proliferation and the genesis of new cells with intermediate PI (intercalation) (French et al. 1976; Bryant et al. 1981). As the blastema forms, the AEC continues to produce FGFs (Han et al. 2001; Christensen et al. 2002), and thus, blastema cells migrate distally during outgrowth of the blastema, comparable to what is observed in the developing limb bud (Li and Muneoka 1999). This model characterizes the blastema (in particular the apical region) as a very dynamic environment, within which cells are continuously altering their interactions with other cells via FGF-induced apical migration and by periodic mitotic divisions (Figure 5.3) (McCusker et al. 2015b). Essentially, nothing is known about any of these cellular behaviors within the regenerating limb blastema (particularly in the apical region) and thus the need and challenge to better understand the dynamics of blastema cell–interactions.

One prediction of this model, which is based on the PCM (Bryant and Gardiner 2016), is that proximal amputations will regenerate faster than most distal amputations, which is what is observed *in vivo* (Iten and Bryant 1973, 1976; Vincent et al. 2015). Given that cell–cell interactions drive growth (Bryant and Gardiner 2016), the probability that the interaction of two cells will stimulate entry into the cell cycle is a function of the probability that the two cells will have different PI. Proximal level amputations will by necessity generate blastema cells with great PI diversity and disparities (i.e. more PI lost requires that more PI needs to be replaced). Conversely, smaller initial disparities (more distal amputations), or in the later stages of resolving large disparities, positional differences between neighbors will be fewer, and thus the probability that interactions between neighboring cells that have different positional information will progressively decrease. The finding that the growth fraction of late blastemas from distal amputations is about half that from proximal amputation (Vincent et al. 2015) is predicted by this model.

This model requires a feedback loop such that so long as cells remain undifferentiated and proliferative, the AEC will continue to produce FGFs (the chemoattractant). Since cells stop migrating in order to divide, dividing cells do not simultaneously migrate, and if cells are either migrating or dividing, they will not differentiate. If migrating cells encounter a cell with PI that is not what they would normally experience (having disparate PI), then they will stop migrating and divide (intercalation). If the cells that are encountered have the same or normally adjacent PI, then the cells are not stimulated to divide, and they continue to migrate distally. So long as the AEC is producing FGF, the cells will remain undifferentiated. A number of observations are consistent with this model, especially the fact that the apical and basal regions of later-stage blastemas are very different (McCusker et al. 2015a). In the case of the apical region, the dynamics of blastema cell behaviors is much like (or the same as) the early blastema and is as described earlier (under the influence of nerve signaling). In contrast, cells in the basal region begin to differentiate, presumably because the range of apical FGF signaling is distally restricted, and the cells are no longer under the influence of the FGF signal. At the same time that undifferentiated apical cells are involved in outgrowth of the blastema, the basal cells are reforming the basal lamina (Neufeld and Day 1996) and expressing markers for differentiation of the dermis (e.g., *Twist*; Satoh et al. 2008a). These differences between apical late-bud/early-bud blastemas and basal late-bud blastemas are also reflected in the observation that the cells have either labile PI (the former) or stabilized PI (the latter) (McCusker and Gardiner 2013, 2014).

Changes in the state of innervation of the WE/AEC are correlated with these changes in the behaviors of the blastema mesenchymal cells. As the severed nerve regenerates and interacts with the WE, there is extensive nerve branching, resulting in the WE/AEC becoming hyperinnervated (Satoh et al. 2008b). As regeneration progresses, the AEC remains hyperinnervated, whereas the more proximal epithelial cells do not. At the end of regeneration, the density of nerves innervating the epidermis returns to normal, and thus, it maybe that the density of nerves interacting with the AEC determines whether or not the AEC functions to direct blastema outgrowth. The challenge is to understand what turns off apical FGF signaling at the end of regeneration. Presumably, this is a consequence of restoring the lost structures such that cell–cell interactions no long stimulate proliferation (i.e., AEC function is downstream of PI and intercalation).

On a side note, this model implicates nerves in the regulation of cancer. The similarity between blastema cells and cancer cells has long been noted (mass of undifferentiated, proliferating cells). The critical role of nerves in the regulation of regeneration may indicate that they have similar, conserved functions in the regulation of the behavior of other cell types. With regard to tumor cells, it maybe that the feedback loop controlling migration, proliferation, and differentiation has become disrupted, specifically that the cells are maintained in an undifferentiated proliferative state either by failing to response to PI cues or by a failure to turn off nerve signaling. This later possibility raises the question of whether nerves always have these pro-regenerative signaling properties or whether they are activated in response to injury, for example, injury-induced expression of *KGF/Fgf-7* (Satoh et al. 2008b).

In contrast to what might be a nerve-mediated, hyper-regenerative behavior associated with cancer, there are wounds that would normally heal but fail to do so, for example, diabetic foot ulcers. These skin wounds fail to heal and are associated with diabetic neuropathies. In regeneration models, nerves are associated with wound healing in addition to blastema formation. For example, denervated limbs form a dense connective tissue cap distally and fail to regenerate (Salley and Tassava 1981), whereas these wounds would normally regenerate without forming a scar. Thus, it appears that at least a low level of nerve signaling is required for scar-free wound healing, even though it may not be quantitatively sufficient to induce blastema formation (Endo et al. 2004). Wound healing in salamanders has begun to attract renewed research attention (e.g., Seifert et al. 2012) and will presumably test the hypothesis that connective tissue fibroblasts exhibit graded responses to variation in the level of innervation, ranging from scar formation to scar-free wound healing, and eventually to blastema formation and regeneration.

5.5 Implications for regenerative engineering

Regardless of what the endogenous nerve signals that control regeneration are, entire limbs can be induced to regenerate in response to mammalian growth factors (Satoh et al. 2015; Makanae et al. 2013, 2014). These discoveries are important for regenerative engineering because they demonstrate that pro-regenerative nerve signals are not a consequence of special, salamander-specific molecules. Since the signals are conserved, the challenge for discovering how to induce a regenerative response in humans is to figure out how to deliver the signals in the appropriate spatial and temporal patterns.

The concept that the key to regeneration is to learn how to control the expression of conserved signaling molecule has been discussed previously, and recently, it has been demonstrated directly genetically (Kang et al. 2016). An enhancer regulatory element that is activated in regenerating hearts and fins of zebrafish can drive injury-dependent expression from minimal promoters in injured neonatal mouse tissues. These findings, along with the identification of comparable enhancer elements associated with activation of BMP5 expression in bone fracture repair in mice, provide evidence for *tissue regeneration enhancer elements* (TREEs) that trigger gene expression of highly conserved regeneration effector molecules. These findings emphasize the importance of discovering the spatial and temporal patterns of regulation of the conserved signals in order to engineer the regenerative response in a normally nonregenerating wound in a human. The conserved nerve signals are among the most important.

There is a major challenge to identify the specific dose/timing of regenerative signaling (nerves in particular), because the stepwise progression of regeneration is dependent on establishing and maintaining the appropriate spatial patterns of cell–cell and cell–ECM interactions. These patterns are lost when the blastema is dissociated, and blastema cells no longer behave *in vitro* as they do *in vivo*. As discussed earlier, the only viable approach to solving this problem at this time is to use OSCs to maintain *in vivo* blastema cell behaviors for a few days *in vitro*. Eventually, it could be feasible to take an engineering approach and recreate the blastema architecture via 3D printing of a pro-regenerative ECM based on discoveries from OSC experiments.

Finally, the main goal of studies of nerve signaling is to identify the processes that are targeted and orchestrated by the signals in the salamander model. One way to achieve the desired outcome (regeneration) is to exactly recapitulate the signals; however, that may not be necessary or even desirable for a mammal. Knowing the targets and processes being regulated will allow for alternative strategies for controlling regeneration in a human. Although the results to date indicate that the factors and pathways are conserved (good news), there maybe a better or more appropriate way to target the critical downstream effectors controlling pro-regenerative cellular behaviors. Whether or not we will need to use the same factors as in the axolotl, we will need to regulate the same processes. Nevertheless, the place to start with is the nerve in a regenerating salamander limb.

Acknowledgments

I am thankful to the present and former members of the Bryant/Gardiner laboratory for discussion of these ideas, especially Drs. Tetsuya Endo, Akira Satoh, Ken Muneoka, Catherine McCusker, and Susan Bryant. The model based on the control of migration being the primary function of nerves is a consequence of discussions over several years with Drs. Ken Muneoka and Susan Bryant.

References

Aguilar, C., and D. M. Gardiner. 2015. DNA methylation dynamics regulate the formation of a regenerative wound epithelium during axolotl limb regeneration. *Plos One* 10 (8): e0134791. doi: 10.1371/journal.pone.0134791.

Bryant, S. V., and D. M. Gardiner. 1992. Retinoic acid, local cell-cell interactions, and pattern formation in vertebrate limbs. *Developmental Biology* 152: 1–25.

Bryant, S. V., and D. M. Gardiner. 2016. The relationship between growth and pattern formation. *Regeneration* 3 (2): 103–122. doi: 10.1002/reg2.55.

Bryant, S. V., T. Endo, and D. M. Gardiner. 2002. Vertebrate limb regeneration and the origin of limb stem cells. The *International Journal of Developmental Biology* 46 (7): 887–896.

Bryant, S. V., V. French, and P. J. Bryant. 1981. Distal regeneration and symmetry. *Science* 212: 993–1002.

Christensen, R. N., M. Weinstein, and R. A. Tassava. 2002. Expression of fibroblast growth factors 4, 8, and 10 in limbs, flanks, and blastemas of ambystoma. *Developmental Dynamics* 223 (2): 193–203. doi: 10.1002/dvdy.10049.

Endo, T., S. V. Bryant, and D. M. Gardiner. 2004. A stepwise model system for limb regeneration. *Developmental Biology* 270 (1): 135–145.

Farkas, J. E., P. D. Freitas, D. M. Bryant, J. L. Whited, and J. R. Monaghan. 2016. Neuregulin-1 signaling is essential for nerve-dependent axolotl limb regeneration. *Development* 143 (15): 2724–2731. doi: 10.1242/dev.133363.

Ferris, D. R., A. Satoh, B. Mandefro, G. M. Cummings, D. M. Gardiner, and E. L. Rugg. 2010. Ex vivo generation of a functional and regenerative wound epithelium from axolotl (Ambystoma mexicanum) skin. *Development, Growth & Differentiation* 52 (8): 715–724. doi: 10.1111/j.1440-169X.2010.01208.x.

French, V., P. J. Bryant, and S. V. Bryant. 1976. Pattern regulation in epimorphic fields. *Science* 193: 969–981.

Gardiner, D. M., K. Muneoka, and S. V. Bryant. 1986. The migration of dermal cells during blastema formation in axolotls. *Developmental Biology* 118: 488–493.

Gilbert, S. F. 2013. *Developmental Biology.* 10th ed. Sunderland, MA: Sinauer Associates, Inc.

Godwin, J. W., A. R. Pinto, and N. A. Rosenthal. 2013. Macrophages are required for adult salamander limb regeneration. *Proceedings of the National Academy of Sciences of the United States of America* 110 (23): 9415–9420. doi: 10.1073/pnas.1300290110.

Groell, A., D. M. Gardiner, and S. V. Bryant. 1993. Stability of positional identity of axolotl blastema cells in vitro. *Roux's Archives of Developmental Biology* 202: 170–175.

Han, M. J., J. Y. An, and W. S. Kim. 2001. Expression patterns of Fgf8 during development and limb regeneration in the axolotl. *Developmental Dynamics* 220: 40–48.

Han, M., X. Yang, G. Taylor, C. A. Burdsal, R. A. Anderson, and K. Muneoka. 2005. Limb regeneration in higher vertebrates: Developing a roadmap. *Anatomical Record. Part B, New Anatomist* 287 (1): 14–24. doi: 10.1002/ar.b.20082.

Hay, E. D., and D. A. Fischman. 1961. Origin of the blastema in regenerating limbs of the newt triturus viridescens. An autoradiographic study using tritiated thymidine to follow cell proliferation and migration. *Developmental Biology* 3 (February): 26–59.

Iten, L., and S. V. Bryant. 1973. Forelimb regeneration from different levels of amputation in the newt, Notophthalmus viridescens. *Wilhelm Roux Archiv fur Entwicklungsmechanik der Organismen* 173: 263–282.

Iten, L. E., and S. V. Bryant. 1976. Regeneration from different levels along the tail of the newt, Notophthalmus viridescens. *The Journal of Experimental Zoology* 196 (3): 293–306.

Kang, J., J. Hu, R.Karra, A. L. Dickson, V. A. Tornini, G. Nachtrab, M. Gemberling, J. A.Goldman, B. L.Black, and K. D. Poss. 2016. Modulation of tissue repair by regeneration enhancer elements. *Nature* 532 (7598): 201–206. doi:10.1038/nature17644.

Knox, S. M., I. M. A. Lombaert, C. L. Haddox, S. R. Abrams, A. Cotrim, A. J. Wilson, and M. P. Hoffman. 2013. Parasympathetic stimulation improves epithelial organ regeneration. *Nature Communications* 4: 1494. doi: 10.1038/ncomms2493.

Knox, S. M., I. M. A. Lombaert, X. Reed, L. Vitale-Cross, J. S. Gutkind, and M. P. Hoffman. 2010. Parasympathetic innervation maintains epithelial progenitor cells during salivary organogenesis. *Science* 329 (5999): 1645–1647. doi:10.1126/science.1192046.

Kumar, A., and J. P. Brockes. 2012. Nerve dependence in tissue, organ, and appendage regeneration. *Trends in Neurosciences* 35 (11): 691–699. doi: 10.1016/j.tins.2012.08.003.

Kumar, A., J. W. Godwin, P. B. Gates, A. A. Garza-Garcia, and J. P. Brockes. 2007. Molecular basis for the nerve dependence of limb regeneration in an adult vertebrate. *Science* 318 (5851): 772–777.

Kumar, A., G. Nevill, J. P. Brockes, and A. Forge. 2010. A comparative study of gland cells implicated in the nerve dependence of salamander limb regeneration. *Journal of Anatomy* 217 (1): 16–25. doi: 10.1111/j.1469-7580.2010.01239.x.

Laufer, E., C. E. Nelson, R. L. Johnson, B. A. Morgan, and C. Tabin. 1994. Sonic hedgehog and Fgf-4 act through a signaling cascade and feedback loop to integrate growth and patterning of the developing limb bud. *Cell* 79 (6): 993–1003.

Lehrberg, J., and D. M. Gardiner. 2015. Regulation of axolotl (Ambystoma mexicanum) limb blastema cell proliferation by nerves and BMP2 in organotypic slice culture. *Plos One* 10 (4): e0123186. doi: 10.1371/journal.pone.0123186.

Li, S., and K. Muneoka. 1999. Cell migration and chick limb development: Chemotactic action of FGF-4 and the AER. *Developmental Biology* 211 (2): 335–347.

Makanae, A., A. Hirata, Y. Honjo, K. Mitogawa, and A. Satoh. 2013. Nerve independent limb induction in axolotls. *Developmental Biology* 381 (1): 213–226. doi: 10.1016/j.ydbio.2013.05.010.

Makanae, A., K. Mitogawa, and A. Satoh. 2014. Co-operative Bmp- and Fgf-signaling inputs convert skin wound healing to limb formation in urodele amphibians. *Developmental Biology* 396 (1): 57–66. doi: 10.1016/j.ydbio.2014.09.021.

Makanae, A., K. Mitogawa, and A. Satoh. 2016. Cooperative inputs of Bmp and Fgf signaling induce tail regeneration in urodele amphibians. *Developmental Biology* 410 (1): 45–55. doi: 10.1016/j.ydbio.2015.12.012.

Mariani, F. V., and G. R. Martin. 2003. Deciphering skeletal patterning: Clues from the limb. *Nature* 423 (6937): 319–325. doi:10.1038/nature01655.

McCusker, C., D. M. Gardiner, and S. V. Bryant. 2015a. The axolotl limb blastema: Cellular and molecular mechanisms driving blastema formation and limb regeneration in tetrapods. *Regeneration* 2: 54–71. doi: 10.1002/reg2.32.

McCusker, C. D., A. Athippozhy, C. Diaz-Castillo, C. Fowlkes, D. M. Gardiner, and S. R. Voss. 2015b. Positional plasticity in regenerating amybstoma mexicanum limbs is associated with cell proliferation and pathways of cellular differentiation. *BMC Developmental Biology* 15 (1): 45. doi: 10.1186/s12861-015-0095-4.

McCusker, C. D., and D. M. Gardiner. 2013. Positional information is reprogrammed in blastema cells of the regenerating limb of the axolotl (Ambystoma mexicanum). *Plos One* 8 (9): e77064. doi: 10.1371/journal.pone.0077064.

McCusker, C. D., and D. M. Gardiner. 2014. Understanding positional cues in salamander limb regeneration: Implications for optimizing cell-based regenerative therapies. *Disease Models & Mechanisms* 7 (6): 593–599. doi: 10.1242/dmm.013359.

McCusker, C., J. Lehrberg, and D. Gardiner. 2014. Position-specific induction of ectopic limbs in non-regenerating blastemas on axolotl forelimbs. *Regeneration* 1 (1): 27–34. doi: 10.1002/reg2.10.

Mescher, A. L. 1996. The cellular basis of limb regeneration in urodeles. *The International Journal of Developmental Biology* 40: 785–795.

Mescher, A. L., A. W. Neff, and M. W. King. 2013. Changes in the inflammatory response to injury and its resolution during the loss of regenerative capacity in developing xenopus limbs. *Plos One* 8 (11): e80477. doi: 10.1371/journal.pone.0080477.

Monaghan, J. R., A. Athippozhy, A. W. Seifert, S. Putta, A. J. Stromberg, M. Maden, D. M. Gardiner, and S. R. Voss. 2012. Gene expression patterns specific to the regenerating limb of the Mexican axolotl. *Biology Open* 1 (10): 937–948. doi: 10.1242/bio.20121594.

Monaghan, J. R., L. G. Epp, S. Putta, R. B. Page, J. A. Walker, C. K. Beachy, W. Zhu, et al. 2009. Microarray and cDNA sequence analysis of transcription during nerve-dependent limb regeneration. *BMC Biology* 7: 1.

Mullen, L., S. V. Bryant, M. A. Torok, B. Blumberg, and D. M. Gardiner. 1996. Nerve dependency of regeneration: The role of distal-less and FGF signaling in amphibian limb regeneration. *Development* 122 (11): 3487–3497.

Muneoka, K., W. Fox, and S. V. Bryant. 1986. Cellular contribution from dermis and cartilage to the regenerating limb blastema in axolotls. *Developmental Biology* 116: 256–260.

Nacu, E., E. Gromberg, C. R. Oliveira, D. Drechsel, and E. M. Tanaka. 2016. FGF8 and SHH substitute for anterior-posterior tissue interactions to induce limb regeneration. *Nature* 533 (7603): 407–410. doi:10.1038/nature17972.

Nedvetsky, P. I., E. Emmerson, J. K. Finley, A. Ettinger, N. Cruz-Pacheco, J. Prochazka, C. L. Haddox, *et al.* 2014. Parasympathetic innervation regulates tubulogenesis in the developing salivary gland. *Developmental Cell* 30 (4): 449–462. doi: 10.1016/j.devcel.2014.06.012.

Neufeld, D. A., and F. A. Day. 1996. Perspective: A suggested role for basement membrane structures during newt limb regeneration. The *Anatomical Record* 246 (2): 155–161.

Niswander, L., S. Jeffrey, G. R. Martin, and C. Tickle. 1994. A positive feedback loop coordinates growth and patterning in the vertebrate limb. *Nature* 371 (6498): 609–612.

Phan, A. Q., J. Lee, M. Oei, C. Flath, C. Hwe, R. Mariano, T. Vu, et al. 2015. Heparan sulfates mediate positional information by position-specific growth factor regulation during axolotl (*Ambystoma mexicanum*) limb regeneration. *Regeneration* 2: 182–201. doi: 10.1002/reg2.40.

Pirotte, N., N. Leynen, T. Artois, and K. Smeets. 2016. Do you have the nerves to regenerate? The importance of neural signalling in the regeneration process. *Developmental Biology* 409 (1): 4–15. doi: 10.1016/j.ydbio.2015.09.025.

Rapraeger, A. C., A. Krufka, and B. B. Olwin. 1991. Requirement of heparan sulfate for bFGF-mediated fibroblast growth and myoblast differentiation. *Science* 252 (5013): 1705–1708.

Salley, J. D., and R. A. Tassava. 1981. Responses of denervated adult newt limb stumps to reinnervation and reinjury. *The Journal of Experimental Zoology* 215 (2). 183–289. doi: 10.1002/jez.1402150208.

Salley-Guydon, J. D, and R. A. Tassava. 2006. Timing the commitment to a wound-healing response of denervated limb stumps in the adult newt, Notophthalmus viridescens. *Wound Repair and Regeneration* 14 (4). 479–483. doi: 10.1111/j.1743-6109.2006.00154.x.

Sarrazin, S., W. C. Lamanna, and J. D. Esko. 2011. Heparan sulfate proteoglycans. *Cold Spring Harbor Perspectives in Biology* 3 (7): a004952–52. doi: 10.1101/cshperspect.a004952.

Satoh, A., S. V. Bryant, and D. M. Gardiner. 2008a. Regulation of dermal fibroblast dedifferentiation and redifferentiation during wound healing and limb regeneration in the axolotl. *Development, Growth & Differentiation* 50 (9): 743–754.

Satoh, A., S. V. Bryant, and D. M. Gardiner. 2012. Nerve signaling regulates basal keratinocyte proliferation in the blastema apical epithelial cap in the axolotl (Ambystoma mexicanum). *Developmental Biology* 366 (2): 374–381. doi: 10.1016/j.ydbio.2012.03.022.

Satoh, A., G. M. Cummings, S. V. Bryant, and D. M. Gardiner. 2010. Neurotrophic regulation of fibroblast dedifferentiation during limb skeletal regeneration in the axolotl (Ambystoma mexicanum). *Developmental Biology* 337 (2): 444–457.

Satoh, A., D. M. Gardiner, S. V. Bryant, and T. Endo. 2007. Nerve-induced ectopic limb blastemas in the axolotl are equivalent to amputation-induced blastemas. *Developmental Biology* 312 (1): 231–244.

Satoh, A., G. M. Graham, S. V. Bryant, and D. M. Gardiner. 2008b. Neurotrophic regulation of epidermal dedifferentiation during wound healing and limb regeneration in the axolotl (Ambystoma mexicanum). *Developmental Biology* 319 (2): 321–335.

Satoh, A., K. Mitogawa, and A. Makanae. 2015. Regeneration inducers in limb regeneration. *Development, Growth & Differentiation* 57 (6): 421–429. doi: 10.1111/dgd.12230.

Seifert, A. W, J. R Monaghan, S. R. Voss, and M. Maden. 2012. Skin regeneration in adult axolotls: A blueprint for scar-free healing in vertebrates. *Plos One* 7 (4): e32875. doi: 10.1371/journal.pone.0032875.

Singer, M. 1974. Neurotrophic control of limb regeneration in the newt. *Annals of the New York Academy of Sciences* 228: 308–321.

Singer, M. 1978. On the nature of the neurotrophic phenomenon in urodele limb regeneration. *American Zoologist* 18: 829–841.

Stocum, D. L. 1968. The urodele limb regeneration blastema: A self-organizing system. I. Differentiation in vitro. *Developmental Biology* 18 (5): 441–456.

Suzuki, M., A. Satoh, H. Ide, and K. Tamura. 2005. Nerve-dependent and -independent events in blastema formation during xenopus froglet limb regeneration. *Developmental Biology* 286 (1): 361–375.

Takeda, M., Y. Suzuki, N. Obara, and Y. Nagai. 1996. Apoptosis in mouse taste buds after denervation. *Cell and Tissue Research* 286 (1): 55–62.

Tanner, K., D. R. Ferris, L. Lanzano, B. Mandefro, W. W. Mantulin, D. M. Gardiner, E. L. Rugg, and E. *Gratton*. 2009. Coherent movement of cell layers during wound healing by image correlation spectroscopy. *Biophysical Journal* 97 (7): 2098–2106.

Tassava, R. A., and C. L. Olsen. 1982. Higher vertebrates do not regenerate digits and legs because the wound epidermis is not functional. A hypothesis. *Differentiation; Research in Biological Diversity* 22 (3): 151–155.

Vincent, C. D, F. Rost, W. Masselink, L. Brusch, and E. M. Tanaka. 2015. Cellular dynamics underlying regeneration of appropriate segment number during axolotl tail regeneration. *BMC Developmental Biology* 15 (1): 48. doi: 10.1186/s12861-015-0098-1.

Voss, S. R., A. Palumbo, R. Nagarajan, D. M. Gardiner, K. Muneoka, A. J. Stromberg, *and A. T. Athippozhy*. 2015. Gene expression during the first 28 days of axolotl limb regeneration I: Experimental design and global analysis of gene expression. *Regeneration* 2 (3): 120–136. doi: 10.1002/reg2.37.

Wallace, H. 1981. *Vertebrate Limb Regeneration*. Chichester, UK: John Wiley and Sons.

Wanek, N., D. M. Gardiner, K. Muneoka, and S. V. Bryant. 1991. Conversion by retinoic acid of anterior cells into ZPA cells in the chick wing bud. *Nature* 350: 81–83.

Wolpert, L., C. Tickle, and A. M. Arias. 2015. *Principles of Development*. Oxford: Oxford University Press.

Yang, E. V., and S. V. Bryant. 1994. Developmental regulation of a matrix metalloproteinase during regeneration of axolotl appendages. *Developmental Biology* 166: 696–703.

Yayon, A., M. Klagsbrun, J. D. Esko, P. Leder, and D. M. Ornitz. 1991. Cell surface, heparin-like molecules are required for binding of basic fibroblast growth factor to its high affinity receptor. *Cell* 64 (4): 841–848.

Yntema, C. L. 1959. Regeneration in sparsely innervated and aneurogenic forelimbs of amblystoma larvae. *The Journal of Experimental Zoology* 140: 101–123.

Chapter 6 Physiological aspects of blastema formation in mice

Ken Muneoka

Contents

Key concepts

a. Regeneration is a stepwise process that can be segregated into two distinct transformations: (1) wound-site transforms into a blastema, and (2) blastema transforms into differentiated replacement structures.

b. Mammals, including humans, possess the ability to regenerate complex structures, but regenerative ability is limited to select structures such as the digit tip.

c. Regeneration in mice can be induced by treating the wound site with specific growth factors such as bone morphogenetic proteins (BMPs) and Wnts.

d. Although many aspects of the mammalian regenerative response are distinct from regeneration in lower vertebrates, such as salamanders and fish, the characteristics of the blastema appear to be generally conserved.

e. In mammalian regeneration, blastema formation involves the controlled histolysis of stump tissues that is impacted by wound closure, revascularization, and oxygen availability.

6.1 Regeneration and regenerative failure: Two sides of the same coin

Regeneration occurs in response to injury that causes a loss of specific and substantial anatomical structures such as a limb or a digit. In animals that possess regenerative capabilities, the injury response involves a transition from mature tissue through a transient undifferentiated proliferative phase and culminates with the re-differentiation of the replacement structures. In the case of the regenerating limb of salamanders or the digit tip of mice, the transient structure that mediates the regenerative response is called a blastema, and the formation of the blastema distinguishes regeneration from a healing response that simply closes the wound site and culminates in the deposition of an abnormally organized fibrous matrix that is remodeled into scar tissue. At its core, the process of regeneration can be segregated into two important and distinct transformations: (1) the transformation from an injury site composed of mature tissues into the blastema (a transient developmental structure), and (2) the transformation of the blastema to the differentiated replacement structures. Without question, regeneration requires the successful completion of both these transformations; without the first transformation, the ability to complete the second transformation is mute, and without the second transformation, there is no regeneration, regardless of whether or not the first transformation is successful. Herein lies a critical concept for thinking about regeneration: it is a stepwise series of inter-connected and inter-dependent processes that must be firing on all cylinders to be successful. Like an automobile, an airplane, or the space shuttle, it is an amazing engineering feat, and regenerative capability has evolved by natural selection over millions of years to be a well-honed and beautifully crafted mechanism to replace the loss of a functional part of the body. One major goal of regenerative biology is to understand the cellular and molecular bases that drive this regenerative engine, and a second major goal is to understand what components of this engine are non-functional in injury responses that fail to regenerate. In mammals such as the mouse and humans, injury responses typically heal without regeneration and scar formation is the norm; however, there are specific injury models that have retained regenerative capabilities that share similarities with lower vertebrates. In this chapter, we focus on the regenerating mouse digit tip as a model to probe how an amputation wound transforms into a blastema, how the digit blastema transforms by differentiating into the replacement digit tip, and how we can utilize an understanding of this process to enhance regeneration of normally non-regenerative injuries.

The regeneration of fingertips in children was the first indication that mammals possessed any regenerative capabilities (Illingworth 1974[*]). It is worth noting that this discovery was made when a mistake was made in a busy emergency room and a patient with a fingertip amputation did not

---

[*] Regeneration of fingertip in children is described in a large cohort of patients.

receive standard care (closing the amputation wound by suturing) but was sent home after simply cleaning the wound. When the patient returned, the wound had healed and eventually regenerated a functional fingertip. Based on these clinical observations, experimental studies conducted on the mouse digit tip showed a similar regenerative response (Borgens 1982). Subsequent studies on mice demonstrated that this regenerative response is mediated by the formation of a blastema of undifferentiated proliferative cells that undergo morphogenesis and redifferentiation to reform the missing digit structures (Fernando et al. 2011; Han et al. 2008). In mice, this response is restricted to the distal part of the digit tip, but it is a robust response that occurs at all developmental stages, including adults. Amputation at more proximal levels of the digit or at any level of the limb results in regenerative failure. The general strategy of using this regeneration model to identify key elements by experimentally inhibiting the response (either genetically or using specific antagonists), and demonstrating that regeneration can be stimulated at normally non-regenerative amputations has proven to be a powerful approach to uncover critical regenerative signaling pathways (Takeo et al. 2013[*]; Yu et al. 2010). Recent studies on the role of bone morphogenetic protein (BMP) signaling have demonstrated enhanced position-specific regenerative responses from amputations at different levels of the mouse digit and limb, thus providing a proof of concept that significant mammalian regeneration can be stimulated (Yu et al. 2010, 2012[†]). These studies involving spatiotemporal targeting of BMP2 treatment provide definitive evidence that mammalian limb tissues possess an unrealized regenerative potential that can be activated by specific modification of the wound-healing environment. Since a similar BMP2 treatment elicits different position-specific responses that correspond to the level of amputation, the results demonstrate that cells involved in amputation wound healing can activate patterning programs that are critical for an appropriate regenerative response. Despite a poor understanding of the positional information system, it is encouraging that the system remains functional and can be activated in mammalian regeneration. This approach of studying endogenous regeneration to identify key regulators and testing potential therapeutic strategies to enhance regeneration presents a clear experimental path to discovering a solution to mammalian regeneration. For this to happen, there is a need to expand our understanding of non-regenerative healing, and a recent study characterizing non-regenerative amputations supports the conclusion that the healing response is dynamic and that multiple tissues within the healing stump display characteristics consistent with an attempted regenerative response (Dawson et al. 2016). From an engineering prospective, uncovering the molecular underpinnings of regenerative potential associated with regenerative failure will prove to be as important as understanding the regeneration process itself, because it represents the interface that needs to be bridged to achieve a functional outcome.

_____

[*] Regeneration is induced by Wnt signaling.
[†] Regeneration is induced by BMP signaling.

6.2 Overview of mouse digit-tip regeneration

The mouse digit tip is one of a handful of mammalian injury models that undergo a regenerative response mediated by the formation of a blastema (Muneoka et al. 2008; Simkin et al. 2015a). The terminal phalanx of the mouse digit (P3) articulates with the second phalanx (P2), forming the P2/P3 joint. From a lateral perspective, the P3 element has a triangular profile that is wide at the P2/P3 joint and tapering to a distal tip; thus, the bulk of the bone volume is associated with the proximal region of the bone (Figure 6.1a–c). The bone marrow cavity is located proximally and is remarkable by comparison with the P2 marrow because of its very high

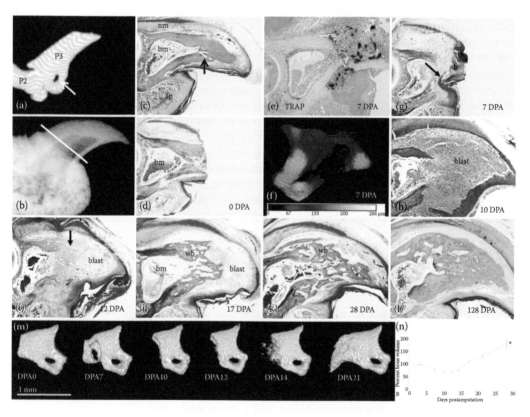

Figure 6.1 Stages of digit regeneration. (a) Three-dimensional microCT rendering of the bone structure of the mouse digit, showing the terminal phalanx (P3), second phalanx (P2), and the proximal–ventral os hole of P3 (arrow). (b) Three-dimensional microCT of the digit shown in (a) that was rendered in BoneJ, using a threshold that allows visualization of soft tissues. Skeletal elements are shown as negative space, and the amputation plane is indicated with the line. (c) Sagittal section through an unamputated digit tip, showing bone marrow cavity (bm) localized in the proximal region. The nail matrix (nm) is at the proximal base of the nail organ, and the fat pad (fp) is a prominent ventral structure. The P3 bone contains a foramen that connects the bone marrow region to the connective tissue (arrow). (d) Sagittal section of the digit tip immediately after amputation. Amputation is through the distal cortical bone and nail bed but leaves the bone marrow (bm) cavity intact. *(Continued)*

vascular content. Most prominent within the P3 marrow is a very large, thin-walled vein that fills much of the cavity space and contrasts two arteries lined with perivascular cells (Figure 6.1c and d). There are a number of foramina in the P3 element, including the os holes (prominent lateral openings that connect the marrow cavity to the proximal connective tissue), and vessel-containing canals that bridge the marrow cavity to the distal connective tissue (Figure 6.1a and c). The P3 element is encased within the nail organ on its dorsal and lateral surface, and a thin layer of connective tissue separates the bone from the nail epidermis. On the ventral surface, the epidermis is contiguous with the thickened epidermis of the fat pad (Figure 6.1c). These structural elements are supported by nerves and vasculature that can be found throughout the terminal phalanx.

Only the distal portion of the P3 element is normally regenerative; amputation in the proximal region of P3 or at any level of P2 fails to regenerate (Dawson et al. 2016; Neufeld and Zhao 1995). We have established a regeneration model that involves amputation of the digit tip without damaging the fat pad (Figure 6.1b and d). This simple amputation removes the distal portion of the P3 element, leaving both the fat pad and the P3 marrow region undamaged (Fernando et al. 2011; Han et al. 2008). In adult mice, amputation triggers a complex series of events in which the bone stump is degraded proximally to the level of the distal bone marrow, and this results in an injury-induced secondary amputation that truncates the digit tip at a more proximal level (Fernando et al. 2011). Bone degradation is mediated by a dramatic rise in osteoclasts that target the stump bone, completely eroding bone tissue surrounding the distal marrow region to effect a secondary amputation (Figure 6.1e and f). This bone-degradation phase takes approximately 9 days to complete,

Figure 6.1 (Continued) Stages of digit regeneration. (e) Tartrate-resistant acid phosphatase (TRAP) staining of regenerates 7 days postamputation (7 DPA) identifies numerous osteoclasts localized to the amputated stump. (f) MicroCT scan of 7 DPA regenerate pseudo-colored according to trabecular thickness to show bone degradation associated with the regenerative response. Color changes indicate bone thickness in micrometers. (g) Histological section of a 7 DPA regenerate showing epidermal migration (arrow) through a region of eroded stump bone. This mode of wound closure causes a re-amputation and sloughing of the distal bone (b). (h) At 10 DPA, a blastema (blast) of undifferentiated cells forms distal to the P3 stump. (i) At 12 DPA, ossification (arrow) is first observed at the interface between the P3 stump and the distal blastema (blast). (j) At 17 DPA, regenerated woven bone (wb) caps the distal region of the bone marrow cavity (bm) and is proximal to the digit blastema (blast). (k) At 28 DPA, the woven bone histology of the regenerated P3 element is regenerated, with an interlacing network of woven bone (wb) that is histologically distinct from the original cortical bone. (l) The 128 DPA–regenerated digit-tip bone has maintained its woven bone appearance. (m) 3D renderings of a regenerating digit (distal is to the left) showing the initial bone degradation response, followed by the ossification of the replacement structure. (n) Quantification of bone volume changes showing the degradation phase, followed by bone regrowth. There is a consistent overshoot in regenerated bone volume (*). (Modified from c–e, g, i–l: Fernando, W.A. et al., *Dev. Biol.*, 350, 301–310, 2011; f: Sammarco, M.C. et al., *PLoS One*, 10, e0140156, 2015; m, n: Sammarco, M.C. et al., *J. Bone Miner. Res.*, 30, 393, 2014.)

and during this phase, the epidermis forms an attachment to the periosteal surface of the bone stump and appears to be inhibited from closing by the presence of the stump bone. Once bone degradation is complete, the wound epidermis can be observed migrating through regions of degraded bone to close the wound (Figure 6.1g). Shortly after the completion of wound closure, a blastema of mesenchymal cells encased within the wound epidermis forms and the first transformative phase of regeneration is complete (Figure 6.1h).

The second transformation involves the differentiation of the replacement digit tip. This phase involves the redifferentiation of blastema cells to reform the distal structures lost by amputation and stump degradation, and there are differences that distinguish the regenerating digits from other appendage regeneration models. During limb development, appendicular skeletal structures differentiate by endochondral ossification, a process that involves forming a chondrogenic template of the skeletal structure that undergoes hypertrophy and is replaced by osteoblasts that differentiate into osteocytes of the forming bone. In salamander limb regeneration, this developmental program is re-utilized, and this portion of the regeneration response has been termed re-development (Bryant et al. 2002). Like other appendicular bones, the P3 element of the mouse forms during development by endochondral ossification, but during postnatal development, the elongation of the P3 element occurs by appositional ossification, a form of direct ossification that forms lamellar bone (Han et al. 2008; Muneoka et al. 2008). During P3 regeneration, newly formed bone does not undergo endochondral ossification but differentiates woven bone, which is another form of direct ossification that involves the coalescing of osteoblasts to form bone aggregates (Fernando et al. 2011) (Figure 6.1i–l). Similar to other regeneration models, the sequence of redifferentiation occurs from proximal to distal, with ossification initiating at the stump–blastema interface. In this way, the regenerated structure is contiguous with the stump. The amputated cortical bone is replaced by woven bone that is characterized by numerous trabecular spaces, thus the regenerated bone is not a perfect replica of the original bone (Figure 6.1l). We have used micro-computed tomography (microCT) as a non-invasive tool to characterize changes in the P3 element during the regenerative response, and both degradation and redifferentiation can be qualitatively and quantitatively profiled (Figure 6.1m and n). Bone volume measurements show that the amount of bone that regenerates is significantly greater than that of the original digit, thus identifying a second characteristic of the regenerative response that is imperfect. These imperfections of the regenerated bone likely reflect the fact that the regenerative process is, in this case, not an example of re-development but is a response that successfully replaces physiological function at the expense of anatomical perfection.

It is relevant to note that when digit regeneration is induced by treatment with BMP2, the induced response forms new bone by establishing a novel endochondral ossification center at the wound site, and this center organizes the formation of newly regenerated bone (Yu et al. 2010, 2012). Thus, the differentiation phase of endogenous regeneration does not involve a re-development response, despite the fact that stump cells retain the capacity to affect a re-development response. This observation suggests that this phase of the regenerative response evolved a completely novel mechanism for bone formation that may reflect inherent limitations imposed by the availability of progenitor cells in the stump wound. For example, a re-development response would require the availability of chondrogenic progenitor cells to form an endochondral ossification center, but only osteogenic progenitor cells appear to be available at the amputation wound. The BMP2-induced regenerative response involves the secondary activation of an SDF1/CXCR4-mediated migration response that functions in the recruitment of progenitor cells, and SDF1 is not expressed at the non-regenerative amputation wound (Lee et al. 2013). The SDF1/CXCR4 signaling is functional in the endogenous regenerative response, so it makes sense that only osteogenic progenitor cells are responsive to this recruitment signal.

Compared to lower-vertebrate regeneration, the overt histolytic degradation of bone tissue, the delay in wound closure of the pre-blastema phase, and the novel mechanism of ossification during the redifferentiation phase are unique features of the mammalian regenerative response. However, the blastema itself shares a number of characteristics that appear to be conserved among vertebrates. First, the blastema consists of a mass of mesenchymal cells covered by a wound epithelium, and there is evidence that the epidermis is required for regenerative outgrowth (Campbell et al. 2011; Poss et al. 2003; Takeo et al. 2013). The wound epidermis is lineage-restricted in that labeling studies fail to show epidermal cells participating in the mesenchymal portion of the blastema during normal regeneration (Rinkevich et al. 2011; Tu and Johnson 2011). During regeneration, the wound epidermis transitions to a signaling tissue, and nerve innervation is required for this transition (Satoh et al. 2012). The mesenchymal component of the blastema is known to be heterogeneous, with cells derived from a number of different tissues present in the stump (Kragl et al. 2009; Rinkevich et al. 2011). These tissues include skeletal, muscle, interstitial, nervous, and vascular. Interstitial contribution is generally thought to be fibroblastic; however, a lack of definitive differentiation markers has made it difficult to carry out detailed lineage analyses (Kragl et al. 2009; Wu et al. 2013). Although there is circumstantial evidence that fibroblasts represent a multipotent cell type in salamander limb regeneration (Dunis and Namenwirth 1977; Kragl et al. 2009; Namenwirth 1974), there is direct evidence that other cell types contribute to the regenerative response in a lineage-restricted manner (Kragl et al. 2009; Rinkevich et al. 2011). The observation that some cell types are lineage-restricted raises a key unanswered question about how cells within the blastema become spatially organized to reform the different tissue types of the regenerate. It seems likely that the

145

blastema is composed of both lineage-restricted and multipotent cells, with multipotent cells functioning to establish boundaries between tissue types and facilitate their integration. In this regard, it is perhaps not surprising that interstitial cells may be serving this function.

Second, enhanced cell proliferation is a characteristic of the blastema, and a number of common mitogenic factors have been identified. Studies on salamander limb mitogens associated with limb regeneration have largely focused on identifying the neurotrophic factor(s) associated with the inhibitory effect caused by denervation (Pirotte et al. 2016; Simoes et al. 2014). Of the numerous factors identified, there are a couple of signaling pathways that have been identified in multiple blastema models. In the mouse digit blastema, BMP signaling is required for both regeneration and direct stimulation of cell proliferation during induced regeneration (Yu et al. 2010, 2012). The Wnt signaling is also required for digit regeneration, and there is evidence that Wnt signaling regulates a nerve-related mitogenic activity linked to fibroblast growth factor (FGF) signaling (Takeo et al. 2013). In salamander limb regeneration, the combined stimulation by FGF and BMP can replace the neurotrophic requirement for accessory blastema induction (Makanae et al. 2014). In fish fin regeneration, FGF signaling promotes cell proliferation and is required for regeneration (Wills et al. 2008). Thus, regeneration is dependent on the stimulation of blastema cell proliferation, and some blastema cell mitogens are conserved among regenerating vertebrates.

Third, there is evidence that cell recruitment is critical for bringing cells to the wound site to form the blastema (Bouzaffour et al. 2009; Gardiner et al. 1986; Lee et al. 2013). The recruitment of blastema cells in fins and digit-tip regeneration utilizes the SDF1/CXCR4 signaling pathway, a chemotactic signaling pathway that is well characterized and has been linked to a number of different injury responses (Cantley 2005; Lee et al. 2013; Ting et al. 2008). In digit regeneration, CXCR4-expressing cells form the blastema and display a dose-dependent migration response to SDF1 *in vitro*, which is antagonized by the CXCR4-specific antagonist AMD3100 (Figure 6.2a–c). In the blastema, *Sdf1* is expressed by the wound epidermis in both fish and mice and also by endothelial cells present within the blastema mesenchyme in the mouse digit (Figure 6.2d). In BMP2-induced digit regeneration, *Sdf1* expression is induced in endothelial cells in the wound bed, and *Sdf1* expression by cultured human vascular endothelial cells is stimulated in a dose-dependent manner by BMP2. Antagonizing systemic activity of SDF1/CXCR4 signaling with AMD3100 causes a significant reduction in the regeneration response, and implanting *Sdf1*-expressing cells into the amputation wound of P2 amputation induces a partial regenerative response in the absence of BMP2 (Figure 6.2e and f). These studies provide both-loss-of function and gain-of-function evidence that SDF1/CXCR4 signaling plays a crucial role in organizing the regenerative response by actively recruiting the stem/progenitor cells that make up the blastema (Lee et al. 2013).

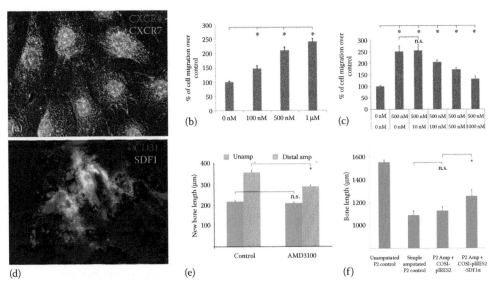

Figure 6.2 Blastema cell recruitment involves SDF1α/CXCR4 signaling. (a) Two known receptors for SDF1α are expressed by blastema cells: CXCR4 (red) and CXCR7 (green). Nuclei were stained with stained with 4',6-diamidino-2-phenylindole (DAPI, blue). (b) Blastema cells display a dose-dependent migration response to SDF1α in the Boyden chamber transwell assay. (c) Inhibition of SDF1α-induced blastema cell migration by AMD3100, a CXCR4-specific antagonist, is dose-dependent under conditions that maintain a constant level of SDF1α (500 nM). (d) In the blastema, double immunohistochemical staining with the endothelial marker, CD31, and SDF1α identify endothelial cells of the blastema as one source of SDF1α during digit-tip regeneration. (e) Daily treatment with the CXCR4-specific antagonist, AMD3100, did not affect neonatal bone growth (blue bars) but had an inhibitory effect on bone regeneration (orange bars). (f) Implantation of COS1 cells expressing SDF1α enhances regeneration after P2 amputation in neonates. P value was calculated by Student's t-test. *: $P < 0.05$; n.s.: no significant difference. Bars on the graphs indicate standard error. (Modified from Lee, J. et al., *Dev. Biol.*, 382, 98–109, 2013.)

To summarize, a comparative assessment of the blastema identifies phylogenetic similarities in cell contribution, how contributing cells are recruited to form the blastema, and the mitogenic requirement of the blastema cells.

This overview of mammalian regeneration highlights the conclusion that only the blastema phase of the vertebrate regenerative response is conserved over phylogenetically distinct groups and that both pre-blastema wound healing and post-blastema differentiation events appear to be dramatically different. As such, the evidence supports the conclusion that mammalian digit regeneration represents a conserved response that has undergone significant phase-specific modifications. An alternative conclusion is that digit-tip regeneration is a completely novel characteristic that has secondarily evolved from a non-regenerative pre-condition and is not a conserved response at all. This latter conclusion seems unlikely, given the complexity of the response; however, we note that there is a mammalian regeneration model that has evolved de novo—deer antler regeneration (Goss 1969). Why is this conclusion important? With respect to vertebrate appendage regeneration in general, the possibility that different models of appendage regeneration present examples of regenerative injury responses that have evolved independently and are potentially largely unrelated would suggest that similarities displayed by the blastema phase represent examples of a type of convergent evolution. Nevertheless, independent of whether mammalian regeneration represents a highly modified conserved response or a newly evolved response, it is clear that understanding how the blastema is built and the microenvironment created by the blastema provide the best avenue for uncovering critical processes, dictating whether or not a regenerative response is successful. In other words, all roads leading to regeneration must pass through the blastema, and because characteristics of blastema are shared among vertebrates, they are likely to be highly relevant for human regeneration.

6.3 Blastema: Dynamic revascularization

In characterizing the digit blastema, we first noted that the blastema microenvironment is largely devoid of vasculature; however, individual endothelial cells are observed scattered throughout the blastemal mesenchyme (Fernando et al. 2011; Mescher 1996). Histological characterization of the salamander limb blastema noted that it is similarly avascular (Mescher 1996). In mice, the endothelial cells of the blastema express the stem cell marker, SCA1, suggesting that they represent a population of endothelial progenitor cells, and cell lineage studies have determined that endothelial cells remain lineage-restricted during digit-tip regeneration (Fernando et al. 2011; Rinkevich et al. 2011). It is likely that these endothelial cells participate in revascularizing the regenerate; however, how this occurs has not yet been studied. The overall role that revascularization plays in digit-tip regeneration has been explored in some detail, with the surprising finding that precocious angiogenesis is inhibitory for the regeneration response (Yu et al. 2014).

The importance of vascular control in any type of injury response is intuitively clear; revascularization is essential for tissue survival, so it seems obvious that this response needs to be promoted during any successful repair response. Thus, the avascularity of the blastema is anomalous and worthy of investigation. Revascularization during the healing of full-thickness skin wounds in mice has been intensively studied. During wound healing, granulation tissue is a transient structure composed of inflammatory cells and fibroblasts and is proposed to be the blastema's equivalent in a non-regenerating injury response (Yu et al. 2014). Revascularization of granulation tissue is stimulated by vascular endothelial growth factor (VEGF), and *Vegfa* expression is regulated by the activity of hypoxia-inducing factor 1α (HIF1α), a transcriptional regulator that is responsive to oxygen availability (Chen and DiPietro 2014).

This transcriptional regulator is constitutively expressed, and when oxygen is available, it is modified by hydrolases, such as propyl hydrolase, that target HIF1α for E3 ubiquitin ligase-dependent degradation. However, under hypoxic conditions, HIF1α is not degraded and is translocated to the nucleus, where it dimerizes with HIF1β and regulates a battery of genes linked to oxygen use, which includes *Vegfa*. It is in the context of this conserved wound-healing response that the avascular character of the blastema is considered.

The first evidence that the blastema differed from a non-regenerative wound was the observation that *pigment epithelium-derived factor (pedf)*, a potent anti-angiogenic factor, is specifically expressed in the blastema during digit regeneration but is absent during non-regenerative amputations (Figure 6.3a) (Muneoka et al. 2008). In addition, in studies where digit regeneration is successfully stimulated by treatment with BMP2 or BMP7, *Pedf* expression is upregulated during the induction

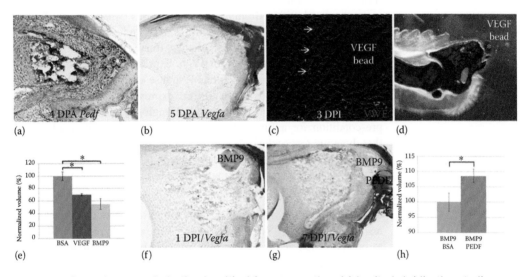

Figure 6.3 Dynamic revascularization is critical for regeneration. (a) In situ hybridization studies show that transcripts for the anti-angiogenic factor, *pedf* (arrows), are upregulated in the digit blastema at 4 days postamputation (4 DPA). (b) At a similar stage of regeneration, *Vegfa* transcripts are not detected in the early blastema or digit stump. (c) Immunohistochemical staining for von Willebrand Factor (VWF, red) 3 days postimplantation (3 DPI) of a VEGF bead shows an enhanced number of endothelial cells (arrows) associated with the VEGF bead in the blastema. (d) Alizarin red whole mount staining of 14 DPI digits are inhibited from regeneration when treated with a VEGF containing bead (*). (e) Graph showing normalized volume measurements from microCT analyses of control BSA-treated (blue), VEGF-treated (red) and BMP9-treated (green) digits at 14 DPI. (f) In situ hybridization studies show that *Vegfa* transcripts are upregulated within 1 day after BMP9 treatment. (g) In PEDF-rescued BMP9-inhibited digits, *Vegfa* expression is downregulated and is similar to untreated regenerates at a similar stage. (h) Bone volume analyzed by microCT imaging is significantly increased by PEDF treatment. Data are normalized to the BMP9-inhibited BSA control digits, *P* < 0.01 (*). (Modified from Yu, L. et al., *Regeneration*, 1, 33–46, 2014.)

of a transient blastema phase associated with induced regeneration (Yu et al. 2010, 2012). Expression studies of the *Vegf* genes (*Vegfa, b, c*) during digit-tip regeneration showed that *Vegfa* was the only *Vegf* family member that can be detected and that *Vegfa* expression was low in the blastema when *Pedf* expression was high (Figure 6.3b) but increased during ossification stages of regeneration (Yu et al. 2014). These observations suggested that the avascular character of the blastema is specifically regulated by the induction of PEDF and the inhibition of VEGFA. To test this hypothesis, digit-tip amputations were treated with purified VEGF, using a microcarrier bead as a vehicle. This single-targeted treatment induced an increase in endothelial cells associated with VEGF bead (Figure 6.3c) and completely inhibited the regenerative response (Figure 6.3d and e). It is of interest that a blastema-like structure formed, so it appears that the avascular microenvironment is not required for the accumulation of the blastemal mesenchymal cell population. To test the role of PEDF in this process, we took advantage of our observation that treatment with BMP9 inhibited regeneration (Figure 6.3e) and that this was associated with a dramatic and precocious enhancement of *Vegfa* expression by blastema cells (Figure 6.3f). Studies in which regenerates were treated with BMP9 and then treated with PEDF 24 hours later caused a downregulation of *Vegfa* expression in the blastema and completely rescued the regenerative response (Figure 6.3g and h). These studies provide the evidence that PEDF expression during blastema formation functions to specifically inhibit angiogenesis and that this is a pre-condition for a functional blastema. The important conclusion is that dynamic revascularization is essential for a successful regenerative response (Yu et al. 2014), and the data support the hypothesis that controlling revascularization distinguishes the regeneration blastema from wound granulation tissue.

Why is restricting revascularization important for the regenerative response? In adults, vasculature allows virtually all parts of the body to be connected and the molecules secreted into the vasculature allow cells from different parts of the body to be interactive. There are many well-studied examples of systemic regulatory signals that are critical for physiological function and homeostasis (e.g., insulin production by the pancreas and hormone production by the pituitary gland). On the other hand, the developing embryonic environment is initially avascular, and in this case, local signaling between cells plays a critical role in regulating how tissues become patterned and undergo morphogenesis. There is considerable evidence that regenerative mechanisms rely on the re-utilization of developmental mechanisms; thus, it is reasonable to conclude that successful regeneration within the context of the adult body is dependent on establishing an environment in which the cells of the blastema are isolated from the rest of the body so that local cellular interactions can prevail over systemic influences. Based on this reasoning, the restriction of revascularization allows the blastema to exist as a temporarily isolated developmental structure in an otherwise mature adult environment, and this condition, which is necessary for regeneration, is created by the localized production of an anti-angiogenetic factor (PEDF) that functions to counteract hypoxia pathways that stimulate angiogenesis.

6.4 Blastema: Hypoxic microenvironment

The requirement for an avascular blastema microenvironment suggests that the mesenchymal cells of the blastema would depend largely on diffusion for the acquisition of nutrients and the removal of waste products. At the same time, these cells are known to be actively involved in a wide range of cellular processes, including the inflammatory response, cell migration, and cell proliferation, and we can predict that this level of high metabolic demand coupled with limitations on the rate of nutrient exchange would limit the rate and possibly the extent of the regenerative response. To begin to explore this possibility, we have focused on oxygen availability within the blastema, primarily because it plays a central and dominant role in ATP production. Currently, there are technical limits in our ability to establish a spatiotemporal map of changing oxygen levels within the blastema and during its formation. To characterize oxygen levels *in vivo*, we used a commercially available reagent, pimonidazole (Hypoxyprobe™), which when introduced systemically forms adducts with thiol-containing proteins at oxygen levels below 1.3%. Using immunohistochemical staining, we were able to use this reagent to identify cells that are exposed to reduced levels of oxygen. We define hypoxyprobe-positive regions as hypoxic, even though we recognize that 2% oxygen is often used to establish hypoxic conditions in *in vitro* studies (Knowles 2015[*]). Thus, our *in vivo* estimates of hypoxia during regeneration are conservative. We have also used immunohistochemical methods to identify regions of higher oxygen levels by immunolocalization of a protein, FBXL5, which is constitutively expressed and degraded but stabilized at oxygen concentrations above 6% (Chollangi et al. 2012). We define FBXL5-immunostained regions as hyperoxic; however, we recognize that 6% oxygen is considered within the normoxic range in many tissues. Using these reagents, we have characterized temporal and spatial changes in oxygen availability during the regenerative response (Figure 6.4a–d) and made some important observations (Sammarco et al. 2014). First, oxygen levels during regeneration are dynamic and highly reproducible; second, hyperoxic regions are typically associated with cells associated with the vascular supply; third, the primary hypoxic region during regeneration is the avascular blastema; and fourth, the hypoxic state of the blastema is transient, and the return to normoxia correlates with revascularization and differentiation. These observations provide a link between the avascular character of the blastema and the hypoxic state of the blastema.

---

[*] An excellent review of hypoxia regulation of osteoclast activity.

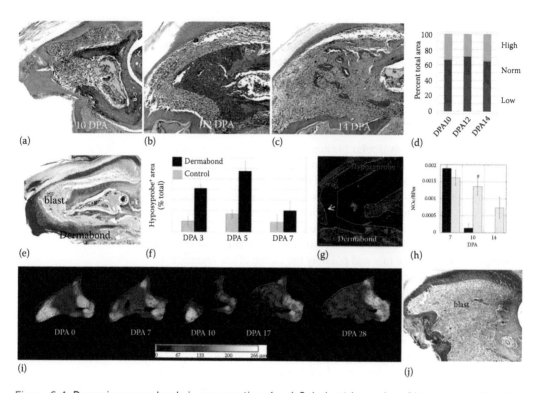

Figure 6.4 Dynamic oxygen levels in regeneration. (a–c) Colorimetric overlay of hypoxyprobe (pink) and FBXL5 (green) as compared with normoxic (purple) areas at DPA 10 (a), 12 (b), and 14 (c). (d) The anti-hypoxyprobe (pink) and anti-FBXL5 (green) stained cells were selected and plotted as cell counts versus DAPI staining to quantitate changing oxygen levels during blastema formation. (e) Dermabond-treated digits at 6 DPA show that wound closure is complete and a small blastema forms at the distal tip of the bone stump that has a reduced level of osteoclast resorption pits. (f) Oxygen profiling of the Dermabond-treated regenerates using Hypoxyprobe indicates a significant (#) increase in hypoxia that is specific to the epidermal microenvironments at 3 and 5 DPA, with a return to control levels after wound closure is complete (7 DPA). (g) Dermabond-treated regenerate at DPA 3 stained for Hypoxyprobe shows that the distal epidermis and wound epidermis (arrow) are hypoxic, whereas the underlying tissues are not hypoxic. The dotted outline delineates epidermis and the underlying bone or connective tissue. (h) Quantification of daily HBO treatment on osteoclast numbers shows that at 7 DPA, control (black bars) and HBO-treated (gray bars) regenerates are similar, but at 10 DPA, control levels decline whereas HBO levels remain high. Results are expressed as mean ± SEM; $^{\#}$ $P < 0.05$; NOc/BPm: number of osteoclasts/bone perimeter. (i) MicroCT scans of an HBO-treated regenerating digit pseudo-colored according to trabecular thickness to show enhanced bone degradation, followed by regeneration from a proximal level. Color changes indicate bone thickness in micrometers. (j) Hyperbaric oxygen–treated regenerate at 14 DPA showing a large proximal blastema (blast) that formed at the base of the digit. (Modified from a–d: Sammarco, M.C. et al., *J. Bone Miner. Res.*, 30, 393, 2014; e–g: Simkin et al. 2015b; h–j: Sammarco, M.C. et al., *PLoS One*, 10, e0140156, 2015.)

To begin to test whether the hypoxic state of the blastema is essential for blastema formation and the regenerative response, we used pulses of hyperbaric oxygen exposure to ameliorate the hypoxic microenvironment within the blastema. We used a protocol similar to the clinically accepted protocol of hyperbaric oxygen (HBO) treatment of wounds in humans that consisted of exposure to 100% oxygen at 2.4 atmospheres of pressure for a 90-minute period (Sammarco et al. 2014). The HBO treatment forces high levels of oxygen directly into the plasma, thereby bypassing limitations of oxygen delivery by red blood cells. By providing periodic HBO in 90-minute pulses during the regenerative response, cells are exposed to oscillating cycles of high oxygen, followed by a gradual return to normoxia. During digit regeneration, a single treatment of HBO inhibits blastema hypoxia based on hypoxyprobe immunostaining, and the hypoxic blastema environment does not completely return for a period of 24–48 hours. We have used HBO treatment in a temporally targeted way on blastema stages or in a continuous way during the entire regenerative process, and under both experimental regimes, the regeneration response is significantly modified but not inhibited (Sammarco et al. 2014, 2015*). Based on these studies, it appears that a continuous state of hypoxia is not a requirement for blastema formation or the regenerative response. This suggests either that vascular inhibition of regeneration is not related to the role that vasculature plays in regulating oxygen availability or that multiple factors that might include hypoxia combine to create a blastema microenvironment that is conducive for a successful regenerative response. We favor the latter of these two possibilities because of the HBO effects outlined as follows.

Although HBO treatment did not inhibit regeneration, there are a number of remarkable oxygen-related effects that modify the regenerative response. Continuous HBO treatment significantly delayed the wound healing response after amputation. The time to complete epidermal wound closure increased by approximately 50%, and this delayed the timing of blastema formation. This was an unexpected effect, because previous wound-healing studies showed that HBO treatment enhanced wound closure in chronic nonhealing diabetic skin wounds (Sander et al. 2009). One explanation for this difference focuses on the nail epidermis and the pause in epidermal closure that is associated with the degradation of the stump bone. Studies aimed at enhancing the timing of epidermal closure after digit amputation showed that the use of a cyanoacrylic wound dressing, Dermabond®, resulted in closure time that was half that of controls (Simkin et al. 2015b†). After application, Dermabond and other cyanoacrylic dressings polymerize across the wound surface, encasing the wound with a flexible skin adhesive (Singer et al. 2008). When used as a wound dressing for digit-tip amputations, Dermabond

---

* Enhanced bone degradation leads to the formation of enlarged blastemas and regeneration from non-regenerative digit levels.
† Rapid wound closure inhibits histolysis and results in smaller blastemas.

promotes epidermal migration across the amputated stump, reduces the bone degradation phase that typically precedes blastema formation, and results in the precocious formation of the blastema (Figure 6.4e). When probed with Hypoxyprobe, Dermabond specifically induces hypoxia in epidermal cells, without modifying oxygen levels in other stump tissues (Figure 6.4f and g). The conclusion that hypoxia enhances epidermal cell migration *in vivo* is in line with the *in vitro* results showing that epidermal cell migration increases when subjected to reduced oxygen levels (Zimmermann et al. 2014). Consistent with the conclusion that the epidermal closure response is directly linked to oxygen availability, we found that HBO treatment of Dermabond-treated amputations delayed the rate of epidermal closure. The evidence suggests that Dermabond acts as an occlusive wound dressing that restricts atmospheric oxygen and enhances epidermal migration and thus the rate of wound closure, but this can be reversed by increasing plasma oxygen availability. In this way, epidermal cells appear to be responsive to levels of oxygen availability, and because wound closure plays a key role in the regenerative response, this represents an example of how oxygen can influence regeneration in a tissue-specific manner.

6.5 Blastema: The role of osteoclasts

A second remarkable effect of HBO treatment on digit regeneration is an enhancement of bone degradation during the pre-blastema phase of regeneration (Sammarco et al. 2014, 2015). Bone degradation is an extreme example of tissue histolysis that remodels the amputation wound before blastema formation. It is reasonable to assume that all tissues involved in the regenerative response undergo some form of histolysis to establish a functional interface between the mature tissues of the stump and the newly developing tissues of the regenerate. Bone degradation involves the activity of highly specialized multinucleated cells called osteoclasts, which are derived from monocytes and attack mineralized bone. Osteoclasts attach to bone and create a focal resorptive compartment that establishes an acidic microenvironment that releases bone minerals and exposes the organic matrix of the bone for proteolytic digestion by cathepsin K, an acid protease secreted by osteoclasts. The activity of osteoclasts has been studied in the context of normal and abnormal bone turnovers, particularly in the context of bone diseases such as osteoporosis and osteopetrosis (Charles and Aliprantis 2014). During normal bone turnover, osteoclast degradation of bone is coupled with the formation of new bone tissue by osteoblasts; thus, these two cell types interact to coordinate the progressive resorption and laying down of new bone tissue in adults. Osteoclastogenesis is regulated in part by the activation of the RANK cell surface receptor (receptor activator of nuclear factor $\kappa\beta$) expressed by osteoclasts and its ligand, RANKL. RANK/RANKL signaling is negatively regulated by a secreted RANK decoy receptor called osteoprotegerin (OPG), which competitively binds RANKL (Honma et al. 2014).

One *in vivo* effect of continuous or targeted HBO treatment on blastema formation is to extend the period of bone degradation without changing the size of the osteoclast population (Figure 6.4h) (Sammarco et al. 2015). The increased period of bone degradation results in an increase in the amount of bone loss, quantified by microCT imaging, effectively driving the degradation-induced secondary re-amputation to more proximal levels (Figure 6.4i). Indeed, bone degradation extends into the P2/P3 joint region in some samples. Regardless of the proximal extent of this degradation response, once this phase terminates, a blastema forms and the digit-tip regenerates; however, in many cases, the proximal extent of degradation spanned into normally non-regenerative parts of the digit. The size of the blastema that forms in HBO-treated regenerates is proportional to the size of the digit stump and is visibly larger than the blastemas that form in control regenerates (Figure 6.4j). These studies suggest that HBO treatment does not enhance osteoclastogenesis but influences bone degradation by delaying the termination of osteoclast activity that precedes wound closure and blastema formation. Although the details of this effect of the regeneration response have not yet been elucidated, it is significant that a recent study on the oxygen-dependent modification of hypoxia-inducing factors (HIFs) by propyl hydroxylases (Phd 1,2,3) in osteoblasts provides a molecular mechanism whereby oxygen availability regulates OPG production via HIF2α (Wu et al. 2015[*]). In conjunction with our HBO studies, the evidence suggests that the termination of osteoclast activity, and thus the bone degradation phase of digit regeneration, is controlled by osteoblasts production of OPG in response to hypoxia.

The bone-degrading action of osteoclasts in the regenerative response is similar to their actions in bone turnover; however, the overall outcome is significantly different. In bone turnover, the general structure of the bone is maintained, while small portions of bone are degraded by osteoclasts and replaced by differentiating osteoblasts. This is a minimally destructive process that is spatiotemporally regulated and isolated by a bone remodeling compartment canopy that confines the remodeling surface (Jensen et al. 2015). In regeneration, osteoclasts are active on both endosteal and periosteal surfaces of the P3 element, and under normal conditions, they completely degrade the entire distal part of the bone (about 50% of the P3 bone volume). By the time osteogenesis begins in the proximal region of the blastema, osteoclasts are no longer present, so there is a complete spatiotemporal separation of osteoclast and osteoblast activity. One outcome of HBO-enhanced degradation is that enhanced osteoclast activity causes bone degradation into the proximal regions of the P3 element that are normally non-regenerative. Prior studies have shown that modification of

---

[*] This study provides a mechanism for how osteoclast activity is regulated by oxygen levels.

BMP or Wnt signaling can induce regeneration from proximal P3 amputations (Takeo et al. 2013; Yu et al. 2010, 2012), so it is clear that cells in this region of the digit have the potential to mount a regenerative response when induced. In the case of HBO treatment, excessive histolytic degradation of stump tissue resulted in blastema formation at non-regenerative levels and in the regeneration of newly patterned bone. Since Wnt signaling and BMP signaling are necessary for regeneration and can induce a regenerative response, it would appear that HBO-mediated bone degradation into non-regenerative regions must be associated with the maintenance of both signaling pathways. This is a novel way to indirectly enhance a regenerative response and highlights the important role of histolytic modification of the wound site to promote blastema formation and regeneration. The role that histolysis plays in regeneration remains largely unstudied and represents a new area of research. Proteolytic activity during salamander limb regeneration has been documented (Yang and Bryant 1994; Yang et al. 1999), and there is evidence that the regeneration process can be modified by inhibiting proteolytic activity (Vinarsky et al. 2005). In mammals, proteolytic digestion of mature tissues results in the release of peptides that possess mitogenic or chemotactic activity (Agrawal et al. 2010), and enhanced regeneration has been reported after repeated treatments with matrix metalloproteinases (MMPs) (Mu et al. 2013). In the case of bone tissue, BMP2 is known to be present and can be isolated from the bone matrix (Urist et al. 1984), so BMP2 release during the bone-degradative response could play a key role in organizing the regeneration response. Histolysis may also function to release stem/progenitor cells and/or establish a regenerative scaffold within the blastema.

6.6 Summary

As we learn more about the regenerative mechanisms in mammals, it is becoming increasingly clear how regenerative biology can impact regenerative engineering. Multiple examples of induced regenerative responses that are appropriately patterned are now available, and this supports the general conclusion that regenerative failure is not the absence of a regenerative process but the failure to complete the process. Conceptually, it is important to recognize that all injury responses begin with initiating a regenerative process, and how this proceeds is highly variable, not only in a phylogenetic sense but also between distinct parts of the body (e.g., the fingertip vs. the limb). In regenerative engineering, implementation of this concept means that to accomplish a regenerative outcome, the field must go beyond solving the problem of engineering a structural replacement and must also solve the problem of establishing a functional interface with the mature host tissue after engraftment. To do the latter, it is critical to consider the injury response of the implantation site and how this process can be managed by providing extrinsic influences such as growth factors treatment, controlling revascularization, and manipulating oxygen availability. These factors influence endogenous and induced regenerative responses and represent obvious targets that must be integrated into the regenerative therapy.

Another important concept that can be derived from regeneration studies is recognizing the importance of specific cellular strategies that have evolved and are necessary to effect a successful regenerative response. These include the following: (1) Identifying the cell types important for regeneration and establishing which cell types are lineage-restricted versus those that are more plastic. In this regard, it is important to note that our current understanding of this problem has been studied only *in vivo* during endogenous regeneration and not in experimental situations, where the availability of other cell types are restricted, or *in vitro*, where cells are tested in isolation and other cell types are absent. It is also critical to remember that although induced pluripotency has been demonstrated (Takahashi et al. 2007), the pluripotent state is probably not well suited for a regenerative response. Instead, a more effective approach will be to identify and expand a cell type or cell types that are better suited for a specific regenerative response. (2) Understanding the importance of histolysis in establishing a functional interface between the mature stump tissues and the regenerated structure and also potentially aiding in the release and/or activation of stem/progenitor cells from mature tissue, so that they can participate in the response. (3) Identifying the chemotactic signaling pathways necessary to recruit stem/progenitor cells to the wound site, where they can participate in the regeneration response. (4) Identifying the mitogenic signals necessary for the expansion of stem/progenitor cell types and understanding that mechanisms for transiently inhibiting differentiation are necessary for the cell expansion process. In this regard, there is considerable evidence that oxygen availability has been linked to the maintenance of stemness in *in vitro* studies (Mohyeldin et al. 2010), so manipulating oxygen availability extrinsically maybe a useful strategy, in combination with approaches that involve the engraftment of stem/progenitor cells in regenerative engineering. (5) Understanding that the revascularization process in regeneration is not static but dynamic and that although revascularization in general is essential for regeneration, controlling the timing of revascularization is equally important for a successful regeneration event.

References

Agrawal, V., Johnson, S.A., Reing, J., Zhang, L., Tottey, S., Wang, G., Hirschi, K.K., Braunhut, S., Gudas, L. J., Badylak, S.F., 2010. Epimorphic regeneration approach to tissue replacement in adult mammals. *Proc Natl Acad Sci U S A* 107, 3351–3355.

Borgens, R.B., 1982. Mice regrow the tips of their foretoes. *Science* 217, 747–750.

Bouzaffour, M., Dufourcq, P., Lecaudey, V., Haas, P., Vriz, S., 2009. Fgf and Sdf-1 pathways interact during zebrafish fin regeneration. *PLoS One* 4, e5824.

Bryant, S.V., Endo, T., Gardiner, D.M., 2002. Vertebrate limb regeneration and the origin of limb stem cells. *Int J Dev Biol* 46, 887–896.

Campbell, L.J., Suarez-Castillo, E.C., Ortiz-Zuazaga, H., Knapp, D., Tanaka, E.M., Crews, C.M., 2011. Gene expression profile of the regeneration epithelium during axolotl limb regeneration. *Dev Dyn* 240, 1826–1840.

Cantley, L.G., 2005. Adult stem cells in the repair of the injured renal tubule. *Nat Clin Pract Nephrol* 1, 22–32.

Charles, J.F., Aliprantis, A.O., 2014. Osteoclasts: More than 'bone eaters.' *Trends Mol Med* 20, 449–459.

Chen, L., DiPietro, L.A., 2014. Production and function of pigment epithelium-derived factor in isolated skin keratinocytes. *Exp Dermatol* 23, 436–438.

Chollangi, S., Thompson, J.W., Ruiz, J.C., Gardner, K.H., Bruick, R.K., 2012. Hemerythrin-like domain within F-box and leucine-rich repeat protein 5 (FBXL5) communicates cellular iron and oxygen availability by distinct mechanisms. *J Biol Chem* 287, 23710–23717.

Dawson, L.A., Simkin, J., Sauque, M., Pela, M., Palkowski, T., Muneoka, K., 2016. Analogous cellular contribution and healing mechanisms following digit amputaton and phalangeal fracture in mice. *Regeneration* 3, 13.

Dunis, D.A., Namenwirth, M., 1977. The role of grafted skin in the regeneration of x-irradiated axolotl limbs. *Dev Biol* 56, 97–109.

Fernando, W.A., Leininger, E., Simkin, J., Li, N., Malcom, C.A., Sathyamoorthi, S., Han, M., Muneoka, K., 2011. Wound healing and blastema formation in regenerating digit tips of adult mice. *Dev Biol* 350, 301–310.

Gardiner, D.M., Muneoka, K., Bryant, S.V., 1986. The migration of dermal cells during blastema formation in axolotls. *Dev Biol* 118, 488–493.

Goss, R.J., 1969. *Principles of Regeneration*. Academic Press, New York.

Han, M., Yang, X., Lee, J., Allan, C.H., Muneoka, K., 2008. Development and regeneration of the neonatal digit tip in mice. *Dev Biol* 315, 125–135.

Honma, M., Ikebuchi, Y., Kariya, Y., Suzuki, H., 2014. Regulatory mechanisms of RANKL presentation to osteoclast precursors. *Curr Osteoporos Rep* 12, 115–120.

Illingworth, C.M., 1974. Trapped fingers and amputated finger tips in children. *J Pediatr Surg* 9, 853–858.

Jensen, P.R., Andersen, T.L., Hauge, E.M., Bollerslev, J., Delaisse, J.M., 2015. A joined role of canopy and reversal cells in bone remodeling—Lessons from glucocorticoid-induced osteoporosis. *Bone* 73, 16–23.

Knowles, H.J., 2015. Hypoxic regulation of osteoclast differentiation and bone resorption activity. *Hypoxia* 3, 10.

Kragl, M., Knapp, D., Nacu, E., Khattak, S., Maden, M., Epperlein, H.H., Tanaka, E.M., 2009. Cells keep a memory of their tissue origin during axolotl limb regeneration. *Nature* 460, 60–65.

Lee, J., Marrero, L., Yu, L., Dawson, L.A., Muneoka, K., Han, M., 2013. SDF-1alpha/CXCR4 signaling mediates digit tip regeneration promoted by BMP-2. *Dev Biol* 382, 98–109.

Makanae, A., Mitogawa, K., Satoh, A., 2014. Co-operative Bmp- and Fgf-signaling inputs convert skin wound healing to limb formation in urodele amphibians. *Dev Biol* 396, 57–66.

Mescher, A.L., 1996. The cellular basis of limb regeneration in urodeles. *Int J Dev Biol* 40, 785–795.

Mohyeldin, A., Garzon-Muvdi, T., Quinones-Hinojosa, A., 2010. Oxygen in stem cell biology: A critical component of the stem cell niche. *Cell Stem Cell* 7, 150–161.

Mu, X., Bellayr, I., Pan, H., Choi, Y., Li, Y., 2013. Regeneration of soft tissues is promoted by MMP1 treatment after digit amputation in mice. *PLoS One* 8, e59105.

Muneoka, K., Allan, C.H., Yang, X., Lee, J., Han, M., 2008. Mammalian regeneration and regenerative medicine. *Birth Defects Res C Embryo Today* 84, 265–280.

Namenwirth, M., 1974. The inheritance of cell differentiation during limb regeneration in the axolotl. *Dev Biol* 41, 42–56.

Neufeld, D.A., Zhao, W., 1995. Bone regrowth after digit tip amputation in mice is equivalent in adults and neonates. *Wound Repair Regen* 3, 461–466.

Pirotte, N., Leynen, N., Artois, T., Smeets, K., 2016. Do you have the nerves to regenerate? The importance of neural signalling in the regeneration process. *Dev Biol* 409, 4–15.

Poss, K.D., Keating, M.T., Nechiporuk, A., 2003. Tales of regeneration in zebrafish. *Dev Dyn* 226, 202–210.

Rinkevich, Y., Lindau, P., Ueno, H., Longaker, M.T., Weissman, I.L., 2011. Germ-layer and lineage-restricted stem/progenitors regenerate the mouse digit tip. *Nature* 476, 409–413.

Sammarco, M.C., Simkin, J., Cammack, A.J., Fassler, D., Gossmann, A., Marrero, L., Lacey, M., Van Meter, K., Muneoka, K., 2015. Hyperbaric oxygen promotes proximal bone regeneration and organized collagen composition during digit regeneration. *PLoS One* 10, e0140156.

Sammarco, M.C., Simkin, J., Fassler, D., Cammack, A.J., Wilson, A., Van Meter, K., Muneoka, K., 2014. Endogenous bone regeneration is dependent upon a dynamic oxygen event. *J Bone Miner Res* 11, 2336–2345.

Sander, A.L., Henrich, D., Muth, C.M., Marzi, I., Barker, J.H., Frank, J.M., 2009. In vivo effect of hyperbaric oxygen on wound angiogenesis and epithelialization. *Wound Repair Regen* 17, 179–184.

Satoh, A., Bryant, S.V., Gardiner, D.M., 2012. Nerve signaling regulates basal keratinocyte proliferation in the blastema apical epithelial cap in the axolotl (Ambystoma mexicanum). *Dev Biol* 366, 374–381.

Simkin, J., Sammarco, M.C., Dawson, L.A., Schanes, P.P., Yu, L., Muneoka, K., 2015a. The mammalian blastema: regeneration at our fingertips. *Regeneration* 2, 93–105.

Simkin, J., Sammarco, M.C., Dawson, L.A., Tucker, C., Taylor, L.J., Van Meter, K., Muneoka, K., 2015b. Epidermal closure regulates histolysis during mammalian (Mus) digit regeneration. *Regeneration* 2, 106–119.

Simoes, M.G., Bensimon-Brito, A., Fonseca, M., Farinho, A., Valerio, F., Sousa, S., Afonso, N., Kumar, A., Jacinto, A., 2014. Denervation impairs regeneration of amputated zebrafish fins. *BMC Dev Biol* 14, 49.

Singer, A.J., Quinn, J.V., Hollander, J.E., 2008. The cyanoacrylate topical skin adhesives. *Am J Emerg Med* 26, 490–496.

Takahashi, K., Tanabe, K., Ohnuki, M., Narita, M., Ichisaka, T., Tomoda, K., Yamanaka, S., 2007. Induction of pluripotent stem cells from adult human fibroblasts by defined factors. *Cell* 131, 861–872.

Takeo, M., Chou, W.C., Sun, Q., Lee, W., Rabbani, P., Loomis, C., Taketo, M.M., Ito, M., 2013. Wnt activation in nail epithelium couples nail growth to digit regeneration. *Nature* 499, 228–232.

Ting, A.E., Mays, R.W., Frey, M.R., Hof, W.V., Medicetty, S., Deans, R., 2008. Therapeutic pathways of adult stem cell repair. *Crit Rev Oncol Hematol* 65, 81–93.

Tu, S., Johnson, S.L., 2011. Fate restriction in the growing and regenerating zebrafish fin. *Dev Cell* 20, 725–732.

Urist, M.R., Huo, Y.K., Brownell, A.G., Hohl, W.M., Buyske, J., Lietze, A., Tempst, P., Hunkapiller, M., DeLange, R.J., 1984. Purification of bovine bone morphogenetic protein by hydroxyapatite chromatography. *Proc Natl Acad Sci U S A* 81, 371–375.

Vinarsky, V., Atkinson, D.L., Stevenson, T.J., Keating, M.T., Odelberg, S.J., 2005. Normal newt limb regeneration requires matrix metalloproteinase function. *Dev Biol* 279, 86–98.

Wills, A.A., Kidd, A.R., 3rd, Lepilina, A., Poss, K.D., 2008. Fgfs control homeostatic regeneration in adult zebrafish fins. *Development* 135, 3063–3070.

Wu, C., Rankin, E.B., Castellini, L., Alcudia, J.F., LaGory, E.L., Andersen, R., Rhodes, S.D. et al., 2015. Oxygen-sensing PHDs regulate bone homeostasis through the modulation of osteoprotegerin. *Genes Dev* 29, 817–831.

Wu, Y., Wang, K., Karapetyan, A., Fernando, W.A., Simkin, J., Han, M., Rugg, E.L., Muneoka, K., 2013. Connective tissue fibroblast properties are position-dependent during mouse digit tip regeneration. *PLoS One* 8, e54764.

Yang, E.V., Bryant, S.V., 1994. Developmental regulation of a matrix metalloproteinase during regeneration of axolotl appendages. *Dev Biol* 166, 696–703.

Yang, E.V., Gardiner, D.M., Carlson, M.R., Nugas, C.A., Bryant, S.V., 1999. Expression of Mmp-9 and related matrix metalloproteinase genes during axolotl limb regeneration. *Dev Dyn* 216, 2–9.

Yu, L., Han, M., Yan, M., Lee, E.C., Lee, J., Muneoka, K., 2010. BMP signaling induces digit regeneration in neonatal mice. *Development* 137, 551–559.

Yu, L., Han, M., Yan, M., Lee, J., Muneoka, K., 2012. BMP2 induces segment-specific skeletal regeneration from digit and limb amputations by establishing a new endochondral ossification center. *Dev Biol* 372, 263–273.

Yu, L., Yan, M., Simkin, J., Ketcham, P.D., Leininger, E., Han, M., Muneoka, K., 2014. Angiogenesis is inhibitory for mammalian digit regeneration. *Regeneration* 1, 33–46.

Zimmermann, A.S., Morrison, S.D., Hu, M.S., Li, S., Nauta, A., Sorkin, M., Meyer, N.P. et al., 2014. Epidermal or dermal specific knockout of PHD-2 enhances wound healing and minimizes ischemic injury. *PLoS One* 9, e93373.

Section II

How cells communicate to remake the pattern and restore function

The early events in regeneration are about recruiting regeneration-competent cells from the tissues that remain after injury in order to form a blastema. However, it takes more than just blastema cells to regenerate the new structure. These cells have to communicate and coordinate their behaviors to end up in the right place and make the right structures. This is accomplished by cells communicating with each other and the surrounding environment to rebuild the complex pattern of tissues of the lost structure.

As is the case during both embryonic and regenerative development, the functional unit of regenerative engineering is the cell. There are many volumes published on the biology of cells, and thus we have not addressed those topics, but rather refer readers who are interested in those references. The focus here is on selected aspects of cellular properties and behaviors that are relevant to regeneration. Specifically, the topics address the issues of how cellular behaviors are orchestrated, so as to achieve a specific and desired regenerative outcome. These behaviors include expansion of the populations of regeneration-competent cells through proliferation, the migration of cells to the right location, and the changes in cell shape associated with morphogenesis and differentiation. Although the behavioral repertoire of cells is limited (proliferate, migrate, and differentiate/dedifferentiate), the orchestration of these behaviors is complex, resulting in the formation of all of our body parts. Ultimately, building body parts in the embryo and regenerating them in the adult are controlled by the processes of pattern formation, resulting in the different tissues becoming assembled in association with each other to yield organ function.

Chapter 7 Retinoic acid and the genetics of positional information

Malcolm Maden,
David Chambers,
and James Monaghan

Contents

Key concepts

    a. Knowledge of a cell's position is encoded on the surface of
       regenerating cells in the blastema.
    b. Connective tissue–derived cells in the blastema are respon-
       sible for assessing positional information.
    c. Positional information on blastemal cells can be changed by
       precise concentrations of retinoic acid.
    d. Retinoic acid acts on the nucleus of blastemal cells to upreg-
       ulate or downregulate many genes during this positional
       information change.

> e. One of those genes is called *Prod1*, and the protein that it encodes is located on the cell surface.
>
> f. Changing PROD1 changes the positional information of blastemal cells.

7.1 Introduction

7.1.1 Positional information in development—gradients

The concept of positional information, introduced by Wolpert (1969), was designed to explain how genetic information gives rise to specific spatial patterns of cellular differentiation, both in the embryo and in regenerating systems. How in the developing neural plate, for example, do neural cells at the rostral end know that they should make a forebrain and those at the caudal end know that they should make a spinal cord? How in the developing limb bud do cells at the anterior side know that they should make digits 1 and 2 and cells at the posterior side know that they should make digits 4 and 5? The regenerating limb of the axolotl or newt provides a superb example of the evocation of positional information—amputation of the limb through the upper arm results in the regeneration of exactly those elements that were lost: the distal humerus, radius, and ulna and the hand, whereas amputation through the lower arm results in the regeneration of just the distal radius and ulna and the hand (Figure 7.1a). Clearly, cells at the amputation plane must know where they are in the proximodistal axis of the limb, in order to know what is missing and thus what to replace.

How is this almost mystical knowledge of position acted upon by cells, where in the cell is it assessed, how is information transferred from cell to cell? The simplest mechanism for specifying positional identity involves an extracellular gradient of a substance (a morphogen), which decreases from one end of the field of cells to the other, and the concentration of that substance specifies the position and thus the positional identity of that cell, the mechanism that was originally proposed by Wolpert (1969). Although details of the release, diffusional spread, mechanics of entry into cells, nuclear interpretation of different concentrations, and so on have been the topics of much debate, there is no doubt that there are many well-established molecules in developing embryos, which ultimately act in such a graded fashion, such as sonic hedgehog (SHH), bone morphogenetic proteins, transforming growth factor family members, fibroblast growth factors (FGFs), and WNTs.

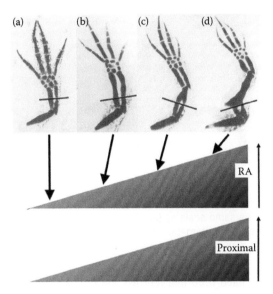

Figure 7.1 Upper panel, Victoria blue–stained axolotl limbs showing the regenerated cartilages after varying treatments with RA. Red lines mark the amputation plane. (a) Control limb amputated through the mid-radius and ulna, which regenerated precisely those elements that were removed. (b) After a low dose of RA, represented by the arrow pointing to the RA concentration graph, the radius and ulna, which have regenerated, are far longer than expected. (c) After a medium dose of RA, the regenerated elements now include the distal part of the humerus, the elbow, and the complete lower arm. (d) After a high dose of RA, the regenerate consists of a complete limb from the shoulder level. Lower graphs represent an increasing concentration of RA (upper graph), paralleling the increasing degree of proximalization of the regenerating limbs (lower graph).

7.1.2 Positional information in regeneration—on the cell surface

In regenerating systems such as the amputated salamander limb, the molecular progress that has been made in understanding the mechanics of the graded nature of positional information has been strikingly absent. Although *Shh* (Imokawa and Yoshizato 1997; Torok et al. 1999; Nacu et al. 2016), *Wnts* (Ghosh et al. 2008), *Bmp*-2 and *Tgfβ*-1 (Levesque et al. 2007; Guimond et al. 2010), and FGFs (Boilly et al. 1991; Mullen et al. 1996; Christensen et al. 2001; Han et al. 2001; Nacu et al. 2016) have been shown to be expressed during limb regeneration, there has been little progress in elaborating whether gradients of these proteins exist in the regeneration blastema, the conical-shaped outgrowth

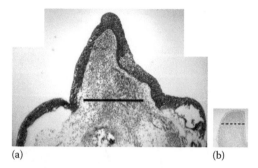

(a)   (b)

Figure 7.2 (a) Masson's trichrome–stained section through a limb regeneration blastema on a mid-sized axolotl larva. Black scale bar = 1 mm. (b) At the same scale as (a), there is a section through a chick limb bud showing the dramatic difference in size. Above the hatched line is the progress zone, where positional information is being generated.

from the amputated limb that regrows the missing limb (Figure 7.2a). The most likely explanation for this apparent dearth of knowledge is that positional information is *not* elaborated by gradients of extracellular molecules but is encoded by a completely different mechanism, perhaps being stably expressed on the surface of blastemal cells. There is considerable experimental evidence for this surface recognition of positional differences. For example, when the blastema is cut off from the limb, rotated 180°, and stuck back on to the same stump frequently, the blastema will de-rotate and end up in exactly the same position that it started (Maden 1978). When a wrist blastema is cut off from the limb and moved proximally to the upper-arm level, and then, the limb is amputated through the upper-arm level, the grafted blastema will move distally as the limb regenerates and will then cease moving and integrate at the precise level from which it originated (Crawford and Stocum 1988). A blastema from a proximal amputation level will engulf a blastema from a distal amputation level when the two are placed next to each other in culture (Nardi and Stocum 1983), a phenomenon attributed to the presence of a gradient of adhesivity along the limb. Furthermore, there is a matter of scale to consider, since developing systems are far smaller than regenerating systems. Extracellular gradients are thought to operate over a field size of fewer than 100 cells and a maximum distance of 1 mm for a low-molecular-weight morphogen (Crick 1970). Figure 7.2b shows a developing chick limb bud at the same scale as the regenerating blastema in Figure 7.2a (the scale bar in Figure 2a = 1 mm). The area at the tip of the limb bud above the hatched line in Figure 7.2b is known as the progress zone, where positional interactions leading to patterning take place, and has an approximate volume of 7×10^6 μm^3. In contrast, the regeneration blastema shown in Figure 7.2a, which is still relatively small and on adult axolotl limbs can be four times this size, has an approximate volume of 440×10^6 μm^3, far larger and containing far more cells than the limb bud. For these and other reasons, different conceptual models of regenerating systems have been described, such as the Polar Coordinate Model (French et al. 1976) and the Boundary Model (Meinhardt 1983), both of which depend on the recognition of positional differences between cells rather than reading extracellular concentrations and both of which have been developed from studies of regenerating systems, not developing ones.

7.2 The effect of retinoic acid on positional information

Despite the powerfully predictive nature of the Polar Coordinate Model and experimental observations such as those described earlier, directing us to investigate the cell surface, it is still the case that we know almost nothing about the nature of positional information in the regenerating limb. Therefore, it was of major importance when it was discovered that positional information could be specifically and predictably altered by the application of precise concentrations of the developmental signaling molecule, retinoic acid (RA) (Maden 1982, 1983). When the limb is amputated through the middle of the forearm and a particular concentration of RA is applied in the water, either by intraperitoneal (IP) injection or by application directly to the blastema, then instead of regenerating the lower arm as controls do (Figure 7.1a), a complete limb is regenerated, starting from the proximal end of the humerus (Figure 7.1d). If a lower concentration of RA is used, then the reduplication is not so extreme and the reduplicated level is only from the distal end of the humerus rather than from the proximal end (Figure 7.1c), and if the concentration is even lower, then the regenerated radius and ulna are longer than normal. The most extreme effect is seen when the limb is amputated through the hand, a relatively high concentration of RA is added, and a complete limb is regenerated from the level of the carpals (Figure 7.3b). Strikingly, the difference in concentration between a minor effect and the maximal effect is only about 2.5 fold and the effect is graded—higher levels of RA give higher degrees of proximalization (lower panels of Figure 7.1).

In addition to the effects on the axolotl (*Ambystoma mexicanum*) limbs shown here (Figure 7.1), all other species of urodeles that are routinely used for limb regeneration studies (Thoms and Stocum 1984; Niazi et al. 1985; Lheureux et al. 1986; Ju and Kim 1994) and anurans (Niazi and Saxena 1978; Scadding and Maden 1986) show this proximalization of the limb by RA. Synthetic agonists (see later) are invariably more potent than RA (Kim and Stocum 1986a; Keeble and Maden 1989; Maden et al. 1991). The dramatic effects of RA are seen not only when limbs are amputated but also when tadpole tails are amputated. Normally, the frog tadpole regenerates its tail rapidly and perfectly from any level of amputation (Figure 7.3c). When RA is applied, up to nine hind limbs, including the pelvic girdle, can be regenerated from the cut ends of the tail (Figure 7.3d) (Mohanty-Hejmadi 1992; Maden 1993). Presumably,

167

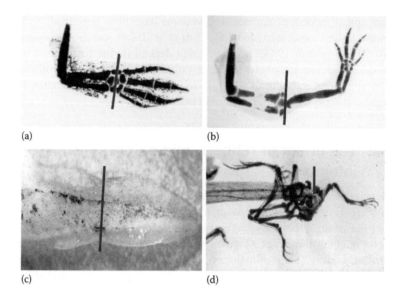

(a)　　　　　　　　　　　　　(b)

(c)　　　　　　　　　　　　　(d)

Figure 7.3 Victoria blue–stained axolotl limbs and tails; red line marks the amputation plane. (a) Control limb amputated through the hand showing perfect regeneration of the elements removed. (b) Limb amputated through the hand and treated with a high concentration of RA. A complete limb starting at the shoulder has regenerated from the hand level. (c) A control *Rana temporaria* tadpole tail, which was amputated at the level of the red line and regenerated exactly what was removed. (d) *Rana* tadpole tail amputated and treated with RA. Instead of tail tissue regenerating, two pairs of hind limbs plus two pelvic girdles have regenerated.

the positional information of tail cells is shifted proximally to the level of the pelvic girdle. These effects are not just restricted to amphibians, because when mouse embryos are treated amazingly early in development, at the egg cylinder stage (4.5–5.5 dpc), which is pre-gastrulation (before the mesoderm is formed), they can develop ectopic pelvic girdles and limbs (Rutledge et al. 1994; Neiderreither et al. 1996).

7.3 Endogenous retinoic acid in the regenerating limb and its mechanism of action

This effect could have been some strange phenomenon induced by a particularly potent chemical taken from the laboratory shelves and not necessarily a reflection of an underlying process. However, RA is indeed an endogenous molecule used both in development and regeneration, and in the maintenance of many of our body systems (Maden 2007), because it is the biologically active molecule through which vitamin A acts. Vitamin A is absorbed from the diet, stored in the liver, and transported around the body in the bloodstream as retinol bound to a plasma-binding protein. Retinol enters cells via the cell surface receptor STRA6 and is converted into RA in the cytoplasm by two specific enzymes: retinol dehydrogenase and retinaldehyde dehydrogenase (Figure 7.4). If RA is acting in an autocrine fashion (on the cell in which it is synthesized), then it enters the nucleus bound to another cytoplasmic-binding protein, cellular RA-binding protein (CRABP), and there, it activates the nuclear RA transcription factors, the RA receptors (RARs). If RA is acting in a paracrine fashion as a signaling molecule, then it will be released from the cytoplasm of the cell in order to act on an adjacent cell (upper arrow in Figure 7.4) and enter the nucleus of that cell. The mechanisms of RA release and transport between cells are completely unknown.

The RARs are present in the nucleus of cells as heterodimers, with the retinoid X receptors (RXRs) bound to a DNA sequence known as an RA response element (RARE) (Figure 7.4). There are three RARs known as RARα, RARβ, and RARγ (and multiple isoforms of each subtype) and three RXRs (RXRα, RXRβ, and RXRγ and their multiple isoforms), and so, there are multiple heterodimeric combinations of a RAR and a RXR

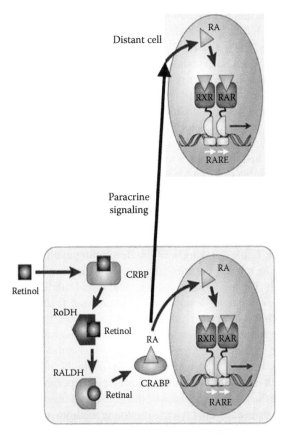

Figure 7.4 Diagram showing the production of RA and its signaling components within a cell (lower section) or how it acts as a signaling molecule on adjacent cells (upper section).

169

that can exist in the nucleus. In the newt, there are at least six isoforms of the RARs: α_1, α_2, δ_{1a}, δ_{1b}, δ_2 (the newt δ is equivalent to the mammalian γ), and β_2. When RA binds the heterodimer composed of one RAR and one RXR, it is transcriptionally activated. Only the RAR/RXR heterodimer is shown in Figure 7.4 for simplicity, but there is a multi-protein corepressor complex that disassociates on RA binding and a coactivator complex that is recruited in order to acetylate and methylate histones to decompact the chromatin (Carrier and Rochette-Egly 2015; Wei 2015).

All of the components of this molecular machinery are present in the regenerating limb. Retinoic acid is readily detectable by high-performance liquid chromatography (HPLC) (Scadding and Maden 1994), and the use of an RA-responsive reporter suggests that endogenous levels of RA are 3.5-fold higher at proximal levels than at distal levels (Brockes 1992); however, the HPLC studies obtained the opposite result—more RA at distal levels than at proximal levels. The enzymes that synthesize RA from retinol, including Rdh10 (Monaghan et al. 2012), Raldh1 (Knapp et al. 2013), Raldh3 (Monaghan et al. 2012), and the CRABP are present in the cytoplasm (Keeble and Maden 1986; McCormick et al. 1988). In the nucleus, RARα (Ragsdale et al. 1989), RARβ (Giguère et al. 1989; Carter et al. 2011), and RARγ (Ragsdale et al. 1989; Hill et al. 1993) are present. When RA synthesis is inhibited by pharmacological inhibitors of the enzymes, limb regeneration is inhibited, confirming that it is required for regeneration (Maden 1998; Scadding 2000). In order to more precisely detect the sites of action of RA in the limb, we generated an RA reporter axolotl by using the enhanced green fluorescent protein (*Egfp*) gene fused to a multimerized RARE sequence from the RARβ gene and generated a transgenic axolotl (Monaghan and Maden 2012). Not only could we identify the site(s) of RA action in the regenerating limb, but we could also examine the site(s) of action *of the same limb* during its development to ask whether the same cells were responding. In the developing axolotl forelimb bud, the mesenchymal cells were uniformly expressing EGFP, whereas the epidermis was negative (Figure 7.5a). Amazingly, in the developing axolotl hind limb bud, which develops much later than the forelimb, there was no expression of EGFP at all (Figure 7.5b), suggesting that hind limbs do not utilize RA for their development, a surprising phenomenon first described in the mouse (Neiderreither et al. 2002; Sandell et al. 2007). However, during the regeneration of both fore *and* hind limbs, expression of EGFP was present not in the mesenchyme but in the apical ectodermal cap (AEC), an outgrowth-controlling region at the tip of the blastema (Figure 7.5c and d). This is remarkably similar to previous experiments that demonstrated that the wound epidermis of the limb regenerate could synthesize RA from retinol and could release it into the medium in culture (Viviano et al. 1995). This comparison between the developing and regenerating limb implies that, at least for RA signaling, limb development and regeneration are very different and the latter is not simply a recapitulation of the former.

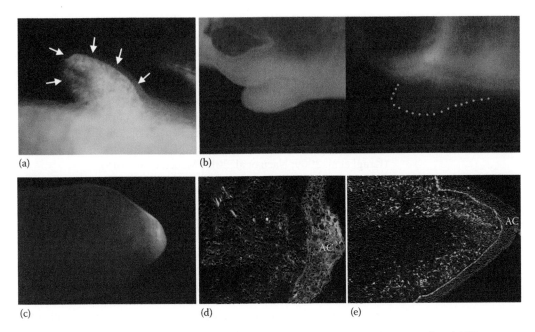

(a)   (b)

(c)   (d)   (e)

Figure 7.5 RARE-EGFP axolotl reporter expression during development and regeneration. (a) The developing limb bud shows GFP throughout the developing forelimb bud in the mesenchyme and not in the overlying ectoderm (white arrows). (b) The developing hind limb bud shows no GFP expression. The left panel shows a light image of the hind limb bud, and the same image under fluorescence is shown in the right panel, with the hind limb bud outlined in white dots. There is fluorescence in the adjacent trunk but not in the limb bud itself. (c) A regenerating forelimb shows GFP expression at the tip of the blastema. (d) A section through a blastema such as (c) shows localized fluorescence in the epidermis of the apical epithelial cap (AEC) and not in the mesenchyme, except for an occasional Schwann cell (top left). (e) When a blastema such as in (c) and (d) is treated with a proximalizing dose of RA, then the cells of the blastema that begin to report are no longer the apical cap cells but the fibroblasts of the blastema in a shell below the ectoderm. Reproduced with permission from Monaghan and Maden (2012).

7.4 The cells that assess positional information during regeneration

We have also used these RA reporter axolotls to ask what happens to the reporter construct in the regenerating limb when these axolotls are treated with a dose of RA that will result in a change of positional information and proximodistal respecification (Figure 7.1). Under these conditions, mesenchymal cells now begin to report EGFP (Figure 7.5e), precisely those cells that we know are responsible for the effects of RA (the mesenchyme rather than the ectoderm; Maden 1984). Furthermore, it is clearly not all the mesenchymal cells that begin to report but a subset of these cells that resemble fibroblasts because of their superficial position in the blastema. This cell type provides an overabundance of cells to the regenerate (Muneoka et al. 1986) and is the *stem cell* of the blastemal mesenchyme in that it can generate multiple connective tissue cell types (Kragl et al. 2009). It is also the cell type that is surely responsible for elaborating positional information and thus the one that responds to administered RA; however, there are currently no specific cell markers for fibroblasts that we could use to confirm this assumption. By contrast, other cell types that contribute to the blastema such as satellite cells and myonuclei from the muscles or Schwann cells from the nerve do not measure positional information or obey the laws of positional information, since they can move proximodistally throughout the limb with no regard to position and do not express positional markers such as *Meis* (Kragl et al. 2009; Nacu et al. 2013; Maden et al. 2015).

7.5 The nuclear retinoic acid receptors that act in positional information

Therefore, we have a cell type (the fibroblast) and a molecular pathway (RA acting in the nucleus) for positional information assessment. We know that there are three classes of RAR present in the nucleus (α, β, and γ/δ), and therefore, we can ask which one is involved in transducing the effects of RA on positional information as a prelude to discovering which genes are the targets of the positional information pathway.

This question was initially approached in a series of experiments involving domain swaps in the RAR genes, whereby the ligand-binding domain (binding RA) of each of the receptors was replaced with the ligand-binding domain from the thyroid hormone receptor, so that the DNA-binding specificity was retained, but they now became responsive to thyroid hormone instead of RA. When the RARα1 receptor was so modified, the proliferation of blastemal cells in culture (a well-established effect of RA is on proliferation of cells) became responsive to thyroid hormone, suggesting that this was at least one function of this RAR (Schilthius et al. 1993). When the RARδ1 receptor was so modified, then a marker of secretory differentiation (another well-established effect of RA) was induced in the wound epidermis by thyroid hormone (Pecorino et al. 1994). Finally, when the RARδ2 isoform was so modified and transfected into the blastemal cells, then after thyroid hormone, the transfected cells moved proximally during regeneration, suggesting that this receptor mediates the ability of RA to proximalize the blastema (Pecorino et al. 1996).

Another technique to answer the same question is to examine the effects of receptor-selective agonists on the regenerating limb to determine whether only RARγ (the mammalian equivalent of the newt RARδ) agonists affect positional information. Although this experiment has not yet been performed, we can certainly say that it is clear that the RARγ agonist, CD1530, has a potent effect at inducing proximodistal reduplications in the regenerating axolotl limb (Nguyen et al., 2017).

7.6 The gene targets used to assess positional information

Now, we are in a position to identify the targets of RARδ2 in the nucleus of blastemal cells to reveal the genetic mechanism of positional information assessment. Since there are specific agonists of the RARs, the obvious question to ask is: what are the target genes induced by an RARγ agonist during the proximalization of blastemal cells? This has not yet been performed during limb regeneration, but we know from other experiments on the neural stem cells of the mouse brain that this is a perfectly feasible question to ask, because we have shown that the three RARs have a small number of overlapping targets and a large number of unique gene targets.

To do this, we treated adult mice with an RARα agonist, an RARβ agonist, or an RARγ agonist, and after 5 days of treatment, we isolated the neural progenitor cells (NPCs) from the sub-ventricular zone of the brain (a known target of the action of RA) and performed a microarray analysis on the RNA from these cells.

These data revealed that each compound elicited a robust and highly individual transcriptional response in NPCs (Figure 7.6). In the group treated with the RARα agonist, 3135 genes in the NPCs were responsive to the treatment. Similarly, 2533 and 1290 genes were identified as being regulated by β and γ treatments, respectively. A high proportion of the NPC-regulated genes was unique to each intervention (α, 1439; β, 1182; and γ, 378), whereas others were in common between treatment groups—899 genes were similarly altered by α and β agonists and only 337 genes were regulated by all three compounds. Furthermore, RARα and β agonists are most related in output, and this response is anti-correlated to that of the RARγ agonist. These data show that each of the RARs elicits specific responses in adult NPCs and that this is a valuable

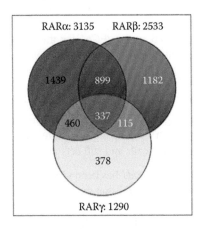

Figure 7.6 Venn diagram of the gene expression changes elicited by RARα, RARβ, and RARγ agonists on neural stem cells in the mouse brain.

methodology for identifying receptor-specific targets that could readily be applied to the regenerating limb.

The equivalent experiment in the regenerating limb awaits to be performed, but in the mean time, gene analysis experiments have been done by using RA, which activates all three of the RARs. The first experiment of this kind was performed by Morais Da Silva et al. (2002), who treated newt distal blastemas with RA in order to proximally respecify them and then performed a differential screen between normal and RA-treated distal blastemas. By applying stringent criteria for candidate genes—proximal blastemas higher than distal blastemas and RA induced in distal blastemas and expressed on the cell surface—one gene was identified, called *Prod1*, a glycosylphosphatidylinisotol (GPI)-linked cell-surface protein. In both the normal limb and in blastemas, there was a 1.7-fold higher level proximally than distally, similar to the 2.5-fold difference in the concentration of RA required to change distal positional levels into proximal levels (Figure 7.1). *Prod1* is not only expressed in blastemal cells but also in the normal adult limb in somewhat of a graded fashion, with proximal levels 2.5-fold higher than distal levels (Kumar et al. 2007a). When the *Prod1* gene was over-expressed in blastemal cells that would normally form distal elements such as the hand and digits, then their positional identity was changed to more proximal cells contributing to the radius and ulna and tissues extending back to the elbow (Echeverri and Tanaka 2005). Thus, *Prod1* is a target of RA, involved in positional specification and expressed on the cell surface exactly as we would have expected (see Introduction). Most interestingly (Shaikh et al. 2011), the *Prod1* promoter has two binding sites for specific transcription factors, the *Meis* genes, which have already been shown to be upregulated by RA and involved in the specification of proximal identity in the developing limb (Mercarder et al. 2000; Yashiro et al. 2004).

The molecular target of newt *Prod1* should tell us something about how positional information is realized. One target of *Prod1* is matrix metallopro-teinase 9 (MMP9), a matrix metalloprotease that degrades the extracellular matrix (ECM) and is necessary for regeneration (Vinarsky et al. 2005), again pointing us toward the cell surface/ECM for the realization of positional cues, but it is difficult to see how such a molecule could be positionally related. To upregulate MMP9, *Prod1* activates the epidermal growth factor (EGF) receptor on the same cell that then signals across the cell membrane to activate ERK1/2 by phosphorylation (Blassberg et al. 2010). Surprisingly, the axolotl orthologue of PROD1 is secreted and not GPI-anchored, yet it only upregulates MMP9 in the cells that are expressing it and not in a para-crine fashion in the adjacent cells and still acts via EGFR interaction.

Another binding partner of *Prod1* has been identified as a secreted protein, which is a blastemal growth factor that the nervous system provides, and it is known as newt anterior gradient (nAG) (Kumar et al. 2007b). The nAG protein is produced by the Schwann cells of the nerve and subsequently by the apical epidermis, but it plays no role in specifying position and acts merely as a proliferative agent. However, the fascinating aspect of these two molecules is that they bring together positional information and proliferation that are obviously linked—when cells of different positional identity are brought together, cell division (intercalation) is stimulated. In the presence of a uniform amount of the proliferative factor nAG, cells with more PROD1 on their surfaces will react differently to those with less PROD1 on their surfaces.

In a similar experimental design, but with far less stringency, using custom *Ambystoma mexicanum* Affymetrix GeneChips to identify all the genes upregulated or downregulated after RA treatment of distal blastemas, a large group of genes were identified as being regulated by RA (Nguyen et al., 2017). These were divided into clusters according to their known function. For example, one group of genes upregulated by RA in distal blastemas ($n = 101$) is known to be expressed in proximal developing limb buds in other vertebrates (e.g., *Meis1*, *Meis2*, *Pbx1*, *Arid5b*, *Fibin*, *Epha7*, *Nrip1*, and *Rnd3*) or is required for proper limb development (e.g., *Mia3*, *Rac1*, *Asph*, *Neo1*, *Cyp26B1*, *Flrt2*, *Rarγ*, *Rbp1*, *KIAA1217*, *Apcdd1*, *Zfn638*, *Stat3*, and *Tsh2*). The association of these genes with limb patterning in other vertebrates supports the idea that RA reprograms the distal cells to resemble a proximal limb cell fate and that proximodistal (PD) respecification involves many genes. Another much smaller cluster included 14 genes that were expressed at lower levels in RA-treated limbs compared with those treated with dimethyl sulfoxide (DMSO). Some of these are known to be expressed in the distal portion of the developing or regenerating vertebrate limb, including *Lhx9*, *Zic5*, *Lmo1*, *Lhx2*, *Spry1*, *Msx2*, and *HoxA13*, and are required for distal identity in developing mouse limbs. This suggests that distal-identity genes are silenced in limbs undergoing PD duplication, in addition to the transcriptional activation of proximal-identity genes during PD duplication.

It is important to note that *HoxA13* was silenced in this experiment as *Hox* genes are the classical targets of RA in development (Marshall et al. 1996), and in the developing limb, genes of the *HoxA* cluster are considered responsible for determining segmental and positional identity. The same is true in the regenerating limb (Gardiner et al. 1995), and when distal blastemal cells are treated with RA, they repress expression of *HoxA13* (Roensch et al. 2013). Clearly, *Hox* genes are crucially involved in the translation of an RA signal from the nucleus to the cell surface. In terms of speed of induction by RA, *Meis* genes are rapidly induced and *Hox* genes are more temporally delayed (Mercarder et al. 2000), suggesting a sequence of inductive events occurring in the nucleus.

This experiment described earlier (Nguyen et al., 2017) also revealed that *RARγ* expression was induced 1.62-fold higher in RA-treated blastemas, supporting the proposition that RARγ is the receptor involved in positional information (see Section 7.5). In addition, 23 extracellular molecules and 11 genes involved in the regulation of cell adhesion were identified, supporting the idea that positional respecification is realized as a cell surface phenomenon. However, surprisingly, an upregulation of the axolotl *Prod1* after RA treatment was not observed.

7.7 Summary and outstanding questions to be answered

We now have a beginning of an understanding of the gene pathways, cell type, and cellular location of the genetics of positional information during limb regeneration that is drawn together in Figure 7.7. It is the fibroblasts, the stem cells of the blastema, that are the effectors of positional

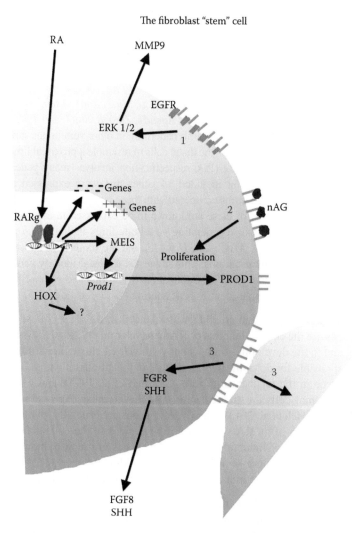

Figure 7.7 Summary diagram of the downstream effects of RA on the fibroblast cell, which is the purveyor of positional information in the blastema. Retinoic acid enters the nucleus, binds to the RARγ heterodimer, activates a host of genes (+ + + +), downregulates a host of genes (− − − −), and upregulates *Meis* and *Hox* genes, which are transcription factors.

(Continued)

information in the blastema. Retinoic acid, perhaps in a concentration gradient deriving from the apical cap of the blastema, acts in the nucleus of blastemal cells via RARγ to upregulate and downregulate many genes with a variety of cellular functions. One group of genes comprises the *Hox* genes, which have classically been associated with positional specification, but few downstream targets, which could play a role in the realization of positional specification, have been identified. Perhaps, their targets are more transcription factors. On the other hand, the transcription factor MEIS has been shown to act on the *Prod1* enhancer to upregulate it, and the PROD1 protein translocates to the cell membrane, where it is present in an amount depending on the location of the cell in the proximodistal axis of the limb. PROD1 interacts with EGFR in the cell membrane to signal back into the cytoplasm via ERK1/2 to upregulate MMP9 that is secreted into the extracellular space (signal 1). PROD1 also binds to nAG, the neurotrophic factor in the extracellular environment, and that interaction signals back into the cytoplasm to stimulate proliferation of the cell (signal 2). PROD1 also interacts with PROD1 on the surface of the adjacent cell, and the detection of concentration differences (perhaps by receptor occupancy Morais da Silva et al. 2002) results in a signal back into the cytoplasm of both cells to respond to positional differences (signal 3). These latter signals are likely to result in the upregulation of developmental signaling molecules such as FGF8 and SHH (Nacu et al. 2016), which are at least two of the intercellular signals responding to the detection of positional differences.

This is clearly a very superficial outline of what is happening in the regenerating limb blastema, and many details need to be filled in. For example, does RA need to be graded or is it simply an on/off signal that is translated into a graded distribution of PROD1; how many other gene pathways are there that are up- or downregulated by the RA signal; are there differences in PROD1 distribution on the surface of single blastemal cells and what are the differences between cells; how do differences in PROD1 signal back to the blastemal cell and how widespread is the response—there must be changes in motility, integrin composition, and cytoskeletal

Figure 7.7 (Continued) There are no *Hox* gene targets that have been identified in blastemal cells, but MEIS binds to the *Prod1* promoter. PROD1 protein translocates to the cell surface (green lines). At the cell surface, PROD1 undergoes at least three interactions. Interaction 1 (red 1) involves reacting with the cell surface EGFR, which induces ERK1/2 signaling, resulting in the upregulation and secretion of MMP9 into the extracellular environment. Interaction 2 (red 2) involves the binding of PROD1 to nAG (purple spheres) released by Schwann cells and the wound epithelium into the blastema. This interaction stimulates cell proliferation of the fibroblast. Interaction 3 (red 3) involves the assessment of differences in PROD1 levels between adjacent fibroblasts, resulting in the upregulation of signaling molecules such as FGF8 and SHH to induce the outgrowth of the blastema in response to positional differences. This latter interaction must somehow feedback to carefully regulate proliferation, such that when positional differences are resolved, regeneration ceases.

activity in addition to the classical signaling molecules? Are the developmental signaling molecules such as FGF8 and SHH simply proliferation agents (Seifert et al. 2010) rather than positional agents, and is the primary positional molecule PROD1? What is the relationship between PROD1 and the composition of the ECM that is considered to have pro-regenerative properties and in particular with regard to the heparan sulfates as potential ECM candidate molecules for positional information (Phan et al. 2015)? Finally, it is important to remember that we have concentrated on only one axis of a three-dimensional blastema and RA has effects on the other two axes of the limb (Kim and Stocum 1986b; Ludolph et al. 1990). How the axes are integrated and coordinated will only triple the level of complexity, whose surface we are just beginning to scratch in the proximodistal axis.

References

Blassberg, R.A., Garza-Garcia, A., Janmohamed, A., Gates, P.B., and Brockes J.P. 2010. Functional convergence of signaling by GPI-anchored and anchorless forms of a salamander protein implicated in limb regeneration. *J Cell Sci* 124:47–56.

Boilly, B., Cavanaugh, K.P., Thomas, D., Hondermarck, H., Bryant, S.V., and Bradshaw, R.A. 1991. Acidic fibroblast growth factor is present in regenerating limb blastemas of axolotls and binds specifically to blastema tissues. *Dev Biol* 145:302–310.

Brockes, J.P. 1992. Introduction of a retinoid reporter gene into the Urodele limb blastema. *Proc Natl Acad Sci USA* 89:11386–11390.

Carrier, M. and Rochette-Egly, C. 2015. Control of gene expression by nuclear retinoic acid receptors: Post-translational and epidgenetic regulatory mechanisms. In The Retinoids: Biology, Biochemistry, and Disease, Eds Dolle, P. and Niederreither, K., pp. 93–116. Hoboken, NJ: Wiley & Sons Inc.

Carter, C., Clark, A., Spencer, G., and Carlone, R. 2011. Cloning and expression of a retinoic acid receptor β2 subtype from the adult newt: Evidence for an early role in tail and caudal spinal cord regeneration. *Dev Dynam* 240:2613–2625.

Christensen, R.N., Weinstein, M., and Tassava, R.A. 2001. Expression of fibroblast growth factors in regenerating limbs of Ambystoma: Cloning and semi-quantitative RT-PCR expression studies. *J Exp Zool* 290:529–540.

Crawford, K. and Stocum, D.L. 1988. Retinoic acid co-ordinately proximalizes regenerate pattern and blastema differential affinity in axolotl limbs. *Development* 102: 687–698.

Crick, F. 1970. Diffusion in embryogenesis. *Nature* 225:420–421.

Echeverri, K. and Tanaka, E.M. 2005. Proximodistal patterning during limb regeneration. *Dev Biol* 279:391–401.

French, V., Bryant, P.J., and Bryant, S.V. 1976. Pattern regulation in epimorphic fields. *Science* 193:969–981.

Gardiner, D.M., Blumberg, B., Komine, Y., and Bryant, S.V. 1995. Regulation of HoxA expression in developing and regenerating axolotl limbs. *Development* 121:1731–1741.

Ghosh, S., Roy, S., Séguin, C., Bryant, S.V., and Gardiner, D.M. 2008. Analysis of the expression and function of Wnt-5a and Wnt-5b in developing and regenerating axolotl (Ambystoma mexicanum) limbs. *Dev Growth Differ* 50:289–297.

Giguère, V., Ong, E.S., Evans, R.M., and Tabin, C.J. 1989. Spatial and temporal expression of the retinoic acid receptor in the regenerating amphibian limb. *Nature* 337:566–569.

Guimond, J.C., Lévesque, M., Michaud, P.L., Berdugo, J., Finnson, K., Philip, A., and Roy, S. 2010. BMP-2 functions independently of SHH signaling and triggers cell condensation and apoptosis in regenerating axolotl limbs. *BMC Dev Biol* 10:15.

Han, M.J., An, J.Y., and Kim, W.S. 2001. Expression patterns of Fgf-8 during development and limb regeneration of the axolotl. *Dev Dynam* 220:40–48.

Hill, D.S., Ragsdale, C.W., Jr., and Brockes, J.P. 1993. Isoform-specific immunological detection of newt retinoic acid receptor delta 1 in normal and regenerating limbs. *Development* 117:937–945.

Imokawa, Y. and Yoshizato, K. 1997. Expression of Sonic hedgehog gene in regenerating newt limb blastemas recapitulates that in developing limb buds. *Proc Natl Acad Sci USA* 94:9159–9164.

Ju, B.-G. and Kim, W.-S. 1994. Pattern duplication by retinoic acid treatment in the regenerating limbs of Korean salamander larvae, Hynobius leechii, correlates well with the extend of dedifferentiation. *Dev Dynam* 199:253–267.

Keeble, S. and Maden, M. 1986. Retinoic acid-binding protein in the axolotl: Distribution in mature tissues and time of appearance during limb regeneration. *Dev Biol* 117:435–441.

Keeble, S. and Maden, M. 1989. The relationship among retinoid structure, affinity for retinoic acid-binding protein, and ability to respecify pattern in the regenerating limb. *Dev Biol* 132:26–34.

Kim, W.-S. and Stocum, D.L. 1986a. Effects of retinoids on regenerating limbs: Comparison of retinoic acid and arotinoid at different amputation levels. *Roux's Arch Dev Biol* 195:455–463.

Kim, W.-S. and Stocum, D.L. 1986b. Retinoic acid modifies positional memory in the anteroposterior axis of regenerating axolotl limbs. *Dev Biol* 114:170–179.

Knapp, D., Schulz, H., Rascon, C.A. et al. 2013. Comparative transcriptional profiling of the axolotl limb identifies a tripartite regeneration-specific gene program. *PLoS ONE* 8:e61352.

Kragl, M., Knapp, D., Nacu, E., Khattak, S., Maden, M., Epperlein, H.H., and Tanaka, E.M. 2009. Cells keep a memory of their tissue origin during axolotl limb regeneration. *Nature* 460:60–65.

Kumar, A., Gates, P.B., and Brockes, J.P. 2007a. Positional identity of adult stem cells in salamander limb regeneration. *C.R. Biologies* 330:485–490.

Kumar, A., Godwin, J.W., Gates, P.B., Garza-Garcia, A.A., and Brockes, J.P. 2007b. Molecular basis for the nerve depenence of limb regeneration in an adult vertrbrate. *Science* 303:540–543.

Levesque, M., Gatien, S., Finnson, K., Desmeules, S., Villard, E., Pilote, M., Philip, A., and Roy, S. 2007. Transforming growth factor-β signaling is essential for limb regeneration in axolotls. *PLoS ONE* 2:e1277.

Lheureux, E., Thoms, S.D., and Carey, F. 1986. The effects of two retinoids on limb regeneration in Pleurodeles waltl and Triturus vulgaris. *J Embryol Exp Morph* 92:165–182.

Ludolph, D.C., Cameron, J.A., and Stocum, D.L. 1990. The effect of retinoic acid on positional memory in the dorsoventral axis of regenerating axolotl limbs. *Dev Biol* 140:41–52.

Maden, M. 1978. Supernumerary limbs in the axolotl. *Nature* 273:232–235.

Maden, M. 1982. Vitamin A and pattern formation in the regenerating limb. *Nature* 295:672–675.

Maden, M. 1983. The effect of vitamin A on the regenerating axolotl limb. *J Embryol Exp Morph* 77:273–295.

Maden, M. 1984. Does vitamin A act on pattern formation via the epidermis or the mesenchyme? *J Exp Zool* 230:387–392.

Maden, M. 1993. The homeotic transformation of tails into limbs in Rana temporaria by retinoids. *Dev Biol* 159:379–391.

Maden, M. 1998. Retinoids as endogenous components of the regenerating limb and tail. *Wound Repair Regen* 6:358–365.

Maden, M. 2007. Retinoic acid in the development, regeneration and maintenance of the nervous system. *Nature Rev Neurosci* 8:755–765.

Maden, M., Avila, D., Roy, M., and Seifert, A.W. 2015. Tissue-specific reactions to positional discontinuities in the regeneratign axolotl limb. *Regeneration* 2:137–147.

Maden, M., Summerbell, D., Maignan, J., Darmon, M., and Shroot, B. 1991. The respecification of limb pattern by new synthetic retinoids and their interaction with cellular retinoic acid-binding protein. *Differentiation* 47:49–55.

Marshall, H., Morrison, A., Studer, M., Pöpperl, H., and Krumlauf, R. 1996. Retinoids and Hox genes. *FASEB J* 10:969–978.

McCormick, A.M., Shubeita, H.E., and Stocum, D.L. 1988. Cellular retinoic acid binding protein: Detection and quantitation in regenerating axolotl limbs. *J Exp Zool* 245:270–276.

Meinhardt, H. 1983. A boundary model for pattern formation in vertebrate limbs. *J Embryol Exp Morph* 76:115–137.

Mercarder, N., Leonardo, E., Peidra, M.E., Martinez, A.C., Ros, M.A., and Torres, M. 2000. Opposing RA and FGF signals control proximodistal vertebrate limb development through regulation of Meis genes. *Development* 127:3961–3970.

Mohanty-Hejmadi P., Dutta, S.K., and Mahapatra, P. 1992. Limbs generated at site of tail amputation in marbled baloon frog after vitamin A treatment. *Nature* 355:352–353.

Monaghan, J.R., Athippozhy, A., Seifert, A.W., Putta, S., Stromberg, A.J., Maden, M., Gardiner, D.M., and Voss, S.R. 2012. Gene expression patterns specific to the regenerating limb of the Mexican axolotl. *Biol Open* 1:937–948.

Monaghan, J.R., and Maden, M. 2012. Visualization of retinoic acid signaling in transgenic axolotls during limb development and regeneration. *Dev Biol* 368:63–75.

Morais da Silva, S., Gates, P.B., and Brockes, J.P. 2002. The newt ortholog of CD59 is implicated in proximodistal identity during amphibian limb regeneration. *Dev Cell* 3:547–555.

Mullen, L.M., Bryant, S.V., Torok, M.A., Blumberg, B., and Gardiner, D.M. 1996. Nerve dependency of regeneration: The role of Distal-less and FGF signaling in amphibian limb regeneration. *Development* 122:3487–3497.

Muneoka, K., Fox, W.F., and Bryant S.V. 1986. Celular contribution from dermis and cartilage to the regenerating limb blastema in axolotl. *Dev Biol* 116:256–260.

Nacu, E., Glausch, M., Le, H.Q., Damanik, F.F.R., Schuez, M., Knapp, D., Khattak, S., Richter, T., and Tanaka, E.M. 2013. Connective tissue cells, but not muscle cells, are involved in establishing the proximo-distal outcome of limb regeneration in the axolotl. *Development* 140:513–518.

Nacu, E., Groberg, E., Olivera, C.R., and Tanaka, E.M. 2016. FGF8 and SHH substitute for anterior-posterior tissue interactions to induce limb regeneration. *Nature*. doi: 10.1038/nature17972.

Nardi, J.B. and Stocum, D.L. 1983. Surface properties of regenerating limb cells: Evidence for gradation along the proximodistal axis. *Differentiation* 25:27–31.

Neiderreither K., Vermot, J., Schuhbaur B., Chambon P., and Dolle P. 2002. Embryonic retinoic acid synthesis is required for forelimb growth and antero-posterior patterning in the mouse. *Development* 129:3563–3574.

Neiderreither K., Ward S.J., Dolle P., and Chambon, P. 1996. Morphological and molecular characterization of retinoic acid-induced limb duplications in mice. *Dev Biol* 176:185–198.

Niazi, I.A., Pescitelli, M.J., and Stocum, D.L. 1985. Stage dependent effects of retinoic acid on regenerating limbs. *Wilhelm Roux Arch Devl Biol* 194:355–363.

Niazi, I.A. and Saxena, S. 1978. Abnormal hundlimb regeneration in tadpoles of the toad, Bufo andersonii, exposed to axcess vitamin A. *Folia Biol (Krakow)* 26:3–8.

Nguyen, M., Singhal, P., Piet, J., Shefelbine S.J., Maden, M., Voss, S.R., and Monaghan, J.R. 2017. Retinoic acid receptor regulation of epimorphic and homeostatic regeneration in the axolotl. *Development* 144:601–611.

Pecorino, L.T., Entwistle, A., and Brockes, J.P. 1996. Activation of a single retinoic acid receptor isoform mediates proximodistal respecification. *Current Biol* 6:563–569.

Pecorino, L.T., Lo, D.C., and Brockes, J.P. 1994. Isoform-specific induction of a retinoid-responsive antigen after biolistic transfection of chimeric retinoic acid/thyroid hormone receptors into a regenerating limb. *Development* 120:325–333.

Phan, A.Q., Lee, J., Oei, M. et al. 2015. Positional information in axolotl and mouse limb extracellular matrix is mediated via heparin sulfate and fibroblast growth factor during limb regeneration in the axolotl (Ambystoma mexicanum). *Regeneration* 2:182–201.

Ragsdale, C.W., Jr., Petkovich, M., Gates, P.B., Chambon, P., and Brockes, J.P. 1989. Identification of a novel retinoic acid receptor in regenerative tissues of the newt. *Nature* 341:654–657.

Roensch, K., Tazaki, A., Chara, O., and Tanaka, E.M. 2013. Progressive specification rather than intercalation of segments during limb regeneration. *Science* 342:1375–1379.

Rutledge, J.C., Shourbaji, A.G., Hughes, L.A., Polif ka, J.E., Cruz, Y.P., Bishop, J.B., and Generoso, W.M. 1994. Limb and lower-body duplications induced by retinoic acid in mice. *Proc Natl Acad Sci USA* 91:5436–5440.

Sandell, L.L., Sanderson, B.W., Moiseyev, G., Johnson, T., Mushegian, A., Young, K., Rey, J.P., Ma, J.X., Staehling-Hampton, K., and Trainor, P.A. 2007. RDH10 is essential for synthesis of embyronic retinoic acid and is required for limb, craniofacial, and organ development. *Genes & Dev* 21:1113–1124.

Scadding, S.R. 2000. Citral, an inhibitor of retinoic acid synthesis, modifies pattern formation during limb regeneration in the axolotl Ambystoma mexicanum. *Canad J Zool* 77:1835–1837.

Scadding, S.R. and Maden, M. 1986. Comparison of the effects of vitamin A on limb development and regeneration in tadpoles of Xenopus laevis. *J Embryol Exp Morph* 91:35–53.

Scadding, S.R. and Maden, M. 1994. Retinoic acid gradients during limb regeneration. *Dev Biol* 162:608–617.

Schilthius, J.G., Gann, A.A., and Brockes, J.P. 1993. Chimeric retinoic acid/thyroid hormone receptors implicate RAR-alpha 1 as mediating growth inhibition by retinoic acid. *EMBO J* 12:3459–3466.

Seifert, A.W., Zheng, Z., Ormerod, B.K., and Cohn, M.J. 2010. Sonic hedgehog controls growth of external genitalia by regulating cell cycle kinetics. *Nat Commun* 1:23. doi: 10.1038/ncomms1020.

Shaikh, N., Gates, P.B., and Brockes, J.P. 2011. The Meis homeoprotein regulates the axolotl Prod1 promoter during limb regeneration. *Gene* 484:69–74.

Thoms, S.D., and Stocum, D.L. 1984. Retinoic acid-induced pattern duplication in regenerating urodele limbs. *Dev Biol* 103:319–328.

Torok, M.A., Gardiner, D.M., Izpisua-Belmonte, J.-P., and Bryant, S.V. 1999. Sonic hedgehog (shh) expression in developing and regenerating Axolotl limbs. *J Exp Zool* 284:197–206.

Vinarsky, V., Atkinson, D.L., Stevenson, T.J., Keating M.L., and Oldenberg, S.J. 2001. Normal newt limb regeneration required matrix metalloprotease function. *Dev Biol* 279:86–98.

Viviano, C.M., Horton, C.E., Maden, M., and Brockes, J.P. 1995. Synthesis and release of 9-cis retinoic acid by the urodele wound epidermis. *Development* 121:3753–3762.

Wei, L.-N. 2015. Retinoic acid receptor coregulators in epigenetic regulation of target genes. In The Retinoids: Biology, Biochemistry, and Disease, Eds Dolle, P. and Niederreither, K., pp. 117–130. Hoboken, NJ: Wiley & Sons Inc.

Wolpert, L. 1969. Positional information and the spatial pattern of cellular differentiation. *J Theor Biol* 25:1–47.

Yashiro, K., Zhao, X., Uehara, M., Yamashita, K., Nishijima, M., Nishino, J., Saijoh, Y., Sakai, Y., Hamada, H. 2004. Regulation of retinoic acid distribution is required for proximodistal patterning and outgrowth of the developing mouse limb. *Dev Cell* 6:411–422.

Chapter 8 MicroRNA signaling during regeneration

Keith Sabin and
Karen Echeverri

Contents

Key concepts

a. MicroRNAs (miRNAs) are small non-coding RNAs that bind to the 3′ untranslated regions of messenger RNAs (mRNAs) to regulate their expression. MicroRNAs bind to 6–8 nucleotide seed sequences, therefore allowing a single miRNA to potentially regulate the expression of thousands of genes.

b. Regeneration of complex tissue requires dynamically coordinated gene expression in multiple different cell types over large periods of time.

c. MicroRNAs play a key role in acting as high-level effectors of gene expression during regeneration in multiple species.

d. MicroRNAs are highly conserved and hence maybe a valuable therapeutic tool in translating regenerative abilities cross-species via their ability to regulate multiple gene pathways.

e. Regenerative engineering approaches focused on the development of biocompatible scaffolds for the effective delivery of microRNAs could represent a novel therapeutic strategy to promote a regenerative response after injury in mammals.

8.1 Introduction

Complex tissue regeneration is a widespread phenomenon across diverse phyla and ranges from whole body regeneration to organ regeneration. However, the extent and functional outcome of regeneration after injury vary drastically between species. Although mammals regenerate very poorly after injury or amputation, other vertebrates, including fish and salamanders, are able to functionally regenerate diverse tissue types after injury (Bryant 1970, Sanchez Alvarado and Tsonis 2006a, b, Tanaka and Ferretti 2009, Tanaka and Reddien 2011, Diaz Quiroz and Echeverri 2013, Gemberling et al. 2013, Reddien 2013). Therefore, these organisms represent a unique opportunity for scientists, clinicians, and engineers to study the biophysical, cellular, and molecular signals required for functional regeneration and translate these findings to mammalian systems. This interdisciplinary approach will allow basic science findings to be transitioned into the clinic, with the explicit goal of therapeutically enhancing the endogenous mammalian injury response to stimulate functional regeneration.

Extensive research efforts have begun to elucidate the molecular and cellular responses necessary to promote regeneration. Interestingly, many of the molecules involved with the development of a certain structure (limb or spinal cord) play critical roles in directing regeneration of the adult structure (Schnapp et al. 2005, Reimer et al. 2009, Singh et al. 2015, Briona et al. 2015). How these signaling pathways are regulated in a spatiotemporal manner to direct faithful regeneration is not well known. Within the last 15–20 years, an increased appreciation of the role of microRNAs (miRNAs) in regulating various cell functions during homeostasis and disease has prompted investigators to study the role of miRNAs during regeneration.

The combination of recent advances in the fields of regeneration, stem cell biology, and bioengineering is now allowing for an unprecedented application of biomaterials and scaffolds to the treatment of human injuries and diseases. In this chapter, we will review the current body of knowledge regarding miRNAs during natural regeneration and contemplate on how this knowledge could be exploited from a regenerative engineering standpoint to manipulate miRNA signaling to promote mammalian regeneration.

8.2 MicroRNAs biogenesis

MicroRNAs (miRNAs) were first identified in *C. elegans* in 1993 (Lee et al. 1993) and have proven to be potent regulators of post-transcriptional gene expression. Most miRNA genes are transcribed by RNA polymerase II to form a primary miRNA that is capped, polyadenylated, and further processed in the nucleus by a muli-enzyme complex, called the microprocessor, to form pre-miRNAs, which are subsequently exported

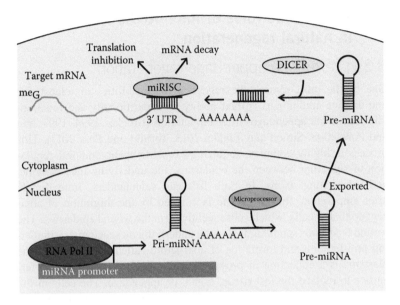

Figure 8.1 Biogenesis and function of miRNAs. MicroRNA gene is transcribed by RNA polymerase II and is polyadenylated. The primary miRNA (pri-miRNA) transcript is processed by a multi-enzyme complex, microprocessor, to generate a pre-miRNA that is subsequently exported to the cytoplasm. After nuclear export, the pre-miRNA is cleaved to ~21 nucleotides by Dicer, and the guide strand is incorporated into miRNA-induced signaling complex (miRISC). Finally, the guide miRNA will anneal to a 6–8 nucleotide seed sequence in the 3′ untranslated region (UTR) of target mRNAs to promote transcript degradation or inhibition of translation.

from the nucleus to the cytoplasm (Yi et al. 2003, Denli et al. 2004, Gregory et al. 2004, Lee et al. 2004, Lund et al. 2004) (Figure 8.1). Once exported into the cytoplasm, the pre-miRNAs are processed further by an RNase III enzyme called Dicer to produce ~22-nucleotide RNA molecule that is incorporated into the miRNA-induced signaling complex, or miRISC (Bernstein et al. 2001, Chendrimada et al. 2005, Park et al. 2011). Once the miRISC is assembled, the guide strand of RNA directs target specificity by binding to complementary seed sequences in the 3′ untranslated region (UTR) of mature mRNAs (Bartel 2009, Shukla et al. 2011). Mature miRNAs regulate gene expression by annealing to the complementary seed sequence in the 3′ UTR of target transcripts to promote mRNA degradation and inhibition of translation (Filipowicz et al. 2008). Considering that a single miRNA can regulate the expression of hundreds or thousands of individual mRNAs and that a single mRNA can be efficiently regulated by multiple miRNAs, therapeutic manipulation of a single miRNA could have profound effects on the cellular response to injury. Indeed, recent evidence supports the hypothesis that differential regulation of miRNA expression after injury is critical for promoting organ and tissue regeneration.

8.3 The diverse roles of microRNAs in natural regeneration

8.3.1 Vertebrate appendage regeneration

One of the most remarkable regenerative capabilities of teleost fish and urodele amphibians is their ability to proportionally and functionally regenerate appendages after amputation (Brockes 1994, 1997, Yin and Poss 2008, Simon and Tanaka 2013, Tornini and Poss 2014). This process is highly complex, requiring exquisite spatiotemporal regulation of signaling between the epidermis, the underlying mesenchyme, and innervating axons. In both fish and salamanders, immediately after amputation, the injury site is covered by the migration of adjacent epithelial cells, which subsequently form the wound epidermis. The wound epidermis sends signals to the underlying mesenchyme, stimulating proliferation and formation of a structure called the blastema. The blastema is a collection of progenitor cells that proliferate and differentiate to replace the lost tissue, organ, or appendage. The cells in the blastema proliferate until a critical mass is reached, at which point an undefined signal(s) promotes cellular differentiation and patterning of the regenerated appendage. Utilizing transcriptional profiling, forward genetic screens, and gain- and loss-of-function experiments, many signaling pathways, including fibroblast growth factor, Wnt, and sonic hedgehog signaling, have been identified as being indispensible for blastema formation, regenerative outgrowth, and patterning of the regenerated appendage (Bryant 1970, Gardiner et al. 1993, 1995, Gardiner and Bryant 1996, Mullen et al. 1996, Brockes 1996, 1997, Carlson et al. 1998, Torok et al. 1999, Raballo et al. 2000, Roy et al. 2000, Bryant et al. 2002, Campbell and Crews 2008, Campbell et al. 2011). These genes, which are re-expressed after injury, are known to play an important role in appendage development but are minimally expressed in the adult tissue. How these genes are regulated after injury to promote appendage regeneration is largely unknown.

Although it is well appreciated that miRNAs are essential for organismal development (Bernstein et al. 2003, Wienholds et al. 2003, Kloosterman and Plasterk 2006), stem cell maintenance (Hatfield et al. 2005), and cellular differentiation (Kanellopoulou et al. 2005), the role of miRNAs during appendage regeneration is not well understood. Zebrafish is a popular model for regeneration due to its fast generation time and availability of genetic tools. The zebrafish caudal fin is composed of bony rays consisting of hemirays that form a protective barrier to the underlying nerves, blood vessels, and mesenchymal cells, and it is often used as a model for understanding bone regeneration. One of the first studies of miRNAs in zebrafish fin regeneration utilized a knockdown of Dicer, the cytoplasmic enzyme complex required for miRNA function, which led to an overall block of zebrafish fin regeneration (Thatcher et al. 2008). However, this blunt approach, though establishing the need for miRNAs during regeneration, did not provide insight into the specific miRNAs and their targets that regulate fin regeneration. Additional studies from the Patton laboratory and others identified several miRNAs that are critical in regulating different stages of zebrafish fin regeneration (Thatcher et al. 2008, Yin et al. 2008a).

During regeneration, many genes that are known to carry out essential roles during development are differentially regulated. For example, upregulation of fibroblast growth factor (FGF) signaling after fin amputation is an essential signal to promote blastema formation during zebrafish fin regeneration (Whitehead et al. 2005). However, the downstream effects of FGF signaling that promote blastema formation were not clear. Interestingly, FGF signaling after fin amputation leads to significant downregulation of miR-133 (Yin et al. 2008a). If FGF signaling during fin regeneration is inhibited by overexpression of dominant negative form of FGF receptor 1, thereby maintaining high levels of miR-133, then cells in the blastema fail to proliferate and fin regeneration does not occur. However, if miR-133 function is inhibited in fish that overexpresses dominant-negative FGF receptor 1 (dnFGFR1), this largely rescues blastema cell proliferation and fin regeneration. This suggests that a major function of FGF signaling after fin amputation is to decrease expression of miR-133 to allow for blastema cell proliferation and regenerative outgrowth. Consistent with this idea, the kinase mps1, which is involved with spindle checkpoint and cell cycle progression (Fisk and Winey 2004, Liu and Winey 2012), was identified as a major target of miR-133 during zebrafish fin regeneration. Downregulation of miR-133 after fin amputation allows for accumulation of mps1 protein in blastema cells and subsequent increase in cell cycle progression.

Another key signaling pathway that is essential for zebrafish fin regeneration is Wnt signaling (Kawakami et al. 2006, Stoick-Cooper et al. 2007). During zebrafish fin regeneration, the Wnt effector gene *lef1* is specifically upregulated in the wound epidermis after amputation and may play a role during scleroblast alignment during fin regeneration (Poss et al. 2000). Interestingly, there are two conserved seed sequences for miR-203 in the 3′ UTR of *lef1*, and microarray-based transcriptional profiling approaches identified significant downregulation of miR-203 after fin amputation (Thatcher et al. 2008). Overexpression or inhibition of miR-203 after fin amputation leads to a marked decrease or increase, respectively, in the protein levels of lef1. Overexpression of miR-203 largely blocked fin regeneration, and subsequent experiments confirmed that co-electroporation of the miR-203 mimic and *lef1* lacking a 3′ UTR was sufficient to rescue regenerative defects. Surprisingly, prolonged inhibition of miR-203 by electroporation of complementary antisense morpholinos leads to excessive regenerative outgrowth. The authors postulated that re-establishment of miR-203 expression to pre-lesion levels could be important to halt regeneration (Thatcher et al. 2008).

Although functional and mechanistic studies examining the role of miRNAs during zebrafish fin regeneration are underway, the role of miRNAs in regulating urodele limb regeneration is still in its infancy. Urodeles can fully and functionally regenerate their fore and hind limbs throughout life. Urodele limbs are structurally remarkably similar to mammalian limbs, and regeneration involves not only the regrowth of lost tissue but also patterning to ensure the correct structure of the arm and hand elements. A recent publication used a microarray-based transcriptional profiling approach to identify differentially expressed miRNAs during axolotl limb regeneration (Holman et al. 2012). This analysis identified the differential expression of dozens of miRNAs in 17 days postamputation (DPA) blastema compared with uninjured tissue. Interestingly, the highly conserved members of the let7 miRNA family were uniformly downregulated in 17 DPA compared with uninjured limb. Let7 has been proposed to be a marker of differentiated somatic cells and inhibits the pluripotent state (Melton et al. 2010). The uniform downregulation of let7 family members in the blastema of 17-DPA limbs would support the idea that those cells have acquired a less differentiated and more progenitor cell-like state. Furthermore, miR-21 was highly upregulated in the blastema of 17-DPA limbs compared with uninjured tissue. Using in silico and *in vitro* luciferase approaches, the Notch ligand, Jagged1, was identified as a target of axolotl miR-21. *Jag1* is essential for limb development in mice and is highly expressed in distal limb bud mesenchyme of human embryos (Crosnier et al. 2000, McGlinn et al. 2005). These potentially regulatory interactions between miR-21 and Jag1 during limb regeneration highlight a possible important signaling node in regulating the regenerative response to amputation. However, a direct interaction between miR-21 and Jag1 *in vivo* during limb regeneration was not determined. Moving forward, it will be important to verify the ability of miR-21 to modulate Jag1 expression during regeneration and to determine if this is sufficient to induce regenerative defects. It will also be interesting to compare the miRNA data from zebrafish caudal fin and axolotl limb to determine how conserved these differentially regulated microRNAs are cross-species when regenerating similar tissue types.

8.3.2 Role of microRNAs during vertebrate nervous system regeneration

Traumatic spinal cord injury (SCI) in mammals results in extensive neuronal cell death, axonal degeneration, and deposition of scar tissue, leading to the loss of sensory and motor function below the injury site. While mammals are unable to regenerate damaged nervous tissue to restore sensory and motor functions, other vertebrates, such as zebrafish and salamanders, are able to functionally regenerate (Tanaka and Ferretti 2009, Diaz Quiroz and Echeverri 2013, Gemberling et al. 2013, Becker and Becker 2015). The molecular mechanisms that promote functional nervous tissue regeneration are still being elucidated, but recent data from our laboratory and others have firmly established an essential role for miRNAs in regulating faithful nervous system regeneration.

After tail amputation in salamanders, the regenerating spinal cord forms a transient structure called the ependymal bulb, which protrudes into the blastema and migrates posteriorly throughout regeneration. Spinal cord outgrowth is dependent on the proliferation and differentiation of $Sox2^+$ ependymoglial cells, which act as neural stem cells in the regenerating spinal cord (Echeverri and Tanaka 2002, Chernoff et al. 2003, Schnapp et al. 2005, McHedlishvili et al. 2007, Fei et al. 2014). The adult spinal cord maintains expression of dorsoventral markers, including Pax7, Pax6, and Msx1/2, which are downregulated after tail amputation within a 500-μm zone anterior to the injury (Schnapp et al. 2005, McHedlishvili et al. 2007). How these dorsoventral (DV) expression domains are differentially regulated between the adult and regenerating spinal cord is not well understood.

Given that these same signaling domains are employed during development and re-accessed during regeneration, an investigation started about the role of miRNAs during spinal cord regeneration after tail amputation (Sehm et al. 2009). Taking a microarray approach comparing miRNA expression in uninjured tail tissue with 3-day post-injury blastema, we identified miR-196 as being highly upregulated after injury. Further analysis showed that miR-196 was specifically expressed in the ependymoglial cells within the 500-μm regeneration zone after injury. Inhibition of miR-196 resulted in overall defective tail regeneration, inhibition of ependymal bulb formation, and blockage of spinal cord regeneration potentially by inhibiting blastema cell proliferation after amputation. Interestingly, inhibition of miR-196 caused an expansion and maintenance of the dorsal Pax7 expression domain in the regenerating spinal cord. Furthermore, miR-196 inhibition increased expression of other dorsal domain proteins, such as BMP4, Msx1, and Meis2, but the expression of the ventral specifying molecule, sonic hedgehog, was not affected, suggesting that miR-196 is essential for appropriate spinal cord patterning during tail regeneration. Interestingly, miR-196 is also involved in regulating Hox genes during limb development and is differentially regulated during axolotl limb regeneration; however, its exact role has yet to be determined in limb regeneration (Hornstein et al. 2005, Holman et al. 2012).

A more comprehensive analysis of miRNA regulation of salamander tail regeneration has been recently reported, taking an unbiased deep-sequencing approach comparing expression levels of all miRNAs in 3-day post-injury blastemas with uninjured tail tissue (Gearhart et al. 2015). This approach identified members of the miR-1 and let-7 families as well as miR-21 and miR-206 families as being significantly differentially regulated after injury compared with uninjured tissue. This is consistent with other reports suggesting important regulatory roles of these miRNAs during regeneration of various organs (Ramachandran et al. 2010,

Holman et al. 2012, Aguirre et al. 2014, Lepp and Carlone 2015). Of interest was the identification and functional validation of putative novel miRNAs that did not share significant homology with any known vertebrate miRNA. This suggests the existence of salamander-specific miRNAs that could represent a species-specific difference in regenerative capabilities compared with mammals. This study highlights the importance of unbiased approaches for identifying species-specific factors that promote tissue regeneration. As bioinformatic and sequencing technologies continue to become more sophisticated, more species-specific factors that could be therapeutically manipulated to enhance mammalian regeneration could be identified.

Although it is clear that miRNAs are essential for functional spinal cord regeneration, the signaling pathways that regulate their expression after injury are less well studied. Recently, the retinoic acid receptor β2 was cloned from the newt and was shown to be upregulated in ependymoglial cells 7 days after tail amputation in newts (Carter et al. 2011). When the retinoic acid receptor β2 (RARβ2) function was pharmacologically blocked, this inhibited spinal cord regeneration. Retinoic acid signaling is a well-established regulator of tissue regeneration and nervous system development, but the downstream-signaling network during spinal cord regeneration in an adult salamander is largely unknown (Maden 2007, Duester 2008, Monaghan and Maden 2012). It was recently shown that the repression of miR-133a expression after tail amputation was necessary for upregulation of RARβ2 and that miR-133a directly targets RARβ2 via a conserved seed sequence in the 3′ UTR (Lepp and Carlone 2014). When miR-133a was overexpressed during tail regeneration, this resulted in decreased levels of RARβ2 and inhibited tail regeneration, phenocopying the effect of pharmacologically inhibiting RARβ2 (Carter et al. 2011, Lepp and Carlone 2014). The present study did not address the signal necessary for miR-133a repression after injury. However, FGF signaling represses miR-133 expression after fin amputation in zebrafish (Yin et al. 2008a), and FGF signaling is critical for ependymoglial cell proliferation and migration after SCI in newts; therefore, FGF family members are likely candidates (Zhang et al. 2000, 2002).

Further transcriptional analysis comparing miRNA expression between regenerating tails treated with a RARβ inhibitor and untreated controls identified a cohort of differentially expressed miRNAs (Lepp and Carlone 2015). These data suggest that one function of RA signaling after tail amputation is to regulate the expression of key miRNAs involved in functional spinal cord regeneration. Of interest, miR-133a, miR-1, and let-7c were all upregulated after RARβ inhibition, suggesting a transcriptionally repressive role for RA signaling at these loci. Furthermore, miR-1 and let-7 family members were identified by deep sequencing of small RNAs during axolotl tail regeneration, suggesting a potential role for RA signaling during tail regeneration in these closely related species (Gearhart et al. 2015). Additional analysis confirmed that the expression pattern of miR-1 mirrored that of miR-133a, and it was also specifically expressed in the ependymoglial cells during regenerative spinal cord outgrowth. Remarkably, there was a conserved miR-1 seed sequence in the 3′ UTR of the newt RARβ2 transcript, further potentiating a negative feedback loop between miR-133a, miR-1, and RARβ2 signaling during spinal cord regeneration.

Apart from tail amputation, salamanders and fish also can regenerate after spinal cord transection injuries (Monaghan et al. 2007, Reimer et al. 2009, 2013, Sabin et al. 2015, Zukor et al. 2011). This process requires the proliferation and migration of ependymoglial cells to reconnect the rostral and caudal ends of the injury and subsequent axon regeneration to restore sensory and motor functions (Butler and Ward 1965, 1967, Zukor et al. 2011, Goldshmit et al. 2012). Interestingly, miR-133b, which is downregulated after mammalian central nervous system (CNS) damage, was shown to be upregulated in certain brain regions after spinal cord transection in zebrafish (Yu et al. 2011). In situ hybridization showed that it was specifically upregulated in neurons of the medial longitudinal fasciculus (MLF) 7 days after spinal cord transection. This nucleus is known to project to the spinal cord in fish (Becker et al. 1997); therefore, increased miR-133b expression could regulate functional axon regeneration in this neuronal population. A known target of miR-133 in mammalian neurons is the small GTPase RhoA that inhibits axon regeneration after mammalian SCI (Dergham et al. 2002, Chiba et al. 2009, Holtje et al. 2009). Functional inhibition of miR-133b after SCI in zebrafish MLF increased RhoA expression and inhibited functional recovery and overall locomotor activity. Furthermore, in miR-133b morpholino-treated animals, biocytin applied caudal to the initial lesion was not retrogradely transported to the MLF, unlike in control fish, suggesting that axon regeneration was inhibited.

A study looking at differentially regulated miRNAs after SCI in the axolotl and rat has provided efficacy for comparative approaches between species (Diaz Quiroz et al. 2014). Using microarray-based transcriptional profiling, it was determined that after SCI in the axolotl, miR-125b was significantly downregulated in ependymoglial cells within 1 day of injury. After a complete spinal cord transection in rats, there was a similar decrease in miR-125b expression at 7 days post-injury; however, the overall levels of miR-125b in rat were six times lower than those in the axolotl. Overexpression of miR-125b in axolotl ependymoglial cells caused random and disorganized axonal sprouting, and functional inhibition of miR-125b led to inhibition of axon regeneration across the lesion. These results suggest that miR-125b could regulate axon guidance molecules during axolotl spinal cord regeneration. A bioinformatic approach identified Sema4D as a potential direct target of miR-125b. Sema4D is an axon-repulsive molecule that is upregulated after mammalian SCI (Moreau-Fauvarque et al. 2003). Subsequent cloning of the rat and axolotl Sema4D 3′ UTR identified a conserved seed sequence for miR-125b, and 3′ UTR luciferase assays confirmed that miR-125b did target that sequence. These data suggest that precise downregulation of miR-125b after SCI in the axolotl allowed for expression of Sema4D, which functioned to faithfully guide regenerating axons along the spinal cord.

The authors went on to show that experimentally increasing the expression of miR-125b *in vivo* in rat after complete spinal cord transection led to improved functional recovery. To increase miR-125b expression, they took a biomaterials approach and mixed the mature form of rat miR-125b into pluronic gel, a biodegradable gel that promotes the uptake of siRNA in the spinal cord *in vivo* (Cronin et al. 2006), and injected this mixture into the injury site. The animals were observed over a span of 8 weeks, and differences in locomotor activity were determined using the Besso, Beattie, and Bresnahan (BBB) test. Remarkably, miR-125b-treated animals had significantly higher BBB scores than controls, and subsequent microarray profiling confirmed that miR-125b treatment caused a decreased expression of glial scar-related genes, including *Gfap*, *Cspg4*, and *Col6A1*. These data suggest that modulation of miR-125b levels inhibits pathways involved in reactive gliosis and glial scar formation and overall leads to the formation of a more permissive regeneration environment in rat. Importantly, these data provide efficacy for comparative studies in lower vertebrates, aimed at elucidating molecular targets that promote regeneration for subsequent manipulation in mammalian regeneration.

8.3.3 Vertebrate cardiac regeneration

After myocardial infarction in mammals, there is extensive cell death of cardiomyocytes that fail to be replaced, leading to cardiac dysfunction and inhibited cardiac output. In addition, fibrotic scarring and extensive deposition of collagens exacerbate cardiac dysfunction. This is in stark contrast to what occurs in fish and salamanders. After cardiac injury in these organisms, cardiomyocytes upregulate expression of cell cycle–related genes and re-enter the cell cycle (Garcia-Gonzalez and Morrison 2014). While a fibrin clot and excessive extracellular matrix components are distributed at the injury site, these structures are remodeled so that approximately 60 days post-injury, scarless regeneration has occurred and cardiac function is restored.

As with limb regeneration, multiple signaling pathways involved with heart development have been identified that are essential to promote functional cardiac regeneration. This highlights a possible role for miRNA-mediated fine-tuning of these signaling pathways in exquisitely directing heart regeneration. Indeed, several studies have identified miRNAs that play important roles in promoting cardiac regeneration (Yin et al. 2012, Witman et al. 2013, Aguirre et al. 2014, Beauchemin et al. 2015). An exciting new study identified miR-101a as regulating multiple steps of zebrafish heart regeneration (Beauchemin et al. 2015). Interestingly, miR-101a expression exhibits a biphasic expression pattern: it is downregulated by 6 hours post-injury and significantly upregulated at 14 days post-injury, a time point at which extracellular matrix (ECM) remodeling occurs. If miR-101a expression is experimentally increased within the first 6 hours of injury, using a heat shock promoter to drive transgene expression, cardiomyocyte proliferation after injury is inhibited. Conversely, if miR-101a function is inhibited at 14 days post-injury, using complementary locked nucleic acid (LNA) oligonucleotides, there is a failure of ECM remodeling and persistence of scar deposition. Although many miR-101a target genes are upregulated within 6 hours post-injury, when miR-101a expression is lowest, only one target gene is also downregulated at 14 days post-injury, when miR-101a expression is highest. This gene was *fosab,* the fish homologue of the mammalian transcription factor c-Fos. While morpholino-mediated knockdown of *fosab* largely ablated the proliferative response of cardiomyocytes to injury, the co-knockdown of miR-101a and *fosab* at 14 days post-injury led to decreased scarring. This suggests that an early increase in *fosab* expression is necessary to stimulate cardiomyocyte proliferation after injury but that *fosab* expression must be decreased at later time points to promote scar tissue remodeling.

In addition to miR-101a, miR-133 also plays an important role in regulating cardiomyocyte proliferation after injury (Yin et al. 2012). After ventricular amputation, miR-133 expression is downregulated but returns to baseline levels by 30 days post-injury. If miR-133 expression is experimentally increased after injury, cardiomyocyte proliferation is blocked. Interestingly, if miR-133 expression is experimentally maintained at high levels past 30 days post-injury, then cardiomyocyte proliferation remains elevated and gaps are observed in the regenerated myocardial wall. This suggests that altered expression of cell-cell junction proteins could be affected by prolonged miR-133 expression during heart regeneration. Consistent with this hypothesis, microarray analysis identified several genes encoding gap junction and tight junction proteins that were downregulated after miR-133 overexpression. Subsequent *in vivo* gene reporter assays confirmed that miR-133 directly targets the 3′ UTR of connexin 43 and that inhibition of gap junction signaling during late stages of heart regeneration phenocopied miR-133 overexpression. Interestingly, connexin 43 dysregulation is associated with a number of human cardiomyopathies, suggesting a possible role of miR-133 in human heart disease. Although these studies provide valuable insight into complex regulatory networks guiding heart regeneration, the signals that regulate miR-101a and miR-133 expression after injury are not clear.

It remains unclear if changes in miRNA signaling during zebrafish heart regeneration are translatable to mammalian cardiac injury. However, in a recent report, researchers took a comparative approach, looking at changes in miRNA expression after cardiac injury in zebrafish to

193

determine if a similar program could induce mammalian cardiomyocyte dedifferentiation, proliferation, and heart regeneration (Aguirre et al. 2014). The authors identified miR-99/100 and let-7a/c as being significantly downregulated during regeneration, and if they are overexpressed, there is decreased cardiomyocyte proliferation and failure of heart regeneration. As mentioned previously, let-7 family members are highly upregulated in differentiated somatic cells and partially function to inhibit cell cycle re-entry. In addition, miR-99a/let-7c regulates mammalian cardiomyogenesis, thus highlighting a possible regulatory role for these family members during mammalian heart regeneration (Coppola et al. 2014). Farnesyl-transferase-beta (Fntβ) and SWI/SNF-related matrix-associated actin-dependent regulator of chromatin subfamily A member 5 (smarca5), were identified as direct targets of miR-99/100 during zebrafish heart regeneration. Primary adult murine cardiomyocytes express very high levels of miR-99/100 or let-7a/c and almost undetectable levels of Smarca5 and Fntβ compared with embryonic murine cardiomyocytes. Interestingly, when miR-99/100 or let-7a/c was inhibited in primary adult murine cardiomyocytes *in vitro*, there was an increase in expression of Smarca5 and Fntβ. More importantly, there was an increase in GATA4 expression, a marker of cardiomyocyte dedifferentiation (Kikuchi et al. 2010), and an increase in adult cardiomyocyte proliferation. Furthermore, inhibition of miR-99/100 or let-7a/c leads to increased dedifferentiation and proliferation of cardiomyocytes in murine cardiac slice organotypic cultures compared with controls. Finally, in an *in vivo* method of myocardial infarction, inhibition of miR-99/100 or let-7a/c by lentiviral or adeno-associated viral delivery of anti-miRs resulted in functional cardiac regeneration and decreased scarring after ligation of the left anterior descending artery. Consistent with the hypothesis that decreased miR-99/100 and let-7a/c expressions after injury promote cardiomyocyte dedifferentiation, proliferation, and heart regeneration by alleviating repression of Smarca5 and Fntβ, the forced overexpression of Smarca5 and Fntβ in cardiomyocytes *in vivo* also promoted functional heart regeneration after myocardial infarction (MI). These findings suggest that an evolutionarily conserved miRNA program is able to promote heart regeneration but is simply not activated after mammalian cardiac injury, unlike in fish. What remains to be discovered is the identity of the upstream signaling cascade that represses miR-99/100 and let-7a/c expression after injury in fish. Furthermore, it remains to be seen whether or not this signaling mechanism is similarly conserved in mammals and just not activated after injury or if a different pathway regulates miR-99/100 and/or let-7a/c expression in mammals.

In addition to zebrafish, adult salamanders exhibit the remarkable ability to regenerate their heart after resection or crush injury (Becker et al. 1974, Oberpriller and Oberpriller 1974, Cano-Martinez et al. 2010, Witman et al. 2011, Piatkowski et al. 2013). Similar to zebrafish heart regeneration, this process involves the dedifferentiation and proliferation of adult cardiomyocytes, and transient deposition of extracellular matrix components that are later remodeled to allow scar-free regeneration (Flink 2002, Cano-Martinez et al. 2010, Witman et al. 2011, Piatkowski et al. 2013). Based on a microarray platform of conserved vertebrate miRNAs, 37 miRNAs are differentially expressed at 7 dpi and 21 dpi compared with uninjured hearts (Witman et al. 2013). Interestingly, miR-128 was most significantly upregulated, up to 8-fold increase, at the 21 dpi time point, when cardiomyocyte proliferation was maximal (Witman et al. 2011, 2013), and was maintained between 30 dpi and 60 dpi when scar tissue remodeling occurred. This suggests that miR-128 could play a role in regulating cardiomyocyte proliferation and tissue remodeling during vertebrate heart regeneration. In situ hybridization revealed that miR-128 is expressed by cardiomyocytes and non-cardiomyocytes within the regeneration zone of 21 dpi hearts. Previous studies have implicated miR-128 as a potential tumor-suppressor gene (Qian et al. 2012). Consistent with this hypothesis, miR-128 inhibition led to significantly more BrdU$^+$ cells compared with controls. The BrdU$^+$ cells did not co-stain with the cardiomyocyte marker myosin heavy chain, suggesting that miR-128 antagonism enhances proliferation of non-cardiomyocyte cell population during newt heart regeneration. However, in situ hybridization confirmed that miR-128 is expressed in cardiomyocytes; therefore, miR-128 could be regulating other processes. Taking a bioinformatic approach, a highly conserved miR-128 seed sequence was identified in the 3′ UTR of the cardiac progenitor transcription factor Islet1 (Isl1). This conserved seed sequence is present in the targets newt *Isl1* transcript, and 3′ UTR luciferase assays confirmed that miR-128 targets newt Islet1. Furthermore, Western blot analysis showed that Isl1 expression is decreased at 21 dpi compared with uninjured heart, and inhibition of miR-128 function causes an increased level of Isl1 protein. Collectively, these data present a role for miR-128 in regulating non-cardiomyocyte proliferation after injury and a potential role in downregulating the cardiac progenitor cell marker Isl1, perhaps allowing for differentiation of cardiomyocytes and re-establishment of a functional myocardium.

Although studies on the role of microRNAs in vertebrates that have the ability to functionally regenerate are still in their infancy, these studies illustrate the importance of these conserved regulatory pathways and begin to identify key miRNAs that play pivotal roles in multiple types of regeneration in various species (Table 8.1). In addition, some interesting trends are arising when profiling data of miRNAs from different types of regeneration are compared, for example, in the heart, spinal cord, lens, and limb regeneration, let-7 is differentially regulated in the early stages of regeneration in all cases; however, the exact role that it plays in all is unclear; in lens, it is important in the dedifferentiation of terminally differentiated cells (Makarev et al. 2006, Tsonis et al. 2007, Yin et al. 2008b, Sehm et al. 2009). Let7 has been shown to play multiple important roles in development, including regulating stem cell differentiation, glucose metabolism, and tumorigenesis (Gaudet and Brisson 2015, Hirschi et al. 2015, Jeker and Marone 2015, Nouraee

Table 8.1 Key MicroRNAs that Play Pivotal Roles in Multiple Types of Regeneration in Various Species

MicroRNA	Type of Regeneration	miRNA Expression After Injury	Species	Known Target Gene	Reference Number
miR-133	Fin	Downregulated	Zebrafish	*Mps1*	Yin et al. 2012
miR-203	Fin and retina	Downregulated	Zebrafish	*Lef1*	Thatcher et al. 2008
miR-21	Limb	Downregulated	Axolotl	*Jagged1*	Holman et al. 2012
miR-196	Spinal cord	Upregulated	Axolotl	*Pax7*	Sehm et al. 2009
miR-133a	Spinal cord	Upregulated	Newt	*RARß2*	Lepp and Carlone 2014
miR-125b	Spinal cord	Downregulated	Axolotl	*Sema4D*	Diaz Quiroz et al. 2014
miR-133b	Spinal cord	Upregulated	Zebrafish	*RhoA*	Yu et al. 2011
miR-101a	Heart	Bimodal	Zebrafish	*fosab*	Beauchemin, Smith and Yin 2015
miR-133	Heart	Downregulated	Zebrafish	*Connexin 43*	Yin et al. 2012
miR-99/100	Heart	Downregulated	Zebrafish	*Smarca5 and Fnt-β*	Aguirre et al. 2014
miR-128	Heart	Upregulated	Newt	*Islet1*	Witman et al. 2013

and Mowla 2015, Orellana and Kasinski 2015, Peng et al. 2015, Ruiz and Russell 2015, Saito et al. 2015, Simonson and Das 2015, Sun et al. 2015, Tessitore et al. 2015, Kuninty et al. 2016, Batty et al. 2016). High levels of let-7 maintain cells in a differentiated state and hence their downregulation in the early phases of regeneration, suggesting that it may be critical for promoting formation of a progenitor pool to drive the regeneration.

8.4 MicroRNAs and regenerative engineering

Regenerative medicine aims to restore structure and function of damaged organs, to replace dying cells, and to regrow organs and limbs one day. The therapeutic modulation of miRNA expression within a tissue after injury is an interesting and a very attractive clinical strategy within regenerative medicine. MicroRNAs have the capability to simultaneously regulate the expression of dozens to hundreds of gene products by binding, with high specificity, to predictable target sites in the 3′ UTR of a transcript, therefore minimizing the potential for off-target effects. Therefore, development of miRNA-based therapeutic reagents could provide clinically desirable outcomes for regulating the complex reaction of tissues to traumatic injury.

One of the keys to effectively utilizing miRNAs in regenerative medicine may lie in developing successful delivery techniques. MicroRNAs have been successfully delivered to multiple different cell types by using various techniques, including viral-, lipid-, and polymer-based delivery systems (Gaudet and Brisson 2015, Hirschi et al. 2015, Jeker and Marone 2015, Nouraee and Mowla 2015, Orellana and Kasinski 2015, Peng et al. 2015, Ruiz and Russell 2015, Saito et al. 2015, Simonson and Das 2015, Sun et al. 2015, Tessitore et al. 2015, Batty et al. 2016, Kuninty et al. 2016).

In the earliest experiments, miRNAs were also delivered systemically via intravenous injections; this approach resulted in the non-specific delivery of miRNAs to the liver and to other organs, leading to multiple effects that are not always beneficial. More recently, with the advance of biomaterials, a more promising option is to deliver miRNA inhibitors or mimics *in vivo* directly to the cells or tissues that you are targeting. Alternatively, patient's autologous cells could be grown *ex vivo* to modulate the miRNA expression and then deliver the manipulated cells embedded in a biodegradable scaffold directly to the site of injury. To date, there have been several interesting uses of scaffold for modulating miRNAs in the mammalian context. For example, in an *in vitro* model of mammalian wound healing, miR-29 has been delivered via a collagen scaffold to treat scarring during wound healing. Delivery of this miR-29 was shown to alter the ratio of type 1 versus type III, an important factor that contributes to scarring, and additional effects were seen on the extracellular remodeling factors MMP8 and TIMP1 (Erler and Monaghan 2014). It will be interesting in the future to determine if this scaffold can also be used in an *in vivo* model of wound healing. In addition to collagen, hydrogels have also been proven a successful mode of delivery for miRNAs. A recent study from Li et al. has shown successful *in vivo* delivery of miR-26 in bone marrow-derived macrophages via hydrogel in mouse bone (Li et al. 2013). The cells were first transfected with the miR-26 inhibitors or mimics and then seeded into the biodegradable hydrogel. This method resulted in an 8-fold increase in miR-26 24 hours post-application and increased to 70 folds by 11 days post-application. Ultimately, this resulted in enhancing vascularization and osteogenesis, which could be a valuable therapy for osteoporosis (Li et al. 2013). However, in moving toward therapies, it may be beneficial to be able to modulate miRNA expression in the endogenous cells at the injury site. Another publication this year has demonstrated enhanced regeneration of critical-sized bone defects via in vivo modulation of miR-26a in mouse (Zhang et al. 2016). This paper used a hyperbranched polymer in which short polyethylene glycol chains and a low-molecular-weight cationic polyethlenimine are attached to the outer shell of a hyperbranched hydrophobic molecular core. When the miRNA is added, it catalyzes further self-assembly into nano-sized polyplexes to directly and efficiently deliver the miRNA. This allows direct, sustained delivery of the miRNA to the endogenous cells, and in bone, this resulted in activation of osteogenic activity in endogenous stem cells (Zhang et al. 2016). There are a limited number of examples of the wide variety of different biomaterials that are currently being tested as a means of delivering miRNA mimics, inhibitors, or cells to different tissue or organs *in vivo* to promote a better response to injury.

To date, little progress has been made in the area of directly translating findings from regeneration-permissive organisms to non-regenerative mammals. One example of this approach has been the work on miRNAs in the axolotl. Diaz et al. identified miR-125b to be an important regulator

197

of functional spinal cord regeneration in the axolotl (Diaz Quiroz et al. 2014). They then modulated the levels of miR-125b after complete transection of the spinal cord in rats and delivered an miR-125b embedded in pluronic gel, a biodegradable gel that is endocytosed by the cells. This approach resulted in a significant increase in the miRNA levels weeks after injury and improved the locomotive recovery of the animals; however, the effect plateaued after 4 weeks (Diaz Quiroz et al. 2014). This is a very promising result, illustrating the potential for translating molecular finding cross-species to enhance functional recovery in mammals. However, optimization of delivery techniques to different tissues and types of injuries maybe the key to developing successful miRNA-based therapeutics in the future, and it offers a potential opening for connecting the diverse field of regeneration biology and bioengineering in the future.

Acknowledgments

Keith Sabin was funded by the National Institutes of Health (NIH) training grant T32 GM113846.

References

Aguirre, A., N. Montserrat, S. Zacchigna, E. Nivet, T. Hishida, M. N. Krause, and L. Kurian et al. 2014. In vivo activation of a conserved microRNA program induces mammalian heart regeneration. *Cell Stem Cell* 15 (5):589–604. doi:10.1016/j.stem.2014.10.003.

Bartel, D. P. 2009. MicroRNAs: Target recognition and regulatory functions. *Cell* 136 (2):215–233. doi:10.1016/j.cell.2009.01.002.

Batty, J. A., J. A. Lima, Jr., and V. Kunadian. 2016. Direct cellular reprogramming for cardiac repair and regeneration. *Eur J Heart Fail* 18 (2):145–156. doi:10.1002/ejhf.446.

Beauchemin, M., A. Smith, and V. P. Yin. 2015. Dynamic microRNA-101a and Fosab expression controls zebrafish heart regeneration. *Development* 142 (23):4026–4037. doi:10.1242/dev.126649.

Becker, C. G., and T. Becker. 2015. Neuronal regeneration from ependymo-radial glial cells: Cook, little pot, cook! *Dev Cell* 32 (4):516–527. doi:10.1016/j.devcel.2015.01.001.

Becker, R. O., S. Chapin, and R. Sherry. 1974. Regeneration of the ventricular myocardium in amphibians. *Nature* 248 (5444):145–147.

Becker, T., M. F. Wullimann, C. G. Becker, R. R. Bernhardt, and M. Schachner. 1997. Axonal regrowth after spinal cord transection in adult zebrafish. *J Comp Neurol* 377 (4):577–595.

Bernstein, E., A. A. Caudy, S. M. Hammond, and G. J. Hannon. 2001. Role for a bidentate ribonuclease in the initiation step of RNA interference. *Nature* 409 (6818):363–366. doi:10.1038/35053110.

Bernstein, E., S. Y. Kim, M. A. Carmell, E. P. Murchison, H. Alcorn, M. Z. Li, A. A. Mills, S. J. Elledge, K. V. Anderson, and G. J. Hannon. 2003. Dicer is essential for mouse development. *Nat Genet* 35 (3):215–217. doi:10.1038/ng1253.

Briona, L. K., F. E. Poulain, C. Mosimann, and R. I. Dorsky. 2015. Wnt/ss-catenin signaling is required for radial glial neurogenesis following spinal cord injury. *Dev Biol* 403 (1):15–21. doi:10.1016/j.ydbio.2015.03.025.

Brockes, J. P. 1994. New approaches to amphibian limb regeneration. *Trends Genet* 10 (5):169–173.

Brockes, J. P. 1996. Retinoid signalling and retinoid receptors in amphibian limb regeneration. *Biochem Soc Symp* 62:137–142.

Brockes, J. P. 1997. Amphibian limb regeneration: Rebuilding a complex structure. *Science* 276 (5309):81–87.

Bryant, S. V. 1970. Regeneration in amphibians and reptiles. *Endeavour* 29 (106):12–17.

Bryant, S. V., T. Endo, and D. M. Gardiner. 2002. Vertebrate limb regeneration and the origin of limb stem cells. *Int J Dev Biol* 46 (7):887–896.

Butler, E. G., and M. B. Ward. 1965. Reconstitution of the spinal cord following ablation in urodele larvae. *J Exp Zool* 160 (1):47–65.

Butler, E. G., and M. B. Ward. 1967. Reconstitution of the spinal cord after ablation in adult Triturus. *Dev Biol* 15 (5):464–486.

Campbell, L. J., and C. M. Crews. 2008. Wound epidermis formation and function in urodele amphibian limb regeneration. *Cell Mol Life Sci* 65 (1):73–79. doi:10.1007/s00018-007-7433-z.

Campbell, L. J., E. C. Suarez-Castillo, H. Ortiz-Zuazaga, D. Knapp, E. M. Tanaka, and C. M. Crews. 2011. Gene expression profile of the regeneration epithelium during axolotl limb regeneration. *Dev Dyn* 240 (7):1826–1840. doi:10.1002/dvdy.22669.

Cano-Martinez, A., A. Vargas-Gonzalez, V. Guarner-Lans, E. Prado-Zayago, M. Leon-Oleda, and B. Nieto-Lima. 2010. Functional and structural regeneration in the axolotl heart (Ambystoma mexicanum) after partial ventricular amputation. *Arch Cardiol Mex* 80 (2):79–86.

Carlson, M. R., S. V. Bryant, and D. M. Gardiner. 1998. Expression of Msx-2 during development, regeneration, and wound healing in axolotl limbs. *J Exp Zool* 282 (6):715–723.

Carter, C., A. Clark, G. Spencer, and R. Carlone. 2011. Cloning and expression of a retinoic acid receptor beta2 subtype from the adult newt: Evidence for an early role in tail and caudal spinal cord regeneration. *Dev Dyn* 240 (12):2613–2625. doi:10.1002/dvdy.22769.

Chendrimada, T. P., R. I. Gregory, E. Kumaraswamy, J. Norman, N. Cooch, K. Nishikura, and R. Shiekhattar. 2005. TRBP recruits the Dicer complex to Ago2 for microRNA processing and gene silencing. *Nature* 436 (7051):740–744. doi:10.1038/nature03868.

Chernoff, E. A., D. L. Stocum, H. L. Nye, and J. A. Cameron. 2003. Urodele spinal cord regeneration and related processes. *Dev Dyn* 226 (2):295–307. doi:10.1002/dvdy.10240.

Chiba, Y., M. Tanabe, K. Goto, H. Sakai, and M. Misawa. 2009. Down-regulation of miR-133a contributes to up-regulation of Rhoa in bronchial smooth muscle cells. *Am J Respir Crit Care Med* 180 (8):713–719. doi:10.1164/rccm.200903-0325OC.

Coppola, A., A. Romito, C. Borel, C. Gehrig, M. Gagnebin, E. Falconnet, and A. Izzo et al. 2014. Cardiomyogenesis is controlled by the miR-99a/let-7c cluster and epigenetic modifications. *Stem Cell Res* 12 (2):323–337. doi:10.1016/j.scr.2013.11.008.

Cronin, M., P. N. Anderson, C. R. Green, and D. L. Becker. 2006. Antisense delivery and protein knockdown within the intact central nervous system. *Front Biosci* 11:2967–2975.

Crosnier, C., T. Attie-Bitach, F. Encha-Razavi, S. Audollent, F. Soudy, M. Hadchouel, M. Meunier-Rotival, and M. Vekemans. 2000. JAGGED1 gene expression during human embryogenesis elucidates the wide phenotypic spectrum of Alagille syndrome. *Hepatology* 32 (3):574–581. doi:10.1053/jhep.2000.16600.

Denli, A. M., B. B. Tops, R. H. Plasterk, R. F. Ketting, and G. J. Hannon. 2004. Processing of primary microRNAs by the Microprocessor complex. *Nature* 432 (7014):231–235. doi:10.1038/nature03049.

199

Dergham, P., B. Ellezam, C. Essagian, H. Avedissian, W. D. Lubell, and L. McKerracher. 2002. Rho signaling pathway targeted to promote spinal cord repair. *J Neurosci* 22 (15):6570–6577. doi:20026637.

Diaz Quiroz, J. F., and K. Echeverri. 2013. Spinal cord regeneration: Where fish, frogs and salamanders lead the way, can we follow? *Biochem J* 451 (3):353–364. doi:10.1042/bj20121807.

Diaz Quiroz, J. F., E. Tsai, M. Coyle, T. Sehm, and K. Echeverri. 2014. Precise control of miR-125b levels is required to create a regeneration-permissive environment after spinal cord injury: A cross-species comparison between salamander and rat. *Dis Model Mech* 7 (6):601–611. doi:10.1242/dmm.014837.

Duester, G. 2008. Retinoic acid synthesis and signaling during early organogenesis. *Cell* 134 (6):921–931. doi:10.1016/j.cell.2008.09.002.

Echeverri, K., and E. M. Tanaka. 2002. Ectoderm to mesoderm lineage switching during axolotl tail regeneration. *Science* 298 (5600):1993–1996.

Erler, P., and J. R. Monaghan. 2014. The link between injury-induced stress and regenerative phenomena: A cellular and genetic synopsis. *Biochim Biophys Acta*. doi:10.1016/j.bbagrm.2014.07.021.

Fei, J. F., M. Schuez, A. Tazaki, Y. Taniguchi, K. Roensch, and E. M. Tanaka. 2014. CRISPR-mediated genomic deletion of Sox2 in the axolotl shows a requirement in spinal cord neural stem cell amplification during tail regeneration. *Stem Cell Reports* 3 (3):444–459. doi:10.1016/j.stemcr.2014.06.018.

Filipowicz, W., S. N. Bhattacharyya, and N. Sonenberg. 2008. Mechanisms of post-transcriptional regulation by microRNAs: Are the answers in sight? *Nat Rev Genet* 9 (2):102–114. doi:10.1038/nrg2290.

Fisk, H. A., and M. Winey. 2004. Spindle regulation: Mps1 flies into new areas. *Curr Biol* 14 (24):R1058–R1060. doi:10.1016/j.cub.2004.11.047.

Flink, I. L. 2002. Cell cycle reentry of ventricular and atrial cardiomyocytes and cells within the epicardium following amputation of the ventricular apex in the axolotl, Amblystoma mexicanum: Confocal microscopic immunofluorescent image analysis of bromodeoxyuridine-labeled nuclei. *Anat Embryol* 205 (3):235–244. doi:10.1007/s00429-002-0249-6.

Garcia-Gonzalez, C., and J. I. Morrison. 2014. Cardiac regeneration in non-mammalian vertebrates. *Exp Cell Res* 321 (1):58–63. doi:10.1016/j.yexcr.2013.08.001.

Gardiner, D. M., B. Blumberg, and S. V. Bryant. 1993. Expression of homeobox genes in limb regeneration. *Prog Clin Biol Res* 383A:31–40.

Gardiner, D. M., B. Blumberg, Y. Komine, and S. V. Bryant. 1995. Regulation of HoxA expression in developing and regenerating axolotl limbs. *Development* 121 (6):1731–1741.

Gardiner, D. M., and S. V. Bryant. 1996. Molecular mechanisms in the control of limb regeneration: The role of homeobox genes. *Int J Dev Biol* 40 (4):797–805.

Gaudet, D., and D. Brisson. 2015. Gene-based therapies in lipidology: Current status and future challenges. *Curr Opin Lipidol* 26 (6):553–565. doi:10.1097/mol.0000000000000240.

Gearhart, M. D., J. R. Erickson, A. Walsh, and K. Echeverri. 2015. Identification of conserved and novel microRNAs during tail regeneration in the Mexican Axolotl. *Int J Mol Sci* 16 (9):22046–22061. doi:10.3390/ijms160922046.

Gemberling, M., T. J. Bailey, D. R. Hyde, and K. D. Poss. 2013. The zebrafish as a model for complex tissue regeneration. *Trends Genet* 29 (11):611–620. doi:10.1016/j.tig.2013.07.003.

Goldshmit, Y., T. E. Sztal, P. R. Jusuf, T. E. Hall, M. Nguyen-Chi, and P. D. Currie. 2012. Fgf-dependent glial cell bridges facilitate spinal cord regeneration in zebrafish. *J Neurosci* 32 (22):7477–7492. doi:10.1523/JNEUROSCI.0758-12.2012.

Gregory, R. I., K. P. Yan, G. Amuthan, T. Chendrimada, B. Doratotaj, N. Cooch, and R. Shiekhattar. 2004. The Microprocessor complex mediates the genesis of microRNAs. *Nature* 432 (7014):235–240. doi:10.1038/nature03120.

Hatfield, S. D., H. R. Shcherbata, K. A. Fischer, K. Nakahara, R. W. Carthew, and H. Ruohola-Baker. 2005. Stem cell division is regulated by the microRNA pathway. *Nature* 435 (7044):974–978. doi:10.1038/nature03816.

Hirschi, K. D., G. J. Pruss, and V. Vance. 2015. Dietary delivery: A new avenue for microRNA therapeutics? *Trends Biotechnol* 33 (8):431–432. doi:10.1016/j.tibtech.2015.06.003.

Holman, E. C., L. J. Campbell, J. Hines, and C. M. Crews. 2012. Microarray analysis of microRNA expression during axolotl limb regeneration. *PLoS One* 7 (9):e41804. doi:10.1371/journal.pone.0041804.

Holtje, M., S. Djalali, F. Hofmann, A. Munster-Wandowski, S. Hendrix, F. Boato, and S. C. Dreger et al. 2009. A 29-amino acid fragment of Clostridium botulinum C3 protein enhances neuronal outgrowth, connectivity, and reinnervation. *Faseb J* 23 (4):1115–1126. doi:10.1096/fj.08-116855.

Hornstein, E., J. H. Mansfield, S. Yekta, J. K. Hu, B. D. Harfe, M. T. McManus, S. Baskerville, D. P. Bartel, and C. J. Tabin. 2005. The microRNA miR-196 acts upstream of Hoxb8 and Shh in limb development. *Nature* 438 (7068):671–674. doi:10.1038/nature04138.

Jeker, L. T., and R. Marone. 2015. Targeting microRNAs for immunomodulation. *Curr Opin Pharmacol* 23:25–31. doi:10.1016/j.coph.2015.05.004.

Kanellopoulou, C., S. A. Muljo, A. L. Kung, S. Ganesan, R. Drapkin, T. Jenuwein, D. M. Livingston, and K. Rajewsky. 2005. Dicer-deficient mouse embryonic stem cells are defective in differentiation and centromeric silencing. *Genes Dev* 19 (4):489–501. doi:10.1101/gad.1248505.

Kawakami, Y., C. Rodriguez Esteban, M. Raya, H. Kawakami, M. Marti, I. Dubova, and J. C. Izpisua Belmonte. 2006. Wnt/beta-catenin signaling regulates vertebrate limb regeneration. *Genes Dev* 20 (23):3232–3237. doi:10.1101/gad.1475106.

Kikuchi, K., J. E. Holdway, A. A. Werdich, R. M. Anderson, Y. Fang, G. F. Egnaczyk, T. Evans, C. A. Macrae, D. Y. Stainier, and K. D. Poss. 2010. Primary contribution to zebrafish heart regeneration by gata4(+) cardiomyocytes. *Nature* 464 (7288):601–605. doi:10.1038/nature08804.

Kloosterman, W. P., and R. H. Plasterk. 2006. The diverse functions of microRNAs in animal development and disease. *Dev Cell* 11 (4):441–450. doi:10.1016/j.devcel.2006.09.009.

Kuninty, P. R., J. Schnittert, G. Storm, and J. Prakash. 2016. MicroRNA targeting to modulate tumor microenvironment. *Front Oncol* 6:3. doi:10.3389/fonc.2016.00003.

Lee, R. C., R. L. Feinbaum, and V. Ambros. 1993. The C. elegans heterochronic gene lin-4 encodes small RNAs with antisense complementarity to lin-14. *Cell* 75 (5):843–854.

Lee, Y., M. Kim, J. Han, K. H. Yeom, S. Lee, S. H. Baek, and V. N. Kim. 2004. MicroRNA genes are transcribed by RNA polymerase II. *Embo J* 23 (20):4051–4060. doi:10.1038/sj.emboj.7600385.

Lepp, A. C., and R. L. Carlone. 2014. RARbeta2 expression is induced by the downregulation of microRNA 133a during caudal spinal cord regeneration in the adult newt. *Dev Dyn* 243 (12):1581–1590. doi:10.1002/dvdy.24210.

Lepp, A. C., and R. L. Carlone. 2015. MicroRNA dysregulation in response to RARbeta2 inhibition reveals a negative feedback loop between microRNAs 1, 133a, and RARbeta2 during tail and spinal cord regeneration in the adult newt. *Dev Dyn* 244 (12):1519–1537. doi:10.1002/dvdy.24342.

Li, Y., L. Fan, S. Liu, W. Liu, H. Zhang, T. Zhou, and D. Wu et al. 2013. The promotion of bone regeneration through positive regulation of angiogenic-osteogenic coupling using microRNA-26a. *Biomaterials* 34 (21):5048–5058. doi:10.1016/j.biomaterials.2013.03.052.

Liu, X., and M. Winey. 2012. The MPS1 family of protein kinases. *Annu Rev Biochem* 81:561–585. doi:10.1146/annurev-biochem-061611-090435.

Lund, E., S. Guttinger, A. Calado, J. E. Dahlberg, and U. Kutay. 2004. Nuclear export of microRNA precursors. *Science* 303 (5654):95-98. doi:10.1126/science.1090599.

Maden, M. 2007. Retinoic acid in the development, regeneration and maintenance of the nervous system. *Nat Rev Neurosci* 8 (10):755–765. doi:10.1038/nrn2212.

Makarev, E., J. R. Spence, K. DelRio-Tsonis, and P. A. Tsonis. 2006. Identification of microRNAs and other small RNAs from the adult newt eye. *Mol Vis* 12:1386–1391.

McGlinn, E., K. L. van Bueren, S. Fiorenza, R. Mo, A. M. Poh, A. Forrest, and M. B. Soares et al. 2005. Pax9 and Jagged1 act downstream of Gli3 in vertebrate limb development. *Mech Dev* 122 (11):1218–1233. doi:10.1016/j.mod.2005.06.012.

McHedlishvili, L., H. H. Epperlein, A. Telzerow, and E. M. Tanaka. 2007. A clonal analysis of neural progenitors during axolotl spinal cord regeneration reveals evidence for both spatially restricted and multipotent progenitors. *Development* 134 (11):2083–2093. doi:10.1242/dev.02852.

Melton, C., R. L. Judson, and R. Blelloch. 2010. Opposing microRNA families regulate self-renewal in mouse embryonic stem cells. *Nature* 463 (7281):621–626. doi:10.1038/nature08725.

Monaghan, J. R., and M. Maden. 2012. Visualization of retinoic acid signaling in transgenic axolotls during limb development and regeneration. *Dev Biol* 368 (1):63–75. doi:10.1016/j.ydbio.2012.05.015.

Monaghan, J. R., J. A. Walker, R. B. Page, S. Putta, C. K. Beachy, and S. R. Voss. 2007. Early gene expression during natural spinal cord regeneration in the salamander Ambystoma mexicanum. *J Neurochem* 101 (1):27–40. doi:10.1111/j.1471-4159.2006.04344.x.

Moreau-Fauvarque, C., A. Kumanogoh, E. Camand, C. Jaillard, G. Barbin, I. Boquet, and C. Love et al. 2003. The transmembrane semaphorin Sema4D/CD100, an inhibitor of axonal growth, is expressed on oligodendrocytes and upregulated after CNS lesion. *J Neurosci* 23 (27):9229–9239.

Mullen, L. M., S. V. Bryant, M. A. Torok, B. Blumberg, and D. M. Gardiner. 1996. Nerve dependency of regeneration: The role of Distal-less and FGF signaling in amphibian limb regeneration. *Development* 122 (11):3487–3497.

Nouraee, N., and S. J. Mowla. 2015. miRNA therapeutics in cardiovascular diseases: Promises and problems. *Front Genet* 6:232. doi:10.3389/fgene.2015.00232.

Oberpriller, J. O., and J. C. Oberpriller. 1974. Response of the adult newt ventricle to injury. *J Exp Zool* 187 (2):249–253. doi:10.1002/jez.1401870208.

Orellana, E. A., and A. L. Kasinski. 2015. MicroRNAs in cancer: A historical perspective on the path from discovery to therapy. *Cancers (Basel)* 7 (3):1388–1405. doi:10.3390/cancers7030842.

Park, J. E., I. Heo, Y. Tian, D. K. Simanshu, H. Chang, D. Jee, D. J. Patel, and V. N. Kim. 2011. Dicer recognizes the 5′ end of RNA for efficient and accurate processing. *Nature* 475 (7355):201–205. doi:10.1038/nature10198.

Peng, B., Y. Chen, and K. W. Leong. 2015. MicroRNA delivery for regenerative medicine. *Adv Drug Deliv Rev* 88:108–122. doi:10.1016/j.addr.2015.05.014.

Piatkowski, T., C. Muhlfeld, T. Borchardt, and T. Braun. 2013. Reconstitution of the myocardium in regenerating newt hearts is preceded by transient deposition of extracellular matrix components. *Stem Cells Dev* 22 (13):1921–1931. doi:10.1089/scd.2012.0575.

Poss, K. D., J. Shen, and M. T. Keating. 2000. Induction of lef1 during zebrafish fin regeneration. *Dev Dyn* 219 (2):282–286. doi:10.1002/1097-0177(2000)9999:9999<::AID-DVDY1045>3.0.CO;2-C.

Qian, P., A. Banerjee, Z. S. Wu, X. Zhang, H. Wang, V. Pandey, and W. J. Zhang et al. 2012. Loss of SNAIL regulated miR-128-2on chromosome 3p22.3 targets multiple stem cell factors to promote transformation of mammary epithelial cells. *Cancer Res* 72 (22):6036–6050. doi:10.1158/0008-5472.CAN-12-1507.

Raballo, R., J. Rhee, R. Lyn-Cook, J. F. Leckman, M. L. Schwartz, and F. M. Vaccarino. 2000. Basic fibroblast growth factor (FGF2) is necessary for cell prolferation and neurogenesis in the developing cerebal cortex. *J. Neuroscience* 20:5012–5023.

Ramachandran, R., B. V. Fausett, and D. Goldman. 2010. Ascl1a regulates Muller glia dedifferentiation and retinal regeneration through a Lin-28-dependent, let-7 microRNA signalling pathway. *Nat Cell Biol* 12 (11):1101–1107. doi:10.1038/ncb2115.

Reddien, P. W. 2013. Specialized progenitors and regeneration. *Development* 140 (5):951–957. doi:10.1242/dev.080499.

Reimer, M. M., V. Kuscha, C. Wyatt, I. Sorensen, R. E. Frank, M. Knuwer, T. Becker, and C. G. Becker. 2009. Sonic hedgehog is a polarized signal for motor neuron regeneration in adult zebrafish. *J Neurosci* 29 (48):15073–15082. doi:10.1523/JNEUROSCI.4748-09.2009.

Reimer, M. M., A. Norris, J. Ohnmacht, R. Patani, Z. Zhong, T. B. Dias, and V. Kuscha et al. 2013. Dopamine from the brain promotes spinal motor neuron generation during development and adult regeneration. *Dev Cell* 25 (5):478–491. doi:10.1016/j.devcel.2013.04.012.

Roy, S., D. M. Gardiner, and S. V. Bryant. 2000. Vaccinia as a tool for functional analysis in regenerating limbs: Ectopic expression of Shh. *Dev Biol* 218 (2):199–205.

Ruiz, A. J., and S. J. Russell. 2015. MicroRNAs and oncolytic viruses. *Curr Opin Virol* 13:40–48. doi:10.1016/j.coviro.2015.03.007.

Sabin, K., T. Santos-Ferreira, J. Essig, S. Rudasill, and K. Echeverri. 2015. Dynamic membrane depolarization is an early regulator of ependymoglial cell response to spinal cord injury in axolotl. *Dev Biol* 408 (1):14–25. doi:10.1016/j.ydbio.2015.10.012.

Saito, Y., T. Nakaoka, and H. Saito. 2015. microRNA-34a as a Therapeutic Agent against Human Cancer. *J Clin Med* 4 (11):1951–1959. doi:10.3390/jcm4111951.

Sanchez Alvarado, A., and P. A. Tsonis. 2006a. Bridging the regeneration gap: Genetic insights from diverse animal models. *Nat Rev Genet* 7 (11):873–884. doi:10.1038/nrg1923.

Sanchez Alvarado, A., and P. A. Tsonis. 2006b. Bridging the regeneration gap: Genetic insights from diverse animal models. *Nat Rev Genet.* 11:873–884.

Schnapp, E., M. Kragl, L. Rubin, and E. M. Tanaka. 2005. Hedgehog signaling controls dorsoventral patterning, blastema cell proliferation and cartilage induction during axolotl tail regeneration. *Development* 132 (14):3243–3253. doi:10.1242/dev.01906.

Sehm, T., C. Sachse, C. Frenzel, and K. Echeverri. 2009. miR-196 is an essential early-stage regulator of tail regeneration, upstream of key spinal cord patterning events. *Dev Biol* 334 (2):468–480. doi:10.1016/j.ydbio.2009.08.008.

Shukla, G. C., J. Singh, and S. Barik. 2011. MicroRNAs: Processing, maturation, target recognition and regulatory functions. *Mol Cell Pharmacol* 3 (3):83–92.

Simon, A., and E. M. Tanaka. 2013. Limb regeneration. *Wiley Interdiscip Rev Dev Biol* 2 (2):291–300. doi:10.1002/wdev.73.

Simonson, B., and S. Das. 2015. MicroRNA therapeutics: The next magic bullet? Mini Rev Med Chem 15 (6):467–474.

Singh, B. N., N. Koyano-Nakagawa, A. Donaldson, C. V. Weaver, M. G. Garry, and D. J. Garry. 2015. Hedgehog signaling during appendage development and regeneration. *Genes (Basel)* 6 (2):417–435. doi:10.3390/genes6020417.

Stoick-Cooper, C. L., G. Weidinger, K. J. Riehle, C. Hubbert, M. B. Major, N. Fausto, and R. T. Moon. 2007. Distinct Wnt signaling pathways have opposing roles in appendage regeneration. *Development* 134 (3):479–489. doi:10.1242/dev.001123.

Sun, R., J. K. Shen, E. Choy, Z. Yu, F. J. Hornicek, and Z. Duan. 2015. The emerging roles and therapeutic potential of microRNAs (miRs) in liposarcoma. *Discov Med* 20 (111):311–324.

Tanaka, E. M., and P. Ferretti. 2009. Considering the evolution of regeneration in the central nervous system. *Nat Rev Neurosci* 10 (10):713–723. doi:10.1038/nrn2707.

Tanaka, E. M., and P. W. Reddien. 2011. The cellular basis for animal regeneration. *Dev Cell* 21 (1):172–185. doi:10.1016/j.devcel.2011.06.016.

Tessitore, A., G. Cicciarelli, V. Mastroiaco, F. D. Vecchio, D. Capece, D. Verzella, M. Fischietti, D. Vecchiotti, F. Zazzeroni, and E. Alesse. 2015. Therapeutic use of microRNAs in cancer. *Anticancer Agents Med Chem* 16 (1):7–19.

Thatcher, E. J., I. Paydar, K. K. Anderson, and J. G. Patton. 2008. Regulation of zebrafish fin regeneration by microRNAs. *Proc Natl Acad Sci U S A* 105 (47):18384–18389. doi:10.1073/pnas.0803713105.

Tornini, V. A., and K. D. Poss. 2014. Keeping at arm's length during regeneration. *Dev Cell* 29 (2):139–145. doi:10.1016/j.devcel.2014.04.007.

Torok, M. A., D. M. Gardiner, J. C. Izpisua-Belmonte, and S. V. Bryant. 1999. Sonic hedgehog (shh) expression in developing and regenerating axolotl limbs. *J Exp Zool* 284 (2):197–206.

Tsonis, P. A., M. K. Call, M. W. Grogg, M. A. Sartor, R. R. Taylor, A. Forge, R. Fyffe, R. Goldenberg, R. Cowper-Sal-lari, and C. R. Tomlinson. 2007. MicroRNAs and regeneration: Let-7 members as potential regulators of dedifferentiation in lens and inner ear hair cell regeneration of the adult newt. *Biochem Biophys Res Commun* 362 (4):940–945. doi:10.1016/j.bbrc.2007.08.077.

Whitehead, G. G., S. Makino, C. L. Lien, and M. T. Keating. 2005. fgf20 is essential for initiating zebrafish fin regeneration. *Science* 310 (5756):1957–1960. doi:10.1126/science.1117637.

Wienholds, E., M. J. Koudijs, F. J. van Eeden, E. Cuppen, and R. H. Plasterk. 2003. The microRNA-producing enzyme Dicer1 is essential for zebrafish development. *Nat Genet* 35 (3):217–218. doi:10.1038/ng1251.

Witman, N., J. Heigwer, B. Thaler, W. O. Lui, and J. I. Morrison. 2013. miR-128 regulates non-myocyte hyperplasia, deposition of extracellular matrix and Islet1 expression during newt cardiac regeneration. *Dev Biol* 383 (2):253–263. doi:10.1016/j.ydbio.2013.09.011.

Witman, N., B. Murtuza, B. Davis, A. Arner, and J. I. Morrison. 2011. Recapitulation of developmental cardiogenesis governs the morphological and functional regeneration of adult newt hearts following injury. *Dev Biol* 354 (1):67–76. doi:10.1016/j.ydbio.2011.03.021.

Yi, R., Y. Qin, I. G. Macara, and B. R. Cullen. 2003. Exportin-5 mediates the nuclear export of pre-microRNAs and short hairpin RNAs. *Genes Dev* 17 (24):3011–3016. doi:10.1101/gad.1158803.

Yin, V. P., A. Lepilina, A. Smith, and K. D. Poss. 2012. Regulation of zebrafish heart regeneration by miR-133. *Dev Biol* 365 (2):319–327. doi:10.1016/j.ydbio.2012.02.018.

Yin, V. P., and K. D. Poss. 2008. New regulators of vertebrate appendage regeneration. *Curr Opin Genet Dev* 18 (4):381–386. doi:10.1016/j.gde.2008.06.008.

Yin, V. P., J. M. Thomson, R. Thummel, D. R. Hyde, S. M. Hammond, and K. D. Poss. 2008a. Fgf-dependent depletion of microRNA-133 promotes appendage regeneration in zebrafish. *Genes Dev* 22 (6):728–733. doi:10.1101/gad.1641808.

Yin, V. P., J. M. Thomson, R. Thummel, D. R. Hyde, S. M. Hammond, and K. D. Poss. 2008b. Fgf-dependent depletion of microRNA-133 promotes appendage regeneration in zebrafish. *Genes Dev* 22:728–733.

Yu, Y. M., K. M. Gibbs, J. Davila, N. Campbell, S. Sung, T. I. Todorova, S. Otsuka, H. E. Sabaawy, R. P. Hart, and M. Schachner. 2011. MicroRNA miR-133b is essential for functional recovery after spinal cord injury in adult zebrafish. *Eur J Neurosci* 33 (9):1587–1597. doi:10.1111/j.1460-9568.2011.07643.x.

Zhang, F., J. D. Clarke, and P. Ferretti. 2000. FGF-2 Up-regulation and proliferation of neural progenitors in the regenerating amphibian spinal cord in vivo. *Dev Biol* 225 (2):381–391. doi:10.1006/dbio.2000.9843.

Zhang, F., J. D. Clarke, L. Santos-Ruiz, and P. Ferretti. 2002. Differential regulation of fibroblast growth factor receptors in the regenerating amphibian spinal cord in vivo. *Neuroscience* 114 (4):837–848.

Zhang, X., Y. Li, Y. E. Chen, J. Chen, and P. X. Ma. 2016. Cell-free 3D scaffold with two-stage delivery of miRNA-26a to regenerate critical-sized bone defects. *Nat Commun* 7:10376. doi:10.1038/ncomms10376.

Zukor, K. A., D. T. Kent, and S. J. Odelberg. 2011. Meningeal cells and glia establish a permissive environment for axon regeneration after spinal cord injury in newts. *Neural Dev* 6:1. doi:10.1186/1749-8104-6-1.

Chapter 9 Recovering what was lost: Can morphogens scale to enable regeneration?

Caroline Beck

Contents

> **Key concepts**
>
> a. Morphogens are a set of diffusible molecules that can direct the development of cells along different fates by establishing gradients across fields of cells.
> b. During regeneration of a complex appendage such as a limb, re-creation of these patterning mechanisms would require scaling up of the system to cover the much larger distances involved.
> c. We are beginning to understand that regulation of these morphogens and their gradients is complex in real life, indicating that there are many possibilities for strategic engineering in order to improve regenerative outcomes for adult organisms.

9.1 Introduction

Since 2001, we humans have had access to our complete genomic DNA sequence (Lander et al. 2001). The human genome is often referred to as the instructions, or blueprints, for building a human; however, this gives the impression that one might, faced with the problem of regenerating a missing body part, reach in, pull out the relevant chapter, and follow the recipe therein. Our genome is not an instruction manual for the human form, but it simply allows each cell access to the information required to make (differentiate into) one of around 200 different cell types that make up the tissues, organs, and systems of our bodies, to enable them to perform their different functions, to replicate themselves when required, and to sacrifice themselves if it is appropriate to do so. Each cell can also influence and respond to its environment, which makes development flexible, and the cells are more capable to cope with minor errors that crop up as typographical errors, or mutations, in the genome sequence.

Genes code for proteins; our genome codes for around 20,500 such proteins (Clamp et al. 2007). Variations in our genetic sequence alter the properties of these proteins, so that we all appear a little different, but our basic form is constructed just once in our lifetime: during the process of development. During development, a single cell replicates, and the daughter cells begin to exhibit subtle differences that result in different developmental programs and behaviors over time: this can be thought of as the *developmental program*. However, once the process of development is concluded, this program shuts down, and although organs and tissues continue to grow, cells are replaced, DNA damage is repaired, diseases are fought, and minor wounds heal, loss or significant damage of an organ or appendage does not normally result in its replacement.

There are, scattered within the animal kingdom, some examples of species with a remarkable ability to regenerate lost body parts; presumably, these species have the ability to re-activate parts of the dormant developmental program (for review, see Carlson 2007). The trigger for this program seems to be wounding—the loss of, or damage to, the body part. Assuming that it is not lethal produces a response in some of the surviving cells that can lead to a spontaneous and spectacular intrinsic response. This response leads to the breakdown of tissue structure and partial or complete reprogramming of some cells toward a developmentally naive state to produce a pool of progenitor cells, as is discussed in earlier chapters.

Here, we will consider the role of a small number of molecules that can influence the behavior of cells. Many of these molecules are peptides, which were first identified as *growth factors* due to their role in encouraging cells to divide, but they also have a unique role to play in shaping body plans due to their specific properties. These are the morphogens: molecules that provide instructions to naive cells, generating organized patterns and structures, and although they are an elegant solution to developing organisms, and therefore, a logical starting place for rebuilding lost structures, they may also be part of the problem when it comes to regeneration. In this chapter, morphogens will be introduced, and mechanisms, limits, pitfalls, problems, and possible solutions for potentially engineering them to work better in driving regeneration will be discussed.

9.1.1 Morphogens: The form producers

In a complex organism, such as ourselves, the major communication system is a network of electrical signals directed via nerves to a central processing unit (CPU) (the brain). The embryo is built and patterned before this system is in place and relies on a rather different mechanism of communication between cells. This is nicely illustrated in the development of the limbs. In the embryos of tetrapod vertebrates, these structures bud out from the main head to tail axis of the embryo under the control of their own intrinsic co-ordinate system and gradually acquire pattern in three dimensions, driven by the differentiation of limb bud cells into first cartilage and then bone of the skeletal elements that give each the recognizable tetrapod limb structure (reviewed in Towers and Tickle 2009).

We were first introduced to the term *morphogen*, meaning *form producer*, by the mathematician Alan Turing. Turing postulated a simple mechanism by which patterns might form in nature (Turing 1952). In Turing's *reaction-diffusion* model, all cells begin equal, and all of them produce two or more diffusible morphogens. If the morphogens have different diffusion rates and can interact with each other and/or feedback on their own production, creating feedback loops, the formerly homogeneous cells will start to spontaneously form self-organizing *patterns* within a tissue (Turing 1952). The reaction-diffusion model is now known to work biologically only with two further conditions: the second morphogen must be an antagonist of the first and must have a longer range of activity, whereas the more local morphogen must feedback positively, enhancing its own production (Gierer and Meinhardt 1972; Meinhardt 2012, Figure 9.1). While Turing mechanisms can be

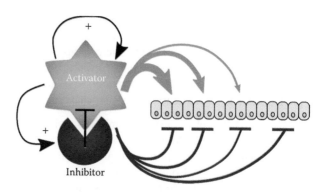

Figure 9.1 Reaction-diffusion models depend on the interactions between and properties of the morphogen activator and inhibitor. Schematic of a biological reaction-diffusion *Turing* mechanism that can generate pattern in an embryonic field, under constraints, first envisaged by Gierer and Meinhardt (1972). A short-range activator (green star) can, via positive feedback, enhance its own production, by binding to its own gene locus, but it also binds to a second locus to activate production of the inhibitor (red pacman). Inhibitor–activator interactions are direct and take place outside of the cells, and the inhibitor has a long range and is relatively stable. This results in a gradient of available activator, which can influence cell identity. (Courtesy of Caroline Beck.)

used to model everything from a zebra's skin stripes (two dimensions) to the cartilage condensations of a vertebrate limb (three dimensions), they do not account for size scaling (Newman et al. 2008; Werner et al. 2015). Neither can they cope with large fields: should the field of cells be larger than the range of the morphogens, the patterns begin to oscillate or repeat, resulting in patches, stripes, or spots (Meinhardt 2012).

There may be good reasons for the restraint on Turing mechanisms: In developing chicken limb buds, the size of organ field can itself influence pattern. It has been demonstrated that alterations to the anterior to posterior width of the autopod field (future hand/foot) can result in changes in digit number, with a narrower autopod producing fewer digits (reviewed in Newman et al. 2008). In this case, scaling is clearly not occurring, suggesting that digit formation depends on a reaction-diffusion (Turing) mechanism to produce the correct number of cartilage condensations for the autopod field size. Oscillating Turing mechanisms have also been proposed to account for the self-organizing properties of cartilage condensations in vertebrate limbs (Cooper 2015).

In Turing's original model, the theoretical morphogens were envisaged as being produced by all cells in a field. However, observations of embryonic development have identified localized morphogen sources, such as *Spemann's organizer*, which regulates vertebrate dorsal-to-ventral patterning, and Saunders' and Gassling's zone of polarizing activity (ZPA) (Saunders and Gasseling 1959), which confers identity to the digits of our limbs. These discoveries led Francis Crick to propose his source/sink model of morphogen gradient formation (Crick 1970). Crick's model again depends on diffusion to disperse the morphogen, but this time, it is made at one area of the cell field and destroyed at the opposite end, creating a stable, linear gradient. Although a localized sink is now thought to be dispensable to the system, there are some examples of source/sink systems in nature, such as retinoic acid (RA) (morphogen, produced primarily by the rate limiting Raldh1-3 enzymes) and cyp26a/b/c (RA-degrading enzyme) (for review, see Pennimpede et al. 2010).

9.1.2 How do morphogens work?

Morphogens can create differential responses across fields of equivalent cells (reviewed in Rogers and Schier 2011). They act to direct, or organize, the cells around them, but they tend to be limited, both in range and in life span. On account of this transience, morphogens are likely to generate pre-patterned fields of cells, which can then self-organize, rather than providing permanent positional information within a 3D organ or tissue. Morphogens are usually small proteins which are synthesized in cells and targeted to the plasma membrane for secretion. One notable exception is RA, a metabolite of vitamin A, which acts by binding to nuclear receptors, which then interact directly with DNA via binding to conserved sequence elements. Once they have passed outside of the cell, they can, presumably, move freely due to their small size. Each is specifically recognized by a receptor: a protein lodged in the plasma membrane of a cell, so that the outside part recognizes the morphogen and the inside part interacts differently with cell machinery, depending on whether the morphogen is bound or not. Each cell, in turn, will have a number of different receptors, and the overall occupancy and combination of these receptors will determine the cell's nuclear response: whether and how it will divide; die; change its shape, adhesiveness, or orientation; differentiate; or release a responding signal. Morphogens generally belong to just a few families of related peptides: hedgehogs (HH), transforming growth factor beta (TGFβ) superfamily (bone morphogenetic proteins [BMPs], growth and differentiation factors (GDFs), activins, nodals, decapentaplegic [Dpp], and so on), wingless (Wg)/int1 (Wnts), epidermal growth factors (EGFs), and fibroblast growth factors (FGFs). These *usual suspects* not only seem to pattern every structure in the human body but are fundamental to all metazoan (multicellular animal) life (Erwin 2009; Slack 2014). They form one part of the developmental toolkit, providing signals to cells that tell them how to behave and differentiate, depending on the concentration and length of exposure time and cellular context.

9.1.3 Context is important for both development and regeneration

As the same developmental toolkit genes shape all our organs and tissues, context is clearly going to be very important in determining outcome. Pluripotent stem cells resemble the very early embryo cells in that they retain

211

the potential to develop into any cell type. In recent years, generation of pluripotent stem cells from somatic cells of individuals has become very achievable; nevertheless, a handful of stem cells in a dish will not spontaneously form a functional limb, unless the unique 3D environment that supports this can be recreated. In other words, cells need to have both potential *and* context to reactivate the programs that drive organ and appendage regeneration. There is evidence that where there are endogenous cases of regeneration of limbs, such as the salamanders, the cells that reform the limb will not be truly pluripotent; rather, they will retain some bias or cellular memory (Kragl et al. 2009). Cellular memory maybe linked to the epigenome, which is also a mechanism by which context is determined. Simply put, the epigenome is another layer of context coded into the DNA and its histones via modifications such as acetylation and methylation. In any cell that is not fully pluripotent, all DNA is not equally primed to respond to morphogen signals. Therefore, cells that used to be muscle might respond differently to those that used to be cartilage. Certain chemicals, such as the histone deacetylase inhibitor valproic acid, can level the playing field, by facilitating removal of epigenetic marks and can be used to encourage cells back into a pluripotent state (Huangfu et al. 2008). The extracellular matrix (ECM) is also certainly critical for providing cellular context and physical support. Many tissue-engineering strategies employ matrix-mimicking scaffolds, and there are many examples where cells are used to seed and re-colonize existing ECM scaffolds, following de-cellularization (Gilbert et al. 2006; Macchiarini et al. 2008; Crapo et al. 2011; Song and Ott 2011). In model systems that can regenerate, such as the salamanders, patterning information from the ECM is a likely source of the positional information (Phan et al. 2015).

9.1.4 Running the program: Gradients, thresholds, switches, and feedback loops

9.1.4.1 Gradients and thresholds: Turning one signal into multiple responses As outlined previously, the idea of morphogens and how they can instruct the cells of developing multicellular animals to develop along different fates has evolved from a synthesis of experimental observations of embryonic induction in amphibians and insects (Stumpf 1966; Spemann and Mangold 2001), patterns of regeneration in planarians (flatworms of the phylum Platyhelminthes (Morgan 1901), and mathematical modeling of molecular gradients (Turing 1952; Crick 1970), combined with the idea of thresholds (Dalcq 1938) for gene activation (for a recent review, see Rogers and Schier 2011). In the late 1960s, Lewis Wolpert cleverly combined these concepts into the much more accessible French Flag model (Wolpert 1969 and Figure 9.2a). In Wolpert's model, a diffusible morphogen is produced from a source cell (or cells), forming a gradient. The nearby cells are exposed to a particular level of morphogen signal, depending on its position relative to the morphogen source. Each cell can therefore read its position relative to the morphogen source.

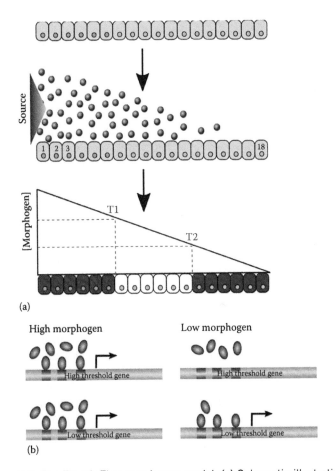

(a)

(b)

Figure 9.2 The French Flag morphogen model. (a) Schematic illustrating Lewis Wolpert's *French Flag* model for how cells determine their position within a field (Wolpert 1969). Cells in a field are initially equivalent (grey). A morphogen produced from a source (brown triangle) will diffuse in the extracellular space, forming a gradient outside the cells (brown spheres). Cells can read their position relative to the morphogen source (1, 2, 3...18) by detecting the morphogen levels locally. The gradient of information is converted from analogue (continuous) to digital (discrete), because cells respond to a certain threshold of morphogen, with cells above threshold T1 adopting one fate (blue), cells above T2 but below T1 white, and cells below T2 red, creating the three equal-sized zones of the *tricolor* flag. (b) One possibility for how cells interpret gradients to generate thresholds. Note: In typical cases, morphogen information is relayed across the cell's plasma and nuclear membranes, omitted here for simplicity. At high levels of morphogen (represented by brown ovals), regulatory binding sites on genes with both high (blue) and low (red) binding thresholds are occupied and are shown here as activated. When the concentration of morphogen is lower, only the red binding sites on the low-threshold gene are occupied, and so, only this gene is activated. (Courtesy of Caroline Beck.)

Above a certain threshold, cells adopt one fate (blue), if they are below this threshold but above a second threshold, they adopt a different fate (white), and below the second threshold, they adopt the third fate (red) (Figure 9.2a). A pattern of cells emerges, resembling the three stripes of the French Flag.

9.1.4.2 The developmental toolkit: Transcription factors and thresholds
Transcription factors are proteins that bind to specific DNA sequences called enhancers, which interact with the promoter regions of nearby genes, turning them on or off. The first clues to how morphogen gradients are interpreted by cells came from studies in fruit flies. Due to a rather unconventional early development, in which the embryo contains no cell membranes, the fruit fly transcription factor bicoid can act like a morphogen. It forms a gradient from anterior to posterior (head to tail) and activates anterior genes at high concentrations and posterior genes at low concentrations. Bicoid is not a typical morphogen, since it does not act outside cells, but it does demonstrate how cell fate can be altered through the differential activation of transcription factors. Then, transcription factors can act as the morphogen readout, changing cell fate accordingly through their interactions with gene targets (Figure 9.2b). The developmental toolkit of animals with more typical, cellularized development also contains a number of transcription factor families, which, though have no direct physical contact with the external morphogen, can respond differently at different concentration gradient thresholds. This is possible because the morphogens interact with cell surface receptors, proteins that sit in the plasma membranes of cells and have an external binding domain and a cytoplasmic effector domain. Binding of the receptor to morphogen ligand initiates a conformational change in the protein, which is relayed to the inside of the cell, often triggering a cascade of cytoplasmic proteins, ultimately resulting in interaction with a transcription factor (reviewed in Ashe and Briscoe 2006; Christian 2011).

The real power of transcription factors as morphogen readouts lies in their ability to act as genetic *switches* (see Meinhardt 2015 for a comprehensive review). Although a morphogen gradient might only be transiently produced and maintained, the effects on a cell can last a lifetime. The simplest type of genetic switch is illustrated in Figure 9.3a: a particular gene is activated by exposure of a cell to a particular threshold level of morphogen. If this gene codes for a transcription factor, the resulting protein product could activate other genetic loci, including its own enhancer, so that the gene remains active even when the morphogen trigger has been removed. In some cases, the positive feedback maybe indirect, or even be provided by inhibition of an inhibitor, as occurs during dorsal–ventral axis patterning in vertebrates by Spemann's organizer. Stable switches like this are key to cellular differentiation: how cells adopt and maintain different fates or roles in the organism; they

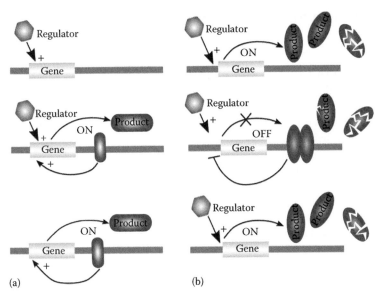

Figure 9.3 Genetic switches and feedback loops. (a) The simplest type of stable switch mechanism is shown here. A regulator (brown hexagon) activates a gene, which results in the formation of (via mRNA intermediates) many copies of a *blue* protein product. If this protein can then activate its own gene by binding to regulatory sites, then a stable, self-sustaining loop is formed, and the regulator is no longer needed to keep the gene active. Note that a blue product that turns off a gene that normally represses the blue product's own gene can also form a positive loop. (b) Negative feedback loops are also common. In the example illustrated here, a regulator turns on a gene, which results in the production of multiple copies of the red protein product. In this case, the red product turns off its own gene by binding to regulatory sites nearby, creating a negative feedback loop, where the regulator cannot activate the gene for long. If the red product is also unstable, and quickly degraded (as shown by cracked red products) by the cell, then an oscillator will result. As long as the regulator is present, the gene will turn on, then off, and then on again at regular intervals. (Courtesy of Caroline Beck.)

ensure that cells remember previous instructions, adding to the context that drives subsequent decisions.

Negative feedback loops are another common mechanism for morphogen action. In this case, the morphogen activates a gene, but the product of this gene acts to turn the gene off, so that only a short pulse of product is synthesized (Figure 9.3b). If this gene product is rapidly degraded and its messenger RNA intermediate is unstable, the repression of transcription by the product will be short-lived, and this can lead to an oscillation effect, where the gene is turned on and off in cycles, as long as the morphogen regulator continues to be present.

9.2 Size matters: The problem of scale

Morphogens, then, are highly potent and powerful agents for change. As such, they are regulated in a number of ways. The amount of morphogen that reaches a particular cell depends on the initial concentration, binding to or manipulation by the ECM, uptake or binding by other cells along the way, the distance from the source, the stability of the molecule, and interactions with receptors and inhibitors. Morphogens, through their ability to activate different cellular outcomes at different concentrations, have given multicellular life pattern in all its forms, from articulated joints of the tetrapod limb to the intricate color patterns of a butterfly's wings. Ultimately, these patterns or forms depend on the concentration and time of exposure of different cells to one or more morphogens. Recall that there are only a handful of these morphogens, and they are used to build every part of our bodies during development.

However, there are limits to this system, as the morphogens cannot diffuse indefinitely at the same rate; therefore, there are limits to their effective range. At the time Lewis Wolpert published his appealing French Flag model, explaining how morphogen gradients can generate positional information and pattern across fields of equivalent cells, the only known systems that employed this were very small: patterning the vertebrate limb bud, for example, would only require morphogens to create a gradient across about 50 cells in any direction from the source (Wolpert 1969).

Biological morphogens seem to be quite limited in their range, beyond what we might expect from the mathematical models. Why is this a problem for regenerative medicine? When pattern is based on a non-scalable system, it cannot be recreated in an adult organism. The distances that a morphogen must travel to do its work in a developing embryo limb bud, shaping it into a tiny limb that later scales in size with the organism, are clearly achievable. The distance across a gaping wound created by the severing of an arm mid-humerus could simply be too large for the morphogen model to work. The program to build an arm does not scale with size. However, all is not lost. Nature has at least managed to solve this problem, in salamanders, which can regenerate limbs even as adults (Brockes and Gates 2014). Somehow, the salamander recreates the embryonic environment within the blastema to rebuild a new limb. If we can reverse-engineer this process, we should be able to learn why we humans are so comparatively limited.

9.2.1 Strategies for scaling morphogens

We know that observable morphogens can only pattern small fields of cells in development, which may explain the poor regenerative response of most post-embryonic metazoans. Yet, salamanders can overcome this, and it is difficult to imagine that they do so by inventing a whole new mechanism for patterning cells. If morphogens can scale, but normally do not, we might expect to see evidence of this in nature, and use this to devise strategies for engineering our morphogens to work over longer distances (reviewed in Akiyama and Gibson 2015).

There are at least three mechanisms by which morphogen-based patterning systems might be able to scale perfectly with size. These are reviewed and discussed in a recent paper (Umulis and Othmer 2013), and the interested reader is directed to this paper for an explanation of the mathematical basis of these models. Here, we will instead focus on how molecules involved in patterning might be regulated (or engineered by modification) in order to work along the predictions of the models.

To begin with, properties of the morphogens themselves can be altered to produce different-gradient profiles (Figure 9.4a), for example, by increasing the rate of production, changing the mobility and range (dispersal rate, D), and altering the stability of the morphogen (degradation rate, k). As described in Umulis and Othmer (2013), scaling of the pattern

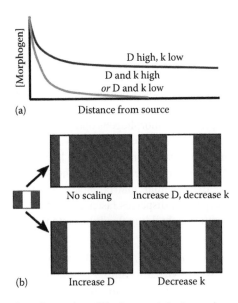

Figure 9.4 Scaling of reaction-diffusion models depends on the properties of the morphogen. (a) The conditions for morphogen scaling can be met by having an inhibitor with a fast dispersal rate (D) and low rate of turnover or destruction (k) and an activator that has either high D and high k or, more commonly observed, a low D and low k, giving it an effectively shorter range than the inhibitor. (b) If we imagine a situation where the field of cells that must be patterned is 3-fold larger in all directions, the system as it stands would not scale the pattern. Altering the properties of the molecules or their ability to disperse unimpeded changes the shape of the gradient and results in a change to the pattern. However, perfect scaling is possible theoretically, within the biological limits of D and k manipulation for a particular morphogen. (Courtesy of Caroline Beck.)

can be achieved if an increase in D is accompanied by a corresponding decrease in k (Figure 9.4b). Examples of mechanisms for regulating D and k will be discussed later.

9.2.1.1 Morphogen production/dispersal There are several plausible competing theories for how morphogens disperse to establish gradients (for review, see Rogers and Schier 2011). Morphogen gradients can be difficult to measure directly, but those that we can observe seem to form exponential gradients, with higher levels nearest the source. The simplest explanation for this pattern is that morphogen molecules are dispersed by diffusion (Crick 1970). Molecules can be removed from activity gradient by inhibitors, ECM binding, and/or endocytosis or may degrade as a natural consequence of their stability (Lander et al. 2009). Rather than having a discrete *sink*, as envisaged by Crick, it seems likely that clearance of the molecules occurs all along the gradient. The range and consequences of the morphogen therefore depend on how fast it disperses and how fast it is cleared. We can mimic a morphogen source by inserting a TGFβ protein, an FGF protein, or an RA-soaked bead into the tissue, a strategy that has been used to attempt to enhance the regenerative properties of the *Xenopus* froglet forelimb (*see later*). However, other morphogens (Wg/Wnt and HH) require post-translational modification for activity, and small-molecule mimics might be more effective. However we do this, adding ectopic morphogen is likely to cause irregular outputs. Since the activity of a typical morphogen depends on binding to its receptor, most morphogens upregulate their own receptors, so in theory, *more morphogen = higher response*. However, there are physiological limits to this: at some point, the lower threshold response will be lost. If the morphogen level is high enough to saturate normal dispersal-limiting extracellular matrix molecules, the dynamics will change, as the morphogen disperses faster and is not cleared. Feedback loops can assist with this, building flexibility into the system.

9.2.1.2 Morphogen stability/half-life Morphogens are often modified to alter their bioactivity after initial production. TGFβ superfamily proteins first form dimers and are then proteolytically cleaved to remove the pro-domain in order to create an active ligand. This might lead to the impression that the pro-domain is unimportant; however, this less conserved region can contain regions that alter the lifespan of the protein, as illustrated in the developing zebrafish embryo, where different TGFβ family morphogens can exert their effects over very different ranges. The TGFβ family member cyclops has a short range of activity compared with its long-range relative squint. Unlike squint, cyclops contains a sequence that targets it for destruction by lysosomes (Tian et al. 2008). If this sequence is deleted, cyclops can extend its activity range, so it becomes as effective as squint over long distances. Regulating the efficiency of pro-domain cleavage also alters the half-life of another TGFβ family member, BMP4 (Cui et al. 2001; Degnin et al. 2004; Goldman et al. 2006; Sopory et al. 2006), again due to the presence of a lysosomal degradation targeting sequence. In the fruit fly *Drosophila*, the BMP5/6/7-like proteins, glass bottom boat (Gbb) and screw (Scw), can be cleaved at a novel site to generate a larger-than-normal ligand. While the longer version of Gbb has both stronger signaling activity and longer range, Scw cleaved in this way is inactive (Akiyama et al. 2012).

9.2.1.3 Dispersal/diffusion Lefty, a third member of nodal-related TGFβ family that includes cyclops and squint, has an even-longer range in zebrafish embryos and acts as an inhibitor of other nodals (Chen and Schier 2001, 2002). Müller et al. (2012) have shown that, in addition to having a longer half-life, lefty has a much higher diffusion coefficient than other nodals, suggesting rapid long-range disposal. Together, the different properties of the nodal-related peptides form a reaction diffusion mechanism, like the one illustrated in Figure 9.1a.

9.2.1.4 Influence of the extracellular matrix There is much evidence that heparan sulfate proteoglycans (HSPGs) in the ECM can dramatically modulate morphogen gradients outside of cells (for review, see Yan and Lin 2009). Heparan sulfate proteoglycans are a group of cell surface and ECM macromolecules, composed of a protein core to which chains of (mainly) heparan sulfate (HS) are added to external serine residues. The biochemistry and synthesis/modification, which require several classes of enzyme, is reviewed in (Yan and Lin 2009). In *Xenopus*, BMP4 dispersal is regulated by interaction with HS through direct interaction with a conserved basic region of the N-terminal part of the protein. The HS-containing HSPGs in the ECM retard the progress of BMP4, and loss of the heparin-binding domain enables BMP4 to act over a longer range. Bone morphogenetic protein 4 can also disperse further in embryos treated with HS-removing enzymes (Ohkawara et al. 2002). Injection of heparin into zebrafish also extends the range of another HSPG-binding morphogen, FGF8, presumably by outcompeting the ECM-bound heparin, freeing the ligand to move (Müller et al. 2013). Similarly, ectopic FGF8 diffuses much faster than endogenous FGF8 (at a rate of 53 μm/second), suggesting that if the ECM heparin gets all bound up, any excess FGF can diffuse freely (Yu et al. 2009), as predicted by Francis Crick's model (Crick 1970).

More evidence for the role of HSPGs comes from mutant studies in the fruit fly, *Drosophila melanogaster* (reviewed in Wartlick et al. 2011). In *Drosophila* embryos, three morphogens, HH, Dpp, and Wg, instruct the formation of wings from the larval wing-imaginal discs. In addition to holding morphogens back from dispersal HSPGs can also facilitate morphogen dispersal (see Section 9.2.2). Carrier of wingless (Cow) is a secreted, diffusible HSPG. When bound to Cow, the morphogen Wg is more mobile, as Cow facilitates diffusion of the morphogen (Chang and Sun 2014). There is a vertebrate homologue of Cow, testican-2. Although no testicans have yet been associated with morphogen function, testican-2 binds Wnt5a outside cells, suggesting possible conservation of this mechanism for controlling Wg/Wnt morphogen dispersal (Chang and Sun 2014).

The glypican Dlp (Dally-like protein) can both inhibit and enhance Wg signaling, depending on the context (Yan and Lin 2009). Glypicans are attached to the cell via a glycosylphosphatidylinositol (GPI) anchor, which can be cleaved, releasing the glypican and potentially enabling it to act as an antagonist of Wg signaling.

Members of the HH family of morphogens also interact with HSPGs (reviewed in Yan and Lin 2009). Before HH can interact with HSPG, it needs to be lipid-modified. Ihog, an Ig/fibronectin superfamily protein, is needed for HH presentation to its receptor. Ihog can bind heparin, which induces it to dimerize and interact better with HH (McLellan et al. 2006). In the vertebrate ventral spinal cord, modification of HSPGs with the endosulfatase enzyme sulf1 restricts diffusion of the co-expressed SHH ligand. Loss of sulf1 activity results in a shallower/gentler SHH gradient and abnormal patterning, suggesting that sulf1 is needed to restrict SHH to the ventral part of the neural tube (Ramsbottom et al. 2014). Interestingly, sulfs are expressed differently in developing and regenerating *Xenopus* hind limbs, suggesting different modulation of HSPGs in the two processes (Wang and Beck 2015), and *sulf1* null mice display defective ability to heal corneal wounds (Maltseva et al. 2013). Sulfs are of particular interest for regeneration, since they are HSPG *editing* enzymes, capable of remodeling the ECM extracellularly, potentially acting to release morphogen ligands from their HSPG tethers. For a further review of HSPG roles in morphogen-based patterning, please see reviews by (Yan and Lin 2009; Christian 2011).

9.2.2 Scaling mechanisms that compensate for size

One attractive mathematical model for perfect scaling calls for the addition of a molecule known as an *expander* (Ben-Zvi and Barkai 2010) that is sensitive to the size of the growing organ field (Figure 9.5). The morphogen forms a gradient from source, but where levels of morphogen are sufficiently low, a stable, rapidly dispersed expander is transcribed. The expander is negatively regulated (at the level of transcription) by the morphogen, so that it is only produced in cells where the morphogen level is very low or absent. Furthermore, the expander molecules are secreted fast-moving molecules that can increase the range of the morphogen by facilitating its spread across cells. As the tissue grows, the expander will extend the range of the morphogen, but new expander molecules will only be made in cells outside the now-extended range, so that the morphogen can subsequently reach these more distant cells, and the gradient will scale until no more new cells can synthesize the expander, leaving only residual expander molecules

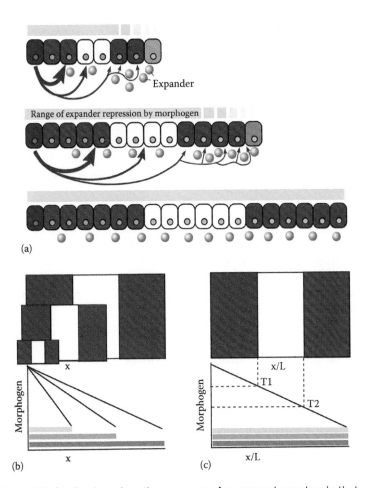

Figure 9.5 Scaling based on the presence of an expander molecule that is sensitive to the size of an expanding field. (a) Schematic of model proposed by Ben-Zvi and Barkai (2010). A morphogen is produced from a source (brown cell) and forms a gradient outside the cells (brown arrows; thicker line indicates more morphogen). Only cells that experience very low or absent levels of morphogen produce a highly diffusible expander molecule (orange spheres). The expander gene is normally silenced in the presence of morphogen. The expander extends the range of the morphogen, and so, the cells producing expander will eventually stop, but as the field expands, more cells find themselves outside of the morphogen range and begin to produce expander. The system will scale pattern until no further cells can produce expander—the morphogen reaches all cells (b and c). Scaling demonstrated using the French Flag model. Perfect scaling requires that the morphogen gradient expand proportionally to the length of the field (L). (b) Three gradient lengths are shown as green shaded boxes. The steeper the gradient, the smaller the x will be, but to accommodate scaling, this gradient needs to respond dynamically, and (c) if x, the distance that the gradient extends from the source, is normalized to L and thresholds remain constant, pattern is maintained. (Courtesy of Caroline Beck.)

and prompting the end to pattern scaling. In the *Drosophila* wing disc, a gradient of Dpp morphogen, originating from a central stripe, directs both patterning and growth. The gradient of the TGFβ family member Dpp has been shown to scale during growth of this organ, and this may be regulated by pentagone (pent), which is only transcribed in cells out of the range of the Dpp morphogen gradient. In the absence of Dpp, Pent is secreted and disperses widely due to its stability. It can alter the Dpp gradient by inducing the endocytosis and degradation of the glypican HSPGs Dally and Dlp (Norman et al. 2016). This loss of dispersal-restricting HSPGs should enable dpp to disperse unhindered, expanding its effective range.

If pent is acting as an expander, we would expect its loss to alter the ability of the Dpp gradient to scale, and consequently, there should be a loss or reduction of part of the normal pattern. Loss of pent function indeed results in the formation of a smaller wing, with one of the distinctive wing veins, L5, missing. Therefore, both patterning and growth are altered, making this a good candidate for a scaling expander molecule (Hamaratoglu et al. 2011).

9.2.3 Limb regeneration in tadpoles

I want to end this chapter by introducing a model system with deficient regeneration: the froglet forelimb. Tadpoles, most commonly of the aquatic genus *Xenopus*, are an interesting example of a halfway house regenerative system. A young tadpole can recover from partial amputation of the limb buds due to an ability to compensate for the loss of morphogen-signaling centers. However, as the animal develops, the ability to recreate a perfect pattern is gradually lost (Dent 1962). By the time the animal completes its metamorphosis into a froglet, limb regeneration is reduced to the production of a hypomorphic spike. The spike contains only a single cartilaginous element; therefore, some elements of the program are retained: the cells of the stump have divided and differentiated (albeit into limited tissues), and appropriate growth in the proximal to distal axis has occurred.

As a *halfway house*, the froglet forelimb model has been extensively adopted to study the later events of regeneration, which would result in perfect replacement of the missing structures in a salamander. As the spike produced is pattern-deficient, it presents an opportunity to study, in isolation, the recreation of skeletal limb pattern *in vivo*. Mechanisms designed to *add back* missing elements of pattern (such as SHH and FGF, which are important morphogens in developing limbs) have led to some success (Figure 9.6). The froglet's poorly patterned limb can be improved slightly by adding morphogens in the form of protein-soaked microbeads or small molecule mimics (for a recent review, see Beck 2015). Perhaps, with a better understanding of the limits of morphogen gradients and how we might be able to engineer these to cope with the

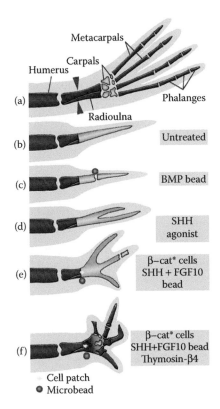

Figure 9.6 The froglet forelimb as an *in vivo* model for testing regenerative strategies. (a) The skeletal elements of a froglet forelimb show clear pattern in both proximal to distal and anterior to posterior planes. Bone is shown red and cartilage blue. Amputation through the radioulna (arrowheads) normally results in the formation of a patternless cartilage spike, (b) a process that takes several weeks. (c) adding a BMP bead can produce a discrete break in the cartilage. (d) a SHH agonist (mimic) can produce simple branched cartilage elements. (e) Combinations of active beta-catenin (β–cat*), SHH and FGF, along with a patch of early limb bud cells, which have intrinsic regenerative potential, can generate both branching and breaks. Adding in the immune suppressant thymosin β4. (f) generates the most complex pattern and also occasionally produced ossified bone elements resembling phalanges (finger bones). * indicates an activated form of beta catenin. (From Beck, C. W.: *Regeneration: Growth Factors in Limb Regeneration*. 2015. Copyright Wiley-VCH Verlag GmbH & Co. KGaA. Reproduced with permission.)

larger distances involved in regenerating an appendage such as the limb, new strategies could be developed and tested in this model system.

If periodic skeletal elements can self-organize, given initial disparities of cells across a field plus some simple behavior rules, then *Xenopus* spikes may fail to generate multiple digits due to a basic deficiency in cell numbers. In support of this, Lin et al. recently produced the most impressive regenerates to

223

date in this system by applying a fibrin patch of young limb bud cells and providing Wnt, SHH, FGF10, and thymosin b4 (Lin et al. 2013; Figure 9.6e and f). Interestingly, this study also showed that early limb bud cells, which have regenerative potential early in limb development, by themselves do very little to rescue the failure of older limb cells to regenerate patterned limbs. If, in the future, we are to develop meaningful strategies to enable us to regrow our own limbs, then we must find better ways to enable morphogens to behave as we want them to, in order to give context to stem cells.

References

Akiyama, T., and M. C. Gibson. 2015. Morphogen transport: Theoretical and experimental controversies. *Wiley Interdisciplinary Reviews: Developmental Biology* 4 (2): 99–112. doi:10.1002/wdev.167.

Akiyama, T., G. Marqués, and K. A. Wharton. 2012. A large bioactive BMP ligand with distinct signaling properties is produced by alternative proconvertase processing. *Science Signaling* 5 (218): ra28–ra28. doi:10.1126/scisignal.2002549.s

Ashe, H. L., and J. Briscoe. 2006. The interpretation of morphogen gradients. *Development* 133 (3): 385–394. doi:10.1242/dev.02238.

Beck, C. W. 2015. *Regeneration: Growth Factors in Limb Regeneration*. Chichester, UK: John Wiley & Sons. doi:10.1002/9780470015902.a0001104.pub2.

Ben-Zvi, D., and N. Barkai. 2010. Scaling of morphogen gradients by an expansion-repression integral feedback control. *Proceedings of the National Academy of Sciences of the United States of America* 107 (15): 6924–6929. doi:10.1073/pnas.0912734107.

Brockes, J. P., and P. B. Gates. 2014. Mechanisms underlying vertebrate limb regeneration: Lessons from the salamander. *Biochemical Society Transactions* 42 (3): 625–630. doi:10.1042/BST20140002.

Carlson, B. M. 2007. *Principles of Regenerative Biology*. San Diego, CA: Academic Press.

Chang, Y.-H., and Y. H. Sun. 2014. Carrier of wingless (cow), a secreted heparan sulfate proteoglycan, promotes extracellular transport of wingless. *PLoS one* 9 (10): 1–17. doi:10.1371/journal.pone.0111573.

Chen, Y., and A. F. Schier. 2001. The zebrafish nodal signal squint functions as a morphogen. *Nature* 411 (6837): 607–610. doi:10.1038/35079121.

Chen, Y., and A. F. Schier. 2002. Lefty proteins are long-range inhibitors of squint-mediated nodal signaling. *Current Biology* 12 (24): 2124–2128.

Christian, J. L. 2011. Morphogen gradients in development: From form to dunction. *Wiley Interdisciplinary Reviews: Developmental Biology* 1 (1): 3–15. doi:10.1002/wdev.2.

Clamp, M., B. Fry, M. Kamal, X. Xie, J. Cuff, M. F. Lin, M. Kellis, K. Lindblad-Toh, and E. S. Lander. 2007. Distinguishing protein-coding and noncoding genes in the human genome. *Proceedings of the National Academy of Sciences of the United States of America* 104 (49): 19428–19433. doi:10.1073/pnas.0709013104.

Cooper, K. L. 2015. Self-organization in the limb: A turing mechanism for digit development. *Current Opinion in Genetics & Development* 32 (June): 92–97. doi:10.1016/j.gde.2015.02.001.

Crapo, P. M., T. W. Gilbert, and S. F. Badylak. 2011. An overview of tissue and whole organ decellularization processes. *Biomaterials* 32 (12): 3233–3243. doi:10.1016/j.biomaterials.2011.01.057.

Crick, F. 1970. Diffusion in embryogenesis. *Nature* 225 (5231): 420–422.

Cui, Y., R. Hackenmiller, L. Berg, F. Jean, T. Nakayama, G. Thomas, and J. L. Christian. 2001. The activity and signaling range of mature BMP-4 is regulated by sequential cleavage at two sites within the prodomain of the precursor. *Genes & Development* 15 (21): 2797–2802.doi:10.1101/gad.940001.

Dalcq, A. M. 1938. *Form and Causality in Early Development.* London: Cambridge University Press.

Degnin, C., F. Jean, G. Thomas, and J. L. Christian. 2004. Cleavages within the prodomain direct intracellular trafficking and degradation of mature bone morphogenetic protein-4. *Molecular Biology of the Cell* 15 (11): 5012–5020. doi:10.1091/mbc.E04-08-0673.

Dent, J. N. 1962. Limb regeneration in larvae and metamorphosing individuals of the South African Clawed Toad. *Journal of Morphology* 110 (1): 61–77. doi:10.1002/jmor.1051100105.

Erwin, D. H. 2009. Early origin of the bilaterian developmental toolkit. *Philosophical Transactions of the Royal Society of London. Series B, Biological Sciences* 364 (1527): 2253–2261. doi:10.1098/rstb.2009.0038.

Gierer, A., and H. Meinhardt. 1972. A theory of biological pattern formation. *Kybernetik* 12 (1): 30–39. doi:10.1007/BF00289234.

Gilbert, T. W., T. L. Sellaro, and S. F. Badylak. 2006. Decellularization of tissues and organs. *Biomaterials* 27 (19): 3675–3683. doi:10.1016/j.biomaterials.2006.02.014.

Goldman, D. C., R. Hackenmiller, T. Nakayama, S. Sopory, C. Wong, H. Kulessa, and J. L. Christian. 2006. Mutation of an upstream cleavage site in the BMP4 prodomain leads to tissue-specific loss of activity. *Development* 133 (10): 1933–1942. doi:10.1242/dev.02368.

Hamaratoglu, F., A. Morton de Lachapelle, G. Pyrowolakis, S. Bergmann, and M. Affolter. 2011. Dpp signaling activity requires pentagone to scale with tissue size in the growing drosophila wing imaginal disc. *PLoS Biology* 9 (10): e1001182–17. doi:10.1371/journal.pbio.1001182.

Huangfu, D., K. Osafune, R. Maehr, W. Guo, A. Ejkelenboom, S. Chen, W. Muhlestein, and Douglas A Melton. 2008. Induction of pluripotent stem cells from primary human fibroblasts with only Oct4 and Sox2. *Nature Biotechnology* 26 (11): 1269–1275. doi:10.1038/nbt.1502.

Kragl, M., D. Knapp, E. Nacu, S. Khattak, M. Maden, H. H. Epperlein, and E. M. Tanaka. 2009. Cells keep a memory of their tissue origin during axolotl limb regeneration *Nature* 460 (7251): 60–65. doi:10.1038/nature08152.

Lander, A. D., W.-C. Lo, Q. Nie, and F. Y. M. Wan. 2009. The measure of success: Constraints, objectives, and tradeoffs in morphogen-mediated patterning. *Cold Spring Harbor Perspectives in Biology* 1 (1): 1–22. doi:10.1101/cshperspect.a002022.

Lander, E. S., L. M. Linton, B. Birren, C. Nusbaum, M. C. Zody, J. Baldwin, and K. Devon, et al. 2001. Initial sequencing and analysis of the human genome. *Nature* 409 (6822): 860–921. doi:10.1038/35057062.

Lin, G., Y. Chen, and J. M. W. Slack. 2013. Imparting regenerative capacity to limbs by progenitor cell transplantation. *Developmental Cell* 24 (1): 41–51. doi:10.1016/j.devcel.2012.11.017.

Macchiarini, P., P. Jungebluth, T. Go, M. A. Asnaghi, L. E. Rees, T. A. Cogan, and A. Dodson, et al. 2008. Clinical transplantation of a tissue-engineered airway. *Lancet* 372 (9655): 2023–2030. doi:10.1016/S0140-6736(08)61598-6.

Maltseva, I., M. Chan, I. Kalus, T. Dierks, and S. D. Rosen. 2013. The SULFs, extracellular sulfatases for heparan sulfate, promote the migration of corneal epithelial cells during wound repair. *PLoS ONE* 8 (8): e69642. doi:10.1371/journal.pone.0069642.

McLellan, J. S., S. Yao, X. Zheng, B. V. Geisbrecht, R. Ghirlando, P. A. Beachy, and D. J. Leahy. 2006. Structure of a heparin-dependent complex of hedgehog and ihog. *Proceedings of the National Academy of Sciences* 103 (46): 17208–17213. doi:10.1073/pnas.0606738103.

225

Meinhardt, H. 2012. Turing's Theory of morphogenesis of 1952 and the subsequent discovery of the crucial role of local self-enhancement and long-range inhibition. *Interface Focus* 2 (4): 407–416. doi:10.1098/rsfs.2011.0097.

Meinhardt, H. 2015. Models for patterning primary embryonic body axes: The role of space and time. *Seminars in Cell and Developmental Biology* 42 (June): 103–117. doi:10.1016/j.semcdb.2015.06.005.

Morgan, T. H. 1901. Growth and regeneration in planaria lugubris. *Development Genes and Evolution* 13 (1): 179–212.

Müller, P., K. W. Rogers, B. M. Jordan, J. S. Lee, D. R., S. Ramanathan, and A. F. Schier. 2012. Differential diffusivity of nodal and lefty underlies a reaction-diffusion patterning system. *Science* 336 (6082): 721–724. doi:10.1126/science.1221920.

Müller, P., K. W. Rogers, S. R. Yu, M. Brand, and A. F. Schier. 2013. Morphogen transport. *Development* 140 (8): 1621–1638. doi:10.1242/dev.083519.

Newman, S. A., S. Christley, T. Glimm, H. G. E. Hentschel, B. Kazmierczak, Y. -T. Zhang, J. Zhu, and M. Alber. 2008. Multiscale models for vertebrate limb development. *Current Topics in Developmental Biology* 81: 311–340. doi:10.1016/S0070-2153(07)81011-8.

Norman, M., R. Vuilleumier, A. Springhorn, J. Gawlik, and G. Pyrowolakis. 2016. Pentagone internalises glypicans to fine tune multiple signalling pathways. *Elife* 5: e13301. doi:10.7554/eLife.13301.

Ohkawara, B., S.-I. Iemura, P. Ten Dijke, and N. Ueno. 2002. Action range of BMP is defined by its N-terminal basic amino acid core. *Current Biology* 12 (3): 205–209.

Pennimpede, T., D. A.Cameron, G. A. MacLean, H. Li, S. Abu-Abed, and M. Petkovich. 2010. The role of CYP26 enzymes in defining appropriate retinoic acid exposure during embryogenesis. *Birth Defects Research. Part A, Clinical and Molecular Teratology* 88 (10): 883–894. doi:10.1002/bdra.20709.

Phan, A. Q., J. Lee, M. Oei, C. Flath, C. Hwe, R. Mariano, T. Vu, C. Shu, K. Muneoka, S. V. Bryant, and D. M. Gardiner. 2015. Heparan sulfates mediate positional information by position-specific growth factor regulation during axolotl (*Ambystoma mexicanum*) limb regeneration. *Regeneration* 2: 182–201. doi: 10.1002/reg2.40.

Ramsbottom, S. A., R. J. Maguire, S. W. Fellgett, and M. E. Pownall. 2014. Sulf1 influences the shh morphogen gradient during the dorsal ventral patterning of the neural tube in xenopus tropicalis. *Developmental Biology* 391 (2): 207–218. doi:10.1016/j.ydbio.2014.04.010.

Rogers, K. W., and A. F. Schier. 2011. Morphogen gradients: From generation to interpretation. *Annual Review of Cell and Developmental Biology* 27 (1): 377–407. doi:10.1146/annurev-cellbio-092910-154148.

Saunders, J. W., and M. T. Gasseling. 1959. Effects of reorienting the wing-bud apex in the chick embryo. *The Journal of Experimental Zoology* 142 (October): 553–569.

Slack, J. 2014. Establishment of spatial pattern. *Wiley Interdisciplinary Reviews: Developmental Biology* 3 (6): 379–388. doi:10.1002/wdev.144.

Song, J. J., and H. C. Ott. 2011. Organ engineering based on decellularized matrix scaffolds. *Trends in Molecular Medicine* 17 (8): 424–432. doi:10.1016/j.molmed.2011.03.005.

Sopory, S., S. M. Nelsen, C. Degnin, C. Wong, and J. L. Christian. 2006. Regulation of bone morphogenetic protein-4 activity by sequence elements within the prodomain. *The Journal of Biological Chemistry* 281 (45): 34021–34031. doi:10.1074/jbc. M605330200.

Spemann, H., and H. Mangold. 2001. *Über induktion von embryoanlagen durch implantation artfremder organisatoren. Roux'Arch. Entwicklungsmech 1924* 100: 599–638.

Stumpf, H. F. 1966. Über gefälleabhängige bildungen des insektensegmentes. *Journal of Insect Physiology* 12 (5): 601–608. doi:10.1016/0022-1910(66)90098-9.

Tian, J., B. Andrée, C. M. Jones, and K. Sampath. 2008. The pro-domain of the zebrafish nodal-related protein cyclops regulates its signaling activities. *Development* 135 (15): 2649–2658. doi:10.1242/dev.019794.

Towers, M., and C. Tickle. 2009. Growing models of vertebrate limb development. *Development* 136 (2): 179–190. doi:10.1242/dev.024158.

Turing, A. M. 1952. The chemical asis of morphogenesis. *Philosophical Transactions of the Royal Society B: Biological Sciences* 237 (641): 37–72. doi:10.1098/rstb.1952.0012.

Umulis, D. M., and H. G. Othmer. 2013. Mechanisms of scaling in pattern formation. *Development* 140 (24): 4830–4843. doi:10.1242/dev.100511.

Wang, Y.-H., and C. Beck. 2015. Distinct patterns of endosulfatase gene expression during xenopus laevislimb development and regeneration. *Regeneration* 2 (1): 19–25. doi:10.1002/reg2.27.

Wartlick, O., P. Mumcu, F. Jülicher, and M. Gonzalez-Gaitan. 2011. Understanding morphogenetic growth control—Lessons from flies. *Nature Publishing Group* 12 (9): 594–604. doi:10.1038/nrm3169.

Werner, S., T. Stückemann, M. Beirán Amigo, J. C. Rink, F. Jülicher, and B. M. Friedrich. 2015. Scaling and regeneration of self-organized patterns. *Physical Review Letters* 114 (13): 138101–138105. doi:10.1103/PhysRevLett.114.138101.

Wolpert, L. 1969. Positional information and the spatial pattern of cellular differentiation. *Journal of Theoretical Biology* 25 (1): 1–47.

Yan, D., and X. Lin. 2009. Shaping morphogen gradients by proteoglycans. *Cold Spring Harbor Perspectives in Biology* 1 (3): a002493. doi:10.1101/cshperspect.a002493.

Yu, S. R., M. Burkhardt, M. Nowak, J. Ries, Z. Petrásek, S. Scholpp, P. Schwille, and M. Brand. 2009. Fgf8 morphogen gradient forms by a source-sink mechanism with freely diffusing molecules. *Nature* 461 (7263): 533–536. doi:10.1038/nature08391.

Chapter 10 Environmental factors contribute to skeletal muscle and spinal cord regeneration

Ophelia Ehrlich,
Yona Goldshmit,
and Peter Currie

Contents

Key concepts

a. The extracellular matrix (ECM)–cell interactions and signals are crucial for tissue and organ regeneration.

b. By further understanding how these environmental cues affect regeneration, we can harness these processes and develop therapies that have the potential to reduce scar formation and improve regeneration.

c. When a skeletal muscle injury occurs, the muscle stem cell niche will generally become activated, and, in turn, proliferate, migrate, and differentiate, forming regenerated skeletal muscle.

d. The ECM glycoprotein laminin can form a protein network that provides structural scaffolding for tissues and organs and is linked to cell migration, differentiation, and proliferation.

e. The role of laminins within skeletal muscle regeneration has been increasingly more apparent within the last few decades of research (e.g., in humans, defective laminin within skeletal muscle can lead to a debilitating disease known as merosin-deficient congenital muscular dystrophy type 1A [MDC1A].

f. Exogenous laminin protein therapy is being thoroughly examined as an option to improve dystrophic muscle and muscle regeneration.

g. Unlike humans and other mammals, there are animals (e.g., the zebrafish) that are able to regenerate their spinal cord.

h. As with muscle regeneration, environmental cues surrounding cells are critical for a regenerative capacity, and by comparing zebrafish spinal cord regeneration with mammals, we can identify specific growth factors that help provide a pro-regenerative environment.

10.1 Introduction

During development, cell fate and organ growth are dictated via a combination of cell–cell and extracellular matrix (ECM)–cell interactions and signals. The adult body also relies on similar interactions for homeostatic maintenance and regeneration of damaged tissues in the contexts where this occurs. This chapter will focus on how cells communicate with their surrounding ECM and their associated growth factors and how we can harness these interactions to improve regeneration. Specifically, we will focus on two distinct tissue paradigms during tissue regeneration. First, we will examine skeletal muscle, a tissue that exhibits regeneration in a variety of animal systems, and discuss what we have learnt, particularly about ECM-related cues, by comparing cellular process in different systems. This section will cover the large amount of data that have been generated about ECM cues in mammalian muscle and the progress we have made in using zebrafish biology to dissect these processes. Extracellular matrix appears to play multiple roles in regulating muscle disease and scarring after injury, and exciting new results appear to directly link specific ECM component in the regulation of the muscle stem cell niche. Second, we will cover the process of spinal cord regeneration in zebrafish and compare and contrast the pro-regenerative environment of the adult zebrafish spinal cord to the limited regenerative capacity evident within the mammals. Here, the role of specific growth factors will be highlighted in driving specific pro-regenerative behaviors.

10.2 Skeletal muscle, a composite tissue

Skeletal muscle composition, in its most basic form, includes innervated muscle fibers, a muscle stem cell population for muscle maintenance and regeneration, and the surrounding ECM that contributes to signaling transduction and mechanical stability. Myofibers are multinucleated muscle cells that contract to provide movement within muscle (Grounds 2008; Waite et al. 2012). Interactions of myofibers with the ECM ensure that there is a transfer of contractile force, and when these interactions do not operate properly, diseases such as muscular dystrophy are observed. The ECM also provides the muscle stem cell niche routes for migration and a platform for biochemical stimulation (Melo et al. 1996; Michele and Campbell 2003; Silva-Barbosa et al. 2008; Webster et al. 2016). On muscle damage, muscle stem cells become activated, proliferate, and differentiate to form regenerated muscle. By observing how injured muscle heals, we can observe the role of the ECM in regeneration and fibrosis. The therapeutic promise of this approach is that knowledge that is gained may allow us to manipulate specific cell signals and develop tools that will improve regeneration and reduce scar formation.

10.2.1 Skeletal muscle regeneration requires a balance between stem cell activity and extracellular matrix deposition

Skeletal muscle is a paradigmatic example of a tissue harboring an adult muscle stem cell niche. Skeletal muscle stem cells are mononucleate cells termed satellite cells, as they are located adjacent to muscle fibers underneath the basement membrane surrounding individual fibers (Mauro 1961; Hack et al. 2000; Guyon et al. 2005; Bunnell et al. 2008). Quiescent muscle stem cells become stimulated by muscle damage, and the local microenvironment is crucial in modulating the muscle stem cell activation. Extracellular matrix molecules and secreted factors have multiple interactions with stem cells that regulate cell fate and determine if they are to proliferate, migrate, differentiate, or self-renew (Hynes 1987; Serrano and Muñoz-Cánoves 2010; Dumont et al. 2015a). Activated satellite cells re-enter the cell cycle to generate a population of daughter cells that commit to differentiate to produce myoblasts that will fuse to other fibers (Shimaoka and Springer 2003; Figeac et al. 2007; Keefe et al. 2015). *Pax7* is the most commonly used marker of satellite cells; however, there are many other markers that are used to identify these cells or a subset of them, including *c-met*, *Pax3*, and integrin-$\alpha7$ (Williams et al. 1994; Clark and Brugge 1995; Giancotti and Ruoslahti 1999; Seale et al. 2000; Takada et al. 2007; Morgan and Zammit 2010). On injury, the process of muscle regeneration includes phases of inflammation, fiber renewal, and fibrosis (Pierschbacher and Ruoslahti 1983; Hawke 2001).

Many mouse and zebrafish injury models have been established to study the kinetics of muscle regeneration *in vivo*, including laceration, freezing, and chemical-induced cardiotoxin injuries (d'Albis et al. 1988; Hawke 2001; Takada et al. 2007; Warren et al. 2007; Seger et al. 2011; Gurevich et al. 2016). Of these injury paradigms, laceration-type injuries, which involve the use of a sharp object to cause direct trauma to the muscle, most accurately model muscle–ECM disruption. These injuries involve the disruption of the connective tissue and myofiber necrosis, followed by inflammation and fiber renewal (Burkin and Kaufman 1999; Almekinders 1999; Boppart et al. 2006; Järvinen et al. 2007; Quintero et al. 2009; de Souza and Gottfried 2013). Within a laceration injury in rats, severed myofibers retract and a hematoma is formed within the first 24 hours after injury (Kaufman et al. 1985; Kääriäinen et al. 1998; Burkin and Kaufman 1999). Inflammatory cells will invade the hematoma and begin to digest the blood clot and necrotic myofibers. Muscle stem cells are attracted to the damaged muscle via cytokines released by macrophages (Mayer et al. 1997; Charge and Rudnicki 2004; Grefte et al. 2007). These cells are also able to secrete ECM proteins that will help anchor fibroblasts within the damaged zone (Lehto et al. 1985; Lehto and Järvinen 1985; Hurme et al. 1991; Vachon et al. 1997). Fibroblasts deposit many ECM proteins, including collagen fibrils that will make up the connective scar tissue (Lehto et al. 1985; Vachon et al. 1997). Within rat laceration injury models, by day 3, the hematoma is replaced by scar tissue. Scar tissue provides a supportive role, keeping the retracted fibers together and providing attachment sites for new myofibers (Kääriäinen et al. 2000; Dumont et al. 2015b). From day 5 post-injury, there is evidence of regenerating fibers that perforate scar tissue; however, if scar tissue is excessive, it will impinge on the ability of muscle fibers to grow through the scar (Hurme et al. 1991; Hurme and Kalimo 1992; Dumont et al. 2015b). Therefore, this scar tissue plays an important supportive role, but it can also become competitive with processes involved in fiber renewal (Kääriäinen et al. 2000; Dumont et al. 2015b). By 21 days after injury, size of the scar tissue is reduced and regenerating fibers have matured; however, skeletal muscle never fully recovers in the rat laceration injury model (Hurme et al. 1991; Kääriäinen et al. 2000; Gnocchi et al. 2009). Fibrosis in scarring is characterized by the accumulation of connective tissue, including collagen (Yao et al. 1996; Sato et al. 2003). Studies of laceration injuries performed on zebrafish also recapitulate many of the observed characteristics of muscle regeneration. Within these models, there was evidence of early muscle regeneration, including muscle fiber necrosis, mononuclear cell invasion, and the initiation of fiber renewal, where muscle stem cells become activated, proliferate, and differentiate into myofibers within the first 7 days after injury (Hughes and Blau 1990; Stoiber and Sanger 1996; Rowlerson et al. 1997).

Muscle laceration injuries are important paradigms to examine the balance of regeneration and fibrosis and to identify treatments for muscle injury that enhance regeneration and prevent fibrosis (Menetrey et al. 2000; Sato et al. 2003; Dumont et al. 2015b). These models also help establish the nature of the muscle stem cell niche regeneration response. Studies have revealed that a correlation exists between myofiber number and improved contractile strength (Sage 1982; Giancotti and Ruoslahti 1999; Sato et al. 2003; Menetrey et al. 2000; Postel et al. 2008). Research has also shown that enhancing the number of muscle fibers within the injury reduced fibrosis within the damaged muscle area (Otey et al. 1990; Guan 1997; Giancotti and Ruoslahti 1999; Parsons 2003; Sato et al. 2003; Hannigan et al. 2005; Wegener et al. 2007; Postel et al. 2008; Janoštiak et al. 2014). Stimulating muscle stem cells will in turn accelerate muscle fiber renewal within the injury, which could then reduce scarring that would otherwise hinder skeletal muscle functionality. To further understand how we could manipulate the regenerative process to further reduce scarring and improve muscle regeneration, this chapter will go into further detail of the ECM–cell interactions.

10.2.2 The extracellular matrix of the muscle cell

Many distinct layers of ECM contribute to ultrastructure and connectivity of adult skeletal muscle. Tendons connect skeletal muscle to bone, the epimysium surrounds the muscle, bundles of myofibers are enclosed

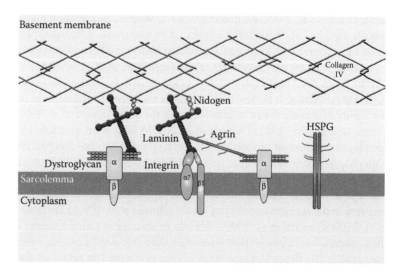

Figure 10.1 The major proteins that make up the basement membrane include non-fibril collagen IV (dark blue), laminin (red), nidogen (green), and heparan sulfate proteoglycans (HSPG) (including agrin, also in orange), which can directly or indirectly interact with membrane-associated ECM receptor proteins, sequentially interacting with the cytoskeleton. (From Goldshmit and Currie, unpublished data. With permission.)

by a perimysium, and individual myofibers are supported within an endomysium. Lying under the endomysium and in direct contact with myofibers is the basement membrane. This highly complex extracellular protein matrix layer not only provides physical support but also contains important signaling factors. The major proteins that make up the basement membrane include non-fibril collagen IV, laminin, nidogen, and proteoglycans, which can directly or indirectly interact with membrane-associated ECM receptor proteins, sequentially interacting with the cytoskeleton (Emery 2002; Jimenez-Mallebrera et al. 2005; Collins and Bönnemann 2010; Frantz et al. 2010; Thorsteinsdóttir et al. 2011). See Figure 10.1 for schematic of major basement membrane proteins.

10.2.2.1 Collagen Extracellular collagen proteins are the primary building blocks of connective tissue (Sage 1982), and collagen fibers are the major component of connective tissue within skeletal muscle, acting to provide crucial structural support, and help transfer mechanical force from muscle to the skeleton. Collagen deposition provides principal tensile elements for tissue strength and scaffolding. Collagen plays additional roles in cell adhesion, migration, and tissue repair (Kadler et al. 2007). Collagen most commonly forms stringent fibrils arising from three polypeptides, self-assembling into supercoils (Brodsky and Persikov 2005). Collagen fibrils (e.g., collagen I, II, and III) make up most of the endomyosin, epimysium, and perimysium of the muscle (Huijing 1997), whereas collagen IV is a self-polymerizing non-fibrillar form of collagen and is a major component of the basement membrane, giving a more flexible network than fibrillar collagen (LeBleu and MacDonald 2007; Khoshnoodi et al. 2008). Collagen IV functions to provide stability through its fibrous network, and molecules are also able to bind collagen, for example, nidogen binds to laminin and collagen IV, and some molecules maybe sequestered within the network itself (Fox et al. 1991; Hudson et al. 1993). Collagen VI is another non-fibrillar collagen, forming tetramers that make bead-filament structures (Kadler et al. 2007). Collagen VI sits adjacent to the basement membrane in the endomysium with other fibril collagens within the skeletal muscle tissue (Groulx et al. 2011). Interestingly, collagen VI plays an important role in cell structure via its interactions with collagen IV, strengthening the links between the endomysium and the basement membrane (Kuo et al. 1997). Collagen VI has also been indicated in matrix-dependent cell signaling roles. It is integral to skeletal muscle ECM, and defects in this isoform leads to muscle disorders (Lampe and Bushby 2005; Bateman et al. 2009; Telfer et al. 2010).

10.2.2.2 Laminin Laminin is the major component of the basement membrane in all tissues (Li et al. 2002). Laminins are cell adhesion molecules and are also implicated in cell signaling pathways (Durbeej 2010). Laminins are heterotrimers, made up of one α, one β, and one γ subunit, forming a cross-like structure (Timpl et al. 1979; Beck et al. 1993). Generally, the α (a), β (b), and γ (c) N-terminals make up a maximum of three short arms. These polypeptides then intertwine to form a coiled-coil domain that forms the long arm, which is followed by a globular structure, named the G domain (Sanes et al. 1990; Tunggal et al. 2000). The G domain is made solely by the α subunit and is critical for laminin–cell interactions (Pegoraro et al. 1998; Tunggal et al. 2000; Allamand and Guicheney 2002). The short arms self-polymerize, forming a laminin network available for other basal lamina components to interact (Colognato et al. 1999; Tunggal et al. 2000).

Laminin isoforms are expressed in specific tissues and at different developmental stages (Engvall 1993). In mammals, five α chains, four β chains, and three γ chains have been described, producing at least 15 different isoforms (Miner and Yurchenco 2004). Names of laminin isoforms are determined by the type(s) of subunits involved: α, β, and/ or γ; for example, laminin-211 represents a trimer formed by the laminin α2, β1, and γ1 chains (Aumailley et al. 2005; McKee et al. 2007). Some laminin isoforms are truncated, for example, *lama3A, lama4, lamb3,* and *lamc2* are chains with truncated N-terminal arms and therefore are unable to polymerize properly, if at all (Miner et al. 2004).

In amniotes and zebrafish, laminin-111 is expressed in embryonic epithelia and has been shown to be crucial for notochord formation in zebrafish (Dziadek and Timpl 1985; Parsons et al. 2002; Li et al. 2002). Mouse mutants that disrupt the interactions of laminin-α1 with laminin-β1 and laminin-γ1 chains were embryonically lethal, highlighting the importance of laminin-111 within the developing embryo. Laminin-111 is essential for the organization of other basement membrane proteins *in vivo* during development (Smyth et al. 1999; Edwards et al. 2010). During murine myogenesis, laminin-α1 expression is observed at the somite boundaries, which later becomes restricted to myotendenous junctions, as the muscle is forming (Tiger and Gullberg 1997). Forming muscle also expresses laminin-α4, laminin-α5, and laminin-α2 chains within the basement membrane (Schuler and Sorokin 1995; Patton et al. 1997, 1998). Adult mouse muscle has diminished expression of the laminin-α4 and -α5 chains, and laminin-α2 is the predominant laminin isoform expressed in adult skeletal mice muscle (Schuler and Sorokin 1995).

In human adult tissue, expression of laminin-111 can be seen in some epithelium and brain blood vessels (Virtanen et al. 2000). Laminin-111 expression function in developing muscle has been extensively examined using *in vitro* studies. *In vitro* experiments using myogenic cell lines have provided valuable insight as to how ECM proteins influence myoblast proliferation, fusion, differentiation, and myotube maintenance (Vachon et al. 1996). Myoblast cell lines express laminin-111, and this protein has an active role in myoblast proliferation, adhesion, and migration (Kühl et al. 1982; Öcalan et al. 1988; Goodman et al. 1989). Primary myoblast lines stimulated by laminin-111 undergo fusion and differentiation (Foster et al. 1987). Time-lapse imaging analysis has suggested that laminin–integrin–α7β1 interactions are required for satellite cell motility and that there is a more-than-a-fourfold increase in myoblast motility when they adhere to laminin compared with other substrates (Siegel et al. 2009). These *in vitro* studies suggest that laminin is important for muscle stem cell survival, proliferation, and migration.

Merosin was the name that was originally used to describe laminin complexes that were composed of an α2 subunit; now, it is most common to name these proteins *laminin* with their subsequent nomenclature (Aumailley et al. 2005). Both laminin-221 and laminin-211 are expressed in the muscle, the heart, and the peripheral nerves. In mammals, laminin-211 is the most predominate laminin isoform in skeletal muscle; however, at myotendious junctions, laminin-221 is seen primarily (Sanes et al. 1990; Patton et al. 1997; Sasaki et al. 2002). As mentioned previously, zebrafish laminin-221 is principally expressed in skeletal muscle (Sztal et al. 2011).

The laminin-α2 chain forms the attachment sites between muscle fibers and the ECM, maintaining the structural integrity of the skeletal muscle and aiding signal transduction (Hall et al. 2007; Laprise et al. 2002). These attachment sites are generally formed by the interaction of the G domain of the α subunit with transmembrane protein integrin or the dystrophin-associated glycoprotein complex (DGC) (Han et al. 2009; Gawlik et al. 2010; Gawlik and Durbeej 2011). It is interesting to note that laminin-α1 and -α2 G domains are most similar, as they both bind to many integrins, including integrin-α7β1, which is the primary integrin isoform in muscle fibers and the DGC (Ehrig et al. 1990; Talts et al. 1999). *In vitro* studies have also identified important roles for laminins in myoblast adhesion and differentiation and in myotube stability and survival (Schuler and Sorokin 1995; Vachon et al. 1996; Kuang et al. 1998). Myotubes lacking *lama2* expression undergo apoptosis (Vachon et al. 1996; Kuang et al. 1998). The importance of laminin-α2 is clearly demonstrated by the fact that laminin-α2 deficiency instigates a severe neuromuscular dystrophy, merosin-deficient congenital muscular dystrophy type 1A (MDC1A) (Engvall 1994; Helbling-Leclerc et al. 1995; Allamand and Guicheney 2002; Hall et al. 2007; Munoz et al. 2010).

10.2.3 Muscle membrane–associated proteins that link to the extracellular matrix

Transmembranous complexes have unique capabilities, as they interact with both extracellular and intracellular proteins, mediating mechanical force and cell signaling (Giancotti and Ruoslahti 1999; Han et al. 2009). In skeletal muscle, both the DGC and integrin-α7β1 have been demonstrated to regulate important cellular activities, including motility, migration, adhesion, proliferation, and apoptosis (Ervasti and Campbell 1993; Yao et al. 1996; Henry and Campbell 1996; Vachon et al. 1997; Michele and Campbell 2003; Weir et al. 2006). See Figure 10.2 for schematic of the DGC and integrin.

10.2.3.1 The dystrophin-associated glycoprotein complex and muscle disease The DGC provides a mechanical link between the ECM and intracellular actin within skeletal, cardiac, and smooth muscle (Ervasti and Campbell 1993; Lapidos et al. 2004). In addition, the DGC complex is expressed in neurons and glial cells, and deficiency within the DGC complex is also associated with a range of brain abnormalities (Waite et al. 2012). This complex is made up of transmembrane proteins: dystroglycan, sarcoglycans, sarcospan, and cytoplasmic proteins: dystrophins, syntrophins, α-dystrobrevins, and neuronal nitric oxide synthase (Michele and Campbell 2003). See Table 10.1 for a summary of the protein complexes and disease phenotypes when they are deficient. Deficiencies of most of the proteins within the DGC have been implicated in muscular dystrophies (Hack et al. 2000; Guyon et al. 2005; Bunnell et al. 2008).

10.2.3.2 Integrin Integrins are transmembrane glycoproteins consisting of one α and one β subunit (Hynes 1987). These heterodimers play a role in cell–ECM and cell–cell interactions, and the α and β subunits interact when activated. A total of 18 α subunits and 8 β subunits have been described in humans (Shimaoka and Springer 2003). Integrins can be regulated by cytoplasmic triggers and therefore are activated by intracellular signals and extracellular ligands, allowing signals to be transmitted from inside of the cell to outside of the cell and vice versa (Williams et al. 1994;

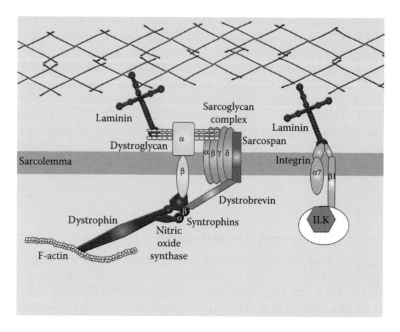

Figure 10.2 The dystrophin-associated glycoprotein complex (DGC) is a mechanical link between the ECM and intracellular actin and is specifically expressed in skeletal, cardiac, and smooth muscle. This complex is made up of transmembrane proteins: dystroglycan, sarcoglycans, and sarcospan, and cytoplasmic proteins: dystrophin, syntrophins, dystrobrevins, and neuronal nitric oxide synthase. A schematic representation of laminin–integrin–ILK is also shown. Laminin binds via its globular domain to integrin. Integrin-linked kinase (ILK) is a downstream effector of integrin signaling. The laminin receptors provide mechanical stability and have also been indicated in cell signaling. (From Goldshmit and Currie, unpublished data. With permission.)

Clark and Brugge 1995; Giancotti and Ruoslahti 1999; Takada et al. 2007). The RGD motif has been used to observe integrins in an activated state. The RGD integrin-ligand-binding motif is derived from a sequence in fibronectin; however, not all integrins will bind to this sequence (Pierschbacher and Ruoslahti 1983). Integrins can be generally grouped into laminin-binding integrins ($\alpha 1\beta 1$, $\alpha 2\beta 1$, $\alpha 3\beta 1$, $\alpha 6\beta 1$, $\alpha 7\beta 1$, and $\alpha 6\beta 4$), collagen-binding integrins, leukocyte integrins, and the RGD-recognizing (fibronectin-binding) integrins (Ruoslahti 1996; Takada et al. 2007). Integrin-$\alpha 7\beta 1$ is the principal integrin expressed in the skeletal and cardiac muscles and is known to interact with laminin-111, -211, and -221 (Burkin and Kaufman 1999; Boppart et al. 2006). Integrin-$\alpha 7\beta 1$ is expressed throughout the sarcolemmal membrane, but there is considerably more integrin-$\alpha 7\beta 1$ at the muscle–tendon junctions (MTJs) and neuromuscular junctions (Kaufman et al. 1985; Burkin and Kaufman 1999). Integrin-$\alpha 7$ null mice exhibit a mild muscular dystrophy phenotype, and in these animals, damage at the site of the MTJs is also evident, indicating that the formation and integrity of MTJs rely on

Table 10.1 A Summary of the DGC Components and Disease Phenotypes When These Proteins Are Deficient

Components of Dystrophin-Associated Glycoprotein Complex	Name of Human Gene	Muscle Disease Phenotypes Due to Deficiency, with Complex	Reference
Dystroglycan	DAG1	Muscular dystrophy-dystroglycanopathy (congenital with brain and eye anomalies), type A, 9	(Geis et al. 2013; Riemersma et al. 2015)
		Muscular dystrophy-dystroglycanopathy (limb-girdle). type C, 9	(Hara et al. 2011; Dong et al. 2015)
Sarcoglycans	α-SGCA	Limb-girdle muscular dystrophy, type 2D	α – (Roberds et al. 1994; Piccolo et al. 1995; Passos Bueno et al. 1995)
	β-SGCB	Limb-girdle muscular dystrophy, type 2E	β – (lim et al. 1995; Bonnemann et al. 1995)
	γ-SGCG	Limb girdle muscular dystrophy type 2C	γ – (Ben Othmane et al. 1992; Noguchi et al. 1995)
	δ-SGCD	Limb girdle muscular dystrophy type 2F	δ – (Nigro et al. 1996; Duggan et al. 1997)
	ε-SGCE	Myoclonus-dystonia syndrome	ε – (Klein et al. 1999; Zimprich et al. 2001)
Sarcospan	SSPN	No known human disease	
Dystrophin	DMD	Duchenne muscular dystrophy	(Bulman et al. 1991, Roberts et al. 1992)
		Becker muscular dystrophy	(Bushby et al. 1991; Boyce et al. 1991)
Syntrophin	SNTA1	No known human disease	
Dystrobrevin	DNTA1	Left ventricular non compaction 1, with or without congenital heart defects	(Ichida et al. 2001)
Nitric oxide synthase	NOS1	No known dystrophies	

Source: From Goldshmit and Currie, unpublished data. With permission.

functional integrin-α7β1 (Mayer et al. 1997). Integrin-α7β1 localization is completely disrupted in laminin-α2-deficient mouse models, indicating its dependency on the laminin-α2 subunit (Vachon et al. 1997). All these experiments reinforce the role of integrin-α7β1 as an integral link to the ECM, specifically regulating laminin linkage to the cytoskeleton of skeletal muscle (Vachon et al. 1997).

10.2.4 Extracellular matrix–muscle linkage components as pro-regenerative tools in muscle disease and injury

Although the role of the complexes within the ECM that we described earlier has been well established in muscle disease and stability, more recently evidence has emerged that they also play direct and pivotal roles in coordinating the regenerative response.

239

10.2.4.1 The dystrophin glycoprotein complex in satellite cell activation   Although the role that the dystrophin-associated complex plays in muscle integrity and muscle disease has been well established, a more recent set of intriguing observation has suggested a new role for this complex in directly regulating satellite cells (Dumont et al. 2015b). Dumont et al. have revealed that, in part, Duchenne muscular dystrophy can also be considered a muscle stem cell disease. The DGC is integral for the regulation of stem cell polarity and apical–basal asymmetric division. By marking the heterogeneous satellite cell population and identifying the different sub-populations, specifically the satellite cells that self-renew and the cells more prone to differentiation, the authors identified that the proportion of committed satellite cells (which differentiate into myofibers) was substantially lower in *mdx* mutant mice than in wildtype (WT) (Dumont et al. 2015b). As a consequence, DGC-deficient satellite cells display impaired regeneration, resulting in fewer differentiated cells within cardiotoxin-injured muscle (Dumont et al. 2015b). This study further revealed that dystrophin is specifically expressed within satellite cells and regulates the asymmetric division and expansion of the satellite cell niche through its binding of the Par complex, which regulates asymmetric process in a number of cell types. This study surprisingly reveals that DGC intracellular dynamics are important for proper regulation of the muscle stem activation (Dumont et al. 2015b). It is currently unresolved as to how the DGC interacts with extracellular cues to drive stem cell activation, and consequently further investigation of the extracellular interactions involving laminin binding with the satellite cell DGC is required to determine how ECM-derived niche factors such as laminins could contribute to the regulation of satellite cell asymmetric division.

10.2.4.2 Integrin signaling and muscle stem cells   Integrin-α7 also reliably marks *Pax7*+ quiescent satellite cells in mice (Gnocchi et al. 2009), and *in vitro* experiments where integrin-α7 was inhibited identified integrin-α7 as a pivotal protein for cell adhesion and cell motility, with laminin-α1 and -α2 as its ligand (Yao et al. 1996). These studies support the theory that on activation, muscle stem cells use the laminin basement membrane to migrate and then regenerate the damaged area (Hughes and Blau 1990; Stoiber and Sanger 1996). Recently, the ability to undertake intervital imaging within living mouse muscle has been developed (Webster et al. 2016). This significant technical breakthrough has documented that activated muscle stem and progenitor cells use the remaining ECM of *ghost fibers* that degenerate after injury, again highlighting the central role that muscle cell matrix plays in guiding and controlling the activated stem cell during the regenerative process (Webster et al. 2016). Extracellular ligands can also activate integrins, changing the intracellular conformation and allowing proteins, such as focal adhesion kinase (FAK) and integrin-linked kinase (ILK), to engage with integrin (Sage 1982; Giancotti and Ruoslahti 1999; Postel et al. 2008). On activation, integrins can either induce downstream signaling cascades or form protein complexes that will in turn interact with actin filaments, maintaining the cell structure (Otey et al. 1990; Guan 1997; Giancotti and Ruoslahti 1999; Parsons 2003; Hannigan et al. 2005; Wegener et al. 2007; Postel et al. 2008; Janoštiak et al. 2014).

10.2.4.3 Laminin treatments for muscular dystrophy   Laminin-111 has been studied as a potential therapeutic tool within skeletal muscle dystrophic and regeneration animal models. Although not the predominate laminin isoform seen in the skeletal muscle, laminin-111 has been shown to interact with the cell receptors, DGC, and integrin-$\alpha7\beta1$ with similar affinities. Historically, laminin expression has been used for the treatment of laminin-$\alpha2$ deficiency. Complete laminin-$\alpha2$ deficiency in humans causes congenital muscular dystrophy (CMD). Congenital muscular dystrophies are a group of neuromuscular disorders that derive from autosomal recessive mutations in a range of genes (Emery 2002; Jimenez-Mallebrera et al. 2005; Collins and Bönnemann 2010). The main clinical features of CMDs are present at birth or within the first few months after birth. Symptoms include hypotonia, muscle weakness, and, often, joint contractures and elevated serum creatine kinase levels (Jimenez-Mallebrera et al. 2005). Merosin-deficient CMD type 1A (MDC1A) is a laminin-$\alpha2$-deficient disease that constitutes ~30% of all European cases of CMD (Allamand and Guicheney 2002). Patients with MDC1A lack independent ambulation, experience respiratory insufficiency, and histologically have increased apoptotic signaling (Hayashi et al. 2001; Allamand and Guicheney 2002). Patient biopsies also indicate variable degrees of muscle regeneration (Jimenez-Mallebrera et al. 2005). Clinically, MDC1A is also identified by white matter changes in the cerebrum of the brain under magnetic resonance imaging (MRI) (de los Angeles Beytía et al. 2014). There is also evidence of patients with milder forms of MDC1A, and this is likely due to having a functional truncated laminin-$\alpha2$ subunit instead of a complete loss of function (Muntoni and Voit 2004; Geranmayeh et al. 2010). Overall, the lack of a laminin-$\alpha2$ subunit, a subunit that is associated with membrane stability and signal transduction, induces the deregulation of a variety of signaling pathways, resulting in apoptosis, fibrosis, chronic inflammation, and failed regeneration (Kuang et al. 1999; Bentzinger et al. 2005; Gawlik and Durbeej 2010; Rooney et al. 2012b; Yamauchi et al. 2013; Mehuron et al. 2014).

Laminin replacement strategies have been explored as a treatment of laminin-$\alpha2$ deficiency in mouse studies. The transgenic expression of laminin-$\alpha1$ was able to rescue neuronal and muscle pathologies in laminin-$\alpha2$-deficient mouse models (Gawlik et al 2006; Gowlik et al. 2010). Another strategy to reintroduce laminin into the laminin-$\alpha2$-deficient models has been to directly inject laminin-111 protein (Rooney et al. 2012b). The intraperitoneal administration of laminin-111 protein increased the life expectancy of $dy^{w-/-}$ mice 3.5-fold and overall improved mobility and muscle force (Rooney et al. 2012b). Laminin-111 protein therapies have also been used to improve skeletal muscle integrity within mice models of dystrophies associated with integrin-α7 and dystrophin deficiencies (Rooney et al. 2009a, b; Goudenege et al. 2010). Dystrophin-deficient mice, animal models of Duchenne muscular dystrophy, treated with systemic delivery of laminin-111 protein, are able to reduce myofiber degeneration, stabilize the sarcolemma, increase the expression of integrin-α7, and protect the muscle from exercise-induced damage (Rooney et al. 2009b; Goudenege et al. 2010). The authors suggest that the laminin protein treatments are able to strengthen the ECM–myofiber interactions through the elevated levels of compensatory proteins, including the laminin–integrin interactions, and improve cell adhesion (Rooney et al. 2009b).

10.2.4.4 Laminin treatments for skeletal muscle regeneration Laminin protein treatment has also been examined in the context of muscle regeneration. Laminin-111 protein has been used to improve regeneration in exercise-induced muscle injuries in wild-type mice (Zou et al. 2014). These studies suggested that laminin-111 protein treatment enhanced the activation of muscle stem cells in response to an exercise-induced muscle injury model (Zou et al. 2014). Laminin-111 directly increased the quantity of *Pax7*+ cells and increased the number of proliferating satellite cells (Zou et al. 2014). There was also increased numbers of new myofibers (Zou et al. 2014). This study confirms that laminin-111 has pro-regenerative qualities within a wild-type animal. They also identified increased expression levels of phosphorylated ILK when mice were treated with laminin-111, suggesting that the ILK activity was involved in the increased number of muscle satellite cells at the site of injury (Zou et al. 2014). The potential to use laminin-111 protein in a variety of muscle pathologies only increases the value of knowledge gained through *in vivo* experiments, providing further information on how exactly laminin-111 protein improves damaged muscle.

To further understand how laminin protein could improve muscle regeneration, interaction of laminins with cell receptors has been examined. The literature suggests that the laminin–integrin interactions are important for muscle regeneration. Gawlik et al. have shown that they can ameliorate the phenotype of dy$^{-/-}$ mice through transgenic overexpression of laminin-α1 (Gawlik et al. 2004). The laminin-α1 globular domain has a separate binding site for dystroglycan and integrin-α7β1. By transgenically expressing laminin-α1 without the dystroglycan binding site, they are able to understand the specific roles that laminin–integrin-α7β1 has in a dystrophic and regeneration context (Gawlik et al. 2010). Cardiotoxin injections were performed into the limb muscles to induce muscle injury, and muscle regeneration was observed (Gawlik et al. 2010). The dy3k$^{-/-}$ mice that transgenically expressed the truncated laminin-α1 protein (does not have a binding domain for dystroglycan, only integrin-α7β1 can interact) had a normal pattern of regeneration that was not morphologically different to its siblings (Gawlik et al. 2010). These results suggest that the regenerative capacity of skeletal muscle depends on the laminin–integrin interactions and that laminin–dystroglycan bonds are not imperative for muscle regeneration but are important for myofiber integrity. This is reiterated by *in vitro* studies that have shown how the laminin–integrin-α7β1 link and its cell signaling pathways contribute to myoblast and myotube survival (Laprise et al. 2002). These studies highlight that laminin–integrin-α7β1 associations are not only crucial for myofiber stability but also important for muscle regeneration capacity.

Recently, laminin-111 treatments have also been administrated to mice lacking wild-type laminin-γ1 expression (Yao et al. 2016). These mice developed a severe skeletal muscle deficit, but the administration of laminin-111 protein directly into the muscle had a positive therapeutic effect. Not only was there an improvement within the muscle on a molecular level, but muscle functions, specifically locomotion and muscle strength, were also significantly improved (Yao et al. 2016).

10.2.5 The potential of exogenous laminin as a therapeutic tool

The data summarized earlier indicate that injections of laminin-111 protein can be used to rescue dystrophic phenotypes and improve the regenerative response, as laminin plays vital roles in myofiber adhesion, survival, myoblast migration, and proliferation. The studies also indicate that laminin–integrin associations are important for stimulating the muscle stem cell niche. However, a detailed knowledge of the downstream signaling pathways of this whole process is lacking. How do laminin interactions with satellite cell receptors lead to myogenic precursor-derived genes being transcribed or repressed? There needs to be a better understanding of this complex signaling process. Identification of the elements involved is essential and could help identify future therapeutic targets. If laminin-111 protein therapy is to be used to improve muscle regeneration, there are also a number of other questions that need to be answered. Should we try to control the localization of the laminin protein only to skeletal muscle? Does a laminin treatment have adverse effects on other organs? Is there a time point during regeneration that would be optimal for a laminin treatment to improve regeneration? Our laboratory has been extensively examining the potentials of laminin-111 protein treatments within zebrafish dystrophic and regenerative models to try to answer these questions.

The zebrafish (*Danio rerio*) is an ideal animal model for studying regeneration and skeletal muscle diseases. Zebrafish lay eggs that are optically clear and allows the visualization and live imaging of spatial and temporal cellular interactions, which is not possible in other animal models (Hall et al. 2007; Chong et al. 2009; Sztal et al. 2011, 2012). For example, laminin-α2-deficient zebrafish (*lama2−/−* mutants) have the capability of being used as a live *in vivo* model (Hall et al. 2007). Despite normal fusion and elongation of skeletal muscle, muscle fiber detachment and retraction from the myotendenous junctions were observed 36 hours post-fertilization, with subsequent apoptosis of these fibers (Hall et al. 2007). Observing skeletal muscle fibers in real time has given vital insight and knowledge of the MDC1A cellular pathology. A way to control localization is to introduce laminin-111 within a biomaterial to localize the laminin to the damaged skeletal muscle. An enzyme-assisted self-assembling 9-fluorenylmethoxycarbonyl-tri-leucine (fmoc-tri-leucine)/laminin hydrogel matrix was microinjected into the *lama2*⁻/⁻ zebrafish and was shown to be stably localized within the skeletal muscle (Figure 10.3; Williams et al. 2011). Transmission electron microscope (TEM) images of the region of injection demonstrate fibrillar nanotubes (the hydrogel) associated with myofiber cell membranes (Figure 10.4; Williams et al. 2011). The laminin (distributed within the hydrogel) was closely localized with the muscle fiber terminations of the myotome 4 days after implantation (Williams et al. 2011). These results suggest that the laminin biomaterial has the potential to play a fundamental role with differentiated myofibers. In addition, our laboratory is now in the process of examining the effects that the laminin biomaterial has on the muscle stem cell niche.

There are a number of zebrafish regeneration paradigms that have been established by using cardiotoxin injection, needle stick injury,

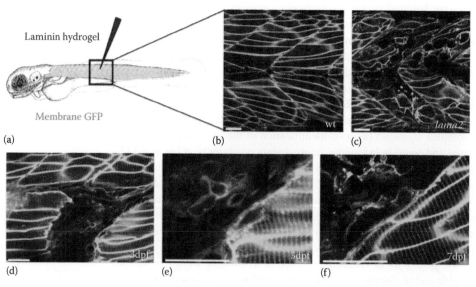

Figure 10.3 Laminin-111 protein delivery to lama2-deficient zebrafish via microinjection of hydrogel matrices. (a) Schematic representation of the site of hydrogel injection in a laminin-deficient zebrafish 3 days post-fertilization (3DPF). (b) Confocal microscope image of intact muscle in wild-type zebrafish, labelled with membrane green fluorescent protein (GFP). (c) Corresponding image of muscle from a laminin-α2-deficient zebrafish. (d and e) Alexa-546-labelled laminin hydrogel (red) implanted into laminin-deficient dystrophic muscle at 3DPF. (f) Alexa-546-labelled laminin hydrogel (red), imaged at 7DPF and 4 days postimplantation (scale bars: 25 µm). (From Williams, R. J. et al., *Biomaterials*, 32, 5304–5310, 2011. With permission.)

Figure 10.4 Transmission electron micrograph shows the structural and spatial stability of the material at the site of implantation. Nanofibers (arrow) on the extracellular side of muscle fiber termini 4 days postimplantation. A sarcomere (white arrowhead) can be seen on the intracellular side of the terminal membrane (black arrowhead) (scale bars: 500 nm). (From Williams, R. J. et al., *Biomaterials*, 32, 5304–5310, 2011. With permission.)

and laser-induced microinjury (Seger et al. 2011; Otten and Abdelilah-Seyfried 2013). It is critical to identify and understand the similarities and the differences between animal models and humans, so we can understand the extent and limits of zebrafish skeletal muscle injury models and how relevant they are to human injuries. These studies have demonstrated that many of the cellular processes involved recapitulate muscle regeneration in amniotes. When comparing mechanical injury paradigms from amniotes and zebrafish, it is clear that similar time frames, molecular markers, and cellular behaviors are involved within the regeneration of skeletal muscle (Siegel et al. 2013). This makes the zebrafish a highly useful tool to examine treatments that could be used as potential therapies for human muscle injuries. Ideally, if this laminin hydrogel is able to rescue the laminin-α2-deficient pathologies and improve muscle regeneration, then this biomaterial could be used as a therapeutic tool for MDC1A and muscle injuries. It is important to be able to control where the laminin protein is delivered *in vivo*, in order to ensure that there are no unforeseen adverse effects. As mentioned previously, the most common cause of death within patients with MDC1A is respiratory insufficiency (Allamand and Guicheney, 2002). If a laminin hydrogel could be localized to the diaphragm, a sheet of skeletal muscles important for respiration function, it could greatly improve MDC1A respiratory problems and these patient's quality of life. Exogenous laminin has also been explored as a co-factor for improving cell therapy in muscular lesions; this has been nicely reviewed in (Riederer et al. 2015).

10.2.6 Summary

To improve muscle regeneration, it will be important to provide an environment for the muscle stem cell niche that enhances proliferation, migration, and survival. Laminin, being an ECM protein, could have an advantageous therapeutic potential, as cells already recognize laminin protein, meaning there is a low probability of inducing a foreign-body immune response. Although exogenous laminin has promising effects, there is still much to determine, including downstream activity of laminin and optimizing therapies, before laminin protein becomes a viable therapeutic tool for human patients with muscular dystrophies or injuries.

10.3 Tissue interactions during spinal cord repair in Zebrafish

Skeletal muscle is an example of a tissue that exhibits regenerative capacity across a broad range of organisms. As we discussed previously, many of the cellular processes and the time frame of regeneration are similar in organisms as phylogenetically diverse as zebrafish and humans. However, by contrast, multiple organ systems that are able to regenerate in zebrafish possess minimal regenerative potential in mammalian systems. One of the most paradigmatic examples of this capacity is spinal cord regeneration, which exhibits an ability to fully regenerate after total resection in zebrafish. An understanding of how these animals are able to regenerate after the spinal cord is damaged and the determination of the growth factors and proteins involved within the spatial and temporal interactions that are required to form a regenerated spinal cord could provide vital insight into how we can improve spinal cord regeneration for human patients. Discovering how cells communicate with surrounding growth factors and ECM within the central nervous system (CNS) of amniotes and zebrafish could facilitate further therapies to improve spinal cord regeneration.

10.3.1 The cellular processes involved in spinal cord regeneration

Damage to the spinal cord produces a lifelong impact on the injured person. Spinal cord injury (SCI) caused by external force most commonly occurs in young adults, meaning that most individuals injured in this manner live with the consequences of their injury for decades. The ability of the CNS to repair itself after injury is highly limited. Breakage of the blood–brain barrier leads to ischemia, edema, and hypoxia that initiate neural cell death, inflammation, and formation of a glial scar. The failure of the spinal cord to regenerate and undergo reconstruction after injury can be a result of an imbalance between factors that prevent regeneration and factors that promote regeneration, in favor of the inhibitory factors. To repair the injured spinal cord, both cellular replacement and axon guidance would be required. Although neurons in the CNS hardly regenerate, glial cells exhibit remarkable self-renewal potential. After injury, astrocytes have a dichotomous role. The injury induces rapid morphological and physiological changes among the different glial cell population at the site of the injury. Astrocytes become reactive and proliferate in order to seal the wound, stabilize the injured tissue, and re-establish homeostasis. However, reactive astrocytes also express high levels of chondroitin sulfate proteoglycans (CSPGs) such as brevican, phosphacan, and neurocan, which are inhibitory for the regeneration process (McKeon et al. 1999; Jones et al. 2003; Tang et al. 2003). On the other hand, glial cells can provide appropriate surfaces for the promotion of axonal elongation and are an important source of trophic factors for the survival of neurons and oligodendrocytes (Noble et al. 1988; Richardson et al. 1988; Patel et al. 1996). A limited number of growth factors have been demonstrated to improve progenitor cell proliferation and migration and therefore regeneration. A cocktail of growth factors is also added in stem cell therapy in order to enhance these cells' survival and proliferation at the lesion site. Growth factors that usually are used include brain-derived neurotrophic factor (BDNF), neurotrophin-3 (NT 3), glial cell-derived neurotrophic factor (GDNF), insulin-like growth factor (IGF), basic fibroblast growth factor (bFGF), epidermal growth factor (EGF), platelet-derived growth factor (PDGF), acidic fibroblast growth factor (aFGF) and hepatocyte growth factor (HGF) (Petter-Puchner et al. 2007; Willerth et al. 2007; Grumbles et al. 2009; Kadoya et al. 2009; Lu et al. 2014a, b; Tuszynski et al. 2014a, b). Delivering stem cells or progenitor cells to the lesion site embedded into fibrin biomatrix, in conjunction with a variety of growth factors, leads to increased survival and integration of the cells into the spinal cord tissue and enhanced axonal regeneration. Understanding the role of each growth factor in the regeneration process will help develop better therapeutic strategies after CNS injury to reduce secondary damage.

10.3.2 Zebrafish models of spinal cord regeneration

During mammalian embryogenesis, radial glial cells are required to direct neurons to their target location and consequently differentiate into astrocytes (Rakic 2003). Axonal regeneration can occur in chick or rat embryonic spinal cord after injury (Shimizu et al. 1990; Hasan et al. 1993; Iwashita et al. 1994). These studies demonstrated that if a lesion is performed at early embryonic stage, approximately day 15 of development in chick or rat, there is axonal regrowth of neurons from brainstem toward the lumbar spinal cord. However, vertebrates such as urodele and tailed amphibians, freshwater turtles, and fish are able to regenerate their spinal cord and recover function even in the adult stage (Chernoff 1996; Becker et al. 1997; Rehermann et al. 2011; Sîrbulescu and Zupanc 2011; Zupanc and Sîrbulescu 2011; Goldshmit et al. 2012). Pro-neuronal transcription factors, such as *Mash1* and *Math3*, play an important role in the maintenance of neural progenitors to ensure a sufficient supply of glial progenitors. MASH-1, for example, is downregulated after development, in both mammals and zebrafish brain; however, it is upregulated after SCI in zebrafish but not in mammals (Ohsawa et al. 2005; Williams et al. 2015). A similar case is with sox11 (Wang et al. 2015), where in zebrafish, upregulation of sox11b mRNA after SCI is mainly localized in ependymal cells lining the central canal and in newly differentiating neuronal precursors or immature neurons in the spinal cord and promotes regeneration (Guo et al. 2011). In mammals, viral infection of sox11 into spinal cord tissue reduced axonal dieback and promoted corticospinal tract regeneration in rat after SCI (Wang et al. 2015). Thus, the mechanisms by which proteins regulate growth-associated genes after injury are not conserved across vertebrates, and this may explain the limited ability for mammalian CNS neurons to regenerate their axons. In zebrafish, disruption of the ependymal layer elicits proliferation predominantly in the radial glia, which persists during their adult life, and this persistence is implied in their ability to regenerate their brain and spinal cord after injury. Therefore, understanding the regeneration process in fish will shed some light on the pro-regenerative pathways in all vertebrates. Toward the end of embryogenesis, most mammalian radial glial cells differentiate into astrocytes (Rakic 2003); however, aquatic vertebrates maintain most of their radial astroglia throughout life, which are located at and around the ventricular zone (VZ) and the central canal (Kaslin et al. 2008, Kroehne et al. 2011). In contrast to the reactive astrocytes in mammals that create the glial scar that prevents any regeneration (Silver and Miller 2004; Cregg et al. 2014), it has been demonstrated in fish that a glial scar fails to form with glial cells, instead creating bridges that promote axonal regeneration (Cohen et al. 1994, Goldshmit et al. 2012). After SCI, glial cells proliferate along the central canal, close to the lesion site, migrate to the lesion site, and elongate along the anteroposterior (AP) axis to develop a bipolar cellular bridge (Goldshmit et al. 2012). Axonal regeneration is dependent on the glial bridge formation, and when bridge formation is impaired, axons fail to cross from one side of the spinal cord to the other. Similar to the process of SCI after brain injury in zebrafish, there is upregulation of the intermediate filamentous proteins glial fibrillary acidic protein (GFAP) and vimentin, swelling of glial processes (hypertrophy), and glial proliferation in the lesioned hemisphere; however, at later time points, gliosis and fibrotic scar formation are not observed, including reactive astrocytes and ECM deposition. Moreover, hypertrophic glia possess a bipolar morphology rather than multipolar morphology, similar to the processes exhibited by spinal cord-derived glia after injury (Kroehne et al. 2011).

During development, at the peak of neurogenesis, all proliferating cells are radial glia that express pax6 and function as progenitors for neurons (Malatesta et al. 2003). Pax6, for example, is expressed in radial glia and in all proliferating progenitors at mid-neurogenesis stages (Gotz et al. 1998). In mammals, Pax6- and nestin-expressing cells are upregulated after injury by dedifferentiated astrocytes at the injury site (Duggal et al. 1997; Sahin Kaya et al. 1999; Goldshmit et al. 2014). On the basis of these findings, radial glia could be a promising source of neural stem cell for the treatment of various diseases such as neurodegenerative diseases and CNS injury. The multiple-potential radial glia-like cells in the zebrafish spinal cord and brain express a variety of glia and neuronal markers, such as olig2 (Park et al. 2007), pax6, sox2, neural/glial antigen 2 (NG2), GFAP, brain lipid-binding protein (BLBP), nestin, vimentin, and GLutamate ASpartate Transporter (GLAST) (Reimer et al. 2008; Goldshmit et al. 2012; Hui et al. 2015) and exist during adulthood in the ependymal cell layer of the ventricular wall. These neuronal markers are upregulated after injury, when these cells initiate proliferation and migrate to the lesion site (Figure 10.5). These cells dedifferentiate, increase their rate of neurogenesis, and regenerate the brain and spinal cord

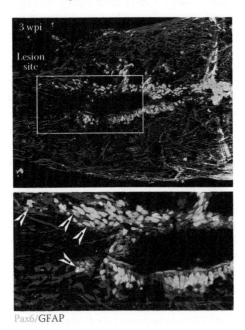

Pax6/GFAP

Figure 10.5 Zebrafish spinal cord 3 weeks post-injury. Glia progenitor cells (radial glia) express GFAP (red) and, along the central canal, co-express pax6 (green). Lesion site is on the left side of the panel. Enlargement of the white box is a confocal image that shows co-expression of pax6 glia progenitor marker with glial marker GFAP. White arrowheads indicate co-expressing cells. (From Goldshmit and Currie, unpublished data. With permission.)

(Kaslin et al. 2008; Reimer et al. 2008; Goldshmit et al. 2012; Briona et al. 2015). Growth factors that are expressed in the VZ of the developing brain have the potential to induce glial cell differentiation to progenitor cells.

10.3.2.1 Fibroblast growth factor   One of the approaches that may reduce glial scar formation is to influence astrocyte morphology and increase progenitor cell marker gene expression. One of the mechanisms that has been suggested to mediate the bipolar morphology of the glial cells and their expression of progenitor markers such as pax6 and sox2 was the signaling pathway of the fibroblast growth factor (FGF). Mammalian FGF family members are expressed not only during embryogenesis but also postnatally and continuing into adulthood. The FGF2 is critical for the maintenance and expression of the precursor cell pool (Palmer et al. 1995; Gritti et al. 1996; Palmer et al. 1999; Zheng et al. 2004), for example, high expression of FGF2 by astrocytes at the subgranular zone (Shetty et al. 2005; Bernal and Peterson 2011). Pringle et al. reported that FGFR3 is expressed in the ependymal zone of the embryonic spinal cord and later becomes restricted to astrocytes in the gray and white matter of the postnatal cord (Pringle et al. 2003). In the amphibian spinal cord, FGF2 is upregulated in ependymal cells at the central canal, retaining radial glia-like morphology 3 days after injury. Its expression is maintained up to 3 weeks post-injury, which may suggest a role of cell differentiation in neurons (Zheng et al. 2004). In the zebrafish spinal cord, *FGFR1–3* are expressed by cells around the central canal (Figure 10.6) and upregulated after spinal cord hemisection a short time after injury. After CNS damage, the levels of FGF ligands and receptors also increase in different cell types at the injury site in the mammalian brain (Gomez-Pinilla and Cotman 1992; Grothe et al. 2001; Yoshimura et al. 2001; Ganat et al. 2002), which makes FGF an attractive candidate to control aspects of glial cell differentiation in mammals. After chronic hypoxia in a rat brain, the number of GFAP-immunoreactive cells decreased throughout the subventricular zone and the proportion of FGF2-GFAP double-positive cells decreased compared with normal conditions; however, the cells in this area re-expressed vimentin, a radial glia marker that co-localized with cells expressing increased levels of FGF receptors and ligands (Ganat et al. 2002). This may suggest that FGF is responsible for glial cell proliferation and dedifferentiation to progenitor cells after injury. The expression of FGF by astrocytes at the injury site has been suggested to be a potent angiogenic and gliogenic mechanism that participates in the wound/repair process (do Carmo Cunha et al. 2007), raising the possibility that FGF acts to protect the brain from pathological events by promoting neuronal survival (Dono 2003) (Figure 10.7).

Inhibition of FGF signaling, by using dominant negative FgfR1 trangenic fish, leads to significant reduction of these cell activities, including reduced proliferation (Kaslin et al. 2009). The FGF3 was previously demonstrated to be upregulated after SCI in fish (Reimer et al. 2009). Goldshmit et al. (2012) demonstrated that FGF8 and FGF3 were highly upregulated on the

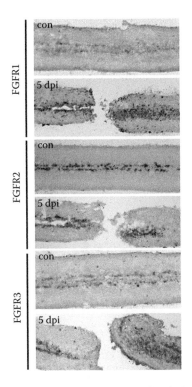

Figure 10.6 In situ hybridization in adult zebrafish in intact spinal cords (con) or injured spinal cords 5 days post-injury (5DPI) for FGFR1–3. FGFR1 and FGFR3 are expressed on radial glial cells at the central canal and upregulated on these cells and on edges of the lesion site after SCI. FGFR2 is mainly expressed on glial cells at the central canal. (From Goldshmit and Currie, unpublished data. With permission.)

glial cells at the lesion site short time after injury. Loss-of-function experiments showed that when knocking down FGFR-1 or using pharmacological inhibitor for the receptor activation after SCI, the bridges were not formed and axonal regeneration was inhibited (Goldshmit et al. 2012).

Exogenous administration of FGF2 in mice after SCI also results in a reduction of glial scarring. The FGF2 treatment not only reduced the number of reactive astrocytes and their processes but also inhibited cytokine and CSPG secretion at the lesion. Moreover, increased FGF2 signaling at the lesion site promoted the formation and propagation of radial/progenitor bipolar glial cells, expressing markers such as pax6 and sox2, which at later stages mediated the formation of GFAP-expressing glial bridges that support regenerating neuronal processes to traverse the lesion (Goldshmit et al. 2014). Furthermore, the increased FGF signaling evident in $spry4^{-/-}$ mice resulted in attenuation of astrocytic gliosis and an increase in neuronal survival and numbers of glia progenitor cells at the lesion site in the mouse SCI hemisection model (Goldshmit et al. 2015). This study also provided evidence for an additional role of FGF in

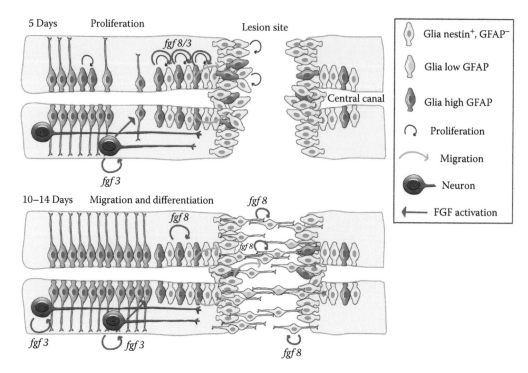

Figure 10.7 Proliferation is an early response to SCI. At 5 days after spinal cord injury (SCI), transection of the spinal cord stimulates secretion of FGF ligands in the central canal, which in turn induce glial cell de-differentiation and proliferation to generate progenitor cells. These have low levels of GFAP and high levels of nestin at the lesion site. The FGF is also released by neuronal cells at the lesion site. Migration and differentiation are later responses to SCI. By 10 days after SCI, nestin-positive cells increase their GFAP levels and maintain their levels of FGF expression. They begin to migrate to and fill the lesion site and elongate into bipolar morphology, a process of differentiation that is also FGF-dependent. FGF expression is also upregulated in neuronal cells upstream of the lesion site. At this stage, axonal regeneration toward, but not through the lesion, occurs. (From Goldshmit, Y. et al., *Journal of Neuroscience*, 32, 7477–7492, 2012. With permission.)

the attenuation of the inflammatory response, which allows better environmental conditions at the lesion site to mediate increased progenitor cell pool.

In addition to the glial-supported axonal regeneration in fish, spinal lesions also trigger local neurogenesis. These neurons, including motor neurons and interneurons, originating from the cells located at the lesion site, generate a short time after SCI (Bernstein and Bernstein 1969; Bernstein and Gelderd 1970; Becker et al. 1997; Reimer et al. 2008, 2009), followed by a complete functional locomotor recovery in the absence of glial scarring several weeks post-SCI. These neurons are generated through asymmetric division from the ependymal layer at the central canal. The FGF2 is associated with this process in urodele amphibians and zebrafish (Zhang et al. 2000). The FGF2 expression is increased in the ependymal layer after tail amputation and later in regenerating neurons. Exogenous FGF2 further increased the proliferation of these *in vivo*, which strongly supports an important role for this factor in spinal cord regeneration.

One clinical study assessed functional outcomes of nerve repair by using acidic FGF1 in patients with cervical SCI. Nine patients in chronic stages received a surgery in which FGF1 was applied to the injury site by using fibrin glue containing the growth factor. The results of this phase I clinical trial showed a significant difference in american spinal injury association scale (ASIA scale) motor and sensory scale scores between the preoperative status and the 6-month postoperative follow-up, suggesting that FGF application may provide some therapeutic benefit in patients with SCI, even in patients with chronic-stage injury (Wu et al. 2008).

10.3.3 Other differences between mammals and zebrafish after spinal cord injury

Besides the differences in progenitor cell pool size and growth factors' re-expression after brain injury or SCI in zebrafish, as opposed to mammals, there are other differences in responses on the molecular and cellular levels after injury between these vertebrates. Understanding the pro-regenerative mechanisms mediating glial cell behavior in the fish after SCI, besides growth factors, will provide a mechanistic basis for which to engender a greater regenerative capacity within the mammalian spinal cord. Exposing mammalian neuronal cells to cells from the fish nervous system or vice versa can distinguish whether differences in the regenerative capacity are attributed to the neuronal cells themselves or to environmental factors only. For example, studies have demonstrated that regenerating axons from both fish and mammals are repelled by mammalian myelin and oligodendrocytes (Fawcett et al. 1989; Bastmeyer et al. 1991). Plating neuronal cells on fish oligodendrocytes, or in the presence of fish condition medium, induced axonal growth (Bastmeyer et al. 1991; Bastmeyer et al. 1993; Schwalb et al. 1995). Thus, it maybe that the fish environmental conditions contain less inhibitory molecules, factors that block inhibitory molecules, or pro-regenerative factors that support axonal regeneration.

The main inhibitory molecules for axonal regeneration after brain injury or SCI that are found in the mammals are myelin debris proteins (myelin-associated glycoprotein [MAG], Nogo-A, and oligodendrocyte myelin glycoprotein [OMgp]), CSPGs, and chemorepulsive guidance molecules (ephrins and semaphorins) (Giger et al. 2010; Rasmussen and Sagasti 2016).

10.3.3.1 Inhibition by myelin debris Nogo-A was first purified as a high-molecular-weight, highly inhibitory membrane protein for spinal cord regeneration (Caroni and Schwab 1988). Application of blocking antibodies (ABs) against the Nogo-A-specific region in spinal cord–lesioned mammals allows axon regeneration and, to some extent, re-establishment of function (Z'Graggen et al. 1998; Schwab 2004), emphasizing its importance. Unlike mammalian CNS myelin, fish

253

myelin and oligodendrocytes did not markedly inhibit axon growth (Bastmeyer et al. 1991). In zebrafish, this protein appears to promote neurite outgrowth, as it is lacking the key inhibitory domain (Diekmann et al. 2005). Abdesselem et al. showed, for example, that zebrafish Nogo-66 sequence is different from the rat sequence in the regions outside the receptor-binding domain. However, zebrafish Nogo-66, in contrast to Rat Nogo-66, increased neurite outgrowth of cultured mouse hippocampal neurons. It activated the receptor but failed to activate the downstream pathway; therefore, it was not inhibitory for axonal regeneration in an optic nerve model (Abdesselem et al. 2009).

10.3.3.2 Inhibition by chondroitin sulfate proteoglycans Another major inhibitory factor for axonal regeneration is the glial scar formation by the reactive astrocytes that create physical and chemical barriers. Reactive astrocytes secrete CSPGs and pro-inflammatory cytokines. It has been demonstrated by Becker et al. that the absence of growth-inhibitory molecules such as CSPGs from the lesion site itself, as they demonstrated in optic nerve crush model, may contribute to axonal regeneration after injury in the CNS of fish (Becker and Becker 2002). It is not that fish are lacking inhibitory molecules in the CNS. The CSPGs and tenascin C are expressed, but they are expressed at the margins of the optic tract, so that they may play a guidance role for axons when they regenerate (Becker and Becker 2002; Becker et al. 2004).

10.3.3.3 Semaphorins Scar tissue formed after traumatic injury in rodent CNS contains fibroblast-like cells expressing high levels of Semaphorin-3A mRNA. These Semaphorin-3A-expressing cells are observed in scars formed after stab lesions in the brain and spinal cord or after optic nerve injury, and levels of expression stay high a few weeks post-injury (Pasterkamp et al. 1999a, b; Shirvan et al. 2002; Nitzan et al. 2006). Interestingly, Rosenzweig et al. found that in goldfish optic nerve, a regenerating CNS system, sema-3A was downregulated in the retina after injury about 2-fold at 3 days after injury. When Semaphorin-3A was injected into the goldfish eye a short time after injury, several events that occur in the regenerative process were affected, including the survival of normally functioning retinal ganglion cells, axonal growth, and myelin clearance from the lesion site by macrophages (Rosenzweig et al. 2010). Another study in lamprey showed that after SCI, semaphorin-positive cells were found in the spinal cord, just rostral and caudal to the scar but not within the scar itself. Sema-4 mRNA expression was downregulated up to 500 μm rostral and caudal to transection in the neurons of the spinal gray matter (Shifman and Selzer 2007).

10.3.4 Summary

We could find that similar molecules are upregulated after CNS injury in zebrafish and higher vertebrates. However, the zebrafish has an intrinsic and impressive ability to regenerate CNS cells and tissues.

It maybe that the timing of the upregulation of these molecules, or their location within the tissue, is a major reason for the ability of zebrafish to regenerate. However, we contend that the main reason for the difference in the regenerative capacity lies in the difference in the differentiated state of resident glial cells of the spinal cord. Zebrafish glial cells are in a differentiation stage comparable to human or rodent cells in the embryonic stage or at the time of birth and retain *radial glia*-like progenitor capacity and consequently possess an ability to respond to regeneration stimuli. By comparison, neural progenitor cells in adult mammalian CNS are very few in number and are located in specific brain regions that are not always available for regeneration. Understanding the growth factors–mediated mechanisms required for the stimulation of these cells that we suggest will enable a better mobilization of these cells within regenerating tissue and lead to better post-injury outcomes.

References

Abdesselem, H., A. Shypitsyna, G. P. Solis, V. Bodrikov, and C. A. Stuermer. 2009. No Nogo66-and NgR-mediated inhibition of regenerating axons in the zebrafish optic nerve. *Journal of Neuroscience* 29 (49):15489–15498. doi:10.1523/JNEUROSCI.3561-09.2009.

Allamand, V., and P. Guicheney. 2002. Merosin-deficient congenital muscular dystrophy, autosomal recessive (MDC1A, MIM|[Num]|156225, LAMA2 gene coding for |[Alpha]|2 chain of laminin). *European Journal of Human Genetics* 10 (2): 91–94. doi:10.1038/sj/ejhg/5200743.

Almekinders, L. C. 1999. Anti-inflammatory treatment of muscular injuries in sport. *Sports Medicine* 28 (6): 383–388.

Aumailley, M., L. Bruckner-Tuderman, W. G. Carter, R. Deutzmann, D. Edgar, P. Ekblom, and J. Engel et al. 2005. A simplified laminin nomenclature. *Matrix Biology* 24 (5): 326–332. doi:10.1016/j.matbio.2005.05.006.

Bastmeyer, M., M. Bahr, and C. A. Stuermer. 1993. Fish optic nerve oligodendrocytes support axonal regeneration of fish and mammalian retinal ganglion cells. *Glia* 8 (1): 1–11. doi:10.1002/glia.440080102.

Bastmeyer, M., M. Beckmann, M. E. Schwab, and C. A. Stuermer. 1991. Growth of regenerating goldfish axons is inhibited by rat oligodendrocytes and CNS myelin but not but not by goldfish optic nerve tract oligodendrocytelike cells and fish CNS myelin. *Journal of Neuroscience* 11 (3): 626–640.

Bateman, J. F., R. P. Boot-Handford, and S. R. Lamandé. 2009. Genetic diseases of connective tissues: Cellular and extracellular effects of ECM mutations. *Nature Reviews Genetics* 10 (3): 173–183. doi:10.1038/nrg2520.

Beck, K., T. W. Dixon, J. Engel, and D. A. D. Parry. 1993. Ionic interactions in the coiled-coil domain of laminin determine the specificity of chain assembly. *Journal of Molecular Biology* 231 (2): 311–323.

Becker, C. G., and T. Becker. 2002. Repellent guidance of regenerating optic axons by chondroitin sulfate glycosaminoglycans in zebrafish. *Journal of Neuroscience* 22 (3): 842–853.

Becker, C. G., J. Schweitzer, J. Feldner, M. Schachner, and T. Becker. 2004. Tenascin-R as a repellent guidance molecule for newly growing and regenerating optic axons in adult zebrafish. *Molecular and Cellular Neuroscience* 26 (3): 376–389. doi:10.1016/j.mcn.2004.03.003.

Becker, T., M. F. Wullimann, C. G. Becker, R. R. Bernhardt, and M. Schachner. 1997. Axonal regrowth after spinal cord transection in adult zebrafish. *The Journal of Comparative Neurology* 377 (4): 577–595.

Bentzinger, C. F., P. Barzaghi, S. Lin, and M. A. Rüegg. 2005. Overexpression of mini-agrin in skeletal muscle increases muscle integrity and regenerative capacity in laminin-A2-deficient mice. *The FASEB Journal* 19 (8): 934–942.

Bernal, G. M., and D. A. Peterson. 2011. Phenotypic and gene expression modification with normal brain aging in GFAP-positive astrocytes and neural stem cells. *Aging Cell* 10 (3): 466–482. doi:10.1111/j.1474-9726.2011.00694.x.

Bernstein, J. J., and M. E. Bernstein. 1969. Ultrastructure of normal regeneration and loss of regenerative capacity following teflon blockage in goldfish spinal cord. *Experimental Neurology* 24 (4): 538–557.

Bernstein, J. J., and J. B. Gelderd. 1970. Regeneration of the long spinal tracts in the goldfish. *Brain Research* 20 (1): 33–38.

Boppart, M. D., D. J. Burkin, and S. J. Kaufman. 2006. Alpha7beta1-Integrin regulates mechanotransduction and prevents skeletal muscle injury. *American Journal of Physiology. Cell Physiology* 290 (6): C1660–C1665. doi:10.1152/ajpcell.00317.2005.

Briona, L. K., F. E. Poulain, C. Mosimann, and R. I. Dorsky. 2015. Wnt/SS-catenin signaling is required for radial glial neurogenesis following spinal cord injury. *Developmental Biology* 403 (1): 15–21. doi:10.1016/j.ydbio.2015.03.025.

Brodsky, B., and A. V. Persikov. 2005. Molecular structure of the collagen triple helix. *Advances in Protein Chemistry* 70: 301–339. doi:10.1016/S0065-3233(05)70009-7.

Bunnell, T. M., M. A. Jaeger, D. P. Fitzsimons, K. W. Prins, and J. M. Ervasti. 2008. Destabilization of the dystrophin-glycoprotein complex without functional deficits in A-dystrobrevin null muscle. *PloS One* 3 (7): e2604. doi:10.1371/journal.pone.0002604.

Burkin, D. J., and S. J. Kaufman. 1999. The alpha7beta1 integrin in muscle development and disease. *Cell and Tissue Research* 296 (1): 183–190.

Caroni, P., and M. E. Schwab. 1988. Antibody against myelin-associated inhibitor of neurite growth neutralizes nonpermissive substrate properties of CNS white matter. *Neuron* 1 (1): 85–96.

Charge, S. B. P., and M. A. Rudnicki. 2004. Cellular and molecular regulation of muscle regeneration. *Physiological Reviews* 84 (1): 209–238. doi:10.1152/physrev.00019.2003.

Chernoff, E. A. 1996. Spinal cord regeneration: A phenomenon unique to urodeles?. *The International Journal of Developmental Biology* 40 (4): 823–831.

Chong, S. W., V. Korzh, and Y. J. Jiang. 2009. Myogenesis and molecules—Insights from zebrafish danio rerio. *Journal of Fish Biology* 74 (8): 1693–1755.

Clark, E. A., and J. S. Brugge. 1995. Integrins and signal transduction pathways: The road taken. *Science* 268 (5208): 233–239.

Cohen, I., T. Sivron, V. Lavie, E. Blaugrund, and M. Schwartz. 1994. Vimentin immunoreactive glial-cells in the fish optic-nerve–Implications for regeneration. *Glia* 10 (1): 16–29. doi:10.1002/glia.440100104.

Collins, J., and C. G. Bönnemann. 2010. Congenital muscular dystrophies: Toward molecular therapeutic interventions. *Current Neurology and Neuroscience Reports* 10 (2): 83–91. doi:10.1007/s11910-010-0092-8.

Colognato, H., D. A. Winkelmann, and P. D. Yurchenco. 1999. Laminin polymerization induces a receptor-cytoskeleton network. *The Journal of Cell Biology* 145 (3): 619–631.

Cregg, J. M., M. A. DePaul, A. R. Filous, B. T. Lang, A. Tran, and J. Silver. 2014. Functional regeneration beyond the glial scar. *Experimental Neurology* 253: 197–207. doi:10.1016/j.expneurol.2013.12.024.

d'Albis, A., R. Couteaux, C. Janmot, A. Roulet, and J. C. Mira. 1988. Regeneration after cardiotoxin injury of innervated and denervated slow and fast muscles of mammals. *European Journal of Biochemistry* 174 (1): 103–110.

de los Angeles Beytía, M., G. Dekomien, S. Hoffjan, V. Haug, C. Anastasopoulos, and J. Kirschner. 2014. High creatine kinase levels and white matter changes: Clinical and genetic spectrum of congenital muscular dystrophies with laminin alpha-2 deficiency. *Molecular and Cellular Probes* 28 (4): 118–122.

de Souza, J., and C. Gottfried. 2013. Muscle injury: Review of experimental models. *Journal of Electromyography and Kinesiology* 23 (6): 1253–1260. doi:10.1016/j.jelekin.2013.07.009.

Diekmann, H., M. Klinger, T. Oertle, D. Heinz, H.-M. Pogoda, M. E. Schwab, and C. A. O. Stuermer. 2005. Analysis of the reticulon gene family demonstrates the absence of the neurite growth inhibitor nogo-a in fish. *Molecular Biology and Evolution* 22 (8): 1635–1648. doi:10.1093/molbev/msi158.

do Carmo Cunha, J., B. de Freitas Azevedo Levy, B. A. de Luca, M. S. de Andrade, V. C. Gomide, and G. Chadi. 2007. Responses of reactive astrocytes containing S100beta protein and fibroblast growth factor-2 in the border and in the adjacent preserved tissue after a contusion injury of the spinal cord in rats: Implications for wound repair and neuroregeneration. *Wound Repair Regen* 15 (1): 134–146. doi:WRR194[pii] 10.1111/j.1524-475X.2006.00194.x.

Dono, R. 2003. Fibroblast growth factors as regulators of central nervous system development and function. *American Journal of Physiology Regulatory, Integrative Comparative Physiology* 284 (4): R867–R881. doi:10.1152/ajpregu.00533.2002 284/4/R867[pii].

Duggal, N., R. Schmidt-Kastner, and A. M. Hakim. 1997. Nestin expression in reactive astrocytes following focal cerebral ischemia in rats. *Brain Research* 768 (1–2): 1–9.

Dumont, N. A., Y. X. Wang, and M. A. Rudnicki. 2015a. Intrinsic and extrinsic mechanisms regulating satellite cell function. *Development* 142 (9): 1572–1581. doi:10.1242/dev.114223.

Dumont, N. A., Y. X. Wang, J. von Maltzahn, A. Pasut, C. F. Bentzinger, C. E. Brun, and M. A. Rudnicki. 2015b. Dystrophin expression in muscle stem cells regulates their polarity and asymmetric division. *Nature Medicine* 21 (12): 1455–1463. doi:10.1038/nm.3990.

Durbeej, M. 2010. Laminins. *Cell and Tissue Research* 339 (1): 259–268. doi:10.1007/s00441-009-0838-2.

Dziadek, M., and R. Timpl. 1985. Expression of nidogen and laminin in basement membranes during mouse embryogenesis and in teratocarcinoma cells. *Developmental Biology* 111 (2): 372–382. doi:10.1016/0012-1606(85)90491-9.

Edwards, M. M., E. Mammadova-Bach, F. Alpy, A. Klein, W. L. Hicks, M. Roux, and P. Simon-Assmann et al. 2010. Mutations in lama1 disrupt retinal vascular development and inner limiting membrane formation. *The Journal of Biological Chemistry* 285 (10): 7697–7711. doi:10.1074/jbc. M109.069575.

Ehrig, K., I. Leivo, and W.S. Argraves. 1990. Merosin, a tissue-specific basement membrane protein, is a laminin-like protein. *The Muscular Dystrophies* 359 (9307): 687–695. doi:10.1016/S0140-6736(02)07815-7.

Emery, A. E. 2002. The muscular dystrophies. *Lancet* 359 (9307): 687–695.

Engvall, E. 1993. Laminin variants: Why, where and when?. *Kidney International* 43: 2–6.

Engvall, E. 1994. Cell adhesion in muscle. *Brazilian Journal of Medical and Biological Research= Revista Brasileira De Pesquisas Medicas E Biologicas/ Sociedade Brasileira De Biofisica...[Et Al.]* 27 (9): 2213–2227.

Ervasti, J. M., and K. P. Campbell. 1993. A role for the dystrophin-glycoprotein complex as a transmembrane linker between laminin and actin. *The Journal of Cell Biology* 122 (4): 809–823.

Fawcett, J. W., J. Rokos, and I. Bakst. 1989. Oligodendrocytes repel axons and cause axonal growth cone collapse. *Journal of Cell Science* 92 (Pt 1): 93–100.

Figeac, N., M. Daczewska, C. Marcelle, and K. Jagla. 2007. Muscle stem cells and model systems for their investigation. *Developmental Dynamics* 236 (12): 3332–3342. doi:10.1002/dvdy.21345.

Foster, R. F., J. M. Thompson, and S. J. Kaufman. 1987. A laminin substrate promotes myogenesis in rat skeletal muscle cultures: Analysis of replication and development using antidesmin and Anti-BrdUrd monoclonal antibodies. *Developmental Biology* 122 (1): 11–20.

Fox, J. W., U. Mayer, R. Nischt, M. Aumailley, D. Reinhardt, H. Wiedemann, K. Mann, R. Timpl, T. Krieg, and J. Engel. 1991. Recombinant nidogen consists of three globular domains and mediates binding of laminin to collagen Type IV. *The EMBO Journal* 10 (11): 3137.

Frantz, C., K. M. Stewart, and V. M. Weaver. 2010. The extracellular matrix at a glance. *Journal of Cell Science* 123 (Pt 24): 4195–4200. doi:10.1242/jcs.023820.

Ganat, Y., S. Soni, M. Chacon, M. L. Schwartz, and F. M. Vaccarino. 2002. Chronic hypoxia up-regulates fibroblast growth factor ligands in the perinatal brain and induces fibroblast growth factor-responsive radial glial cells in the sub-ependymal zone. *Neuroscience* 112 (4): 977–991. doi:S030645220200060X [pii].

Gawlik, K. I., M. Akerlund, V. Carmignac, H. Elamaa, and M. Durbeej. 2010. Distinct roles for laminin globular domains in laminin alpha1 chain mediated rescue of murine laminin alpha2 chain deficiency. *PloS One* 5 (7): e11549. doi:10.1371/journal.pone.0011549.

Gawlik, K. I., and M. Durbeej. 2010. Transgenic overexpression of laminin A1 chain in laminin A2 chain-deficient mice rescues the disease throughout the lifespan. *Muscle & Nerve* 42 (1): 30–37. doi:10.1002/mus.21616.

Gawlik, K. I., and M. Durbeej. 2011. Skeletal muscle laminin and MDC1A: Pathogenesis and treatment strategies. *Skeletal Muscle* 1 (1): 9. doi:10.1186/ 2044-5040-1-9.

Gawlik, K. I., J. Y. Li, and Å. Petersén. 2006. Laminin A1 chain improves laminin A2 chain deficient peripheral neuropathy. *Human Molecular Genetics* 15(18): 2690–2700.

Gawlik, K., Y. Miyagoe-Suzuki, P. Ekblom, S. I. Takeda, and M. Durbeej. 2004. Laminin alpha1 chain reduces muscular dystrophy in laminin alpha2 chain deficient mice. *Human Molecular Genetics* 13 (16): 1775–1784. doi:10.1093/hmg/ddh190.

Geranmayeh, F., E. Clement, L. H. Feng, C. Sewry, J. Pagan, R. Mein, S. Abbs, L. Brueton, A.-M. Childs, and H. Jungbluth. 2010. Genotype–phenotype correlation in a large population of muscular dystrophy patients with LAMA2 mutations. *Neuromuscular Disorders* 20 (4): 241–250.

Giancotti, F. G., and E. Ruoslahti. 1999. Integrin signaling. *Science* 285 (5430): 1028–1033. doi:10.1126/science.285.5430.1028.

Giger, R. J., E. R. Hollis, 2nd, and M. H. Tuszynski. 2010. Guidance molecules in axon regeneration. *Cold Spring Harbor Perspectives in Biology* 2 (7): a001867. doi:10.1101/cshperspect.a001867.

Gnocchi, V. F., R. B. White, Y. Ono, J. A. Ellis, and P. S. Zammit. 2009. Further characterisation of the molecular signature of quiescent and activated mouse muscle satellite cells. *PloS One* 4 (4): e5205.

Goldshmit, Y., F. Frisca, J. Kaslin, A. R. Pinto, J. K. K. Y. Tang, A. Pebay, R. Pinkas-Kramarski, and P. D. Currie. 2015. Decreased anti-regenerative effects after spinal cord injury in Spry4-/-mice. *Neuroscience* 287 (February): 104–112. doi:10.1016/j.neuroscience.2014.12.020.

Goldshmit, Y., F. Frisca, A. R. Pinto, A. Pebay, J. K. K. Y. Tang, A. L. Siegel, J. Kaslin, and P.D. Currie. 2014. Fgf2 improves functional recovery-decreasing gliosis and increasing radial glia and neural progenitor cells after spinal cord injury. *Brain and Behavior* 4 (2): 187–200. doi:10.1002/brb3.172.

Goldshmit, Y., T. E. Sztal, P. R. Jusuf, T. E. Hall, M. Nguyen-Chi, and P. D. Currie. 2012. Fgf-dependent glial cell bridges facilitate spinal cord regeneration in zebrafish. *Journal of Neuroscience* 32 (22): 7477–7492. doi:10.1523/JNEUROSCI.0758-12.2012.

Gomez-Pinilla, F., and C. W. Cotman. 1992. Transient lesion-induced increase of basic fibroblast growth factor and its receptor in layer VIb (subplate cells) of the adult rat cerebral cortex. *Neuroscience* 49 (4): 771–780. doi:0306-4522(92)90355-6 [pii].

Goodman, S. L., G. Risse, and K. von der Mark. 1989. The E8 subfragment of laminin promotes locomotion of myoblasts over extracellular matrix. *Journal of Cell Biology* 109 (2): 799–809.

Gotz, M., A. Stoykova, and P. Gruss. 1998. Pax6 controls radial glia differentiation in the cerebral cortex. *Neuron* 21 (5): 1031–1044.

Goudenege, S., Y. Lamarre, N. Dumont, J. Rousseau, J. Frenette, D. Skuk, and J. P. Tremblay. 2010. Laminin-111: A potential therapeutic agent for duchenne muscular dystrophy. *Molecular Therapy* 18 (12): 2155–2163.

Grefte, S., A. M. Kuijpers-Jagtman, R. Torensma, and J. W. Von den Hoff. 2007. Skeletal muscle development and regeneration. *Stem Cells and Development* 16 (5): 857–868. doi:10.1089/scd.2007.0058.

Gritti, A., E. A. Parati, L. Cova, P. Frolichsthal, R. Galli, E. Wanke, L. Faravelli, D. J. Morassutti, F. Roisen, D. D. Nickel, and A. L. Vescovi. 1996. Multipotential stem cells from the adult mouse brain proliferate and self-renew in response to basic fibroblast growth factor. *Journal of Neuroscience* 16 (3): 1091–1100.

Grothe, C., C. Meisinger, and P. Claus. 2001. In vivo expression and localization of the fibroblast growth factor system in the intact and lesioned rat peripheral nerve and spinal ganglia. *Journal of Comparative Neurology* 434 (3): 342–357.

Groulx, J.-F., D. Gagné, Y. D. Benoit, D. Martel, N. Basora, and J.-F. Beaulieu. 2011. Collagen VI is a basement membrane component that regulates epithelial cell–fibronectin interactions. *Matrix Biology* 30 (3): 195–206.

Grounds, M. D. 2008. Complexity of extracellular matrix and skeletal muscle regeneration. *Advances in Muscle Research* 3: 269–302. doi:10.1007/978-1-4020-6768-6_13.

Grumbles, R. M., S. Sesodia, P. M. Wood, and C. K. Thomas. 2009. Neurotrophic factors improve motoneuron survival and function of muscle reinnervated by embryonic neurons. *Journal of Neuropathology and Experimental Neurology* 68 (7): 736–746. doi:10.1097/NEN.0b013e3181a9360f.

Guan, J.-L. 1997. Role of focal adhesion kinase in integrin signaling. *The International Journal of Biochemistry & Cell Biology* 29 (8–9): 1085–1096. doi:10.1016/S1357-2725(97)00051-4.

Guo, Y., L. Ma, M. Cristofanilli, R. P. Hart, A. Hao, and M. Schachner. 2011. Transcription factor Sox11b is involved in spinal cord regeneration in adult zebrafish. *Neuroscience* 172 (January): 329–341. doi:10.1016/j.neuroscience.2010.10.026.

Gurevich, D. B., P. D. Nguyen, A. L. Siegel, O. V. Ehrlich, C. Sonntag, J. M. N. Phan, and S. Berger et al. 2016. Asymmetric division of clonal muscle stem cells coordinates muscle regeneration in vivo. *Science*, May. doi:10.1126/science.aad9969.

Guyon, J. R., A. N. Mosley, S. J. Jun, F. Montanaro, L. S. Steffen, Y. Zhou, V. Nigro, L. I. Zon, and L. M. Kunkel. 2005. Δ-Sarcoglycan is required for early zebrafish muscle organization. *Experimental Cell Research* 304 (1): 105–115.

Hack, A. A., M. Y. Lam, L. Cordier, D. I. Shoturma, C. T. Ly, M. A. Hadhazy, M. R. Hadhazy, H. L. Sweeney, and E. M. McNally. 2000. Differential requirement for individual sarcoglycans and dystrophin in the assembly and function of the dystrophin-glycoprotein complex. *Journal of Cell Science* 113 (14): 2535–2544.

Hall, T. E., R. J. Bryson-Richardson, S. Berger, A. S. Jacoby, N. J. Cole, G. E. Hollway, J. Berger, and P. D. Currie. 2007. The zebrafish candyfloss mutant implicates extracellular matrix adhesion failure in laminin 2-deficient congenital muscular dystrophy. *Proceedings of the National Academy of Sciences* 104 (17): 7092–7097. doi:10.1073/pnas.0700942104.

Han, R., M. Kanagawa, T. Yoshida-Moriguchi, E. P. Rader, R. A. Ng, D. E. Michele, and D. E. Muirhead et al. 2009. Basal lamina strengthens cell membrane integrity via the laminin G domain-binding motif of A-dystroglycan. *Proceedings of the National Academy of Sciences* 106 (31): 12573–12579. doi:10.1073/pnas.0906545106.

Hannigan, G., A. A. Troussard, and S. Dedhar. 2005. Integrin-linked kinase: A cancer therapeutic target unique among its ILK. *Nature Reviews Cancer* 5 (1): 51–63. doi:10.1038/nrc1524.

Hasan, S. J., H. S. Keirstead, G. D. Muir, and J. D. Steeves. 1993. Axonal regeneration contributes to repair of injured brainstem-spinal neurons in embryonic chick. *The Journal of Neuroscience* 13 (2): 492–507.

Hawke, T. J. 2001. Myogenic satellite cells: Physiology to molecular biology. *Journal of Applied Physiology* 91(2): 534–551.

Hayashi, Y. K., Z. Tezak, T. Momoi, I. Nonaka, C. A. Garcia, E. P. Hoffman, and K. Arahata. 2001. Massive muscle cell degeneration in the early stage of merosin-deficient congenital muscular dystrophy. *Neuromuscular Disorders* 11 (4): 350–359.

Helbling-Leclerc, A., X. Zhang, H. Topaloglu, C. Cruaud, F. Tesson, J. Weissenbach, and F. M. S. Tomé et al. 1995. Mutations in the laminin A2–chain gene (LAMA2) cause merosin–deficient congenital muscular dystrophy. *Nature Genetics* 11 (2): 216–218. doi:10.1038/ng1095-216.

Henry, M. D., and K. P. Campbell. 1996. Dystroglycan: An extracellular matrix receptor linked to the cytoskeleton. Current opinion in cell biology 8 (5): 625–631. doi:10.1016/S0955-0674(96)80103-7.

Hudson, B. G., S. T. Reeders, and K. Tryggvason. 1993. Type IV collagen: Structure, gene organization, and role in human diseases. Molecular basis of goodpasture and alport syndromes and diffuse leiomyomatosis. *The Journal of Biological Chemistry* 268(35): 26033–26036.

Hughes, S. M., and H. M. Blau. 1990. Migration of myoblasts across basal lamina during skeletal muscle development. *Nature* 345 (6273): 350–353.

Hui, S. P., T. C. Nag, and S. Ghosh. 2015. Characterization of proliferating neural progenitors after spinal cord injury in adult zebrafish. *PloS One* 10(12): e0143595.

Huijing, P. A. 1997. Muscle as a collagen fiber reinforced composite: A review of force transmission in muscle and whole limb. *Journal of Biomechanics* 32 (4): 329–345.

Hurme, T., and H. Kalimo. 1992. Adhesion in skeletal muscle during regteneration. *Muscle & Nerve* 15 (4): 482–489.

Hurme, T., H. Kalimo, M. Lehto, and M. Järvinen. 1991. Healing of skeletal muscle injury: An ultrastructural and immunohistochemical study. *Medicine and Science in Sports and Exercise* 23 (7): 801–810.

Hynes, R. O. 1987. Integrins: A family of cell surface receptors. *Cell* 48 (4): 549–554.

Iwashita, Y., S. Kawaguchi, and M. Murata. 1994. Restoration of function by replacement of spinal cord segments in the rat. *Nature* 367 (6459): 167–170. doi:10.1038/367167a0.

Janoštiak, R., A. C. Pataki, J. Brábek, and D. Rösel. 2014. Mechanosensors in integrin signaling: The emerging role of p130Cas. *Cell Migration and Invasion in Physiology and Pathology* 93 (10–12): 445–454.

Järvinen, T. A. H., T. L. N. Järvinen, M. Kääriäinen, V. Äärimaa, S. Vaittinen, H. Kalimo, and M. Järvinen. 2007. Muscle injuries: Optimising recovery. *Best Practice & Research Clinical Rheumatology* 21 (2): 317–331.

Jimenez-Mallebrera, C., S. C. Brown, and C. A. Sewry. 2005. Congenital muscular dystrophy: Molecular and cellular aspects. *Cellular and Molecular* 62(7–8): 809–823.

Jones, L. L., R. U. Margolis, and M. H. Tuszynski. 2003. The chondroitin sulfate proteoglycans neurocan, brevican, phosphacan, and versican are differentially regulated following spinal cord injury. *Experimental Neurology* 182 (2): 399–411.

Kääriäinen, M., J. Kääriäinen, T. L. Järvinen, L. Nissinen, J. Heino, M. Järvinen, and H. Kalimo. 2000. Integrin and dystrophin associated adhesion protein complexes during regeneration of shearing-type muscle injury. *Neuromuscular Disorders* 10 (2): 121–32.

Kääriäinen, M., J. Kääriäinen, T. L. N. Järvinen, H. Sievänen, H. Kalimo, and M. Järvinen. 1998. Correlation between biomechanical and structural changes during the regeneration of skeletal muscle after laceration injury. *Journal of Orthopaedic Research* 16 (2): 197–206. doi:10.1002/jor.1100160207.

Kadler, K. E., C. Baldock, J. Bella, and R. P. Boot-Handford. 2007. Collagens at a glance. *Journal of Cell Science* 120 (12): 1955–1958.

Kadoya, K., S. Tsukada, P. Lu, G. Coppola, D. Geschwind, M. T. Filbin, A. Blesch, and M. H. Tuszynski. 2009. Combined intrinsic and extrinsic neuronal mechanisms facilitate bridging axonal regeneration one year after spinal cord injury. *Neuron* 64 (2): 165–172. doi:10.1016/j.neuron.2009.09.016.

Kaslin, J., J. Ganz, and M. Brand. 2008. Proliferation, neurogenesis and regeneration in the non-mammalian vertebrate brain. *Philosophical Transactions of the Royal Society of London B: Biologicl Sciences* 363 (1489): 101–122. doi:10.1098/rstb.2006.2015.

doi:Kaslin, J., J. Ganz, M. Geffarth, and H. Grandel. 2009. Stem cells in the adult zebrafish cerebellum: Initiation and maintenance of a novel stem cell niche. *The Journal of Neuroscience* 29 (19): 6142–6153.

Kaufman, S. J., R. F. Foster, K. R. Haye, and L. E. Faiman. 1985. Expression of a developmentally regulated antigen on the surface of skeletal and cardiac muscle cells. *Journal of Cell Biology* 100 (6): 1977–1987.

Keefe, A. C., J. A. Lawson, S. D. Flygare, Z. D. Fox, M. P. Colasanto, S. J. Mathew, M. Yandell, and G. Kardon. 2015. Muscle stem cells contribute to myofibres in sedentary adult mice. *Nature Communications* 6: 7087. doi:10.1038/ncomms8087.

Khoshnoodi, J., V. Pedchenko, and B. G. Hudson. 2008. Mammalian collagen IV. *Microscopy Research and Technique* 71 (5): 357–370. doi:10.1002/jemt.20564.

Kroehne, V., D. Freudenreich, S. Hans, J. Kaslin, and M. Brand. 2011. Regeneration of the adult zebrafish brain from neurogenic radial glia-type progenitors. *Development* 138 (22): 4831–4841. doi:10.1242/dev.072587.

Kuang, W., H. Xu, P. H. Vachon, and E. Engvall. 1998. Disruption of thelama2Gene in embryonic stem cells: Laminin A2 is necessary for sustenance of mature muscle cells. *Experimental Cell Research* 241 (1): 117–125.

Kuang, W., H. Xu, J. T. Vilquin, and E. Engvall. 1999. Activation of the lama2 gene in muscle regeneration: Abortive regeneration in laminin Alpha2-Deficiency. *Laboratory Investigation: A Journal of Technical Methods and Pathology* 79 (12): 1601–1613.

Kühl, U., R. Timpl, and K. von der Mark. 1982. Synthesis of Type IV collagen and laminin in cultures of skeletal muscle cells and their assembly on the surface of myotubes. *Developmental Biology* 93 (2): 344–354.

Kuo, H. -J., C. L. Maslen, D. R. Keene, and R. W. Glanville. 1997. Type VI collagen anchors endothelial basement membranes by interacting with Type IV collagen. *Journal of Biological Chemistry* 272 (42): 26522–26529.

Lampe, A. K., and K. M. D. Bushby. 2005. Collagen VI related muscle disorders. *Journal of Medical Genetics* 42 (9): 673–685.

Lapidos, K. A., R. Kakkar, and E. M. McNally. 2004. The dystrophin glycoprotein complex: Signaling strength and integrity for the sarcolemma. *Circulation Research* 94 (8): 1023–1031.

Laprise, P., È. Marie Poirier, A. Vézina, N. Rivard, and P. H. Vachon. 2002. Merosin-integrin promotion of skeletal myofiber cell survival: Differentiation state-distinct involvement of p60Fyn tyrosine kinase and P38α stress-activated MAP kinase. *Journal of Cellular Physiology* 191 (1): 69–81.

LeBleu, V. S., and B. MacDonald. 2007. Structure and function of basement membranes. *Experimental Biology and Medicine* 232(9): 1121–1129.

Lehto, M., and M. Järvinen. 1985. Collagen and glycosaminoglycan synthesis of injured gastrocnemius muscle in rat. *European Surgical Research* 17 (3): 179–185.

Lehto, M., V. C. Duance, and D. Restall. 1985. Collagen and fibronectin in a healing skeletal muscle injury. An immunohistological study of the effects of physical activity on the repair of injured gastrocnemius muscle in the rat. *Journal of Bone & Joint Surgery, British* 67 (5): 820–828.

Li, S., D. Harrison, S. Carbonetto, R. Fassler, N. Smyth, D. Edgar, and P. D. Yurchenco. 2002. Matrix assembly, regulation, and survival functions of laminin and its receptors in embryonic stem cell differentiation. *Journal of Cell Biology* 157 (7): 1279–1290.

Lu, P., L. Graham, Y. Wang, D. Wu, and M. Tuszynski. 2014a. Promotion of survival and differentiation of neural stem cells with fibrin and growth factor cocktails after severe spinal cord injury. *Journal of Visualized Experiments* (89): e50641. doi:10.3791/50641.

Lu, P., G. Woodruff, Y. Wang, L. Graham, M. Hunt, D. Wu, and E. Boehle et al. 2014b. Long-distance axonal growth from human induced pluripotent stem cells after spinal cord injury. *Neuron* 83 (4): 789–796. doi:10.1016/j.neuron.2014.07.014.

Malatesta, P., M. A. Hack, E. Hartfuss, H. Kettenmann, W. Klinkert, F. Kirchhoff, and M. Gotz. 2003. Neuronal or glial progeny: Regional differences in radial glia fate. *Neuron* 37 (5): 751–764.

Mauro, A. 1961. Satellite cell of skeletal muscle fibers. *The Journal of Biophysical and Biochemical Cytology* 9 (February): 493–495.

Mayer, U., G. Saher, R. Fässler, and A. Bornemann. 1997. Absence of integrin A7 causes a novel form of muscular dystrophy. *Nature Genetics* 17 (3): 318–323.

McKee, K. K., D. Harrison, S. Capizzi, and P. D. Yurchenco. 2007. Role of laminin terminal globular domains in basement membrane assembly. *The Journal of Biological Chemistry* 282 (29): 21437–21447. doi:10.1074/jbc. M702963200.

McKeon, R. J., M. J. Jurynec, and C. R. Buck. 1999. The chondroitin sulfate proteoglycans neurocan and phosphacan are expressed by reactive astrocytes in the chronic CNS glial scar. *The Journal of Neuroscience: The Official Journal of the Society for Neuroscience* 19 (24): 10778–10788.

Mehuron, T., A. Kumar, L. Duarte, J. Yamauchi, A. Accorsi, and M. Girgenrath. 2014. Dysregulation of matricellular proteins is an early signature of pathology in laminin-deficient muscular dystrophy. *Skeletal Muscle* 4 (1): 14.

Melo, F., D. J. Carey, and E. Brandan. 1996. Extracellular matrix is required for skeletal muscle differentiation but not myogenin expression. *Journal of Cellular Biochemistry* 62 (2): 227–239. doi:10.1002/(SICI)1097-4644(199608)62:2<227:AID-JCB11>3.0.CO;2-I.

Menetrey, J., C. Kasemkijwattana, C. S. Day, P. Bosch, M. Vogt, F. H. Fu, M. S. Moreland, and J. Huard. 2000. Growth factors improve muscle healing in vivo. *Journal of Bone & Joint Surgery, British* 82 (1): 131–137.

Michele, D. E., and K. P. Campbell. 2003. Dystrophin-glycoprotein complex: Post-translational processing and dystroglycan function. *The Journal of Biological Chemistry* 278 (18): 15457–15460. http://www.jbc.org/content/278/18/15457.short.

Miner, J. H., and P. D. Yurchenco. 2004. Laminin functions in tissue morphogenesis. *Annual Review of Cell and Developmental Biology* 20: 255–284. doi:10.1146/annurev.cellbio.20.010403.094555.

Miner, J. H., C. Li, J. L. Mudd, G. Go, and A. E. Sutherland. 2004. Compositional and structural requirements for laminin and basement membranes during mouse embryo implantation and gastrulation. *Development* 131 (10): 2247–256. doi:10.1242/dev.01112.

Morgan, J. E., and P. S. Zammit. 2010. Direct effects of the pathogenic mutation on satellite cell function in muscular dystrophy. *Experimental Cell Research* 316 (18): 3100–3108. doi:10.1016/j.yexcr.2010.05.014.

Munoz, J., Y. Zhou, and H. W. Jarrett. 2010. LG4–5 domains of laminin-211 binds A-dystroglycan to allow myotube attachment and prevent anoikis. *Journal of Cellular Physiology* 222 (1): 111–119. doi:10.1002/jcp.21927.

Muntoni, F., and T. Voit. 2004. The congenital muscular dystrophies in 2004: A century of exciting progress. *Neuromuscular Disorders* 14 (10): 635–649. doi:10.1016/j.nmd.2004.06.009.

Nitzan, A., P. Kermer, A. Shirvan, M. Bahr, A. Barzilai, and A. S. Solomon. 2006. Examination of cellular and molecular events associated with optic nerve axotomy. *Glia* 54 (6): 545–556. doi:10.1002/glia.20398.

Noble, M., K. Murray, P. Stroobant, M. D. Waterfield, and P. Riddle. 1988. Platelet-derived growth factor promotes division and motility and inhibits premature differentiation of the oligodendrocyte/Type-2 astrocyte progenitor cell. *Nature* 333 (6173): 560–562. doi:10.1038/333560a0.

Öcalan, M., S. L. Goodman, U. Kühl, S. D. Hauschka, and K. Von der Mark. 1988. Laminin alters cell shape and stimulates motility and proliferation of murine skeletal myoblasts. *Developmental Biology* 125 (1): 158–167. doi:10.1016/0012-1606(88)90068-1.

Ohsawa, R., T. Ohtsuka, and R. Kageyama. 2005. Mash1 and Math3 are required for development of branchiomotor neurons and maintenance of neural progenitors. *The Journal of Neuroscience: The Official Journal of the Society for Neuroscience* 25 (25): 5857–5865. doi:10.1523/JNEUROSCI.4621-04.2005.

Otey, C. A., F. M. Pavalko, and K. Burridge. 1990. An interaction between alpha-actinin and the beta 1 integrin subunit in vitro. *Journal of Cell Biology* 111 (2): 721–729.

Otten, C. 233 cile, and S. Abdelilah-Seyfried. 2013. Laser-inflicted injury of zebrafish embryonic skeletal muscle. Journal of Visualized Experiments (71): e4351.

Palmer, T. D., E. A. Markakis, A. R. Willhoite, F. Safar, and F. H. Gage. 1999. Fibroblast growth factor-2 activates a latent neurogenic program in neural stem cells from diverse regions of the adult CNS. *Journal of Neuroscience* 19 (19): 8487–8497.

Palmer, T. D., J. Ray, and F. H. Gage. 1995. FGF-2-responsive neuronal progenitors reside in proliferative and quiescent regions of the adult rodent brain. *Molecular and Cellular Neuroscience* 6 (5): 474–486. doi:S1044-7431(85)71035-4[pii] 10.1006/mcne.1995.1035.

Park, H.-C., J. Shin, R. K. Roberts, and B. Appel. 2007. An Olig2 reporter gene marks oligodendrocyte precursors in the postembryonic spinal cord of zebrafish. *Developmental Dynamics* 236 (12): 3402–3407. doi:10.1002/dvdy.21365.

Parsons, J. T. 2003. Focal adhesion kinase: The first ten years. *Journal of Cell Science* 116 (8): 1409–1416.

Parsons, M. J., S. M. Pollard, L. Saúde, B. Feldman, P. Coutinho, E. M. A. Hirst, and D. L. Stemple. 2002. Zebrafish mutants identify an essential role for laminins in notochord formation. *Development* 129 (13): 3137–3146.

Pasterkamp, R. J., R. J. Giger, M. J. Ruitenberg, A. J. Holtmaat, J. De Wit, F. De Winter, and J. Verhaagen. 1999a. Expression of the gene encoding the chemorepellent semaphorin III is induced in the fibroblast component of neural scar tissue formed following injuries of adult but not neonatal CNS. *Molecular and Cellular Neuroscience* 13 (2): 143–166. doi:10.1006/mcne.1999.0738.

Pasterkamp, R. J., M. J. Ruitenberg, and J. Verhaagen. 1999b. Semaphorins and their receptors in olfactory axon guidance. *Cellular and Molecular Biology (Noisy-le-grand)* 45 (6): 763–779.

Patel, A. J., C. Wickenden, A. Jen, and H. A. R. de Silva. 1996. Glial cell derived neurotrophic factors and alzheimer's disease. *Neurodegeneration* 5(4): 489–496.

Patton, B. L., A. Y. Chiu, and J. R. Sanes. 1998. Synaptic laminin prevents glial entry into the synaptic cleft. *Nature* 393 (6686): 698–701. doi:10.1038/31502.

Patton, B. L., J. H. Miner, A. Y. Chiu, and J. R. Sanes. 1997. Distribution and function of laminins in the neuromuscular system of developing, adult, and mutant mice. *The Journal of Cell Biology* 139 (6): 1507–1521.

Pegoraro, E., H. Marks, C. A. Garcia, T. Crawford, P. Mancias, A. M. Connolly, M. Fanin, F. Martinello, C. P. Trevisan, and C. Angelini. 1998. Laminin A2 muscular dystrophy genotype/phenotype studies of 22 patients. *Neurology* 51 (1): 101–110.

Petter-Puchner, A. H., W. Froetscher, R. Krametter-Froetscher, D. Lorinson, H. Redl, and M. van Griensven. 2007. The long-term neurocompatibility of human fibrin sealant and equine collagen as biomatrices in experimental spinal cord injury. *Experimental and Toxicologic Pathology: Official Journal of the Gesellschaft Für Toxikologische Pathologie* 58 (4): 237–245. doi:10.1016/j.etp.2006.07.004.

Pierschbacher, M. D., and E. Ruoslahti. 1983. Cell attachment activity of fibronectin can be duplicated by small synthetic fragments of the molecule. *Nature* 309 (5963): 30–33.

Postel, R., P. Vakeel, J. Topczewski, R. Knöll, and J. Bakkers. 2008. Zebrafish integrin-linked kinase is required in skeletal muscles for strengthening the integrin-ECM adhesion complex. *Developmental Biology* 318 (1): 92–101. doi:10.1016/j.ydbio.2008.03.024.

Pringle, N. P., W. P. Yu, M. Howell, J. S. Colvin, D. M. Ornitz, and W. D. Richardson. 2003. Fgfr3 expression by astrocytes and their precursors: Evidence that astrocytes and oligodendrocytes originate in distinct neuroepithelial domains. *Development* 130 (1): 93–102. doi:10.1242/dev.00184.

Quintero, A. J., V. J. Wright, F. H. Fu, and J. Huard. 2009. Stem cells for the treatment of skeletal muscle injury. *Clinics in Sports Medicine* 28 (1): 1–11.

Rakic, P. 2003. Elusive radial glial cells: Historical and evolutionary perspective. *Glia* 43 (1): 19–32. doi:10.1002/glia.10244.

Rasmussen, J. P., and A. Sagasti. 2016. Learning to swim, again: Axon regeneration in fish. *Experimental Neurology* 287: 318–330. doi:10.1016/j.expneurol.2016.02.022.

Rehermann, M. I., F. F. Santiñaque, B. López-Carro, R. E. Russo, and O. Trujillo-Cenóz. 2011. Cell proliferation and cytoarchitectural remodeling during spinal cord reconnection in the fresh-water turtle trachemys dorbignyi. *Cell and Tissue Research* 344 (3): 415–433. doi:10.1007/s00441-011-1173-y.

Reimer, M. M., V. Kuscha, C. Wyatt, I. Sorensen, R. E. Frank, M. Knuwer, T. Becker, and C. G. Becker. 2009. Sonic hedgehog is a polarized signal for motor neuron regeneration in adult zebrafish. *The Journal of Neuroscience: The Official Journal of the Society for Neuroscience* 29 (48): 15073–15082. doi:10.1523/JNEUROSCI.4748-09.2009.

Reimer, M. M., I. Sorensen, and V. Kuscha. 2008. Motor neuron regeneration in adult zebrafish. *The Journal of Neuroscience* 28 (34) 8510–8516.

Richardson, W. D., N. Pringle, M. J. Mosley, B. Westermark, and M. Dubois-Dalcq. 1988. A role for platelet-derived growth factor in normal gliogenesis in the central nervous system. *Cell* 53 (2): 309–319.

Riederer, I., A. C. Bonomo, V. Mouly, and W. Savino. 2015. Laminin therapy for the promotion of muscle regeneration. *Febs Letters* 589 (22): 3449–53. doi:10.1016/j.febslet.2015.10.004.

Rooney, J. E., P. B. Gurpur, and D. J. Burkin. 2009a. Laminin-111 protein therapy prevents muscle disease in the mdx mouse model for duchenne muscular dystrophy. *Proceedings of the National Academy of Sciences* 106 (19): 7991–7996. doi:10.1073/pnas.0811599106.

Rooney, J. E., P. B. Gurpur, Z. Yablonka-Reuveni, and D. J. Burkin. 2009b. Laminin-111 restores regenerative capacity in a mouse model for alpha7 integrin congenital myopathy. *The American Journal of Pathology* 174 (1): 256–264. doi:10.2353/ajpath.2009.080522.

Rooney, J. E., J. R. Knapp, B. L. Hodges, R. D. Wuebbles, and D. J. Burkin. 2012. Laminin-111 protein therapy reduces muscle pathology and improves viability of a mouse model of merosin-deficient congenital muscular dystrophy. *The American Journal of Pathology, February.* doi:10.1016/j.ajpath.2011.12.019.

Rosenzweig, S., D. Raz-Prag, A. Nitzan, R. Galron, M. Paz, G. Jeserich, G. Neufeld, A. Barzilai, and A. S. Solomon. 2010. Sema-3A indirectly disrupts the regeneration process of goldfish optic nerve after controlled injury. *Graefe's Archive for Clinical and Experimental Ophthalmology* 248 (10): 1423–1435. doi:10.1007/s00417-010-1377-y.

Rowlerson, A., G. Radaelli, F. Mascarello, and A. Veggetti. 1997. Regeneration of skeletal muscle in two teleost fish: Sparus aurata and brachydanio rerio. *Cell and Tissue Research* 289 (2): 311–22.

Ruoslahti, E. 1996. Rgd and other recognition sequences for integrins. *Annual Review of Cell and Developmental Biology* 12 (1): 697–715. doi:10.1146/annurev.cellbio.12.1.697.

Sage, H. 1982. Collagens of basement membranes. *Journal of Investigative Dermatology* 79 (s 1): 51–59.

Sahin Kaya, S., A. Mahmood, Y. Li, E. Yavuz, and M. Chopp. 1999. Expression of nestin after traumatic brain injury in rat brain. *Brain Research* 840 (1–2): 153–157.

Sanes, J. R., E. Engvall, R. Butkowski, and D. D. Hunter. 1990. Molecular heterogeneity of basal laminae: Isoforms of laminin and collagen IV at the neuromuscular junction and elsewhere. *The Journal of Cell Biology* 111 (4): 1685–1699.

Sasaki, T., K. Mann, J. H. Miner, N. Miosge, and R. Timpl. 2002. Domain IV of mouse laminin B1 and B2 chains. *European Journal of Biochemistry* 269 (2): 431–442.

Sato, K., Y. Li, W. Foster, K. Fukushima, N. Badlani, N. Adachi, A. Usas, F. H. Fu, and J. Huard. 2003. Improvement of muscle healing through enhancement of muscle regeneration and prevention of fibrosis. *Muscle & Nerve* 28 (3): 365–372.

Schuler, F., and L. M. Sorokin. 1995. Expression of laminin isoforms in mouse myogenic cells in vitro and in vivo. *Journal of Cell Science* 108 (12): 3795–3805. http://jcs.biologists.org/content/108/12/3795.short.

Schwab, M. E. 2004. Nogo and axon regeneration. *Current Opinion in Neurobiology* 14 (1): 118–124.

Schwalb, J. M., N. M. Boulis, M. F. Gu, J. Winickoff, P. S. Jackson, N. Irwin, and L. I.Benowitz. 1995. Two factors secreted by the goldfish optic nerve induce retinal ganglion cells to regenerate axons in culture. *Journal of Neuroscience* 15 (8): 5514–5525.

Seale, P., L. A. Sabourin, A. Girgis-Gabardo, A. Mansouri, P. Gruss, and M. A. Rudnicki. 2000. Pax7 is required for the specification of myogenic satellite cells. *Cell* 102 (6): 777–786. doi:10.1016/S0092-8674(00)00066-0.

Seger, C., M. Hargrave, X. Wang, R. J. Chai, S. Elworthy, and P. W. Ingham. 2011. Analysis of Pax7 expressing myogenic cells in zebrafish muscle development, injury, and models of disease. *Developmental Dynamics* 240 (11): 2440–2451. doi:10.1002/dvdy.22745.

Serrano, A. L., and P. Muñoz-Cánoves. 2010. Regulation and dysregulation of fibrosis in skeletal muscle. *Experimental Cell Research* 316 (18): 3050–3058. doi:10.1016/j.yexcr.2010.05.035.

265

Shetty, A. K., B. Hattiangady, and G. A. Shetty. 2005. Stem/progenitor cell proliferation factors FGF-2, IGF-1, and VEGF exhibit early decline during the course of aging in the hippocampus: Role of astrocytes. *Glia* 51 (3):173–186. doi:10.1002/glia.20187.

Shifman, M. I., and M. E. Selzer. 2007. Differential expression of class 3 and 4 semaphorins and netrin in the lamprey spinal cord during regeneration. *Journal of Comparative Neurology* 501 (4): 631–646. doi:10.1002/cne.21283.

Shimaoka, M., and T. A. Springer. 2003. Therapeutic antagonists and conformational regulation of integrin function. *Nature Reviews Drug Discovery* 2 (9): 703–716.

Shimizu, I., R. W. Oppenheim, and M O'Brien. 1990. Anatomical and functional recovery folloing spinal cord transection in the chick embryo. *Journal of neurobiology* 21 (6): 918–937.

Shirvan, A., M. Kimron, V. Holdengreber, I. Ziv, Y. Ben-Shaul, S. Melamed, E. Melamed, A. Barzilai, and A. S. Solomon. 2002. Anti-semaphorin 3A antibodies rescue retinal ganglion cells from cell death following optic nerve axotomy. *Journal of Biological Chemistry* 277 (51): 49799–49807. doi:10.1074/jbc. M204793200.

Siegel, A. L., K. Atchison, K. E. Fisher, and G. E. Davis. 2009. 3D timelapse analysis of muscle satellite cell motility. *Stem Cells* 27 (10): 2527–2538.

Siegel, A. L., D. B. Gurevich, and P. D. Currie. 2013. A myogenic precursor cell that could contribute to regeneration in zebrafish and its similarity to the satellite cell. *FEBS Journal* 280 (17): 4074–4088.

Silva-Barbosa, S. D., G. S. Butler-Browne, W. de Mello, I. Riederer, J. P. Di Santo, W. Savino, and V. Mouly. 2008. Human myoblast engraftment is improved in laminin-enriched microenvironment. *Transplantation* 85 (4): 566–575. doi:10.1097/TP.0b013e31815fee50.

Silver, J., and J. H. Miller. 2004. Regeneration beyond the glial scar. *Nature Reviews Neuroscience* 5 (2): 146–156. doi:10.1038/nrn1326nrn1326[pii].

Sîrbulescu, R. F., and G. K. H. Zupanc. 2011. Spinal cord repair in regeneration-competent vertebrates: Adult teleost fish as a model system. *Brain Research Reviews* 67 (1–2): 73–93. doi:10.1016/j.brainresrev.2010.11.001.

Smyth, N., H. Seda Vatansever, P. Murray, M. Meyer, C. Frie, M. Paulsson, and D. Edgar. 1999. Absence of basement membranes after targeting the LAMC1Gene results in embryonic lethality due to failure of endoderm differentiation. *The Journal of Cell Biology* 144 (1): 151–160. doi:10.1073/pnas.94.19.10189.

Stoiber, W., and A. M. Sanger. 1996. An electron microscopic investigation into the possible source of new muscle fibres in teleost fish. *Anatomy and Embryology* 194 (6): 569–579.

Sztal, T., S. Berger, and P. D. Currie. 2011. Characterization of the laminin gene family and evolution in zebrafish. *Developmental Dynamics* 240 (2): 422–431.

Sztal, T. E., C. Sonntag, T. E. Hall, and P. D. Currie. 2012. Epistatic dissection of laminin-receptor interactions in dystrophic zebrafish muscle. *Human Molecular Genetics* 21 (21): 4718–4731. doi:10.1093/hmg/dds312.

Takada, Y., X. Ye, and S. Simon. 2007. The integrins. *Genome Biology* 8 (5): 215. doi:10.1186/gb-2007-8-5-215.

Talts, J. F., Z. Andac, W. Göhring, A. Brancaccio, and R. Timpl. 1999. Binding of the G domains of laminin A1 and A2 chains and perlecan to heparin, sulfatides, A-Dystroglycan and several extracellular matrix proteins. *The EMBO Journal* 18 (4): 863–870. doi:10.1093/emboj/18.4.863.

Tang, X., J. E. Davies, and S. J. A. Davies. 2003. Changes in distribution, cell associations, and protein expression levels of NG2, neurocan, phosphacan, brevican, Versican V2, and Tenascin-C during acute to chronic maturation of spinal cord scar tissue. *Journal of Neuroscience Research* 71 (3): 427–444. doi:10.1002/jnr.10523.

Telfer, W. R., A. S. Busta, C. G. Bonnemann, E. L. Feldman, and J. J. Dowling. 2010. Zebrafish models of collagen VI-Related myopathies. *Human Molecular Genetics* 19 (12): 2433–2444.

Thorsteinsdóttir, S., M. Deries, and A. S. Cachaĵo. 2011. The extracellular matrix dimension of skeletal muscle development. *Developmental* biology 354 (2): 191–207.

Tiger, C. F., and D. Gullberg. 1997. Absence of laminin A1 chain in the skeletal muscle of dystrophic Dy/Dy mice. *Muscle & Nerve* 20 (12): 1515–1524.

Timpl, R., H. Rohde, P. G. Robey, and S. I. Rennard. 1979. Laminin–a glycoprotein from basement membranes. *Journal of Biological* Chemistry 254 (19): 9933–9937.

Tunggal, P., N. Smyth, and M. Paulsson. 2000. Laminins: Structure and genetic regulation. Microscopy *Research and Technique* 51 (3): 214–227.

Tuszynski, M. H., Y. Wang, L. Graham, M. Gao, and D. Wu. 2014a. Neural stem cell dissemination after grafting to CNS injury sites. *Cell* 156 (3): 388–389.

Tuszynski, M. H., Y. Wang, L. Graham, and K. McHale. 2014b. Neural stem cells in models of spinal cord injury. *Experimental* neurology 261: 494–500.

Vachon, P. H., F. Loechel, H. Xu, U. M. Wewer, and E. Engvall. 1996. Merosin and laminin in myogenesis: Specific requirement for merosin in myotube stability and survival. *The Journal of Cell Biology* 134 (6): 1483–1497. http://gateway. webofknowledge.com/gateway/Gateway.cgi?GWVersion=2&SrcAuth=meke ntosj&SrcApp=Papers&DestLinkType=FullRecord&DestApp=WOS&KeyU T=A1996VJ71900012.

Vachon, P. H., H. Xu, L. Liu, F. Loechel, Y. Hayashi, K. Arahata, J. C. Reed, U. M. Wewer, and E. Engvall. 1997. Integrins (Alpha7beta1) in muscle function and survival. Disrupted expression in merosin-deficient congenital muscular dystrophy. *The Journal of Clinical Investigation* 100 (7): 1870–1881.

Virtanen, I., D. Gullberg, J. Rissanen, E. Kivilaakso, T. Kiviluoto, L. A. Laitinen, V. -P. Lehto, and P. Ekblom. 2000. Laminin A1-chain shows a restricted distribution in epithelial basement membranes of fetal and adult human tissues. *Experimental Cell Research* 257 (2): 298–309.

Waite, A., S. C. Brown, and D. J. Blake. 2012. The dystrophin-glycoprotein complex in brain development and disease. *Trends in Neurosciences* 35 (8): 487–496. doi:10.1016/j.tins.2012.04.004.

Wang, Z., A. Reynolds, A. Kirry, C. Nienhaus, and M. G. Blackmore. 2015. Overexpression of Sox11 promotes corticospinal tract regeneration after spinal injury while interfering with functional recovery. *The Journal of Neuroscience: The Official Journal of the Society for Neuroscience* 35 (7): 3139–3145. doi:10.1523/JNEUROSCI.2832-14.2015.

Warren, G. L., M. Summan, X. Gao, R. Chapman, T. Hulderman, and P. P. Simeonova. 2007. Mechanisms of skeletal muscle injury and repair revealed by gene expression studies in mouse models. *The Journal of Physiology* 582 (2): 825–841.

Webster, M. T., U. Manor, J. Lippincott-Schwartz, and C.-M. Fan. 2016. Intravital imaging reveals ghost fibers as architectural units guiding myogenic progenitors during regeneration. *Cell Stem Cell* 18 (2): 243–252. doi:10.1016/j. stem.2015.11.005.

Wegener, K. L., A. W. Partridge, J. Han, A. R. Pickford, R. C. Liddington, M. H. Ginsberg, and I. D. Campbell. 2007. Structural basis of integrin activation by talin. *Cell* 128 (1): 171–182.

Weir, M. L., M. L. Oppizzi, M. D. Henry, A. Onishi, K. P. Campbell, M. J. Bissell, and J. L. Muschler. 2006. Dystroglycan loss disrupts polarity and beta-casein induction in mammary epithelial cells by perturbing laminin anchoring. *Journal of Cell Science* 119 (19): 4047–4058. doi:10.1242/jcs.03103.

Willerth, S. M., T. E. Faxel, D. I. Gottlieb, and S. E. Sakiyama-Elbert. 2007. The effects of soluble growth factors on embryonic stem cell differentiation inside of fibrin scaffolds. *Stem Cells* 25 (9): 2235–2244. doi:10.1634/ stemcells.2007-0111.

Williams, M. J., P. E. Hughes, T. E. O'Toole, and M. H. Ginsberg. 1994. The inner world of cell adhesion: Integrin cytoplasmic domains. *Trends in Cell Biology* 4 (4): 109–112.

267

Williams, R. J., T. E. Hall, V. Glattauer, J. White, P. J. Pasic, A. B. Sorensen, L. Waddington, K. M. McLean, P. D. Currie, and P. G. Hartley. 2011. The in vivo performance of an enzyme-assisted self-assembled peptide/protein hydrogel. *Biomaterials* 32 (22): 5304–5310. doi:10.1016/j.biomaterials.2011.03.078.

Williams, R. R., I. Venkatesh, D. D. Pearse, A. J. Udvadia, and M. B. Bunge. 2015. MASH1/Ascl1a leads to GAP43 expression and axon regeneration in the adult CNS. *PloS One* 10 (3): e0118918. doi:10.1371/journal.pone.0118918.

Wu, J.-C., W.-C. Huang, Y.-A. Tsai, Y.-C. Chen, and H. Cheng. 2008. Nerve repair using acidic fibroblast growth factor in human cervical spinal cord injury: A preliminary Phase I clinical study. *Journal of Neurosurgery. Spine* 8 (3): 208–214. doi:10.3171/SPI/2008/8/3/208.

Yamauchi, J., A. Kumar, L. Duarte, T. Mehuron, and M. Girgenrath. 2013. Triggering regeneration and tackling apoptosis: A combinatorial approach to treating congenital muscular dystrophy Type 1 A. *Human Molecular Genetics* 22 (21): 4306–4317.

Yao, C. C., B. L. Ziober, A. E. Sutherland, D. L. Mendrick, and R. H. Kramer. 1996. Laminins promote the locomotion of skeletal myoblasts via the Alpha 7 integrin receptor. *Journal of Cell Science* 109 (Pt 13): 3139–3150.

Yao, Y., E. H. Norris, C. E. Mason, and S. Strickland. 2016. Laminin regulates PDGFRβ(+) cell stemness and muscle development. *Nature Communications* 7: 11415. doi:10.1038/ncomms11415.

Yoshimura, S., Y. Takagi, J. Harada, T. Teramoto, S. S. Thomas, C. Waeber, J. C. Bakowska, X. O. Breakefield, and M. A. Moskowitz. 2001. FGF-2 regulation of neurogenesis in adult hippocampus after brain injury. *Proceedings of the National Academy of Sciences* U S A 98 (10): 5874–5879. doi:10.1073/pnas.101034998101034998[pii].

Z'Graggen, W. J., G. A. Metz, G. L. Kartje, M. Thallmair, and M. E. Schwab. 1998. Functional recovery and enhanced corticofugal plasticity after unilateral pyramidal tract lesion and blockade of myelin-associated neurite growth inhibitors in adult rats. *The Journal of Neuroscience* 18 (12): 4744–4757.

Zhang, F., J. D. W. Clarke, and P. Ferretti. 2000. FGF-2 Up-Regulation and proliferation of neural progenitors in the regenerating amphibian spinal cord in vivo. *Developmental Biology* 225 (2): 381–391.

Zheng, W., R. S. Nowakowski, and F. M. Vaccarino. 2004. Fibroblast growth factor 2 is required for maintaining the neural stem cell pool in the mouse brain subventricular zone. *Developmental Neuroscience* 26 (2–4): 181–196. doi:82136[pii] 10.1159/000082136.

Zou, K., M. De Lisio, H. D. Huntsman, Y. Pincu, Z. Mahmassani, M. Miller, D. Olatunbosun, T. Jensen, and M. D. Boppart. 2014. Laminin-111 improves skeletal muscle stem cell quantity and function following eccentric exercise. *Stem Cells Translational Medicine* 3 (9): 1013–1022.

Zupanc, G. K. H., and R. F. Sîrbulescu. 2011. Adult neurogenesis and neuronal regeneration in the central nervous system of teleost fish. *The European Journal of Neuroscience* 34 (6): 917–929. doi:10.1111/j.1460-9568.2011.07854.x.

Chapter 11 Engineered flies for regeneration studies

Florenci Serras

Contents

Key concepts

a. Genetic engineering applied to *Drosophila* has permitted the development of non-invasive tools to study regeneration *in vivo*.

b. Genetically induced apoptosis results in the activation of proliferation to compensate for the missing cells.

c. The combination of genetically engineered constructs allows simultaneous induction of cell death and activation of genetic constructs (e.g., RNA interference [RNAi] of a desired gene) to monitor the effects on regeneration *in vivo*.

> d. Using engineered flies, the Jun N-terminal kinase (JNK), p38, Wnt, and the Janus kinase/signal transducers and activators of transcription (JAK-STAT) signaling pathways have been found to play a key role in regeneration.
> e. Epigenetic control of regeneration includes the inhibition of gene silencing and the activity of specific stress-responding enhancers.
> f. Systemic control of regenerative growth is crucial for maintaining the proportionality of the organs.

11.1 Introduction

Regeneration is classically considered the ability to reconstruct missing parts of the body or even of a whole animal. The organism is constantly exposed to aggressors that can lead to physiological alterations, result in degeneration, and ultimately provoke cell death. Organisms react to such insults by activating mechanisms of cell replacement to restore tissue homeostasis. In a broad sense, the current view of regeneration includes the cellular responses that occur after stimuli ranging from amputation to tissue renewal.

Among the variety of organisms being used in regeneration studies, only a few are as genetically tractable as *Drosophila*. The discovery of gene regulatory networks that respond to chemical or physical insults is crucial for understanding regeneration and engineering tissues for regenerative medicine. Thus, *Drosophila* can be used as an experimental source of information to determine the principles of tissue recovery. In addition, genetically engineered tools can be relatively easily mastered and used to manipulate tissues to determine the importance of certain genes and signaling pathways in regeneration.

Imaginal discs of *Drosophila* are well-characterized tissues in which regeneration can be examined (Bergantiños et al. 2010b; Worley et al. 2012). Imaginal discs are epithelial sacs in which cells proliferate and grow and in which cell fate becomes determined during the larval stages. After metamorphosis, imaginal discs contribute to different adult structures and appendages, such as wings and legs. Thus, by the end of the larval stages, imaginal discs contain all the information required to differentiate into the respective adult structure. After the larval stages, the developing organism is enclosed in the puparium, where only two additional cell divisions occur and the cell cycle is permanently arrested in G1 (Milán et al. 1996a, b).

Two early discoveries were key to identifying imaginal discs as models in regeneration biology. First, cultured disc fragments were found to regenerate after implantation into the abdomen of adult females (Hadorn and Buck 1962; Schubiger 1971). Second, larvae irradiated with doses that caused cell death in about 50% of the disc cells induced compensatory proliferation to rescue their normal size and shape (Haynie and Bryant 1977).

With the advancement of genetic engineering, new tools have been successfully established in *Drosophila*. This chapter aims to introduce the imaginal discs and the genetic designs that can be used for regeneration research, without the need for surgical fragmentation or irradiation. Briefly, the main design is based on the controlled transcriptional activation of pro-apoptotic genes in time and space and the subsequent monitoring of the regenerative capacity of the tissue to restore the dead zones (Smith-Bolton et al. 2009; Bergantiños et al. 2010a). Using these approaches, some relevant genes, signals, and regulatory networks that are activated after genetically induced regeneration are currently being discovered.

11.2 Basic principles of imaginal disc regeneration

Imaginal discs were originally used in regeneration studies by Ernst Hadorn and colleagues (Hadorn and Buck 1962). The main goal at that time was to determine the developmental potential of the discs. The main protocol involved surgically cutting pieces of imaginal discs and implanting them with a micropipette into the abdomen of a fly (see Bergantiños et al. 2010b and Worley et al. 2012 for review). This ectopic environment behaves as a natural culture chamber where the disc can regenerate. In the early days of imaginal disc research, this technique revealed some of the key features of imaginal disc regeneration.

According to the classical definition of Thomas Hunt Morgan (Morgan 1901), regeneration of lost structures in animals can be achieved either by epimorphosis, which implies local stimulation of cell proliferation, or by morphallaxis, which refers to cell re-specification without cell division. Surgical removal of a piece of tissue results in the apposition of new neighbors as a result of wound healing, and therefore, cells that were previously separated will meet and contact. This generates a discontinuity in positional information (Wolpert 1969) that may trigger local cell proliferation (French et al. 1976). A blastema-like structure consisting of mitotic and S-phase cells forms near the wound edges (Dale and Bowes 1980, 1981; Adler 1981; Kiehle and Schubiger 1985; Bryant and O'Brochta 1986; Fain and Alvarez 1987; O'Brochta and Bryant 1987; Bryant and Fraser 1988; Bosch et al. 2005; Lee et al. 2005; Diaz-Garcia and Baonza 2013). These studies revealed some of the basic traits of disc regeneration, including proliferation near the wound, and, therefore, epimorphic and intercalary proliferations to restore the lost pattern and positional continuity (Haynie and Bryant 1977).

Another key trait in disc regeneration is cell plasticity, or the capacity to switch cell fates. This phenomenon was discovered in studies designed to understand the determination state of imaginal disc fragments (Hadorn 1965). Hadorn and collaborators found that fragments of imaginal discs cultured *in vivo* for long periods were able to switch cell fates. Fragments from a particular imaginal disc were cultured for many generations in fly abdomens, where they proliferated but did not differentiate. Eventually, they were transplanted into larvae, where they differentiated after pupation into the cell types of the discs from which they were derived. Occasionally, the imaginal disc fragment differentiated into structures typical of another disc. As the cells of the disc are determined and not fully differentiated, it was considered that discs underwent transdetermination. Moreover, some preferential transdeterminations were described. For example, leg to wing transdetermination is more frequent than wing to leg transdetermination. In addition, during transdetermination, cells switch determination state directly from one state to another, without reverting to an undefined younger state (Hadorn et al. 1970). Transdetermination can also be achieved by ectopic expression of *wingless* (*wg*) (Johnston and Schubiger 1996; Maves and Schubiger 1998, 1995). After fragmentation, *wg* expression localizes near the wound (Gibson and Schubiger 1999; M. Schubiger et al. 2010). Moreover, transdetermination from leg to wing, which is the most frequently studied example (Steiner et al. 1981; McClure and Schubiger 2007), has been successfully reproduced by ectopic expression of *wg* (Maves and Schubiger 1995; Johnston and Schubiger 1996). Thus, transdetermination is a paradigm of cell plasticity during regeneration.

11.3 Genetically engineered flies for regeneration research

Despite these striking observations, interest in imaginal disc regeneration remained low, mainly because of the difficulties in performing microsurgical damage and transplantation in abdomens.

This situation turned around after the implementation of genetically engineered designs that allowed specific zones of the disc to be killed by conditionally activating apoptosis. As discussed later, many of the recent findings have helped to strengthen imaginal discs as a true model regeneration system.

11.3.1 The use of transactivators to induce cell death

One of the major advances in *Drosophila* research has been the introduction of the Gal4/UAS transactivator system to drive the expression of transgenes in particular regions. *Gal4* encodes a transcription factor in *Saccharomyces cerevisiae* (Laughon and Gesteland 1984). To activate transcription, *Gal4* requires the function of two domains: a DNA-binding domain located at the amino terminus and a separate activation domain near the carboxy terminus (Fischer et al. 1988). *Gal4* regulates transcription by binding to a 17-base-pair site (CGG-N_{11}-CCG) known as upstream activation sequence (*UAS*). On *Gal4* binding, the *UAS* acts as an enhancer that drives transcription of the sequences cloned downstream. This Gal4/UAS yeast system can be used in flies to activate genes in particular tissues or organs (Fischer et al. 1988).

In *Drosophila*, the *Gal4* gene can be expressed under the control of a tissue-specific promoter (the driver). *Gal4* is transcribed only in those cells in which the native promoter driver is active. It is precisely in those cells where the Gal4 protein will bind to *UAS* and the expression of the target gene will be forced. It is important to note that *Gal4* has no deleterious effects in flies and that to control expression, *Gal4* and *UAS* are kept in separate strains. To drive expression of a target gene, a fly carrying the promoter-*Gal4* is mated to a fly carrying the *UAS*-target gene, and the resulting progeny will express the target gene in the cells in which the promoter is active.

Since the introduction of the Gal4/UAS system in flies, a wide range of *Gal4* drivers specific for many tissues, organs, and even individual cells have been developed, and most importantly, many are available in public *Drosophila* stock centers. A P-transposable-element-based vector was also generated to clone the target gene under *UAS* control (pUAST). This vector contains an array of five *UAS* sites, a *heat shock protein 70* (*hsp70*) basal promoter, multiple cloning site, an SV40 terminator intron, the polyadenylation signal, and a genetic marker (mini-white) to identify the transgene *UAS* strain (Brand and Perrimon 1993).

11.3.2 The cell killer: Activation of pro-apoptotic genes under *Gal4/UAS* control

Pioneering studies on regeneration after cell death were originally performed in irradiated larvae (Haynie and Bryant 1977; Szabad et al. 1979). Temperature-sensitive cell-autonomous lethal alleles have also been used to genetically induce cell death, in order to study the subsequent regeneration (Brook et al. 1993; Russell et al. 1998). A considerable improvement has been the use of *Gal4* expressed under the control of tissue-specific promoters to drive the expression of pro-apoptotic genes controlled by *UAS*, resulting in the genetic ablation of specific cells.

In *Drosophila*, many signaling pathways converge into one apoptotic program, in which inhibitors of apoptosis and activators interplay to balance the death/live fate of a cell (Bergmann and Steller 2010). Cells have commonly inhibited the apoptotic pathway by the action of the inhibitor of apoptosis (IAP) proteins. The IAPs directly bind to caspases through the baculovirus IAP repeat (BIR) domain, thereby inhibiting their activation (Salvesen and Duckett 2002). *Drosophila* IAP (Diap1) can be inhibited by the family of IAP antagonists Reaper (Rpr), Hid and Grim proteins, initially discovered in flies (Goyal et al. 2000). This family of Diap1 inhibitors contains IAP-binding motifs (IBM) that are required to bind the BIR domains and hamper the anti-apoptotic function of IAP, thereby releasing the caspases from Diap1 (Sandu et al. 2010). Therefore, inducible expression of these Diap1 antagonists, mostly Hid and Rpr, has become a convenient way to induce apoptosis.

Cell death can also be induced after expression of *eiger*, which encodes for *Drosophila* tumor necrosis factor alpha (TNFα) (Igaki et al. 2002; Moreno et al. 2002). Eiger activates its receptor Wengen (Kanda et al. 2002) or Grindelwald (Andersen et al. 2015), both of which induce apoptosis via JNK-dependent activation of caspases.

Activation of an apoptotic inducer using the Gal4/UAS expression system alone would continuously kill cells, and thus, regeneration would be difficult to monitor, because regenerating cells would die as long as they fell into the domain of Gal4/UAS activity. To overcome this caveat, the Gal4 inhibitor Gal80 was introduced to control the activation of the Gal4/UAS system. The Gal80 protein from yeast binds to 30 amino acids of the carboxyl-terminus activation domain of Gal4 and blocks the transcriptional machinery (Ma and Ptashne 1987). In addition, Gal80 from yeast ubiquitously expressed in flies is not poisonous, and as in yeast, it acts as a potent antagonist of Gal4 (Lee and Luo 1999; Zeidler et al. 2004). Therefore, Gal80 has become a powerful tool for conditionally controlling Gal4 activity in *Drosophila* research laboratories (Figure 11.1).

A temperature-sensitive allele of Gal80 has been developed and is applicable to any experimental design in which temperature changes can be imposed (McGuire et al. 2003; Zeidler et al. 2004). The *tubGal80^{TS}* allele, expressed ubiquitously by the *tubulin* 1α promoter, represses the transcriptional activity of Gal4 at 17°C and thus prevents the expression of the target *UAS* transgene. At 29°C, *tubGal80^{TS}* becomes inactive, allowing Gal4 to drive the expression of the UAS transgene (Figure 11.1). Thus, a simple temperature shift that activates Gal80 enables gene expression to be turned on or off and, therefore, activate and inactivate any of *rpr*, *hid*, or *eiger* in a temporal and/or spatial manner to induce genetic ablation (Figure 11.2).

A refinement of this technique has been the combination of two trans-activation systems in which two cell populations can be programmed

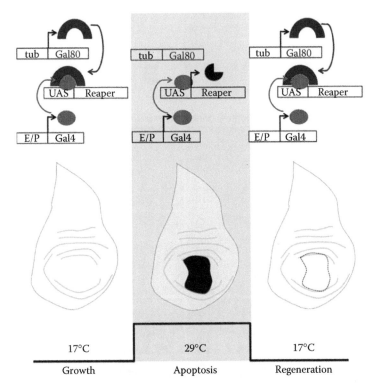

Figure 11.1 Genetic system to induce tissue damage and stimulate regeneration in wing-imaginal discs. At 17°C, the *tub-Gal80^{TS}* temperature-sensitive transgene is ubiquitously transcribed under the *tubulin* regulatory region (*tub*). Gal80 protein (red) binds to the activation domain of Gal4 protein (blue) and inhibits its function. Gal4 is expressed under a tissue-specific enhancer/promoter (E/P). Temperature shift to 29°C (shaded) results in inhibition of Gal80 transcription; Gal4 (blue) bound to DNA can now initiate transcription of the pro-apoptotic gene *reaper* (black). Apoptosis will be induced only in the zone where the tissue-specific E/P is active (black zone). Shifting the temperature back to 17°C will discontinue apoptosis, and regeneration will be completed (dotted line).

to express different transgenes, with some overlapping cells that activate both. One transactivator can drive the pro-apoptotic gene (e.g., *rpr*), whereas the other can express a desired transgene (e.g., a *UAS-RNAi* for a gene of interest). For example, the combination of *LexA* and *Gal4* transactivator systems has improved genetic analysis and functional studies of regeneration.

LexA is a bacterial transcription factor that binds to specific sequences known as the *lexA operator* or *lexO*. An optimized LexA transactivation system has been engineered for *Drosophila* to work in conjunction

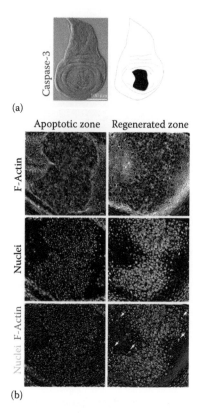

(a)

(b)

Figure 11.2 Genetic ablation of specific zones in the wing-imaginal disc. (a) Using a Gal4 driven by a tissue-specific enhancer for the central zone of the imaginal disc (black zone), apoptosis can be detected by cleaved caspase-3 staining (red). (b) Zoom of one of the regions in which apoptosis has been activated and regeneration has been stimulated. The first column shows a basal section, where the dead cells accumulate and extrude from the disc epithelium. Note that the F-actin is well organized outside the apoptotic zone, whereas the inside shows the existence of cell debris. Nuclei labeling in the apoptotic zone shows DNA fragmentation and condensation. In contrast, the right column shows an apical view of the same disc in which regeneration occurred. Note that actin is well organized, and mitoses are scattered around the zone (white arrows).

with Gal4. Basically, chimeric proteins have been created, in which *lexA* (*L*) contains the activation domain of yeast *Gal4* (*G*), separated by a hinge region (*H*), which allows *lexA-LHG* to be suppressible by Gal80 (Yagi et al. 2010).

Activated *LHG* induces the expression of any transgene under the control of the *lexO* sequences and does not interfere with *Gal4/UAS*. For combined transactivation systems, the *LHG/lexO-rpr* has been used to induce cell death in a specific domain, and the Gal4/UAS system has been used to activate the desired *UAS* transgene (UAS-RNAi, UAS-mutated forms, and so on), both controlled by the thermosensitive *tub-Gal80^{TS}* (Herrera and Morata 2014; Santabárbara-Ruiz et al. 2015). This combined system of two transactivators has allowed the function of several genes to be monitored during regeneration (Figure 11.3).

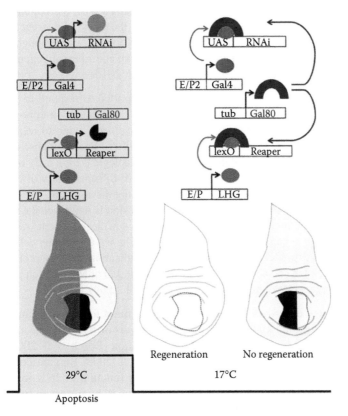

Figure 11.3 Double transactivator system to induce genetic ablation simultaneously with activation of a second transgene. The Gal4/UAS is used to drive expression of a second transgene, which could be any *UAS-RNAi* for diminishing the expression of genes of interest (green zone). The LHG/lexO system is used to drive *reaper* expression (black). At 29°C, *tub-Gal80^{TS}* is inhibited, *LHG* binds to *lexO,* and *reaper* is transcribed. *LHG* is expressed under a tissue-specific enhancer/promoter (E/P) (black). *Gal4,* transcribed under another tissue-specific regulatory region (E/P2), binds to *UAS* to drive expression of the RNAi in the E/P2 acting domain (green). Shifting the temperature to 17°C results in Gal80 expression and inhibition of both Gal4 and LHG. In this phase, regeneration can occur and reconstruct the ablated domain (dotted line; left disc) or interfere with regeneration in the zone where the transgene is acting (right disc). Note that this double transactivator system allows monitoring control and experimental cells on the same tissue.

11.4 Imaginal disc genetic ablation studies: Emerging properties for epithelial recovery

11.4.1 Signaling in regeneration

Two studies initially showed that after genetic ablation using the Gal4/UAS/Gal80 system, proliferation was localized near the damaged tissues. Cell lineage experiments showed that the origin of the regenerated zone is primarily cells outside the ablated domain, even though a few surviving cells within the ablated domain can contribute to regeneration (Smith-Bolton et al. 2009; Bergantiños et al. 2010a).

Several categories of signals have been reported to respond to genetic ablation. Two morphogens, molecules that direct morphogenesis depending on their local concentration, have been found to respond to ablation: *wg*, a fly member of the Wnt family of ligands whose expression is strongly upregulated, and Decapentaplegic (Dpp), a fly member of the transforming growth factor beta (TGF-ß) superfamily of ligands, which expands its expression area surrounding the ablation zone (Smith-Bolton et al. 2009). Interestingly, *wg* expression is reminiscent of developmental expression during the early stages of development, when the proliferative capabilities are exacerbated. The *wg*-expressing cells were found to express E2F transcription factor, indicating that they proliferate, block Notch (N) activity, and express high levels of Myc, which potentiates regenerative growth (Smith-Bolton et al. 2009).

A widely described signaling pathway required for regeneration is the Jun N-terminal kinase (JNK). Elevated levels of JNK have been found in fragmented discs and after genetic ablation (Bosch et al. 2005; Lee et al. 2005; Mattila et al. 2005; Bosch et al. 2008; Bergantiños et al. 2010a; Herrera et al. 2013). However, the function of JNK in regeneration is under debate. Briefly, JNK seems to be necessary in dying cells that activate apoptosis in order to signal for compensatory proliferation, yet regenerating cells also require JNK. For example, if the apoptotic program of dying cells is truncated by ectopically expressing p35, a potent inhibitor of executioner caspases, Rpr or Hid can activate JNK (Ryoo et al. 2004). However, in the absence of p35, JNK has been found to act in regenerating cells (Bergantiños et al. 2010a). This conflict has recently been reconciled, as JNK has been found to act in both dying and regenerating cells; JNK is active in dead cells and induces caspase-mediated apoptosis, perhaps by amplifying a loop mechanism that ensures apoptosis (Shlevkov and Morata 2012). Concomitantly, high levels of JNK in dying cells co-exist with non-poisonous levels of JNK in living tissues that are necessary to stimulate regeneration (Smith-Bolton et al. 2009; Bergantiños et al. 2010a; Herrera et al. 2013; Fan et al. 2014; Santabárbara-Ruiz et al. 2015).

11.4.2 Cell plasticity

Since the early experiments of Ernst Hadorn on the stability of cell fates (Hadorn 1965, 1978), interest in understanding the biological basis of cell plasticity in imaginal disc research has been rising. The transdetermination that disc fragments undergo after long episodes of culture indicates that disc cell fates are labile and are capable of following alternative developmental programs. It is likely that cells remodel the epigenetic marks that eventually fix cell fates. It has been reported that JNK-dependent silencing of the Polycomb group (PcG) genes is key in transdetermination events, which are more conspicuous in heterozygous mutant backgrounds for PcG genes. Moreover, as JNK leads to downregulation of PcG genes, JNK pathway mutants impair transdetermination (Lee et al. 2005). Gene expression profiles of regenerating disc fragments showed upregulation of chromatin remodeling genes (Blanco et al. 2010). This is the case of *ash2*, a co-factor required for trimethylation of histone H3 lysine 4, a conserved epigenetic mark associated with transcriptionally active chromatin regions (Perez-Lluch et al. 2011; Carbonell et al. 2013).

In addition to regenerating disc fragments, epigenetic control of plasticity has been documented after genetic ablation. The transgression of cells through compartmental boundaries rarely occurs, and if it does occur, it is mainly when cell death is activated in the entire compartment (Herrera and Morata 2014). However, in genetic backgrounds heterozygous for the PcG allele Pc^3, the number and size of transgressions are greatly increased. When heterozygous for the trithorax group allele trx^{E2}, transgressions diminish considerably, as occurs after JNK is blocked (Herrera and Morata 2014). In summary, gene silencing is turned off in a JNK-dependent manner to promote cell plasticity.

Other examples of cell plasticity include re-specification of wing veins to regenerate interveins (Repiso et al. 2013) and the contribution of hinge cells to the regeneration of wing tissue (Smith-Bolton et al. 2009; Herrera et al. 2013).

11.4.3 Polarity of growth

Clonal analysis of genetic mosaics allows for the tracing of genetically labeled cells generated in a single mitotic recombination event. If the biological system is undergoing growth, these labeled cells will proliferate and clones of labeled cells will emerge. The technique is very useful for studying cell behavior and relevant properties such as the control of organ size and growth orientation. This technique for clonal analysis commonly requires the use of chromosomes that carry a heat shock–inducible recombinase flipase (*hsFLP*) and the flipase recombination targets (FRT), which are specific recombination sites for FLP.

By combining the recombination and genetic ablation tools, clones of labeled cells (often green fluorescent protein [GFP]-labeled) can be recovered, and their shape, size, and orientation can be compared to discs without cell death. After inducing massive cell death in the whole central zone of the disc, clones grow twice as much as controls, which indicates that the cells undergo additional divisions (Herrera et al. 2013). Clone orientation was studied in relation to the proximal–distal pattern of growth, which is indicative of normal growth orientation in limbs and wings. It was found that after cell death, clones change orientation toward the damaged zone (Repiso et al. 2013). This observation is consistent with regenerative growth that senses the damaged zone and preferentially orients cell divisions toward the missing part.

Mechanisms of growth and size control in normal development also act in regeneration. The transcriptional co-activator found downstream from the Hippo pathway, Yorkie (*yki*), is implicated in size control in regeneration (Grusche et al. 2010; Sun and Irvine 2011). Upstream cell-to-cell contact molecules, such as the *protocadherin Fat*, which is involved in planar cell polarity, and the apical transmembrane protein Crumbs, which is involved in apico-basal polarity, regulate the Hippo pathway for growth control (Bennett and Harvey 2006; Cho et al. 2006; Silva et al. 2006; Chen et al. 2010; Ling et al. 2010; Robinson et al. 2010). Reduction of the levels of Fat and Crumbs disrupts clone orientation (Repiso et al. 2013). Therefore, cell contacts and cell polarity proteins may act as cell-surface sensors in the stimulation of oriented regenerative growth. Although the signals received by surface contacts are not yet known, one cannot rule out the possibility that mechanical (e.g., changes in stretching or tension) or chemical (e.g., oxidative stress) stimuli play a role.

11.4.4 Mechanical forces and oxidative stress

Little is known about mechanical forces in regeneration. However, some studies point to a scenario in which these forces are key in cell division orientation and growth control, suggesting the involvement of physical contacts (Aegerter-Wilmsen et al. 2007; Mao et al. 2013). An exciting proposal is that the existence of differential rates of proliferation in the wing disc epithelium causes a global tension pattern that produces the mechanical forces needed to orient mitotic spindles (Mao et al. 2013).

Recent evidence also emphasizes the role of oxidative stress in regeneration. The production of reactive oxygen species (ROS) upstream in signaling pathways required for regeneration has been reported in *Xenopus* and zebrafish (Niethammer et al. 2009; Love et al. 2013). The production of ROS has also been described in regenerating imaginal discs (Santabárbara-Ruiz et al. 2015; Fogarty et al. 2016). Both induction of cell death and physical damage result in an oxidative burst necessary for JNK and p38 mitogen-activated protein (MAP) kinase activation. Consequently, animals fed with antioxidants before damage regenerate poorly. In addition, JNK and p38 are required for stress-dependent activation of the cytokines Unpaired (Upd), which ultimately participates in controlling regenerative growth (Pastor-Pareja et al. 2008; Santabárbara-Ruiz et al. 2015). Reactive oxygen species were also found to act as an attractant for macrophages, which in turn induced JNK activation in the epithelium (Fogarty et al. 2016).

11.5 Parallels between cancer and regeneration: Systemic responses after damage and cancer

Knowledge about growth control will provide valuable information for regenerative medicine. It has long been assumed that cancer and regeneration share some commonalities, such as the stimulus for growth. Regenerating organs or tissues must have developed mechanisms to avoid overgrowth based on systemic control to maintain organ proportions. *Drosophila* larval development, and in particular the imaginal discs, can shed some light on this matter.

After embryonic development, *Drosophila* goes through three larval stages. The nuclei of most cells of the larvae replicate to increase the ploidy and the cell size. In contrast, the diploid cells of the imaginal discs proliferate during these larval stages and increase in cell number until reaching a fixed disc size. At the end of the third larval stage, production of the steroid hormone ecdysone peaks, and most cell divisions stop and the larva enters metamorphosis. During pupation, most polyploid larval cells will die, whereas the cell derivatives of the imaginal discs will contribute to the adult parts (eyes, legs, wings, and so on).

Remarkably, the duration of the third larval stage adapts to accommodate imaginal disc growth (Shingleton 2010). On tissue damage, the onset of pupation is postponed until the tissue recovers. This delay occurs during regeneration, transdetermination, and tumor growth. The duration of delay increases with the amount of damage (Hussey et al. 1927). A mechanism must exist that allows the damaged tissue to reach its final size, without interference from pupation, and, at the same time, prevents overgrowth of the other unaffected organs. Indeed, when an imaginal disc is damaged, repair and regeneration occur in that disc, but the surrounding organs stop developing until the damaged disc recovers. This delay provides an excellent opportunity to explore the systemic mechanisms that control tissue proportions during regeneration and ultimately to discover mechanisms responsible for allometry, or organ growth scaling.

In the search of genes responsible for the delay in pupation after tumor production, the *Drosophila insulin-like peptide 8* (*dilp8*) emerged as a candidate responsible for growth coordination (Colombani et al. 2012; Garelli et al. 2012). Dilp8 is a 150-amino-acid peptide, containing a predicted secretion domain with a signal peptide, followed by a cleavage site at its N-terminus. *Dilp8* is upregulated in disc tumors and secreted into the hemolymph. The delay in pupation due to tumor growth can be rescued in a *dilp8* mutant background. In the absence of damage, ectopic expression of *dilp8* results in pupation delay. Strikingly, after genetic ablation or gamma irradiation, both of which induce regenerative growth, expression of *dilp8* is upregulated. As with tumors, reduction of *dilp8* levels results in a shorter delay that prevents full regeneration. Thus, dilp8 is a peptide that responds to damage and regulates the phasing of pupation to allow enough time for tissue recovery. Dilp8 targets the relaxin receptor Lgr3 of neurons that belong to a brain circuit that controls coordinated body size (Vallejo et al. 2015).

Despite the longer larval interval, the final size of the individual after damage is similar to that of undamaged controls (Colombani et al. 2012; Garelli et al. 2012). This means that dilp8 is not only involved in the lengthening of developmental timing but also extends its control of growth to the rest of the body. Thus, it appears that while the damaged organ regenerates, the rest of the organs stop growing until the damaged organ reaches its final size. It has been found that the dilp8-dependent growth coordination between regenerating and undamaged tissues requires the activity of nitric oxide synthase (NOS) and that NOS regulates growth in undamaged tissues by reducing ecdysone biosynthesis in the prothoracic gland (Jaszczak et al. 2015). Thus, dilp8 signal to the endocrine gland is mediated by NOS to prevent overgrowth of undamaged discs.

This systemic mechanism prevents the overgrowth of undamaged organs by synchronizing growth of regenerating tissues with peripheral organs. This discovery is key to understanding how a regenerating organ stops growing after regeneration is completed.

11.6 Conclusion: A summary of the evidence for imaginal discs being a true model for regeneration

By definition, imaginal discs are in a state of developmental growth, during which cell determination occurs but cell differentiation does not occur. Therefore, at first glance, it is difficult to discern developmental growth from truly regenerative growth.

In his classical essay, Thomas Hunt Morgan (Morgan 1901, p. 23) stated that "The word regeneration has come to mean, in general usage, not only the replacement of a lost part, but also the development of a new, whole organism, or even a part of an organism, from a piece of an adult, or of an embryo, or of an egg." Later definitions disregarded embryonic regeneration and focused on regrowth and repair of tissues and organs in the adult (Sanchez Alvarado and Tsonis 2006), thus excluding imaginal discs. However, the existence of regeneration-specific features is a strong argument in favor of the distinction between development and regeneration (Vervoort 2011).

Recent lines of evidence indicate that damaged discs provoke regeneration-specific responses. First, JNK is activated in both dying cells (Shlevkov and Morata 2012) and surviving cells near the damaged tissue (Bosch et al. 2005; Lee et al. 2005; Mattila et al. 2005; Smith-Bolton et al. 2009; Bergantiños et al. 2010a; Herrera et al. 2013; Fan et al. 2014). Activation of JNK in surviving cells contributes to regeneration of the missing part (Bosch et al. 2008; Herrera et al. 2013; Fogarty et al. 2016).

Second, MAP kinase p38 is phosphorylated near the damaged zone but not in dead cells (Santabárbara-Ruiz et al. 2015). Discs show basal levels of P-p38 that strongly and rapidly increase on damage.

Third, oxidative stress acts as a physiological signal that triggers regeneration (Santabárbara-Ruiz et al. 2015; Fogarty et al. 2016). Both p38 and JNK activation are ROS-dependent (Santabárbara-Ruiz et al. 2015).

Fourth, cytokine *upd* expression and the Janus kinase/signal transducers and activators of transcription (JAK-STAT) activation are associated with damage in a development-independent manner (Pastor-Pareja et al. 2008; Santabárbara-Ruiz et al. 2015; Katsuyama et al. 2015).

Fifth, cells that are undergoing regeneration can be re-programmed—or re-specified—to new fates (Lee et al. 2005; Smith-Bolton et al. 2009; Repiso et al. 2013; Herrera and Morata 2014; Schuster and Smith-Bolton 2015). Regeneration in imaginal discs does not seem to be driven by stem cells but rather by proliferation and re-programming of specified cells.

Sixth, neural network controls triggered by the production of dilp8 control the balance between the regenerating organ (disc) and the rest of the body (Colombani et al. 2012; Garelli et al. 2012; Katsuyama et al. 2015; Vallejo et al. 2015).

Seventh, a program for *wg* regulation during regeneration, independent of development, has been described. Cells in imaginal discs lose their capacity to regenerate in late larval stages. How this regenerative capacity is blocked while still preserving the developmental program, including some proliferation and differentiation, is an intriguing question. Several studies have shown a decrease in regeneration capacity over time (Smith-Bolton et al. 2009; Halme et al. 2010), perhaps by silencing *regeneration* regulatory regions (Harris et al. 2016). A Wnt regeneration-specific enhancer is epigenetically silenced in late stages, when the regeneration capacity is abolished (Johnston and Schubiger 1996; Schubiger et al. 2010; Harris et al. 2016).

Together, these lines of evidence consolidate the imaginal discs as a true model for the discovery of the genetic basis of epithelial regeneration. In addition, recent observations that resulted from the analysis of genetic ablation, which in many cases confirm previous observations after physical injury, point to imaginal disc regeneration as allowing replacement of the lost part with a mechanism distinct from normal development. One can predict that due to the versatility and amenability of *Drosophila*, engineered constructs will provide valuable information for other regeneration systems.

Acknowledgments

I wish to thank Paula Santabárbara, Elena Vizcaya-Molina, Emili Saló, and Montserrat Corominas for the discussions and help. This work was supported by the Spanish Ministerio de Economía y Competitividad grant BFU2015-67623-P.

References

Adler, P N. 1981. Growth during pattern regulation in imaginal discs. *Dev Biol* 87 (2): 356–373. doi:0012-1606(81)90159-7[pii].

Aegerter-Wilmsen, T, C M Aegerter, E Hafen, and K Basler. 2007. Model for the regulation of size in the wing imaginal disc of drosophila. *Mech Dev* 124 (4): 318–326. doi:S0925-4773(06)00220-6[pii] 10.1016/j.mod.2006.12.005.

Andersen, D S, J Colombani, V Palmerini, K Chakrabandhu, E Boone, M Röthlisberger, J Toggweiler et al. 2015. The drosophila TNF receptor grindelwald couples loss of cell polarity and neoplastic growth. *Nature* 522 (7557): 482–486. doi:10.1038/nature14298.

Bennett, F C, and K F Harvey. 2006. Fat cadherin modulates organ size in drosophila via the salvador/warts/hippo signaling pathway. *Curr Biol* 16 (21): 2101–2110. doi:10.1016/j.cub.2006.09.045.

Bergantiños, C, M Corominas, and F Serras. 2010a. Cell death-induced regeneration in wing imaginal discs requires JNK signalling. *Development* 137 (7): 1169–1179. doi:137/7/1169[pii] 10.1242/dev.045559.

Bergantiños, C, X Vilana, M Corominas, and F Serras. 2010b. Imaginal discs: Renaissance of a model for regenerative biology. *Bioessays* 32 (3): 207–217. doi:10.1002/bies.200900105.

Bergmann, A, and H Steller. 2010. Apoptosis, stem cells, and tissue regeneration. *Sci Signal* 3 (145): re8. doi:10.1126/scisignal.3145re8.

Blanco, E, M Ruiz-Romero, S Beltran, M Bosch, A Punset, F Serras, and M Corominas. 2010. Gene expression following induction of regeneration in drosophila wing imaginal discs. Expression profile of regenerating wing discs. *BMC Dev Biol* 10: 94. doi:1471-213X-10-94[pii] 10.1186/1471-213X-10-94.

Bosch, M, J Baguna, and F Serras. 2008. Origin and proliferation of blastema cells during regeneration of drosophila wing imaginal discs. *Int J Dev Biol* 52 (8): 1043–1050. doi:082608mb [pii] 10.1387/ijdb.082608mb.

Bosch, M, F Serras, E Martin-Blanco, and J Baguna. 2005. JNK signaling pathway required for wound healing in regenerating drosophila wing imaginal discs. *Dev Biol* 280 (1): 73–86. doi:S0012-1606(05)00003-5[pii] 10.1016/j. ydbio.2005.01.002.

Brand, A H, and N Perrimon. 1993. Targeted gene expression as a means of altering cell fates and generating dominant phenotypes. *Development* 118 (2): 401–415.

Brook, W J, L M Ostafichuk, J Piorecky, M D Wilkinson, D J Hodgetts, and M A Russell. 1993. Gene expression during imaginal disc regeneration detected using enhancer-sensitive P-elements. *Development* 117 (4): 1287–1297.

Bryant, P J, and D A O'Brochta. 1986. Growth patterns in drosophila imaginal discs. *Prog Clin Biol Res* 217A: 297–300.

Bryant, P J, and S E Fraser. 1988. Wound healing, cell communication, and DNA synthesis during imaginal disc regeneration in drosophila. *Dev Biol* 127 (1): 197–208. doi:0012-1606(88)90201-1[pii].

Carbonell, A, A Mazo, F Serras, and M Corominas. 2013. Ash2 acts as an ecdysone receptor coactivator by stabilizing the histone methyltransferase trr. *Mol Biol Cell* 24 (3): 361–372. doi:10.1091/mbc. E12-04-0267; 10.1091/mbc. E12-04-0267.

Chen, C L, K M Gajewski, F Hamaratoglu, W Bossuyt, L Sansores-Garcia, C Tao, and G Halder. 2010. The apical-basal cell polarity determinant crumbs regulates hippo signaling in drosophila. *Proc Natl Acad Sci U S A* 107 (36): 15810–15815. doi:10.1073/pnas.1004060107.

Cho, E, Y Feng, C Rauskolb, S Maitra, R Fehon, and K D Irvine. 2006. Delineation of a fat tumor suppressor pathway. *Nat Genet* 38 (10): 1142–1150. doi:10.1038/ng1887.

Colombani, J., D. S. Andersen, and P. Leopold. 2012. Secreted peptide dilp8 coordinates drosophila tissue growth with developmental timing. *Science* 336 (6081): 582–585. doi:10.1126/science.1216689.

Dale, L, and M Bowes. 1980. Is regeneration in drosophila the result of epimorphic regeneration? *Wilhelm Roux's Arch* 189: 91–96.

Dale, L, and M Bowes. 1981. Wound healing and regeneration in the imaginal wing disc of drosophila. *Wilhelm Roux's Arch* 190: 185–190.

Diaz-Garcia, S, and A Baonza. 2013. Pattern reorganization occurs independently of cell division during drosophila wing disc regeneration in situ. *Proc Natl Acad Sci U S A* 110 (32): 13032–13037. doi:10.1073/pnas.1220543110; 10.1073/pnas.1220543110.

Fain, M J, and C Alvarez. 1987. The cell cycle and its relation to growth during pattern regulation in wing discs of drosophila. *J Insect Physiol* 33 (10): 697–705.

Fan, Y, S Wang, J Hernandez, V B Yenigun, G Hertlein, C E Fogarty, J L Lindblad, and A Bergmann. 2014. Genetic models of apoptosis-induced proliferation decipher activation of JNK and identify a requirement of EGFR signaling for tissue regenerative responses in drosophila. *PLoS Genet* 10 (1): e1004131. doi:10.1371/journal.pgen.1004131.

Fischer, J A, E Giniger, T Maniatis, and M Ptashne. 1988. GAL4 activates transcription in drosophila. *Nature* 332 (6167): 853–856. doi:10.1038/332853a0.

Fogarty, C E, N Diwanji, J L Lindblad, M Tare, A Amcheslavsky, K Makhijani, K Brückner, Y Fan, and A Bergmann. 2016. Extracellular reactive oxygen species drive apoptosis-induced proliferation via drosophila macrophages. *Curr Biol* 26 (5): 575–584. doi:10.1016/j.cub.2015.12.064.

French, V, P J Bryant, and S V Bryant. 1976. Pattern regulation in epimorphic fields. *Science* 193 (4257) : 969–981.

Garelli, A, A M Gontijo, V Miguela, E Caparros, and M Dominguez. 2012. Imaginal discs secrete insulin-like peptide 8 to mediate plasticity of growth and maturation. *Science* 336 (6081): 579–582. doi:10.1126/science.1216735.

Gibson, M C, and G Schubiger. 1999. Hedgehog is required for activation of engrailed during regeneration of fragmented drosophila imaginal discs. *Development* 126 (8): 1591–1599.

285

Goyal, L, K McCall, J Agapite, E Hartwieg, and H Steller. 2000. Induction of apoptosis by drosophila reaper, hid and grim through inhibition of IAP function. *The EMBO J* 19 (4): 589–597. doi:10.1093/emboj/19.4.589.

Grusche, F A, H E Richardson, and K F Harvey. 2010. Upstream regulation of the hippo size control pathway. *Curr Biol* 20 (13): R574–R582. doi:10.1016/j.cub.2010.05.023.

Hadorn, E. 1965. Problems of determination and transdetermination. *Genetic Control of Differentiation, Brook Symp Biol* 18: 148–161.

Hadorn, E. 1978. Transdetermination. *The Genetics and Biology of Drosophila*, edited by M Ashburner and T R F Wright, 2c: 555–617. New York: Academic Press.

Hadorn, E, and D Buck. 1962. Ober entwicklungsleistungen transplantierter teilstiicke von flugel-imaginalscheiben von drosophila melanogaster. *Rev Suisse Zool* 69: 302–310.

Hadorn, E, R Gsell, and J Schultz. 1970. Stability of a position-effect variegation in normal and transdetermined larval blastemas from drosophila melanogaster. *Proc Natl Acad Sci U S A* 65 (3): 633–637.

Halme, A, M Cheng, and I K Hariharan. 2010. Retinoids regulate a developmental checkpoint for tissue regeneration in drosophila. *Curr Biol* 20 (5): 458–463. doi:10.1016/j.cub.2010.01.038.

Harris, R E, L Setiawan, J Saul, and I K Hariharan. 2016. Localized epigenetic silencing of a damage-activated WNT enhancer limits regeneration in mature drosophila imaginal discs. *eLife* 5: e11588. doi:10.7554/eLife.11588.

Haynie, J, and P J Bryant. 1977. The effects of X-Rays on the proliferation dynamics of cells in the imaginal wing disc of drosophila melanogaster. *Roux's Arch Dev Biol* 183 (2): 85–100. doi:10.1007/BF00848779.

Herrera, S C, R Martin, and G Morata. 2013. Tissue homeostasis in the wing disc of drosophila melanogaster: Immediate response to massive damage during development. *PLoS Genet* 9 (4): e1003446. doi:10.1371/journal.pgen.1003446; 10.1371/journal.pgen.1003446.

Herrera, S C, and G Morata. 2014. Transgressions of compartment boundaries and cell reprogramming during regeneration in drosophila. *eLife* 3: e01831. doi:10.7554/eLife.01831.

Hussey, R G, W R Thompson, and E T Calhoun. 1927. The influence of x-rays on the development of drosophila larvae. *Science* 66 (1698): 65–66. doi:10.1126/science.66.1698.65.

Igaki, T, H Kanda, Y Yamamoto-Goto, H Kanuka, E Kuranaga, T Aigaki, and M Miura. 2002. Eiger, a TNF superfamily ligand that tTriggers the drosophila JNK pathway. *The EMBO J* 21 (12): 3009–3018. doi:10.1093/emboj/cdf306.

Jaszczak, J S, J B Wolpe, A Q Dao, and A Halme. 2015. Nitric oxide synthase regulates growth coordination during drosophila melanogaster imaginal disc regeneration. *Genetics* 200 (4): 1219–1228. doi:10.1534/genetics.115.178053.

Johnston, L A, and G Schubiger. 1996. Ectopic expression of wingless in imaginal discs interferes with decapentaplegic expression and alters cell determination. *Development* 122 (11): 3519–3529.

Kanda, H, T Igaki, H Kanuka, T Yagi, and M Miura. 2002. Wengen, a member of the drosophila tumor necrosis factor receptor superfamily, is required for eiger signaling. *J Biol Chem* 277 (32): 28372–28375. doi:10.1074/jbc. C200324200.

Katsuyama, T, F Comoglio, M Seimiya, E Cabuy, and R Paro. 2015. During drosophila disc regeneration, JAK/STAT coordinates cell proliferation with dilp8-mediated developmental delay. *Proc Natl Acad Sci U S A* 112 (18): E2327–E2336. doi:10.1073/pnas.1423074112.

Kiehle, C P, and G Schubiger. 1985. Cell proliferation changes during pattern regulation in imaginal leg discs of drosophila melanogaster. *Dev Biol* 109 (2): 336–346.

Laughon, A, and R F Gesteland. 1984. Primary structure of the saccharomyces cerevisiae GAL4 Gene. *Mol Cell Biol* 4 (2): 260–267.

Lee, N, C Maurange, L Ringrose, and R Paro. 2005. Suppression of Polycomb Group Proteins by JNK signalling induces transdetermination in drosophila imaginal discs. *Nature* 438 (7065): 234–237. doi:nature04120[pii] 10.1038/nature04120.

Lee, T, and L Luo. 1999. Mosaic analysis with a repressible cell marker for studies of gene function in neuronal morphogenesis. *Neuron* 22 (3): 451–461. doi:10.1016/S0896-6273(00)80701-1.

Ling, C, Y Zheng, F Yin, J Yu, J Huang, Y Hong, S Wu, and D Pan. 2010. The apical transmembrane protein crumbs functions as a tumor suppressor that regulates hippo signaling by binding to expanded. *Proc Natl Acad Sci U S A* 107 (23): 10532–10537. doi:10.1073/pnas.1004279107.

Love, N R, Y Chen, S Ishibashi, P Kritsiligkou, R Lea, Y Koh, J L Gallop, K Dorey, and E Amaya. 2013. Amputation-induced reactive oxygen species are required for successful xenopus tadpole tail regeneration. *Nat Cell Biol* 15 (2): 222–228. doi:10.1038/ncb2659.

Ma, J, and M Ptashne. 1987. The carboxy-terminal 30 amino acids of GAL4 are recognized by GAL80. *Cell* 50 (1): 137–142. doi:10.1016/0092-8674(87)90670-2.

Mao, Y, A L Tournier, A Hoppe, L Kester, B J Thompson, and N Tapon. 2013. Differential proliferation rates generate patterns of mechanical tension that orient tissue growth. *The EMBO J* 32 (21): 2790–2803. doi:10.1038/emboj.2013.197; 10.1038/emboj.2013.197.

Mattila, J, L Omelyanchuk, S Kyttala, H Turunen, and S Nokkala. 2005. Role of jun N-terminal kinase (JNK) signaling in the wound healing and regeneration of a drosophila melanogaster wing imaginal disc. *Int J Dev Biol* 49 (4): 391–399. doi:052006jm [pii] 10.1387/ijdb.052006jm.

Maves, L, and G Schubiger. 1995. Wingless induces transdetermination in developing drosophila imaginal discs. *Development* 121 (5): 1263–1272.

Maves, L, and G Schubiger. 1998. A molecular basis for transdetermination in drosophila imaginal discs: Interactions between wingless and decapentaplegic signaling. *Development* 125 (1): 115–124.

McClure, K D, and G Schubiger. 2007. Transdetermination: Drosophila imaginal disc cells exhibit stem cell-like potency. *Int J Biochem Cell Biol* 39 (6): 1105–1118. doi:S1357-2725(07)00030-1[pii] 10.1016/j.biocel.2007.01.007.

McGuire, S E, P T Le, A J Osborn, K Matsumoto, and R L Davis. 2003. Spatiotemporal rescue of memory dysfunction in drosophila. *Science* 302 (5651): 1765–1768. doi:10.1126/science.1089035 302/5651/1765[pii].

Milán, M, S Campuzano, and A García-Bellido. 1996a. Cell cycling and patterned cell proliferation in the wing primordium of drosophila. *Proc Natl Acad Sci U S A* 93 (2): 640–645.

Milán, M, S Campuzano, and A García-Bellido 1996b. Cell cycling and patterned cell proliferation in the drosophila wing during metamorphosis. *Proc Natl Acad Sci U S A* 93 (21): 11687–11692.

Moreno, E, K Basler, and G Morata. 2002. Cells compete for decapentaplegic survival factor to prevent apoptosis in drosophila wing development. *Nature* 416 (6882): 755–759. doi:10.1038/416755a 416755a [pii].

Morgan, T H. 1901. *Regeneration*. New York: The Macmillan Company. doi:http://dx.doi.org/10.5962/bhl.title.1114.

Niethammer, P, C Grabher, A T Look, and T J Mitchison. 2009. A tissue-scale gradient of hydrogen peroxide mediates rapid wound detection in zebrafish. *Nature* 459 (7249): 996–999. doi:10.1038/nature08119.

O'Brochta, D A, and P J Bryant. 1987. Distribution of S-phase cells during the regeneration of drosophila imaginal wing discs. *Dev Biol* 119 (1): 137–142. doi:0012-1606(87)90215-6[pii].

287

Pastor-Pareja, J C, M Wu, and T Xu. 2008. An innate immune response of blood cells to tumors and tissue damage in drosophila. *Dis Model Mech* 1 (2–3): 144–154. doi:10.1242/dmm.000950.

Perez-Lluch, S, E Blanco, A Carbonell, D Raha, M Snyder, F Serras, and M Corominas. 2011. Genome-wide chromatin occupancy analysis reveals a role for ASH2 in transcriptional pausing. *Nucleic Acids Res* 39 (11): 4628–4639. doi:10.1093/nar/gkq1322.

Repiso, A, C Bergantinos, and F Serras. 2013. Cell fate respecification and cell division orientation drive intercalary regeneration in drosophila wing discs. *Development* 140 (17): 3541–3551. doi:10.1242/dev.095760; 10.1242/dev.095760.

Robinson, B S, J Huang, Y Hong, and K H Moberg. 2010. Crumbs regulates salvador/warts/hippo signaling in drosophila via the FERM-domain protein expanded. *Curr Biol* 20 (7): 582–590. doi:10.1016/j.cub.2010.03.019.

Russell, M A, L Ostafichuk, and S Scanga. 1998. Lethal P-lacZ insertion lines expressed during pattern respecification in the imaginal discs of drosophila. *Genome* 41 (1): 7–13.

Ryoo, H D, T Gorenc, and H Steller. 2004. Apoptotic cells can induce compensatory cell proliferation through the JNK and the wingless signaling pathways. *Dev Cell* 7 (4): 491–501. doi:S1534580704003247[pii] 10.1016/j.devcel.2004.08.019.

Salvesen, G S, and C S Duckett. 2002. IAP proteins: Blocking the road to death's door. *Nat Rev Mol Cell Biol* 3 (6): 401–410. doi:10.1038/nrm830.

Sanchez Alvarado, A, and P A Tsonis. 2006. Bridging the regeneration gap: Genetic insights from diverse animal models. *Nat Rev Genet* 7 (11): 873–884. doi:nrg1923[pii] 10.1038/nrg1923.

Sandu, C, H Don Ryoo, and H Steller. 2010. Drosophila IAP antagonists form multimeric complexes to promote cell death. *J Cell Biol* 190 (6): 1039–1052. doi:10.1083/jcb.201004086.

Santabárbara-Ruiz, P, M López-Santillán, I Martínez-Rodríguez, A Binagui-Casas, L Pérez, M Milán, M Corominas, and F Serras. 2015. ROS-induced JNK and p38 signaling is required for unpaired cytokine activation during drosophila regeneration. *PLoS Genet* 11 (10): e1005595. doi:10.1371/journal.pgen.1005595.

Schubiger, G. 1971. Regeneration, duplication and transdetermination in fragments of the leg disc of drosophila melanogaster. *Dev Biol* 26 (2): 277–295. doi:0012-1606(71)90127-8[pii].

Schuster, K J, and R K Smith-Bolton. 2015. Taranis protects regenerating tissue from fate changes induced by the wound response in drosophila. *Dev Cell* 34 (1): 119–128. doi:10.1016/j.devcel.2015.04.017.

Schubiger, M, A Sustar, and G Schubiger. 2010. Regeneration and transdetermination: The role of wingless and its regulation. *Dev Biol* 347 (2): 315–324. doi:10.1016/j.ydbio.2010.08.034.

Shingleton, A W. 2010. The regulation of organ size in drosophila: Physiology, plasticity, patterning and physical force. *Organogenesis* 6 (2): 76–87.

Shlevkov, E, and G Morata. 2012. A dp53/JNK-dependant feedback amplification loop is essential for the apoptotic response to stress in drosophila. *Cell Death Differ* 19 (3): 451–460. doi:10.1038/cdd.2011.113.

Silva, E, Y Tsatskis, L Gardano, N Tapon, and H McNeill. 2006. The tumor-suppressor gene fat controls tissue growth upstream of expanded in the hippo signaling pathway. *Curr Biol* 16 (21): 2081–2089. doi:10.1016/j.cub.2006.09.004.

Smith-Bolton, R K, M I Worley, H Kanda, and I K Hariharan. 2009. Regenerative growth in drosophila imaginal discs is regulated by wingless and myc. *Dev Cell* 16 (6): 797–809. doi:10.1016/j.devcel.2009.04.015.

Steiner, E, M Koller-Wiesinger, and R Nöthiger. 1981. Transdetermination in leg imaginal discs of drosophila melanogaster und drosophila nigromelanica. *Wilhelm Roux's Archives of Dev Biol* 190: 156–160.

Sun, G, and K D Irvine. 2011. Regulation of hippo signaling by jun kinase signaling during compensatory cell proliferation and regeneration, and in neoplastic tumors. *Dev Biol* 350 (1): 139–151. doi:10.1016/j.ydbio.2010.11.036.

Szabad, J, P Simpson, and R Nothiger. 1979. Regeneration and compartments in drosophila. *J Embryol Exp Morphol* 49: 229–241.

Vallejo, D M, S Juarez-Carreño, J Bolivar, J Morante, and M Dominguez. 2015. A brain circuit that synchronizes growth and maturation revealed through dilp8 binding to Lgr3. *Science* 350 (6262): 1–16. doi:10.1126/science.aac6767.

Vervoort, M. 2011. Regeneration and development in animals. *Biol Theory* 6: 25–35. doi:10.1007/s13752-011-0005-3.

Wolpert, L. 1969. Positional information and the spatial pattern of cellular differentiation. *J Theor Biol* 25 (1): 1–47. doi:10.1016/S0022-5193(69)80016-0.

Worley, M I, L Setiawan, and I K Hariharan. 2012. Regeneration and transdetermination in drosophila imaginal discs. *Annu Rev Genet* 46: 289–310. doi:10.1146/annurev-genet-110711-155637; 10.1146/annurev-genet-110711-155637.

Yagi, R, F Mayer, and K Basler. 2010. Refined lexA transactivators and their use in combination with the drosophila gal4 system. *Proc Natl Acad Sci U S A* 107 (37): 16166–16171. doi:1005957107[pii] 10.1073/pnas.1005957107.

Zeidler, M P, C Tan, Y Bellaiche, S Cherry, S Hader, U Gayko, and N Perrimon. 2004. Temperature-sensitive control of protein activity by conditionally splicing inteins. *Nat Biotechnol* 22 (7): 871–876. doi:10.1038/nbt979nbt979[pii].

Chapter 12 Positional information in the extracellular matrix
Regulation of pattern formation by heparan sulfate

Anne Q. Phan
and Md. Ferdous
Anower-E-Khuda

Contents

Key concepts

a. In organisms, cells are embedded in an extracellular matrix (ECM) that determines the physical, chemical, and biological properties of the cells and tissues. Together, the components of the ECM create an instructive extracellular environment that regulates cell signaling, proliferation, migration, and differentiation.

b. The Accessory Limb Model (ALM) is an *in vivo* gain-of-function assay that can be used to determine the bioactive components of the ECM that can regulate limb regeneration.

c. Heparan sulfate (HS) is a bioactive component of the ECM that regulates limb regeneration.

d. To engineer therapeutic regenerative matrices, the blueprint of positional information embedded in the ECM needs to be mapped out.

12.1 The extracellular matrix

In organisms, cells are embedded in an extracellular matrix (ECM) that determines the physical, chemical, and biological properties of the cells and tissues. The ECM is composed of fibrous proteins that provide tensile strength and elasticity (collagens and elastins), adhesive glycoproteins (fibronectin, laminin, and tenascin), and proteoglycans (PGs) that interact with other extracellular components. Together, the various ECM components create an instructive extracellular environment that regulates cell proliferation, migration, and differentiation, thereby dictating the physical and biological properties of cells and tissues. The ECM provides essential physical scaffolding for the cellular constituents and is required for cell adhesion, cell polarity, cell migration, cell differentiation, tissue coherence, tissue morphogenesis, and homeostasis. The three-dimensional architecture of the ECM defines tissue boundaries.

The fundamental components of ECM are glycans and proteins. Cells are coated in a dense layer of glycans, also known as the glycocalyx (Figure 12.1), that dominates the biological characteristics of the cell membrane and controls the interface of a cell with its environment. The glycocalyx serves as a platform for morphogen and growth factor binding. All cell–cell interactions are either cell–matrix–cell or cell–matrix communications.

HS glycosaminoglycans (GAGs) are one of the most diverse and heterogeneous macromolecules in the ECM. HS is known to interact with numerous growth factors, morphogens, cytokines, and chemokines. Through specific interactions of HS and growth factors, we have

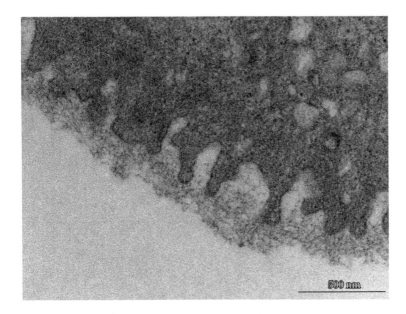

Figure 12.1 The cell surface is coated in a dense layer of glycans, known as the glycocalyx. Transmission electron microscopy (TEM) micrograph of a cell with the glycocalyx visualized as the gray matter coating the outside of the cell. (Courtesy of Miriam Cohen.)

demonstrated that HS regulates pattern formation during axolotl limb regeneration (Phan et al. 2015).

This chapter will briefly discuss the components of the ECM, HS signaling, HS in pattern formation, *in vivo* assays for ECM signaling in regeneration, and our ideas about applying an understanding of the role of the ECM in pattern formation toward engineering therapeutic regenerative matrices.

12.2 Components of the extracellular matrix that can regulate signaling

The ECM is a macromolecular network composed of PGs with long unbranched GAG chains, collagen, elastin, fibronectin, laminin, and several other glycoproteins (Figure 12.2). In most tissues, fibril-forming collagen type I associates with other collagens, ECM proteins, and PGs to assemble large fibrillar structures. Cells can interact with the ECM through specific cell surface receptors such as cell surface PGs, integrins, selectins, and cadherins.

Synthesis and secretion of the ECM are tightly yet dynamically regulated and are unique for each tissue and developmental time, and the composition of ECM is extremely heterogeneous. The components of ECM are organized into two different classes of macromolecules: fibrous proteins

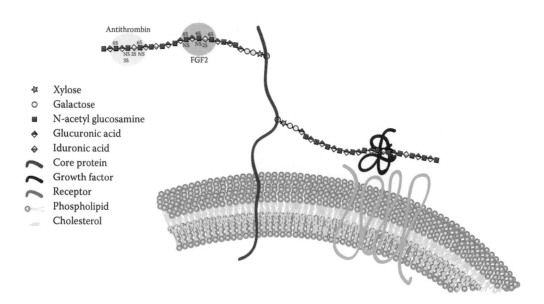

Figure 12.2 Diagram of heparan sulfate and its interaction with growth factors. Heparan sulfate proteoglycans (HSPG) are composed of a core protein, with covalently attached heparan sulfate glycosaminoglycan chains (HS GAG). The HS GAG chains are composed of repeating disaccharide units of N-acetyl glucosamine and glucuronic acid. The HS GAG chains can be modified by epimerization of glucuronic acid to iduronic acid and N, 2-O, 6-O, and 3-O sulfation to generate binding sites for proteins such as antithrombin and FGF2 (depicted). The HSPGs can act as co-receptors for growth factor binding to receptors. (Courtesy of Eillen Tecle and Pascal Gagneux.)

(including collagen and elastin) and glycoproteins (including laminin, fibronectin, and PGs). In addition, proteases (matrix metalloproteinases) and protease inhibitors that regulate ECM turnover also shape the ECM architecture and signaling.

Although a major interest of matrix engineering has focused on the intrinsic mechanical properties of ECM (adhesive affinity, matrix stiffness, fiber alignment, and matrix density), this chapter will focus on the bioactive components, whose composition can directly regulate cell signaling that is important in pattern formation and regeneration.

Collagen is the most abundant ECM protein and provides tensile strength, regulates cell adhesion and migration, and directs tissue development. Collagen fibril formation is strictly regulated by other ECM molecules such as decorin, fibromodulin, and integrins.

Integrins regulate epidermal growth factor receptor (EGFR), focal adhesion kinases (FAK), Src, mitogen-activated protein kinase (MAPK), c-Jun N-terminal kinases (JNK), p38, P13K, and Akt signaling. Collagen receptor integrins are involved in the regulation of cell proliferation and survival. Focal adhesion kinases, Src-family kinases (SFKs), and integrin-linked kinases (ILK) are the integrin signaling molecules.

Elastin and elastin-associated proteins create a large ECM structure and provide essential support to tissues such as large elastic blood vessels, lungs, heart, skin, bladder, and elastic cartilage to recover from continuous stretching. Elastin is composed of single tropoelastin subunits that are cross-linked with an outer layer of fibrillin microfibrils. Fibrillin contains calcium-binding epidermal growth factor (EGF)-like domains and heparin-binding domains that bind HS (Tiedemann et al. 2001), suggesting that fibrillin may directly signal cells, and specific cell surface receptor may regulate the assembly of fibrillin.

Laminins are glycoproteins made up of three subunits α, β, and γ: five α chains, three β chains, and three γ chains. By combination, they assemble more than 14 isoforms. The distribution of laminin isoforms is tissue-specific, suggesting that specific isoforms are required for specific tissue function. Laminin molecules interact with each other and with other ECM molecules to organize ECM structure, promote cell adhesion, angiogenesis, induce neurite growth, and regulate cell proliferation, migration, and differentiation. Laminins play crucial roles in early embryonic development and organogenesis. The biological effects of laminin are mediated by integrin and non-integrin receptors such as dystroglycan and HS.

Fibronectin is a multifunctional molecule that is ubiquitously expressed in the ECM. Fibronectin has a key role in cell adhesion, embryonic development, and wound healing. The molecule exists in two forms: the circulating form is present in plasma and the cellular form is expressed by fibroblasts. Fibronectin binds to cells through integrin receptors and forms binding domains for a variety of proteins and carbohydrates. The integrin-$\alpha5\beta1$ is the primary fibronectin receptor that mediates cell adhesion, migration, and signaling. Cell signaling through fibronectin involves two mechanisms: integrin–ligand interaction and integrin clustering or aggregation. Syndecans, a family of HS PG (HSPG) cell surface receptors, bind to fibronectin through their GAG chains and induce intracellular cell signaling.

Matrix metalloproteinases (MMPs) are the principal mediators of ECM remodeling. The activity of MMPs is largely regulated by $\alpha2$-macroglobulin and tissue inhibitors of metalloproteinases (TIMPs). Extracellular matrix binding can affect the interaction of MMPs with TIMPs and is highly dependent on the sulfation pattern of the GAGs (Troeberg et al. 2014). ADAMs (a disintegrin and metalloproteinase) are a family of zinc-dependent proteases related to the MMPs. ADAMs have been implicated in rheumatoid arthritis and osteoarthritis, wound healing, cardiovascular disease nephritis, cancer, and inflammation. MMPs, TIMPs, and ADAMs help organize growth factors and cytokines in ECM through other ECM molecules, including collagen, cell surface receptors, and sulfated GAGs.

Proteoglycans are composed of a core protein, with one or more GAG chains covalently attached. The GAG side chains are unbranched, long polysaccharides decorated with sulfate residues at different positions and are hence often referred to as sulfated GAGs. The GAGs are highly negatively charged repeating disaccharide units composed of N-acetylated hexosamine and D-/L-hexuronic acid. The GAG chain length and degree of sulfation of PGs are extremely heterogeneous and play critical roles in cellular signaling mechanisms. To date, six types of GAGs have been described: heparin (Hep), heparan sulfate (HS), chondroitin sulfate (CS), dermatan sulfate (DS), keratan sulfate (KS), and hyaluronan (HA). Based on the localization of PGs, four major classes have been identified: intracellular, cell surface, pericellular, and extracellular PGs.

Extracellular PGs, the largest class of PGs, function as bridging molecules connecting various ECM molecules and as mediators of cell signaling. Aggrecan is dominant in cartilage and provides a mechanosensitive feedback to the chondrocytes. Versican regulates cell adhesion, migration, and inflammation. Small leucine-rich PGs (SLRPs) interact with various collagens, bind with receptor tyrosine kinases and innate immune receptors, and modulate various cell signaling pathways. Decorin binds transforming growth factor beta (TGF-β), insulin-like growth factor receptor (IGFR), and vascular endothelial growth factor receptor 2 (VEGFR2). Circulating macrophages secret biglycan as a danger-signaling molecule for Toll-like receptor (TLR 2/4), an innate immunity receptor. Perlecan influences cell adhesion, growth, and survival. Agrin plays an important biological role in organizing basement membrane and clustering acetylcholine receptors. The cell surface PGs, syndecans, and glypicans interact with other ECM components and numerous growth factors mostly through their HS GAG chains. HS is involved in a wide variety of biological functions, too vast to be reviewed here, but briefly summarized in the next section.

12.3 Heparan sulfate composition mediates cell signaling

HS PGs are a major bioactive component of the ECM that can regulate growth factor signaling by (1) sequestering growth factors in the ECM, (2) protecting growth factors against proteolysis, and (3) acting as a necessary co-receptor for signaling.

HS PGs consist of a core protein, with one or more covalently attached HS chains. HS, a type of GAG, are linear polysaccharide chains of repeating disaccharides of N-acetyl-glucosamine and uronic acid (glucuronic acid or iduronic acid). Unlike heparin, HS is larger, less sulfated, and synthesized by all cell types (rather than just mast cells) (Esko et al. 2009). Typical HS GAG chains are extensive (~40–300 sugar residues, 7–80 kD, ~20–150 nm) and therefore tend to dominate the biological activity of HSPGs. These sugar residues are extensively modified by N-deacetylase-N-sulfotransferases (Ndsts), epimerases (Glce), and O-sulfotransferases (Hs2st, Hs6sts, and Hs3sts) to generate unique HS fine structure. HS GAGs are among the most negatively charged biopolymers in nature, and variations in chain number, chain length, and GAG sulfation level lead to extreme diversity in chemical and biological properties. Arrangement of negatively charged sulfate groups and orientation of carboxyl groups can specify ligand-binding sites (e.g., *antithrombin, fibroblast growth factor 2* [FGF2]; Figure 12.2). In addition, clusters of sulfated residues (NS domains) are separated by non-sulfated regions rich in glucuronic acid (NA domains). Patterns of HS GAG sulfation are more specific to cell type rather than to specific core proteins (Kato et al. 1994).

Localization of HS chains, as they are facing the extracellular environment, put them in advantageous positions to interact with abundant ligand molecules. Different gene knockout mice models highlight the importance of HS sulfation on ligand or growth factor binding (Sarrazin et al. 2011). Antithrombin binding to HS is critically dependent on the 3-O-sulfate residues in HS chains. Vascular endothelial growth factor (VEGF), platelet-derived growth factor (PDGF), fibroblast growth factor (FGF), and Wnt binding to HS and downstream cell signaling are dependent on HS sulfation pattern. The VEGF binding is affected by loss of N-sulfation. 2-O-sulfation facilitates FGF2 binding, whereas FGF1 binding is dependent on 6-O-sulfation. Storage of mast cell proteases inside the granules is critically dependent on 6-O-sulfation (Anower-E-Khuda et al. 2013); conversely, loss of 6-O-sulfation is required for Wnt binding and signal activation. Biochemistry of HSPG is reviewed in more detail by Sarrazin and colleagues (Sarrazin et al. 2011).

12.4 Assay for pro-regenerative activity of extracellular matrix components

Regeneration is an intrinsically stepwise process that is initiated in response to injury (Endo et al. 2004). Since any given step must occur successfully in order for the system to progress to the next step, failure of one step results in the failure of all downstream steps to occur. Consequently, it is not possible to identify the downstream steps in order to test whether or not they have the potential to occur (regenerative potential). Thus, one strategy for discovering how to induce regeneration in a non-regenerating animal is to identify the stepwise developmental processes that are required, by studying an animal that can regenerate (e.g., the axolotl), and then determine whether or not those processes can occur in mammals.

A major goal in the field of regeneration is to learn how the salamander regenerates and then use this information to develop therapies to induce a regenerative response in humans. The Accessory Limb Model (ALM) was designed to advance this goal by functioning as an *in vivo* gain-of-function assay for the requirements of axolotl limb regeneration (Endo et al. 2004; Satoh et al. 2007). Basically, the ALM involves the surgical induction of ectopic limb formation as a model to study limb regeneration. The ALM is described in (Endo et al. 2004; Satoh et al. 2007; Lee et al. 2013; Endo et al. 2015).

The ALM demonstrated that there are three discrete steps required for axolotl limb regeneration: (1) formation of a wound epithelium generated by wound healing after injury, (2) neurotrophic signaling between the cut end of a nerve and the wound epithelium to induce dedifferentiation, and formation of a mass of multipotent progenitor cells (blastema formation), and (3) positional information that instructs the blastema cells how to differentiate in order to reform the lost pattern (Endo et al. 2004). In brief summary, the ALM is composed of three surgical steps: (1) lateral skin wound, (2) nerve deviation, and (3) contralateral skin graft.

The ALM allows for the functional testing of candidate signaling factors that are required for each step in regeneration: wound healing, blastema formation, and pattern formation. Although a limb can be induced de novo by surgically deviating a nerve and grafting skin in an axolotl, these techniques are not realistic for human therapies. Therefore, discovering the underlying mechanisms by which the nerve and skin grafts function and how they can be replaced with human viable strategies would be the most important step in moving the field of regeneration toward medical application.

Since 2007, at least 18 papers have been published using the ALM to study regeneration (Satoh et al. 2008; Satoh et al. 2008; Ferris et al. 2010; Hirata et al. 2010; Satoh et al. 2011; Satoh et al. 2012; Makanae and Satoh 2012; Hirata et al. 2013; Makanae et al. 2013; McCusker and Gardiner 2013; Makanae et al. 2014; McCusker et al. 2014; Mitogawa et al. 2014; Satoh and Makanae 2014; Endo et al. 2015; Aguilar and Gardiner 2015; Phan et al. 2015; Nacu et al. 2016; McCusker et al. 2016). Ectopic limbs have been induced on the flank of the axolotl (Hirata et al. 2013). The ALM has also been adapted to *Xenopus* (Mitogawa et al. 2014). Growth-factor-soaked beads have been used to replace the nerve to induce blastema formation (Makanae et al. 2013; Makanae et al. 2014).

Phan et al. found that cell-free ECM, rather than a live skin graft, can be used to induce pattern (Figure 12.3), allowing for testing of non-axolotl–derived

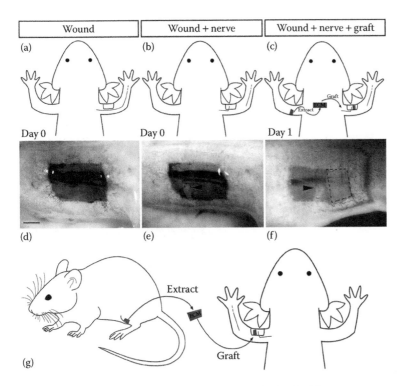

Figure 12.3 Adaptation of the Accessory Limb Model (ALM) as an *in vivo* gain-of-function assay for ECM induction of pattern formation during axolotl limb regeneration. (a, d) A full-thickness skin wound is made on the anterior side of the upper arm. (b, e) The brachial nerve is severed at the elbow and deviated to the wound site (black arrowhead). This induces a blastema to form. (c, f) Grafts of ECM are placed in the wound site to assay for induction of pattern formation. (g) Adaptation of the ALM to assay mouse limb ECM. Scale bar = 1 mm.

ECM. We demonstrated that the pattern-inducing activity can be mimicked by grafts of artificial ECM composed of purified mammalian factors (porcine HS and bovine collagen) as a substitute for a skin graft (Figure 12.4) (Phan et al. 2015). Similarly, the signaling from a deviated nerve can be mimicked by implanting gelatin beads that have been soaked in purified human growth factors (FGF and bone morphogenetic protein [BMP]) into an axolotl wound, which results in blastema formation (Makanae et al. 2013; Makanae et al. 2014). The two steps required for stimulating an axolotl wound to form an ectopic limb (blastema formation and pattern formation) can be induced individually by defined mammalian factors. A combination of these factors is sufficient to induce axolotl limb regeneration without a deviated nerve or skin graft, essentially induction of a *limb from scratch* (Figure 12.5). Extracellular matrix composition, stiffness, graft timing, and growth factor release can all be varied to find the exact combination required for perfect induction of limb regeneration.

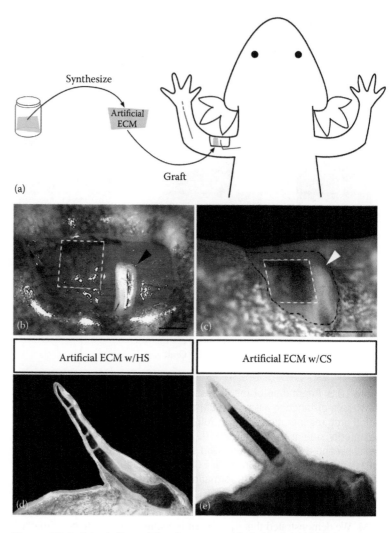

Figure 12.4 Adaptation of the Accessory Limb Model (ALM) to assay synthetic artificial ECM for induction of pattern formation during axolotl limb regeneration. (a) Artificial ECM can be synthesized from defined components and grafted into a wound with a deviated nerve. (b) Example of wound with deviated nerve (white, black arrowhead) and artificial ECM graft (green, outlined with white dashed line). (c) 1 week after surgery, retention of the artificial ECM graft can still be traced with black India ink (black, outlined with white dashed line) and the deviated nerve (white, white arrowhead). (d, e) Whole mount alcian blue/alizarin red skeletal staining of pattern induced by artificial ECM with heparan sulfate (HS) (D) or artificial ECM with chondroitin sulfate (CS) (E). Blue = cartilage and red = ossified bone. Scale bar = 1 mm.

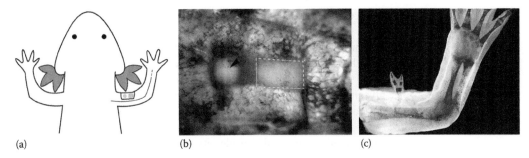

(a)   (b)   (c)

Figure 12.5 Adaptation of the Accessory Limb Model to assay for defined synthetic factors to induce axolotl limb regeneration. (a) A growth factor–soaked bead (substituting a deviated nerve) and artificial ECM (substituting a skin graft) are grafted into a wound to induce blastema and pattern formation during limb regeneration. (b) Example of a wound with a bead (white sphere, black arrowhead) artificial ECM graft (green, outlined with white dashed line). (c) Whole mount alcian blue/ alizarin red skeletal stain of ectopic limb induced by FGF2-, FGF8-, and BMP2-soaked bead and artificial ECM with heparan sulfate.

12.5 Glycans in the extracellular matrix as candidates for mediating positional information

In 1969, Lewis Wolpert outlined the concepts of *positional information* and *pattern formation* (Wolpert 1969). Since then, the best evidence for these phenomena has been intercalation during regeneration (Agata et al. 2003; French et al. 1976; Bryant et al. 1981). Freely diffusible morphogens and growth factors alone cannot account for positional specification and pattern formation; therefore, additional or alterative mechanisms involving interactions between cells are required (Kerszberg and Wolpert 2007; Wolpert 2011).

One area of biology that is rapidly expanding is the field of glycobiology. For decades, the complex and indecipherable sugars attached to every extracellular and secreted protein and coating the surface of all cell membranes has been too difficult to study compared with the less diverse polymers (e.g., nucleic acids and proteins). However, with the development of new tools and the realization that complex biological processes cannot be simply explained with genes and proteins, glycobiology is one of the fastest growing research areas in the natural sciences. Biological cells are coated in a dense layer of glycans (glycocalyx; Figure 12.1) that dominate the biological characteristics of the cell membrane and control the cell's interface with everything else outside (Cohen et al. 2007).

Cells in organisms are embedded in an ECM full of glycans that determine the physical, chemical, and biological properties of the cells and tissue. All cell–cell interactions are either cell–matrix–cell or cell–matrix communications. Therefore, if positional information is a result of cell–cell interactions, glycans in the ECM must be involved. Unlike proteins, glycans do not have to be transcribed and translated and therefore are more dynamically regulated and can be altered at the cell surface. With hundreds of building blocks with thousands of modifications, and some GAG chains being hundreds of residues in length, the possible glycan diversity is orders of magnitudes higher in complexity than genes or proteins. Therefore, sugars that can encode diversely specific cell-regulatory information are dynamically regulated on the outside of cells, function in organismal self-identity, and are in abundance at the cell interface where communication must occur are the most likely suspects when looking for molecules that mediate positional information.

Studies in which posterior mouse limb bud cells were grafted into the chick limb bud suggest that zone of polarizing activity (ZPA) signaling ability (posterior positional information) is maintained in the ECM rather than by the cells and is mediated by HS GAGs in the ECM (Schaller and Muneoka 2001). These studies became the basis for our own experiments to investigate the role of HS in pattern formation during axolotl limb regeneration. Using the ALM, we found that positional information in axolotl limb regeneration is mediated by HS in the ECM (Phan et al. 2015).

12.6 Heparan sulfate in pattern formation and regeneration

HS PGs are important regulators of the extracellular signaling molecules in pattern formation during development (Schaller and Muneoka 2001; Häcker et al. 2005; Wang and Beck 2015; Bause et al. 2016). Interaction of growth factors with HS is dependent on HS composition (Guimond et al. 1993; Olwin and Rapraeger 1992; Chang et al. 2000). Growth factors and artificial matrix components are being engineered based on their interaction with each other (Martino et al. 2014; Murali et al. 2013). The concept of HS stabilizing and shaping morphogen gradients has been demonstrated in a number of systems (Akiyama et al. 2008; Häcker et al. 2005; Takeo et al. 2005; Wang and Beck 2015). Specific chondroitin sulfate and HS sulfation motifs are associated with sites of growth, differentiation, repair, and molecular recognition (Melrose et al. 2012). On account of the specificity of HS growth factor regulation and the ability to identify specific HS-binding residue sequences, an *HS code* has been proposed (Habuchi et al. 2004). There is spatial-temporal regulation of HS sulfotransferases and dynamic expression of HSPGs (Phan et al. 2015; Nogami et al. 2004; Kobayashi et al. 2010). This HS code can be so tightly and rapidly controlled during development that FGF2 signaling during mouse embryonic day 9 can be changed to FGF1 signaling by embryonic day 11 (Nurcombe et al. 1993).

12.7 Positional information is a blueprint of position-specific growth factor regulation embedded in the extracellular matrix

We propose a model in which pattern formation is dictated by position-specific extracellular regulation of morphogens and growth factors, which then transduce signaling into specific cells. We hypothesize that positional information is a large-scale blueprint of signaling instructions encoded in the ECM. These blueprints dictate the body axes (anterior/posterior, dorsal/ventral, and proximal/distal) and gross morphology of organs and tissues. The blueprint directs position-specific growth factor signaling (e.g., FGF2 on right, FGF8 on the right, BMP on top, and Wnt on bottom) (Phan et al. 2015). These growth factors then instruct the cells to migrate, proliferate, or differentiate in a manner orchestrated by the extracellular matrix to eventually integrate all the cells into a functional pattern.

Presumably, pattern is initially formed during embryogenesis, with maternal and zygotic genes establishing the body axis. These growth factors then induce expression and synthesis of their corresponding HS code. The code is maintained in the matrix through adulthood (Figure 12.6). The regulation of HS biosynthesis is an area that needs further study.

In animals that regenerate, this positional information is stored in the connective tissue in the form of an HS code, and specific factors preferentially bound to the code. On traumatic injury, the positional information is used to regenerate the missing part from what is remaining and to recreate the position-specific modifications (Figure 12.7).

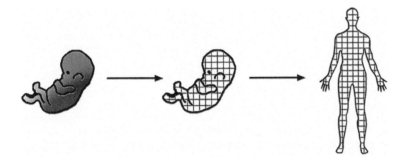

Figure 12.6 Model of establishment of positional information. Initial body axis is established by morphogens and growth factor expression during development. The initial axis information is then encoded into the ECM as a grid of positional information and then retained through development into adulthood.

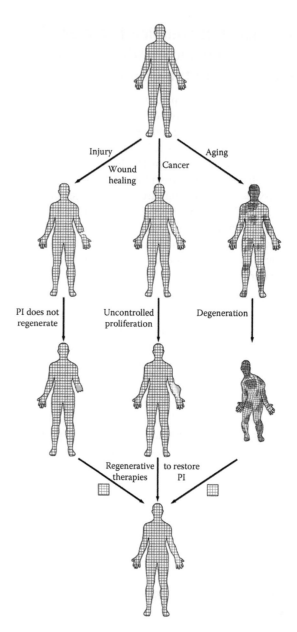

Figure 12.7 Model for application of positional information to development of regenerative therapies for wound healing, cancer, and aging. The grid of positional information (PI) is encoded in the ECM. On injury, the mammalian wound healing response results in loss of the PI required for regeneration. Development of cancer is correlated with degradation of the ECM, causing distortion of the PI grid, resulting in disorganized and unregulated proliferation. Over aging, the grid of PI is corroded, leading to degeneration associated with aging. If regenerative matrices encoding the proper grid of PI can be applied, regeneration could be induced.

12.8 Mapping out extracellular matrix composition to engineer regenerative matrices

One of the first steps to developing regenerative therapies with positional information is to map out the ECM composition across the entire body. In the way that the Polar Coordinate Model assigned arbitrary numbers and letters to denote positional values, ECM composition-based coordinates can be assigned to each position (e.g., high 3-O sulfation, low N-sulfation = posterior forelimb mid-stylopod). The HS composition can be cross-referenced with already described growth factor–binding sites (e.g., FGF2, 2-O, and N) and known areas of signaling (sonic hedgehog [SHH] in the ZPA). Transcriptional analysis of skin fibroblast across the human body has shown site-specific expression patterns (Rinn et al. 2008; Rinn et al. 2006). Similar methods could be used to analyze and map the HS/growth factor code. Once the HS code of positional information is mapped out, regenerative matrices can be engineered and applied to therapeutics for injury, cancer, or aging (Figure 12.7).

References

Agata, K., T. Tanaka, C. Kobayashi, K. Kato, and Y. Saitoh. 2003. Intercalary regeneration in planarians. *Developmental Dynamics* 226 (2): 308–316. doi:10.1002/dvdy.10249.

Aguilar, C. and D. M. Gardiner. 2015. DNA methylation dynamics regulate the formation of a regenerative wound epithelium during axolotl limb regeneration. *PloS One* 10 (8): e0134791. doi:10.1371/journal.pone.0134791.

Akiyama, T., K. Kamimura, C. Firkus, S. Takeo, O. Shimmi, and H. Nakato. 2008. Dally regulates Dpp morphogen gradient formation by stabilizing Dpp on the cell surface. *Developmental Biology* 313 (1): 408–419. doi:10.1016/j.ydbio.2007.10.035.

Anower-E-Khuda, M. F., H. Habuchi, N. Nagai, O. Habuchi, T. Yokochi, and K. Kimata. 2013. Heparan sulfate 6-O-sulfotransferase isoform-dependent regulatory effects of heparin on the activities of various proteases in mast cells and the biosynthesis of 6-O-sulfated heparin. *The Journal of Biological Chemistry* 288 (6): 3705–3717. doi:10.1074/jbc. M112.416651.

Bause, M., R. van der Horst, and F. Rentzsch. 2016. Glypican1/2/4/6 and sulfated glycosaminoglycans regulate the patterning of the primary body axis in the cnidarian Nematostella vectensis. *Developmental Biology* 414: 108–120. doi:10.1016/j.ydbio.2016.04.011.

Bryant, S. V., V. French, and P. J. Bryant. 1981. Distal regeneration and symmetry. *Science* 212 (4498): 993–1002. doi:10.1126/science.212.4498.993.

Chang, Z., K. Meyer, A. C. Rapraeger, and A. Friedl. 2000. Differential ability of heparan sulfate proteoglycans to assemble the fibroblast growth factor receptor complex in situ. *FASEB Journal* 14 (1): 137–44.

Cohen, M., D. Joester, I. Sabanay, L. Addadi, and B. Geiger. 2007. Hyaluronan in the pericellular coat: An additional layer of complexity in early cell adhesion events. *Soft Matter* 3 (3): 327. doi:10.1039/b613770a.

Endo, T., S. V. Bryant, and D. M. Gardiner. 2004. A stepwise model system for limb regeneration. *Developmental Biology* 270 (1): 135–145. doi:10.1016/j.ydbio.2004.02.016.

Endo, T., D. M. Gardiner, A. Makanae, and A. Satoh. 2015. The accessory limb model: An alternative experimental system of limb regeneration. *Methods in Molecular Biology (Clifton, N.J.)* 1290: 101–113. doi:10.1007/978-1-4939-2495-0_8.

Esko, J. D., K. Kimata, and U. Lindahl. 2009. Proteoglycans and sulfated glycosaminoglycans. In *Essentials of Glycobiology*, 2nd ed., edited by A. Varki, R. D. Cummings, and J. D. Esko. New York: Cold Spring Harbor Laboratory Press.

Ferris, D. R, A. Satoh, B. Mandefro, G. M. Cummings, D. M. Gardiner, and E. L. Rugg. 2010. Ex vivo generation of a functional and regenerative wound epithelium from axolotl (Ambystoma mexicanum) skin. *Development, Growth & Differentiation* 52 (8): 715–724. doi:10.1111/j.1440-169X.2010.01208.x.

French, V., P. J. Bryant, and S. V. Bryant. 1976. Pattern regulation in epimorphic fields. *Science* 193 (4257): 969–981.

Guimond, S., M. Maccarana, and B. B. Olwin. 1993. Activating and inhibitory heparin sequences for FGF-2 (Basic FGF). Distinct Requirements for FGF-1, FGF-2, and FGF-4. *Journal of Biological* 268 (32): 23906–23914.

Habuchi, H., O. Habuchi, and K. Kimata. 2004. Sulfation pattern in glycosaminoglycan: Does it have a code? *Glycoconjugate Journal* 21 (1–2): 47–52. doi:10.1023/B:GLYC.0000043747.87325.5e.

Häcker, U., K. Nybakken, and N. Perrimon. 2005. Heparan sulphate proteoglycans: The sweet side of development. *Nature Reviews Molecular Cell Biology* 6 (7): 530–541. doi:10.1038/nrm1681.

Hirata, A., D. M. Gardiner, and A. Satoh. 2010. Dermal fibroblasts contribute to multiple tissues in the accessory limb model. *Development, Growth & Differentiation* 52 (4): 343–350. doi:10.1111/j.1440-169X.2009.01165.x.

Hirata, A., A. Makanae, and A. Satoh. 2013. Accessory limb induction on flank region and its muscle regulation in axolotl. *Developmental Dynamics* 242 (8): 932–940. doi:10.1002/dvdy.23984.

Kato, M., H. Wang, M. Bernfield, J. T. Gallagher, and J. E. Turnbull. 1994. Cell surface syndecan-1 on distinct cell types differs in fine structure and ligand binding of its heparan sulfate chains. *The Journal of Biological Chemistry* 269 (29): 18881–18890.

Kerszberg, M. and L. Wolpert. 2007. Specifying positional information in the embryo: Looking beyond morphogens. *Cell* 130 (2): 205–209. doi:10.1016/j.cell.2007.06.038.

Kobayashi, T., H. Habuchi, K. Nogami, S. Ashikari-Hada, K. Tamura, H. Ide, and K. Kimata. 2010. Functional analysis of chick heparan sulfate 6-O-sulfotransferases in limb bud development. *Development, Growth & Differentiation* 52 (2): 146–156. doi:10.1111/j.1440-169X.2009.01148.x.

Lee, J., C. Aguilar, and D. Gardiner. 2013. Gain-of-function assays in the axolotl (Ambystoma mexicanum) to identify signaling pathways that induce and regulate limb regeneration. *Methods in Molecular Biology* 1037: 401–117. doi:10.1007/978-1-62703-505-7_23.

Makanae, A., A. Hirata, Y. Honjo, K. Mitogawa, and A. Satoh. 2013. Nerve independent limb induction in axolotls. *Developmental Biology* 381 (1): 213–226. doi:10.1016/j.ydbio.2013.05.010.

Makanae, A., K. Mitogawa, and A. Satoh. 2014. Co-operative Bmp- and Fgf-signaling inputs convert skin wound healing to limb formation in urodele amphibians. *Developmental Biology* 396 (1): 57–66. doi:10.1016/j.ydbio.2014.09.021.

Makanae, A. and A. Satoh. 2012. Early regulation of axolotl limb regeneration. *Anatomical Record* 295 (10): 1566–1574. doi:10.1002/ar.22529.

Martino, M. M., P. S. Briquez, E. Güç, F. Tortelli, W. W. Kilarski, S. Metzger, J. J. Rice. et al. 2014. Growth factors engineered for super-affinity to the extracellular matrix enhance tissue healing. *Science* 343 (6173): 885–888. doi:10.1126/science.1247663.

McCusker, C. D. and D. M. Gardiner. 2013. Positional information is reprogrammed in blastema cells of the regenerating limb of the axolotl (Ambystoma mexicanum). *PloS One* 8 (9): e77064. doi:10.1371/journal.pone.0077064.

McCusker, C. D., C. Diaz-Castillo, J. Sosnik, A. Q. Phan, and D. M. Gardiner. 2016. Cartilage and bone cells do not participate in skeletal regeneration in Ambystoma mexicanum limbs. *Developmental Biology* 416 (1): 26–33. doi:10.1016/j.ydbio.2016.05.032.

McCusker, C., J. Lehrberg, and D. Gardiner. 2014. Position-specific induction of ectopic limbs in non-regenerating blastemas on axolotl forelimbs. *Regeneration* 1 (1): 27–34. doi:10.1002/reg2.10.

Melrose, J., M. D. Isaacs, S. M. Smith, C. E. Hughes, C. B. Little, B. Caterson, and A. J. Hayes. 2012. Chondroitin sulphate and heparan sulphate sulphation motifs and their proteoglycans are involved in articular cartilage formation during human foetal knee joint development. *Histochemistry and Cell Biology* 138 (3): 461–475. doi:10.1007/s00418-012-0968-6.

Mitogawa, K., A. Hirata, M. Moriyasu, A. Makanae, S. Miura, T. Endo, and A. Satoh. 2014. Ectopic blastema induction by nerve deviation and skin wounding: A new regeneration model in xenopus laevis. *Regeneration* 1 (2): 26–36. doi:10.1002/reg2.11.

Murali, S., B. Rai, C. Dombrowski, J. L. J. Lee, Z. X. H. Lim, D. S. Bramono, L. Ling. et al. 2013. Affinity-selected heparan sulfate for bone repair. *Biomaterials* 34 (22): 5594–5605. doi:10.1016/j.biomaterials.2013.04.017.

Nacu, E., E. Gromberg, C. R. Oliveira, D. Drechsel, and E. M. Tanaka. 2016. FGF8 and SHH substitute for anterior-posterior tissue interactions to induce limb regeneration. *Nature* 533: 407–410. doi:10.1038/nature17972.

Nogami, K., H. Suzuki, H. Habuchi, N. Ishiguro, H. Iwata, and K. Kimata. 2004. Distinctive expression patterns of heparan sulfate O-sulfotransferases and regional differences in heparan sulfate structure in chick limb buds. *The Journal of Biological Chemistry* 279 (9): 8219–8229. doi:10.1074/jbc.M307304200.

Nurcombe, V., M. D. Ford, J. A. Wildschut, and P. F. Bartlett. 1993. Developmental regulation of neural response to FGF-1 and FGF-2 by heparan sulfate proteoglycan. *Science* 259 (5104): 103–106.

Olwin, B. B. and A. Rapraeger. 1992. Repression of myogenic differentiation by aFGF, bFGF, and K-FGF is dependent on cellular heparan sulfate. *The Journal of Cell Biology* 118 (3): 631–639.

Phan, A. Q., J. Lee, M. Oei, C. Flath, C. Hwe, R. Mariano, T. Vu. et al. 2015. Positional information in axolotl and mouse limb extracellular matrix is mediated via heparan sulfate and fibroblast growth factor during limb regeneration in the axolotl (Ambystoma mexicanum). *Regeneration* 2 (4): 182–201. doi:10.1002/reg2.40.

Rinn, J. L., C. Bondre, H. B. Gladstone, P. O. Brown, and H. Y. Chang. 2006. Anatomic demarcation by positional variation in fibroblast gene expression programs. *PLoS Genetics* 2 (7): e119. doi:10.1371/journal.pgen.0020119.

Rinn, J. L., J. K. Wang, H. Liu, K. Montgomery, M. van de Rijn, and H. Y. Chang. 2008. A systems biology approach to anatomic diversity of skin. *The Journal of Investigative Dermatology* 128 (4): 776–782. doi:10.1038/sj.jid.5700986.

Sarrazin, S., W. C. Lamanna, and J. D. Esko. 2011. Heparan sulfate proteoglycans. *Cold Spring Harbor Perspectives in Biology* 3 (7). doi:10.1101/cshperspect.a004952.

Satoh, A., S. V. Bryant, and D. M. Gardiner. 2008. Regulation of dermal fibroblast dedifferentiation and redifferentiation during wound healing and limb regeneration in the axolotl. *Development, Growth & Differentiation* 50 (9): 743–754. doi:10.1111/j.1440-169X.2008.01072.x.

Satoh, A., D. M. Gardiner, S. V. Bryant, and T. Endo. 2007. Nerve-induced ectopic limb blastemas in the axolotl are equivalent to amputation-induced blastemas. *Developmental Biology* 312 (1): 231–244. doi:10.1016/j.ydbio.2007.09.021.

Satoh, A., G. M. C. Graham, S. V. Bryant, and D. M. Gardiner. 2008. Neurotrophic regulation of epidermal dedifferentiation during wound healing and limb regeneration in the axolotl (Ambystoma mexicanum). *Developmental Biology* 319 (2): 321–335. doi:10.1016/j.ydbio.2008.04.030.

Satoh, A., A. Hirata, and A. Makanae. 2012. Collagen reconstitution is inversely correlated with induction of limb regeneration in Ambystoma mexicanum. *Zoological Science* 29 (3): 191–197. doi:10.2108/zsj.29.191.

Satoh, A. and A. Makanae. 2014. Conservation of position-specific gene expression in axolotl limb skin. *Zoological Science* 31 (1): 6–13. doi:10.2108/zsj.31.6.

Satoh, A., A. Makanae, A. Hirata, and Y. Satou. 2011. Blastema induction in aneu-rogenic state and Prrx-1 regulation by MMPs and FGFs in Ambystoma mexicanum limb regeneration. *Developmental Biology* 355 (2): 263–274. doi:10.1016/j.ydbio.2011.04.017.

Schaller, S. and K. Muneoka. 2001. Inhibition of polarizing activity in the anterior limb bud is regulated by extracellular factors. *Developmental Biology* 240 (2): 443–456. doi:10.1006/dbio.2001.0500.

Takeo, S., T. Akiyama, C. Firkus, T. Aigaki, and H. Nakato. 2005. Expression of a secreted form of dally, a drosophila glypican, induces overgrowth pheno-type by affecting action range of hedgehog. *Developmental Biology* 284 (1): 204–218. doi:10.1016/j.ydbio.2005.05.014.

Tiedemann, K., B. Bätge, P. K. Müller, and D. P. Reinhardt. 2001. Interactions of fibrillin-1 with heparin/heparan sulfate, implications for microfibrillar assem-bly. *The Journal of Biological Chemistry* 276 (38): 36035–36042. doi:10.1074/jbc.M104985200.

Troeberg, L., C. Lazenbatt, M. F. Anower-E-Khuda, C. Freeman, O. Federov, H. Habuchi, O. Habuchi, K. Kimata, and H. Nagase. 2014. Sulfated glycos-aminoglycans control the extracellular trafficking and the activity of the metalloprotease inhibitor TIMP-3. *Chemistry & Biology* 21 (10): 1300–1309. doi:10.1016/j.chembiol.2014.07.014.

Wang, Y.H. and C. Beck. 2015. Distinct patterns of endosulfatase gene expression during xenopus laevis limb development and regeneration. *Regeneration* 2 (1): 19–25. doi:10.1002/reg2.27.

Wolpert, L. 1969. Positional information and the spatial pattern of cellular differen-tiation. *Journal of Theoretical Biology* 25 (1): 1–47.

Wolpert, L. 2011. Positional information and patterning revisited. *Journal of Theoretical Biology* 269 (1): 359–365. doi:10.1016/j.jtbi.2010.10.034.

Chapter 13 Organ shaping by localized signaling centers

Stephanie Tsai,
Randal B. Widelitz,
Alaa Abdelhamid, and
Cheng-Ming Chuong

Contents

Key concepts

a. Successful regenerative medicine requires guiding the competent progenitor cells to form the desired tissue/organ for transplantation. The organ architecture is important for proper function. We use the following examples to illustrate this principle.

b. The number and spatiotemporal distribution of localized growth zones determine beak shape. Beak width, length, and curvature maybe controlled by different gene expression levels and patterns.

c. A series of specific molecules with biological, chemical, and mechanical couplings mediates a chemical gradient that defines the shape and orientation of feathers within a spatiotemporal distribution.
d. Branching may arise from the periodic patterning of budding (e.g., feather) or apoptosis (e.g., mammary gland).
e. Patterns of tooth morphology and their localization in the jaw are dynamically determined by signaling centers.

13.1 Introduction: Shaping organs by localized signaling centers

Regenerative medicine carries the hope to rebuild an injured or lost organ to its full original function. There are three levels of issues that will have to be solved to ease the path of regenerative medicine: (1) acquiring progenitor cells, (2) guiding these progenitor cells to form the tissues/organs that one desires, and (3) having these newly formed tissues/organs integrate with the host without side effects. For the first issue, progenitors may come from resident adult stem cells, de-differentiation of endogenous cells, or exogenously delivered stem cells. The second issue must consider both differentiation and morphogenesis, so the regenerated organ can have the proper function and morphology. For the third issue, one has to be concerned about tumor formation, immune rejection, or loss of the stem cell product. In this chapter, we focus on the morphogenesis of the second issue.

Organ architecture is very important for proper function. The arrangement of cells within each organ depends on cellular and molecular interactions. Although proliferation, migration, differentiation, and/or apoptosis can be judged at the level of single or dissociated cells, tissues and organs should be considered a cell collective, possessing a higher level of organization. The patterned organization and shape of the collective are new properties that emerge only in a cell population. Morphogenesis deals with this level of behavior. There are four scenarios in which morphogenesis-competent fields are established, where a collective of morphogenesis-competent progenitor cells interact to build a new organ: (1) embryonic development, (2) cyclic regeneration under physiological conditions (Chuong et al. 2012), (3) blastema-based regeneration after injury (McCusker and Gardiner 2014), and (4) tissue engineering based on derived stem cells.

In these four scenarios, the sources of cells can be very different, yet the principles of morphogenesis can be similar. Based on this premise, we would like to learn more about how morphogenesis occurs during development and physiological regeneration (scenarios I and II) and try to apply this knowledge to achieve regenerative wound healing and regenerative engineering (scenarios III and IV). Earlier, we have discussed different types of regeneration (Chueh et al. 2013) and how to engineer stem cells into organs with topological transformations (Chuong et al. 2006). Recently, we reviewed the self-organization process in periodic patterning (Chuong et al. 2013) and the overall principles of morphogenesis and integument patterns (Li et al. 2015). In this chapter, we will focus on one of these principles: using localized signaling centers (LSCs) to shape the organ.

At the cellular level, behavior has been described in terms of cell proliferation, cell differentiation, cell apoptosis, cell adhesion, and cell migration. In tissue morphogenesis and regeneration, behavior exists at the level of a cell population. Thus, an LSC is defined as a cluster of cells exhibiting specific molecular information that can guide their own or adjacent cell group's proliferation, differentiation, apoptosis, adhesion, or migration (Figure 13.1b). The activity is transient in space and time; therefore, the relative position, size, and duration of activity can vary and have implications on organ shapes. A combination of these LSCs then works coordinately to sculpt and build an organ's shape. In the following section, we will use four examples to illustrate this perspective in different contexts.

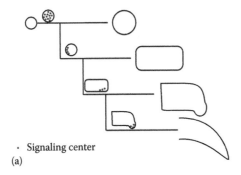

· Signaling center

(a)

Figure 13.1 Conceptual design principles of organ shaping by localized signaling centers (LSCs). (a) Schematic drawing showing how a strategically positioned LSC can mold an organ shape. In this case, a localized proliferation center is shown. Randomly distributed proliferating cells can create a ball-like cellular mass (e.g., tumor). An asymmetrically positioned LSC can convert the cellular mass into an elongated shape (e.g., epiphyseal plate in the long bone) or can create a bent structure, as seen in different types of beaks (Wu et al. 2006). The LSCs can be deployed in a dynamic way, spatially and temporally, during development. They can also be layered upon a previous shape (e.g., articular cartilage head). These multiple steps build up complex shapes step by step to achieve the desired shape and size. They likely represent evolutionary novel steps as well.   (Continued)

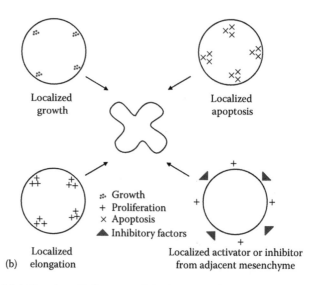

Figure 13.1 (Continued) Conceptual design principles of organ shaping by localized signaling centers (LSCs). (b) Particular shapes (e.g., branches) can be achieved by different LSC mechanisms. The signal can be located in the epithelium or adjacent mesenchyme.

13.2 Case studies

13.2.1 Shapes of beaks: The classic case of function and form

The shaping of the bird beak is a paradigm for evolution. It is not only an evolutionary innovation from reptiles to birds but also represents a specific functional and developmental module that can be modulated to generate a spectrum of shapes (Wagner and Lynch 2010). Studies on series of fossils demonstrate that the beak is not a structure fused directly to the pre-maxillae, since it geometrically differs from those of ancestral archosaurs (Zhang et al. 2002). The upper beak is composed of the frontal nasal mass (FNM), lateral nasal prominences (LNP), and maxillary prominences (MXP); the mandibular prominence (MDP) comprises the lower beak. The mesenchymal core of these prominences is derived from neural crest and mesoderm. Its epithelial covering is derived from the ectoderm and endoderm (Helms and Schneider 2003, Richman and Lee 2003).

The shape of beaks correlates with growth zones during early- to mid-development stages, especially in the mid-sagittal dimension that could be highly controlled by two parameters (length and height; Figure 13.1a). Later, the growth zone disappears, whereas further changes in size are based on uniform cell proliferation. Birds with curved beaks maybe controlled by several other factors (Fritz et al. 2014). Activation of the two frontonasal signaling systems during specific development stages has been hypothesized to generate the facial differences seen between avians and mice: (i) Frontonasal ectodermal zone: Fibroblast growth factor 8 (FGF8), dominant in the frontonasal prominence during early chicken embryonic stages, predicts ventrally located sonic hedgehog (*SHH*) in facial regions (Hu et al. 2003). (ii) Wnt is then activated and required for the formation of proper skeletal elements (Brugmann et al. 2007). Inhibition of Wnt in the frontonasal prominence caused both skeletal phenotypes and altered palatal regions into a more ancestral phenotype (Bhullar et al. 2015).

The number and spatiotemporal distribution of localized growth zones determine beak shape. Studies on duck and quail chimerae demonstrated that beak morphology is determined by neural crest cells instead of dermal cells, even though the ratio of neural crest cells to dermal cells may affect the signaling of recombinant tissues (Noden 1983, Trainor and Krumlauf 2000, Schneider and Helms 2003). The different duration and area of proliferation during development could mold the frontonasal mass into different shapes. Both chickens and ducks have two localized growth zones in the FNM at stage 26 (Hamburger and Hamilton 1951), but the two zones converge centrally to form a single growth zone in chickens at stages 28. The frontonasal ectodermal cells expressing FGF8 and SHH would determine mesenchymal patterning. Persistent FGF8 expression would also contribute to more cartilage growth (Abzhanov and Tabin 2004, Wu et al. 2006). Sonic hedgehog and Wnt are also responsible for many differences in facial width and fusion of facial primordia (Bhullar et al. 2015). However, SHH and *FGF8* showed no significant differences in beak expression between species of Darwin's finches, whereas Bone morphogenetic protein 2 (BMP2) and BMP7 correlate more with the beak size than with shape (Richman and Lee 2003, Abzhanov et al. 2004). During the development of dentary and frontal bones, instead of directly becoming osteoblasts (expressing *Opn*, *BspII*, and *Ihh*), pre-osteoblasts (expressing *Runx2*, *Coll II*, *Coll IX*, *Ptc1*, *Gli1*, *PTHrP-R*, and *Bmp4*) are induced by BMP2 and BMP4 (Abzhanov et al. 2007) to differentiate into chondrocyte-like osteoblasts (expressing *Opn*, *Coll II*, *Coll IX*, *Ihh*, *Ptc1*, *Gli1*, *Bmp4*, and *PTHrP*).

The BMPs work both upstream and parallel to Edn1 in the ventral arch, and several other ventral-specific (*hand2* and *dlx6a*) and dorsal-specific (*jag1b*) genes promote arch patterning after neural crest migration. Later, BMP action on craniofacial skeleton patterning functions independently of Edn1 (Alexander et al. 2011). The BMP4 correlates not only with the localized growth zones in chickens and ducks but also with beak morphology in Darwin's finches (Abzhanov et al. 2004, Wu et al. 2004). Overexpressed BMP4 enhanced cell proliferation and skeletal differentiation, causing larger beaks in embryonic stage 22–23 chickens. These results are consistent with those in which BMP receptors are overexpressed and contrast with those in which noggin (a BMP antagonist) is overexpressed. *Msx1*, a marker of craniofacial patterning, co-localized with *BMP4* during the shift in the growth zone, showing that its activity is regulated by BMP. The above-mentioned BMP4 regulation could also produce beak asymmetry or curvature in the final form. Besides, BMP4-coated beads, though highly site-specific, could partially rescue the growth of a surgically deleted mesenchymal zone (Abzhanov et al. 2004, Wu et al. 2004, 2006). Similar facial patterning results have also shown that ectopic expression of BMP4 could rescue $Msx1^{-/-}$ mice with developmental palate deficiencies (Zhang et al. 2002). Combining noggin and retinoic acid in early avian embryos led to a homeotic transformation in the frontonasal mass (Richman et al. 1997, Plant et al. 2000). Higher levels of *CALM*, a gene modulating calcium signaling, correlate with longer beaks in Darwin's finches. Elevating Ca^{2+} levels by infecting RCAS-CA-CaMKII (*CaMKII*, downstream effector of CaM) into the distal mesenchyme of early stage 24 chick frontonasal processes upregulates proliferation surrounding the pre-nasal cartilage, without directly infecting the skeletal tissues, indicating that multiple mechanisms are involved in generating longer beaks. Beak length is independently controlled from width in finches that strongly express *CALM* during morphogenesis. In finches, beak length is regulated by *CALM* expression during morphogenesis, while the depth and width are controlled by BMP4. This finding suggests that they may function independently in controlling beak shape (Abzhanov et al. 2006).

HMGA2 regulates beak and body size in the Galapagos finches but not beak shape. Two different *HMGA2* alleles control beak and body size: the L allele is present in larger birds that also have large beaks, whereas the S allele is present in smaller birds with corresponding smaller beaks. Heterozygous finches have an intermediate beak and body size (Lamichhaney et al. 2016). In contrast to the *HMGA2* alleles, the Alx1 homeodomain protein regulates beak shape (Lamichhaney et al. 2015). Alx1 regulates neural crest migration and development of craniofacial mesenchyme. Two *Alx1* alleles correspond with the pointed (P) or blunted (B) beaks. In general, pointed beaks were longer and straighter than blunt beaks (Dee et al. 2013, Lamichhaney et al. 2015). Bmp4 and CaM independently regulate the growth of prenasal cartilage. The transforming growth factor βIIr (TGFβIIr) and β-catenin regulate the premaxillary bone length and depth, either directly or via DKK3, Dickkopf3 (Mallarino et al. 2011).

Whether a localized growth zone or universal proliferation configures the mammalian jaw shape remains to be definitively determined; however, directional proliferation is generally accepted as one of the major factors contributing to jaw morphogenesis. Lengthening of the maxilla combines both palatal-directed sutural growth and tuberosity-directed periosteal apposition. Expansion of the jaw along the cranial–caudal axis combines both growth at the sutural articulations of the frontal and zygomatic processes and the periosteal apposition on the alveolus (Björk 1968).

The correlation between form and function in Galapagos finches inspired Darwin's theories of evolution. Here, we can see that these diverse forms were achieved by the simple yet consequential modulation of LSC activity.

13.2.2 Feather polarizing activity: An example of a localized signaling center that translates diffusible signals into tissue elongation

Feathers initiate from dome-shaped epithelial thickenings (placodes) at stage 29, followed by the formation of rounded dermal condensations to form symmetric primordia. The feather primordia elongate to form a symmetric bud. Then, the bud apex shifts to the posterior side to form anterior–posterior asymmetry at the asymmetric short bud stage. The feather continues to elongate, forming the long bud stage, and then, the bud invaginates at the follicle stage (Lin et al. 2006). During this process, buds dramatically change their shape and establish anterior–posterior polarity. Scientists studying limb buds have identified the zone of polarizing activity (ZPA) that produces anterior–posterior polarity through tissue transplantation studies (Balcuns et al. 1970). This is a classic example of identifying the function of an LSC. We took a similar approach to identify the feather zone of polarizing activity (FPA). For the ZPA, limb investigators identified SHH as the molecule responsible for polarizing activity. In the FPA, we found β-catenin, Delta, and non-muscle myosin to be the molecular module that drives polarizing activity. This zone strongly expresses non-muscle myosin IIB (NM IIB), which is the motor protein for directional cell rearrangements and feather elongation. Cells from the posterior part of the dermis are also required to orient this elongation (Li et al. 2013).

The function of each component was tested (Li et al. 2013). Perturbing the spatial distribution of nuclear β-catenin-positive cells, either by misexpression of nuclear β-catenin or by decreasing β-catenin nuclear accumulation, significantly altered feather polarity. The dermal nuclear β-catenin zone induces *Jag1*, which is surrounded by *Notch1*-positive dermal cells. Inhibition of Notch signaling causes the nuclear β-catenin zone to loosen its original spatial configuration, resulting in randomized feather bud orientation. Mathematical modeling simulations show that Notch lateral inhibition is critical to reducing fluctuations and variations in the Wnt7a gradient that maintains the sharp-edged spatial configuration of the β-catenin-positive zone. The nuclear β-catenin also induces expression of NM IIB. Blocking NM IIB disrupts the original directional cell rearrangement and abolishes the elongation of feather buds.

These data identify a novel mechanism that contributes to elongated organ shape with a particular orientation. A series of specific molecules, including biological, chemical, and mechanical couplings, mediates a chemical gradient for defining spatiotemporal limitations to shape and orient the feathers. We also developed a model whereby Notch signaling fine-tunes the spatial configuration of the Wnt responding zone (Figure 13.2). The model shows how positive feedback between Notch and Wnt signaling helps reduce noise (spatial variations) and ultra-sensitizes cells to the Wnt gradient. This mechanism converts a pliable chemical gradient signal into a more precise domain conducting a

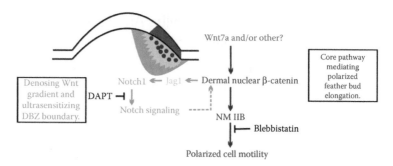

Figure 13.2   Feather polarizing activity as an example of a localized signaling center: a Wnt/Notch/non-muscle myosin module that translates diffusible signals into tissue elongation. Using a transplantation/recombination strategy similar to that for identifying the limb bud zone of polarizing activity (ZPA) (Balcuns et al. 1970), we found a unique zone (about 50 µm in dimension) at the posterior bud with biological activity for directional elongation. Purple: Wnt7a and/or other Wnt molecules from posterior epithelium induce β-catenin nuclear accumulation in dermal cells (red spots). Yellow: Jag1-positive zone. Blue: Notch1-positive zone. Red arrows: Nuclear β-catenin—Notch feedback loop. Green arrows: Nuclear β-catenin-induced directional cell motility. Dashed lines: Unknown mechanism. The region is modulated both chemically and mechanically, defining robust feedback interactions of well-patterned development. The dashed lines indicate the potential mechanisms that are under studying. (Adapted from Li, A. et al., *Proc. Natl. Acad. Sci. U S A*, 110, E1452–E1461, 2013.)

particular mechanical activity, thus defining the position, boundary, and duration of localized morphogenetic activities that are essential to shape organs. Therefore, this is a good example of an LSC, a polarizing activity zone, which is responsible for modulating directional elongation.

13.2.3   Feathers and mammary glands: Building two branched ectodermal organs with opposite mechanisms

Branching expands the surface area of an organ, enabling it to interact more efficiently with its micro- or macro-environment. The surface expansion may have arisen through budding and growth, as seen in glands, or by the apoptotic splitting of the organ into branches, as seen in feathers. Symmetric and asymmetric branches can form to serve different purposes, as can be seen in the morphogenesis of glands (e.g., mammary and salivary), feathers, lung alveoli, and several other tissues and organs.

Feathers show hierarchical branch patterns on different body parts that may differ at different life stages (chicks vs sexually mature adults) or between species (Figure 13.3b). The cylindrical epithelium of the feather filament forms periodic invaginations and segregates into alternating zones of proliferation or death. Each valley becomes a marginal plate,

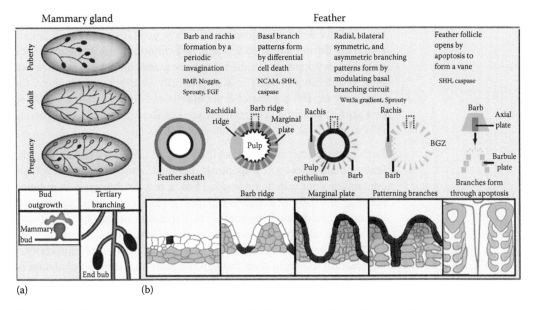

Figure 13.3 Branches in mammary glands and feathers are achieved by opposite mechanisms: Differential growth or differential death. (a) The branching of mammary glands. It depends on differential growth of duct epithelia into fat pad mesenchyme. Secondary and tertiary branchings are added, which are required to complete the adult mammary gland. The growth results from epithelial–mesenchymal interactions, which are also regulated by systemic hormone conditions. (b) The branching of feather barbs. It mainly depends on differential apoptosis. Feathers form an epithelial cylinder first. The basal layer then forms alternating zones of proliferation (barb plate) and death (marginal plate). Epithelia in the marginal plate express Neural Cell Adhesion Molecule (NCAM) and SHH and eventually undergo apoptosis. The branching pattern results from epithelial–mesenchymal interactions. Depending on the activity of BMP and Wnt3a, radially or bilaterally symmetric feather shapes can form. (Adapted from Chuong, C. M. et al., *Cell*, 158, 1212–1212, 2014.)

creating future interbarb space via apoptosis. Each ridge becomes a barb ridge containing barbule plates bilaterally and axial plates centrally. The axial plates will apoptose, enabling the barbs to open. Periodic patterning within the follicle is initiated by activators (e.g., BMPs) and inhibitors (e.g., SHH and caspase-3). Apoptosis also occurs at the follicular level, in the pulp epithelia (internal to the filament cylinder), in feather sheath (enclosing the filament cylinder), and in the barb-generative zone (in the posterior follicle) (Chuong et al. 2014). Thus, apoptosis is the major source of feather branching.

The budding and branching morphogenesis of mammary glands also develop as a self-organized module (Figure 13.3a). Female mammary glands undergo many structural changes from embryonic to adult stages. In order to adapt different functions required during different life stages, mammary gland cells are remodeled by proliferation during pregnancy, differentiation during late-stage pregnancy and early lactation, or apoptosis during involution in response to different signaling. The signaling that initiates and supports the growth of the mammary gland is also

317

similar to the one that participates in hair and dental placode formation: Lef1 (a Wnt target gene) and FGFs (e.g., FGF10 and FGFR2b) are expressed during the placode stage, and some continue to be expressed until bud development. In the mammary gland, branching requires cellular proliferation, which is also generally distributed throughout the whole organ. The pattern of branching is not derived from a localized cell growth zone via concentrated growth factor receptors or signaling (Pispa and Thesleff 2003, Nelson et al. 2006).

Both of these organs are cyclically built and degraded. In feathers, the feather filament is lost by apoptosis, but the lower portion of the follicle containing stem cells is retained. Then, a new feather cycle is initiated (Chen et al. 2016). In mammary glands, pregnancy induces budding and increased proliferation promotes the expansion of the ductal network and alveoli into the surrounding stroma. These then differentiate and produce milk-secreting lobules. Following lactation, involution restores the gland to its virgin state through the Stat3-mediated upregulation of lysosomal proteases Cathepsin B and L (Kreuzaler et al. 2011). Both organ systems also respond to hormones to alter their structures. In birds, the shape, size, and color of feathers are dramatically altered by the onset of hormones accompanying puberty (Chen et al. 2015). Mammary glands respond to hormones produced at the onset of pregnancy and lactogenesis that induce significant organ remodeling (Mehta et al. 2014).

In the *in vitro* cultures of salivary gland, the combination of dissociated submandibular gland epithelial and mesenchymal cells self-organizes into branches (but without a lumen). This occurs through the expression of b1 integrin-dependent migration, E-cadherin-dependent compaction, and differentiation into similar structures that form during normal salivary gland growth (Wei et al. 2007). During embryonic development, all parenchymal cells can proliferate. In adult salivary gland, cells (including the acinar cells and stem cells) that are responsible for regeneration (such as repairing from chronic sialadenitis) have not yet been established, but it is generally believed that regeneration involves proliferation and self-renewal of acinar cells (Aure et al. 2015).

13.2.4 Molding the tooth: Simple or complex, depending on the generation of localized signaling centers

The tooth is another organ with a range of shapes, from simple to complex. As in hair and feather follicles, the signaling centers in tooth are not pinned to specific locations and are dynamically regulated at different developmental stages. In the placode initiation stage, expression of FGF8, BMP2, and BMP4 plus the inhibition of Pax9 and Pitx2 determine the region of the placode signaling center. The FGF induces placode stratification but not invagination by triggering asymmetric cell division, whereas SHH signaling promotes invagination by triggering cell rearrangements

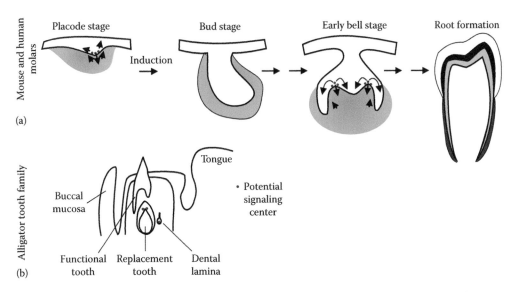

Figure 13.4 The sequential appearance of localized signaling centers shapes the teeth. (a) Shows how the sequential addition of signaling centers can build complex tooth shapes during development and evolution. Early induction leads to the formation of tooth buds that can form conically shaped teeth. Generation of enamel knots (Jernvall et al. 1994) leads to the formation of cusps on the crown surface. Downward migration of Hertwig's epithelial root sheath (HERS) at the apical end of teeth leads to the appearance of roots. (b) Shows a tooth family in the alligator's mandible and how the generation of a new signaling center allows regenerative tooth cycling. When a functional tooth is lost or the root is being resorbed, the replacement tooth starts to grow into a new functional tooth and the dental lamina grows into a new replacement tooth. During this process, a new signaling center forms in the sheath of the growing replacement tooth, allowing the formation of a new dental lamina, which contains stem cells for the next tooth generation. (Modified from Wu, P. et al., *Proc. Natl. Acad. Sci. U S A*, 110, E2009–E2018, 2013.)

(Li et al. 2016). Following budding, the early epithelial signaling center also expresses *BMPs*, *FGFs*, *SHH*, and *Wnts* that shift to the mesenchyme at later stages. From the cap to bell stage during tooth development, proliferation is localized to specific regions of the enamel organ and dental papilla (Figure 13.4a). Even though the primary and secondary enamel knots express many signaling pathways (e.g., SHH and FGFs) similar to those in the placode or bud stages, the signals originating from these signaling centers regulate crown and cusp patterns by instructing local folding and proliferation (Jernvall and Thesleff 2000). In molar crown patterning, the multi-cuspal morphology also depends on BMP4 (one of the enamel knot inducers), downregulating *Foxi3*, which inhibits formation of the enamel knot and cervical loops (Jussila et al. 2015).

Cell proliferation is most active at the cervical loop of the inner dental epithelium and is controlled by Wnts and FGFs. In mammals, tooth roots can be patterned in various numbers and morphologies, which differ between tooth positions (anterior vs posterior and maxillary vs mandibular), individuals (genetic or epigenetic factors), and probably local environments

319

(the asymmetry of specific tooth positions and/or morphology within an individual) during development. The apical papilla (harboring mesenchymal stem cells) coordinates with the Hertwig's epithelial root sheath (HERS) to form a localized signaling loop, leading to the development of roots. Proliferation of the cervical loop is maintained by FGF3 and/or FGF10 during root formation. In continuously growing mouse incisors and vole molars, FGF10 is expressed continuously, whereas prolonged Fgf10 can activate abnormal HERS enlargement (which should maintain the same root size after development) (Huang and Chai 2012). Sonic hedgehog, secreted by the HERS, induces Nfic expression in the dental papilla, which in turn induces HhiP, an SHH attenuator, in the HERS and the dental papilla. Correct levels of SHH activity are required for elongation of the tooth root. Increasing SHH activity in an Nfic knockout mouse that has reduced HhiP activity produces teeth with short roots (Huang et al. 2010, Liu et al. 2015). Furthermore, multiple Wnt ligands and signaling (e.g., Wnt/β-catenin, Wnt10a, Wnt5a, and Wnt3a) in odontoblasts also support root elongation and morphogenesis (Bae et al. 2015).

The placement of tooth buds within the jawbones also requires the combination of several signaling factors expressed with specific spatial and temporal patterns. In cichlid fish, the teeth and taste buds develop in close proximity and might share similar signaling pathways. Inhibiting Wnt, BMP, or SHH at the preplacode stage leads to disrupted tooth placode and taste bud formation. Later, at the placode stage, inhibiting Wnt still suppresses tooth and taste bud formation, whereas inhibiting BMP and SHH inhibits tooth bud formation and increases taste bud density (Bloomquist et al. 2015). In mice, FGF signaling can regulate the position of the first molar. When *FGF* is upregulated, it can lead to the formation of supernumerary teeth. Suppressing FGF signaling with *Spry2/Spry4* leads to the inhibition of SHH and prevents the formation of supernumerary teeth (Lochovska et al. 2015). The FGF also plays a role in epithelial migration during tooth development in the mandible. The migration of FGF8-positive epithelial cells toward SHH is required for tooth formation. Inhibition of both *FGF* and *SHH* interferes with the migration of epithelial cells in early placode formation, blocking tooth initiation (Prochazka et al. 2015). Polyphyodont reptiles, such as alligators, have three teeth located at each tooth position. These teeth comprise a tooth family unit. Each tooth within a family unit is at different stages of development. The position of each family unit and of each tooth within a family unit is patterned, making it more complex than diphyodont teeth (Figure 13.4b). Although *wave stimulus* and *zone of inhibition* theories are widely accepted, neither has been proven to be responsible for polyphyodont patterning. The *wave stimulus* theory considers that proliferation occurs as a wave-like function. The *zone of inhibition* theory proposes that a few inhibitory signals may help restrict tooth formation to specific regions. More recent studies have identified sfrp1 and sfrp2 as molecules that can serve in this inhibitory capacity (Whitlock and Richman 2013, Wu et al. 2013).

13.3 What we can learn about nature's way of building organ architecture

For stem cells to be applied to regenerative medicine in different organs, there are four categories with different difficulty levels. The first level is where cellular differentiation is required but structural order is unnecessary, such as in bone marrow transplantation. This is already in practice. The second level is where a molecular circuit controls the proper secretion of a molecule, as needed to treat diabetes. In this case, insulin-producing cells can be used, yet the challenge for regenerative engineering is in reconstituting the sensor, controlling when to start and stop secretion. The third level is for organs in which architecture is important, such as the skin and skeleton. In this case, scientists have been able to generate chondrocytes and cartilage in culture but do not have good control of its shape. The most challenging level is to rebuild complex connections such as neural circuits. In this case, neurons can now be generated, but rebuilding functional network remains difficult. Here, we aim at the organ's architecture level.

We have learned that cells do not build organ architecture by using a blueprint. While the information is in DNA, there is no direct blueprint. Yet, the information is translated through a series of signaling modules. It is the combinatory activities of these signaling centers, and their interactions, that give rise to the shape. By modulating the relative ratio or duration of each activity, the body can modulate the shape and relative size without too many new efforts. Determining molecules regulating the formation and regeneration of organs with different shapes, or extreme shapes, can reveal how nature regulates these LSCs.

We have earlier dealt with the periodic patterning, using a self-organization approach in feathers and hairs (Jiang et al. 1999, Lee et al. 2011). We can appreciate that if LSCs are deployed properly, we can sculpt and shape an organ. However, how are these signaling centers positioned, and how is their activity duration and level regulated? The answer maybe at the level of genomic control, and indeed, progress in epigenetics has brought possible hope to this area of new research.

In terms of tissue engineering for structures, one may consider using a 3D printing or tissue scaffolding approach directly. However, experiences have shown that a direct scaffold approach may not last, as cells forced to form a certain shape may not stay that way for long. It may be more effective to reprogram cells to a progenitor stage (such as the blastema generated via *endogenous reprogramming*) and then let the cells take over to do self-morphogenesis. Axolotl limb regeneration remains the most elegant example of regeneration in adult vertebrates (Nacu and Tanaka 2011, McCusker and Gardiner 2014). In regenerative engineering, we have to help cells overcome roadblocks to achieve our desired results, but we have to let cells take over as early as possible. We can facilitate the process by providing necessary signals early rather than late in the process. Tissue engineering can be provided extrinsically at key stages and integrated with intrinsic morphogenetic ability of progenitor cells. By doing this, we are more likely to produce organs with shapes that cells want to form, not the organ shape we want the cells to form, which might be a more successful strategy.

13.4 Summary

In regeneration, primordia cells have to be generated, and they progress to tissues and organs. In regenerative engineering, one has to be concerned with where the primordia cells come from and how to guide them to form tissues and organs of the right shape and size. This is a multi-dimensional issue that involves tissue interactions, cell fate specification, morphogenesis, tissue pattern organization, size regulation, and organ shaping. In this perspective, we focused on how organ shapes, simple or complex, can be explained by the topological deployment of LSCs. Cell behavior has been described in terms of cell proliferation, cell differentiation, cell apoptosis, cell adhesion, and cell migration. In regeneration, we deal with cell populations, but LSCs can also be ascribed to specify localized proliferative zones, localized differentiation zones, localized apoptotic zones, localized cellular elongation zones, and localized epithelial–mesenchymal transition (EMT) zones. Their relative position, size, and duration of activity can vary. A combination of these LSCs then works coordinately to sculpt and build an organ shape. By analyzing four integument organs (beak, feather, mammary gland, and tooth), we reveal different aspects of this concept. From these, we hope to learn how nature builds its architecture, which would allow us to apply these principles to guide stem cells toward the bigger goal of regenerative engineering.

Acknowledgment

This work was supported by the National Institute of Arthritis and Musculoskeletal and Skin Diseases of the National Institutes of Health (NIH) under award numbers AR 47364 and AR 60306 of NIH in the United States. It was also supported by the National Plan for Science, Technology and Innovation (MAARIFAH)—King Abdulaziz City for Science and Technology—the Kingdom of Saudi Arabia, award number (ASTP-09). Stephanie Tsai is supported by NIH/NIDCR grant # T90 DE021982. The content is solely the responsibility of the authors and does not necessarily represent the official views of the NIH.

References

Abzhanov, A., W. P. Kuo, C. Hartmann, B. R. Grant, P. R. Grant, and C. J. Tabin. 2006. The calmodulin pathway and evolution of elongated beak morphology in Darwin's finches. *Nature* 442 (7102):563–567. doi:10.1038/nature04843.

Abzhanov, A., M. Protas, B. R. Grant, P. R. Grant, and C. J. Tabin. 2004. Bmp4 and morphological variation of beaks in Darwin's finches. *Science* 305 (5689):1462–1465. doi:10.1126/science.1098095.

Abzhanov, A., S. J. Rodda, A. P. McMahon, and C. J. Tabin. 2007. Regulation of skeletogenic differentiation in cranial dermal bone. *Development* 134 (17):3133–3144. doi:10.1242/dev.002709.

Abzhanov, A. and C. J. Tabin. 2004. Shh and Fgf8 act synergistically to drive cartilage outgrowth during cranial development. *Dev Biol* 273 (1):134–148. doi:10.1016/j.ydbio.2004.05.028.

Alexander, C., E. Zuniga, I. L. Blitz, N. Wada, P. Le Pabic, Y. Javidan, T. Zhang, K. W. Cho, J. G. Crump, and T. F. Schilling. 2011. Combinatorial roles for BMPs and endothelin 1 in patterning the dorsal-ventral axis of the craniofacial skeleton. *Development* 138 (23):5135–5146. doi:10.1242/dev.067801.

Aure, M. H., S. Arany, and C. E. Ovitt. 2015. Salivary glands: Stem cells, self-duplication, or both? *J Dent Res* 94 (11):1502–1507. doi:10.1177/0022034515599770.

Bae, C. H., T. H. Kim, S. O. Ko, J. C. Lee, X. Yang, and E. S. Cho. 2015. Wntless regulates dentin apposition and root elongation in the mandibular molar. *J Dent Res* 94 (3):439–445. doi:10.1177/0022034514567198.

Balcuns, A., M.T. Gasseling, and J.W. Jr. Saunders. 1970. Spatio temporal distribution of a zone that controls antero posterior polarity in the limb bud of the chick and other bird embryos. *Amer Zool* 10 (3):323.

Bhullar, B. A., Z. S. Morris, E. M. Sefton, A. Tok, M. Tokita, B. Namkoong, J. Camacho, D. A. Burnham, and A. Abzhanov. 2015. A molecular mechanism for the origin of a key evolutionary innovation, the bird beak and palate, revealed by an integrative approach to major transitions in vertebrate history. *Evolution* 69 (7):1665–1677. doi:10.1111/evo.12684.

Björk, A. 1968. The use of metallic implants in the study of facial growth in children: Method and application. *Am J Phys Anthropol* 29 (2):243–254.

Bloomquist, R F., N. F. Parnell, K. A. Phillips, T. E. Fowler, T. Y. Yu, P. T. Sharpe, and J. D. Streelman. 2015. Co-evolutionary patterning of teeth and taste buds. *Proc Nat Acad Sci U S A* 112 (44):E5954–E5962.

Brugmann, S. A., L. H. Goodnough, A. Gregorieff, P. Leucht, D. ten Berge, C. Fuerer, H. Clevers, R. Nusse, and J. A. Helms. 2007. Wnt signaling mediates regional specification in the vertebrate face. *Development* 134 (18):3283–3295. doi:10.1242/dev.005132.

Chen, C. C., M. V. Plikus, P. C. Tang, R. B. Widelitz, and C. M. Chuong. 2016. The modulatable stem cell niche: Tissue interactions during hair and feather follicle regeneration. *J Mol Biol* 428 (7):1423–1440. doi:10.1016/j.jmb.2015.07.009.

Chen, C. F., J. Foley, P. C. Tang, A. Li, T. X. Jiang, P. Wu, R. B. Widelitz, and C. M. Chuong. 2015. Development, regeneration, and evolution of feathers. *Annu Rev Anim Biosci* 3:169–195. doi:10.1146/annurev-animal-022513-114127.

Chueh, S. C., S. J. Lin, C. C. Chen, M. Lei, L. M. Wang, R. Widelitz, M. W. Hughes, T. X. Jiang, and C. M. Chuong. 2013. Therapeutic strategy for hair regeneration: Hair cycle activation, niche environment modulation, wound-induced follicle neogenesis, and stem cell engineering. *Expert Opin Biol Ther* 13 (3):377–391. doi:10.1517/14712598.2013.739601.

Chuong, C. M., R. Bhat, R. B. Widelitz, and M. J. Bissell. 2014. SnapShot: Branching morphogenesis. *Cell* 158 (5):1212–1212. doi:10.1016/j.cell.2014.08.019.

Chuong, C. M., V. A. Randall, R. B. Widelitz, P. Wu, and T. X. Jiang. 2012. Physiological regeneration of skin appendages and implications for regenerative medicine. *Physiology* (*Bethesda*) 27 (2):61–72. doi:10.1152/physiol.00028.2011.

Chuong, C. M., P. Wu, M. Plikus, T. X. Jiang, and R. B. Widelitz. 2006. Engineering stem cells into organs: Topobiological transformations demonstrated by beak, feather, and other ectodermal organ morphogenesis. *Cur Top Dev Bio* 72:237–274. doi:10.1016/S0070-2153(05)72005-6.

Chuong, C. M., C. Y. Yeh, T. X. Jiang, and R. Widelitz. 2013. Module-based complexity formation: Periodic patterning in feathers and hairs. *Wiley Interdiscip Rev Dev Biol* 2 (1):97–112. doi:10.1002/wdev.74.

Dee, C. T., C. R. Szymoniuk, P. E. Mills, and T. Takahashi. 2013. Defective neural crest migration revealed by a Zebrafish model of Alx1-related frontonasal dysplasia. *Hum Mol Genet* 22 (2):239–251. doi:10.1093/hmg/dds423.

Fritz, J. A., J. Brancale, M. Tokita, K. J. Burns, M. B. Hawkins, A. Abzhanov, and M. P. Brenner. 2014. Shared developmental programme strongly constrains beak shape diversity in songbirds. *Nat Commun* 5:3700. doi:10.1038/ncomms4700.

Hamburger, V., and H. L. Hamilton. 1951. A series of normal stages in the development of the chick embryo. *J Morphol* 88 (1):49–92.

Helms, J. A. and R. A. Schneider. 2003. Cranial skeletal biology. *Nature* 423 (6937):326–331. doi:10.1038/nature01656.

Hu, D., R. S. Marcucio, and J. A. Helms. 2003. A zone of frontonasal ectoderm regulates patterning and growth in the face. *Development* 130 (9):1749–1758.

Huang, X. F. and Y. Chai. 2012. Molecular regulatory mechanism of tooth root development. *Int J Oral Sci* 4 (4):177–181. doi:10.1038/ijos.2012.61.

Huang, X., X. Xu, P. Bringas, Jr., Y. P. Hung, and Y. Chai. 2010. Smad4-Shh-Nfic signaling cascade-mediated epithelial-mesenchymal interaction is crucial in regulating tooth root development. *J Bone Miner Res* 25 (5):1167–1178. doi:10.1359/jbmr.091103.

Jernvall, J. and I. Thesleff. 2000. Reiterative signaling and patterning during mammalian tooth morphogenesis. *Mech Dev* 92 (1):19–29.

Jernvall, J., P. Kettunen, I. Karavanova, L.B. Martin, and I. Thesleff. 1994. Evidence for the role of the enamel knot as a control center in mammalian tooth cusp formation: Non-dividing cells express growth stimulating Fgf-4 gene. *Int J Dev Biol* 38 (3):463–469.

Jiang, T. X., H.S. Jung, R. B. Widelitz, and C. M. Chuong. 1999. Self-organization of periodic patterns by dissociated feather mesenchymal cells and the regulation of size, number and spacing of primordia. *Development* 126 (22):4997–5009.

Jussila, M., A. J. Aalto, M. Sanz Navarro, V. Shirokova, A. Balic, A. Kallonen, T. Ohyama, A. K. Groves, M. L. Mikkola, and I. Thesleff. 2015. Suppression of epithelial differentiation by Foxi3 is essential for molar crown patterning. *Development* 142 (22):3954–3963. doi:10.1242/dev.124172.

Kreuzaler, P. A., A. D. Staniszewska, W. Li, N. Omidvar, B. Kedjouar, J. Turkson, V. Poli, R. A. Flavell, R. W. Clarkson, and C. J. Watson. 2011. Stat3 controls lysosomal-mediated cell death in vivo. *Nat Cell Biol* 13 (3):303–309. doi:10.1038/ncb2171.

Lamichhaney, S., J. Berglund, M. S. Almen, K. Maqbool, M. Grabherr, A. Martinez-Barrio, M. Promerova. et al. 2015. Evolution of Darwin's finches and their beaks revealed by genome sequencing. *Nature* 518 (7539):371–375. doi:10.1038/nature14181.

Lamichhaney, S., F. Han, J. Berglund, C. Wang, M. S. Almen, M. T. Webster, B. R. Grant, P. R. Grant, and L. Andersson. 2016. A beak size locus in Darwin's finches facilitated character displacement during a drought. *Science* 352 (6284):470–474. doi:10.1126/science.aad8786.

Lee, L. F., T. X. Jiang, W. Garner, and C. M. Chuong. 2011. A simplified procedure to reconstitute hair-producing skin. *Tissue Eng Part C Methods* 17 (4):391–400. doi:10.1089/ten. TEC.2010.0477.

Li, A., M. Chen, T. X. Jiang, P. Wu, Q. Nie, R. Widelitz, and C. M. Chuong. 2013. Shaping organs by a wingless-int/Notch/nonmuscle myosin module which orients feather bud elongation. *Proc Natl Acad Sci U S A* 110 (16):E1452–E1461. doi:10.1073/pnas.1219813110.

Li, A., Y. C. Lai, S. Figueroa, T. Yang, R. B. Widelitz, K. Kobielak, Q. Nie, and C. M. Chuong. 2015. Deciphering principles of morphogenesis from temporal and spatial patterns on the integument. *Dev Dyn* 244 (8):905–920. doi:10.1002/dvdy.24281.

Li, J., L. Chatzeli, E. Panousopoulou, A. S. Tucker, and J. B. Green. 2016. Epithelial stratification and placode invagination are separable functions in early morphogenesis of the molar tooth. *Development* 143 (4):670–681. doi:10.1242/dev.130187.

Lin, C. M., T. X. Jiang, R. B. Widelitz, and C. M. Chuong. 2006. Molecular signaling in feather morphogenesis. *Curr Opin Cell Biol* 18(6):730–741. Epub 2006 Oct 17.

Liu, Y., J. Feng, J. Li, H. Zhao, T. V. Ho, and Y. Chai. 2015. An Nfic-hedgehog signaling cascade regulates tooth root development. *Development* 142 (19):3374–3382. doi:10.1242/dev.127068.

Lochovska, K., R. Peterkova, Z. Pavlikova, and M. Hovorakova. 2015. Sprouty gene dosage influences temporal-spatial dynamics of primary enamel knot formation. *BMC Dev Biol* 15:21. doi:10.1186/s12861-015-0070-0.

Mallarino, R., P. R. Grant, B. R. Grant, A. Herrel, W. P. Kuo, and A. Abzhanov. 2011. Two developmental modules establish 3D beak-shape variation in Darwin's finches. *Proc Natl Acad Sci U S A* 108 (10):4057–4062. doi:10.1073/pnas.1011480108.

McCusker, C. D. and D. M. Gardiner. 2014. Understanding positional cues in salamander limb regeneration: Implications for optimizing cell-based regenerative therapies. *Dis Model Mech* 7 (6):593–599. doi:10.1242/dmm.013359.

Mehta, R. G., M. Hawthorne, R. R. Mehta, K. E. O. Torres, X. J. Peng, D. L. McCormick, and L. Kopelovich. 2014. Differential roles of ER alpha and ER beta in normal and neoplastic development in the mouse mammary gland. *Plos One* 9 (11):e113175. doi:10.1371/journal.pone.0113175.

Nacu, E. and E. M. Tanaka. 2011. Limb regeneration: A new development? *Annu Rev Cell Dev Biol* 27:409–440. doi:10.1146/annurev-cellbio-092910-154115.

Nelson, C. M., M. M. Vanduijn, J. L. Inman, D. A. Fletcher, and M. J. Bissell. 2006. Tissue geometry determines sites of mammary branching morphogenesis in organotypic cultures. *Science* 314 (5797):298–300. doi:10.1126/science.1131000.

Noden, D. M. 1983. The role of the neural crest in patterning of avian cranial skeletal, connective, and muscle tissues. *Dev Biol* 96 (1):144–165.

Pispa, J. and I. Thesleff. 2003. Mechanisms of ectodermal organogenesis. *Dev Biol* 262 (2):195–205.

Plant, M. R., M. E. MacDonald, L. I. Grad, S. J. Ritchie, and J. M. Richman. 2000. Locally released retinoic acid repatterns the first branchial arch cartilages in vivo. *Dev Biol* 222 (1):12–26. doi:10.1006/dbio.2000.9706.

Prochazka, J., M. Prochazkova, W. Du, F. Spoutil, J. Tureckova, R. Hoch, T. Shimogori, R. Sedlacek, J. L. Rubenstein, T. Wittmann, and O. D. Klein. 2015. Migration of founder epithelial cells drives proper molar tooth positioning and morphogenesis. *Dev Cell* 35 (6):713–724. doi:10.1016/j.devcel.2015.11.025.

Richman, J. M., M. Herbert, E. Matovinovic, and J. Walin. 1997. Effect of fibroblast growth factors on outgrowth of facial mesenchyme. *Dev Biol* 189 (1):135–147. doi:10.1006/dbio.1997.8656.

Richman, J. M. and S. H. Lee. 2003. About face: Signals and genes controlling jaw patterning and identity in vertebrates. *Bioessays* 25 (6):554–568. doi:10.1002/bies.10288.

Schneider, R. A. and J. A. Helms. 2003. The cellular and molecular origins of beak morphology. *Science* 299 (5606):565–568. doi:10.1126/science.1077827.

Trainor, P. and R. Krumlauf. 2000. Plasticity in mouse neural crest cells reveals a new patterning role for cranial mesoderm. *Nat Cell Biol* 2 (2):96–102. doi:10.1038/35000051.

Wagner, G. P. and V. J. Lynch. 2010. Evolutionary novelties. *Curr Biol* 20 (2):R48–R52. doi:10.1016/j.cub.2009.11.010.

Weber, E. L. and C. M. Chuong. 2013. Environmental reprogramming and molecular profiling in reconstitution of human hair follicles. *Proc Natl Acad Sci U S A* 110 (49):19658–19659. doi:10.1073/pnas.1319413110.

Wei, C., M. Larsen, M. P. Hoffman, and K. M. Yamada. 2007. Self-organization and branching morphogenesis of primary salivary epithelial cells. *Tissue Eng* 13 (4):721–735. doi:10.1089/ten.2006.0123.

Whitlock, J. A. and J. M. Richman. 2013. Biology of tooth replacement in amniotes. *Int J Oral Sci* 5 (2):66–70. doi:10.1038/ijos.2013.36.

Wu, P., T. X. Jiang, J. Y. Shen, R. B. Widelitz, and C. M. Chuong. 2006. Morphoregulation of avian beaks: Comparative mapping of growth zone activities and morphological evolution. *Dev Dyn* 235 (5):1400–1412. doi:10.1002/dvdy.20825.

Wu, P., T. X. Jiang, S. Suksaweang, R. B. Widelitz, and C. M. Chuong. 2004. Molecular shaping of the beak. *Science* 305 (5689):1465–1466. doi:10.1126/science.1098109.

Wu, P., X. Wu, T. X. Jiang, R. M. Elsey, B. L. Temple, S. J. Divers, T. C. Glenn, K. Yuan, M. H. Chen, R. B. Widelitz, and C. M. Chuong. 2013. Specialized stem cell niche enables repetitive renewal of alligator teeth. *Proc Natl Acad Sci U S A* 110 (22):E2009–E2018. doi:10.1073/pnas.1213202110.

Zhang, Z., Y. Song, X. Zhao, X. Zhang, C. Fermin, and Y. Chen. 2002. Rescue of cleft palate in Msx1-deficient mice by transgenic Bmp4 reveals a network of BMP and Shh signaling in the regulation of mammalian palatogenesis. *Development* 129 (17):4135–4146.

Chapter 14 The positional information grid in development and regeneration

Susan V. Bryant and
David M. Gardiner

Contents

Key concepts

a. There is a subpopulation of cells within the connective tissue
   that has stable information about their position (positional
   information [PI]) within organs (e.g., a limb).
b. These PI cells respond to injury by recognizing when PI
   has been lost, thereby creating a confrontation between normally non-adjacent cells, which stimulates proliferation and
   restoration of the missing pattern by the process of intercalation; in other words, PI cells drive regeneration.

c. The PI cells are distributed as a two-dimensional grid that underlies epithelial sheets (e.g., within the dermis) and surrounds all internal organs (e.g., intestine, bones, muscles, blood vessels, and nerves).

d. Cytonemes are long cellular processes that are likely to be the cellular structures responsible for long-range cell–cell contact-mediated signaling properties of the PI grid.

e. The PI cells encode information in the extracellular matrix (ECM) of the connective tissue during embryonic development and remake PI during regenerative development.

f. The PI grid functions (at least in part) to regulate growth factor (morphogen) signaling, and by controlling proliferation, it also controls pattern formation via cell–cycle dependent gating of expression of critical transcription factors.

g. Success in engineering ways to regenerate and/or rejuvenate the PI grid likely will lead to enhanced regeneration and novel strategies for controlling cancer and aging.

14.1 Introduction

The ability of some animals to completely and accurately regenerate missing parts of appendages has been known and studied for several centuries (Dinsmore 1992; Morgan 1901). Progress in identifying the principles that govern appendage regeneration came from findings from studies of three completely different and phylogenetically unrelated organisms that can regenerate: *Drosophila* (regenerating imaginal discs, the anlage of the future appendages in dipterans), salamanders (regenerating arms and legs), and cockroaches (regenerating legs). By comparing regeneration in these different models, it became evident that there is a set of basic rules that govern regeneration, which led to the development of the Polar Coordinate Model (PCM) for regeneration (for details, see Bryant et al. 1981[*]; French et al. 1976[†]).

Among these rules for regeneration, the most important is that growth and pattern formation during regeneration are controlled by short-range interactions between cells with a property referred to as positional information (PI). This information is equivalent to a postal zip code, such that each cell has a unique identity, based on its location along the proximal-distal axis and around the circumferential axis (Figure 14.1a and b). Injury results in the loss of a subpopulation of cells that have this information, thus creating a PI gap. When PI cells surrounding the wound migrate under the wound epidermis to the center of the wound bed, cells from different parts of the limb circumference come into contact and

---

[*] This paper refines the original Polar Coordinate Model published in 1976.

[†] This is the original Polar Coordinate Model that articulated the view that the mechanisms for pattern formation are conserved across species and involve short-range cell–cell signaling leading to growth and intercalation.

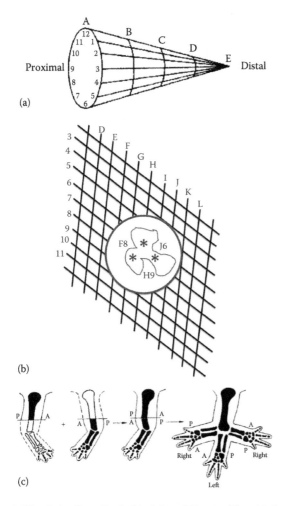

Figure 14.1 The Polar Coordinate Model and the positional information (PI) grid. (a) The positional identity of a cell is specified in relation to the limb circumference (1–12) and to the long axis of the appendage (A–E). (b) The PI grid is composed of cells that are arrayed in a two-dimensional sheet under the epidermis and around internal structures (e.g., muscles, bones, nerves, and blood vessels), such that when there is an injury (e.g., amputation), a hole is created in the grid. When the edges of the hole come together during wound healing, cells with different PI come together and interact (red asterisks). (c) When cells with different PI interact, growth and pattern formation are stimulated (intercalation), resulting in either repair of the grid or the generation of supernumerary structures. An example of the latter is observed when limbs are grafted from left to right (and vice versa). Cells with different positional identities are brought together at the graft–host interface (A next to P), intercalation is stimulated, and two supernumerary limbs are formed (right), in addition to the limb formed from the graft (left). (Courtesy of Susan V. Bryant and David M. Gardiner.)

recognize that there is PI gap between them. These cells normally would not be neighbors and therefore not normally in communication with each other (Figure 14.1b). In response to the PI gap (discontinuity), the cells near the wound margin begin to proliferate, giving rise to new cells that acquire the PI that was lost (or missing). Growth continues until all PI cells have normal neighbors, thereby eliminating any positional discontinuities between adjacent cells, at which point proliferation ceases. This process of restoring the intermediate PI that was lost during injury is referred to as *intercalation* (Bryant et al. 1981; French et al. 1976). Based on the PCM, it is the interaction of cells with PI, leading to intercalation, that is the driving force behind regeneration. The focus of this chapter is on the biology of the PI cells and how they function to control intercalation and regeneration.

14.2 Existence of the positional information grid in vertebrates

Although the molecular basis of PI is yet unknown, the phenomenology of how cells with PI interact with others to control regeneration is well characterized. Cells without PI (e.g., muscle and skeleton) (McCusker and Gardiner 2014; McCusker et al. 2016) do not stimulate intercalary growth and pattern formation when grafted into new locations. By contrast, the PI cells within the connective tissue do stimulate intercalation when they make contact with connective tissue PI cells from different parts of the circumference during wound healing. Healing after simple amputation leads to the formation of a replica of the part that was missing. Healing after grafting to alter the arrangement of PI around the amputation site characteristically generates complex supernumerary regenerates (Bryant et al. 1981; French et al. 1976; McCusker et al. 2015[*]). The ability to stimulate the formation of supernumerary structures, and even entire new limbs, has allowed for (1) the identification of connective tissues cells as having the necessary PI for complete regeneration, and (2) the identification of the quality of the *identity* of the PI that these cells express. For example, grafting cells from one position on the limb to a host wound site at the same position (anterior to anterior) does not stimulate intercalation. In contrast, grafting cells from the posterior of the limb to an anterior wound site induces formation of an entire ectopic limb (Endo et al. 2004[†]; Satoh et al. 2007).

In some situations, such as when a limb blastema of a regenerating axolotl is grafted from a left limb to an amputation site on the right limb, there is no possible way that all the positions around the limb circumference

---

[*] This is a recent review of the mechanisms controlling blastema formation and regeneration, with a focus on the axolotl model.

[†] This paper establishes the Accessory Limb Model as a gain-of-function assay for signals that control limb regeneration

can be matched, resulting in two locations in the graft host junction that are matched and two other locations at which they maximally mismatch. At the site of a graft/host mismatch, the mismatch is eliminated by the growth of a supernumerary limb (Figure 14.1c). It also is possible to change, or reprogram, PI experimentally; for example, treatment with retinoic acid (RA) reprograms the PI of anterior cells to posterior, and these formerly *anterior* cells are now capable of stimulating ectopic pattern when grafted into an anterior wound site (Bryant and Gardiner 1992*; McCusker et al. 2014). Going forward, these assays will play a role in the identification of the specific molecules and signaling pathways involved in PI-mediated interaction (e.g., heparin sulfate–mediated fibroblast growth factor [FGF] signaling; Phan et al. 2015†).

Since the PI cells are localized in the connective tissue, and the connective tissue is distributed throughout and surrounding all the tissues of the body, they form a network, or grid, of cells that corresponds to the shape of each organ and to the body in its entirety. If this network of cells could be visualized apart from all the other cell types of the body, as has been done for other tissues in the Body Worlds exhibits via the technique of plastination, the shape and form of all the tissues would be preserved and would be identical to that in the intact body. Thus, PI cells form an interconnected network that determines the form, size, and location of the biological structures. We refer to this network as the PI grid (Bryant and Gardiner 2016‡). Thus, among the cell types that participate in regeneration, it is the PI grid cells that have the information for rebuilding the lost structures. The non-grid cells (pattern-*following* cells) replace the functional tissues (e.g., muscle, bone, nerves, and blood vessels), as guided by cells of the PI grid (pattern-*forming* cells) that coordinate the spatial organization of the tissues (McCusker et al. 2016). Thus, the PI grid functions to integrate the growth and pattern formation essential to the success of both embryonic development and post-embryonic regenerative development.

14.3 Origin and location of the positional information grid

In vertebrate embryos, constructed from three embryonic germ layers: ectoderm, endoderm, and mesoderm, the PI grid is a property of the mesoderm, which is the middle layer of the body (Figure 14.2a). During early embryonic development, the mesoderm splits into an outer layer and

---

* This essay established the model that RA induces supernumerary pattern formation by reprograming anterior cells rather than by functioning as a diffusible morphogen.
† This study demonstrated the presence of a heparan sulfate code in the ECM of the axolotl and mouse skin that regulates growth factor/morphogen signaling in position-specific and developmental stage–specific manners.
‡ This essay presents a model for how regulation of cell cycle kinetics can function upstream of pattern formation and also states that morphogens function to control the length of the cell cycle.

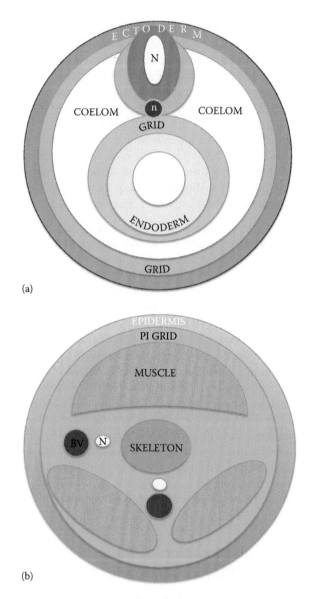

(a)

(b)

Figure 14.2 Distribution of the positional information (PI) grid in the embryo and the limb. (a) A cross-section through a stylized embryo after gastrulation, at which time, the surface ectoderm (blue) is underlain by a layer of mesoderm (green), which also surrounds the developing neural tube and notochord. The endoderm of the gut tube (yellow) is also surrounded by a layer of mesoderm, and the space between the gut- and epidermis-associated mesoderm is the coelom. (b) The limb in an adult tetrapod is surrounded by the epidermis and is composed of multiple tissues of mesodermal origin (e.g., muscle, pink; blood vessels, red; nerves, white; and skeletal tissues, brown). The PI grid is located within the loose connective tissues (green); it underlies the epidermis and surrounds all the internal tissues.

an inner layer, with a cavity, the coelom, in between. The outer layer of the mesodermal mantle associates with the undersurface of the outermost layer, the ectoderm, and together these layers form the skin and the peripheral appendages, including limbs and genitalia, and the epidermal appendages such as hair, feathers, and scales. The inner layer of the mesodermal mantle associates with the endoderm to form the gut and associated endodermal organs, such as the liver and lungs. Although many structures involve the association of mesoderm, either with epidermis or with endoderm, some organs (e.g., kidneys, muscle, and skeleton) are constructed from mesoderm alone. Nevertheless, the development of these mesodermal structures is dependent on the presence of connective tissue cells that contain PI. Thus, tissues and organs develop from the embryonic cells derived from the ectoderm, endoderm, and/or mesoderm that interact with the mesoderm-derived PI grid cells of the future loose connective tissue that control growth and pattern formation (Bryant and Gardiner 2016).

The final arrangement of the embryonic germ layers occurs as a result of the process of gastrulation. The early-cleavage/blastula-stage embryo is a relatively uniform ball of cells, which rearrange during gastrulation through dramatic morphogenetic movements to form the three germ layers. As a consequence, cells from distant positions in the embryo are brought together, and the inner and outer mesodermal layers of cells begin to interact with the adjacent layers of the endoderm and ectoderm, respectively, between which the mesodermal sheet resides. Most of these early-interacting cells are the progenitors of the cells that will build the structures of the embryo and adult. They acquire specialized functions (e.g., nerves, muscles, and blood cells), but they do not have PI (i.e., they are embryonic progenitors of the pattern-following cells). A much smaller subpopulation of cells has PI and is fated to give rise to the pattern-forming cells. It is unclear at what point in development do cells with PI arise; however, as discussed later, differences in cell fates associated with differences in the cell cycle kinetics of subpopulations of cells become evident during early cleavage stages in *Drosophila* (Foe 1989; see Bryant and Gardiner 2016). In summary, the PI needed to drive the development of embryonic structures in the right place, at the right time, and of the right size resides in the mesodermal grid. Gastrulation puts the PI grid cells where they are needed in order to interact with all the non-grid cells to build tissues and organs (Bryant and Gardiner 2016).

14.4 Evidence for the presence and function of the positional information grid comes from studies of regeneration

The most direct evidence for the PI grid comes from experiments on vertebrates capable of regeneration (Bryant et al. 1981; Endo et al. 2004; French et al. 1976; Satoh et al. 2007). As discussed earlier, we presume that the PI grid is established during embryonic development and persists into adulthood, when it can function to control growth and pattern formation during regeneration. For successful limb regeneration in urodeles (salamanders), experiments have shown that two main requirements must be met: (1) the wound edge must include dermal cells from opposite parts of the limb circumference (different PI), and (2) because regeneration is a nerve-dependent process, an adequate nerve supply must be present at the site of the wound (McCusker et al. 2015). Both of these requirements are met when a limb is amputated, since the limb has a nerve supply, and cells from different positions around the limb circumference migrate to the center of the amputation plane to form the early blastema (Gardiner et al. 1986). These migrating cells bring different PI with them as they migrate toward the center of the wound, and as they interact with one another on the wound surface, growth is stimulated and a blastema is formed. Progeny of the cells beneath the apical epidermis of the blastema will form the distal tip of the future regenerate. Within the blastema, interactions between cells from different parts of the circumference continue, generating new pattern along the proximal–distal axis, until all cells have normal neighbors, at which point growth ceases and regeneration is complete.

The requirements for both nerves and PI have been demonstrated repeatedly over the decades. Limbs that have been denervated or had their supply of nerves reduced below a critical threshold level fail to regenerate (McCusker et al. 2015; Singer 1974). In the case of PI, supernumerary limbs can be induced to form by grafting skin or blastemas, so that opposite PI is now present at the wound site. Blastemas only form at positions where cells with significantly different PI are brought together. No supernumerary pattern is formed at sites of interactions between cells with the same PI. The need for diversity of PI for regeneration has been demonstrated directly by the regenerative failure of limbs that are created surgically so as to contain cells with only anterior PI (double-half-anterior limbs) (Holder et al. 1980; McCusker et al. 2015; Wigmore and Holder 1985). Surgically created limb stumps that have both anterior and posterior cells regenerate normally or produce supernumerary limbs (Figure 14.3).

In recent years, regeneration of an entire limb has been achieved by providing the same signals (nerve plus PI) to a skin wound on the side of the arm of an axolotl (Endo et al. 2004; Satoh et al. 2007). A simple skin wound alone will heal without forming supernumerary structures, since there are both a limited nerve supply (subcutaneous sensory nerves) and a lack of diversity of PI. However, if the wound is provided with enhanced nerve signals from a surgically deviated nerve and enhanced PI by grafting cells from the opposite side of the limb, a supernumerary limb is induced to form (Endo et al. 2004; Satoh et al. 2007). The function of the nerve can be replaced by a cocktail of human growth factors (Makanae et al. 2014, 2016), and the requirement for extra PI can be replaced in part by artificial

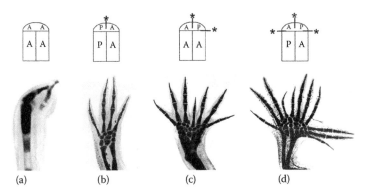

(a)   (b)   (c)   (d)

Figure 14.3 The relationship between the degree of disparity in positional information (PI) and the amount of new pattern generated. As illustrated in Figure 14.1c, two supernumerary limbs are induced to form when limbs are grafted from left to right (d). If a limb is amputated, cells from the anterior (A) and posterior (P) interact in the blastema that forms, and a single limb is formed (b). Regeneration is inhibited if information is not present for both A and P, as observed in (a) in which a surgically created double-anterior limb is created. A double-anterior limb can regenerate if a blastema with both A and P cells is grafted distally (c). The top row of the diagram illustrates the limb types generated by grafts of half limbs and early blastemas, and the bottom row shows the resulting limbs that have been stained to show the skeletal patterns. Positional confrontations between A and P cells are marked by an asterisk. (Courtesy of Susan V. Bryant and David M. Gardiner.)

extracellular matrix (ECM) containing heparan sulfate proteoglycan (HSPG) (Phan et al. 2015). Given that HSPGs function to regulate growth factor signaling, one hypothesis is that the PI grid functions, at least in part, to control the spatial and temporal activity of growth factors (see later).

Although it is well established that cells with PI are localized within the loose connective tissues, the specific identities and characteristics of these cells have yet to be discovered. The limb and regenerating limb stump are composed of cells from multiple lineages, and it is evident that not all of these cells have PI (in fact, most appear not to be having PI) (McCusker and Gardiner 2014; McCusker et al. 2016). Operationally, cells that have PI are defined as those cells that are able to induce supernumerary growth when grafted adjacent to the cells with different PI (McCusker and Gardiner 2013[*]). Conversely, if no supernumerary structures are induced, then the cells either lack PI or have the same PI as the graft site (Bryant et al. 1981; French et al. 1976; McCusker and Gardiner 2013). By using this assay, it has been possible to identify PI as a property of the cells of the loose connective

---

[*] This is an experimental study demonstrating that positional information is both stabile and labile.

tissue (fibroblasts). Since fibroblasts are associated with nearly all tissues, grafts of most tissues will induce formation of supernumerary structures, unless they are depleted of fibroblasts. For example, when the connective tissues surrounding limb skeletal elements are removed before grafting, the graft loses its ability to induce supernumerary outgrowths (McCusker et al. 2016). Conversely, tissues that are enriched in fibroblasts, such as the dermis, induce a stronger supernumerary response than tissues with fewer fibroblasts (skeletal elements with associated peripheral connective tissues) (see Bryant and Gardiner 2016; McCusker et al. 2015). In recent times, the recognition that fibroblasts have a dominant role in the control of growth and pattern formation during regeneration has led to the concept of there being two populations of cells required for regeneration: the pattern-forming cells of the connective tissue and the pattern-following cells that make the functional tissues of the limb based on information derived from the connective tissue (e.g., muscle and skeleton) (McCusker et al. 2015, 2016).

14.5 Growth factors, morphogens, and the positional information grid

The central hypothesis of the PCM is that growth and pattern formation are functionally linked. This link is via the process of intercalation, such that a discontinuity in the pattern stimulates growth that continues until the discontinuity has been eliminated, at which point growth ceases. The molecular basis for intercalation is unclear; however, recent data on the regulation of growth factor signaling by temporal–spatial differences in HSPGs are a strong indication that the PI grid functions at least in part by controlling growth factor activity (Phan et al. 2015). Based on the observations that growth factors function as morphogens (see Bryant and Gardiner 2016) and that growth factors are associated with the PI grid, an integrated view of the coordinate regulation of growth and pattern formation argues for a cause-and-effect relationship between the two. As discussed later, a mechanistic relationship between proliferation and pattern formation has been well characterized in *Drosophila* embryogenesis but not yet in regeneration.

In earlier years, much attention was focused on the signals that control cell proliferation. With the explosive growth of developmental genetics, the focus shifted to the gene regulatory networks that control differentiation and pattern formation, which has led to the recognition of the function of morphogens. Morphogens are molecules that are critically involved in, and essential for, normal development and regeneration. They are made at a localized source and diffuse away from the source, creating a morphogen gradient across a field of cells (see Gilbert 2013; Wolpert et al. 2015). They earned their name as a result of the effects that experimental changes in morphogen levels (e.g., exposure to exogenous factors) have on developing systems—they alter the morphology of the embryo and cause supernumerary structures to form. Thus, morphogens are obvious candidates for the signaling molecules that function as part of the PI grid.

As noted earlier, concomitant with their effect on the structure of the embryos (pattern formation), all morphogens are growth factors (alter cell cycle kinetics) and actually were first identified based on their function as growth factors. We have argued that the importance of the role of morphogens in development is integrally associated with their role in controlling growth (Bryant and Gardiner 2016). As an example of the potential of morphogens for changing pattern, a bead soaked in RA implanted into the anterior edge of a developing chick's wing bud causes a complete duplication of the limb bud, which differentiates as a symmetrical double limb (Tickle et al. 1975). The dramatic effect of RA on developing and regenerating organs can be understood as a consequence of its effect on the cell cycle. Cells of the zone of polarizing activity (ZPA), a localized region of cells at the posterior margin of the limb bud that has a relatively long cell cycle, interact with adjacent (more anterior) cells to cause the development of the limb. Grafting of an RA-soaked bead into the anterior of the limb bud causes anterior cells adjacent to the bead to increase the length of the cell cycle, and in so doing, they acquire a new fate as a second ZPA signaling region (Ohsugi et al. 1997). This induced ZPA (posterior cells) then interacts with non-reprogrammed anterior cells to generate supernumerary digits in a pattern that is mirror symmetrical with the original limb bud (Bryant and Gardiner 1992; Noji et al. 1991; Wanek et al. 1991).

Given that the PI grid can function to control growth factor/morphogen signaling via HSPGs, the next question becomes: how changes in growth can lead to changes in pattern formation? From studies of early *Drosophila* development, it is clear that pattern formation is a consequence of changes in the distribution of the transcription factors (TFs) that control cell fate (see Bryant and Gardiner 2016). In particular, an essential ingredient of successful development is the establishment of fields of cells expressing graded levels of TFs, now known to be essential for pattern formation in all multicellular organisms and not just in *Drosophila*. Since TFs are intracellular, the question then becomes: how an extracellular gradient of a growth factor/morphogen that is regulated by the PI grid leads to intracellular gradients of TFs?

One well-established mechanism for control of transcription by the cell cycle is embodied in the *intron delay hypothesis*, proposed originally by Gubb (1986). The essence of this mechanism is that it takes a finite amount of time to transcribe a gene and that the expression of a gene with a relatively large transcription unit (TU) cannot occur in the time available if the cell cycle is short (the length of the cell cycle gates transcription) (Bryant and Gardiner 2016; Shermoen and O'Farrell 1991).

The fact that cell cycle gating functions to control pattern formation was first demonstrated experimentally for *Ultrabithorax* (*Ubx*) (Shermoen and O'Farrell 1991). During the early cleavage stages of *Drosophila* embryos, the cell cycle is very short, of the order of 8 minutes during the first 13 mitotic cycles. Given the rate of RNA polymerization (variable, but on average approximately 1.4 kb/min), it takes about 55 minutes to completely transcribe the 77-kb *Ubx* TU. Thus, functional Ubx transcripts are not detected until cell cycles 14 and 15, when the duration of the cell cycle becomes long enough, so that it no longer gates *Ubx* transcription (Shermoen and O'Farrell 1991). This phenomenon has subsequently been demonstrated for other genes in *Drosophila* and in other species, including mammals (see Bryant and Gardiner 2016).

Cell cycle gating of transcription likely is a common phenomenon in early embryos, which are characterized by a period of rapid cell division in order to create the cell mass (blastula) needed to begin building the embryo. These early divisions are rapid, and the cells have an abbreviated cell cycle, lacking the variably sized gap phases that are typical of cell cycles at later stages of development, when large genes are being transcribed. During this early period of rapid division, only the shortest of genes can be completely transcribed in the time available before the next cell division begins and transcription is aborted (Bryant and Gardiner 2016; Shermoen and O'Farrell 1991). The question is: can this mechanism also function at later stages of embryonic and regenerative development?

The core concept of this model is that proliferation controls pattern formation via the relative relationship between the length of G1 and the size of the TU (Figure 14.4). As sizes of TU increase as a consequence of increased intron size, corresponding increases in the duration of G1 are needed to allow sufficient time for long transcripts to be made. For vertebrates, the time window needed for synthesis of genes with large TUs will be in terms of hours rather than of minutes, as originally described in *Drosophila* (Gubb 1986; Shermoen and O'Farrell 1991). Unfortunately, the details of cell cycle kinetics are limited for vertebrates, particularly with regard to data with high spatial and temporal resolution (Boehm et al. 2010). When such data have been obtained (developing mouse limb bud), it is evident that there are discrete domains of cells with different cell cycle kinetics—areas of low proliferation and others of high proliferation (Boehm et al. 2010). Obviously, making progress toward better understanding the co-regulation of the cell cycle and gene expression will necessitate utilizing techniques for high temporal and spatial resolution of cell cycle kinetics.

With regard to the importance of cell cycle kinetics for the control of pattern formation, it is noteworthy that many classical signaling centers exhibit unique cell cycle kinetics, specifically a very long G1/G0 phase (Hay and Fischman 1961; Ohsugi et al. 1997; Satoh et al. 2012). These

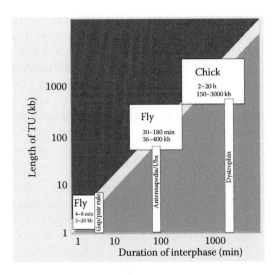

Figure 14.4 The size (length) of the transcription unit (TU) and the length of interphase of the cell cycle interact to control gene expression. The vertical axis is the TU length, and the horizontal axis is the duration of interphase. Completion of transcription will occur in the green region, but transcripts will be aborted in the red region. The yellow zone indicates the situation in which transcription potentially could be gated. This predicted relationship is based on the assumption of a transcription rate between 1.2 kb/minute and 2.4 kb/minute. Since RNA polymerization is restricted to the G1 phase of the cell cycle and nascent transcripts are aborted when the cell progress beyond G1, for a given duration of G1, TUs that are small enough will be able to complete transcription (green zone), but transcription will be aborted for TUs that are too large (red zone). Examples of reported values for the length of G1 and transcription unit sizes that would be predicted to be affected are illustrated in the white boxes. (Courtesy of Susan V. Bryant and David M. Gardiner.)

regions include the apical ectodermal ridge (AER) of both mouse and chick limb buds, the ZPA of the chick limb bud, the apical epithelial cap (AEC) of regenerating axolotl limbs, the notochord and floor plate of the chick embryo, the AEC of regenerating lizard tails and zebrafish fin rays, the enamel knot of developing teeth, and the midbrain–hindbrain boundary (see Bryant and Gardiner 2016). Experimentally, an ectopic signaling center (ZPA) can be induced in developing chick limb buds by reversibly blocking cell cycle progression in anterior cells with aphidicolin (Ohsugi et al. 1997). During the period of experimentally extended interphase, ectopic bone morphogenetic protein (*BMP2*) signaling is induced, resulting in the formation of supernumerary, mirror-image digits comparable to the response to grafting of the ZPA to the anterior, or by reprogramming anterior cells to become ZPA cells with RA (discussed earlier).

In addition to the presence of domains of cells with distinct cell cycle kinetics associated with the control of pattern formation, there are developmentally important genes with relatively large TUs. Although there has been no exhaustive search of genomic databases, several candidate genes that are expected to be susceptible to cell cycle gating (i.e., large TU) have been identified (Bryant and Gardiner 2016). In addition, there are very long antisense transcripts that overlap important developmental regulatory genes, and thus, gating of antisense TUs would indirectly regulate expression of the target sense transcripts, regardless of the size of the targeted gene (Bryant and Gardiner 2016).

Our interpretation of how growth factors/morphogens control pattern formation is based on the fact that these molecules fall into two classes: those that promote cell division and those that inhibit it. Factors that slow the cell cycle will provide more time in the cycle for the synthesis of large as well as small gene products from large TUs. Conversely, in those that speed up the cell cycle, synthesis of long transcripts will be gated by the cell cycle (Figure 14.4).

By this model, the length of the cell cycle is controlled directly by positively and negatively acting growth factors, and pattern formation is controlled indirectly via cell cycle gating of expression of critical TFs. Morphogens can be either positively acting growth factors that accelerate the rate of cell division or negatively acting growth factors that slow the rate (Figure 14.5). These extracellular signaling gradients generate an underlying gradient of cell cycle lengths. The longer the cell cycle, the more the transcripts that can accumulate before the cell enters mitosis and gene expression is terminated, and vice versa. Similarly, the shorter a gene, the more the molecules of its product that can be made before the cell cycle interrupts the process. The interplay

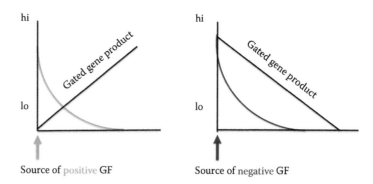

Figure 14.5 A model for how diffusion of extracellular growth factors would gate gene expression. High levels of a growth-stimulating (green) or a growth-inhibiting (red) growth factor would establish a gradient of cell cycle times, which in turn would establish a gradient of intracellular TFs, as predicted by the model for cell cycle gating of transcription (see Figure 14.4). (Courtesy of Susan V. Bryant and David M. Gardiner.)

between the size of a specific TF gene and the length of time available in the cell cycle for its expression creates the conditions needed to generate gradients of TFs on which the development and regeneration of the pattern depend (Figure 14.4). By this view, the PI grid acts directly to create discrete spatial and temporal cellular domains with unique cell cycle kinetics, which in turn result in discrete patterns of expression of key transcript factors via cell cycle gating (Bryant and Gardiner 2016; Foe 1989).

Finally, it is important to recognize that in the context of regenerative engineering, this view of cell cycle gating of transcription places regulation of the cell cycle upstream of the regulation of gene expression and developmental fate, which is the opposite of most views of development and regeneration. Thus, engineering strategies should focus on the spatial and temporal regulation of the cell cycle, which in turn will result in changes in the behavior and fate of the cells (i.e., pattern formation). This is a fundamentally alternative approach to attempts to alter the fate of the cells by the delivery of factors that are part of a signaling cascade associated with expression of a desired (differentiation-inducing) TF.

14.6 Cytonemes as cellular structures that could function as part of the positional information grid

Based on the PCM, it is evident that the PI grid involves cell–cell contacts over long distances (multiple cell diameters) that regulate growth factor/morphogen signaling and pattern formation. In recent years, it has become evident that many cells have specialized filopodia-like extensions (cytonemes) that function to regulate this type of cell–cell signaling (see Kornberg and Roy 2014; Roy and Kornberg 2015; Roy et al. 2014*). In general, filopodia are thin protrusions that extend from cells that have long been recognized as being involved in diverse cellular behaviors. Although their physical properties vary (1–400 μm in length and 0.1–0.4 μm in diameter), all are actin-based, they extend and retract quickly (multiple cell diameters in a few minutes), and their tips can contact other cells. Their diverse shapes and roles have led to a diverse terminology associated with the idea that they have diverse functions, including the early speculation that filopodia function as sensors of patterning information (Gustafson and Wolpert 1961). In 1999, filopodia with long and fragile cellular protrusions were identified in *Drosophila* and were speculated to function by transporting signaling molecules over long distances (Ramírez-Weber and Kornberg 1999). These cell extensions with signaling functions were defined as cytonemes and since

---

* The is the first experimental demonstration that cytonemes function to control short-range, cell–cell contact-mediated growth factor signaling.

then have been identified to function in regulating a number of signaling pathways in a variety of model systems (see Kornberg and Roy 2014).

The discovery and subsequent demonstration of the function of cytonemes are particularly important for understanding the mechanisms underlying pattern formation and the PI grid. Although growth factors/morphogens function in a graded, dose-dependent fashion (see Bryant and Gardiner 2016), the mechanism(s) for establishing such gradients of extracellular factors is/are unclear. Historically, the dominant view of almost every model of pattern-generating gradient formation is that such gradients are established by some version of diffusion. As a mechanism, diffusion is particularly appealing, given the mathematical models that have led to the theoretical generation of a multitude of biologically relevant patterns (Meinhardt, 1982; Turing 1952; Wolpert 1969). The diffusion mechanism is most famously embodied in the French Flag model (Wolpert 1969), which proposed that the positional value of a cell is established by a particular concentration of morphogen and that cells respond to different threshold concentrations of diffusing morphogens. Although appealing in a relatively simple way, the challenge of understanding pattern formation in complex tissues in time and space has led Wolpert to conclude that diffusible morphogen gradients lack the necessary precision and are *too messy* (Wolpert 2009). In contrast, from the early days of regeneration research, diffusion of morphogens is not a mechanism that is consistent with the governing principles as formalized in the PCM. In regeneration, signaling that controls pattern formation is mediated by short-range cell–cell interactions, which mechanistically can be accounted for by the functions of cytonemes.

Although filopodia were observed 100 years ago, their diverse functions have been fully appreciated only recently. This is particularly the case for cytonemes, given their fragile nature due to the small size. Although their structure was observed in some tissue preparations, it was lost in others (see Kornberg and Roy 2014). With advances in tissue preservation, the availability of fluorescent markers, and high-resolution imaging, it is now possible to study their structure and function (Kornberg and Roy 2014). Based on these new and improved technologies, it is now evident that cytonemes are present in many different cell types and that they function to control growth factor/morphogen signaling.

Specific examples of the function of cytonemes include regulation of most of the well-known morphogens, including Wnts, FGFs, decapentaplegic (Dpp) in *Drosophila*, BMP in vertebrates, HHs (hedgehogs), and Notch/Delta. In each case, all the signaling molecules and other molecules associated with each signaling pathway are associated with the cytonemes and the target cells localized at the site of contact with cytonemes. Thus, everything needed for signaling is present in the right place and at the right time to account for the phenomenology and phenotypes associated with each pathway. The more recent findings indicate that the transport mechanism itself is based on the trafficking of extracellular vesicles (exosomes) that have encapsulated the morphogen and are then translocated along the cytoneme (Gradilla et al. 2014). Most importantly, signaling fails if cytoneme structure is disrupted (see Roy and Kornberg 2015); in other words, signaling is dependent on the presence of cytonemes.

The best characterized model of this signaling mechanism is DPP (BMP in vertebrates) and FGF signaling between the *Drosophila* wing-imaginal disc and air sac primordia (Roy et al. 2014). Cells receive DPP by extending cytonemes that directly contact DPP-producing cells. These cytonemes contain the DPP receptor in motile puncta that also contain DPP taken up from DPP-producing cells. In contrast, a different set of cytonemes that contact FGF-producing cells contain the FGF receptor and take up FGF but not DPP. Thus, cytonemes are not homogenous but specialized to receive and regulate signaling associated with specific growth factors/morphogens; in other words, they are ligand-specific. In addition, this signaling is bidirectional, such that the cytonemes of the ligand-receiving cells trigger the release of the ligand from the producing cells (Roy et al. 2014).

Although, to date, much of the work has been carried out in the *Drosophila* wing disc model, there is increasing evidence that cytoneme-mediate cell–cell signaling is a conserved mechanism for the control of pattern formation. Cytonemes have been reported in a number of cells in culture (Kornberg and Roy 2014). In zebrafish, Wnt8a is transported via cytonemes to regulate the signaling required for the specification of neural plate cells (Stanganello and Scholpp 2016). The Wnt receptor Frz is localized to cytonemes, where it functions to transport Wnt proteins during flight muscle formation in *Drosophila* (Huang and Kornberg 2015). During chick development, Frz7 is present on cytonemes associated with somite development (Sagar et al. 2015), and sonic hedgehog (SHH) is present in vesicles that move along specialized filopodia in the limb bud (Sanders et al. 2013).

It is noteworthy that the behavior of cytonemes is ever changing. Assuming that cytonemes are involved in the regulation of PI, such dynamic behaviors could account for the phenomena of plasticity of PI and the PI grid associated with regeneration. PI plasticity could arise from changes in the structural properties of cytonemes (numbers and length) that can change rapidly, as well as their ligand-specific properties (types and concentrations of ligands bound and transported). Perhaps, most important is their ability to regulate release and uptake of signals and to direct signals to a preselected target. Thus, once established, the arrangement of cytonemes could be stabilized but then rapidly reorganized in response to injury, a property that is essential for successful regeneration.

Finally, the distribution and behavior of cytonemes are regulated by the ECM, HSPG in particular (see Roy and Kornberg 2015), which is another feature associated with regeneration (Phan et al. 2015). The idea is that HSPGs are essential for cytoneme growth and/or stability, which in turn mediate signaling between producing and responding cells. The essential role of the HSPGs is to provide a substrate for the cytonemes that track over the surface of the adjacent cells, much like the growth cones of extending neurites are guided by cues in the ECM (see Roy and Kornberg 2015). Therefore, we envision a model for the integrated function of the PI grid, whereby the information controlling cytoneme dynamics is encoded in the PI grid, which in turn regulates the delivery of morphogens temporally and spatially in order to form the pattern of the morphogen-responding cells.

14.7 Cancer and the positional information grid

A remarkable observation about salamanders is that in addition to their excellent regenerative abilities, they are also reported to be resistant to cancer. Attempts to induce cancer in urodeles by the administration of chemicals known to be carcinogenic in mammals lead to increased growth, as expected, but the growth leads to ectopic differentiated structures, not to undifferentiated cancer cells (see Tsonis 1983). It may be that cancer results from a weakness or failure of the PI grid in mammals, resulting in the decoupling of the regulation of growth and pattern formation. In contrast, a highly functional PI grid in urodeles inhibits the development of invasive cancer. This suggests that a possible approach to dealing with cancer in humans is early detection, coupled with treatments that attract cancer cells to a location in the body that would be minimally affected by the cancer cells and where they could be induced to differentiate. If this leads to the formation of supernumerary but differentiated structures, then surgical repairs would be needed, but the cancer would have been stopped in its tracks. An alternate and ultimately preferred remedy for cancer would be to seek treatments to reactivate the PI grid, either locally or globally, in affected individuals. Typically, proliferation of the tumor cells is written about as a downstream consequence of carcinogenesis. Similarly, dedifferentiation associated with carcinogenesis is conceptualized as being upstream of proliferation. In contrast, the cell cycle gating model places proliferation upstream of gene expression that regulates the state of differentiation. Revisiting the cause-and-effect relationship between proliferation and differentiation could provide insights into regulating the state of differentiation of tumor cells.

14.8 Aging and the positional information grid

It is important to recognize that although humans do not regenerate most of our complex, multi-tissue structures (e.g., an entire limb), they are very good at regenerating the individual tissues that make up those structures (e.g., epidermis, vasculature, nerves, and epidermis). Thus, it appears that regenerative failure in humans is not a consequence of the bits and pieces not regenerating but rather of the failure to coordinate and integrate them, so as to reform the more complex structure, the PI. The most notable exceptions in human regeneration are digit tips, which regenerate well in young mammals, including humans, but less well as the individual ages. This phenomenon of ontogenetic decline of regenerative abilities is widely observed among animals and thus appears to be a fundamental property of regeneration (Gardiner 2005). Thus, embryos and larvae can regenerate body parts well, even if the adults do not (see Gardiner 2005; Gardiner and Holmes 2012; McCusker and Gardiner 2011). The most dramatic decline in regenerative abilities is associated with differentiation at the end of embryogenesis. It appears that the remarkable regenerative ability of animals such as salamanders is a consequence of the ability of their cells to revert to a more embryonic state in order to re-access the earlier developmental programs (dedifferentiate; Han et al. 2005; McCusker et al. 2015).

In addition to the loss of regenerative ability associated with the transition from embryonic to post-embryonic development, there is a slower yet continuously progressive loss of regenerative abilities associated with aging. If throughout life there is a balance between the forces that cause damage to body parts and the ability to regenerate and repair that damage, then the phenomenon of aging maybe a consequence of the decline in regenerative abilities, resulting in the accumulation of ever-increasing amounts of damage. Thus, understanding the mechanisms of the ontogenetic decline of regenerative abilities could lead to strategies for reversing this process.

A number of causes for regenerative decline are being actively investigated (e.g., depletion of the pool of adult stem cells and/or changes in the stem cell niche). In the context of this chapter, we note that any age-associated modifications to the PI grid would result in changes in the way the behavior of regeneration-competent cells is regulated during regeneration. One such modification is the accumulation of advanced glycation end products (AGEs) with age, which is accelerated in association with diseases of aging (e.g., diabetes). The AGEs are protein modifications that occur during metabolism by the non-enzymatic glycation resulting from reactions between glucose and the amino groups of proteins. By incorporating the observation of increasing accumulation of AGEs with aging, along with what we know about the role of the ECM in contributing to and regulating the PI grid, it is possible that the active sites of the PI grid are progressively altered over time. Presumably, glycation of the PI grid would lead to progressive regenerative failure, and thus, therapies to reverse glycation would restore PI grid's function. The thinking here is that over time, the PI grid needs to be cleaned up, which would lead to the long-sought goal of rejuvenation.

14.9 Preventing positional information grid failure

As discussed in multiple contexts earlier, a healthy and complete PI grid is required for the development of an embryo, and in adults, deterioration of the integrity of the PI grid leads to cancer and aging. Wounds are either very slow to heal or sometimes do not heal at all (e.g., diabetic foot ulcers), leading to gaps in the PI grid. Scars, the thickened regions of connective tissue that remain in the skin after wounds have healed, are subject to remodeling over time, leading to thinning and eventual disappearance of the scar, but this process is much slower in older people. The prospect that at some point in the future, as we learn more about the PI Grid and its essential function in the structural integrity of the body, we will be in a position to repair or enhance it as needed in humans. This is likely to involve the development of transplantable artificial matrices containing regeneration-stimulating molecules that can be applied to fresh wounds, with the goal of speeding wound healing and reducing scarring. Once this goal has been successfully achieved, the next challenge will be to find ways to manipulate amputation wounds so as to stimulate regeneration of not only tissues but also eventually of organs. This leads us to recognize the importance of discovering strategies to promote PI grid health. Research progress on how to reverse the accumulation of AGEs in many diseases should lead to possible therapies.

The PI grid is located in the connective tissue compartment of our body, within which tissue fluids accumulate and through which they circulate via the lymphatic system. We consider it likely that therapies that alter the physical–chemical properties of the connective space would also impact the PI grid and the cells that respond to the PI grid. Thus, activities such as exercise and massages that result in the purging and refreshing of connective tissue compartment fluids might also be important in terms of maintaining the biological activity of the PI grid.

14.10 Engineering the positional information grid

As discussed earlier, we need to understand whether a defective or damaged PI grid can be repaired. In fact, we have examples of vertebrates with essentially the same body plan as ours (salamanders) that are capable of PI grid regeneration, and they can regenerate just about any part that can be removed, without compromising viability. We raise the possibility that regenerative failure in humans is a consequence of a defect in the PI grid. In salamander, the PI grid has the ability to self-heal in order to restore continuity. This healing can occur in different ways and, depending on how a wound heals, can lead to replacement of the missing part, to extra structures, or no regeneration at all (see McCusker et al. 2015). By this view, the challenge for regenerative engineering is how to induce healing and regeneration of the PI grid.

Among the challenges that mammals, including ourselves, face, one challenge is how to stimulate growth and repair (i.e., regeneration) without also unleashing cancer. One way to minimize unintended consequences for humans is to provide them with grafts of organs developed outside of the body—grown in donor animals or *in vitro*. As with endogenous regeneration, such approaches will require that we learn more about the PI grid and what it needs for optimal function.

We are encouraged by the findings about the relationship between the cells of the PI grid and the ECM, in particular, the specific modification of the ECM that regulates growth factor/morphogen signaling (Phan et al. 2015). This makes the ECM, and sulfated GAGs in particular, amenable to an engineering approach to creating pro-regenerative biomaterials that could function as smart bandages. This approach would lead eventually to engineering the PI grid in such a way as to make it tunable, in terms of both the quality (types of signals transported and delivered) and the quantity (variations in the amount and rate of transport). This physical nature of the PI grid is much more amenable to a regenerative engineering approach than trying to generate multiple complex gradients of diffusing chemical signals in order to achieve a reliable and quantitative mechanism for controlled cell–cell and cell–ECM signaling.

A major challenge to understanding and eventually engineering the PI grid is the lack of molecular markers for the cells that make the PI grid. Given the heterogeneity of cells of the connective tissue, and advances in understanding the role of the ECM in controlling the behavior of other cells, we now appreciate that fibroblasts are not just simple, spindle-shaped glue cells. There is at least one subpopulation of fibroblasts that functions to encode the information of the PI grid. Finally, the connective tissue fibroblasts that play such an important role in regeneration in species that can regenerate (the axolotl) function to make scars in species that cannot regenerate (e.g., humans). Thus, it is likely that by understanding and manipulating the behavior of these cells, we will be able to control scarring and induce regeneration in humans.

Acknowledgments

We wish to thank the members of the Bryant/Gardiner Laboratory past and present for help with and encouragement of the development of the ideas presented.

References

Boehm, B., H. Westerberg, G. Lesnicar-Pucko, S. Raja, M. Rautschka, J. Cotterell, J. Swoger, and J. Sharpe. 2010. The role of spatially controlled cell proliferation in limb bud morphogenesis. *PLoS Biology* 8 (7): e1000420. doi:10.1371/journal.pbio.1000420.

Bryant, S. V., V. French, and P. J. Bryant. 1981. Distal regeneration and symmetry. *Science* 212: 993–1002.

Bryant, S. V., and D. M. Gardiner. 1992. Retinoic acid, local cell-cell interactions, and pattern formation in vertebrate limbs. *Developmental Biology* 152: 1–25.

Bryant, S. V., and D. M. Gardiner. 2016. The relationship between growth and pattern formation. *Regeneration* 3 (2): 103–122. doi:10.1002/reg2.55.

Dinsmore, C. E. 1992. *The Foundations of Contemporary Regeneration Research: Historical Perspectives. Monographs in Developmental Biology.* Vol. 23. Basel, Switzerland: Karger.

Endo, T., S. V. Bryant, and D. M. Gardiner. 2004. A stepwise model system for limb regeneration. *Developmental Biology* 270 (1): 135–145.

Foe, V. E. 1989. Mitotic domains reveal early commitment of cells in drosophila embryos. *Development* 107 (1): 1–22.

French, V., P. J. Bryant, and S. V. Bryant. 1976. Pattern regulation in epimorphic fields. *Science* 193: 969–981.

Gardiner, D. M. 2005. Ontogenetic decline of regenerative ability and the stimulation of human regeneration. *Rejuvenation Research* 8 (3): 141–153.

Gardiner, D. M., and L. B. Holmes. 2012. Hypothesis: Terminal transverse limb defects with 'nubbins' represent a regenerative process during limb development in human fetuses. *Birth Defects Research. Part a, Clinical and Molecular Teratology* 94 (3): 129–133. doi:10.1002/bdra.22876.

Gardiner, D. M., K. Muneoka, and S. V. Bryant. 1986. The migration of dermal cells during blastema formation in axolotls. *Developmental Biology* 118: 488–493.

Gilbert, S. F. 2013. *Developmental Biology.* 10th ed. Sunderland, MA: Sinauer Associates, Inc.

Gradilla, A., E. González, I. Seijo, G. Andrés, M. Bischoff, L. González-Méndez, V. Sánchez et al. 2014. Exosomes as hedgehog carriers in cytoneme-mediated transport and secretion. *Nature Communications* 5 (December): 5649. doi:10.1038/ncomms6649.

Gubb, D. 1986. Intron-delay and the precision of expression of homoeotic gene products in Drosophila. *Developmental Genetics* 7 (3): 119–131. doi:10.1002/dvg.1020070302.

Gustafson, T., and L. Wolpert. 1961. Studies on the cellular basis of morphogenesis in the sea urchin embryo. Gastrulation in vegetalized larvae. *Experimental Cell Research* 22 (January): 437–449.

Han, M., X. Yang, G. Taylor, C. A. Burdsal, R. A. Anderson, and K. Muneoka. 2005. Limb regeneration in higher vertebrates: Developing a roadmap. *Anatomical Record. Part B, New Anatomist* 287 (1): 14–24. doi:10.1002/ar.b.20082.

Hay, E. D., and D. A. Fischman. 1961. Origin of the blastema in regenerating limbs of the newt triturus viridescens. An autoradiographic study using tritiated thymidine to follow cell proliferation and migration. *Developmental Biology* 3 (February): 26–59.

Holder, N., P. W. Tank, and S. V. Bryant. 1980. Regeneration of symmetrical forelimbs in the axolotl, Ambystoma mexicanum. *Developmental Biology* 74 (2): 302–314.

Huang, H., and T. B. Kornberg. 2015. Myoblast cytonemes mediate Wg signaling from the wing imaginal disc and Delta-Notch signaling to the air sac primordium. *eLife* 4 (May): e06114. doi:10.7554/eLife.06114.

Kornberg, T. B., and S. Roy. 2014. Cytonemes as specialized signaling filopodia. *Development* 141 (4): 729–736. doi:10.1242/dev.086223.

Makanae, A., K. Mitogawa, and A. Satoh. 2014. Co-operative Bmp- and Fgf-signaling inputs convert skin wound healing to limb formation in urodele amphibians. *Developmental Biology* 396 (1): 57–66. doi:10.1016/j.ydbio.2014.09.021.

Makanae, A., K. Mitogawa, and A. Satoh. 2016. Cooperative inputs of Bmp and Fgf signaling induce tail regeneration in urodele amphibians. *Developmental Biology* 410 (1): 45–55. doi:10.1016/j.ydbio.2015.12.012.

McCusker, C., and D. M. Gardiner. 2011. The axolotl model for regeneration and aging research: A mini-review. *Gerontology* 57 (6): 565–571. doi:10.1159/000323761.

McCusker, C., D. M. Gardiner, and S. V. Bryant. 2015. The axolotl limb blastema: Cellular and molecular mechanisms driving blastema formation and limb regeneration in tetrapods. *Regeneration* 2: 54–71. doi:10.1002/reg2.32.

McCusker, C. D., and D. M. Gardiner. 2013. Positional information is reprogrammed in blastema cells of the regenerating limb of the axolotl (Ambystoma mexicanum). *Plos One* 8 (9): e77064. doi:10.1371/journal.pone.0077064.

McCusker, C. D., and D. M. Gardiner. 2014. Understanding positional cues in salamander limb regeneration: Implications for optimizing cell-based regenerative therapies. *Disease Models & Mechanisms* 7 (6): 593–599. doi:10.1242/dmm.013359.

McCusker, C. D., C. Diaz-Castillo, J. Sosnik, A. Q. Phan, and D. M. Gardiner. 2016. Cartilage and bone cells do not participate in skeletal regeneration in Ambystoma mexicanum limbs. *Developmental Biology* 416 (1): 26–33. doi:10.1016/j.ydbio.2016.05.032.

McCusker, C., J. Lehrberg, and D. M. Gardiner. 2014. Position-specific induction of ectopic limbs in non-regenerating blastemas on axolotl forelimbs. *Regeneration* 1 (1): 27–34. doi:10.1002/reg2.10.

Meinhardt, H. 1982. *Models of Biological Pattern Formation*. London: Academic Press.

Morgan, T. H. 1901. *Regeneration*. New York: The Macmillan Company.

Noji, S., T. Nohno, E. Koyama, K. Muto, K. Ohyama, Y. Aoki, K. Tamura et al. 1991. Retinoic acid induces polarizing activity but is unlikely to be a morphogen in the chick limb bud. *Nature* 350: 83–86.

Ohsugi, K., D. M. Gardiner, and S. V. Bryant. 1997. Cell cycle length affects gene expression and pattern formation in limbs. *Developmental Biology* 189: 13–21.

Phan, A. Q., J. Lee, M. Oei, C. Flath, C. Hwe, R. Mariano, T. Vu et al. 2015. Heparan sulfates mediate positional information by position-specific growth factor regulation during axolotl (*Ambystoma mexicanum*) limb regeneration. *Regeneration* 2: 182–201. doi:10.1002/reg2.40.

Ramírez-Weber, F. A., and T. B. Kornberg. 1999. Cytonemes: Cellular processes that project to the principal signaling center in Drosophila imaginal discs. *Cell* 97 (5): 599–607.

Roy, S., H. Huang, S. Liu, and T. B. Kornberg. 2014. Cytoneme-mediated contact-dependent transport of the Drosophila decapentaplegic signaling protein. *Science* 343 (6173): 1244624. doi:10.1126/science.1244624.

Roy, S., and T. B. Kornberg. 2015. Paracrine signaling mediated at cell-cell contacts. *BioEssays* 37 (1): 25–33. doi:10.1002/bies.201400122.

Sagar, F. P., C. Wiegreffe, and M. Scaal. 2015. Communication between distant epithelial cells by filopodia-like protrusions during embryonic development. *Development* 142 (4): 665–671. doi:10.1242/dev.115964.

Sanders, T. A., E. Llagostera, and M. Barna. 2013. Specialized filopodia direct long-range transport of SHH during vertebrate tissue patterning. *Nature* 497 (7451): 628–632. doi:10.1038/nature12157.

Satoh, A., S. V. Bryant, and D. M. Gardiner. 2012. Nerve signaling regulates basal keratinocyte proliferation in the blastema apical epithelial cap in the axolotl (Ambystoma mexicanum). *Developmental Biology* 366 (2): 374–381. doi:10.1016/j.ydbio.2012.03.022.

Satoh, A., D. M. Gardiner, S. V. Bryant, and T. Endo. 2007. Nerve-induced ectopic limb blastemas in the axolotl are equivalent to amputation-induced blastemas. *Developmental Biology* 312 (1): 231–244.

Shermoen, A. W., and P. H. O'Farrell. 1991. Progression of the cell cycle through mitosis leads to abortion of nascent transcripts. *Cell* 67 (2): 303–310.

Singer, M. 1974. Neurotrophic control of limb regeneration in the newt. *Annals of the New York Academy of Sciences* 228: 308–321.

Stanganello, E., and S. Scholpp. 2016. Role of cytonemes in Wnt transport. *Journal of Cell Science* 129 (4): 665–672. doi:10.1242/jcs.182469.

Tickle, C., D. Summerbell, and L. Wolpert. 1975. Positional signalling and specification of digits in chick limb morphogenesis. *Nature* 254 (5497): 199–202.

Tsonis, P. A. 1983. Effects of carcinogens on regenerating and non-regenerating limbs in amphibia (review). *Anticancer Research* 3 (3): 195–202.

Turing, A. M. 1952. The chemical basis of morphogenesis. *Philosophical Transactions of the Royal Society of London B: Biological Sciences* B237: 37–72.

Wanek, N., D. M. Gardiner, K. Muneoka, and S. V. Bryant. 1991. Conversion by retinoic acid of anterior cells into ZPA cells in the chick wing bud. *Nature* 350: 81–83.

Wigmore, P., and N. Holder. 1985. Regeneration from isolated half limbs in the upper arm of the axolotl. *Journal of Embryology and Experimental Morphology* 89 (October): 333–347.

Wolpert, L. 1969. Positional information and the spatial pattern of cellular differentiation. *Journal of Theoretical Biology* 25 (1): 1–47.

Wolpert, L. 2009. Diffusible gradients are out - An interview with Lewis Wolpert. Interviewed by Richardson, Michael K. *International Journal of Developmental Biology* 53 (5–6): 659–562. doi:10.1387/ijdb.072559mr.

Wolpert, L., C. Tickle and A. Martinez Arias. 2015. *Principles of Development*. Oxford: Oxford University Press.

Chapter 15 Theorizing about gene expression heterogeneity patterns after cell dedifferentiation and their potential value for regenerative engineering

Carlos Díaz-Castillo

Contents

It is not just that noise is a bad signal. And it is not just that we want to actively reduce noise. We all want cleaner images and more reliable data lines and more soundproof walls and windows. We want more: We want to eliminate noise. We want to wipe noise out of digital existence. We want to win the war on noise through total annihilation.

But we never will.

Bart Kosko
(Kosko 2006)

In short, variation is an endless source of challenging questions.

Ernst Mayr
(Hallgrímsson and Hall 2011)

Key concepts

a. Cell dedifferentiation, the process by which specialized cells are reverted to less specialized, more proliferative, and more pluripotent stages, is important for the regeneration of complex structures, both in plants and in animals.

b. Cell dedifferentiation might result in cell convergence; in other words, dedifferentiated cells would be more similar among themselves than the specialized cells they originated from were.

c. Information theory-based measures for transcriptome specialization and diversity from mouse and human tumors and the organ they originate from reflect the cell convergence that follows cell dedifferentiation.

d. Cell dedifferentiation commonly concurs with a general decompaction of chromatin across genomes.

e. Since chromatin compaction is an important enhancer of gene expression noise, it would be expected that gene expression became generally less noisy after cell dedifferentiation.

f. Transcriptome convergence and gene expression noise reduction after cell dedifferentiation should be manifested as a general reduction of gene expression heterogeneity, which might be very useful for the characterization of naturally occurring regenerative processes and the inception of regenerative engineering approaches.

15.1 Introduction

For ages, variation in living organisms has been a constant source of inspiration for biologists. However, it is humbling to acknowledge how little we know about the molecular and cellular bases of biological variation and its integration with deterministic mechanisms to explain life or to find applied solutions to human troubles (Hallgrímsson and Hall 2011). Gene expression, the concatenation of steps toward the production of active RNAs and proteins from instructions encoded in the genome, is particularly suitable to study the molecular and cellular bases of biological variation. Gene expression can vary due to genetic changes in elements that participate in its regulation, in response to environmental changes and/or developmental processes, and also in the absence of any of these inputs (Kaern et al.

2005*, Raser and O'Shea 2005*, Raj and van Oudenaarden 2008*, Kilfoil et al. 2009*). In addition, gene expression variation results in a phenotypic variation that can be manifested as heterogeneity at different levels, from cells within a tissue to individuals within a population, with an important effect on normal and pathological development or the evolution of species (Kaern et al. 2005, Raser and O'Shea 2005, Raj and van Oudenaarden 2008, Kilfoil et al. 2009). Thus, the study of gene expression variation might permit disentangling the factors contributing to biological variation and how these factors might influence the course of developmental, pathological, or evolutionary processes.

Regenerative engineering represents a brave attempt of scientific and technological convergence, with concrete goals toward the regeneration of complex structures (Laurencin and Nair 2015). The completion of regeneration of human joints and whole limbs has been scheduled for 2022 and 2030, respectively (Khademhosseini 2015), forcing us to explore even the most unlikely avenue of potential convergence toward these goals. Traditionally, some of the main areas contributing to regenerative engineering have been heavily invested toward the study of conserved principles—developmental biology—or even dismissive of the utility of manifestations of variation such as noise—engineering. In the present essay, I reflect on two particular aspects associated with cell dedifferentiation to suggest that gene expression heterogeneity across genomes will decrease in events that course with cell dedifferentiation: plant somatic embryogenesis, cancer, or regenerative development. I kept this piece simple and concise because it is considerably speculative and it is aimed to elicit a biological variation-friendly mindset in regenerative engineers, while pointing to basic researchers the utility that the study of regenerative development might have for the understanding of the molecular and cellular bases of biological variation. A manuscript reporting preliminary analyses that confirm the central hypothesis drawn here is currently in preparation and will be published separately.

15.2 Cell dedifferentiation

From a cellular point of view, the development of multicellular organisms is a process characterized by cell differentiation: the progressive and coordinated specialization and loss of proliferative potential of initially pluripotent cells. The study of cell differentiation is one of the main pillars of developmental biology. Far less studied, cell dedifferentiation, or the cases in which specialized cells are reverted to less specialized, more proliferative, and more pluripotent stages, is important for the formation of naturally occurring calli and tumors or the regeneration of complex structures in both plants and animals (King and Newmark 2012,

---

* This is a great review on stochastic biological variation—the variation observed in biological systems, even in the absence of genetic or environmental cues.

Ikeuchi et al. 2013, Friedmann-Morvinski and Verma 2014, Varga et al. 2014, Yamada et al. 2014, Campos-Sanchez and Cobaleda 2015, Grafi and Barak 2015, Li et al. 2015, Sugiyama 2015). In addition, cell dedifferentiation is important for induced somatic embryogenesis in plants or human pluripotent stem cells (induced pluripotent stem cells [iPSCs]) (Jopling et al. 2011, Kami and Gojo 2014, El-Badawy and El-Badri 2015, Feher 2015, Krause et al. 2015, Sugiyama 2015). Thus, the understanding of dedifferentiation dynamics is of broad interest for plant biotechnology or medicine and especially for regenerative engineering.

Two features that are characteristic of dedifferentiating cells permit speculating that a general reduction in gene expression heterogeneity might be observable for events that course with cell dedifferentiation. These two features are the cellular convergence that cell dedifferentiation represents and the general reduction in chromatin compaction characteristic of cell dedifferentiation that might cause a reduction in gene expression noise.

15.3 Cell dedifferentiation-based convergence

Since all specialized cells within an organism ultimately originate from a single cell, the zygote, the mere idea that specialized cells can be reverted to previous pluripotent stages implies that events that course with cell dedifferentiation could be conceptualized as cases of cell convergence; *in other words*, dedifferentiated cells would be more similar among themselves than the specialized cells they originated from are. Whether all cell dedifferentiations reach the same level of undifferentiated pluripotency is still contentious. For example, iPSCs and tumor cells are supposed to be reverted to a more ancestral level of pluripotency than dedifferentiated cells in salamander limb regeneration, which seem to maintain some memory of the differentiated lineage from which they come (Kragl et al. 2009, Eguizabal et al. 2013, Yamada et al. 2014, Campos-Sanchez and Cobaleda 2015). Notwithstanding these differences, if cell dedifferentiation truly embodied a convergence of cells toward undifferentiated stages, this convergence could be detected as a reduction in heterogeneity for cellular features.

Interestingly, the putative convergence of dedifferentiated cells might be detectable at the transcriptomic level. In two different articles, Martínez el al. used information theory to derive two measures for specialization and diversity of transcriptomes from different tissues and applied them to the study of transcriptomes obtained from different human and murine cancers and the tissues from which they originated (Martinez and Reyes-Valdes 2008, Martinez et al. 2010*). These authors showed

---

* This and the previous reference use information theory to define measures for transcriptome specialization and divergence and apply them to the study of transcriptomes from mouse and human tumors and the organs from which they originated. Their results are consistent with the hypothesized cellular convergence associated with cell dedifferentiation.

that despite their different independent origins, the transcriptomes of most cancers show a very similar reduction in the level of expression for tissue-specific genes and a homogenization of the level of expression for expressed genes when compared with the tissues from which they originated (Martinez et al. 2010). Furthermore, cancer transcriptomes seem to be very similar in terms of specialization and divergence to the transcriptome of undifferentiated embryonic stem cells, underscoring the potential convergence putatively associated with cell dedifferentiation (Martinez et al. 2010). Whether similar trends are visible in all events coursing with cell dedifferentiation is still unknown.

15.4 Cell dedifferentiation-based reduction in gene expression noise

Recently, the study of gene expression in populations of clonal cells maintained in the same environment has suggested that gene expression is prone to vary even in the absence of genetic, environmental, or developmental cues (Kaern et al. 2005, Raser and O'Shea 2005, Raj and van Oudenaarden 2008, Kilfoil et al. 2009). Like many other intracellular processes, gene expression relies on a few copies of many intervening elements acting amidst very congested intracellular contexts, and therefore, it is susceptible to stochastic fluctuations of both required and contextual molecules (Kaern et al. 2005, Raser and O'Shea 2005, Raj and van Oudenaarden 2008, Kilfoil et al. 2009). The variation in gene expression that ultimately depends on molecular stochasticity is commonly referred to as stochastic variation in gene expression, or gene expression noise. The characterization of gene expression noise is still in its infancy, and many aspects are yet to be clarified. For example, it is unclear how many cells within a population could be considered in comparable microenvironments and developmental/cell cycle states, so the variation they show can be unambiguously deemed as stochastic or if stochastic variation caused at any of the steps of gene expression is mitigated in subsequent steps or propagated through biological systems, resulting in phenotypic noise (Battich et al. 2015).

One interesting aspect for gene expression noise is that it might be variable itself, being such a variation of great value to understand nuclear dynamics of cells involved in certain processes. One of the factors that contribute to the variation in gene expression noise is chromatin compaction (Kaern et al. 2005, Raser and O'Shea 2005, Raj and van Oudenaarden 2008). Slow dynamics for the transition between compacted and open chromatin states for regions with highly compacted chromatin or heterochromatin make the expression of genes located there particularly noisy (Kaern et al. 2005, Raser and O'Shea 2005, Raj and van Oudenaarden 2008). Although the study of gene expression

355

noise in clonal cells is recent (Kaern et al. 2005, Raser and O'Shea 2005, Raj and van Oudenaarden 2008), an extensive literature exist on stochastic patterns of gene expression variation for genes located close to or within heterochromatin (reviewed in Elgin and Reuter 2013*). Position-effect variegation (PEV) was first discovered in *Drosophila*, associated with the relocation of a gene close to heterochromatin, caused by a chromosome inversion, and since then, it has been extended to other species for genes located close to or within heterochromatin (Elgin and Reuter 2013). The expression of these genes show signs of stochastic variation with regard to their state of activation/repression, their level of expression for cells of the same individual, and their expression patterning between individuals of the same progeny (Elgin and Reuter 2013). Since chromatin compaction for many genes can change during the cell cycle, between cell types, or in response to biotic and abiotic signals, it would be expected that chromatin compaction–dependent gene expression noise across genomes was also variable. In fact, it is known that factors that affect chromatin compaction, such as temperature, genetic variation in heterochromatin-forming elements, and the direction of chromosome inheritance, result in the modification of stochastic patterns in gene expression, such as PEV (Maggert and Golic 2002, Elgin and Reuter 2013). Furthermore, recently, it has been suggested that large differences in junk DNA genomic content between individuals of different sexes in the same species can cause a variation in gene expression noise, with an important effect on the expression of phenotypes and species dynamics (Diaz-Castillo 2015†).

Interestingly, both naturally occurring and induced dedifferentiation events are characterized by a general opening of the chromatin (Jiang et al. 2013, El-Badawy and El-Badri 2015, Feher 2015, Grafi and Barak 2015, Jiang et al. 2015, Krause et al. 2015, Lee et al. 2015). Such changes would result in many genes to become more accessible to the basic machinery and regulatory elements of transcription. In fact, the early expression of otherwise-silent transposable elements in dedifferentiating cells is consistent with the derived accessibility of many loci in the genome that are associated with the general opening of the chromatin (Wang and Wang 2012, Zhu et al. 2012, Macia et al. 2015). Considering that chromatin compaction is an important enhancer for gene expression noise (Kaern et al. 2005, Raj and van Oudenaarden 2008, Raser and O'Shea 2005), it could be inferred that the extensive reduction in chromatin compaction across dedifferentiating nuclei would cause a general

---

* This review is a great segue into the extensive literature on the phenomenon referred to as position-effect variegation (PEV)—the stochastic phenotypic variation derived from genes being located close to or within chromosome regions with highly compacted chromatin or heterochromatin.

† This article shows that in metazoans, gene expression noise might be generally larger for heterogametic individuals than for homogametic individuals and that such sexual dimorphism might be dependent on chromatin formation early in embryogenesis, in the presence or absence of junk DNA–enriched sex-specific chromosomes.

reduction in gene expression noise. Whether events that course with cell dedifferentiation are truly characterized by a general decrease in gene expression noise is also unknown.

15.5 Summary and perspectives

In this brief essay, I hypothesized that processes that concur with cell dedifferentiation would do so with a general decrease in gene expression heterogeneity because of the convergence that cell dedifferentiation represents itself and the general reduction in gene expression noise caused by the generalized opening of dedifferentiated cells chromatin. To the best of my knowledge, no literature exists that directly addresses gene expression heterogeneity dynamics along processes that concur with cell dedifferentiation. Limb regeneration in salamanders might be a very suitable model system to test the validity of this hypothesis and explore its value further for regenerative engineering. In recent years, a number of studies analyzed transcriptome dynamics along limb regeneration in *Ambystoma mexicanum* or Mexican axolotl (Monaghan et al. 2009, Monaghan et al. 2012, Stewart et al. 2013, Wu et al. 2013, McCusker et al. 2015, Voss et al. 2015). Although most of these studies have low levels of biological replication, which would be an inconvenience for the study of gene expression heterogeneity, they can be used to make preliminary observations and better design-specific studies in the future. Moreover, despite the difficulty that entails sequencing large genomes cluttered with large amounts of repetitive DNA, the genome of the axolotl is currently being sequenced, and soon, it would be feasible to relate gene expression heterogeneity dynamics with other aspects of the nuclear architecture (Keinath et al. 2015). Although it is still too soon to appreciate the value of the putative reduction in gene expression heterogeneity along processes that concur with cell dedifferentiation for regenerative engineering, the study of gene expression heterogeneity along limb regeneration might help to better characterize gene expression dynamics along processes where cells dedifferentiate, to identify elements that might be important for the regulation of these processes, or to locate chromosome domains with particular gene expression heterogeneity dynamics that can be used as target for the insertion of reporter genes to help better monitor the correct progress of regenerative processes by using non-invasive methodologies.

Acknowledgments

I want to express my deepest gratitude to Raquel Chamorro-García for her unfailing support and to David Gardiner for thinking that there is room within the regenerative engineering arena for thoughts on biological variation.

References

Battich, N., T. Stoeger, and L. Pelkmans. 2015. Control of transcript variability in single mammalian cells. *Cell* 163 (7):1596–1610. doi:10.1016/j.cell.2015.11.018.

Campos-Sanchez, E. and C. Cobaleda. 2015. Tumoral reprogramming: Plasticity takes a walk on the wild side. *Biochim Biophys Acta* 1849 (4):436–447. doi:10.1016/j.bbagrm.2014.07.003.

Diaz-Castillo, C. 2015. Evidence for a sexual dimorphism in gene expression noise in metazoan species. *PeerJ* 3:e750. doi:10.7717/peerj.750.

Eguizabal, C., N. Montserrat, A. Veiga, and J. C. I. Belmonte. 2013. Dedifferentiation, transdifferentiation, and reprogramming: Future directions in regenerative medicine. *Semin Reprod Med* 31 (1):82–94. doi:10.1055/s-0032-1331802.

El-Badawy, A. and N. El-Badri. 2015. Regulators of pluripotency and their implications in regenerative medicine. *Stem Cells Cloning* 8:67–80. doi:10.2147/SCCAA.S80157.

Elgin, S. C. and G. Reuter. 2013. Position-effect variegation, heterochromatin formation, and gene silencing in Drosophila. *Cold Spring Harb Perspect Biol* 5 (8):a017780. doi:10.1101/cshperspect.a017780.

Feher, A. 2015. Somatic embryogenesis - Stress-induced remodeling of plant cell fate. *Biochim Biophys Acta* 1849 (4):385–402. doi:10.1016/j.bbagrm.2014.07.005.

Friedmann-Morvinski, D. and I. M. Verma. 2014. Dedifferentiation and reprogramming: Origins of cancer stem cells. *EMBO Rep* 15 (3):244–253. doi:10.1002/embr.201338254.

Grafi, G. and S. Barak. 2015. Stress induces cell dedifferentiation in plants. *Biochim Biophys Acta* 1849 (4):378–384. doi:10.1016/j.bbagrm.2014.07.015.

Hallgrímsson, B. and B. K. Hall. 2011. *Variation: A Central Concept in Biology*. Academic Press: New York.

Ikeuchi, M., K. Sugimoto, and A. Iwase. 2013. Plant callus: Mechanisms of induction and repression. *Plant Cell* 25 (9):3159–3173. doi:10.1105/tpc.113.116053.

Jiang, F., Z. Feng, H. Liu, and J. Zhu. 2015. Involvement of plant stem cells or stem cell-like cells in dedifferentiation. *Front Plant Sci* 6:1028. doi:10.3389/fpls.2015.01028.

Jiang, F., J. Zhu, and H. L. Liu. 2013. Protoplasts: A useful research system for plant cell biology, especially dedifferentiation. *Protoplasma* 250 (6):1231–1238. doi:10.1007/s00709-013-0513-z.

Jopling, C., S. Boue, and J. C. I. Belmonte. 2011. Dedifferentiation, transdifferentiation and reprogramming: Three routes to regeneration. *Nat Rev Mol Cell Biol* 12 (2):79–89. doi:10.1038/nrm3043.

Kaern, M., T. C. Elston, W. J. Blake, and J. J. Collins. 2005. Stochasticity in gene expression: From theories to phenotypes. *Nat Rev Genet* 6 (6):451–464. doi:10.1038/nrg1615.

Kami, D. and S. Gojo. 2014. Tuning cell fate: From insights to vertebrate regeneration. *Organogenesis* 10 (2):231–240. doi:10.4161/org.28816.

Keinath, M. C., V. A. Timoshevskiy, N. Y. Timoshevskaya, P. A. Tsonis, S. R. Voss, and J. J. Smith. 2015. Initial characterization of the large genome of the salamander Ambystoma mexicanum using shotgun and laser capture chromosome sequencing. *Sci Rep* 5:16413. doi:10.1038/srep16413.

Khademhosseini, A. 2015. HEAL project aims to regenerate human limbs by 2030. *Regen Eng Transl Med* 1 (1–4):50–57. doi:10.1007/s40883-015-0007-y.

Kilfoil, M. L., P. Lasko, and E. Abouheif. 2009. Stochastic variation: From single cells to superorganisms. *HFSP J* 3 (6):379–385. doi:10.2976/1.3223356.

King, R. S. and P. A. Newmark. 2012. The cell biology of regeneration. *J Cell Biol* 196 (5):553–562. doi:10.1083/jcb.201105099.

Kosko, B. 2006. *Noise*. Viking: New York.

Kragl, M., D. Knapp, E. Nacu, S. Khattak, M. Maden, H. H. Epperlein, and E. M. Tanaka. 2009. Cells keep a memory of their tissue origin during axolotl limb regeneration. *Nature* 460 (7251):60–65. doi:10.1038/nature08152.

Krause, M. N., I. Sancho-Martinez, and J. C. I. Belmonte. 2015. Understanding the molecular mechanisms of reprogramming. *Biochem Biophys Res Commun* 473 (3): 693–697. doi:10.1016/j.bbrc.2015.11.120.

Laurencin, C. T. and L. S. Nair. 2015. Regenerative engineering: Approaches to limb regeneration and other grand challenges. *Regen Eng Transl Med* 1 (1–4):1–3. doi:10.1007/s40883-015-0006-z.

Lee, K., O. S. Park, S. J. Jung, and P. J. Seo. 2015. Histone deacetylation-mediated cellular dedifferentiation in Arabidopsis. *J Plant Physiol* 191:95–100. doi:10.1016/j.jplph.2015.12.006.

Li, Q., H. Yang, and T. P. Zhong. 2015. Regeneration across metazoan phylogeny: Lessons from model organisms. *J Genet Genomics* 42 (2):57–70. doi:10.1016/j.jgg.2014.12.002.

Macia, A., E. Blanco-Jimenez, and J. L. Garcia-Perez. 2015. Retrotransposons in pluripotent cells: Impact and new roles in cellular plasticity. *Biochim Biophys Acta* 1849 (4):417–426. doi:10.1016/j.bbagrm.2014.07.007.

Maggert, K. A. and K. G. Golic. 2002. The Y chromosome of Drosophila melanogaster exhibits chromosome-wide imprinting. *Genetics* 162 (3):1245–1258.

Martinez, O. and M. H. Reyes-Valdes. 2008. Defining diversity, specialization, and gene specificity in transcriptomes through information theory. *Proc Natl Acad Sci USA* 105 (28):9709–9714. doi:10.1073/pnas.0803479105.

Martinez, O., M. H. Reyes-Valdes, and L. Herrera-Estrella. 2010. Cancer reduces transcriptome specialization. *PLoS One* 5 (5):e10398. doi:10.1371/journal.pone.0010398.

McCusker, C. D., A. Athippozhy, C. Diaz-Castillo, C. Fowlkes, D. M. Gardiner, and S. R. Voss. 2015. Positional plasticity in regenerating Amybstoma mexicanum limbs is associated with cell proliferation and pathways of cellular differentiation. *BMC Dev Biol* 1–17. doi:10.1186/s12861-015-0095-4.

Monaghan, J. R., A. Athippozhy, A. W. Seifert, S. Putta, A. J. Stromberg, M. Maden, D. M. Gardiner, and S. R. Voss. 2012. Gene expression patterns specific to the regenerating limb of the Mexican axolotl. *Biol Open* 1 (10):937–948. doi:10.1242/bio.20121594.

Monaghan, J R., L. G. Epp, S. Putta, R. B. Page, J. A. Walker, C. K. Beachy, W. Zhu et al. 2009. Microarray and cDNA sequence analysis of transcription during nerve-dependent limb regeneration. *BMC Biol* 7 (1):1. doi:10.1186/1741-7007-7-1.

Raj, A. and A. van Oudenaarden. 2008. Nature, nurture, or chance: Stochastic gene expression and its consequences. *Cell* 135 (2):216–226. doi:10.1016/j.cell.2008.09.050.

Raser, J. M. and E. K. O'Shea. 2005. Noise in gene expression: Origins, consequences, and control. *Science* 309 (5743):2010–2013. doi:10.1126/science.1105891.

Stewart, R., C. A. Rascón, S. Tian, J. Nie, C. Barry, L.-F. Chu, H. Ardalani et al. 2013. Comparative RNA-seq analysis in the unsequenced axolotl: The oncogene burst highlights early gene expression in the blastema. *PLoS Comp Biol* 9 (3):e1002936. doi:10.1371/journal.pcbi.1002936.

Sugiyama, M. 2015. Historical review of research on plant cell dedifferentiation. *J Plant Res* 128 (3):349–359. doi:10.1007/s10265-015-0706-y.

Varga, J., T. D. Oliveira, and F. R. Greten. 2014. The architect who never sleeps: Tumor-induced plasticity. *FEBS Lett* 588 (15):2422–2427. doi:10.1016/j.febslet.2014.06.019.

Voss, S. R., A. Palumbo, R. Nagarajan, D. M. Gardiner, K. Muneoka, A. J. Stromberg, and A. T. Athippozhy. 2015. Gene expression during the first 28 days of axolotl limb regeneration I: Experimental design and global analysis of gene expression. *Regeneration* 2 (3):120–136. doi:10.1002/reg2.37.

Wang, Q. M. and L. Wang. 2012. An evolutionary view of plant tissue culture: Somaclonal variation and selection. *Plant Cell Rep* 31 (9):1535–1547. doi:10.1007/s00299-012-1281-5.

Wu, C-H., M.-H. Tsai, C.-C. Ho, C.-Y. Chen, and H.-S. Lee. 2013. De novo transcriptome sequencing of axolotl blastema for identification of differentially expressed genes during limb regeneration. *BMC Genomics* 14 (1):1. doi:10.1186/1471-2164-14-434.

Yamada, Y., H. Haga, and Y. Yamada. 2014. Concise review: Dedifferentiation meets cancer development: Proof of concept for epigenetic cancer. *Stem Cells Transl Med* 3 (10):1182–1187. doi:10.5966/sctm.2014-0090.

Zhu, W., D. Kuo, J. Nathanson, A. Satoh, G. M. Pao, G. W. Yeo, S. V. Bryant, S. R. Voss, D. M. Gardiner, and T. Hunter. 2012. Retrotransposon long interspersed nucleotide element-1 (LINE-1) is activated during salamander limb regeneration. *Dev Growth Differ* 54 (7):673–685. doi:10.1111/j.1440-169X.2012.01368.x.

Section III

Integration of new structures with the old

In order for a newly formed body part to function, it needs to integrate structurally with the rest of the body that was not lost to injury. This is an issue for regeneration that does not occur for embryogenesis. In the embryo, all the tissues are at comparable stages of development, and their development is coordinately regulated such that each piece is formed in the right place relative to all the other pieces. With regeneration (and thus with regenerative engineering), the new part is more comparable to an embryonic structure; whereas, the uninjured parts of the tissue or organ are already differentiated and functional. Thus the new part at some point in its development needs to become integrated with the preexisting host tissues. Based on our current understanding, this will require controlling the state of differentiation of both the engineered cells and the host cells at the interface between the two. Specifically, there needs to be a continuity in the positional information (PI) between the host and regenerated cells, and this will require learning how to control the plasticity of PI. By establishing continuity of PI, the cells collectively can make the bridge between the new and the old.

Dedifferentiation is important at both the beginning (formation of the early blastema by fibroblasts with PI) and the end (integration of regenerated and uninjured structures) of regeneration. As cells of the adult must be reprogrammed to make new cells to replace the missing parts during regeneration, there necessarily will be epigenetic modifications that are essential for regulating differentiation and dedifferentiation. In this section of this book, we explore the hypothesis that the injured host cells are induced to dedifferentiate, and then interact with developmentally plastic cells at the interface with the blastema, leading to integration. If this is the case, then it will be necessary to discover how to engineer the processes controlling the state of differentiation.

Chapter 16 Directed differentiation of pluripotent stem cells *in vitro*

Diane L. Carlisle

Contents

Key concepts

a. Multiple approaches to directed differentiation *in vitro* are available, but they are not interchangeable; the research goals often dictate the most appropriate differentiation method.

b. Modifying published differentiation protocols to best answer the research question requires a solid understanding of developmental biology.

c. Moving from enriched monolayer culture to multicellular tissues necessitates incorporation of structural supports and modification of growth factors to allow multiple germ layers to develop simultaneously.

d. To demonstrate that the differentiated cells are the desired type, the gold-standard evidence is functionality, in addition to mRNA and protein profile.

16.1 Introduction

The generation of pluripotent stem cells (PSCs), which includes embryonic stem cells (ESCs) and induced pluripotent stem cells (iPSCs), gives us the ability to differentiate cell *in vitro* into any cell type in the body. Regenerative medicine research investigates utilization of PSCs *in vitro* for transplantation to restore physiologic function. Personalized medicine researchers use differentiated PSCs to develop patient-specific drugs and to investigate disease pathogenesis, and developmental biologists can use them to understand cell fate decision-making and disruptions in normal development, especially human development, in ways not previously possible. Although there is significant enthusiasm for the use of differentiated PSCs for all these purposes, designing the ideal differentiation protocol is not always straightforward. In this chapter, I will guide readers through some of the key factors that influence experimental design. In addition, I will discuss the advantages and challenges inherent to differentiation paradigms. Some of the key points to consider are the renewability of the source of research material and the need for a highly enriched single cell type or a multicellular tissue for the experimental question. However, the central point is the nature of the research question itself. If we need only differentiated cells as a source material, we might use a different method than the one if we are investigating the differentiation process. In this chapter, I will focus on these key factors and touch on the most important practical points to consider when designing a directed differentiation protocol.

16.2 Choosing a differentiation method

Many methods are available for differentiating cells into specific cell fates, and many more methods are under development. Each method has specific advantages and disadvantages that need to be considered before deciding on the optimal choice. Key considerations include the renewability of the source cell population, how closely the process needs to mimic developmental biology, how enriched the final population needs to be, and the specific experimental paradigm to be tested.

16.2.1 Developmental modeling through a pluripotent stem cell stage

The most well-established differentiation paradigms are those that start with a PSC stage. In this model, either ESCs are used or source cells are converted to pluripotency by using established iPSC derivation methods. Pluripotent stem cells are then differentiated into the desired cell type by adding cytokines based on known developmental signaling. This method has two major strengths. A key defining feature of PSCs is their immortality (Mitalipova et al. 2003). If cultured meticulously, PSCs provide an unlimited clonal source material for differentiation. This provides the ability to expand an otherwise-limited resource.

For example, patient-specific fibroblasts from an individual patient that are converted to iPSCs turned a very limited resource into an unlimited resource, because somatic cells have limited life spans, whereas iPSCs are immortal. Furthermore, when the same starting material is used for multiple experiments, it can decrease experimental variability.

The second key advantage of using a PSC in an approach that mimics embryonic development is the ability to build on decades of core developmental biological knowledge. The key signaling molecules that lead to each of the three germ layers are well known and are reviewed in detail in Irion et al. (2008). Some protocols rely on this knowledge to try to tightly control differentiation at each decision-making step, which also allows quality-control analysis during the protocol. For example, when differentiating to lung, the protocol allows for the analysis of, first, definitive endoderm markers, then anterior foregut, followed by lung progenitor, and then finally, mature lung markers (Green et al. 2011, Longmire et al. 2012, Wong et al. 2012, Wong and Rossant 2013, Huang et al. 2015). If the hypothesis to be tested investigates perturbations of the developmental pathway, this type of differentiation is essential, because each cell-fate step can be individually tested. In no system has this been more extensively optimized than in neurons, where cells are quickly moved along the developmental pathway by using small molecules to produce mature neurons in about 3 weeks (Maury et al. 2015). Conversely, if a large amount of a highly enriched cell population is not needed, researchers can rely significantly on spontaneous differentiation of PSCs, which results in a mixture of all cell types that can be either used directly as resource material or enriched for particular cells by using sorting or other methods (Clark et al. 2004). Furthermore, the foundational knowledge of developmental biology allows researchers to make educated alterations to differentiation protocols. Certain signaling molecules can be added, omitted, or manipulated in differentiation protocols if the goal of the project is to investigate the downstream effects of that manipulation.

Despite these advantages, going through a pluripotent stage is not always required, and it does have disadvantages. First, either we must obtain an ESC line, or somatic cells must be reprogrammed to generate iPSCs. Some funding sources limit the use of human ESCs, and human ESCs are not available for non-genetic diseases, since non-genetic diseases cannot be identified using pre-implantation genetic diagnosis of the source embryos. The iPSC lines can be made from somatic cells of adults with known disease, genetic or epigenetic; however, it is unlikely that cells retain their epigenetic status through reprogramming to pluripotency and redifferentiation into somatic cells (reviewed in Liang and Zhang 2013). Furthermore, differentiation using developmental paradigms takes significantly longer than other differentiation protocols. Finally, there is increased likelihood of variability in the end result, because there are multiple steps to the protocols over time. Thus, for some research questions, other methods such as transdifferentiation maybe preferred.

16.2.2 Transdifferentiation

Transdifferentiation between cell types can be done either by converting a somatic cell directly into another somatic cell or by going through a common progenitor (defined as having the ability to divide and redifferentiate into a limited number of cell types). This method was first validated for the generation of neurons (Son et al. 2011) and is under development for other cell types (Sekiya and Suzuki 2011, Yang et al. 2014). Reprogramming a somatic cell directly into another somatic cell type is significantly faster than going through a PSC intermediate. Although no one has, as yet, characterized the epigenetics of these cells, it is possible that epigenetic markers are more likely to be retained, since the reprogramming is faster and with fewer steps (one reprogramming step instead of reprogramming to PSC and then reprogramming again to a differentiated cell type). For research questions with unlimited source material that need differentiated cells only as an end product, this type of protocol maybe ideal. Disadvantage of transdifferentiation is that the biological processes involved do not reflect classic developmental biology and thus are not well understood. Thus, this method is not appropriate for answering questions about developmental cell-fate decision-making. In addition, transdifferentiation converts all the cells in each sample to the new cell type. Since the starting and ending somatic cell types have limited life spans, over time, the starting material will be exhausted.

Two intermediate protocols are in use to try to overcome the disadvantages of transdifferentiation. Some transdifferentiation protocols have been developed to go through a progenitor stage, where the progenitor can be expanded to at least a limited extent, to increase the material available for research (Thier et al. 2012). In other protocols, transdifferentiation factors are applied to PSCs, to provide unlimited source material. Somatic cells are differentiated from PSCs more quickly than using multistep developmentally based protocols; however, other disadvantages such as loss of epigenetic markers due to the use of pluripotency reprogramming and inability to answer basic developmental biology questions still apply.

16.3 Harnessing developmental principles for directed differentiation

16.3.1 Know your starting cells

Once a protocol is chosen, it is time to think about the details of differentiation, so that we can obtain useful, reproducible, reliable results. The starting point is your source cells. Quality-controlled PSCs express standard pluripotency markers, and they have been demonstrated to produce all three germ layers in a teratoma assay (Navara et al. 2007). However, not all PSCs behave in the same way in directed differentiation protocols. This is not surprising, since PSC lines are made from genetically diverse source tissues. Thus, we expect to see differences in lines based on normal genetic variation (International Stem Cell et al. 2011). In addition, some iPSCs are made with integrating virus, some are made with episomal virus, and others are made using other epigenetic methods. When designing our experiments, we need to balance experimental reproducibility (decreasing variation by using a single line) with questions of physiologic relevance (making sure that results are generalizable and not a quirk of a single genetic feature of an individual line).

Pluripotent stem cells are immortal; however, they can change over time (Mitalipova et al. 2005). Routine quality-control measures include karyotyping and pluripotency confirmation, at least annually. A common irony in the field is that if we believe we have suddenly become experts in growing our PSCs because they are growing easily, then we should check the karyotype, because PSCs that are suddenly easy to grow have likely become aneuploid; aneuploidy was demonstrated to occur over time (Mitalipova et al. 2005) (International Stem Cell et al. 2011). The pluripotency of the source cells should be fully characterized by markers and teratoma formation before any experiments. However, established PSCs can spontaneously differentiate during routine culture (Liang and Zhang 2013). If so, downstream experiments will obviously be affected; thus, PSCs should be checked for routine pluripotency markers such as OCT4, NANOG, and others. This is easily done using live-cell antibody-based stains every few months.

When using PSCs for experiments, it is important to remember that an immortal pluripotent cell state is an artifact of culture. Many products are commercially available in which cells can be grown (International Stem Cell Initiative et al. 2010, Chen et al. 2011, Rao 2011). Most of them work, but they are not equivalent and do not produce equivalent cells. Since they are all artificial, no system is better or worse than the other, but they are different. In my own laboratory, we found that a differentiation protocol that we used to produce epithelium was more likely to include keratinocytes if the starting PSCs were grown on mouse embryonic fibroblasts rather than on Matrigel. Clearly, although it was the same human ESC line, with the same pluripotency marker-staining profile, the starting cells were not equivalent when grown on different matrices. With this in mind, if we are having trouble with a differentiation protocol, it might be worth changing the culture of our source cells to see if it helps in differentiation. Conversely, if we have a protocol that is working, we do not switch PSC culture reagents simply because a less expensive product comes along, unless we are willing to commit to extensive quality-control experiments to ensure that all our differentiation protocols still work.

16.3.2 Early cell-fate specification

The signaling networks that lead to specification of the three germ layers, endoderm, mesoderm, and ectoderm, are extensively researched

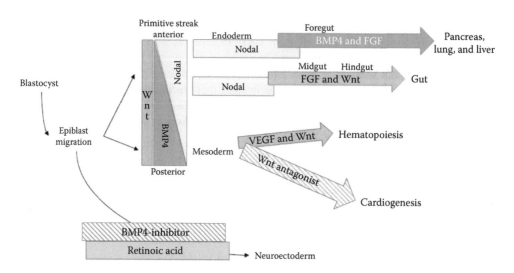

Figure 16.1 Specific cytokines specify cell-fate decision-making. Signaling molecules that were identified as key factors, specifying that embryonic development can be used to differentiate pluripotent stem cells *in vitro*.

(reviewed by Irion et al. 2008). Gradients and stepwise expression of bone morphogenetic protein 4 (BMP4), Nodal, Wnt, retinoic acid (RA), and other molecules control the development of these tissues (Figure 16.1). To obtain enriched cell types *in vitro*, we mimic these gradients and their stepwise signaling by adding recombinant cytokines or small molecule mimic and inhibitors to our differentiating cells in a stepwise manner. The benefit of understanding the developmental pathways behind the differentiation protocol is that it allows us to modify our protocol to ask specific developmental biology questions. For example, if we want to investigate the effect of an environmental exposure on endogenous Wnt signaling, we cannot add endogenous Wnt or Wnt inhibitors to our culture medium, because that would override the environmental effect. In order to make this adjustment, we have to understand what each small molecule included in our culture is (including the components of prepackaged differentiation kits) and how it works. Similarly, if we want to see the effect of a genetic mutation known to be associated with a specific disease, a fully controlled differentiation protocol may override the developmental basis of the pathology, because the exogenous factors added may compensate for the genetic modifier. As with the previous environmental modifier example, the differentiation protocol may need to be modified to assess the genetic change.

16.4 Monolayer-enriched culture or multicellular tissue?

Very detailed differentiation methods are useful for reliably producing highly enriched monolayer cultures of somatic cells. Culture on plastic plates does not support the three-dimensional growth of cells, limiting cellular crosstalk and structure formation. The most common strategy for overcoming this challenge is the use of structural supports to allow three-dimensional cellular differentiation. Common supports include Matrigel and Transwell filters and bioreactors. Use of these supports allows cells to form spheres or more complex structures, depending on the type of support.

However, what if our experimental question includes analysis of multiple cell types generated from different germ layers? For example, when we study lung development, we may want to assess foregut the interaction of endoderm and lateral plate mesoderm (Morrisey and Hogan 2010, Morrisey et al. 2013). Endoderm generation requires high nodal/low BMP4 signaling as a first step, whereas mesoderm requires low nodal signaling and high BPM4 signaling (Irion et al. 2008). There are a few protocols that allow multicellular tissue generation, and those that do exist often come from the same germ layer; for example, the generation of cerebral organoids containing neurons and glia, which arise from a common neural progenitor cell in the neuroectoderm, has been published (Lancaster and Knoblich 2014). Current work modifies exposure to BMP4/Nodal/Wnt/(RA) mimics and inhibitors in combination with structural supports to allow the cells to develop as multicellular three-dimensional tissues (Dye et al. 2015). By allowing cells to endogenously produce cytokines and influence each other, as would happen in an embryo, we believe that complex tissues may be obtained by using fewer exogenous molecules added to the culture, as compared with highly controlled monolayers of a single cell type.

16.5 Characterizing differentiated cells

Now that we have followed our differentiation protocols, how do we know if we were successful? The simplest analyses are mRNA and protein expression studies to determine if our cells express markers consistent with the cell type that we expect and do not express markers of non-related cells. Although straightforward in theory, these analyses can be complicated, because even the most well-established protocols result in enriched, but not pure, populations of the cell type of interest. Contaminating cells that veered off the desired differentiation path are always present in the culture. Thus, the use of positive and negative markers must be interpreted with this in mind. If the desired cells can be selected, for example, by laser capture microdissection or by fluorescence-activated cell sorting (FACS), characterization is more definitive (Pruszak et al. 2007). However, the gold standard of characterization is always the demonstration of a functional cell phenotype, for example, an invoked action potential in a neuron.

16.6 Upcoming challenges

There are many challenges in directed differentiation that will be addressed as research moves forward. These include the development of protocols for three-dimensional differentiation of homogenous cells and increased development of tissues and organoids with multiple cell types arising from different germ layers. These will be greatly enhanced by the use of bioengineered structures. Currently, one major challenge in the development and analysis of tissues generated from PSCs is that they lack an axis of organization. Current protocols that add stepwise cytokines would not address this issue. In the developing embryo, the body axis is organized by cytokine gradients, among other signals. Bioengineering solutions to this challenge will greatly enhance our ability to obtain useful tissues for regeneration research.

Acknowledgments

I would like to thank Minnu Suresh and Marielle Siebert for their editorial assistance. I must also acknowledge all the current and former faculty members at the MBL Frontiers in Stem Cells and Regeneration course for sharing their knowledge and ideas. There are too many to name individually that I would like to acknowledge; however, they have collectively shaped my scientific perspective, in terms of both specific protocols for differentiation and my overall perception of the role of *in vitro* pluripotency research in developmental and regeneration biology.

References

Chen, G., D. R. Gulbranson, Z. Hou, J. M. Bolin, V. Ruotti, M. D. Probasco, K. Smuga-Otto et al. 2011. Chemically defined conditions for human ipsc derivation and culture. *Nat Methods* 8(5): 424–429.

Clark, A. T., M. S. Bodnar, M. Fox, R. T. Rodriquez, M. J. Abeyta, M. T. Firpo, and R. A. Pera. 2004. Spontaneous differentiation of germ cells from human embryonic stem cells *in vitro*. *Hum Mol Genet* 13(7): 727–739.

Dye, B. R., D. R. Hill, M. A. Ferguson, Y. H. Tsai, M. S. Nagy, R. Dyal, J. M. Wells et al. 2015. *In vitro* generation of human pluripotent stem cell derived lung organoids. *Elife* 4.

Green, M. D., A. Chen, M. C. Nostro, S. L. d'Souza, C. Schaniel, I. R. Lemischka, V. Gouon-Evans et al. 2011. Generation of anterior foregut endoderm from human embryonic and induced pluripotent stem cells. *Nat Biotechnol* 29(3): 267–272.

Huang, S. X., M. D. Green, A. T. de Carvalho, M. Mumau, Y. W. Chen, S. L. D'Souza, and H. W. Snoeck. 2015. The *in vitro* generation of lung and airway progenitor cells from human pluripotent stem cells. *Nat Protoc* 10(3): 413–425.

International Stem Cell, Initiative, K. Amps, P. W. Andrews, G. Anyfantis, L. Armstrong, S. Avery, H. Baharvand et al. 2011. Screening ethnically diverse human embryonic stem cells identifies a chromosome 20 minimal amplicon conferring growth advantage. *Nat Biotechnol* 29(12): 1132–1144.

International Stem Cell Initiative, Consortium, V. Akopian, P. W. Andrews, S. Beil, N. Benvenisty, J. Brehm, M. Christie et al. 2010. Comparison of defined culture systems for feeder cell free propagation of human embryonic stem cells. *In vitro Cell Dev Biol Anim* 46(3–4): 247–258.

Irion, S., M. C. Nostro, S. J. Kattman, and G. M. Keller. 2008. Directed differentiation of pluripotent stem cells: From developmental biology to therapeutic applications. *Cold Spring Harb Symp Quant Biol* 73: 101–110.

Lancaster, M. A. and J. A. Knoblich. 2014. Generation of cerebral organoids from human pluripotent stem cells. *Nat Protoc* 9(10): 2329–2340.

Liang, G. and Y. Zhang. 2013. Genetic and epigenetic variations in ipscs: Potential causes and implications for application. *Cell Stem Cell* 13(2): 149–159.

Longmire, T. A., L. Ikonomou, F. Hawkins, C. Christodoulou, Y. Cao, J. C. Jean, L. W. Kwok et al. 2012. Efficient derivation of purified lung and thyroid progenitors from embryonic stem cells. *Cell Stem Cell* 10(4): 398–411.

Maury, Y., J. Come, R. A. Piskorowski, N. Salah-Mohellibi, V. Chevaleyre, M. Peschanski, C. Martinat, and S. Nedelec. 2015. Combinatorial analysis of developmental cues efficiently converts human pluripotent stem cells into multiple neuronal subtypes. *Nat Biotechnol* 33(1): 89–96.

Mitalipova, M., J. Calhoun, S. Shin, D. Wininger, T. Schulz, S. Noggle, A. Venable et al. 2003. Human embryonic stem cell lines derived from discarded embryos. [In eng]. *Stem Cells* 21(5): 521–526.

Mitalipova, M. M., R. R. Rao, D. M. Hoyer, J. A. Johnson, L. F. Meisner, K. L. Jones, S. Dalton, and S. L. Stice. 2005. Preserving the genetic integrity of human embryonic stem cells. [In eng]. *Nat Biotechnol* 23(1): 19–20.

Morrisey, E. E., W. V. Cardoso, R. H. Lane, M. Rabinovitch, S. H. Abman, X. Ai, K. H. Albertine et al. 2013. Molecular determinants of lung development. *Ann Am Thorac Soc* 10(2): S12–S16.

Morrisey, E. E. and B. L. Hogan. 2010. Preparing for the first breath: Genetic and cellular mechanisms in lung development. *Dev Cell* 18(1): 8–23.

Navara, C. S., C. Redinger, J. Mich-Basso, S. Oliver, A. Ben-Yehudah, C. Castro, and C. Simerly. 2007. Derivation and characterization of nonhuman primate embryonic stem cells. [In eng]. *Curr Protoc Stem Cell Biol* Chapter 1: Unit 1A 1.

Pruszak, J., K. C. Sonntag, M. H. Aung, R. Sanchez-Pernaute, and O. Isacson. 2007. Markers and methods for cell sorting of human embryonic stem cell-derived neural cell populations. [In eng]. *Stem Cells* 25(9): 2257–2268.

Rao, M. 2011. Keeping things simple. *Nat Methods* 8(5): 389–390.

Sekiya, S. and A. Suzuki. 2011. Direct conversion of mouse fibroblasts to hepatocyte-like cells by defined factors. *Nature* 475(7356): 390–393.

Son, E. Y., J. K. Ichida, B. J. Wainger, J. S. Toma, V. F. Rafuse, C. J. Woolf, and K. Eggan. 2011. Conversion of mouse and human fibroblasts into functional spinal motor neurons. *Cell Stem Cell* 9(3): 205–218.

Thier, M., P. Worsdorfer, Y. B. Lakes, R. Gorris, S. Herms, T. Opitz, D. Seiferling et al. 2012. Direct conversion of fibroblasts into stably expandable neural stem cells. *Cell Stem Cell* 10(4): 473–479.

Wong, A. P., C. E. Bear, S. Chin, P. Pasceri, T. O. Thompson, L. J. Huan, F. Ratjen, J. Ellis, and J. Rossant. 2012. Directed differentiation of human pluripotent stem cells into mature airway epithelia expressing functional cftr protein. *Nat Biotechnol* 30(9): 876–882.

Wong, A. P. and J. Rossant. 2013. Generation of lung epithelium from pluripotent stem cells. *Curr Pathobiol Rep* 1(2): 137–145.

Yang, R., Y. Zheng, L. Li, S. Liu, M. Burrows, Z. Wei, A. Nace et al. 2014. Direct conversion of mouse and human fibroblasts to functional melanocytes by defined factors. *Nat Commun* 5: 5807.

Chapter 17 Dedifferentiation as a cell source for organ regeneration

José E. García-Arrarás

Contents

Key concepts

a. Many animal species, particularly deuterostomes with high regenerative capacities, undergo cell dedifferentiation before organ regeneration.
b. These dedifferentiated cells usually undergo active proliferation and become the main source of cells for the formation of new organ.
c. Dedifferentiated cells maintain a limited differentiation potential, often giving rise to cells of the same type as those of the original cell.

> d. Dedifferentiation can also be observed at the molecular level, where a limited re-programming of the cell's gene expression takes place.
> e. Maintenance of the differentiated state involves both intracellular and extracellular factors.

17.1 Introduction

In animal species that are able to regenerate organs after injury, the cells for the formation of the new organ originate via one of two different processes: (1) proliferation and differentiation of uni-, pluri-, or multipotent stem cells or (2) the dedifferentiation of mature cells, followed by their proliferation and differentiation. The former will be discussed in depth by other authors, whereas we will focus on the latter. Those readers interested in particular topics that have been reviewed by other authors are directed to some excellent reviews that have appeared in recent years (Tanaka and Reddien 2011, Baddour et al. 2012, Brockes and Gates 2014, Mashanov et al. 2014a, Sanchez Alvarado and Yamanaka 2014, Krause et al. 2015, Takahashi and Yamanaka 2015, Wabik and Jones 2015).

The dedifferentiation process, as defined by Hall (2005), is "a process in which a cell loses its specialized morphology, function and biochemistry to initiate cell division and reverts to a less differentiated state in order to redifferentiate again." Although cellular dedifferentiation has been known to be involved in regeneration processes for several decades, it is only in recent years that it has received serious attention. Part of the problem might have been that the early researches were done in regenerating amphibian limbs, and thus, investigators were reticent about ascribing a similar process to mammals. However, the main problem most likely stems from the fact that cellular dedifferentiation goes against the reigning dogma that views embryological development as a linear, irreversible process, where cell progenitors move from a less differentiated phenotype to a more differentiated phenotype. This implies that cells acquire their morphological, biochemical, and physiological properties in a stepwise manner controlled by the activation of a specific genetic profile. This differentiation pathway is usually accompanied by a loss of cellular plasticity; therefore, as cells acquire a specific phenotype, concomitantly, they are thought to lose the possibility of acquiring other alternative phenotypes (Figure 17.1). An example of this is the cells from the ectoderm embryological layer that can initially give rise to epidermis or nervous system if they take the nervous system route, which, later in development, will become either neurons or glia, and if they choose the neuronal route, later, they will have to choose what type of neuron to form (e.g., cholinergic vs. catecholaminergic). The outcome of the differentiation process is the formation of organs made up of cells that have reached their final differentiated state, which, in many cases, is considered irreversible.

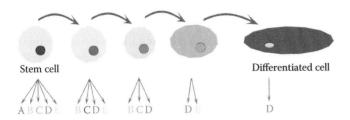

Figure 17.1 Cell differentiation during embryogenesis, or in whole body regeneration, is considered a lineal process, where the original cell (a stem cell or the zygote) is totipotent and can give rise to any cell of the organism. As development or regeneration advances, cells undergo a step-by-step process of differentiation. At the same time, they lose their potential, becoming pluripotent (can give rise to a few cell types) and, at their most mature stage, unipotent, where they are able to give rise to only one cell type.

What is now becoming obvious is that the differentiated state can be (unstable) tenuous, and in some cases, cells can backtrack into less differentiated states (Figure 17.2). Some cancers are example of this dedifferentiation, where the differentiated state of cell is lost and cells revert to *less mature* or *more embryological* phenotypes that are often less restricted in their proliferation or differentiation potential when compared with the mature, differentiated cells that gave rise to these cells (Daley 2008, Yamada et al. 2014, Campos-Sanchez and Corbaleda 2015). In fact, there are many parallelisms between onco-genesis and regenerative processes, but the topic is too extensive to be discussed in this review. For those interested readers, we suggest reviews by Oviedo and Beane (2009), Trusolino (2010), Kalluri and Weinberg 2009, Le Guen (2015), Hong et al. (2014), Matsumoto and Ema (2014), among others.

Figure 17.2 Mature dedifferentiated cells can lose their differentiated state and advance backward to a more *embryological* or *less differentiated* state. It remains unclear as to how far into the dedifferentiated state do cells move and what are the restrictions (if any) in this dedifferentiation process.

375

17.2 Natural dedifferentiation

Cellular dedifferentiation usually takes place during regenerative events in many animals that are known to possess extraordinary regenerative abilities. This is particularly true in the deuterostome group—animals considered higher in the evolutionary scale—which includes vertebrates. Thus, dedifferentiation of mature cells is thought to be the main mechanism by which some organs are able to regenerate in echinoderms, fish, amphibians, and neonate mammals.

The dedifferentiation process as part of a regenerative event was first described during newt limb regeneration (Thornton 1949, 1950, Hay 1959). Multinucleated muscle fibers were observed to form mononucleated cells that became part of the regenerating structure. These observations were done at the light and electron microscopy levels, and it took several additional decades for the phenomenon to be corroborated using other techniques. Extensive studies on limb regeneration have now been done, and these provide strong evidence that not only the muscle cells of the amphibian limb but also other cells such as cartilage, dermis, and glia can dedifferentiate (Brockes 1998, Brockes and Kumar 2002, Brockes and Gates 2014). As part of this work, it was shown by lineage tracing studies that labeled myofibers transplanted from culture to regenerating blastemas formed mononucleated cells that became part of the regenerated limb (Lo et al. 1993, Kumar et al. 2000). In similar experiments in the axolotl tail, when a single muscle fiber was labeled and then followed upon regeneration, the fiber was observed to dedifferentiate after amputation and the dedifferentiated cells proliferated as part of the regenerating blastema (Echeverri and Tanaka 2002).

Amphibians were also used as model systems to study another particular case of dedifferentiation, called *transdifferentiation*. During this process, cells lose their mature phenotype and undergo a somewhat-direct conversion to another phenotype (Tsonis 2002). The best-studied case for transdifferentiation is the regeneration of the lens in the newt's eye (Eguchi 1988, Tsonis 2006). When the lens of an animal is removed, the dorsal iris cells lose their phenotype and form lens cells, without necessarily passing through an *undifferentiated cell* stage. This process can be clearly followed as cells lose their pigmented granules and enter the cell cycle.

Many additional experiments have shown that cellular dedifferentiation is not limited to urodele amphibians. Dedifferentiation has been shown to be the main mechanism during muscle regeneration in echinoderms (Garcia-Arraras and Dolmatov 2010). Dedifferentiation of myocytes and myoepithelial cells in echinoderms is characterized by a simplification of cellular structures and by nuclear activation. The process has been well described at the electron and light microscope levels (Garcia-Arraras and Greenberg 2001, Candelaria et al. 2006, San Miguel and Garcia-Arraras 2007, Garcia-Arraras et al. 2011). Muscle cells have been shown to form spindle-like structures (SLS), where the contractile apparatus of muscle cells is packaged and eliminated into the coelomic cavity and the adjacent connective tissues. The cell nucleus with a small amount of cytoplasm remains. These cells can now undergo proliferation and are eventually incorporated into the regenerated structure, where they eventually redifferentiate (Murray and García-Arrarás 2004).

Cell dedifferentiation has also been shown to take place during certain regenerative events in tunicates, marine chordate invertebrates (Kawamura et al. 2008). Specifically, epicardial cells in tunicates have been shown to discard portions of their cytoplasm during the dedifferentiation process before regeneration (Kawamura et al. 2008).

Zebrafish, which has become a model organism for regeneration studies, also shows that cells can lose their differentiated state, undergo proliferation, and contribute to the regenerated organ. In particular, two cases highlight the process of cell dedifferentiation in zebrafish and its association with regenerative processes, the regeneration of the heart, and the regeneration of the radial fin. Zebrafish can regenerate parts of their heart that have been damaged by injury (Poss et al. 2002). It was shown that after removal of part of the ventricle and before entering the cell cycle, cardiomyocytes close to the injury lost the cell-to-cell contacts and underwent changes in their cytoskeleton, mainly observed by the disassembly of their contractile apparatus before they enter a proliferative state. Once regeneration was completed, the cells redifferentiate to mature cardiomyocytes (Odelberg 2005, Jopling et al. 2010, Kikuchi et al. 2011). In fact, the process of heart regeneration in fish follows a similar pattern to that shown years ago for heart regeneration in newts (Oberpriller and Oberpriller 1974).

In fin regeneration in zebrafish, dedifferentiation of various cell types has been documented (Stewart and Stankunas 2012). These include the osteoblasts, osteoclasts, and scleroclasts (Knopf et al. 2011, Sousa et al. 2011, Stewar and Stankunas 2012). Moreover, recent experiments have shown that such dedifferentiation is not limited to fin regeneration, but it also takes place in other type of injuries such as traumatic injury to the skull (Geurtzen et al. 2014).

Cases of cellular dedifferentiation and reprogramming have also been documented in mammals. Dedifferentiation occurs during the mouse digit-tip regenerative process (Lehoczky et al. 2011 and Rinkevich et al. 2011, 2015). Similarly, during heart regeneration in postnatal mice, cardiomyocytes regress to a more embryonic stage, expressing particular genes, loss of cell adhesion, and partial disassembly of sarcomeres (D'Uva et al. 2015).

Not to convey the idea that dedifferentiation is the sole process for organ regeneration in deuterostomes, there are other examples of alternative processes used. For example, skeletal muscle regenerates from specialized, unipotential cells called *satellite cells* that lie close to the mature muscle fibers (Carlson 2007); deer antler regeneration takes place by the proliferation of a particular cell precursor (Li et al. 2014); and tunicates possess multipotential stem cells that can give rise to complete animals (Rinkevich et al. 1995, 2010, Kürn et al. 2011). A surprising finding was the recent discovery that two amphibian species well known for their regeneration abilities utilize fundamental different regeneration programs for muscle regeneration. While muscle fibers in the newt, as described earlier, undergo dedifferentiation and proliferation to give rise to the new muscle fibers, in the axolotl, muscle progenitors similar to satellite cells form the new muscle fibers, with no contribution from the pre-existing myofibers (Sandoval-Guzman et al. 2014). Even more surprising was the subsequent finding that newts undergo a developmental change in their muscle regeneration mechanisms, where larval newts regenerate their muscle by using stem cells, whereas post-metamorphic newts undergo muscle dedifferentiation to provide the muscle cells for the new limb (Tanaka et al. 2016).

17.3 Formation of stem cells by dedifferentiation

Interestingly, in some tissues, there is a link between cell dedifferentiation and stem cell formation that takes place after massive injury. These cases have been documented in both mammals and insects when toxins and/or genetic tools are used to eliminate the endogenous tissue stem cells. In this scenario, the remaining cells need to be activated to give rise to a new structure.

What follows stem cells' disappearance is that their immediate progeny, which are cells that are already routed in their differentiation process, backtrack and undergo a partial dedifferentiation, themselves becoming the new stem cells and repopulating the tissue by converting to fully functional stem cells. Eliminating the stem cells from the intestinal luminal epithelium showed the best example of this differentiation reversal. In this tissue, all cell types originate from Lgr5-labeled stem cells within the intestinal crypts (Barker et al. 2008). These cells give rise to transiently dividing cells that migrate toward the tip of the intestinal villi or to the base of the crypt and eventually differentiate into 4–5 cell types. The migration to the tip of the villi is in a linear pathway, as cells remain attached to the epithelium basement membrane. However, if the stem cells are depleted by using genetic methods or toxins, then cells that are advanced in their differentiation path reprogram and become the new stem cells (Li and Clevers 2010, Tian et al. 2011, Roth et al. 2012, Tetteh et al. 2015). In other cases, it has been shown that even fully differentiated cells can dedifferentiate and become stem cells (Desai and Krasnow 2013).

A similar process has been documented in the luminal epithelium of the airways (trachea) in mammals on the elimination of the existing stem cells

(Tata et al. 2013). Here, secretory cells show the ability to dedifferentiate and take a step back in their initial differentiation process, becoming the basal stem cells of the trachea. In this particular case, the dedifferentiation capability is restricted to the cells that are not fully mature, and it appears that once full maturation is reached, the cells lose their capacity to become stem cells.

It might be a general process of some animal tissues that partially differentiated cells can backtrack and give rise to stem cells under particular circumstances; this is shown by a similar finding in *Drosophila*, in which, on the elimination of the germ cells, some of their immediate progenies undergo partial dedifferentiation and become the new germ cells of the gonads (Brawley and Matunis 2004). Nonetheless, some tissues, such as the hair follicle, are refractive to this type of dedifferentiation, and once the cells have begun their differentiation process, they cannot be turned back to form new stem cells (Tata et al. 2013).

Summarizing, we observe that dedifferentiation is an important process to form the cells that are needed for organ regeneration. This dedifferentiation can give rise to precursor cells of the organ directly or to organ stem cells that eventually give rise to the organ's cellular component. It is thus plausible that, by activating the dedifferentiation process, organ regeneration can be achieved. In fact, induced dedifferentiation with particular genes is a potential therapy for organ regeneration. One of the best examples of this was shown by inducing dedifferentiation and proliferation in adult mice cardiomyocytes with a constitutively active ERBB2 receptor for neuregulin-1 (D'Uva et al. 2015). The partial dedifferentiation allowed the cells to re-enter the cell cycle and proliferate, producing new cardiomyocytes, a process that does not occur in mature cardiac muscle cells.

17.4 Potency of dedifferentiated cells

The initial view of the dedifferentiation process was that it provided a homogenous pool of undifferentiated cells that formed the blastema, from which the regenerating organ with all its cell types formed. Thus, it was assumed that all cells within the blastema were similar, no matter their origin (e.g., whether they originated from dedifferentiated skeletal muscle or from dedifferentiated cartilage), and that all of these cells were able to give rise to any of the cell types in the regenerated structure. In recent years, experimental results have greatly modified this view by showing that in fish and amphibians, cells in the blastema are not homogenous and that they are lineage-restricted. Most of these studies were done using lineage tracers that could be followed up through dedifferentiation and redifferentiation. For example, in the axolotl, green fluorescent protein (GFP)-labeled cells were tracked after limb transection to show that the dedifferentiated cells are highly restricted in their potential: muscle formed muscle, cartilage formed cartilage, and glia formed glia; only dermal cells appeared to be able to cross lineages and form cartilage (Kragl et al. 2009). Similarly, in zebrafish, different cell types were followed after fin transection and none of the dedifferentiated cells crossed from one lineage to another. This includes studies in which osteoblasts were shown to dedifferentiate to an osteoblast progenitor-like stage that allowed for proliferation and a further increase in the number of osteoblasts (Knopf et al. 2011, Stewart and Stankunas 2012, Geurtzen et al. 2014). Finally, experiments in the skull of zebrafish also showed that mature osteoblasts downregulated the expression of differentiation markers and upregulate pre-osteoblast markers and proliferated (Geurtzen et al. 2014).

The lineage restriction of dedifferentiated cells has also been shown in mammals. In the mouse digit-tip model, the regenerated epidermal cells were found to originate from dedifferentiated keratinocytes, whereas osteoblasts originated from dedifferentiated osteoblasts (Lehoczky et al. 2011, Rinkevich et al. 2011).

The implication of these findings is that the reversion observed during dedifferentiation is not to an embryonic or stem-like state but to a tissue- or cell-specific progenitor state (Sugimoto et al. 2011).

The status of the lineage restriction in non-vertebrate deuterostomes still remains to be determined. As to the potential of the echinoderm dedifferentiated cells, it is clear that in some cases, the cells give rise to the same cell type or a similar derivative of the original cell (e.g., dedifferentiated muscle giving rise to new muscle cells in the intestine and dedifferentiated glia giving rise to glia or neurons in the radial nerve cord) (Garcia-Arraras and Dolmatov 2010, Garcia-Arraras et al. 2011, Mashanov et al. 2013, 2015). However, in other cases, the possibility that they give rise to a broader range of cell types is present. Such is the case of the regeneration of the luminal epithelium of the gut in a sea cucumber model system that originates from dedifferentiated mesothelium. Therefore, in this case, cells that have a mesodermal origin give rise to endoderm derivatives (Mashanov et al. 2005).

Transdifferentiation might be viewed as an exception to the lineage-restriction finding, in view that cells undergo a partial dedifferentiation in order to adopt the destiny of another cell type. However, in these cases, it cannot be excluded that the two cell types share a common lineage or are not that far apart in their embryological lineage. Such is the case of ependymal cells becoming neurons after spinal cord injury in amphibians (Walder et al. 2003), dedifferentiated radial glia becoming neurons in the radial nerve cord of echinoderms (Mashanov et al. 2013), and Muller cells in the retina giving rise to other retinal types (Bernardos et al. 2007, Qin et al. 2009) after some degree of dedifferentiation.

The restrictions in potential of dedifferentiated cells offer both pros and cons as sources of progenitor cells for regenerative medicine and organ engineering. On the one hand, these cells represent partially committed precursors that might lack the plasticity to become cells other than those from where they originated (muscle-dedifferentiated cells will become muscle, cartilage-dedifferentiated cells will become cartilage, and so on). However, on the other hand, these same restrictions might serve to restrain the redifferentiation of cells to specific cells and tissues, eliminating one of the main problems of stem-like cells: the possibility that they differentiate into the wrong cell type or that there is loss of the tissue mechanisms that control replication and differentiation and the cells become teratomas.

17.5 Induced dedifferentiation

It is evident that molecular events, particularly changes in gene expression, must accompany the dramatic morphological and biochemical changes observed during dedifferentiation, as viewed during depigmentation of dorsal iris cells or the cellularization of muscle fibers. Thus, as with many other biological processes, the study of dedifferentiation has extended to the molecular level. It is a common occurrence that during regenerative processes, one observes diminishing expression of certain gene products associated with fully differentiated cells, whereas new gene markers, usually associated with less differentiated cells, appear in the dedifferentiated cells (Sugimoto et al. 2011). Thus, the process of natural cell dedifferentiation can be compared with the process termed *cellular or nuclear reprogramming*, which is associated with changes in gene expression that take place during induced cell dedifferentiation. In its beginning, the induction of dedifferentiation was pioneered by Gurdon's nuclei-transfer experiments, showing that the nuclei of a mature cell could be reprogrammed to an embryological state (Gurdon and Wilmut 2011). Follow-up of these experiments gave rise to the successful cloning of animals by nuclear transfer from adult (differentiated) cells, such as shown by the cloning of the sheep Dolly. These experiments gave rise to studies directed at identifying the factors responsible for the reprogramming response. Once again, the link with cancer is evident, since oncogenesis is viewed as a disease in which cells are reprogrammed to an embryonic-like state (Halley-Stott et al. 2013).

A decade ago, experimental results from Yamanaka's laboratory shook the field of cell biology by demonstrating that fully differentiated mammalian cells could be reprogrammed to become stem cells. In the original experiments, fibroblasts were reprogrammed by the introduction of four transcription factors (Oct3/4, Sox2, c-Myc, and Klf4). The reprogrammed cells were named induced pluripotent stem cells (iPSCs) and were shown to share many characteristics (molecular and physiological) with embryonic stem cells (Takahashi and Yamanaka 2006). The results tore down the dogma that fully differentiated cells had arrived at a final stage that was irreversible and could not go back in their developmental progress to an earlier stage. Moreover, the experiments demonstrated that cellular dedifferentiation (and reprogramming) could be easily induced by the expression of particular genes, in this case, a set of four transcription factors. Since then, the findings have been not only repeated but also expanded to many other cell types and to alternative routes of reprogramming, where cells can be partially dedifferentiated or transdifferentiated to specific cell types by using specific combinations of transcription factors (Vierbuchen and Wernig 2012).

17.6 Molecular basis of dedifferentiation activation and reprogramming

After the discovery of iPSCs, the most obvious gene candidates for inducing dedifferentiation during natural regenerative events were the transcription factors shown to reprogram fibroblasts (and other cells) into iPSCs (Takahashi and Yamanaka 2006) (Figure 17.3). However, in several studies that have been done using different model systems, the results have been disappointingly negative or inconclusive (Maki et al. 2009, Christen et al. 2010, Perry et al. 2013, Bhavsar and Tsonis 2014). For example, there is no corresponding increase in the expression of the Yamanaka pluripotency factors in either lens or limb regeneration in newts (Maki et al. 2009). Two of the factors (Klf4 and Sox2) increase in early stages of lens regeneration and myc increases at later stages. Oct4 was not expressed either before or after lens regeneration (Maki et al. 2009). Moreover, when Oct4 was artificially overexpressed during lens regeneration, the process was inhibited instead of being activated (Bhavsar and Tsonis 2014). Similar results were found in other regenerative models: in zebrafish fin regeneration, none of the factors showed upregulation (Christen et al. 2010). In axolotl limb regeneration, Klf2/4 was highly induced, and a small, but late, increase in c-myc

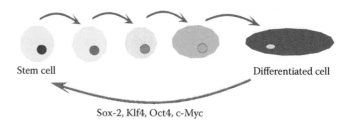

Stem cell   Differentiated cell

Sox-2, Klf4, Oct4, c-Myc

Figure 17.3 Mature differentiated cells, such as fibroblasts, have been shown to be able to reprogram their gene expression and become dedifferentiated pluripotent cells. This process relies on the reprogramming of genetic expression by the addition of four different transcription factors. The induced pluripotent stem cells (iPSCs) can then be made to differentiate into various cell types, depending on the factors added to the cell culture.

was observed; no significant changes were observed in Oct4, Sox2, or Nanog (Knapp et al. 2013). Similarly, in *Holothuria glaberrima*, only one of the Yamanaka factors (cmyc) is overexpressed during intestinal or radial nerve regeneration (Mashanov et al. 2014b, 2015a,b). These studies suggest that the natural induction of dedifferentiation or reprogramming does not follow the same gene activation profile as the artificial induction that leads to the formation of iPSC.

Nonetheless, one should not forget that in many models, there is an underlying expression of some of these factors, even in non-injured animals (Mashanov et al. 2015), suggesting that although the factors might not be overexpressed during regeneration, they might still be needed for the dedifferentiation process to occur. Experiments such as those performed in zebrafish, where Sox2 was knocked down and as a result fin regeneration was inhibited (Christen et al. 2010), provide strong support for this idea. Other experiments also suggest that some of these factors might play additional roles during the regenerative process (Neff et al. 2011).

On the other hand, the finding that the profile of gene expression changes does not correlate with those observed during induced dedifferentiation should not be surprising, particularly when one takes into account that in most animal models where natural dedifferentiation has been studied, the dedifferentiated cells have a restricted potential or plasticity. The dedifferentiated cells are usually restricted in their potential to become cells other than their previous mature selves or at most become cells that lie within the same embryological lineage. Therefore, it should not be expected that the changes in gene expression that lead to a multipotent induced stem cell from a mature differentiated cell should be the same as those that occur in natural dedifferentiation. It could be argued that in the latter, the product of the dedifferentiation process would be a cell with restricted potential, whose main role is to re-enter the cell cycle and form the progenitors for the same cell type within the regenerated tissue or organ (Figure 17.4).

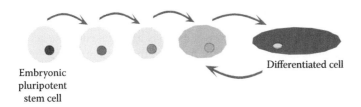

Embryonic
pluripotent
stem cell

Differentiated cell

Figure 17.4 Mature cells that dedifferentiate during regenerative processes appear to have a restricted differentiation potential. Thus, their dedifferentiation does not involve the extensive reprogramming observed in iPSCs but is limited to a late precursor with limited or unipotent capacities.

Other experiments have been done to try to establish the gene activation pathway for cellular dedifferentiation; however, these experiments are confounded by various issues. First, the process of dedifferentiation occurs concomitantly with other cellular processes, making gene expression profiling difficult to tease apart as to which genes are involved in which process. In this respect, wound healing and cell proliferation processes, which are usually parts of the regenerative response, also cause large changes in gene expression, which are difficult to separate from the dedifferentiation event. In other words, many genes have been associated with the regeneration process; however, few genes, if any, have been shown to be directly associated with the cellular dedifferentiation that takes place during regeneration. For example, expression of Msx1 was initially correlated to the muscle cellularization observed in regenerating axolotl limbs (Kumar et al. 2004). However, further experiments where the expression of Msx1 was modulated concluded that there was little probability of Msx1 involvement in muscle fragmentation during regeneration (Schnapp and Tanaka 2005). Second, it has been difficult to find a common thread in different experimental data using various animal models, different types of dedifferentiating cells, and a focus on different gene candidates. Third, most of the evidence that has been found is directed mainly at the signaling pathways that activate the dedifferentiation process rather than at the intracellular gene expression pathways that are responsible for the process to take place. Nonetheless, it is important to go over the available information as an introduction to future experiments—to see what is known and what remains to be described.

17.6.1 Amphibians

The molecular basis of natural cell dedifferentiation during regeneration has been mostly studied using the urodele (newt) limb regeneration model. In this system, dedifferentiation is closely associated with cell proliferation. Therefore, as muscle fibers undergo *cellularization*, where the multinucleated skeletal muscle fibers fragment into mononucleated cells, these mononucleated cells also re-enter the cell cycle (Lo et al. 1993, Brockes 1998). However, experimental evidence suggests that cellular dedifferentiation can take place in the absence of proliferation activation (see review by Straube and Tanaka 2006).

Dedifferentiation of muscle myotubes has been linked to the activation of thrombin, suggesting that the presence of thrombin during the initial stage of wound healing is the detonator for the cascade of events, in particular cellular dedifferentiation, of the regenerative process (Tanaka et al. 1999). Further experiment showed that thrombin proteolysis acts on a serum factor, which then activates and induces cell dedifferentiation and proliferation (Echeverri and Tanaka 2002). Along these lines, it was reported that muscle cell re-entry into the cell cycle was activated by some unknown factor that is probably available or released in the blastema (Brockes 1998). This re-entry has been duplicated in culture conditions by a serum factor. Thrombin activation has also been shown to be of critical importance in lens regeneration (Imokawas et al. 2003, 2004, Godwin et al. 2010).

17.6.2 Zebrafish

It has been suggested that the difference between the regenerating zebrafish heart cells and those of mammals that are reactive to regeneration is the capacity of adult cells to proliferate and that this might be due to the presence of a mitogen-activated protein (MAP) kinase that is activated after injury in the former (Knapp and Tanaka 2013).

17.6.3 Tunicates

Several molecules have been associated with regeneration in tunicates; however, it is difficult to determine whether these molecules have a direct effect on the dedifferentiation process. Most interesting is the finding of a serine protease (tunicate retinoic acid–inducible modular protease [TRAMP]) that is produced by the coelomic cells and that induces dedifferentiation in the atrial epithelium (Hara et al. 1992, Kawamura et al. 2008).

It should be noted that thrombin is also a serine protease. Moreover, in echinoderms, our group has shown that serine proteases might play a role in the intestinal regenerative process (Pasten et al. 2012). Thus, there is a growing evidence that serine proteases might be involved in the initial regenerative events associated with several animal species. Now, whether this role is specific to the dedifferentiation process or to other processes during regeneration remains to be determined.

17.7 Role of extracellular matrix in maintenance of the differentiated state

Although many extracellular molecules such as hormones and growth factors or cell contacts can affect the dedifferentiation/differentiation status of a cell, there is a component that is usually overseen or little studied: the extracellular matrix (ECM). The ECM is a dynamic network made up of molecules secreted by adjacent cells. During embryonic development, the ECM is known to provide instructive cues that help cells attain and maintain their differentiated status (Lin and Bissell 1993, Hansen et al. 2015). Similarly, it has been shown that during regenerative events, the ECM undergoes dramatic changes in its content and organization. These changes in ECM are thought to be critical for creating the tissue environmental conditions that promote the regeneration process. In fact, activation of matrix metalloproteinases, the enzymes responsible for the ECM remodeling, is a common event that follows injury in animals with large regenerative abilities, as has been shown during limb regeneration in amphibians and intestinal regeneration in echinoderms (Park and Kim 1998, Yang et al. 1999, Quiñones et al. 2002, Bellayr et al. 2009).

The role of the ECM in dedifferentiation has been probed directly in *in vitro* studies (Calve et al. 2010). By growing cells in different ECM components, these investigators showed that events associated with dedifferentiation, such as myotube fragmentation and DNA synthesis, occurred more often in ECM components, such as hyaluronic acid, which are upregulated during regeneration, than in components such as collagen, which are associated with the mature (uninjured) organ. Moreover, they also showed that the ECM components that mediate the regeneration of the limb blastema had similar effects on mammalian cells, suggesting that the information coded by these molecules can also direct mammalian regenerative processes (Calve et al. 2010).

The critical information provided by the molecules in the ECM in the differentiation state of cells is best ascertained in experiments where organs are decellularized and then re-seeded with isolated cells (Swinehart and Badylak 2016, Katari et al. 2014, Scarritt et al. 2015, Swinehart et al. 2015). In these experiments, all the cellular components are eliminated and what is left is a scaffold of the original organ made up by the ECM and cellular debris. If this organ scaffold is seeded with isolated cells, the cells can re-localize within the tissue, form a cellularized structure, and function as if in the normal organ (Soto-Gutierrez et al. 2010). Thus, the ECM scaffold helps maintain the differentiation status of the cells. Liver cells (hepatocytes) exemplify the importance of the ECM (Baptista et al. 2014, Rudnick 2014). When isolated from the organ and cultured *in vitro*, hepatocytes soon loose morphological and biochemical properties that correspond to the liver functions (Fujita et al. 1986). The ECM of the cultures can be manipulated to increase the expression of liver-specific molecules. Moreover, if hepatocytes are seeded into a decellularized liver scaffold, they are associated with the ECM and maintain their normal physiological functions (Baptista et al. 2014).

Recent experiments have shown that in some cases, when stem cells or iPSCs or other type of undifferentiated cells are perfused into this scaffold, the remaining ECM directs their differentiation toward the cell types found in the original organ (Zhou et al. 2014, Hoshiba et al. 2016).

The message that arises from these experiments is that the ECM must be taken into account in any manipulation that aims at restoring a tissue or organ to its original function. Moreover, changes that occur in the ECM influence the process of dedifferentiation and differentiation. Therefore, the ECM plays important roles in moving cells toward a particular differentiated phenotype, maintaining cells in a specific differentiated phenotype or inducing dedifferentiation in a specific cellular population.

17.8 Epithelial–mesenchymal transition of dedifferentiated cells

The morphological changes that take place during dedifferentiation are usually accompanied by alterations in cellular organization. In many cases, in addition to the morphological changes, cellular dedifferentiation has been associated with a decrease in cell adherence.

In some cases, the changes are rather dramatic, and cells might leave their tissue of origin and incorporate or participate in the formation of a new tissue. When the cellular changes include cells leaving an epithelial tissue, losing their attachment to the basement membrane, and passing to an adjacent connective tissue, the process is termed an epithelial–mesenchymal transition (EMT).

We have documented this process during intestinal regeneration in the holothurian *H. glaberrima* (Garcia-Arraras et al. 1998, 2011, Garcia-Arraras and Greenberg 2001, Candelaria et al. 2006, Mashanov and Garcia-Arraras 2011). In this case, some of the epithelial cells that surround the connective tissue of the regenerating intestinal rudiment and that probably originate from the dedifferentiated cells in the mesentery lose their attachment to other epithelial cells and ingress into the underlying connective tissue. The cells then acquire a mesenchymal morphology and become part of the new intestinal submucosa. The EMT after injury has also been documented in various vertebrates tissues, including mammalian heart (Smart et al. 2011, Zhou and Pu 2011), kidney (Swetha et al. 2011, Johansson 2014), and lung (Kalluri and Weinberg 2009, Moheimani et al. 2015), among others. The process has also been documented in non-mammalian vertebrates. For example, in zebrafish heart regeneration, the cells of the endocardium dedifferentiate, separate from other epithelial cells, and migrate to form new cardiomyocytes (Lepilina et al. 2006). The EMT is also observed during zebrafish fin regeneration, where epithelial cells enter the blastema and form part of the blastemal cell population (Stewart and Stankunas 2012). The EMT mechanism associated with regeneration appears to be well conserved during evolution since recent studies describe EMT mechanisms in sponges, a basal group in the animal evolutionary tree. (Borisenko et al. 2015).

We have previously conjectured that EMT plays an important role during visceral regeneration, which takes place within the animal cavities that are sterile, and thus, the overlying epithelial tissues can provide cells to form the new organ (Garcia-Arraras et al. 2011). In contrast, the epithelial tissues (skin) of limbs and other appendages that are in contact with the outside environment are not amenable for the EMT in view that they might introduce bacteria or other pathogenic organisms from the environment.

17.9 Concluding remarks in the application of dedifferentiation to regenerative engineering

It is evident from the information provided earlier that animals that show strong regeneration potential rely on cellular dedifferentiation as a mechanism to provide the cells for the new organ or tissue. It can be said that in natural regeneration, at least in those cases that have been described up to present, the differentiation process takes the cell to the state of a tissue progenitor with limited differentiation potential. If one could harness this dedifferentiation mechanism to be used in animals with less regeneration potential, then dedifferentiation could provide a source of cells for regenerative engineering of specific organs. This process has been named *reverse engineering a regeneration blastema* (Gardiner et al. 2013). For this, mature cells of the organ to be regenerated would be induced to dedifferentiate and proliferate. Following this, they would be used to form a regeneration blastema composed of the dedifferentiated cell types of tissues that make the original organ.

There are possible additional advantages to this approach. First, these cells would originate from the same individual in whom the organ will be regenerated, thus eliminating possible problems of immune rejection. In addition, dedifferentiated cells would probably be more amenable to the same proliferation and differentiation controls as the *normal* cells of the organisms, thus diminishing the possibility of uncontrolled growth or oncogenetic proliferation. Nevertheless, to achieve these goals, it is first necessary to find the answers to various questions:

First, what are the injury-activated mechanisms that trigger the dedifferentiation response?

Second, what intracellular signaling pathways and the modulation of which genes are involved in the dedifferentiation process?

Third, what are the factors that play key roles in the restriction of plasticity and inducing or maintaining lineage restriction?

Fourth, what molecules direct the progenitor/dedifferentiated cells toward the correct differentiation within the regenerated organ?

Deciphering these and other events will be key to the possibility of engineering organs from cells and tissues.

References

Baddour JA, Sousounis K, Tsonis PA (2012) Organ repair and regeneration: An overview. *Birth Def Res* (Part C) 96:1–29.

Baptista PM, Moran E, Vyas D, Shupe T, Soker S (2014) Liver regeneration and bioengineering: The role of liver extra-cellular matrix and human stem/progenitor cells. In: *Regenerative Medicine Applications in Organ Transplantation*. Orlando G, Lerut J, Soker S, Stratta RJ (eds.), pp. 391–400, Elsevier, Amsterdam, the Netherlands.

Barker N, vande Wetering M, Clevers H (2008) The intestinal stem cell. *Genes Dev* 22:1856–1864.

Bellayr IH, Mu X, Li Y (2009) Biochemical insights into the role of matrix metallo-protienases in regeneration: Challenges and recent developments. *Future Med Chem* 1:1095–1111.

Bernardos RL, Barthel LK, Meyers JR, Raymond PA (2007) Late-stage neuronal progenitors in the retina are radial Müller glia that function as retinal stem cells. *J Neurosci* 27:7028–7040.

Bhavsar RA, Tosnis PA (2014) Exogenous Oct-4 inhibits lens transdifferentiation in the newt *Notophthalmus viridescens*. *PLoS One* 9:e102510.

Borisenko IE, Adamska M, Tokina DB, Ereskovsky AV (2015) Transdifferentiation is a driving force of regeneration in *Halisarca dujardini* (Demospongiae, Porifera). *Peer J* 3:e1211.

Brawley C, Matunis E (2004) Regeneration of male germline stem cells by sper-matogonial dedifferentiation in vivo. *Science* 304:1331–1334.

Brockes JP (1998) Progenitor cells for regeneration: Origin by reversal of the dif-ferentiated state. In *Cellular and Molecular Basis of Regeneration*. Ferreti P, Géraudie J (eds.), pp. 63–77, John Wiley & Sons, West Sussex, UK.

Brockes JP, Gates PB (2014) Mechanisms underlying vertebrate limb regeneration: Lessons from the salamander. *Bichem Soc Trans* 42:625–630.

Brockes JP, Kumar A (2002) Plasticity and reprogramming of differentiated cells in amphibian regeneration. *Nat Rev Mol Cell Biol* 3:566–574.

Calve S, Odelberg SJ, Simon H-G (2010) A transitional extracelular matrix instructs cell behavior during muscle regeneration. *Dev Biol* 344:259–271.

Campos-Sánchez E, Cobaleda C (2015) Tumoral reprogramming: Plasticity takes a walk on the wild side. *Biochim Byophys Acta* 1849:436–447.

Candelaria AG, Murray G, File SK, García-Arrarás JE (2006) Contribution of mes-enterial muscle de-differentiation to intestine regeneration in the sea cucumber *Holothuria glaberrima*. *Cell Tissue Res* 325:55–65.

Carlson BM (2007) *Principles of Regenerative Biology*. Elsevier, Amsterdam, the Netherlands.

Christen B, Robles V, Raya M, Paramonov I, Izpisua Belmonte JC (2010) Regeneration and reprogramming compared. *BMC Biology* 8:5.

Daley GQ (2008) Common themes of dedifferentiation in somatic cell reprogram-ming and cancer. *Cold Spring Harb Symp Quant Biol* 73:171–174.

Desai TJ, Krasnow MA (2013) Differentiated cells in a back-up role. *Nature* 503:204–205.

D'Uva G, Aharonov A, Lauriola M, Kain D, Yahalom-Ronen Y et al. (2015) ERBB2 triggers mammalian heart regeneration by promoting cardiomyocyte dedif-ferentiation an proliferation. *Nature Cell Biol* 17:627–638.

Echeverri K, Tanaka EM (2002) Mechanisms of muscle dedifferentiation during regeneration. *Semin Cell Dev Biol* 13:353–360.

Eguchi G (1988) Cellular and molecular background of wolffian lens regeneration. *Cell Differ Dev* 25 Suppl:147–158.

Faulk DM, Badylak SF (2014) Natural biomaterials for regenerative medicine appli-cations. In: *Regenerative Medicine Applications in Organ Transplantation*. Orlando G, Lerut J, Soker S, Stratta RJ (eds.), pp. 101–112, Elsevier, Amsterdam, the Netherlands.

Fujita M, Spray DC, Choi H, Saez J, Jefferson DM, Hertzberg E, Rosenberg LC, Reid LM (1986) Extracellular matrix regulation of cell-cell communication and tissue-specific gene expression in primary liver cultures. *Prog Clin Biol Res* 226:333–360.

Garcia-Arrarás JE, Estrada-Rodgers L, Santiago R, Torres II, Díaz-Miranda L, Torres-Avillán I (1998) Cellular mechanisms of intestine regeneration in the sea cucumber, *Holothuria glaberrima*. *J Exp Zool* 281:288–304.

389

Garcia-Arrarás JE, Dolmatov IY (2010) Echinoderms; potential model systems for studies on muscle regeneration. *Curr Pharm Design* 16:942–955.

Garcia-Arrarás JE, Greenberg MJ (2001) Visceral regeneration in holothurians. *Microsc Res Tech* 55:438–451.

Garcia-Arrarás JE, Valentín G, Flores J, Rosa R, Rivera-Cruz A, San Miguel-Ruiz JE, Tossas K (2011) Cell de-differentiation and epithelial to mesenchymal transitions during intestinal regeneration in *H. glaberrima. BMC Dev Biol* 11:61.

Gardiner DM, Bryant SV, Muneoka K (2013) Engineering limb regeneration; Lessons from animals that can regenerate. In: *Regenerative Engineering.* Laurencin CT, Kahn Y (eds.), pp. 387–404, CRC Press, Boca Raton, FL.

Geurtzen K, Knopf F, Wehner D, Huitema LFA, Schulte-Merker S, Weidinger G (2014) Mature osteoblasts dedifferentiate in response to traumatic injury in the zebrafish fin and skull. *Stem Cells Reg* 141:2225–2234.

Godwin JW, Liem KF, Brockes JP (2010) Tissue factor expression in newt iris coincides with thrombin activation and lens regeneration. *Mech Dev* 127:321–328.

Gurdon JB, Wilmut I (2011) Nuclear transfer to eggs and oocytes. *Cold Spring Harb Perspect Biol* 3:pii:a002659.

Hall BK (2005) *Bones and Cartilage: Developmental and Evolutionary Skeletal Biology.* San Diego, CA: Academic Press, p. 166.

Halley-Stott RP, Pasque V, Gurdon JB (2013) Nuclear reprogramming. *Development* 140:2468–2471.

Hansen NUB, Genovese F, Leeming DJ, Karsdal MA (2015) The importance of extracellular matrix for cell function and in vivo likeness. *Exp Mol Pathol* 98:286–294.

Hara K, Fujiwara S, Kawamura K (1992) Retinoic acid can induce a secondary axis in developing buds of a colonial ascidian, *Polyandrocarpa misakiensis. Dev Growth Differ* 34:437–445.

Hay ED (1959) Microscopic observations of muscle dedifferentiation in regenerating *Amblystoma* limbs. *Dev Biol* 1:555–585.

Hong L, Cai Y, Jiang M, Zhou D, Chen L (2014) The Hippo signaling pathway in liver regeneration and tumorigenesis. *Acta Biochim Biophys Sin* (Shanghai) 47:46–52.

Hoshiba T, Chen G, Endo C, Maruyama H, Wakui M, Nemoto E, Kawazoe N, Tanake M (2016) Decellularized extracelular matrix as an in vitro model to study the comprehensive roles of the ECM in stem cell differentiation. *Stem Cells Int* 2016;6397820.

Imokawas Y, Brockes J (2003) Selective activation of thrombin is a critical determinant for vertebrate lens regeneration. *Curr Biol* 13:877–881.

Imokawas Y, Simon A, Brockes J (2004) A critical role for thrombin in vertebrate lens regeneration. *Phil Trans R Soc Lond B* 359:765–776.

Johansson ME (2014) Tubular regeneration: When can the kidney regenerate from injury and what turns failure into success? *Nephron Exp Nephrol* 126:76–81.

Jopling C, Sleep E, Raya M, Marti M, Raya A, Izpisua-Belmonte JC (2010) Zebrafish heart regeneration occurs by cardiomyocyte dedifferentiation and proliferation. *Nature* 464:606–609.

Kalluri R, Weinberg RA (2009) The basics of epithelial-mesenchymal transition. *J Clin Inv* 119:1420–1428.

Katari R, Peloso A, Zambon JP, Soker S, Stratta RJ, Atala A, Orlando G (2014) Renal bioengineering with scaffolds generated from human kidneys. *Nephron Exp Nephrol* 126:119.

Kawamura K, Sugino Y, Sunanaga T, Fujiwara S (2008) Multipotent epithelial cells in the process of regeneration and asexual reproduction in colonial tunicates. *Dev Growth Differ* 50:1–11.

Kikuchi K, Holdway JE, Major RJ, Blum N, Dahn RD, Begemann G, Poss KD (2011) Retinoic acid production by endocardium and epicardium is an injury response essential for zebrafish heart regeneration. *Dev Cell* 20:397–404.

Knapp D, Tanaka E (2012) Regeneration and reprogramming. *Curr Opin Genet Dev* 22:485–493.

Knapp D, Schulz H, Rascon CA, Volkmer M, Scholz J et al. (2013) Comparative transcriptional profiling of the axolotl limb identifies a tripartite regeneration-specific gene program. *PLoS ONE* 8(5):e61352.

Knopf F, Hammond C, Chekuru A, Kurth T, Hans S, Weber CW, Mahatma G, Fisher S, Brand M, Schulte-Merker S, Weidinger G (2011) Bone regenerates via dedifferentiation of osteoblasts in the zebrafish fin. *Dev Cell* 20:713–724.

Kragl M, Knapp D, Nacu E, Khattak S, Maden M, Epperlein HH, Tanaka EM (2009) Cells keep a memory of their tissue origin during axolotl limb regeneration. *Nature* 460:60–67.

Krause MN, Sancho-Martinez I, Izpisua Belmonte JC (2015) Understanding the molecular mechanisms of reprogramming. *Biochem Biophys Res Commun* 473:693–697.

Kumar A, Velloso CP, Imokawa Y, Brockes JP (2000) Plasticity of retrovirus-labelled myotubes in the newt limb regeneration blastema. *Dev Biol* 218:125–136.

Kumar A, Velloso CP, Imokawa Y, Brockes JP (2004) The regenerative plasticity of isolated urodele myofibers and its dependence on MSX1. *PLoS Biol* 2:E218.

Kürn U, Rendulic S, Tiozzo S, Lauzon RJ (2011) Asexual propagation and regeneration in colonial ascidians. *Biol Bull* 221:43–61.

Le Guen L, Marchal S, Faure S, de Santa Barbara P (2015) Mesenchymal-epithelial interactions during digestive tract development and epithelial stem cell regeneration. *Cell Mol Life Sci* 72:3883–3896.

Lehoczky JA, Robert B, Tabina CJ (2011) Mouse digit tip regeneration is mediated by fate-restricted progenitor cells. *PNAS* 108:20609–20614.

Lepilina A, Coon AN, Kikuchi K, Holdway JE, Roberts RW, Burns CG, Poss KD (2006) A dynamic epicardial injury response supports progenitor cell activity during zebrafish heart regeneration. *Cell* 127:607–19.

Li C, Zhao H, Liu Z, McMahon C (2014) Deer antler- a novel model for studying organ regeneration in mammals. *Int J Biochem Biol* 56:111–122.

Li L, Clevers H. (2010) Coexistence of quiescent and active adult stem cells in mammals. *Science* 327:542–545.

Lin CQ, Bissell MJ (1993) Multifaceted regulation of cell differentiation by extracellular matrix. *FASEB J* 7:737–743.

Lo DC, Allen F, Brockes JP (1993) Reversal of muscle differentiation during urodele limb regeneration. *Proc Natl Acad Sci U S A* 90:7230–7234.

Maki N, Suetsugu-Maki R, Tarui H, Agata K, Del Rio-Tsonis K et al. (2009) Expression of stem cell pluripotency factors during regeneration in newts. *Dev Dyn* 238:1613–1616.

Mashanov VS, Dolmatov IY, Heinzeller T (2005) Transdifferentiation in holothurian gut regeneration. *Biol Bull* 209:184–193.

Mashanov VS, Zueva OR, García-Arrarás JE (2013) Echinoderms employ phylogenetically conserved mechanisms of neural regeneration through activation of the radial glia. *BMC Biol* 11:49.

Mashanov VS, García-Arrarás JE (2011) Gut regeneration in holothurians: A snapshot of recent developments. *Biol Bull* 221:93–109.

Mashanov VS, Zueva OR, Garcia-Arraras JE (2014a) Post-embryonic visceral organogenesis: Why does it occur in worms and sea cucumbers and fail in humans? *Curr Topics Dev Biol* 108:185–215.

Mashanov VS, Zueva OR, Garcia-Arraras JE (2014b) Transcriptomic changes during regeneration of the central nervous system in an echinoderm. *BMC Genomics* 15:357.

Mashanov VS, Zueva OR, Garcia-Arraras JE (2015a) Myc regulates programmed cell death and radial glia dedifferentiation after neural injury in an echinoderm. *BMC Dev Biol* 15:24.

Mashanov VS, Zueva OR, Garcia-Arraras JE (2015b) Expression of pluripotency factors in echinoderm regeneration. *Cell Tissue Res* 359:521–536.

Matsumoto K, Ema M (2014) Roles of VEGF-A signaling in development, regeneration, and tumours. *J Biochem* 156(1):1–10.

Moheimani F, Roth HM, Cross J, Reid AT, Shaheen F et al. (2015) Disruption of β-catenin/CBP signaling inhibits human airway epithelial-mesenchymal transition and repair. *Int J Biochem Cell Biol* 68:59–69.

Murray G, García-Arrarás JE (2004) Myogenesis during holothurian intestinal regeneration. *Cell Tissue Res* 318:515–524.

Neff AW, King MW, Mescher AL (2011) Dedifferentiation and the role of Sall4 in reprogramming and patterning during amphibian limb regeneration. *Dev Dyn* 240:979–989.

Oberpriller JO, Oberpriller JC (1974) Response of the adult newt ventricle to injury. *J Exp Zool* 187:249–253.

Odelberg S (2005) Cellular plasticity in vertebrate regeneration. *Anat Rec* 287B:25–35.

Oviedo NJ, Beane WS (2009) Regeneration: The origin of cancer or a possible cure? *Semin Cell Dev Biol* 20:557–564.

Park IS, Kim WS (1998) Modulation of gelatinase activity correlates with the dedifferentiation profile of regenerating salamander limbs. *Mol Cells* 9:119–126.

Pasten C, Rosa R, Ortiz S, García-Arrarás JE (2012) Characterization of proteolytic activities during intestinal regeneration of the sea cucumber, *Holothuria glaberrima*. *Intl J Dev Biol* 56:681–691.

Poss KD, Wilson LG, Keating MT (2002) Heart regeneration in zebrafish. *Science* 298:2188–2190.

Qin Z, Barthel LK, Raymond PA (2009) Genetic evidence for shared mechanisms of epimorphic regeneration in zebrafish. *PNAS* 106:9310–9315.

Quiñones JL, Rosa R, Ruiz DL, García-Arrarás JE (2002) Extracellular matrix remodeling involvement during intestine regeneration in the sea cucumber *Holothuria glaberrima*. *Dev Biol* 250:181–197.

Rinkevich B, Shlemberg Z, Fishelson L. (1995) Whole-body protochordate regeneration from totipotent blood cells. *Proc Natl Acad Sci USA* 92:7695–7699.

Rinkevich Y, Lindau P, Ueno H, Longaker MT, Weissman IL (2011) Germ-layer and lineage-restricted stem/progenitors regenerate the mouse digit tip. *Nature* 476:409–413.

Rinkevich Y, Maan ZN, Walmsely G, Sen SK (2015) Injuries to appendage extremities and digit tips: A clinical and cellular update. *Dev Dyn* 244:641–650.

Rinkevich Y, Rosner A, Rabinowitz C, Lapidot Z, Moiseeva E, Rinkevich B (2010) Piwi positive cells that line the vasculature epithelium, underlie whole body regeneration in a basal chordate. *Dev Biol* 345:94–104.

Roth S, Franken P, Sacchetti A, Kremer A, Anderson K, Sansom O, Fodde R (2012) Paneth cells in intestinal homeostasis and tissue injury. *PLoS One* 7:e38965.

Rudnick DA (2014) Liver regeneration: The developmental biology approach. In: *Regenerative Medicine Applications in Organ Transplantation*. Orlando G, Lerut J, Soker S, Stratta RJ (eds.), pp. 353–374, Elsevier, Amsterdam, the Netherlands.

Sanchez Alvarado A, Yamanaka S (2014) Rethinking differentiation. *Cell* 157:110–119.

San Miguel-Ruiz JE, García-Arrarás JE (2007) Common cellular events occur during wound healing and organ regeneration in the sea cucumber *Holothuria glaberrima*. *BMC Dev Biol* 7:115.

Sandoval-Guzman T, Wng H, Khattak S, Schuez M, Roensch K, Nacu E, Tazaki A, Joven A, Tanaka EM, Simon A (2014) Fundamental differences in dedifferentiation and stem cell recruitment during skeletal muscle regeneration in two salamander species. *Cell Stem Cell* 14:174–187.

Scarritt ME, Pashos NC, Bunnell BA (2015) A review of cellularization strategies for tissue engineering of whole organs. *Front Bioeng Biotech* 3:1–17.

Schnapp E, Tanaka EM (2005) Quantitative evaluation of morpholino-mediated protein knock-down of GFP, MSx1 and Pax6 during tail regeneration in *Ambystoma mexicanum*. *Dev Dyn* 232:162–170.

Smart N, Bollini S, Dube KN, Vieira JM, Zhou B et al. (2011) *De novo* cardiomyocytes from within the activated adult heart after injury *Nature* 474:640–644.

Soto-Gutierrez A, Yagi H, Uygun BE, Navarro-Alvarez N, Uygun K, Kobayashi N, Yang YG, Yarmush ML (2010) Cell delivery: From cell transplantation to organ engineering. *Cell Transpl* 19:655–665.

Sousa S, Alfonso N, Bensimon-Brito A, Fonseca M, Simoes M, Leon J, Roehl H, Cancela ML, Jacinto A (2011) Differential skeletal cells contribute to blastema formation during zebrafish fin regeneration. *Development* 138:3897–3905.

Stewart S, Stankunas K (2012) Limited dedifferentiation provides replacement tissue during zebrafish fin regeneration. *Dev Biol* 365:339–349.

Straube WL, Tanaka EM (2006) Reversibility of the differentiated state: Regeneration in amphibians. *Artif Organs* 30:743–755.

Sugimoto K, Gordon SP, Meyerowitz EM (2011) Regeneration in plants and animals: Dedifferentiation, transdifferentiation, or just differentiation. *Trends Cell Biol* 21:212–28.

Swetha G, Chandra V, Phadnis S, Bhonde R (2011) Glomerular parietal epithelial cells of adult murine kidney undergo EMT to generate cells with traits of renal progenitors. *J Cell Mol Med* 15:396–413.

Swinehart IT, Badylak SF (2016) Extracellular matrix bioscaffolds in tissue remodeling and morphogenesis. *Dev Dyn* 245:351–360.

Takahashi K, Yamanaka S (2006) Induction of pluripotent stem cells from mouse embryonic and adult fibroblast cultures by defined factors. *Cell* 126:663–676.

Takahashi K, Yamanaka S (2015) A developmental framework for induced pluripotency. *Development* 142(19):3274–3285.

Tanaka EM, Drechsel DN, Brockes JP (1999) Thrombin regulates S-phase re-entry by cultured newt myocytes. *Curr Biol* 9:792–799.

Tanaka E, Reddien PW (2011) The cellular basis for animal regeneration. *Dev Cell* 21:172–185.

Tanaka HV, Ng NCY, Yu ZY, Casco-Robles MM, Maruo F, Tsonis PA, Chiba C (2016) A developmentally regulated switch from stem cells to dedifferentiation for limb muscle regeneration in newts. *Nat Commun* 7:11069.

Tata PR, Mou H, Pardo-Saganta A, Zhao R, Prabhu M et al. (2013) Dedifferentiation of committed epithelial cells into stem cells in vivo. *Nature* 503:218–223.

Tetteh PW, Farin HF, Clevers H (2015) Plasticity within stem cell hierarchies in mammalian epithelia. *Trends Cell Biol* 25:100–108.

Thornton CS (1949) Beryllium inhibition of regeneration. I. Morphological effects of beryllium on amputated forelimbs of larval *Ambystoma*. *J Morphol* 84:459–494.

Thornton CS (1950) Beryllium inhibition of regeneration. II. Localization of beryllium effect in amputated limbs of larval *Ambystoma*. *J Exp Zool* 114: 305–333.

Tian H, Biehs B, Warming S, Leong KG, Rangell L, Klein OD, de Sauvage FJ (2011) A reserve stem cell population in small intestine renders Lgr5-positive cells dispensable. *Nature* 478:255–259.

Trusolino L1, Bertotti A, Comoglio PM. (2010) MET signaling: Principles and functions in development, organ regeneration and cancer. *Nat Rev Mol Cell Biol* 11:834–848.

Tsonis PA (2002) Regenerative biology: The emerging field of tissue repair and restoration. *Differentiation* 70:397–409.

Tsonis PA (2006) How to build and rebuild a lens. *J Anat* 209:433–437.

Vierbuchen T, Wernig M (2012) Molecular roadblocks for cellular reprogramming. *Mol Cell* 47:827–838.

Wabik A, Jones PH (2015) Switching roles: The functional plasticity of adult tissue stem cells. *EMBO J* 34:1164–1179.

Walder S, Zhang F, Ferretti P (2003) Up-regulation of neural stem cell markers suggests the occurrence of dedifferentiation in regenerating spinal cord. *Dev Genes Evol* 213:625–630.

Yamada Y, Haga H, Yamada Y (2014) Concise review: Dedifferentiation meets cancer development: Proof of concept for epigenetic cancer. *Stem Cell Transl Med* 3:1182–1187.

Yang EV, Gardiner DM, Carlson MRJ, Nugas CA, Bryant SV (1999) Expression of MMP-9 and related matrix metalloproteinase genes during axolotl limb regeneration. *Dev Dyn* 216:2–9.

Zhou B, Pu WT (2011) Epicardial epithelial-to-mesenchymal transition in injured heart. *J Cell Mol Med* 15:2781–2783.

Zhou Q, Ye X, Sun R, Mastsumoto Y, Moriyama M, Asano Y, Ajioka Y, Saijo Y (2014) Differentiation of mouse induced pluripotent stem cells into alveolar epithelial cells in vitro for use in vivo. *Stem Cells Translat Med* 3:675–685.

Chapter 18 Epigenetic control of cell fate and behavior

Cristian Aguilar

Contents

Key concepts

  a. The initiation of regeneration involves a rapid change in the potency and behavior of cells located at the site of injury.

  b. Mechanisms that are able to affect multiple signaling pathways at once are likely needed to illicit large-scale and rapid changes in gene expression.

  c. The regenerative environment is heterogeneous and is composed of multiple tissues with varying involvement in healing and regeneration.

  d. Therapies must be able to target specific tissues in a directed fashion, within the context of natural and functional relationships with surrounding tissues.

18.1 Introduction

The process of regeneration involves a rapid and large-scale change in the behavior of cells located at the site of injury or tissue loss. Although much focus has been placed on identifying specific signaling factors responsible for initiating and controlling regeneration, the complexity of regeneration as a cellular program likely necessitates the use of mechanisms that can broadly and quickly affect states of cell fate and behavior. Epigenetic mechanisms may fill this role, as recent inquiries are beginning to uncover the important changes to epigenetic patterns underlying the initiation, progress, and resolution of the replacement of lost structures and function. DNA modifications, a wide array of small RNAs, and histone modifications have recently been determined to play critical roles throughout regeneration by influencing the expression or silencing of large sets of genes and, perhaps, most specifically, by influencing states of cellular differentiation (Maatouk et al. 2006, Epsztejn-Litman et al. 2008, Hu et al. 2010, Challen et al. 2011). The control of cell fate *in vivo* is arguably the most important aspect of regeneration and maybe the critical first step in reproducing a regeneration program in organisms without one. In the context of vertebrate regeneration models, fully differentiated cells at the site of injury must revert to a more stem-like state through a process known as dedifferentiation (Satoh et al. 2008). It is this reversion that presumably allows cells to proliferate once again, in order to replace the lost tissues. After proliferation, cells must redifferentiate into the proper cell types in the appropriate locations and integrate functionally in order for a structure to be considered properly regenerated (Satoh et al. 2007). Each of these steps involves a multitude of genes being switched on or off, through positive and negative controls, in order to occur rapidly and within defined windows of time. Fortunately, these changes in cell fate and behavior are not completely exclusive to regeneration. The control of stem cells during embryogenesis and development and the formation of cancerous growths in healthy individuals also involve the changing of cell fates and controlling cellular proliferation. Indeed, the fields of stem cell and cancer biology have much in common with regeneration biology, and epigenetic mechanisms uncovered in any of these three fields can provide important inroads to understanding the others. When attempting to engineer a regenerative response *de novo*, an additional layer of complexity arises when we consider that the regeneration environment is heterogeneous (Kragl et al. 2009). Not only is there a need to influence epigenetic patterns within cells, but we must also devise ways to do so in a targeted fashion. Cells of epithelial tissues will need to be influenced differently from nerves, muscles, and connective tissue. Furthermore, any epigenetic control will also need to be elicited in a specific temporal fashion, for which the sequence is yet to be fully understood. Ultimately, the identification of regeneration-specific epigenetic patterns and the specific induction of such patterns may likely prove to be the foundation on which an engineered regenerative therapy involving a patient's own cells and tissues is built.

18.2 History of epigenetic mechanisms across models

The overall process of regeneration is somewhat unique when considering its starting and endpoints. From the time of injury, a symphony of cellular events takes place, allowing for the creation of a regeneration-permissible environment while simultaneously priming tissues to restart developmental pathways and maintain the memory of lost structures. In the final act, new cells cooperate to restore pattern and integrate with uninjured tissues nearby (McCusker et al. 2015). However, this arrangement of cellular behaviors that we call regeneration is ultimately a precise organization of individual movements, which themselves are not solely specific to regeneration. Therefore, previous studies of these movements in other models can, and should, be utilized to inform the ongoing analysis of regeneration.

18.2.1 Discovery of epigenetic control

British developmental biologist Conrad Waddington first coined the term *epigenetics* in 1942 (Waddington 1942). Literally translating to *above genetics*, this term was initially used to refer to a model through which genes could interact with their environment in order to bring about a cellular phenotype, specifically within the context of cellular differentiation from a totipotent state during embryonic development of organisms. At that time, the true composition of genes as DNA had not been identified, but Waddington appreciated that the progression of differentiation that a cell undergoes is actually the culmination of many decisions resulting in slight changes to phenotype until the cell's final fate is determined. He likened this series of decisions to a ball rolling down an irregular hillside (Goldberg et al. 2007), as shown in Figure 18.1. Each turn down a ravine or gully, influenced by the peaks and valleys of the hillside, represented a different pattern of gene expression that a cell would undertake. As the ball approaches the bottom of the hillside, the cell has become further differentiated and its epigenetic pattern of gene expression has become more specific to its ultimate function as a cell within a tissue, comprising the fully patterned body of an organism.

Today, we understand that these epigenetic patterns of expression are controlled through the mechanisms of DNA methylation, histone modifications, and several types of non-coding RNAs. DNA can be methylated at cytosine bases in eukaryotes, changing the identity of the base to 5-methylcytosine (Chen and Li 2004). This modification is typically observed in cytosine bases that are followed by a guanine base, known as CpG dinucleotide, as illustrated in Figure 18.2. DNA methylation is generally associated with transcriptional silencing; however, there is some evidence of transcriptional activation resulting from methylation (Wu et al. 2010). During the process of DNA methylation, methyl groups are transferred from the co-substrate S-adenosyl-L-methionine by various

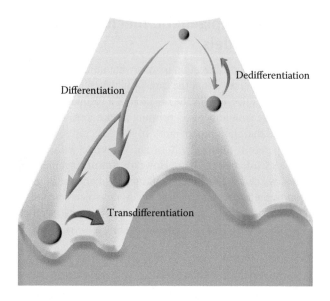

Figure 18.1 Epigenetic landscape of cellular differentiation. At the top of the slope, a cell maintains its highest level of plasticity. Progressing downward though the valley, each turn and ravine represent a more determined state of differentiation. We now recognize that cells are not necessarily terminally differentiated but can undergo the process of transdifferentiation into a closely related cell type or dedifferentiation back toward a more stem-like state. Dedifferentiation is a common mechanism observed during regeneration in higher-order organisms. (Adapted from Yamanaka, S., *Nature*, 460, 49–52, 2009.)

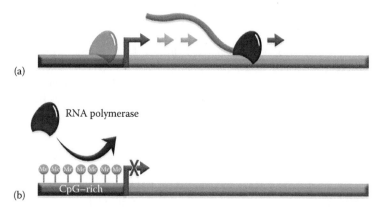

Figure 18.2 DNA methylation inhibits transcription. Methylation of DNA is predominantly found on cytosine bases that neighbor guanosine bases in what is known as a CpG dinucleotide. A promoter sequence allows for the recruitment of transcriptional machinery (a), but the presence of methylation in areas rich with CpG dinucleotides upstream of promoter sequences prevents transcription by RNA polymerase (b). In this way, gene expression patterns can be modified from their original state.

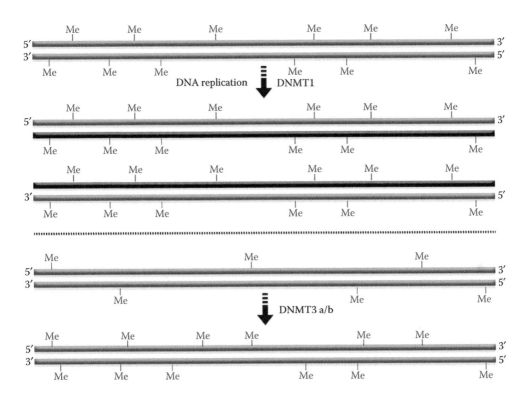

Figure 18.3 Mechanisms of DNA methylation. The maintenance methyltransferase DNMT1 is active during DNA replication, where it copies the methylation pattern of a parent strand (green) in order to reproduce it on the newly formed daughter strands (blue). The DNMT3 class of methyltransferases does not depend on DNA replication but instead introduces methyl groups where they had not been found previously.

enzymes known as DNA methyltransferases (DNMTs). Figure 18.3 illustrates the activity of two classes of DNMTs. The DNMT1 class of methyltransferases is responsible for the maintenance of methylation patterns in DNA. In the process of cell division, DNMT1 follows the DNA replication machinery, reading the parent strand's pattern of methylation and copying it faithfully onto the newly created daughter strand (Chen and Li 2004). In this way, a cell's pattern of DNA methylation can be passed onto cellular progeny, in order to maintain specific patterns of gene expression. This is particularly important during the growth and homeostasis of tissues and organs. As cells turn over, new cells can be generated to perfectly replace lost cells in identity and function.

A second class of DNMTs, the DNMT3 class, is responsible for the modification of cytosine bases that had not been previously methylated in a process termed *de novo* methylation (Okano et al. 1999). By creating new patterns of DNA methylation, the activity of these enzymes would result in the creation of gene expression profiles that are different from those of the parent cells. Methyl groups on cytosine bases can also be

399

modified and removed, adding further complexity to the effects of this modification on gene expression. The ten-eleven translocation (TET) family of enzymes can oxidize methyl groups on cytosine bases, resulting in products such as 5-hydroxymethyl cytosine. Modified methylcytosines have unique interactions with various transcription factors and can also be further oxidized, which may lead to their active removal from DNA through base excision repair mechanisms (Tahiliani et al. 2009).

Besides covalent modifications of nucleic acids, DNA-associated proteins can also be modified to affect gene expression. Histone proteins aid in the compaction and structure of chromatin, but they can be modified through various mechanisms in order to influence their level of association with DNA and their ability to interact with transcription factors. Mechanisms of histone modification include phosphorylation, ADP-ribosylation, ubiquitylation, and sumoylation, but the most characterized mechanisms are methylation and acetylation (Jenuwein and Allis 2001). Methylation of histones occurs primarily on lysine and arginine residues and can result in either gene silencing or activation, depending on the residue modified. Figure 18.4 depicts DNA associated with histones containing various modifications. For example, trimethylation of histone 3 at lysine position 4 (H3K4me3) is found to be enriched at promoters of transcriptionally active genes. This modification recruits chromatin-remodeling factors and positively regulating transcription factors resulting in gene expression (Tomazou and Meissner 2010). On the other hand, trimethylation of histone 3 at lysine position 27 (H3K27me3) is tightly associated with inactive gene promoters (Meissner 2010). Generally speaking, H3K4 trimethylation and H3K27 trimethylation are opposed to one another; however, several genes have been discovered to have

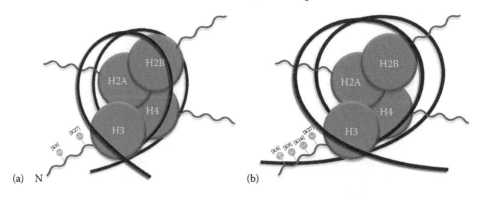

Figure 18.4 Histone modifications affect gene expression. DNA is organized into nucleosome structures composed of two each of four different histone proteins: H2A, H2B, H3, and H4. Modifications to the N-terminal tails of histone proteins, histone 3, can cause DNA to be more tightly associated in the nucleosome to result in a transcriptionally inactive state (a) or to become more loosely associated to facilitate transcription (b). Common modifications of histone 3 that lead to changes in nucleosome structure are depicted.

both modifications, also known as marks, at their promoters. This results in a sort of primed state of bivalency, in which a gene is silenced but it carries marks for activation. Presumably, the removal of the repressive mark would then allow the already-present active mark to induce gene expression more rapidly (Bernstein et al. 2006).

Acetylation of histones is nearly always associated with genes that are actively transcribed. The most commonly studied sites of acetylation are on lysine residues of histones 3 and 4. Lysine is a positively charged amino acid, which facilitates binding with negatively charged DNA. The acetylation of lysine effectively neutralizes the positive charge and weakens the affinity of DNA for the histone (Struhl 1998). In doing so, the DNA becomes more accessible to positively regulating transcription factors and, in some cases, recruits the transcription factors directly. Enzymes known as histone acetyltransferases (HATs) perform the placement of acetyl groups on histones, and the removal of acetyl groups is performed by histone deacetylases (HDACs).

Recently, small RNAs have been described to play large roles in mediating states of cellular differentiation through modulating gene expression states, generally through silencing mechanisms. Small double-stranded RNA molecules, called pre-microRNAs (pre-miRNA), are produced from target genes and processed by the RNA-induced silencing complex (RISC) containing the Dicer protein (Krol et al. 2004). Dicer cleaves the double-stranded structure into two single strands, one of which is the complement of the target gene transcript. The single-stranded mature miRNA then associates with a member of the Argonaute family of proteins, which facilitates base pairing between the miRNA and the target messenger RNA (mRNA). The formation of the new double-stranded structure results in the degradation of the target mRNA, which can no longer be translated to produce a functional protein (Valencia-Sanzhez et al. 2006). The complement of miRNAs within a cell differs based on cell identity and appears to be important in directing the proper development of tissues within an organism (Sayed and Abdellatif 2011). Taken together, the myriad of epigenetic control mechanisms can largely influence states of cellular differentiation, a process that we must hope to control within the context of directed regeneration of complex structures.

18.2.2 Stem cell potency

During the progressive development of organisms from the embryonic stage, cells begin to differentiate into various tissues and eventual organs, while forming fully patterned, complex structures of the adult. Embryonic stem cells are observed to contain chromatin that exists in a more accessible state than the chromatin of the cells that have progressed further through differentiation (Meshorer et al. 2006). Embryonic stem cells in which all

DNA methylytransferase activity was removed exhibited expression of tissue-specific transcription factors and signaling molecules, suggesting a state of differentiation in contrast to stemness (Tsumura et al. 2006). The manipulation of other chromatin modifiers, such as the Polycomb group complex, in embryonic stem cells through experimental means has led to the inability to differentiate or even the death of treated cells (Surface et al. 2010). The Polycomb group complex is implicated in the methylation of lysine 27 of histone 3, previously described as a mark of silenced genes. When examining subsets of genes known to control differentiation in the developing embryo, many carry the H3K27me3 repressive mark and the H3K4me3 activating mark deposited by the Trithorax group of histone methyltransferases (Bernstein et al. 2006). This bivalency underscores the importance of activating developmentally relevant genes in a rapid manner, mediated by epigenetic histone modifications. Furthermore, transcription factors demonstrated to be critical to maintaining the self-renewal capability of stem cells, such as Oct4, function in part by recruiting a histone 3 lysine 9 demethylase known as Jumonji domain–containing protein 1C (Jmjd1c). This also leads to the inhibition of DNA methylation activity by DNMT3a (Shakya et al. 2015). The promoter region of the gene encoding Nanog, a second transcription factor contributing to the self-renewal capability of stem cells, has been demonstrated to be bound by Oct4 in conjunction with Jmjd1C.

Outside of developing embryonic stem cells, the process of somatic cell reprogramming (returning differentiated cells to a point of stem cell potency *in vitro*) is also aided by experimental manipulation of epigenetic factors. Classically, the creation of induced pluripotent stem cells could be accomplished by the expression of a handful of transcription factors such as Oct4, Sox2, Klf4, and c-myc (Takahashi and Yamanaka 2006), but the efficiency of reprogramming is often poor. In an effort to further understand the ability of this small subset of factors to induce pluripotency, the interactome of Oct4 was analyzed. Again, various histone modifiers were identified, such as the lysine demethylases of the switch/sucrose non-fermentable (SWI/SNF) group and HDACs of the nucleosome remodeling and histone deacetylase (NuRD) complex (van den Berg et al. 2010, Pardo et al. 2010). In situations where efficiency of reprogramming is poor or incomplete, DNMT inhibitors have been added to the media, allowing cells to reach a more complete level of dedifferentiation. It is apparent that the process of controlling states of cellular differentiation, whether in embryonic stem cells or through somatic cell reprogramming, requires the precise control of epigenetic modifiers. In the event that these modifiers become misregulated, abnormal cell activity may arise.

18.2.3 Cancer cells and proliferation

Cancer cells as a model demonstrate the effects of aberrant cell behavior and activity; however, through studying the disease, we can learn much about how normally functioning processes are connected to one another and ideally about the ways in which we can return that normal function to the cell, so that it may resume its original role in the body. The development of cancer has long been understood to be the product of genetic mutation accumulations that reduce the normal function of proteins preventing cell growth (known as tumor suppressors) or that cause a change in proteins such that they become drivers of cell division (known as oncogenes). One mechanism driving the silencing of tumor suppressors is hypermethylation of the promoter regions of the DNA that encodes them (Esteller et al. 2000). In addition, levels of methylated cytosine bases outside of tumor suppressor genes are significantly lower in cancer cells compared with the non-cancerous tissue from which they arose (Wilson et al. 2007). Beyond the changes to DNA methylation levels, the expression of patterns of the enzymes responsible for these modifications are also modulated. Cancers such as breast, prostate, and ovarian cancers exhibit higher-than-normal levels of DNMTs. It is interesting to examine an increase in methyltransferase expression while simultaneously observing lower levels of methylation outside of key regions. This highlights the importance of identifying specific targets of methylation and discovering how the recruitment of DNA-modifying enzymes to precise sequences is accomplished.

DNA methyltransferases have not been described to be able to identify specific DNA sequences themselves in order to modify cytosine bases. Therefore, the targeting of DNMTs must require either the direction of cofactors that are able to recognize specific DNA sequences or the addition of separate modifications to DNA that can, in turn, be recognized by DNMTs. Components of the PIWI-interacting RNA pathway have been shown to be required for site-directed DNA methylation of the promoter of the *Rasgrf1* gene in mice (Watanabe et al. 2011). Elsewhere, some previously described histone modifications have been found to correlate with DNA methylation, or lack thereof, within specific regions of the genome. Methylation of histone 3 at the lysine 4 position is associated with active gene expression. Conversely, regions of DNA lacking this epigenetic mark are often silenced and found to have a higher distribution of DNA methylation (Meissner et al. 2008). The increase in DNA methylation within regions of DNA associated with unmethylated histone tails appears to be mediated by a second factor known as DNMT3l. DNMT3l is a regulatory cofactor of DNMT3a, unable to perform methylation itself for lack of the necessary enzymatic domains. Yet this cofactor can bind to DNMT3a and associate with histones, in this case with unmethylated histone tails (Ooi et al. 2007). Furthermore, DNA methyltransferase 3l can associate with histone deacetylase enzymes to mediate transcriptional repression (Dhayalan et al. 2010). Finally, DNMT3a has been described to directly recognize the methylation of lysine 36 of histone 3, itself a mark of gene silencing. These interactions can allow for the formulation of strategies to precisely add epigenetic modifications in the aim of controlling gene expression during regeneration.

Although the specific methylation of key promoters can lead to silencing of tumor-suppressor genes in cancer, the global hypomethylation must

403

also be considered. By removing suppressive methylation marks, the cell is presumably able to express a larger repertoire of gene products, some of which may be outside of the program for that particular source cell type, which could lead to the adoption of different cell identities. Again, the specific role of the target gene is ultimately the most important factor. Demethylation may allow genes to be expressed more readily, but the newly expressed gene may further serve to activate, or repress, its target pathway.

Histone-modifying enzyme activity also becomes altered under, or leads to, cancerous conditions. Enhancer of zeste homolog 2 (EZH2), the catalytic subunit of the Polycomb repressive complex, has been observed to be overexpressed in cancers such as prostate cancer, ovarian cancer, brain cancer, and melanoma, leading to increased levels of the repressive histone 3 lysine 27 trimethylation mark (Yamaguchi and Hung 2014). The SWI/SNF chromatin-remodeling complex typically plays a role of tumor suppressor in normal cells, but its function is inactivated within many cancer types. It functions through somewhat unknown mechanisms but has been determined to shift the position of nucleosomes within a region of DNA (Whitehouse et al. 1999). Approximately 20% of cancers contain mutations in at least one subunit of the SWI/SNF complex, highlighting the importance of epigenetic modifiers in controlling cellular proliferation and differentiation. Overall, the most common aberrant histone modifications described in cancerous states are the methylation of lysine 4, 9, and 27 on histone 3 and the deacetylation of lysine 16.

Taken together, epigenetic modifications such as DNA methylation, histone modifications, and small RNAs play large roles in influencing the proliferation and differentiation of cells in both disease and normal development. The similarities between regeneration and each of these processes warrant particular attention during the course of developing clinically relevant therapies for injury.

18.3 Epigenetics in regeneration

In the aim of elucidating natural mechanisms of regeneration, we must utilize the multitude of models available to us. Some of the most well-studied and relevant models are those of the planarian flatworm, zebrafish, and salamander. Although each system utilizes unique mechanisms, the behavior of cells appears to be similar. Each has the ability to control and, in a sense, reprogram its identity *in vivo*, responding to cues and information found within the wound environment. The role of epigenetic modifications as a means through which this cellular behavior is controlled has only recently begun to be examined. Here, we will discuss these studies within three regenerative models.

18.3.1 Planaria

The planarian flatworm is a relatively simple invertebrate organism composed of the three basic germ layers organized into a simple digestive system and primitive nervous system. No circulatory or respiratory system is present, as the organism is small enough to utilize simple diffusion for the distribution of nutrients and gasses. However, this simple organism is a wonderful model of regeneration, utilizing a population of adult stem cells termed neoblasts. Neoblasts are the only proliferative cells in the organism, comprising about 20% of the total cell count. In response to injury, the neoblasts become activated and participate in the formation of blastema, the regenerative structure of the worm. Cells of the blastema, along with cells of the area surrounding the injury site, interact in order to adopt new pattern and replace lost structures through a morphallactic regeneration mechanism. Recently, studies have begun to investigate the role of epigenetics in mediating this regenerative response. As planarians are fairly primitive organisms, it is important to identify epigenetic modifiers that are well conserved.

Two components of the NuRD complex have been identified within the planarian species *Schmidtea mediterranea* and *Dugesia japonica* (Scimone et al. 2010). The NuRD complex is implicated in controlling cell fates in various tissues and contains more than seven proteins in mammals. A third component of the NuRD complex, methyl-CpG-binding domain 2/3 (mbd2/3), was also identified in *S. mediterranea*, and knockout studies determined that mbd2/3 is required for the correct differentiation of neoblasts into newly patterned tissues during regeneration (Jaber-Hijazi et al. 2013). Additional microarray analysis has identified a heterochromatin protein 1–like protein (HP1-1) in *Planaria*, which when knocked down leads to a decrease in histone H3 lysine 9 trimethylation levels and results in a loss of self-renewal capability in neoblasts (Zeng et al. 2013). Premature differentiation into committed cells during early blastema formation was also observed. Finally, six members of the SET1/MLL family of histone methyltransferases were identified in *S. mediterranea* through an extensive tblastn database search using the known SET domain protein sequence from humans to search for homologous proteins in *Planaria* (Hubert et al. 2013). The SET1/MLL family of proteins positively regulates gene expression through methylation of lysine 4 on histone H3. RNA interference knockdown of the planarian homologs led to a depletion of neoblasts in uninjured worms and a reduced blastema growth after amputation. Although histone modifications are beginning to be described in planarian regeneration, it is interesting to note that DNA methylation is not detected across the genome, and no definitive DNMTs have been described. This suggests that control of cell fate through DNA methylation arose later in evolutionary history for eukaryotes (Jaber-Hijazi et al. 2013).

18.3.2 Zebrafish

Among vertebrate models of regeneration, zebrafish have a well-characterized genome, allowing for complex and specific analysis of chromatin modifications. During regeneration, developmentally related genes must be reactivated in order to direct the replacement of lost tissues. The dynamics of reactivating silenced genes have been found to involve epigenetic switches in the form of bivalent chromatin. Bivalent chromatin is not transcriptionally active due to the presence of the repressive histone H3 lysine 27 trimethylation mark. However, the developmental genes also carry methylation on histone H3 lysine 4, a mark of active transcription. In order for the genes to be activated at the initiation of regeneration, the repressive mark must be removed by a histone demethylase, Kdm6b.1 (Stewart et al. 2009). The experimental knockdown of Kdm6b.1 results in blocking the initiation of regeneration of zebrafish caudal fins by preventing developmental genes, such as distal-less homeobox 4, from being expressed.

Although histone modifications of promoter regions can control genes immediately downstream, DNA sequences known as enhancers can be located thousands of base pairs away from their target gene, and they may also control the expression of many genes at once (Reeder et al. 2015). Enhancers are characterized as regions of open chromatin, associated with histones marked by acetylation of lysine 27 on histone 3. Recently, a specific enhancer region upstream of the leptin b (*lepb*) gene was found to become active in response to injury of the zebrafish caudal fin and cardiac tissue (Kang et al. 2016). By introducing the zebrafish *lepb*-linked enhancer region to mouse tissue, it was discovered that the enhancer has a sufficient level of conservation to be able to act upon the genes of other species. Furthermore, regenerative programs became initiated in mouse wounds that would not regenerate otherwise. In this case, epigenetic marks are demonstrated to control regulatory gene elements in a conserved fashion, suggesting that regeneration mechanisms of classic model organisms could be translated into regeneration-deficient organisms such as mammals.

As regeneration progresses in the zebrafish, new tissues must receive adequate oxygenation and nutrients. After amputation of the caudal fin, initial regenerating structures lack vasculature, which results in a hypoxic environment that leads to the expression of a histone deacetylase HDAC6. While HDACs are named for their activity on histones, some also interact with non-histone proteins to remove acetyl groups. Histone deacetylase 6 has been found to deacetylate the actin-remodeling protein cortactin in endothelial cells, allowing them to migrate into the blastema and sprout new blood vessels (Kaluza et al. 2011). This discovery highlights the importance of studying the role of epigenetic modifiers in regeneration beyond their traditional activity associated with chromatin. The range of epigenetic studies in zebrafish highlights the broad scope of control that these mechanisms can elicit on cellular behavior.

18.3.3 Newts and salamanders

The newt and salamander are powerful models of regeneration due to their anatomical similarities to humans and their wide range of regeneration that includes structures such as the lens of the eye, heart, limbs, and spinal cord. However, the current state of genomic research in these organisms is fairly limited. Most studies investigating epigenetic mechanisms of regeneration have focused on miRNAs. During cardiac regeneration in the newt, a specific miRNA was observed to increase in expression through a microarray screen. miR-128 has been determined to inhibit proliferation of non-cardiomyocytes surrounding the site of injury (Witman et al. 2013). Furthermore, newts with depleted levels of miR-128 showed evidence of persistent scarring, evidenced by fibrotic tissue in the wound site. This phenotype is in contrast to the remodeling of scar tissue that precedes normal newt cardiac regeneration.

In addition to cardiac regeneration, miRNAs are implicated in creating a regeneration-permissive environment for spinal cord repair. After severing of the spinal cord, a regeneration program involves the crossing of the transected region by axons in order to re-establish connection. In the axolotl salamander, miR-125b is expressed and leads to the downregulation of Sema4D, a protein that normally blocks axon growth across the injury plane (Diaz Quiroz et al. 2014). In this manner, the activity of miR-125b allows axons to continue to grow after injury. Interestingly, mammals also utilize this same pathway but fail to regulate it in a way that could lead to regeneration. Experimental induction of miR-125b expression in rat spinal cord injuries led to the downregulation of Sema4D and enhanced the functional recovery of the animal. This study highlights the importance of identifying evolutionarily conserved epigenetic mechanisms that can be more easily translated into functional treatments for mammalian injuries. To that end, a wider assessment of miRNA expression has been carried out, leading to the identification of 4564 miRNA families present in the regenerating axolotl tail that are also conserved across a multitude of vertebrates (Gearhart et al. 2015).

Beyond miRNAs, histone modifications and DNA methylation have been investigated only to a very limited extent. After removal of the newt eye lens, cells of the iris dedifferentiate and migrate in order to reform the lost structure. Within this context, an array of histone modifications was observed to change in distribution (Maki et al. 2010). The active transcription marks of histone H3 lysine 4 trimethylation and acetylated histone H4 increased in pigmented epithelial cells of the iris, whereas the histone H3 lysine 9 acetylation mark decreased. Within the ventral portion of the iris, the repressive mark of histone H3 lysine 27 trimethylation was seen to increase. Clearly, the complexity of histone modification regulation warrants further examination, in order to more completely understand the most important dynamics necessary to initiate a regenerative response.

Recently, studies of DNA methylation in the regenerating axolotl limb have uncovered an important role for DNA methylation in controlling the formation of a regeneration-capable wound epithelium. After amputation of a limb, one of the earliest events is the healing of the epithelium. The cells of the wound epithelium must be able to respond to signals from an underlying nerve in order to begin the formation of a blastema structure (Endo et al. 2004). Superficial wounding of the limb epithelium alone does not result in cells capable of responding to nerves. When comparing the expression of epigenetic modifiers between superficial wounds and regenerating wounds, the *de novo* DNA methyltransferase DNMT3a was expressed at a relatively low level in regenerating wounds (Aguilar and Gardiner 2015). Furthermore, treatment of non-regenerating wounds with a small molecule DNMT inhibitor to mimic the regenerative condition resulted in the successful participation of non-regenerating wounds in the formation of a blastema. This research continues to suggest that epigenetic mechanisms may serve as critical switches utilized by cells to activate regeneration-specific pathways. As we continue to glean important insights about conserved epigenetic mechanisms from an array of model organisms, we will approach the point at which we are prepared to construct methods to precisely manipulate the epigenetic pattern of cells in mammalian injuries to improve wound repair and even replace complex structures.

Acknowledgments

Thank you to Elijah Schaffer of Azusa Pacific University for his talent and expertise in the generation of the figures in this chapter.

References

Aguilar C, Gardiner DM. 2015. DNA Methylation dynamics regulate the formation of a regenerative wound epithelium during axolotl limb regeneration. *PLoS One* 10(8):e0134791.

Bernstein BE, Mikkelsen TS, Xie X et al. 2006. A bivalent chromatin structure marks key developmental genes in embryonic stem cells. *Cell* 125(2):315–326.

Challen GA, Sun D, Jeong M et al. 2011. Dnmt3a is essential for hematopoietic stem cell differentiation. *Nat Genet* 44(1):23–31.

Chen T, Li E. 2004. Structure and function of eukaryotic DNA methyltransferases. *Curr Top Dev Biol* 60:55–89.

Dhayalan A, Rajavelu A, Rathert P et al. 2010. The Dnmt3a PWWP domain reads histone 3 lysine 36 trimethylation and guides DNA methylation. *J Biol Chem* 285(34):26114–26120.

Diaz Quiroz JF, Tsai E, Coyle M, Sehm T, Echeverri K. 2014. Precise control of miR-125b levels is required to create a regeneration-permissive environment after spinal cord injury: A cross-species comparison between salamander and rat. *Dis Model Mech* 7(6):601–611.

Endo T, Bryant SV, Gardiner DM. 2004. A stepwise model system for limb regeneration. *Dev Biol* 270(1):135–45.

Epsztejn-Litman S, Feldman N, Abu-Remaileh M et al. 2008. De novo DNA methylation promoted by G9a prevents reprogramming of embryonically silenced genes. *Nat Struct Mol Biol* 15(11):1176–1183.

Esteller M, Silva JM, Dominguez G et al. 2000. Promoter hypermethylation and BRCA1 inactivation in sporadic breast and ovarian tumors. *J Natl Cancer Inst* 92:564–9.

Gearhart MD, Erickson JR, Walsh A, Echeverri K. 2015. Identification of conserved and novel MicroRNAs during tail regeneration in the mexican axolotl. *Int J Mol Sci* 16(9):22046–22061.

Goldberg AD, Allis CD, Bernstein E. 2007. Epigenetics: A landscape takes shape. *Cell* 128:635–638.

Hu B, Gharaee-Kermani M, Wu Z, Phan SH. 2010. Epigenetic regulation of myofibroblast differentiation by DNA methylation. *Am J Pathol* 177(1):21–28.

Hubert A, Henderson JM, Ross KG, Cowles MW, Torres J, Zayas RM. 2013. Epigenetic regulation of planarian stem cells by the SET1/MLL family of histone methyltransferases. *Epigenetics* 8(1):79–91.

Jaber-Hijazi F, Lo PJ, Mihaylova Y et al. 2013. Planarian MBD2/3 is required for adult stem cell pluripotency independently of DNA methylation. *Dev Biol* 384(1):141–153.

Jenuwein T, Allis CD. 2001. Translating the histone code. *Science* 293(5532): 1074–1080.

Kaluza D, Kroll J, Gesierich S et al. 2011. Class IIb HDAC6 regulates endothelial cell migration and angiogenesis by deacetylation of cortactin. *EMBO J* 30(20):4142–4156.

Kang J, Hu J, Karra R et al. 2016. Modulation of tissue repair by regeneration enhancer elements. *Nature* 532(7598):201–206.

Kragl M, Knapp D, Nacu E et al. 2009. Cells keep a memory of their tissue origin during axolotl limb regeneration. *Nature* 460(7251):60–65.

Krol J, Sobczak K, Wilczynska U et al. 2004. Structural features of microRNA (miRNA) precursors and their relevance to miRNA biogenesis and small interfering RNA/short hairpin RNA design. *J Biol Chem* 279(40):42230–9.

Maatouk DM, Kellam LD, Mann MR et al. 2006. DNA methylation is a primary mechanism for silencing postmigratory primordial germ cell genes in both germ cell and somatic cell lineages. *Development* 133(17):3411–3418.

Maki N, Tsonis PA, Agata K. 2010. Changes in global histone modifications during dedifferentiation in newt lens regeneration. *Mol Vis* 16:1893–1897.

McCusker CD, Athippozhy A, Diaz-Castillo C, Fowlkes C, Gardiner DM, Voss SR. 2015. Positional plasticity in regenerating Amybstoma mexicanum limbs is associated with cell proliferation and pathways of cellular differentiation. *BMC Dev Biol* 15:45.

Meissner A. 2010. Epigenetic modifications in pluripotent and differentiated cells. *Nat Biotechnol* 28(10):1079–1088.

Meissner A, Mikkelsen TS, Gu H et al. 2008. Genome-scale DNA methylation maps of pluripotent and differentiated cells. *Nature* 454(7205):766–770.

Meshorer E, Yellajoshula D, George E, Scambler PJ, Brown DT, Misteli T. 2006. Hyperdynamic plasticity of chromatin proteins in pluripotent embryonic stem cells. *Dev Cell* 10(1):105–116.

Okano M, Bell DW, Haber DA, Li E. 1999. DNA methyltransferases Dnmt3a and Dnmt3b are essential for de novo methylation and mammalian development. *Cell* 99(3):247–257.

Ooi SK, Qiu C, Bernstein E et al. 2007. DNMT3L connects unmethylated lysine 4 of histone H3 to de novo methylation of DNA. *Nature* 448(7154):714–717.

Pardo M, Lang B, Yu L et al. 2010. An expanded Oct4 interaction network: Implications for stem cell biology, development, and disease. *Cell Stem Cell* 6(4):382–395.

Reeder C, Closser M, Poh HM, Sandhu K, Wichterle H, Gifford D. 2015. High resolution mapping of enhancer-promoter interactions. *PLoS One* 10(5):e0122420.

Satoh A, Gardiner DM, Bryant SV, Endo T. 2007. Nerve-induced ectopic limb blastemas in the Axolotl are equivalent to amputation-induced blastemas. *Dev Biol* 312(1):231–244.

Satoh A, Graham GMC, Bryant SV, Gardiner DM. 2008. Neurotrophic regulation of epidermal dedifferentiation during wound healing and limb regeneration in the axolotl (*Ambystoma mexicanum*). *Dev Biol* 319(2):321–335.

Sayed D, Abdellatif M. 2011. MicroRNAs in development and disease. *Physiol Rev* 91(3):827–887.

Scimone ML, Meisel J, Reddien PW. 2010. The Mi-2-like Smed-CHD4 gene is required for stem cell differentiation in the planarian Schmidtea mediterranea. Development 137:1231–1241.

Shakya A, Callister C, Goren A et al. 2015. Pluripotency transcription factor Oct4 mediates stepwise nucleosome demethylation and depletion. *Mol Cell Biol* 35(6):1014–1025.

Stewart S, Tsun ZY, Izpisua Belmonte JC. 2009. A histone demethylase is necessary for regeneration in zebrafish. *Proc Natl Acad Sci U S A* 106(47):19889–19894.

Struhl K. 1998. Histone acetylation and transcriptional regulatory mechanisms. *Genes & Dev* 12(5):599–606.

Surface LE, Thornton SR, Boyer LA. 2010. Polycomb group proteins set the stage for early lineage commitment. *Cell Stem Cell* 7(3):288–298.

Tahiliani M, Koh KP, Shen Y, Pastor WA, Bandukwala H, Brudno Y et al. 2009. Conversion of 5-methylcytosine to 5-hydroxymethylcytosine in mammalian DNA by MLL partner TET1. *Science* 324(5929):930–935.

Takahashi K, Yamanaka S. 2006. Induction of pluripotent stem cells from mouse embryonic and adult fibroblast cultures by defined factors. *Cell* 126(4):663–676.

Tomazou EM, Meissner A. 2010. Epigenetic regulation of pluripotency. *Adv Exp Med Biol* 695:26–40.

Tsumura A, Hayakawa T, Kumaki Y et al. 2006. Maintenance of self-renewal ability of mouse embryonic stem cells in the absence of DNA methyltransferases Dnmt1, Dnmt3a and Dnmt3b. *Genes Cells* 11(7):805–814.

Valencia-Sanchez MA, Liu J, Hannon GJ, Parker R. 2006. Control of translation and mRNA degradation by miRNAs and siRNAs. *Genes Dev* 20(5):515–524.

van den Berg DL, Snoek T, Mullin NP et al. 2010. An Oct4-centered protein interaction network in embryonic stem cells. *Cell Stem Cell* 6(4):369–381.

Waddington CH. 1942. The epigenotype. *Endeavour* 1:18–20.

Watanabe T, Tomizawa S, Mitsuya K et al. 2011. Role for piRNAs and noncoding RNA in de novo DNA methylation of the imprinted mouse Rasgrf1 locus. *Science* 332(6031):848–852.

Whitehouse I, Flaus A, Cairns BR et al. 1999. Nucleosome mobilization catalysed by the yeast SWI/SNF complex. *Nature* 400(6746):784–787.

Wilson AS, Power BE, Molloy PL. 2007. DNA hypomethylation and human diseases. *Biochim Biophys Acta* 1775:138–162.

Witman N, Heigwer J, Thaler B, Lui WO, Morrison JI. 2013. miR-128 regulates non-myocyte hyperplasia, deposition of extracellular matrix and Islet1 expression during newt cardiac regeneration. *Dev Biol* 383(2):253–263.

Wu H, Coskun V, Tao J et al. 2010. Dnmt3a-dependent nonpromoter DNA methylation facilitates transcription of neurogenic genes. *Science* 329(5990):444–448.

Yamaguchi H, Hung MC. 2014. Regulation and role of EZH2 in cancer. *Cancer Res Treat* 46:209–222.

Yamanaka S. 2009. Elite and stochastic models for induced pluripotent stem cell generation. *Nature* 460:49–52.

Zeng A, Li YQ, Wang C et al. 2013. Heterochromatin protein 1 promotes self-renewal and triggers regenerative proliferation in adult stem cells. J Cell Biol 201(3):409–425.

Chapter 19 Developmental plasticity and tissue integration

Warren A. Vieira and
Catherine D. McCusker

Contents

Key concepts

a. Generation of cells exhibiting a pluripotent nature may not be necessary for successful regeneration, as natural systems, such as the regenerating *Ambystoma mexicanum* limb, do not make use of such cells.

b. Within the blastema, all cells contribute to the new regenerate; however, some are responsible for establishing the pattern of the missing structures in the limb regenerate (*pattern forming*), whereas other types respond to positional cues from the pattern-forming cells (*pattern following*).

c. Distinct cell types have differing abilities to contribute to the regenerated structure.

d. The positional information present at the site of injury is imperative to establishing the pattern of the regenerated limb structure.

e. Tissue integration is a key step in the generation of a functional biological structure.

f. Tissue regeneration and integration do not appear to be necessarily coupled processes, but both of them are dependent on the property of positional information.

g. Plasticity of positional information promotes the integration of new tissue (regenerated or grafted) into the surrounding host environment.

19.1 Introduction

The future vision of many cell-based regenerative therapies is to repair or replace damaged tissues by grafting in healthy cell populations. In some cases, this requires that the grafted population integrate with the host tissues to generate a fully functional structure. Research on mammalian and amphibian model systems suggests that the ability of a graft to integrate into the host site is dependent on the positional information in both tissues. In this chapter, we will explain the differences in the positional information of the cell types that contribute to regenerating mammalian and amphibian organs and how they impact the process of tissue integration. Last, we dissect the potential causes of integration defects in cell-based regenerate therapies and present methods that could be utilized to overcome these challenges.

19.2 Developmental plasticity

Building an organ, whether it is during embryogenesis or a regeneration event in adults, requires that the building blocks of the organ (cells and extracellular matrix) are assembled according to a specific blueprint or pattern. The regenerative process has the additional challenge of constructing organs from previously patterned adult tissues. In tetrapods, the regenerated tissue is composed of cells of adult stem cell (ASC) and differentiated cell origins. In this section, we will describe what is currently known about the characteristics of the cell types that contribute to regenerating structures in amphibians and mammals and how this knowledge can change the way in which we promote the integration of cell-based regenerative therapies.

19.2.1 Developmental plasticity in tissue identity

Urodele amphibian species (salamanders and newts) exhibit robust regenerative capabilities throughout their lives. Although several different structures within their bodies are able to regenerate, limb regeneration has been most extensively studied. In this context, wound healing begins immediately after limb amputation, whereby the site of injury is rapidly covered by a wound epithelium derived from stump keratinocytes (Hay and Fischman, 1961; Ferris et al., 2010). The wound epithelium is then converted to an apical epithelial cap (AEC) in response to innervation by the underlying nerves (Singer and Inoue, 1964; Singer, 1974), and it acts as a signaling structure. Apical epithelial cap–derived factors, including fibroblastic growth factor (Mullen et al., 1996), bone morphogenetic proteins, and Wnt signaling proteins (Kawakami et al., 2006), are required to stimulate the establishment and expansion of an underlying cell mass, the blastema. This proliferative population of regeneration-competent cells is used to reconstruct the missing portion of the limb and is considered an equivalent to the limb bud, required for embryonic limb development (Bryant et al., 2002). However, unlike its embryonic counterpart, the blastema is formed from cells that arise from mature tissues.

In response to amputation, cells derived from the muscle, loose connective tissue, and Schwann cells migrate and accumulate in the space between the AEC and the amputation plane to generate the blastema (Kragl et al., 2009[*]). However, not all stump cell types contribute equally to this regenerative structure; substantially, more blastema cells are of connective tissue rather than of muscle origin, and there is no apparent contribution from underlying skeletal elements (bone and cartilage) (Muneoka et al., 1985; Tank, 1987; McCusker et al., 2016). As the blastema grows over time, the more proximal cells differentiate in order to pattern the new tissue; however, these cells are

---

[*] The authors performed a series of grafting experiments using tissues (skin, skeletal, muscle, and Schwann cells) from transgenic axolotl expressing green fluorescent protein (GFP) to determine their potential to contribute to the tissues in the regenerated limb. They observed that each tissue type contributed to specific tissues in the regenerate. This study revealed that blastema cells were not dedifferentiated to pluripotency, as was previously hypothesized. Rather, they are a heterogeneous population of cells with restricted potential, based on their original tissue identities, when contributing to the regenerate.

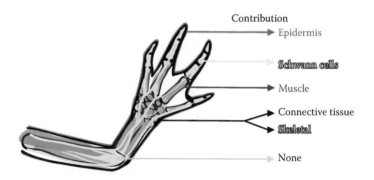

Figure 19.1 Developmental plasticity in tissue identity. The limb is composed of a variety of tissue types, including epidermis (green), muscle (red), neural (yellow), connective (blue), and skeletal (cartilage and bone) (grey). The cells from each tissue contribute differently to the regenerating blastema and have differing potentials to differentiate into the tissues of the regenerated limb. Epidermal cells from the mature limb tissue contribute to the wound epithelium region of limb blastema and eventually differentiate into the mature epidermis in the complete regenerate. Either satellite cells or dedifferentiated mononucleated myocytes from the muscle contribute only to regenerated muscle (Kragl et al., 2009). Schwann cells contribute only to Schwann cells in the regenerate. With the exception of bone/cartilage, which do not contribute to the regenerate (McCusker et al., 2016), connective tissue cells overcontribute disproportionality to the blastema (greater than 50%) and differentiate into skeletal and other connective tissues (Muneoka et al., 1985; Kragl et al., 2009; Nacu et al., 2013; McCusker et al., 2016). Thus, connective tissue–derived cells have greater plasticity in terms of tissue identity in the regenerated structures.

restricted in terms of their differentiation potential according to their original uninjured tissue identities (Kragl et al., 2009; Nacu et al., 2013). As depicted in Figure 19.1, muscle and Schwann-derived cells exhibit little plasticity, contributing to only muscle and nerve fibers, respectively, whereas connective tissue cells have the greatest plasticity and contribute to cartilage, connective tissue, and tendons in the regenerated limb (Kragl et al., 2009; Nacu et al., 2013; McCusker et al., 2016). Thus, the plasticity of tissue identity is largely dependent on the type of tissue that contributes to the limb regenerate.

Unlike urodele amphibians, which can faithfully regenerate a structure after amputation, higher vertebrates, such as mammals, are highly restricted in their regenerative capabilities. However, the digit tips of neonate and adult mice offer a unique system for scientific exploration, as they are capable of regeneration (Han et al., 2008; Rinkevich et al., 2011*;

---

* This study used genetic fate mapping and clonal analysis to test whether the mouse digit-tip blastema were composed of mature cells that had been dedifferentiated to a pluripotent state. The authors observed that the blastema cells arise from the resident tissues and are lineage-restricted, similar to what has been previously observed in amphibian limb regenerates.

Wu et al., 2013). Amputation of the distal digit tip results in cellular responses that are similar to those documented in the amphibian limb; a wound epithelium forms over the site of injury, followed by the generation of a stump-derived, proliferating progenitor cell mass, the blastema. As in the amphibian, these latter cells contribute to the new structure in a germ-line linage-restricted manner (Rinkevich et al., 2011). Taken together, these data suggest that epimorphic regeneration is an evolutionary conserved mechanism, even though it is limited in higher vertebrates. Furthermore, cells with limited plasticity, rather than having a pluripotent nature, are sufficient for successful regeneration. This is an important concept within the field of regenerative medicine, where a large effort has been on the establishment of pluripotent stem cell lines for therapeutic use. As natural systems do not even make use of such cells, therapeutic strategies could be simplified by taking advantage of cells with limited differentiation plasticity to facilitate successful regeneration (Kragl et al., 2009).

19.2.2 Developmental plasticity in positional identity

The cells that contribute to the amphibian limb regenerate are heterogeneous in terms of their ability to retain memory of their positional identity. Although the concept of positional identity is well described in the developing embryo, a complete understanding of its contribution and functional relevance in adult cells is still outstanding. Current experimental data in amphibians and mammals suggest that positional information is retained throughout adult life by cells within the connective tissues, with adult fibroblasts being the most extensively studied cell type in this regard (Chang et al., 2002*; Endo et al., 2004†; Nacu et al., 2013; Rinn et al., 2006). These cells exhibit specific gene-expression profiles that correspond to their anatomical origins that were established during embryogenesis (Chang et al., 2002). As we will describe in the following sections, there are cells that are responsible for establishing the pattern of the missing structures in the limb regenerate; thus, these are a *pattern-forming* population of cells. On the other hand, other cell types

---

* This study performed a large-scale transcriptional analysis on fibroblast populations that had been isolated from different locations throughout the human body. The authors discovered that each population of fibroblasts expressed a different and characteristic set of transcripts, including members of the *Hox* family. They observed that the expression of *Hox* genes that were established during embryogenesis were maintained in the fibroblasts that were located in the corresponding adult tissues, suggesting that these human cells were retaining memory of their positional identity.

† This paper describes the Accessory Limb Model. The authors showed that the formation of an ectopic blastema at a wound site requires nerve input (in the form of a deviated nerve to the wound site). If a contralateral graft of dermal cells (to establish positional discontinuity at the wound site) is put in the wound site, the blastema forms a supernumerary limb. In addition to establishing that regeneration is a stepwise process, the Accessory Limb Model that was generated in this study serves as a simplified gain-of-function assay to study the basic components of the regenerative process.

such as muscle, Schwann, and epidermal cells do not retain memory of their positional origin but respond to positional cues from the cells that do retain memory; thus these are *pattern-following* cell types (Kragl et al., 2009; Nacu et al., 2013; reviewed in McCusker et al., 2015a). Thus, all of the above-listed cell populations contribute to the formation of the amphibian and mammalian blastema; however, each has a differential role (direct vs indirect) in establishing the pattern of the regenerating structure (Figure 19.2).

Since the *pattern-forming* cells establish a regenerate that has new (i.e., distal) positional information, the positional identity of these cells or their progeny must be somewhat plastic to adopt the new positional program. Evidence from regenerating salamander limbs reveals that positional plasticity is a property that is retained by undifferentiated blastema cells and that this property is maintained by signaling downstream of nerves (McCusker and Gardiner, 2013*). The positional program of these plastic cells can be reprogrammed by either changing the positional identity (through grafting manipulations) of the surrounding mature tissues that have stabilized positional information (McCusker and Gardiner, 2013) or via exposure to ectopic levels of retinoic acid (RA) (Maden, 1983; McCusker et al., 2014), a molecule that is instrumental in embryonic and regenerating limb pattern formation. It is speculated that

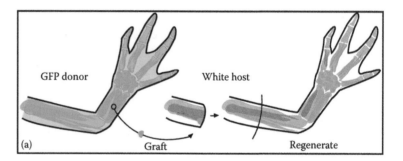

(a)   GFP donor   White host   Graft   Regenerate

Figure 19.2 Developmental plasticity in positional identity. The cell populations that contribute to the regenerate have differential roles (direct and indirect) in establishing the pattern in the regenerating structure. (a) Some cell progenitors are *pattern-following* cells that do not retain positional memory but respond to positional cues from cells that do retain positional memory. The cartoon in (a) illustrates how grafts of muscle from transgenic green fluorescent protein (GFP)+ animals follow the patterning information in the new host location (Kragl et al., 2009).   (*Continued*)

---

* This article evaluates the stability of positional information in cells of the blastema and the stump during regeneration, using the axolotl model. Cells derived from the early blastema and those constituting the tip of the late blastema are liable in terms of their positional identity; however, cells derived from the basal region of the late-stage blastema and those from the stump tissue are stable in this regard.

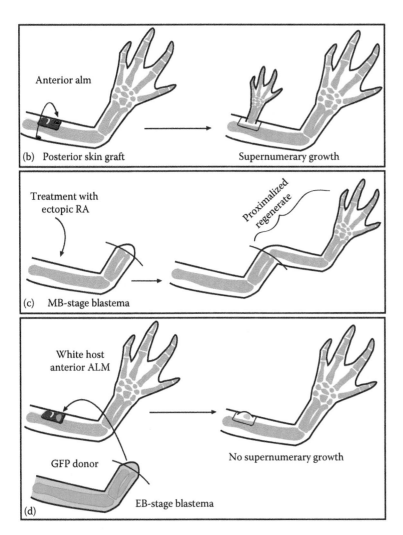

Figure 19.2 (Continued) Developmental plasticity in positional identity.
(b) Other cells, which arise from the connective tissues, retain positional
memory and use this memory to generate the pattern of the missing limb
structures. For example, grafts of dermal cells from the posterior side of
the limb result in the generation of supernumerary limbs when grafted
into an innervated wound on the anterior side of the limb (the Accessory
Limb Model [ALM]) (Endo, 2004). (c and d) Blastema cells that arise
from connective tissues are positionally plastic in the undifferentiated
blastema tissue. For example, (c) illustrates that distal blastemas are
reprogrammed to have a proximal positional identity on treatment with
ectopic retinoid acid (Maden, 1983). Panel (d) illustrates that early
blastema grafts from a GFP+ animal into ALMs do not induce the
formation of supernumerary limb structures (McCusker, 2013).

positional plasticity is required to ensure that the regenerate structure is completely and appropriately patterned before cell differentiation (McCusker and Gardiner, 2013).

19.3 The role of positional identity in injured and uninjured tissues

19.3.1 Positional identity in the context of tissue plasticity

In the axolotl, there is a correlation between whether a blastema cell originates from a *pattern-forming* or *pattern-following* cell type and the degree of plasticity of its tissue identity in the regenerate (Kragl et al., 2009; Nacu et al., 2013; McCusker et al., 2016). For example, *pattern-following* muscle-, Schwann-, and epidermal-derived blastema cells do not retain positional memory and contribute exclusively to the same type of tissue in the regenerate as their original identity in the mature tissue (Kragl et al., 2009; Nacu et al., 2013). However, *pattern-forming* dermal and periosteum cells (both connective tissue cell types) exhibit the greatest plasticity, contributing to the formation of several tissue types, including skeletal tissue, connective tissue, dermis, and vasculature (Kragl et al., 2009; Nacu et al., 2013; McCusker et al., 2016). Bone and cartilage, which appear to lack positional memory and do not contribute to the blastema, are generated from dermal and periosteum cells that have positional memory (Kragl et al., 2009; McCusker et al., 2016). A corollary of this is that positional memory must be erased in connective tissue–derived blastema cells on differentiation into cartilage and bone. How positional memory is lost during skeletal tissue differentiation and whether it plays a physiological role in tissue generation and homeostasis are uncertain and require further exploration.

Restricted plasticity of tissue identity during regeneration is a phenomenon documented in other regenerative model systems as well; however, the plasticity of connective tissue–derived cells has not been thoroughly explored in these systems. In the zebrafish caudal fin (Tu and Johnson, 2011), anuran tadpole tail (Gargioli and Slack, 2004), and murine digit tip (Rinkevich et al., 2011), the blastema generated in response to amputation is derived from a mixture of underlying mature tissues, which contribute to the regenerate in a lineage-restricted manner. In all three of these models, *position-following* cells (e.g., nerve, bone, and muscle) contribute exclusively to the same cell types in the regenerate (Gargioli and Slack, 2004; Rinkevich et al., 2011; Tu and Johnson, 2011). The fate of distinct *pattern-forming* cell populations (e.g., fibroblasts) in the regenerating tadpole tail and mouse digit tip has not been considered; however, in the zebrafish, these cells show restricted plasticity and contribute to only the same cell type in the regenerate (Tu and Johnson, 2011). This deviation from the axolotl system may be accounted for by technical issues, with the lineage-marking strategy used by Tu and colleagues (2011) possibly excluding the identification of all the fibroblast populations within the organism and, therefore, limiting the observation of cell-fate plasticity (as reviewed by Tanaka and Reddien, 2011).

19.3.2 Positional identity in the context of regeneration

Understanding the factors required for the initiation and progression of regeneration is complicated in the context of limb amputation, as this mode of injury results in excessive trauma and associated cellular processes, including inflammation and necrosis. The Accessory Limb Model (ALM) circumvents this problem by using a simple skin wound on the limb, in conjunction with a tissue graft from a different position on the limb and a nerve deviated to the site, to generate an ectopic limb by molecular mechanisms comparable to amputation-driven regeneration (Endo et al., 2004). However, ectopic outgrowth occurs only if normally non-adjacent cells of mesodermal origin (in other words, cells with differing positional information) are brought into contact with one another at the site of injury (French et al., 1976[*]; Bryant et al., 1981[†]; Endo et al., 2004). A skin graft, contralateral to the site of interest, is sufficient for this purpose; however, skin derived from unrelated sites, such as the head relative to a wound site on the limb, fails to support ectopic outgrowth (Tank, 1987; Satoh et al., 2007). These findings suggest that there may be a threshold of the positional discrepancy that can stimulate, and therefore be resolved, by regeneration (reviewed in McCusker and Gardiner, 2014[‡]). In addition, the requirement of positional information for regeneration to occur may,

---

[*] This paper presents the PCM of regeneration. The model was based on observations in regenerating insect and amphibian appendages and has been a powerful predictor of the pattern of the regenerated structures. The PCM is based on the idea that limb cells from different positions in the limb induce new limb pattern to form when coming in contact with cells from a different limb location. Regeneration at a wound or grafting site is generated through a process termed intercalation, whereby new cells are generated and these exhibit an intermediate positional identity to that of its adjacent cells. It is suggested that this process occurs by the shortest possible route and that distal outgrowth depends on a complete circle of positional information.

[†] This paper is a revision of the PCM of regeneration. In particular, the authors broaden the *complete circle* rule of the PCM, which states that distal outgrowth (the establishment of the pattern of the missing limb structure) will occur only if a *complete circle* of radial positional information is present in the regenerating limb environment. The thesis of this paper is that distal outgrowth occurs if a positional discontinuity is present. The authors predict that *incomplete circles* of positional information result in symmetrical or distally incomplete outgrowths.

[‡] This paper discusses the concept of tissue integration, in the context of axolotl regeneration, and how this could explain integration defects that have been observed in current regenerative medicine strategies.

in part, explain the high proportion of connective tissue cells contributing to the formation of the blastema (Muneoka et al., 1986[*]). As the cells of the blastema generate a new, integrated structure, a sufficient amount of local positional identity and disparity thereof would be required to guide the process. Connective tissue cells, as opposed to other cell types that do not retain positional memory, would therefore be a suitable source of this information, with more initial input dictating a greater likelihood of generating a complete regenerate (Muneoka et al., 1986).

The Polar Coordinate Model (PCM) of regeneration was established from observations in amphibian and insect models in the 1970s and thus far has been the most powerful predictive model to describe the pattern that forms when tissues with varying positional identities are allowed to interact in a permissive environment (French et al., 1976; Bryant et al., 1981). According to the PCM, discrepancies in the positional information of blastema cells that arise from different locations in the mature stump are resolved by a process called intercalation, where new cells are generated with the intermediate (missing) positional identities (French et al., 1976; Bryant et al., 1981). Although the molecular mechanisms of the intercalary response have not yet been identified, there is a strong correlation between the amount of sustained proliferation in the blastema and the complexity of the pattern that is regenerated in amphibian systems (Endo et al., 2004; McCusker et al., 2015b). Thus, the positional information present at the site of injury is imperative to establish the pattern of the regenerated limb structure, and an understanding of the mechanisms by which this occurs requires further investigation.

19.3.3 Positional identity in the context of normal tissue turnover

Although studies of positional identity are still in their infancy, experimental data suggest that positional identity maintains homeostasis during normal tissue turnover. Fibroblasts derived from the terminal murine phalanx, naturally associated with the nail bed, stimulated a phenotype in co-cultured epithelial progenitor cells that is reminiscent of a nail matrix; however, fibroblasts from the sub-terminal phalangeal element, natively covered by skin, stimulated keratinocyte differentiation instead (Wu et al., 2013). Furthermore, appropriate distal epithelial pathways are not induced in keratinocytes when co-cultured with human fibroblasts that

---

[*] This paper focuses on the contribution and role of the dermal and skeletal tissues in pattern formation during limb regeneration. The authors used triploid cell markers to lineage-trace the grafted dermal and skeletal populations in the host environment and observed that almost half of the blastema population was composed of cells from dermal origin, whereas only 2% of skeletal tissue contributed to the regenerate. The authors point out that the contribution of these tissues correlate with their effect on pattern formation: dermal cells having a large influence on the pattern of the regenerate and skeletal tissues having no effect.

do not express HoxA13, a distal location marker, relative to HoxA13-expressing counterparts (Rinn et al., 2008[*]). As epithelial cells do not carry positional information and undergo constant renewal, these data support the hypothesis that fibroblasts (and possibly other cells in the connective tissue that exhibit positional memory) dictate site-specific fates to their neighboring cells (Rinn et al., 2008; Wu et al., 2013; Sriram et al., 2015). Additional support for such a hypothesis can be found in planarians. β-catenin, an important regulator of gene transcription, exhibits graded expression along the anterior–posterior axis of the animal's body, which is highest at the proximal end. This is believed to maintain positional identity within planarians, the importance of which was clearly demonstrated when knocking down the expression of this protein resulted in the ectopic formation of anterior tissue at the proximal end of the animal during normal tissue turnover (Reuter et al., 2015).

19.4 Engineering the process of tissue integration

19.4.1 Overview and definition of integration

Integration is the process by which the pattern (or blueprint) of the newly regenerated or recently grafted tissues aligns seamlessly with the existing pattern in the surrounding tissues. Although tissue integration is a key step in the generation of a functional biological structure, it does not always happen during endogenous regeneration events under some surgically manipulated circumstances (Figure 19.3). For example, the proximal skeletal elements of an ectopic limb that have formed as a result of a nerve deviation and a contralateral tissue graft into a lateral limb wound do not integrate with the existing skeletal element in the limb proper (Endo et al., 2004). This simple observation suggests that tissue regeneration and integration are not necessarily coupled processes (reviewed in McCusker and Gardiner, 2014). Intriguingly, both processes are dependent on the property of positional information; the positional information in the remaining tissue generates the missing limb pattern during regeneration, and the positional information in both the regenerated and existing tissues controls their integration (reviewed in McCusker and Gardiner, 2014). In the following text, we will explain how the property of positional plasticity could also play a role in tissue integration.

---

[*] The focus of this study was to determine whether the Hox program in dermal fibroblasts contributes to skin positional identity. The authors discovered that dermal fibroblasts from the human foot or thigh stably express *Hox* genes after multiple passages. It was additionally observed that dermal fibroblasts from the foot use Wnt5A to induce the expression of K9, a *distal* marker in epidermal cells.

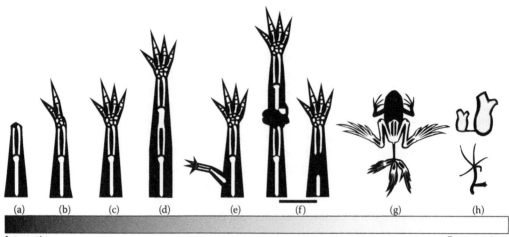

Integration   Regeneration

Figure 19.3 Integration phenotypes in animals that regenerate complex body structures. (A, B, C, D, F) Phenotypes from urodele salamanders that have been amputated at the wrist. (a) Denervation of a urodele limb early during the process of regeneration can completely inhibit the formation of new structures (Singer, 1946; Mullen et al., 1996; Suzuki et al., 2005). (b) Denervation of a limb at later stages, when the regenerate has initiated tissue specification, results in a hypomorphic regenerate with truncated distal structures. The same phenotype occurs when a late-stage regenerate is exposed to ectopic retinoic acid (Niazi et al., 1985). (c) A non-manipulated regenerate forms a normally patterned limb, where the new structures are completely integrated into the old structures. (d) Treatment of a mid- to late-stage regenerate with ectopic retinoic acid results in duplication of structures along the proximal/distal axis, which fuse and integrate with the old tissues. (e) An ectopic limb can be induced by innervating a wound site with a tissue graft from the opposite axis of the limb. The skeletal structures of the ectopic limb do not integrate with the skeleton in the stump, whereas the soft tissues have limited integration (Endo, 2004; Satoh, 2007; McCusker, 2013). (f) Increased duration of exposure of a mid-stage regenerate with retinoic acid often results in the formation of an amorphous tissue mass between the stump tissue and the regenerate, resulting in diminished integration with the stump tissue (Maden, 1982). (g) If retinoic acid signaling has been disrupted in some anuran species before metamorphosis, ectopic limbs form from the tail bud. These ectopic limbs are sometimes attached to the main body axis by a thin piece of tissue, exhibiting minimal integration (Mohanty-Hejmadi and Crawford, 2003). (h) Some organisms that reproduce asexually, such as ascidians and hydra, appear to be able to control the level of integration of the newly generated daughter offspring (Otto and Campbell, 1977; Manni et al., 2007).

19.4.2 The impact of cell plasticity on integration

We propose that plasticity of positional information promotes the integration of new tissue (regenerated or grafted) into the surrounding host environment. In amphibians, the integration of the grafted tissues is dependent on the positional information in both the grafted and host-site tissues (Figure 19.4) (reviewed in McCusker and Gardiner, 2014). This is exemplified by the following observations. If the grafted tissue is positionally plastic (e.g., early stage regenerate) or pattern-following (e.g., muscle tissue), it adopts the positional information of the host location and integrates seamlessly (Kragl et al., 2009; McCusker and Gardiner,

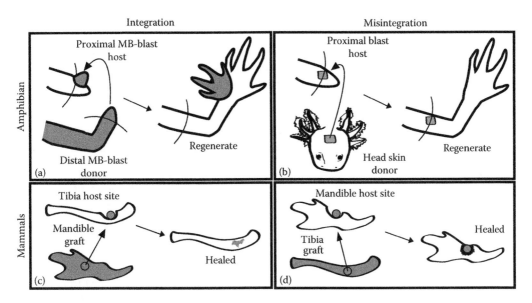

Figure 19.4 The relationship between positional plasticity and tissue integration. (a) In regenerating salamanders, grafts of midbud-staged blastemas onto host blastemas result in the integration of grafted cells with the host cells with the same positional identity, such that distal blastemas integrate into the distal location of the host blastema (Crawford and Stocum, 1988). (b) In contrast, grafts of head skin onto the salamander limb regenerate do not integrate into the host environment (Tank, 1987; Satoh et al., 2007). (c) In mammals, grafts of mandible tissue into the tibia alter their *Hox* code to match, and integrate, into the host tissue (Leucht et al., 2008). (d) Tibia grafts into the mandible retain the position-specific code of the tibia and fail to integrate into the host tissue (Leucht et al., 2008).

2013; Nacu et al., 2013; McCusker et al., 2014). If the positional information of the grafted tissue is not positionally plastic, it either integrates with the host by generating a new (intermediate) pattern or does not integrate (Figure 19.4a and b) (Tank, 1987; Satoh et al., 2007; McCusker and Gardiner, 2013). Positional plasticity also appears to be connected with tissue integration in mammals (Figure 19.4c and d). In mice, tibia skeletal progenitor cells retain positional markers consistent with their anatomical location, and when grafted into the mandible, these cells fail to integrate into the host location (Leucht et al., 2008). In contrast, skeletal progenitor cells derived from the mandible exhibit positional plasticity and are able to adopt the positional program of the host location when grafted into the tibia, thereby integrating seamlessly into this new environment (Leucht et al., 2008). Thus, the relationship between positional plasticity and tissue integration appears to be conserved among tetrapods.

19.4.3 Challenges of integration of grafted populations in cell-based therapies

Given that the goal of many cell-based regenerative therapies is to synthesize new tissues outside the body and then integrate them with the damaged host

423

environment, we will likely need to learn more about the positional information of both the graft and host tissues to facilitate this process. For example, how will we promote the integration of grafts into the existing structures that are composed of diverse populations of cells arranged in a specific way, without causing harm to the overall structure and/or function thereof, for example, in the heart (reviewed in Yang et al., 2014)? Implantation of a multi-cell-type graft, as opposed to a single-cell-type unit, is likely to be more biologically permissive, but the generation of such a structure is complicated and will require positional appropriate integration.

Moreover, will we be able to avoid the potential issues, such as the induction of ectopic growth (by means of an inappropriate intercalary response) and the presentation of aberrant positional cues to *pattern-following* cells, that may result from grafting in populations that have positional memory distinct to that of the host environment (reviewed in McCusker and Gardiner, 2014)? It is unknown whether the positional memory is erased or rendered plastic in grafted populations derived from ASC or induced pluripotent cell (iPSCs) lines established from connective tissue cells that stably retain positional information (Chang et al., 2002). Future work should focus on whether positional interrelations are responsible for graft/host pathologies such as tumor formation.

Lastly, can we control the integration of tissue grafts in order to prevent the common problem of *overintegration* of the grafted populations, such that they disappear from the host environment over time (reviewed in McCusker and Gardiner, 2014)? As we have described previously, cells with positional information appear to play an important role in tissue homeostasis by providing positional cues to cell types that do not retain positional information. Thus, the use of cells that lack positional information (e.g., muscle and nerves) in therapeutic strategies does not necessarily mitigate positional-related problems, because these cells still need to integrate with already-patterned tissues by following the positional cues dictated by the host environment, which could be difficult to control (McCusker and Gardiner, 2014). Further research is required to determine how such cell lines respond to pre-established positional information, so as to allow for functionally appropriate integration into the host site.

19.4.4 Potential solutions to promote integration of cell-based regenerative therapies

As we have described earlier, different biological challenges arise, depending on whether the cell type used for engraftment retains positional memory or not. As such, the solutions to promote tissue integration of engrafted populations will likely be different, depending on the status of this cellular property. In the following text, we will consider possible methods to facilitate the proper integration of grafted tissues by either (1) promoting positional plasticity in *pattern-forming* cells or (2) modulating positional cues to *pattern-following* cells by the host environment.

19.4.4.1 Promoting the integration of pattern-forming cells The ability to integrate grafted populations into a host environment is linked to the positional plasticity of the grafted population during endogenous regeneration events in amphibians and mammals. Thus, this phenomenon could be exploited to promote the integration of grafts that retain positional memory into humans (Figure 19.5). Since the discovery of the Yamanaka factors in 2006, a number of molecules have been shown to promote the plasticity of tissue identity, and it is possible that some of these factors also promote positional

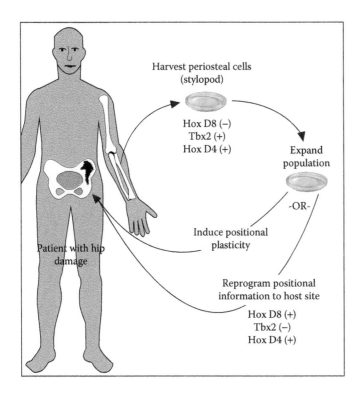

Figure 19.5 Incorporating positional plasticity in cell-based regenerative therapies to promote the integration of grafted tissues. We propose that altering the positional information in the grafted populations to match the host environment will facilitate tissue integration in future cell-based therapies. For example, human dermal fibroblasts from the forearm are *Hox D8* (–), *Tbx2* (+), and *Hox D4* (+), whereas fibroblasts from the hip area are *Hox D8* (+), *Tbx2* (–), and *Hox D4* (+) (Rinn, 2008). Thus, if periskeletal tissues were to be harvested from the forearm to generate a population of cells to graft and repair a damaged iliac crest of the hip, the positional information of the grafted population could be reprogrammed to match the host location: *Hox D8* (+), *Tbx2* (–), and *Hox D4* (+). Alternately, positional plasticity could be induced in the grafted population such that it adopts the positional identity of the host location endogenously, as described in (Leucht et al., 2008).

425

plasticity (Takahashi and Yamanaka, 2006). In addition, the induction of the wound-healing response also appears to induce dedifferentiation in the wound environment. This is exemplified by similar large-scale, structural modifications observed in the chromatin of cells situated throughout a wound site and those being reprogrammed to form iPSCs (Hay, 1959). Future research will be required to determine whether the wound-healing response also promotes positional plasticity.

Under some circumstances, it may also be beneficial to reprogram the positional information in grafted populations, so that it matches the new host location, and thus, the tissue can efficiently integrate at the host site (Figure 19.5). This will require a thorough examination of the positional program in the host location and the identification of methods to stably reprogram the positional information in the grafted tissue to match this program. Positional reprogramming has been accomplished in amphibian and mammalian embryos and adult tissues by treatment with ectopic RA (Bryant and Gardiner, 1992*; Manohar et al., 1996; Chen et al., 2007; Voss et al., 2009). Thus, the use of RA, or other molecules that are capable of reprogramming positional information, could be an invaluable tool in the development of some cell-based therapies.

19.4.4.2 Promoting the integration of pattern-following tissue grafts
Cell-based therapies that depend on the engraftment of cell types that do not retain positional information pose different challenges to integration. Since these cell types are *pattern-following*, their integration depends on their ability to receive positional cues from the cells in the host environment. Evidence from regenerating amphibian limbs suggests that growth factor signaling that is downstream of wound epithelium/nerve interactions induces an environment where *pattern-following* cell types integrate with the host site (McCusker and Gardiner, 2013). Therefore, one possible solution to promote integration of these cell types could be through the coordinated application of specific growth factors or other signaling molecules. However, continued integration would also not be desirable, as the *pattern-following*/grafted cells would eventually disappear from the original graft site. Thus, overintegration could be prevented by regulating the positional cues coming from the host environment. For example, evidence suggests that *pattern-forming* dermal cells use Wnt molecules to provide positional cue to *pattern-following* epithelia cells (Rinn et al., 2008). In this example, one could control the communication of positional cues between cell types by applying Wnt signaling inhibitors.

---

* This paper addresses the different effects of ectopic RA and local cell–cell interactions on patterning during regeneration. The authors propose that RA reprograms the positional information in the regenerating limb tissue to a singular identity on the limb, the most Posterior/Ventral/Proximal identity (PVPr).

Aside from overintegration, another reason why *pattern-following* cells disappear from the host site is that they require the coexistence of adult stem cell populations to replenish their population over time. Thus, there are multiple factors in addition to positional identity that could be harnessed to overcome the potential challenges of tissue integration in cell-based therapies.

References

Bryant, S. V., Endo, T. and Gardiner, D. M. (2002). Vertebrate limb regeneration and the origin of limb stem cells. *Int. J. Dev. Biol.* 46, 887–896.

Bryant, S. V. and Gardiner, D. M. (1992). Retinoic acid, local cell-cell interactions, and pattern formation in vertebrate limbs. *Dev. Biol.* 152, 1–25.

Bryant, S. V., French, V. and Bryant, P. J. (1981). Distal regeneration and symmetry. *Science* 212, 993–1002.

Chang, H. Y., Chi, J.-T., Dudoit, S., Bondre, C., van de Rijn, M., Botstein, D. and Brown, P. O. (2002). Diversity, topographic differentiation, and positional memory in human fibroblasts. *Proc. Natl. Acad. Sci. U.S.A.* 99, 12877–12882.

Chen, H., Zhang, H., Lee, J., Liang, X., Wu, X., Zhu, T., Lo, P.-K., Zhang, X. and Sukumar, S. (2007). HOXA5 acts directly downstream of retinoic acid receptor beta and contributes to retinoic acid-induced apoptosis and growth inhibition. *Cancer Res.* 67, 8007–8013.

Crawford, K. and Stocum, D. L. (1988). Retinoic acid coordinately proximalizes regenerate pattern and blastema differential affinity in axolotl limbs. *Development* 102, 687–698.

Endo, T., Bryant, S. V. and Gardiner, D. M. (2004). A stepwise model system for limb regeneration. *Dev. Biol.* 270, 135–145.

Ferris, D. R., Satoh, A., Mandefro, B., Cummings, G. M., Gardiner, D. M. and Rugg, E. L. (2010). *Ex vivo* generation of a functional and regenerative wound epithelium from axolotl (*Ambystoma mexicanum*) skin. *Dev. Growth Differ.* 52, 715–724.

French, V., Bryant, P. J. and Bryant, S. V. (1976). Pattern regulation in epimorphic fields. *Science* 193, 969–981.

Gargioli, C. and Slack, J. M. W. (2004). Cell lineage tracing during *Xenopus* tail regeneration. *Development* 131, 2669–2679.

Han, M., Yang, X., Lee, J., Allan, C. H. and Muneoka, K. (2008). Development and regeneration of the neonatal digit tip in mice. *Dev. Biol.* 315, 125–135.

Hay, E. D. (1959). Electron microscopic observations of muscle dedifferentiation in regenerating *Ambystoma* limbs. *Dev. Biol.* 1, 555–585.

Hay, E. D. and Fischman, D. A. (1961). Origin of the blastema in regenerating limbs of the newt *Triturus viridescens*. *Dev. Biol.* 3, 26–59.

Kawakami, Y., Rodriguez Esteban, C., Raya, M., Kawakami, H., Marti, M., Dubova, I. and Izpisúa-Belmonte, J. C. (2006). Wnt/beta-catenin signaling regulates vertebrate limb regeneration. *Genes Dev.* 20, 3232–3237.

Kragl, M., Knapp, D., Nacu, E., Khattak, S., Maden, M., Epperlein, H. H. and Tanaka, E. M. (2009). Cells keep a memory of their tissue origin during axolotl limb regeneration. *Nature* 460, 60–65.

Leucht, P., Kim, J.-B., Amasha, R., James, A. W., Girod, S. and Helms, J. A. (2008). Embryonic origin and Hox status determine progenitor cell fate during adult bone regeneration. *Development* 135, 2845–2854.

Maden, M. (1982). Vitamin A and pattern formation in the regenerating limb. *Nature* 295, 672–675.

Maden, M. (1983). The effect of vitamin A on the regenerating axolotl limb. *J. Embryol. exp. Morph.* 77, 273–295.

Manni, L., Zaniolo, G., Cima, F., Burighel, P. and Ballarin, L. (2007). *Botryllus schlosseri*: A model ascidian for the study of asexual reproduction. *Dev. Dyn.* 236, 335–352.

Manohar, C. F., Salwen, H. R., Furtado, M. R. and Cohn, S. L. (1996). Up-regulation of HOXC6, HOXD1, and HOXD8 homeobox gene expression in human neuroblastoma cells following chemical induction of differentiation. *Tumour Biol.* 17, 34–47.

McCusker, C., Athippozhy, A., Diaz-Castillo, C., Fowlkes, C., Gardiner, D. M. and Voss, S. R. (2015a). Positional plasticity in regenerating *Ambystoma mexicanum* limbs is associated with cell proliferation and pathways of cellular differentiation. *BMC Dev. Biol.* 15, 45–62.

McCusker, C., Bryant, S. V. and Gardiner, D. M. (2015b). The axolotl limb blastema: Cellular and molecular mechanisms driving blastema formation and limb regeneration in tetrapods. *Regeneration* 2, 54–71.

McCusker, C., Lehrberg, J. and Gardiner, D. (2014). Position-specific induction of ectopic limbs in non-regenerating blastemas on axolotl forelimbs. *Regeneration* 1, 27–34.

McCusker, C. D., Diaz-Castillo, C., Sosnik, J., Phan, A. and Gardiner, D. M. (2016). Cartilage and bone cells do not participate in skeletal regeneration in *Ambystoma mexicanum* limbs. *Dev. Biol.* 416, 26–33.

McCusker, C. D. and Gardiner, D. M. (2013). Positional information is reprogrammed in blastema cells of the regenerating limb of the axolotl (*Ambystoma mexicanum*). *PLoS ONE* 8, e77064.

McCusker, C. D. and Gardiner, D. M. (2014). Understanding positional cues in salamander limb regeneration: Implications for optimizing cell-based regenerative therapies. *Dis Model Mech.* 7, 593–599.

Mohanty-Hejmadi, P. and Crawford, M. J. (2003). Vitamin A, regeneration and homeotic transformation in anurans. *Proc. Indian Nat. Sci. Acad. B* 69, 673–690.

Mullen, L. M., Bryant, S. V., Torok, M. A., Blumberg, B. and Gardiner, D. M. (1996). Nerve dependency of regeneration: The role of distal-less and FGF signaling in amphibian limb regeneration. *Development* 122, 3487–3497.

Muneoka, K., Fox, W. F. and Bryant, S. V. (1986). Cellular contribution from dermis and cartilage to the regenerating limb blastema in axolotls. *Dev. Biol.* 116, 256–260.

Muneoka, K., Holler-Dinsmore, G. V. and Bryant, S. V. (1985). A quantitative analysis of regeneration from chimaeric limb stumps in the axolotl. *J. Embryol. Exp. Morphol.* 90, 1–12.

Nacu, E., Glausch, M., Le, H. Q., Damanik, F. F. R., Schuez, M., Knapp, D., Khattak, S., Richter, T. and Tanaka, E. M. (2013). Connective tissue cells, but not muscle cells, are involved in establishing the proximo-distal outcome of limb regeneration in the axolotl. *Development* 140, 513–518.

Niazi, I., Pescitelli, M. and Stocum, D. (1985). Stage-dependent effects of retinoic acid on regenerating urodele limbs. *Roux Arch. Dev. Biol.* 194, 355–363.

Otto, J. J. and Campbell, R. D. (1977). Budding in hydra attenuata: Bud stages and fate map. *J. Exp. Zool.* 200, 417–428.

Reuter, H., März, M., Vogg, M. C., Eccles, D., Grífol-Boldú, L., Wehner, D., Owlarn, S., Adell, T., Weidinger, G. and Bartscherer, K. (2015). B-catenin-dependent control of positional information along the AP body axis in planarians involves a teashirt family member. *Cell Rep.* 10, 253–265.

Rinkevich, Y., Lindau, P., Ueno, H., Longaker, M. T. and Weissman, I. L. (2011). Germ-layer and lineage-restricted stem/progenitors regenerate the mouse digit tip. *Nature* 476, 409–413.

Rinn, J. L., Bondre, C., Gladstone, H. B., Brown, P. O. and Chang, H. Y. (2007). Anatomic demarcation by positional variation in fibroblast gene expression programs. *PLoS Genet* 2, e119.

Rinn, J. L., Wang, J. K., Allen, N., Brugmann, S. A., Mikels, A. J., Liu, H., Ridky, T. W. et al. (2008). A dermal HOX transcriptional program regulates site-specific epidermal fate. *Genes Dev.* 22, 303–307.

Satoh, A., Gardiner, D. M., Bryant, S. V. and Endo, T. (2007). Nerve-induced ectopic limb blastemas in the axolotl are equivalent to amputation-induced blastemas. *Dev. Biol.* 312, 231–244.

Singer, M. (1946). The nervous system and regeneration of the forelimb of adult Triturus. V. The influence of number of nerve fibers, including a quantitative study of limb innervation. *J. Exp. Zool.* 101, 299–337.

Singer, M. (1974). Trophic functions of the neuron. VI. Other trophic systems. Neurotrophic control of limb regeneration in the newt. *Ann. N Y Acad. Sci.* 228, 308–322.

Singer, M. and Inoue, S. (1964). The nerve and the epidermal apical cap in regeneration of the forelimb of adult Triturus. *J. Exp. Zool.* 155, 105–115.

Sriram, G., Bigliardi, P. L. and Bigliardi-Qi, M. (2015). Fibroblast heterogeneity and its implications for engineering organotypic skin models in vitro. *Eur. J. Cell Biol.* 94, 483–512.

Suzuki, M., Satoh, A., Ide, H. and Tamura, K. (2005). Nerve-dependent and -independent events in blastema formation during *Xenopus* froglet limb regeneration. *Dev. Biol.* 286, 361–375.

Takahashi, K. and Yamanaka, S. (2006). Induction of pluripotent stem cells from mouse embryonic and adult fibroblast cultures by defined factors. *Cell* 126, 663–676.

Tanaka, E. M. and Reddien, P. W. (2011). The cellular basis for animal regeneration. *Dev. Cell* 21, 172–185.

Tank, P. W. (1987). The effect of nonlimb tissues on forelimb regeneration in the axolotl, *Ambystoma-Mexicanum*. *J. Exp. Zool.* 244, 409–423.

Tu, S. and Johnson, S. L. (2011). Fate restriction in the growing and regenerating zebrafish fin. *Dev. Cell* 20, 725–732.

Voss, A. K., Collin, C., Dixon, M. P. and Thomas, T. (2009). Moz and retinoic acid coordinately regulate H3K9 acetylation, hox gene expression, and segment identity. *Dev. Cell* 17, 674–686.

Wu, Y., Wang, K., Karapetyan, A., Fernando, W. A., Simkin, J., Han, M., Rugg, E. L. and Muneoka, K. (2013). Connective tissue fibroblast properties are position-dependent during mouse digit tip regeneration. *PLoS ONE* 8, e54764.

Yang, X., Pabon, L. and Murry, C. E. (2014). Engineering adolescence: Maturation of human pluripotent stem cell-derived cardiomyocytes. *Circ. Res.* 114, 511–523.

Section IV
Principles of organ development and regeneration

Efforts to induce tissue repair and regeneration have been ongoing since the early discoveries that regeneration can occur naturally in some animals. In recent years, remarkable progress has been made in understanding how to go about regenerating body parts, and this last section of this book focuses on some examples of what is possible for the future. These advances are evidence that we should be encouraged about the potential for regenerative engineering, not only in replaced tissue defects resulting from injury, but also those that are a consequence of aging. If the function of an organ can be enhanced or restored as we age, regenerative engineering will prove to be the long sought-after fountain of youth.

In part, these advances have resulted from a paradigm shift in thinking about inducing human regeneration. For over a century, the regenerative ability of animals such as worms, fishes, and salamanders was taken for granted, yet there was a presumption that it would never occur for humans. Over the past few decades this attitude has changed, and it is time to consider regeneration as a fundamental biological property that is shared to varying degrees by all animals. There are increasing examples of where advances in regeneration are being made, and based on the successes and failures to date, it is evident that it will be easier for some tissues and organs, and harder for others. Regardless, it was not very long ago that the question was whether or not human regeneration was possible; whereas, today the question is how long will it take.

Chapter 20 Functional ectodermal organ regeneration based on epithelial and mesenchymal interactions

Masamitsu Oshima
and Takashi Tsuji

Contents

Key concepts

a. We have developed a bioengineering technology for generating a three-dimensional organ germ that is induced by reciprocal epithelial–mesenchymal interactions in the developing embryo.

b. We have represented a fully functional bioengineered ectodermal organ replacement after the transplantation of respective ectodermal bioengineered organ germ. Our technology demonstrates the potential for bioengineered organ replacement in future regenerative therapy.

c. We have demonstrated a next-generation important concept in the development of a bioengineered 3D integumentary organ system from induced pluripotent stem (iPS) cells by using an *in vivo* clustering-dependent embryoid body (CDB) transplantation model.

d. Our bioengineered integumentary organ system was fully functional after transplantation *in vivo* and could be a proper connection to the host's surrounding tissues without tumorigenesis.

20.1 Introduction

Organogenesis is accomplished through a complex process comprising tissue self-organization, cell-to-cell interactions, spatiotemporal gene expressions, and cell movement (Takeichi 2011, Sasai 2013a, 2013b). During embryonic development, organ-forming fields are generated in an orderly, specific process known as embryonic pattern formation (Sasai 2013b). Almost all organs originate from their respective organ germs through reciprocal epithelial–mesenchymal interactions between the immature epithelium and the mesenchyme in each organ-forming field (Pispa and Thesleff 2003, Thesleff 2003, Tucker and Sharpe 2004, Sharpe and Young 2005). Organ development, which relies on inductive properties such as regional and genetic specificity, is regulated by a developmental mechanism based on epithelial–mesenchymal interactions, based on signaling molecules and transcription factor pathways (Gilbert 2013). Ectodermal organs such as the teeth, salivary glands, lacrimal glands, and hair follicles demonstrate extremely similar developmental processes and develop from their respective germ layers through reciprocal interactions based on the epithelium and mesenchyme in the developing embryo. The principal interactions of ectodermal organ development are common to those of other organs. They allow for the organization of a 3D structure consisting of various tissues and cell populations that coordinate with surrounding tissues such as blood vessels and peripheral nerves to achieve the respective physiological organ functions (Jahoda et al. 2003, Pispa and Thesleff 2003, Thesleff 2003, Nishimura et al. 2005, Schechter et al. 2010, Tucker and Miletich 2010).

Current advances in future regenerative therapies have been inspired by many previous research fields such as embryonic development, stem cell biology, and tissue engineering technology (Langer and Vacanti 1999, Atala 2005, Brockes and Kumar 2005, Madeira et al. 2015). As an attractive regenerative concept, stem cell transplantation using tissue-derived stem cells, embryonic stem (ES) cells, or induced pluripotent stem (iPS) cells has been attempted for repairing the damaged tissues underlying structural and functional diseases (Addis and Epstein 2013, Takebe et al. 2013, Trounson et al. 2013, Kamao et al. 2014). Regenerative therapy can develop fully functional bioengineered tissues/organs that can replace lost or damaged organs after disease, injury, or aging (Ikeda and Tsuji 2008). Cell-sheet–based therapy allows tissue reconstruction from stem cells grown on a sheet. This regenerative approach can regenerate a broad range of damaged tissues, such as those from burns and cardiac dysfunction, through cell-sheet transplantation (Yang et al. 2005, Miyahara et al. 2006). In addition, an organoid model, which replicates 3D structures, such as parts of organs, has been generated for several organs, including the intestine, pancreas, and liver. Organoids can be derived from both isolated tissue-specific stem cells and isolated tissue fragments from the corresponding organs and would be available for regenerative therapies through the replication of their tissue-specific stem cell niches (Sato et al. 2009, Greggio et al. 2013, Huch et al. 2013a, Huch et al. 2013b, Tan and Barker 2014). Organ replacement regenerative therapy offers great potential for the replacement of dysfunctional organs with fully functional bioengineered organs that are reconstructed by *in vitro* 3D cell manipulation using candidate stem cells (Atala 2005, Purnell 2008). Many attempts to generate fully functional substitute organs that can replace lost or damaged tissues have been reported. This concept also includes fully artificial organs, which are made from mechanical and chemical devices with computer chips, to reproduce the physiological organ functions of organs, including heart, eyes, and kidneys (Wolf 1952, Copeland et al. 2004). Another tissue engineering strategy that has been used is to create bio-artificial organs composed of living cells and natural or artificial polymers, such as those that can reproduce the biochemical organ functions in the liver and pancreas (Colton 1995, Fort et al. 2008). However, current artificial organs consisting of various functional cells and artificial materials cannot achieve full functionality and thus are not available for long-term organ replacement therapy *in vivo* (Uygun et al. 2010). Further advances in biotechnology are truly required for achieving the functional replacement of lost or damaged tissues and organs by using organ regenerative therapy.

Pluripotent stem cells, including ES and iPS cells, can be induced to differentiate into specific somatic cell lineages by using cytokines or oxygen concentrations that reproduce the environments and the distinctive patterning and positioning signals during embryogenesis (Cohen and Melton 2011). The patterning signals indicating the body axis and the organ-forming fields are strictly regulated by signaling centers based on the embryonic body plan (Walck-Shannon and Hardin 2014). These signals regarding complex pattern formation in the local sites of the body may lack centralized organizing signals, as can be observed in the teratomas, which generate disorganized neural tissues, muscle, cartilage, and bronchial epithelia (Sasai 2013b). Although pluripotent stem cells have a self-organization ability, it can be difficult to regulate the development of specific types of compartmentalized tissue. Recently, the generation of neuroectodermal and endodermal organs has been demonstrated through the regulation of complex patterning signals during embryogenesis and the self-formation of pluripotent stem cells in 3D stem cell culture (Tavassoll and Devilee 2003, Eiraku et al. 2011, Suga et al. 2011, Nakano et al. 2012, Koehler et al. 2013, Sasai 2013a, 2013b, Watson et al. 2014). These bioengineered organs may contribute to the development of next-generation regenerative approaches; however, ectodermal organs such as the skin, hair follicles, teeth, and exocrine organs cannot be sufficiently reproduced from pluripotent stem cells (Nakano et al. 2012, Sasai 2013a). It is thus anticipated that future bioengineering technology will eventually be able to reproduce the fully functional regenerated organs through the proper application of pluripotent stem cells by using an *in vitro* 3D culture (Figure 20.1).

In this chapter, we will describe a bioengineering technology for the generation of a 3D germ layer, using completely dissociated epithelial and mesenchymal cells (Nakao et al. 2007). This technology established the fully functional organ regeneration in teeth; exocrine organs, including the salivary glands/lacrimal glands; and hair follicles. In addition, we recently generated a bioengineered 3D integumentary organ system from iPS cells that includes appendage organs such as the hair follicles and sebaceous glands (Takagi et al. 2016). These bioengineering technologies can provide substantial advances in future organ replacement regenerative therapy.

20.2 A novel three-dimensional cell manipulation technology: The "organ germ method"

The ultimate goal of regenerative therapy is to completely restore lost or damaged tissues by using a fully functioning bioengineered organ (Oshima and Tsuji 2014). Almost all organs, including ectodermal organs, arise from their respective organ germs, which are induced by

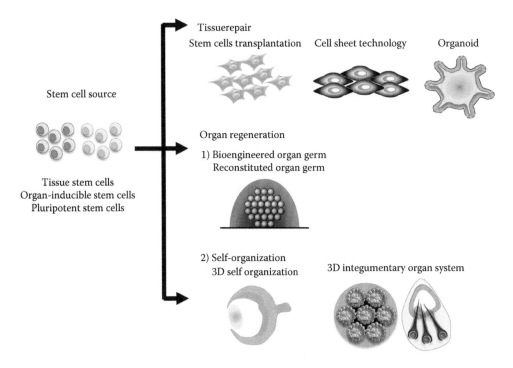

Figure 20.1 **Regenerative concept for tissue repair and organ regeneration.** As attractive regenerative concepts for tissue repair, stem cell transplantation and cell-sheet technology using tissue-derived stem cells and pluripotent stem cells have been attempted for repairing damaged tissues. In addition, next-generation bioengineering technologies are applicable for complex organ regeneration by using organoid models, the bioengineered organ germ method, 3D self-organization culture, and 3D integumentary organ system.

reciprocal epithelial–mesenchymal interactions during early embryonic development (Figure 20.2). The bioengineering technology for regenerating 3D organs has progressed to the replication of organogenesis based on epithelial–mesenchymal interactions, thereby enabling the development of fully functional bioengineered organs by using bioengineered organ germs that are generated from immature stem cells through 3D cell manipulation *in vitro* (Ikeda and Tsuji 2008, Oshima and Tsuji 2014).

To achieve the precise replication of the developmental processes in organogenesis, an *in vitro* 3D cell manipulation method designated as an *organ germ method* has recently been established by using a tooth-and-hair-follicle-developmental model (Figure 20.3a) (Nakao et al. 2007). We demonstrated the possibility of generating a bioengineered tooth germ by using completely dissociated single epithelial and mesenchymal cells derived from tooth germs in ED14.5 mice. The most important breakthrough in our organ germ method is the achievement of the 3D cell compartmentalization of epithelial and mesenchymal cells at a high cell density in a collagen gel. This bioengineered tooth germ

437

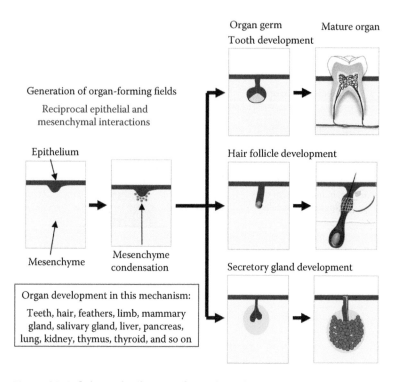

Figure 20.2 Schematic diagram of ectodermal organ development. Organogenesis occurs through reciprocal epithelial–mesenchymal interactions. Almost all organs, including ectodermal organs, arise from their respective organ germs, which are induced by reciprocal epithelial–mesenchymal interactions during early embryonic development. Ectodermal organs such as the teeth, hair follicles, salivary glands, and lacrimal glands develop by similar processes.

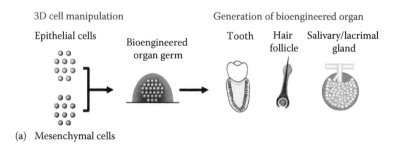

Figure 20.3 The organ germ method: a novel 3D cell-processing system. (a) Strategy for functional organ regeneration using the organ germ method. Functional organs such as teeth, hair follicles, salivary glands, and lacrimal glands could be regenerated by the transplantation of bioengineered organ germs reconstituted from epithelial and mesenchymal cells through the organ germ method.   *(Continued)*

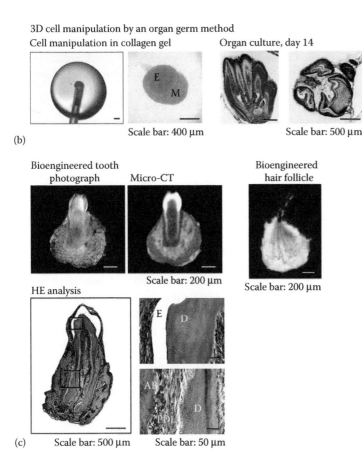

3D cell manipulation by an organ germ method

Cell manipulation in collagen gel   Organ culture, day 14

Scale bar: 400 μm   Scale bar: 500 μm

(b)

Bioengineered tooth photograph   Micro-CT   Bioengineered hair follicle

Scale bar: 200 μm   Scale bar: 200 μm

HE analysis

Scale bar: 500 μm   Scale bar: 50 μm

(c)

Figure 20.3 (Continued) The organ germ method: a novel 3D cell-processing system. (b) Tooth germ–derived mesenchymal cells at a high density are injected into the center of a collagen drop. Subsequently, tooth germ?derived epithelial cells are injected into the drop adjacent to the mesenchymal cell aggregate. Within 1 day of the organ culture, the formation of bioengineered tooth germ, with appropriate compartmentalization between the epithelial and mesenchymal cells and cell-to-cell compaction, was observed (left panels). Bioengineered incisor and molar tooth germs, which were reconstituted using dissociated cells isolated from respective tooth germs, developed in organ culture in 14 days (right panels). (c) Transplantation of this bioengineered tooth germ into a subrenal capsule for 30 days (left-upper panels). A bioengineered tooth unit comprising a mature tooth, PDL, and alveolar bone, with the correct tooth tissue structures such as enamel (E), dentin (D), the PDL, and alveolar bone (left-lower panels). The bioengineered hair follicle is also generated by the organ germ method (right panels).

can achieve initial tooth development, with the appropriate cell-to-cell compaction between epithelial and mesenchymal cells, in an *in vitro* organ culture. The development of bioengineering tooth germ, which is adopted in our method, successfully replicates the multicellular assembly, including ameloblasts, odontoblasts, pulp cells, and dental follicle cells, underlying the epithelial–mesenchymal interactions and natural tooth development (Figure 20.3b). The bioengineered tooth germ generates a structurally correct tooth after transplantation into a subrenal capsule *in vivo* and in organ culture *in vitro* (Figure 20.3c) (Nakao et al. 2007). The tooth morphology is defined by not only the tooth length and crown size as macromorphological features but also the cusp numbers/ positions as micro-morphological features. These morphological properties are regulated by specific gene expressions at the boundary surface of the immature oral epithelium and neural crest-derived mesenchyme in the embryonic jaw. Our bioengineered tooth germ could be reconstructed by adjusting the cell-to-cell contact length between the epithelial and mesenchymal cell layers, and the crown width and cusp number of the bioengineered tooth are dependent on the contact length of the epithelial and mesenchymal cell layers (Ishida et al. 2011). Our cell manipulation technology can also be used for the regeneration of hair follicles (Figure 20.3c), and it offers the possibility of generating many types of bioengineered organ germs (Nakao et al. 2007).

20.3 Functional tooth regeneration

Teeth have important oral functions such as mastication, pronunciation, and facial aesthetics and thus have a critical influence on the local/ general health and quality of life (Dawson 2006). These tooth-related biological functions are established with the dentition, masticatory muscles, and the temporomandibular joint under the control of the central nervous system (Dawson 2006). To restore the occlusal function after tooth loss, several dental therapies that replace the tooth with artificial materials, such as fixed dental bridges, removable dentures, and dental implants, have been widely performed as conventional dental treatments (Brenemark and Zarb 1985, Burns et al. 2003, Pokorny et al. 2008, Rosenstiel et al. 2015). Although these artificial therapies have been widely applied in the rehabilitation of tooth loss, further technological improvements based on biological findings are expected to restore physiological functions of the tooth (Huang et al. 2008).

To represent a successful tooth replacement therapy, a bioengineered tooth must be capable of being grafted into the tooth-loss region under the adult oral environment and achieving full functionality, including sufficient occlusal performance, biological cooperation with the periodontal ligament (PDL), and afferent responsiveness to noxious stimuli in the maxillofacial region (Wise et al. 2002, Ohazama et al. 2004, Wise and King 2008). We have recently demonstrated that a bioengineered tooth germ can develop with the correct tooth structure and completely erupt in an oral cavity 49 days after transplantation (Figure 20.4a) (Nakao et al. 2007, Ikeda et al. 2009). The erupted bioengineered tooth not only reached the occlusal plane but also performed an occlusal function with the opposing natural tooth (Figure 20.4a) (Ikeda et al. 2009). As an another tooth regenerative concept, the most critical consideration is whether a bioengineered tooth unit composed of mature tooth, PDL, and alveolar bone and engrafted into the tooth-loss region can exert the immediate organ performance, with its full functions *in vivo* (Gridelli and Remuzzi 2000). We have demonstrated the successful engraftment of a bioengineered tooth unit, with bone integration

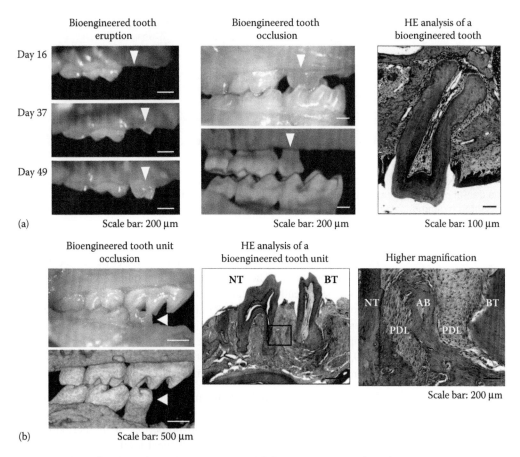

Figure 20.4 Functional whole-tooth regeneration. (a) The transplanted bioengineered tooth germ erupted (arrowheads, left panels) and reached the occlusal plane with the opposing lower first molar at 49 days after transplantation (center panels). The bioengineered tooth also formed a correct tooth structure comprising enamel, dentin, dental pulp, and periodontal tissue (right panels). (b) The bioengineered tooth unit was engrafted through bone integration and reached the occlusal plane with the opposing upper first molar at 40 days after transplantation (arrowheads, left panels). The engrafted bioengineered tooth unit also had the correct tooth structure in the recipient jaw. NT, natural tooth; BT, bioengineered tooth; AB, alveolar bone; PDL, periodontal ligament (center and right panels). (*Continued*)

441

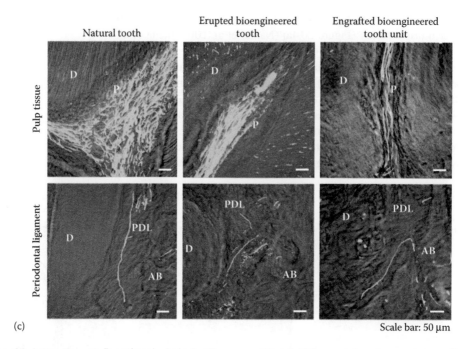

Natural tooth

Erupted bioengineered tooth

Engrafted bioengineered tooth unit

Pulp tissue

Periodontal ligament

(c)

Scale bar: 50 μm

Figure 20.4 (Continued) Functional whole-tooth regeneration. (c) The peripheral nerve innervation of the dental pulp and periodontal ligament area in the natural and bioengineered teeth were represented by immunostaining of neurofilament-H. D, dentin; P, pulp; PDL, periodontal ligament; AB, alveolar bone.

between the alveolar bone of the tooth unit and that of the recipient 40 days after transplantation. The PDL that originated in the bioengineered tooth unit was sufficiently maintained during the bone integration and regeneration (Figure 20.4b) (Oshima et al. 2011). Furthermore, the hardness of the enamel and dentin in the bioengineered tooth was equivalent to that of the natural teeth, as measured using the Knoop hardness test (Ikeda et al. 2009, Oshima et al. 2011). These tooth regenerative approaches indicate the potential for successfully restoring masticatory performance.

For the realization of tooth regenerative therapy, the bioengineered tooth must have the correct tooth structure and appropriate biological functionality *in vivo*. The structural properties of the periodontal tissue are important to exert the physiological tooth functions, including the absorption of occlusal loading, the maintenance of the alveolar bone height, and orthodontic tooth movement through bone remodeling (Avery 2002, Fukumoto and Yamada 2005). Autologous tooth transplantation studies have indicated that the remaining healthy PDLs around the tooth root could successfully restore the physiological tooth functions, including bone remodeling and the prevention of ankylosis (Tsukiboshi 1993). Thus, the functional cooperation between the teeth and the maxillofacial region through the biological connection of the PDL is essential for the exertion of oral functions (Avery 2002, Lindhe et al. 2008). We have demonstrated that bioengineered teeth are able to successfully undergo functional tooth movement equivalent to that of natural teeth, based on the proper localization of osteoclasts and osteoblasts in response to mechanical stress (Ikeda et al. 2009, Oshima et al. 2011). These findings indicate that a bioengineered tooth can replicate the PDL function and lead to the re-establishment of functional teeth within the maxillofacial region (Ikeda et al. 2009, Oshima et al. 2011). The teeth and periodontal tissue are peripherally governed by the sensory trigeminal and sympathetic nerves. The afferent nervous system in the maxillofacial region plays important roles in the regulation of the physiological functions of the tooth and the perception of noxious stimuli (Luukko et al. 2005). It is thus anticipated that neuronal regeneration, including the re-entry of nerve fibers after the engraftment of the bioengineered tooth, is required for regenerating the perceptive potential against noxious stimuli in future tooth regenerative therapy (Luukko et al. 2005). We have demonstrated that sensory and sympathetic nerve fibers can innervate both the pulp and the PDL region in the engrafted bioengineered tooth (Figure 20.4c) (Ikeda et al. 2009, Oshima et al. 2011). Bioengineered teeth can receive pain stimulation due to pulp injury and orthodontic movement and have the ability to properly transduce these peripheral stimulations to the central nervous system via c-Fos immunoreactive neurons (Ikeda et al. 2009, Oshima et al. 2011). These findings indicate that bioengineered teeth can restore the proprioceptive responses to noxious stimuli within the maxillofacial region.

Regenerated teeth developed from bioengineered germs or transplanted bioengineered mature tooth units have successfully demonstrated physiological functions of the tooth in cooperation with the maxillofacial region such as sufficient masticatory performance, biological connections with periodontal tissues, and afferent responsiveness to noxious stimuli. Our bioengineering technology for fully functional tooth regeneration will contribute to the substantial advances in whole-tooth replacement regenerative therapy in the future (Nakao et al. 2007, Ikeda et al. 2009, Oshima et al. 2011).

20.4 Functional secretory gland regeneration

Secretory glands, including salivary glands and lacrimal glands, are important for the protection of appendage organs and the maintenance of physiological functions in the microenvironment of the oral and ocular surfaces. Salivary glands play essential roles in the normal functioning of the upper gastrointestinal tract and oral health via saliva production. There are three major salivary glands: the parotid gland (PG), submandibular gland (SMG), and sublingual gland (SLG). There is also a minor salivary gland. Serous saliva that is produced from the PG and SMG primarily contains amylase proteins for the digestion of foods, whereas mucous saliva that is produced from the SLG contains glycoproteins and mucin proteins for dryness protection in the oral cavity (Avery 2002, Edgar et al. 2004, Tucker and Miletich 2010). Lacrimal glands play multilateral roles in the protection of a healthy ocular surface epithelium through the production of tears from the impairment of air exposure. The lacrimal gland comprises a main gland that secretes aqueous tears and some small accessory glands (Schechter et al. 2010). These mature glands are organized in accordance with a tubuloalveolar scheme: the acini that carry the fluid to the mucosal surface through the lacrimal duct, the myoepithelial cells that envelop the acini, and the early duct elements (Melnick et al. 2009, Schechter et al. 2010). The aqueous layer of the tear is formed by secretion from the lacrimal glands. It contains water and many tear proteins, including lactoferrin, for the expression of biological functions such as moisturizing the ocular surface and antimicrobial activity (Ohashi et al. 2006, Hirayama et al. 2013b).

Secretory gland impairment causes acinar cell damage and dysfunction of the fluid secretion, which lead to xerostomia (dry mouth syndrome) and dry eye disease, respectively. These diseases can be caused by various physiological disorders such as aging, injury, Sjögren's syndrome, and complications from surgery and radiation therapy (Ship et al. 2002, Gary 2007). Xerostomia exacerbates some oral problems, including dental decay, periodontal disease, bacterial infection, swallowing dysfunction, and dysgeusia (Atkinson et al. 2005), whereas dry eye disease causes ocular discomfort, fatigue, and visual impairment that can interfere with daily activities (Miljanovic et al. 2007). Thus, secretory gland dysfunctions are recognized as important public health problems that may decrease the daily activity and quality of life. Conventional clinical therapies for xerostomia and dry eye disease are mainly palliative approaches utilizing the artificial substitutes (Stephen 2007). Therefore, a novel curative treatment based on a biological approach is required for the restoration of the salivary gland and lacrimal gland functions (Kagami et al. 2008, Zoukhri 2010).

To recover secretory gland dysfunction, we tried to regenerate secretory glands, including the salivary and lacrimal glands, as an organ replacement regenerative approach for xerostomia and dry eye disease. We investigated whether our organ germ method could regenerate bioengineered salivary and lacrimal glands. The bioengineered gland germs successfully underwent branching morphogenesis, followed by stalk elongation and cleft formation. Thus, bioengineered salivary and lacrimal gland germs could be generated by our organ germ method. A functional duct association between the host salivary/lacrimal gland ducts and the bioengineered salivary/lacrimal gland germ is essential for the differentiation of the acinar formation and the oral/ocular physiological functions. It is important to establish a bioengineered salivary/lacrimal gland transplantation method to achieve fluid secretion into the oral cavity or ocular surface from the regenerated glands. We therefore developed gland-defect models in mice in which we resected all three major salivary glands or lacrimal glands for the demonstration of fluid secretion from the bioengineered glands.

The bioengineered salivary/lacrimal gland germ has been transplanted into defect mice using an inter-epithelial tissue-connecting plastic method (Toyoshima et al. 2012). The duct connection between the host duct and bioengineered salivary/lacrimal gland epithelial cells could be achieved, and then, the bioengineered salivary/lacrimal glands were engrafted into adult mice at 30 days after transplantation (Figure 20.5a). The engrafted bioengineered glands had a correct acinar structure, including the expression of aquaporin-5, which was a membrane channel involved in the secretion of water, and were properly enveloped as an acinar structure by myoepithelial cells. The engrafted glands also received correct nerve invasion (Figure 20.5b). These findings indicate that the bioengineered glands

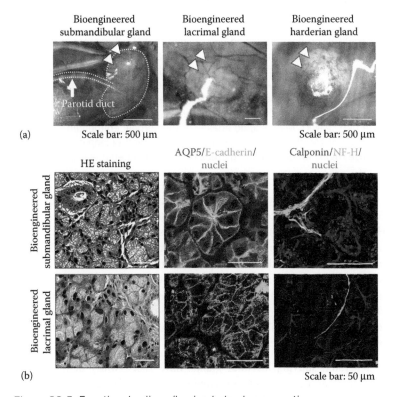

Figure 20.5 Functional salivary/lacrimal gland regeneration.
(a) Photographs of the bioengineered submandibular gland germ, lacrimal gland germ, and harderian gland germ after transplantation. At 30 days after transplantation, the bioengineered submandibular gland, bioengineered lacrimal gland, and bioengineered harderian gland were engrafted. (b) Histological and immunohistochemical analyses of the submandibular gland and bioengineered lacrimal gland after transplantation. Hematoxylin and eosin staining revealed that the bioengineered lacrimal gland achieved a mature secretory gland structure, including acini and a duct (left panels). Aquaporin-5 is red and E-cadherin is green in the center panel. Calponin is red and E-cadherin is green in the center-right panel. Calponin is red, neurofilament-H (NF-H) is green, and DAPI is blue in the right panel. *(Continued)*

445

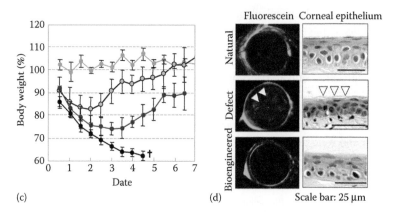

Figure 20.5 (Continued) Functional salivary/lacrimal gland regeneration. (c) Measurement of body weight every 0.5 days after transplantation in normal mice (gray dots), salivary-gland–defect mice (black dots), salivary-gland–engrafted mice (red dots), and salivary-gland–defect mice that were given high-viscosity water (green dots). All salivary-gland–defect mice died within 5 days (†) after the removal of all major salivary glands. (d) Representative images of the corneal surface of a natural lacrimal gland (left, upper), a lacrimal-gland–defect mouse (left, center), and a bioengineered lacrimal-gland–engrafted mouse (left, lower). Microscopic images of the corneal epithelium, including a natural mouse (right, upper), lacrimal-gland–defect mouse (right, center), and bioengineered lacrimal-gland–transplanted mouse (right, lower). Impaired areas of the corneal epithelium are highlighted by arrowheads.

could successfully develop both the correct histological structures and functional performance required for fluid secretion (Ogawa et al. 2013, Hirayama et al. 2013a). Our bioengineered salivary gland could mitigate disorders resulting from salivary gland hypofunction, including dryness, bacterial infection, and swallowing dysfunction (Figure 20.5c). Similarly, our bioengineered lacrimal gland could prevent ocular surface impairment resulting from dry eye disease (Ogawa et al. 2013, Hirayama et al. 2013a) (Figure 20.5d). These results indicate that functional replacements using bioengineered salivary/lacrimal glands would be an attractive therapeutic strategy for severe dry syndromes.

Regenerated salivary/lacrimal glands have successfully demonstrated physiological secretory functions under afferent and efferent neuron network control, in cooperation with the central nervous system. Our bioengineering technology for fully functional secretory gland regeneration showed a proof of concept for bioengineered secretory organ replacement therapy in the future.

20.5 Functional hair follicle regeneration

The hair organ plays biologically important roles in thermoregulation, physical insulation from ultraviolet (UV) radiation, waterproofing, tactile sensation, protection against noxious stimuli, camouflage, and social communication (Chuong 1998). The hair follicle consists of a permanent upper region, including the infundibulum and isthmus, and a variable lower region containing the hair matrix, differentiated epithelial cells, and dermal papilla (DP) cells that can generate dermal cell populations (Fuchs 2007, Yamao et al. 2010). Hair contains various stem cell origins such as epithelial cells in the follicle stem cell niche of the bulge region, multipotent mesenchymal precursors in the DP cells (Oshima et al. 2001, Blanpain et al. 2004), neural crest–derived melanocyte progenitors of the sub-bulge region (Nishimura et al. 2002), and follicle epithelial cells of the bulge region to complement the connection to the arrector pili muscle (Fujiwara et al. 2011). Reciprocal interactions between the epithelial stem cells and mesenchymal precursor cells promote the hair cycle that depends on the cell activation in the telogen–anagen transition and the anagen, catagen, and telogen phases (Greco et al. 2009).

Hair-loss disorders, such as alopecia areata and androgenetic alopecia, cause emotional distress to the patient and have negative effects on the quality of life in both men and women of all ages (Springer et al. 2003, Mounsey 2009). Current pharmacological treatments do not achieve a sufficient effect against hair loss, even in common conditions such as androgenetic alopecia and alopecia areata (Springer et al. 2003, Mounsey 2009). Although hair replacement surgery has experienced a rapid evolution since the inception of the large-scale punch graft many decades ago, follicular unit transplantation (FUT) is the culmination of the refinement of hair transplantation techniques. It is considered that FUT achieves superior results, with minimized morbidity, and is thus the state-of-the-art method for the treatment of both male and female pattern alopecia (Rousso and Presti 2008). These approaches for conventional hair treatments are very effective in the clinic, but from a biological perspective, it is hoped that the development of hair organ regenerative technology will widely enable future regenerative therapy for hair loss.

Critical issues to be considered in hair regenerative therapy include whether the bioengineered hair follicles can regenerate a proper hair structure and maintain full physiological functions according to their fate determination (Hardy 1992, Kutzner et al. 2006). To achieve the fully functional regeneration of hair follicles, many bioengineering technologies have been developed, including reconstructing the variable region of the hair follicle (Jahoda et al. 1984, Jahoda et al. 1996) and *de novo* folliculogenesis via the self-assembly of epithelial and mesenchymal cells that are isolated from the skin and hair follicles (Weinberg et al. 1993). In a previous report, we demonstrated that a bioengineered hair follicle germ, through the reconstruction of embryonic follicle germ-derived epithelial and mesenchymal cells by using our organ germ method, could regenerate functional bioengineered hair follicle and hair shaft (Nakao et al. 2007). To establish the proper connection between the epithelium of host skin and the bioengineered hair follicle germ, we developed an interepithelial tissue-connecting plastic device that uses a nylon thread as an artificial guide for the correct directional formation

447

of the infundibulum. We also indicated that bioengineered hair follicles such as pelage and vibrissae could successfully erupt from skin surface, grow at a high frequency, and re-establish successful connections with the recipient skin in adult mice (Toyoshima et al. 2012). Thus, the bioengineered pelage and vibrissae follicle germs could regenerate structurally correct hair follicles and shafts after intracutaneous transplantation (Figure 20.6a).

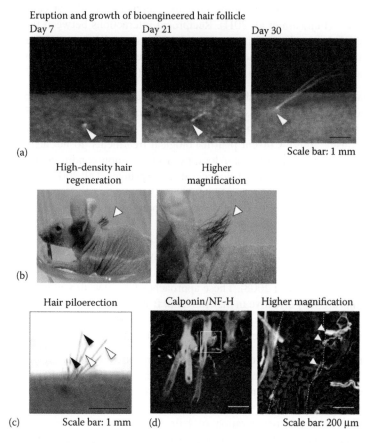

Figure 20.6 Functional hair follicle regeneration. (a) Macromorphological observations of the bioengineered hairs (arrowhead), reconstituted from the epithelial and mesenchymal cells from adult hair follicles at day 7, day 21, and day 30 after transplantation. (b) Macromorphological observations of high-density intracutaneous transplantation of the bioengineered follicle germs. (c) Successful piloerection was observed in bioengineered hair follicles after intradermal injection of acetylcholine (Ach). White, before intradermal injection; and black, after intradermal injection. (d) Successful nerve fiber innervations (arrowhead) between the bioengineered hair follicle and the arrector pili muscles were observed by immunohistochemical staining. Red, calponin for smooth muscle; white, neurofilament-H (NF-H); and broken lines, outermost limit of the hair follicle.

The organ-inductive epithelial and mesenchymal stem cells derived from the hair follicle can provide a cell source of differentiated hair follicle cells that enable hair cycling to occur over an individual's lifetime (Lee and Chuong 2009). It is practically anticipated that these epithelial/mesenchymal stem cells isolated from the adult tissue of patients can be utilized for hair follicle regenerative therapy (Weinberg et al. 1993, Claudinot et al. 2005, Lee and Chuong 2009, Bonfanti et al. 2010). We thus demonstrated that a human bioengineered hair follicle germ, which is composed of dissociated epithelial cells derived from bulge region and scalp hair follicle–derived intact DPs from a patient with androgenetic alopecia patient, grew a pigmented hair shaft on intracutaneous transplantation into the back skin of nude mice (Figure 20.6a) (Toyoshima et al. 2012). In addition, the bioengineered pelage follicle germs could generate all types of pelage hair, including awl/auchene and zigzag, in accordance with their follicle fate, as determined during embryonic development (Toyoshima et al. 2012). The hair features such as hair color could also be regulated by the addition of cells from the sub-bulge region that contains melanoblasts (Toyoshima et al. 2012). Furthermore, we demonstrated the high-density transplantation of bioengineered hair follicles for future clinical applications (Figure 20.6b). It is essential for rearrangements of various hair follicle stem cells and their niches in the bioengineered follicle to reproduce permanent hair cycles (Lee and Chuong 2009). We showed that the bioengineered pelage and vibrissae could repeat hair cycles properly, in accordance with the cell origins that were maintained into a stem cell niche (Toyoshima et al. 2012). Thus, bioengineered follicles could maintain stem cells via the reconstitution of hair follicle niches for epithelial stem cells and DP cells. In the hair follicle, the piloerection of pelage and vibrissae can achieve contraction of the surrounding arrector pili muscles through the efferent stimulation of the sympathetic nerves (Fujiwara et al. 2011). We demonstrated that the bioengineered follicles could transduce tactile sensations from the erected hair follicle to the connecting muscles and peripheral nerve fibers (Figure 20.6c and d). We further demonstrated that mature bioengineered hair follicles, which were generated by ectopic transplantation *in vivo*, could be replaced by the orthotopic hair follicle unit transplantation therapy (Sato et al. 2012). This therapeutic concept could achieve the clinical restoration of the proper hair appearance by controlling the aesthetic and functional requirements, including hair type, hair density, hair stream, and physiological piloerection through the surgical transplantation of hair follicles (Sato et al. 2012).

The bioengineered hair follicle germ and the mature bioengineered hair follicular unit could recover natural hair function and re-establish cooperation with the surrounding recipient muscles and nerve fibers. These approaches are potentially available in future hair regenerative therapy for hair loss caused by injury or by diseases such as alopecia areata and androgenic alopecia.

20.6 Functional three-dimensional integumentary organ system from induced pluripotent stem cells, utilizing the clustering-dependent embryoid body method

The integumentary organ system is composed of the skin and its append-ages, including the hair, sebaceous glands, sweat glands, feathers, and nails, and this organ system contributes to biological functions such as cushioning, waterproofing, protecting subcutaneous tissues, excreting waste, and thermoregulation (Jiang et al. 2004). The integumentary organs arise from their respective organ germs, with epithelial and mesenchymal interactions in the skin field (Pispa and Thesleff 2003). In embryonic development, the skin field is determined by a regulated process of pattern formation, and then, its appendage organs are induced in that field by the epithelial and mesenchymal interactions based on a Turing model of expression of activators and inhibitors (Jiang et al. 2004). The generation of the integumentary system could contribute to regenerative therapies for patients with burn, scar, and alopecia and could be available as an *in vitro* assay system for non-animal safety testing of cosmetics and quasi-drugs (Fukano et al. 2006, Sun et al. 2014). However, it is difficult to generate the complex bioengineered 3D integumentary organ system by using *in vitro* stem cell culture method or *in vivo* transplantation models and to restore the physiological skin functions through bioengineered 3D integumentary organ transplantation.

In previous studies, pluripotent stem cells, including ES and iPS cells, have represented the generation of whole embryos by tetraploid complementation in blastocysts (Suga et al. 2011). Moreover, pluripotent stem cells can generate tumors, commonly referred to as teratomas, that include multiple organized tissues derived from three germ layers, such as neural tissue (ectodermal), cartilage and muscle (mesenchymal), and gut and bronchial epithelia (endodermal), by an *in vivo* transplantation (Takahashi and Yamanaka 2006). It is well known that several cutaneous types of mature teratomas, such as cystic dermoids in the ovary and orbital region, are able to generate ectodermal organs, including the teeth and hair follicles, at a high frequency; however, the distinctive molecular mechanisms of the onset of these dermoids are not known (Suga et al. 2011). Furthermore, at present, the generation of bioengineered 3D tissues in 3D stem cell culture is limited to neuroectodermal and endodermal tissues (Sasai 2013b). Recently, we successfully generated a bioengineered 3D integumentary organ system from iPS cells that includes appendicular organs such as hair follicles and sebaceous glands. We established a novel *in vivo* transplantation model designated as a *clustering-dependent embryoid body (CDB) transplantation method* by using iPS cells that formed ideal connections to the surrounding epidermis, dermis, arrector pili muscles, and subcutaneous adipose tissues (Takagi et al. 2016). In an *in vitro* culture of iPS cells,

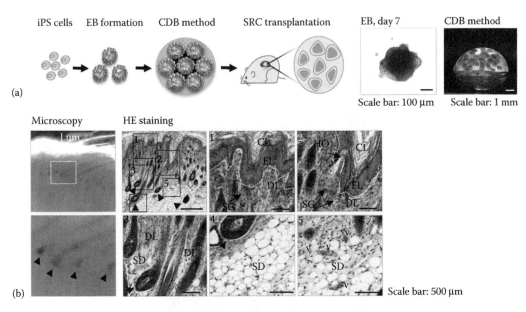

Figure 20.7 Functional 3D integumentary organ system from induced pluripotent stem (iPS) cells. (a) Schematic illustration of EB cultures and the CDB transplantation method (left panels). Phase-contrast images of iPS cells and the formation of EBs, which were cultured in non-adherent plastic plates for 7 days. Macromorphological observations of multiple EBs in a collagen gel before transplantation (right panels). (b) Macromorphological and histological observations of the hair follicles and their surrounding tissues in iPS-cell–derived bioengineered 3D integumentary organ system. Macroscopic observation of isolated cystic structures with hair follicles (left panels). In the hematoxylin and eosin (HE) analyses, the boxed areas in the low-magnification macroscopic views were shown at a higher magnification in the other panels. CL, cystic lumen; EL, epidermal layer; DL, dermal layer; SD, subdermal tissue; SG, sebaceous gland; HO, hair opening; v, vessel. *(Continued)*

the embryoid bodies (EBs) were formed in 3 days and were then allowed to grow for 1 week (Figure 20.7a). We transplanted these EBs in various culture conditions into the subrenal capsule of severe combined immunodeficient mice *in vivo*. Both single iPS cells and single EB transplants formed teratoma-like tissues that included neural tissue, cartilage, muscle, and bronchial epithelia, similar to the previous report (Okita et al. 2008). In contrast, in the combined multiple EB transplants in the CDB method, which contained more than 30 EBs cultured for 7 days under non-adhesive conditions, distinctive cystic epithelia could be observed. The tissue weight of the CDB transplants was heavy compared with that of the single EB transplants, and the cystic tissue area proportion was significantly larger than that in explants of single iPS and single EB cells. In general, the cultured EBs form an outer-layer epithelial progenitor (Metallo et al. 2008, Kadoshima et al. 2013). The epithelium induction in our method could be established by the autonomous connection with the outer-layer epithelium of each EB and the self-formation in the explants (Figure 20.7b). The frequency of hair follicle generation with

451

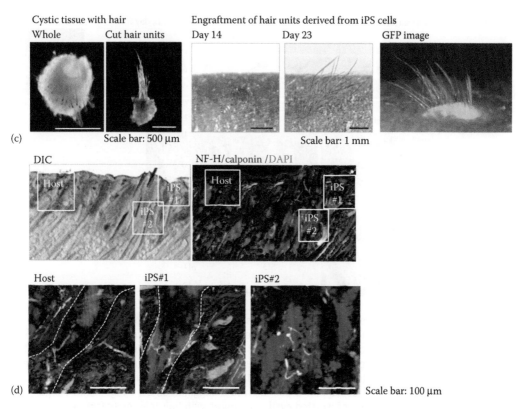

Cystic tissue with hair
Whole   Cut hair units

Engraftment of hair units derived from iPS cells
Day 14   Day 23   GFP image

(c)   Scale bar: 500 µm   Scale bar: 1 mm

DIC   NF-H/calponin/DAPI

Host   iPS#1   iPS#2

(d)   Scale bar: 100 µm

Figure 20.7 (Continued) Functional 3D integumentary organ system from induced pluripotent stem (iPS) cells. (c) Photographs of the methods for the transplantation of iPS-cell–derived hair follicles. Cystic tissue with hair follicles was isolated and divided into small pieces, including 10–20 hair follicles (left panels). Macromorphological observations of the engraftments into the dorsoventral skin of nude mice showing the eruption of skin and growth of iPS-cell–derived hair follicles (center and right panels). Functional engraftment is indicated by the GFP-labeled transplants (right panels). (d) Immunohistological analysis of the iPS-cell–derived hair follicles and their surrounding tissues. Host pelage (left) and iPS-cell–derived hair follicles (right #1, #2) were stained by neurofilament-H (NF-H) and calponin.

sebaceous glands in the explants of our bioengineered 3D integumentary organ system could be regulated by Wnt signaling (Takagi et al. 2016).

As a regenerative therapy for patients with burns, a cultured epithelial sheet is available for the replacement of the skin structure and physiological skin functions, including the protection of subcutaneous tissues, waterproofing, and thermoregulation (Sun et al. 2014). However, it is reported that current treatments using cultured epithelium tissues suffer from fundamental problems such as the non-excretion of sweat and lipids from exocrine organs and dissatisfaction with the aesthetic aspects (Niemann and Watt 2002). Artificial skins, which consist of cultured epidermis and dermis without appendage organs or pores, are commonly utilized for safety testing during the newly developed cosmetics and quasi-drugs, but these skins have critical limitations related to the permeability of cosmetics and drugs and the reliability of the physiological responses (Kandyba et al. 2008). It is therefore expected that a bioengineered 3D integumentary organ system will be developed to overcome these issues (Bellas et al. 2012). We next investigated whether the iPS-cell–derived integumentary organ system generated with our CDB transplantation method could be transplanted into the cutaneous environment of adult nude mice. We prepared small specimens containing 10–20 follicular units derived from cystic tissues in explant and transplanted them onto the backs of nude mice by using a follicular unit transplantation method (Figure 20.7c). We observed the successful eruption and complete growth of black hair shafts at 14 days after transplantation (Figure 20.7c). The bioengineered hair follicles were successfully engrafted and established full functionality, with proper connections to the surrounding tissues such as epithelium, dermis, adipose tissues, arrector pili muscles, and nerve fibers (Figure 20.7d). Our findings demonstrate the potential clinical applications of our bioengineered 3D integumentary organ system as a novel non-animal assay system for cosmetics and quasi-drugs (Takagi et al. 2016).

Our current study presents the development of a functional bioengineered 3D integumentary organ system and the realization of organ replacement therapy using iPS cells. These data will substantially contribute to the development of bioengineering technologies that will provide future regenerative therapies for patients with burns, scars, and alopecia (Takagi et al. 2016).

20.7 Future perspectives for organ replacement regenerative therapy

The progress made in our regenerative technology, including the bioengineered organ germ method and bioengineered 3D integumentary organ system, is remarkable for possible future organ replacement regenerative strategies (Figure 20.8), and many patients can benefit from the practical applications of organ replacement regenerative therapy. To address the future clinical use of organ replacement regenerative therapy based on replicating the epithelial–mesenchymal interactions, one of the major research hurdles remaining is the identification of appropriate cell sources (Ikeda and Tsuji 2008). Recent studies of stem cell biology have led to the identification of candidate cell sources in humans for tooth-tissue repair (Huang et al. 2009, Egusa et al. 2012), salivary gland tissue repair or saliva secretion recovery (Sonis et al. 2000, Dörr et al. 2001, Ihrler et al. 2002, Baum et al. 2004, Nakamura et al. 2004, Man et al. 2011, Lombaert and Hoffman 2013), and lacrimal grand tissue regeneration (Zoukhri 2010, You et al. 2011, You et al. 2012). Although these cell lineages would be available for tissue repair, the organ-inducible stem cells, which are able to replicate epithelial–mesenchymal interactions in

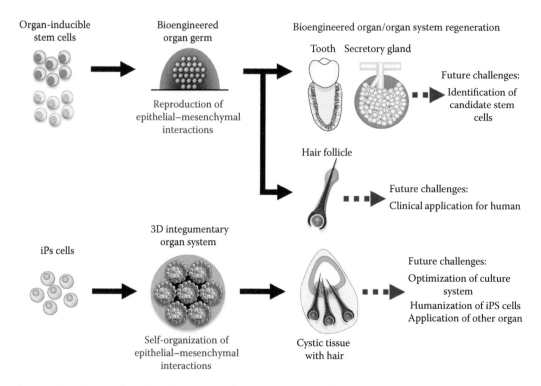

Organ-inducible stem cells

Bioengineered organ germ

Reproduction of epithelial–mesenchymal interactions

Bioengineered organ/organ system regeneration

Tooth   Secretory gland

Future challenges:
Identification of candidate stem cells

Hair follicle

Future challenges:
Clinical application for human

iPs cells

3D integumentary organ system

Self-organization of epithelial–mesenchymal interactions

Cystic tissue with hair

Future challenges:
Optimization of culture system
Humanization of iPS cells
Application of other organ

Figure 20.8 Future directions for organ replacement regenerative therapy. Our bioengineered organ germ method and bioengineered 3D integumentary organ system are expected to lead to future organ replacement regenerative strategies based on the reproduction of the epithelial–mesenchymal interaction. To address the future clinical use of organ replacement regenerative therapy, further research challenges must be addressed.

their respective organogenesis, are anticipated to lead to future whole-organ regeneration (Egusa et al. 2012, Egusa et al. 2013, Ogawa and Tsuji 2015). At present, the hair follicles in the adult human, which undergo hair cycle and hair growth phases throughout the lifetime, are able to maintain various stem cell types for hair organ regeneration (Jahoda and Reynolds 2001, Jahoda et al. 2003, Blanpain et al. 2004, Chuong et al. 2006, Yamao et al. 2010). Our technology demonstrates fully functional bioengineered hair organ regeneration by utilizing human stem cells derived from a patient with androgenetic alopecia (Toyoshima et al. 2012). This study is considered to provide an important model of organ replacement regenerative therapy for practical use in human dermatological medicine. The iPS cells are candidate cell sources for organ regeneration that can differentiate into endodermal, ectodermal, and mesodermal cell lineages. They also represent the potential to differentiate into dental epithelial or mesenchymal cells for the reconstitution of bioengineered germ (Yan et al. 2010, Egusa et al. 2010, Arakaki et al. 2012, Otsu et al. 2012). Notably, the self-organization system, including the optic cup and adenohypophysis, and our bioengineered 3D integumentary organ system demonstrate the feasibility of the available bioengineering technology and the realization of whole-organ replacement therapy using iPS cells (Eiraku et al. 2011, Suga et al. 2011, Sasai 2013b, Takagi et al. 2016). Therefore, ectodermal organ regenerative therapy is now regarded as a viable model for studying future organ replacement regenerative therapies that can be applied to several complex organs, and it will contribute substantially to the knowledge and technology in all of organ regeneration (Volponi et al. 2010, Oshima and Tsuji 2014).

Acknowledgments

This work was supported by Health and Labour Sciences Research Grants from the Ministry of Health, Labour and Welfare (No. 21040101) awarded to Akira Yamaguchi and a Grant-in-Aid for Scientific Research (A) (No. 20249078) awarded to Takashi Tsuji (2008–2010) and a Grant-in-Aid for Young Scientists (B) (No. 22791941) awarded to Masamitsu Oshima from the Ministry of Education, Culture, Sports, Science and Technology (MEXT). This work was partially supported by a collaboration grant from Organ Technologies Inc.

References

Addis, R. C., Epstein, J. A. 2013. Induced regeneration—the progress and promise of direct reprogramming for heart repair. *Nature Medicine* 19:829–836.

Arakaki, M., Ishikawa, M., Nakamura, T. et al. 2012. Role of epithelial-stem cell interactions during dental cell differentiation. *Journal of Biological Chemistry* 287:10590–10601.

Atala, A. 2005. Tissue engineering, stem cells and cloning: Current concepts and changing trends. *Expert Opinion on Biological Therapy* 5:879–892.

Atkinson, J. C., Grisius, M., Massey, W. 2005. Salivary hypofunction and xerostomia: Diagnosis and treatment. *Dental Clinics of North America* 49:309–326.

Avery, J. K. 2002. *Oral Development and Histology.* New York: Thieme Press.

Baum, B. J., Voutetakis, A., Wang, J. 2004. Salivary glands: Novel target sites for gene therapeutics. *Trends in Molecular Medicine* 10(12):585–590.

Bellas, E., Seiberg, M., Garlick, J., Kaplan, D. L. 2012. In vitro 3D full-thickness skin-equivalent tissue model using silk and collagen biomaterials. *Macromolecular Bioscience* 12(12):1627–1636.

Blanpain, C., Lowry, W. E., Geoghegan, A. et al. 2004. Self-renewal, multipotency, and the existence of two cell populations within an epithelial stem cell niche. *Cell* 118:635–648.

Bonfanti, P., Claudinot, S., Amici, A. W. et al. 2010. Microenvironmental reprogramming of thymic epithelial cells to skin multipotent stem cells. *Nature* 466:978–982.

Brenemark, P. I., Zarb, G. A. 1985. Tissue-integrated prostheses. In *Osseointegration in Clinical Dentistry.* Albrektsson T., Ed. Berlin, Germany: Quintessence Publishing Co Press, pp. 211–232.

Brockes, J. P., Kumar, A. 2005. Appendage regeneration in adult vertebrates and implications for regenerative medicine. *Science* 310:1919–1923.

Burns, D. R., Beck, D. A., Nelson, S. K. 2003. A review of selected dental literature on contemporary provisional fixed prosthodontic treatment: Report of
the Committee on Research in Fixed Prosthodontics of the Academy of Fixed
Prosthodontics. *Journal of Prosthetic Dentistry* 90:474–497.

Chuong, C. M. 1998. *Molecular Basis of Epithelial Appendage Morphogenesis.*
Austin, TX: R.G. Landes Company.

Chuong, C. M., Wu, P., Plikus, M., Jiang, T. X., Bruce, W. R. 2006. Engineering
stem cells into organs: Topobiological transformations demonstrated by
beak, feather, and other ectodermal organ morphogenesis. *Current Topics in
Developmental Biology* 72:237–274.

Claudinot, S., Nicolas, M., Oshima, H. et al. 2005. Long-term renewal of hair follicles
from clonogenic multipotent stem cells. *Proceedings of the National Academy
of Sciences of the United States of America* 102:14677–14682.

Cohen, D. E., Melton, D. 2011. Turning straw into gold: Directing cell fate for regenerative medicine. *Nature Reviews Genetics* 12(4):243–252.

Colton, C. K. 1995. Implantable biohybrid artificial organs. *Cell Transplantation*
4:415–436.

Copeland, J. G., Smith, R. G., Arabia, F. A. et al. 2004. Cardiac replacement with a
total artificial heart as a bridge to transplantation. *The New England Journal
of Medicine* 351:859–867.

Dawson, P. E. 2006. *Functional Occlusion: From TMJ to Smile Design.* St. Louis,
MO: Mosby.

Dörr, W., Noack, R., Spekl, K., Farrell, C. L. 2001. Modification of oral mucositis by
keratinocyte growth factor: Single radiation exposure. *International Journal
of Radiation Biology* 77(3):341–347.

Edgar, M., Dawes, C., Mullane, O. D. 2004. *Saliva and Oral Health.* 3rd ed.
Duns Tew: Stephen Hancocks Limited.

Egusa, H., Okita, K., Kayashima, H., Yu, G., Fukuyasu, S., Saeki, M., Matsumoto, T.,
Yamanaka, S., Yatani, H. 2010. Gingival fibroblasts as a promising source of
induced pluripotent stem cells. *PLoS One* 5:e12743.

Egusa, H., Sonoyama, W., Nishimura, M., Atsuta, I., Akiyama, K. 2012. Stem cells
in dentistry—part I: Stem cell sources. *Journal of Prosthodontic Research*
56(3):151–165.

Egusa, H., Sonoyama, W., Nishimura, M., Atsuta, I., Akiyama, K. 2013. Stem
cells in dentistry—part II: Clinical applications. *Journal of Prosthodontic
Research* 56(4):229248.

Eiraku, M., Takata, N., Ishibashi, H. et al. 2011. Self-organizing optic-cup morphogenesis in three-dimensional culture. *Nature* 472(7341):51–56.

Fort, A., Fort, N., Ricordi, C. et al. 2008. Biohybrid devices and encapsulation
technologies for engineering a bioartificial pancreas. *Cell Transplantation*
17:997–1003.

Fuchs, E. 2007. Scratching the surface of skin development. *Nature* 445:834–842.

Fujiwara, H., Ferreira, M., Donati, G. et al. 2011. The basement membrane of hair
follicle stem cells is a muscle cell niche. *Cell* 144:577–589.

Fukano, Y., Knowles, N. G., Usui, M. L. et al. 2006. Characterization of an in vitro
model for evaluating the interface between skin and percutaneous biomaterials. *Wound Repair and Regeneration* 14(4):484–491.

Fukumoto, S., Yamada, Y. 2005. Review: Extracellular matrix regulates tooth morphogenesis. *Connective Tissue Research* 46:220–226.

Gary, N. F. 2007. The epidemiology of dry eye disease: Report of the Epidemiology
Subcommittee of the International Dry Eye WorkShop. *The Ocular Surface*
5:93–107.

Gilbert, S. F. 2013. *Developmental Biology.* 10th ed. Sunderland, MA: Sinauer
Associates.

Greco, V., Chen, T., Rendl, M. et al. 2009. A two-step mechanism for stem cell activation during hair regeneration. *Cell Stem Cell* 4:155–169.

Greggio, C., De Franceschi, F., Figueiredo-Larsen, M. et al. 2013. Artificial three-dimensional niches deconstruct pancreas development in vitro. *Development* 140:4452–4462.

Gridelli, B., Remuzzi, G. 2000. Strategies for making more organs available for transplantation. *The New England Journal of Medicine* 343:404–410.

Hardy, M. H. 1992. The secret life of the hair follicle. *Trends in Genetics* 8:55–61.

Hirayama, M., Ogawa, M., Oshima, M. et al. 2013a. Functional lacrimal gland regeneration by transplantation of a bioengineered organ germ. *Nature Communications* 4:2497.

Hirayama, M., Oshima, M., Tsuji, T. 2013b. Development and prospects of organ replacement regenerative therapy. *Cornea* 32 Suppl 1:S13–S21.

Huang, G. T., Gronthos, S., Shi, S. 2009. Mesenchymal stem cells derived from dental tissues vs. those from other sources: their biology and role in regenerative medicine. *Journal of Dental Research* 88:792–806.

Huang, G. T., Sonoyama, W., Liu, Y. et al. 2008. The hidden treasure in apical papilla: the potential role in pulp/dentin regeneration and bioroot engineering. *Journal of Endodontics* 34:645–651.

Huch, M., Bonfanti, P., Boj, S. F. et al. 2013a. Unlimited in vitro expansion of adult bi-potent pancreas progenitors through the Lgr5/R-spondin axis. *EMBO Journal* 32:2708–2721.

Huch, M., Dorrell, C., Boj, S. F. et al. 2013b. In vitro expansion of single Lgr5+ liver stem cells induced by Wnt-driven regeneration. *Nature* 494(7436):247–250.

Ihrler, S., Zietz, C., Sendelhofert, A. et al. 2002. A morphogenetic concept of salivary duct regeneration and metaplasia. *Virchows Archiv* 440(5):519–526.

Ikeda, E., Morita, R., Nakao, K. et al. 2009. Fully functional bioengineered tooth replacement as an organ replacement therapy. *Proceedings of the National Academy of Sciences of the United States of America* 106:13475–13480.

Ikeda, E., Tsuji, T. 2008. Growing bioengineered teeth from single cells: Potential for dental regenerative medicine. *Expert Opinion on Biological Therapy* 8:735–744.

Ishida, K., Murofushi, M., Nakao, K. et al. 2011. The regulation of tooth morphogenesis is associated with epithelial cell proliferation and the expression of Sonic hedgehog through epithelial-mesenchymal interactions. *Biochemical and Biophysical Research Communications* 405:455–461.

Jahoda, C. A., Reynolds, A. J. 2001. Hair follicle dermal sheath cells: Unsung participants in wound healing. *The Lancet* 358:1445–1448.

Jahoda, C. A., Horne, K. A., Oliver, R. F. 1984. Induction of hair growth by implantation of cultured dermal papilla cells. *Nature* 311:560–562.

Jahoda, C. A., Oliver, R. F., Reynolds, A. J. et al. 1996. Human hair follicle regeneration following amputation and grafting into the nude mouse. *Journal of Investigative Dermatology* 107:804–807.

Jahoda, C. A., Whitehouse, J., Reynolds, A. J. et al. 2003. Hair follicle dermal cells differentiate into adipogenic and osteogenic lineages. *Experimental Dermatology* 12:849–859.

Jiang, T. X., Widelitz, R. B., Shen, W. M. et al. 2004. Integument pattern formation involves genetic and epigenetic controls: Feather arrays simulated by digital hormone models. *The International Journal of Developmental Biology* 48(2–3):117–135.

Kadoshima, T., Sakaguchi, H., Nakano, T. et al. 2013. Self-organization of axial polarity, inside-out layer pattern, and species-specific progenitor dynamics in human ES cell-derived neocortex. *Proceedings of the National Academy of Sciences of the United States of America* 110(50):20284–20289.

Kagami, H., Wang, S., Hai, B. 2008. Restoring the function of salivary glands. *Oral Diseases* 14:15–24.

Kamao, H., Mandai, M., Okamoto, S. et al. 2014. Characterization of human induced pluripotent stem cell-derived retinal pigment epithelium cell sheets aiming for clinical application. *Stem Cell Reports* 2:205–218.

Kandyba, E. E., Hodgins, M. B., Martin, P. E. 2008. A murine living skin equivalent amenable to live-cell imaging: Analysis of the roles of connexins in the epidermis. *Journal of Investigative Dermatology* 128(4):1039–1049.

Koehler, K. R., Mikosz, A. M., Molosh, A. I., Patel, D., Hashino, E. 2013. Generation of inner ear sensory epithelia from pluripotent stem cells in 3D culture. *Nature* 500(7461):217–221.

Kutzner, H., Requena, L., Rutten, A. et al. 2006. Spindle cell predominant trichodiscoma: A fibrofolliculoma/trichodiscoma variant considered formerly to be a neurofollicular hamartoma: a clinicopathological and immunohistochemical analysis of 17 cases. *American Journal of Clinical Dermatology* 28:1–8.

Langer, R. S., Vacanti, J. P. 1999. Tissue engineering: The challenges ahead. *Scientific American* 280:86–89.

Lee, L. F., Chuong, C. M. 2009. Building complex tissues: High-throughput screening for molecules required in hair engineering. *Journal of Investigative Dermatology* 129:815–817.

Lindhe, J., Lang, N. P., Karring, T. 2008. *Clinical Periodontology and Implant Dentistry*, 5th ed. Hoboken, NJ: Blackwell.

Lombaert, I. M., Hoffman, M. P. 2013. *Stem Cells in Salivary Gland Development and Regeneration. Stem Cells in Craniofacial Development and Regeneration.* Hoboken, NJ: Wiley-Blackwell, pp. 271–284.

Luukko, K., Kvinnsland, I. H., Kettunen, P. 2005. Tissue interactions in the regulation of axon pathfinding during tooth morphogenesis. *Developmental Dynamics* 234:482–488.

Madeira, C., Santhagunam, A., Salgueiro, J. B., Cabral, J. M. S. 2015. Advanced cell therapies for articular cartilage regeneration. *Trends in Biotechnology* 33:35–42.

Man, Y. G., Ball, W. D., Marchetti, L., Hand, A. R. 2011. Contributions of intercalated duct cells to the normal parenchyma of submandibular glands of adult rats. *The Anatomical Record* 263(2):202–214.

Melnick, M., Phair, R. D., Lapidot, S. A. et al. 2009. Salivary gland branching morphogenesis: a quantitative systems analysis of the Eda/Edar/NFkappaB paradigm. *BMC Developmental Biology* 9:32.

Metallo, C. M., Ji, L., de Pablo, J. J., Palecek, S. P. 2008. Retinoic acid and bone morphogenetic protein signaling synergize to efficiently direct epithelial differentiation of human embryonic stem cells. *Stem Cells* 26(2):372–380.

Miljanovic, B., Dana, R., Sullivan, D. A. et al. 2007. Impact of dry eye syndrome on vision-related quality of life. *American Journal of Ophthalmology* 143:409–415.

Miyahara, Y., Nagaya, N., Kataoka, M. et al. 2006. Monolayered mesenchymal stem cells repair scarred myocardium after myocardial infarction. *Nature Medicine* 12:459–465.

Mounsey, A. L. 2009. Diagnosis and treating hair loss. *American Family Physician* 80(4):356–362.

Nakamura, T., Matsui, M., Uchida, K. et al. 2004. M3 muscarinic acetylcholine receptor plays a critical role in parasympathetic control of salivation in mice. *The Journal of Physiology* 558:561–575.

Nakano, T., Ando, S., Takata, N. et al. 2012. Self-formation of optic cups and storable stratified neural retina from human ESCs. *Cell Stem Cell* 10(6):771–785.

Nakao, K., Morita, R., Saji, Y. et al. 2007. The development of a bioengineered organ germ method. *Nature Methods* 4:227–230.

Niemann, C., Watt, F. M. 2002. Designer skin: Lineage commitment in postnatal epidermis. *Trends in Cell Biology* 12(4):185–192.

Nishimura, E. K., Granter, S. R., Fisher, D. E. 2005. Mechanisms of hair graying: incomplete melanocyte stem cell maintenance in the niche. *Science* 307:720–724.

Nishimura, E. K., Jordan, S. A., Oshima, H. et al. 2002. Dominant role of the niche in melanocyte stem-cell fate determination. *Nature* 416:854–860.

Ogawa, M., Oshima, M., Imamura, A. et al. 2013. Functional salivary gland regeneration by transplantation of a bioengineered organ germ. *Nature Communications* 4:2498.

Ogawa, M., Tsuji, T. 2015. Functional salivary gland regeneration as the next generation of organ replacement regenerative therapy. *Odontology* 103(3):248–257.

Ohashi, Y., Dogru, M., Tsubota, K. 2006. Laboratory findings in tear fluid analysis. *Clinica Chimica Acta* 369:17–28.

Ohazama, A., Modino, S. A., Miletich, I., Sharpe, P. T. 2004. Stem-cell-based tissue engineering of murine teeth. *Journal of Dental Research* 83:518–522.

Okita, K., Nakagawa, M., Hyenjong, H., Ichisaka, T., Yamanaka, S. 2008. Generation of mouse induced pluripotent stem cells without viral vectors. *Science* 322(5903):949–953.

Oshima, H., Rochat, A., Kedzia, C. et al. 2001. Morphogenesis and renewal of hair follicles from adult multipotent stem cells. *Cell* 104:233–245.

Oshima, M., Mizuno, M., Imamura, A. et al. 2011. Functional tooth regeneration using a bioengineered tooth unit as a mature organ replacement regenerative therapy. *PLoS One* 6:e21531.

Oshima, M., Tsuji, T. 2014. Functional tooth regenerative therapy: Tooth tissue regeneration and whole-tooth replacement. *Odontology* 102:123–136.

Otsu, K., Kishigami, R., Oikawa-Sasaki, A. et al. 2012. Differentiation of induced pluripotent stem cells into dental mesenchymal cells. *Stem Cells and Development* 21:1156–1164.

Pispa, J. and Thesleff, I. 2003. Mechanisms of ectodermal organogenesis. *Developmental Biology* 262(2):195–205.

Pokorny, P. H., Wiens, J. P., Litvak, H. 2008. Occlusion for fixed prosthodontics: A historical perspective of the gnathological influence. *Journal of Prosthetic Dentistry* 99:299–313.

Purnell, B. 2008. New release: The complete guide to organ repair. *Science* 322:1489.

Rosenstiel, S. F., Land, M. F., Fujimoto, J. 2015. *Contemporary Fixed Prosthodontics.* Missouri, MO: Mosby Press, pp. 209–366.

Rousso, D. E., Presti, P. M. 2008. Follicular unit transplantation. *Facial Plastic Surgery* 24(4):381–388.

Sasai, Y. 2013a. Cytosystems dynamics in self-organization of tissue architecture. *Nature* 493(7432):318–326.

Sasai, Y. 2013b. Next-generation regenerative medicine: Organogenesis from stem cells in 3D culture. *Cell Stem Cell* 12:520–529.

Sato, A., Toyoshima, K., Toki, H. et al. 2012. Single follicular unit transplantation reconstructs arrector pili muscle and nerve connections and restores functional hair follicle piloerection. *The Journal of Dermatology* 39:682–687.

Sato, T., Robert, G., Vries, Hugo J. et al. 2009. Single Lgr5 stem cells build crypt-villus structures in vitro without a mesenchymal niche. *Nature* 459(7244):262–265.

Schechter, J. E., Warren, D. W., Mircheff, A. K. 2010. A lacrimal gland is a lacrimal gland, but rodent's and rabbit's are not human. *The Ocular Surface* 8:111–134.

Sharpe, P. T., Young, C. S. 2005. Test-tube teeth. *Scientific American* 293:34–41.

Ship, J. A., Pillemer, S. R., Baum, B. J. 2002. Xerostomia and the geriatric patient. *Journal of the American Geriatrics Society* 50:535–543.

Sonis, S. T., Peterson, R. L., Edwards, L. J. et al. 2000. Defining mechanisms of action of interleukin-11 on the progression of radiation-induced oral mucositis in hamsters. *Oral Oncolgy* 36(4):373–381.

Springer, K., Brown, M., Stulberg, S. et al. 2003. Common hair loss disorders. *American Family Physician* 68(1):93–102.

Stephen, C. 2007. Management and therapy of dry eye disease. report of the Management and Therapy Subcommittee of the International Dry Eye WorkShop. *The Ocular Surface* 5:163–178.

Suga, H., Kadoshima, T., Minaguchi, M. et al. 2011. Self-formation of functional adenohypophysis in three-dimensional culture. *Nature* 480(7375):57–62.

Sun, B. K., Siprashvili, Z., Khavari, P. A. 2014. Advances in skin grafting and treatment of cutaneous wounds. *Science* 346(6212):941–945.

Takagi, R., Ishimaru, J., Sugawara, A. et al. 2016. Bioengineering a 3D integumentary organ system from iPS cells using an in vivo transplantation model. *Science Advances* 2(4):e1500887.

Takahashi, K., Yamanaka, S. 2006. Induction of pluripotent stem cells from mouse embryonic and adult fibroblast cultures by defined factors. *Cell* 126(4):663–676.

Takebe, T., Sekine, K., Enomura, M. et al. 2013. Vascularized and functional human liver from an iPSC-derived organ bud transplant. *Nature* 499:481–484.

Takeichi, M. 2011. Self-Organization of Animal Tissues: Cadherin-Mediated Processes. *Development Cell* 21(1):24–26.

Tan, D. W., Barker, N., 2014. Intestinal stem cells and their defining niche. *Current Topics in Development Biology* 107:77–107.

Tavassoll, F. A., Devilee, P. (ed.). 2003. *Pathology and Genetics of Tumours of the Breast and Female Genital Organs*. World Health Organization Classification of Tumours. Lyon, France: IARC Press.

Thesleff, I. 2003. Epithelial-mesenchymal signalling regulating tooth morphogenesis. *Journal of Cell Science* 116:1647–1648.

Toyoshima, K. E., Asakawa, K., Ishibashi, N. et al. 2012. Fully functional hair follicle regeneration through the rearrangement of stem cells and their niches. *Nature Communications* 3:784.

Trounson, A., Daley, G. Q., Pasque, V., Plath, K. 2013. A new route to human embryonic stem cells. *Nature Medicine* 19:820–821.

Tsukiboshi, M. 1993. Autogenous tooth transplantation: A reevaluation. *The International Journal of Periodontics and Restorative Dentistry* 13:120–149.

Tucker, A., Sharpe, P. 2004. The cutting-edge of mammalian development; how the embryo makes teeth. *Nature Reviews Genetics* 5:499–508.

Tucker, A. S., Miletich, I. 2010. *Salivary Glands; Development, Adaptations, and Disease*. London, UK: Karger.

Uygun, B. E., Soto-Gutierrez, A., Yagi, H. et al. 2010. Organ reengineering through development of a transplantable recellularized liver graft using decellularized liver matrix. *Nature Medicine* 16:814–820.

Volponi, A. A., Pang, Y., Sharpe, P. T. 2010. Stem cell-based biological tooth repair and regeneration. *Trends in Cell Biology* 20:715–722.

Walck-Shannon, E., Hardin, J. 2014. Cell intercalation from top to bottom. *Nature Reviews Molecular Cell Biology* 15(1):34–48.

Watson, C. L., Mahe, M. M., Munera, J. et al. 2014. An in vivo model of human small intestine using pluripotent stem cells. *Nature Medicine* 20(11):1310–1314.

Weinberg, W. C., Goodman, L. V., George, C. et al. 1993. Reconstitution of hair follicle development in vivo: Determination of follicle formation, hair growth, and hair quality by dermal cells. *Journal of Investigative Dermatology* 100:229–236.

Wise, G. E., Frazier-Bowers, S., D'Souza, R. N. 2002. Cellular, molecular, and genetic determinants of tooth eruption. *Critical Reviews in Oral Biology & Medicine* 13:323–334.

Wise, G. E., King, G. J. 2008. Mechanisms of tooth eruption and orthodontic tooth movement. *Journal of Dental Research* 87:414–434.

Wolf, A. V. 1952. The artificial kidney. *Science* 115:193–199.

Yamao, M., Inamatsu, M., Ogawa, Y. et al. 2010. Contact between dermal papilla cells and dermal sheath cells enhances the ability of DPCs to induce hair growth. *Journal of Investigative Dermatology* 130:2707–2718.

Yan, X., Qin, H., Qu, C. et al. 2010. iPS cells reprogrammed from human mesenchymal-like stem/progenitor cells of dental tissue origin. *Stem Cells and Development* 19(4):469–480.

Yang, J., Yamato, M., Kohno, C. et al. 2005. Cell sheet engineering: Recreating tissues without biodegradable scaffolds. *Biomaterials* 26(33):6415–6422.

You, S., Avidan, O., Tarig, A. et al. 2012. Role of epithelial-mesenchymal transition in repair of the lacrimal gland after experimentally induced injury. *Investigative Ophthalmology & Visual Science* 53:126–135.

You, S., Kublin, C. L., Avidan, O., Miyasaki, D., Zoukhri, D. 2011. Isolation and propagation of mesenchymal stem cells from the lacrimal gland. *Investigative Ophthalmology & Visual Science* 52:2087–2094.

Zoukhri, D. 2010. Mechanisms involved in injury and repair of the murine lacrimal gland: Role of programmed cell death and mesenchymal stem cells. *Ocular Surface* 8:60–69.

Chapter 21 Blastema formation in mammalian digit-tip regeneration

Makoto Takeo and
Mayumi Ito

Contents

Key concepts

a. Although mammals have limited regenerative abilities, they exhibit simultaneous multiple tissue regeneration when regenerating a digit tip.

b. Mammalian digit tips regenerate via blastema formation, which is the mechanism of epimorphic regeneration used in lower vertebrates to regenerate entire limbs.

> c. The digit-tip blastema is composed of multiple types of cells that originate from different cells, and nail organ has a critical role in the blastema formation and subsequent digit-tip regeneration.
> d. There are both intrinsic mechanisms that control cell fate and extrinsic mechanisms associated with the extracellular environment that influence whether or not cells contribute to regeneration of the digit-tip structure.
> e. The biochemical characteristics (signaling pathways and gene regulatory networks) have been well characterized for the regeneration of mammalian digit tip.
> f. The challenge is to overcome the regenerative limitations that prevent the intrinsic regenerative potential of mammals.

21.1 Introduction

The mammalian fingertip is composed of the nail plate, nail epithelium, skin, tendon, nerves, blood vessels, and terminal phalanx, which is the most distal bone of the limb. Defects in digit tip due to congenital anomalies or amputation negatively affect the quality of patients' life. Digit replantation is currently utilized as the treatment for digit amputation, with the digit survival rate being between 57% and 90%, depending on the method of analysis and country (Sebastin and Chung 2011, Waikakul et al. 2000). Prospective survey research by Waikakul et al. (2000) showed that 946 of 1018 digits (92.9%) were successfully engrafted. Sebastin et al. systemically reviewed 30 studies of digit replantation worldwide and reported that the mean survival rate was 86%. However, recently, Futa et al. claimed that the survival rate is 57% at two academic level-I trauma hospitals in the United States. The survival rates of crush-cut amputations were significantly lower compared with clean-cut amputations (Sebastin and Chung 2011). Moreover, replantation cannot be performed on the crush or burst amputations, in which the amputated part is seriously smashed or broken into pieces and undergoes prolonged ischemia time (Shieh and Cheng 2015, Chen and Chen 1991). Reconstitution of digital tissue using the regenerative approach based on the knowledge of developmental biology, stem cell biology, and tissue engineering is one of the alternative options for these amputees (Badylak 2007, Howard et al. 2008). Regeneration of single targeted tissue, such as skin, cartilage, bone, blood vessels, nerves, tendon, and skeletal muscle, which are the basic components of digit, can occur (Shieh and Cheng 2015). However, there is no way to assemble all these tissues and regenerate entire digit or limb, and new approaches to achieve simultaneous regeneration of different types of tissues are eagerly anticipated.

Mammals are generally considered to have poor regenerative ability; however, the digit tip of humans and rodents can regenerate on amputation within the part of the digit distal to the interphalangeal joint between the middle and distal phalanges (Murai et al. 1997). On amputation of the digit tip, epithelial cells migrate to re-epithelialize the wound, followed by accumulation of undifferentiated mesenchymal cells, traditionally referred to as the blastema (Neufeld 1980), from which all components of digit tip, except the epithelium, coordinately regenerate. Mechanisms underlying digit-tip regeneration, especially how the blastema forms and differentiates, may provide insight into developing new strategies in regenerative medicine (Muneoka et al. 2008, Muller et al. 1999, Han et al. 2005).

21.2 Regional limitation in digit regeneration in mammals

The regeneration of the mammalian digit tip was first reported in humans as clinical observations and was later shown to also occur in experimental animals, such as rodent, opossum, and squirrel monkey (Scharf 1961, Reginelli et al. 1995, Borgens 1982, Neufeld and Zhao 1995, Han et al. 2008, Mizell 1968, Singer et al. 1987). A series of studies has shown that digit-tip regeneration occurs in both juveniles and adults of both humans and rodents, and this regeneration response is restricted to the levels distal to the nail matrix (Douglas 1972, Lee et al. 1995). For example, in humans, Illingworth reported that complete regrowth of the fingertip is observed even fingers amputated at the level between skin creases and nail cuticle, but more a proximal amputation fails to regeneration (Illingworth 1974).

This amputation level–dependent ability of digit regeneration is later extensively examined in mice (Neufeld and Zhao 1993, Zhao and Neufeld 1995, Han et al. 2008). Pioneering studies by Neufeld and Zhao revealed that amputation through the distal-most 30% of the terminal phalanx (distal amputation) resulted in the significant distal elongation and distal tapering of the terminal phalanx (Figure 21.1) (Neufeld and Zhao 1993, 1995). In contrast, amputations through the proximal 20% of the distal phalanx (proximal amputation) precluded nail plate regrowth, and the bone grew minimally or regressed (Figure 21.1). Amputation through the intermediate 40% of the distal phalanx yielded variable responses, and both bone growth and lack of bone growth were observed. Importantly, they noticed that nail regrowth also was variable, and bone regrowth was always correlated with nail regrowth, suggesting the role of the nail organ in digit-tip regeneration.

To examine this relationship between nail and digit regeneration, they surgically removed nail organ before distal amputation and found that this treatment inhibited digit regeneration after distal amputation, in which digit regeneration normally occurred (Zhao and Neufeld 1995). Conversely, when digit was amputated from the ventral surface of

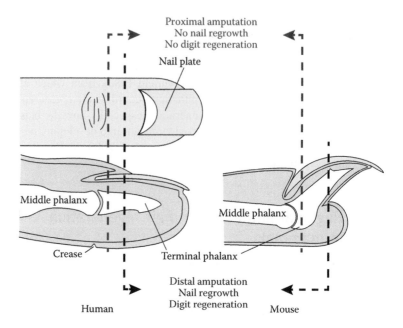

Figure 21.1 Schematic illustration of digit morphology and amputation level–dependent regeneration capacity in human and mouse.

the digit to remove bone but preserve nail and matrix, bone regrowth occurred from a more proximal level of amputation (Zhao and Neufeld 1995). Similar results were also reported in a human. Vidal and Dickson reported the case of a 9-year-old boy who had an L-shaped laceration on the radial side of the dorsum of the terminal phalanx, but the nail and nail bed appeared to be intact (Vidal and Dickson 1993). Radiograph taken on admission disclosed the complete absence of the terminal phalanx, with some tiny shreds of periosteal tissue. At sixth month after suture of wound, they found that the injured digit regrew with a nearly normal appearance. X-ray examination confirmed regeneration of the terminal phalanx that had almost normal morphology, except the absence of an epiphysis. These observations suggested that there is a relationship between nail organ and the level-specific regeneration of mammalian digit tip.

21.3 The mechanisms underlying digit-tip regeneration in mammals

21.3.1 Bone morphogenetic protein signaling

Reginelli et al. reported that the proximal limit of the regenerative capacity in fetal and neonatal mouse digits corresponds to the proximal extent of the *Msx1* expression domain. They hypothesized that the expression of *Msx* genes is important for digit cells to participate in a regenerative response (Reginelli et al. 1995). This hypothesis was further examined by Han et al. from the same laboratory, using *Msx1* or *Msx2* null mice (Han et al. 2003). Using an organ culture system, they demonstrated that *Msx1*, but not *Msx2*, null fatal mice display a regenerative defect. In *Msx1* null mice, they also found that bone morphogenetic protein 4 (*BMP4*) is downregulated in mutant digits, and treatment with exogenous *BMP4* rescues digit regeneration in a dose-dependent manner. The involvement of BMP signaling in digit-tip regeneration was further examined in neonatal mice by the same group (Yu et al. 2010). Yu et al. demonstrated that *BMP2* and *BMP7* are upregulated in the blastema, and implantation of Noggin-soaked beads inhibit digit-tip regeneration. These results demonstrate critical roles of *Msx1* and BMP during digit-tip regeneration.

21.3.2 Wnt signaling

Studies have revealed that nail stem cells are located at the proximal part of the nail matrix and differentiate into distal part of the nail matrix, nail bed, and nail plate in a Wnt-dependent manner (Takeo et al. 2013, 2016). Lehoczky and Tabin (2015) demonstrated that one of the Wnt agonists, Lgr6, is expressed in the nail stem cells at distal regions of the proximal matrix, and loss of Lgr6 expression results in the failure of digit regeneration. Moreover, when Wnt activation in the nail epithelium is suppressed by removing β-catenin, a key component of Wnt signaling, or Wntless (Wls), which is required for Wnt ligand secretion, nail stem cells fail to differentiate and nail growth is inhibited (Takeo et al. 2013) (Figure 21.2).

In addition to the inhibition of nail growth and digit bone maintenance in the homeostatic condition, depletion of β-catenin in the epithelium results in decreased innervation at the amputation site, which is very important for amphibian limb regeneration. Moreover, there was a lack of fibroblast growth factor 2 (FGF2) expression in the nail epithelium that overlies the wound site on loss of β-catenin or Wls in epithelial cells (Takeo et al. 2013). These changes were associated with the failure of blastema growth. These observations clearly suggest that a Wnt-active epithelium is essential to promote digit-tip regeneration through induction of blastema development and growth.

21.3.3 Innervation

Denervation before a distal amputation results in failure in blastema growth (Mohammad and Neufeld 2000, Rinkevich et al. 2014), and subsequent digit-tip regeneration is abrogated in mice (Takeo et al. 2013). Interestingly, the digits with denervation treatment lack FGF2 expression in nail epithelium. Implantation of FGF2-soaked beads in denervated digit after distal amputation overcomes the effect of denervation and rescues blastema growth in mice (Takeo et al. 2013). The FGF2 also

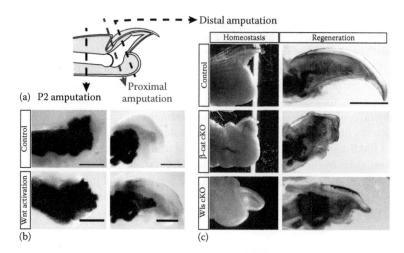

Figure 21.2 Wnt activation in nail epithelium is important for digit regeneration. (a) Schematic illustration of amputation level. (b) Whole mount alizarin red staining on control and K14-CreER;β-catenin[fl(ex3)/+] mouse, in which epithelial cells forced activate Wnt signaling, mutant digit at 5 weeks after amputation at proximal and P2 level. (c) Gross appearance of non-amputated digit at second month after gene modification (left) and regenerated digit at 5 weeks after amputation (right) of control and epithelial Wnt-deficient mouse models (K14-CreER;β-catenin[fl/fl], K14-CreER;Wls[fl/fl]). Bars = 500 μm. (Modified from Takeo, M. et al., *Nature*, 499, 228–232, 2013.)

shows potent effects in promoting proliferation of blastema cells cultured *in vitro* (Takeo et al. 2013). These observations maybe consistent with the observation by Zhao and Neufeld (1996) that bone regeneration after distal amputation is stimulated by FGF2 *in vitro*. In addition, a recent study by Johnston et al. unveils the important role of Schwann cell precursors (SCPs), which secret growth factors such as oncostatin M and platelet-derived growth factor-AA, in digit regeneration in adult mice (Johnston et al. 2016). Denervation before digit amputation results in the depilation of SCPs in the regenerating blastemal area, leading to an impaired digit regeneration that can be rescued by transplantation of exogenous SCPs.

21.4 Overcoming the regional limitation in regeneration

Series of works have provided the evidence that activation of endogenous regeneration mechanism can overcome the regenerative limitation after proximal digit amputation that does not normally induce regeneration. It was shown that BMP7-soaked beads induce *Msx1* upregulation in the dorsal region of the blastema after proximal amputation. In addition, bone regrowth was also seen in neonatal mice that received implantation of BMP2-soaked beads or stromal cell-derived factor (SDF)-1a-expressing cells at 4 days after proximal amputation (Yu et al. 2012, Lee et al. 2013). Another study showed that proximal amputation in normal mice resulted in the removal of most Wls-expressing distal matrix cells, and Wnt activation in the nail epithelium did not occur, leading to a failure of innervation in the amputation site. These observations demonstrated that Wnt activation in nail epithelium is required for nail growth and digit-tip regeneration; digit amputation at proximal level results in the removal of almost all Wls-expressing cells, and Wnt activation does not occur. Thus, regeneration of the nail and digit is coupled, and a level-dependent ability of digit regeneration exists in postnatal mice (Figure 21.2). In contrast, forced Wnt activation by stabilizing β-catenin in the epithelium induces innervation, FGF2 expression, and blastema growth after proximal amputations. However, forced Wnt activation in epithelial cells after complete removal of the nail organ is not capable of inducing digit regeneration (Figure 21.2).

21.5 Partial regeneration from non-regenerative limb/digit amputations

Although complete regeneration, other than the digit tip, has not been reported in mammals, blastema formation and subsequent partial regeneration occur after amputation of limbs or digits proximal to the terminal phalanx, either naturally or inductively.

Partial regeneration after limb amputation was first described by Selye (1934). In this work, he amputated the right limb of 12- to 15-day-old rats via a transverse section above the junction cartilage of the femur, resulting in the removal of the entire growth zone. A few days after the surgery, he found that the end of the stump was closed by proliferating spindle cells showing all the morphological characteristics of osteoblasts. After 19–25 days, the proximal part of the bone showed the typical structure of the growth zone, and the bone regrew in length. However, whether the osteoblasts seen around the amputation stump represent blastema cells or not was not mentioned in this work.

Blastema formation in mammalian limb regeneration was first reported in neonatal opossums (Mizell 1968). After amputation at the level through the distal part of the radius and ulna, forelimbs showed no external indication of a regenerative response; however, histological examination revealed that the underlying tissue underwent some dedifferentiation and cartilaginous callus formation around the radius. In contrast to the forelimb, hind limb displayed a regenerative response, and Mizell reported a blastema-like structure in one case out of 94 controls. Moreover, Mizell also reported that when nerve tissue from the forebrain of the opossum

469

was implanted in the hind limb before amputation, a regenerative response was seen in 8 out of 30 cases, and basal portion of three digits emerged at 38 days after amputation in the best regenerates. A subsequent study based on fate mapping of the developing opossum hind limb did not observe a regenerative response (Fleming and Tassava, 1981).

In adult mammals, partial limb regeneration can also be stimulated by a variety of methods, such as NaCl treatment, muscle flapping, and modulating electrical environment of amputation site; however, regeneration does not occur naturally (Neufeld 1980, Scharf 1961, Becker 1972, Becker and Spadaro 1972). Among such experiments, Becker and Spadaro first mentioned the effect of electrical stimulation on the formation of blastema after limb amputation. They amputated the forelimb at the level of mid-humerus of 21-day-old rat and implanted bimetallic electrogenic devices, which could control electrical current at the amputation site. They reported that the electrical stimulation resulted in the formation of a larger blastema than that in controls and the restoration of the missing portion of the humeral shaft at 7th day or less after amputation (Becker 1972, Becker and Spadaro 1972).

Although the formation of a blastema in response to digit amputation was mentioned in limb regeneration in fetal and adult mammals, these studies lack detailed analysis of this structure, and whether this structure has characteristics of so-called *blastema* similar to those in amphibians was under debate. To address this question, Neufeld (1980) performed a detailed analysis. In normal situations, blastema formation never occurred after amputation through the proximal phalanx of adult rats. To induce blastema formation after this level of amputation, Neufeld removed the scar pad on the 27th and 33rd days and treated with NaCl solution for 15 days, starting from 27th day after amputation. This treatment induced the formation of thickened wound epithelium and a large mass of irregularly oriented sub-epithelial cells at the 42nd day; however, complete blastema formation was not achieved. Closer examination of the sub-epithelial cells of treated animals revealed that these cells are more numerous, less regularly oriented, and more mitotically active, as indicated by the enhanced incorporation of [3H] thymidine. All these characteristics are seen in amphibian blastema during early stage of regeneration, and therefore, Neufeld concluded that the mammalian blastema that is formed after digit amputation could resemble that of amphibians.

Although Neufeld did not observe further regenerative response beyond the early stage of blastema formation in this work, Neufeld's laboratory later reported partial bone regeneration (Mohammad et al. 1999). In this study, they amputated the middle digit of the hind limb of 7- to 21-day-old rat through the proximal phalanx and implanted one-quarter of the nail plate, including the nail matrix, into the bone stump of amputated proximal phalanx. Among more than 50 digits that received nail transplantation, 6 transplants successfully engrafted and formed a nail bed and keratinized nail at the amputation site. In these successful transplants, bone growth toward nail organ was confirmed by X-ray examinations and histological analysis.

21.6 The characteristics of blastema cells and their origins

Classically, a blastema is considered a homogeneous population of mesenchymal cells based on the histological studies. However, studies using transplantation and genetic lineage tracing show that the blastema is composed of several different types of cells in amphibians and mice (Kragl et al. 2009, Rinkevich et al. 2011, Lehoczky et al. 2011). Using several lineage-specific Cre driver mice, such as Sp7-tTA:tetO-Cre mice for osteoblast lineage and Msx1-CreER for mesenchymal lineage, Lehoczky et al. (2011) demonstrated that the blastema is composed of cells derived from osteoblast and mesenchymal lineage but not epithelium and differentiated bone lineage. They also demonstrated that blastema cells derived from osteocyte lineage contribute to bone and periosteum regeneration and progeny of Msx1-positive cells give rise to bone and dermis. Rinkevic et al. (2011) also reached the same conclusion that blastema cells differentiate in a lineage-specific manner; however, they used a different mouse model. These studies demonstrate that the sources of blastema cells in regenerates are lineage-restricted.

Blastema cells that formed after distal amputation express *Msx1*, *Msx2*, and *BMP4* in fetal mice (Han et al. 2003). Postnatally, *BMP4* and *BMP2* are expressed in the distal and proximal regions of the blastema, whereas the expression of *Msx1* and *Msx2* is not detectable (Han et al. 2008, Yu et al. 2010). The majority of blastemal cells also co-express the stem cell marker Sca1 and the endothelial marker CD31 (Table 21.1) (Fernando et al. 2011). The accumulation of Sca1-positive cells is also found in the amputation

Table 21.1 Marker Expression of Mesenchymal Cells in Amputation Site

Amputation Level	Marker or Cre Driver	Reference
Distal amputation	BMP2	Han et al. 2003. Yu et al. 2010
	BMP4	Han et al. 2003. Yu et al. 2010
	Sca1/CD31/Lin–	Fernando et al. 2011
	Sca1+/CD29+/V-CAM+/ CD45–/C-kit–/CD49f–/ CD104–/CD34–/BMP4+	Wu et al. 2013
	CXCR4	Yu et al. 2012
P2 amputation	Sca1+/Sox2+/CD90+/CD133–/c-kit–	Agrawal et al. 2012
	Sca1+/CD29+/V-CAM+/ CD45–/C-kit–/CD49f–/ CD104–/CD34–	Wu et al. 2013
	CXCR4	Lee et al. 2013

site after P2 amputation in mice; however, blastema formation does not occur after this level of amputation (Table 21.1) (Agrawal et al. 2010, 2011, 2012, Mu et al. 2013). Agrawal et al. showed that Sca1+ cells that did not express epithelial or hematopoietic progenitor cell marker CD133 or c-kit but did express Sox2 and the periosteal mesenchymal stem cell marker CD90 accumulated at the amputation site after P2 amputation in adult mice (Biernaskie et al. 2009, Spangrude et al. 1988, Zhang et al. 2005, Agrawal et al. 2010, 2012). When an amputated digit was treated with a peptide that is identical to a string of amino acids from the C-terminal telopeptide of collagen III alpha, the number of Sca1+ cells increased, and these cells proliferated actively. They also showed that these cells were capable of differentiating into mesenchymal lineage cells, such as neuroectoderm, adipocyte, and osteoblast, *in vitro*. Moreover, Mu et al. (2013) demonstrated that treatment with matrix metalloproteinase 1 (MMP1) promoted the accumulation of Sca1+ cells at the amputation site and the formation of CD31+ capillary vessels but not bone regeneration.

Wu et al. (2013) isolated mesenchymal cells, which are positive for Sca-1, CD29, and V-CAM and negative for CD45, C-kit, CD49f, CD104, and CD34, from both the digit tip and the area distal and proximal to the interphalangeal joint (P2 area) (Table 21.1). They found that isolated mesenchymal cells only from the digit tip expressed BMP4, which is normally expressed in the blastema during digit-tip regeneration. In contrast, this expression was not detectable in the mesenchymal cells isolated from P2 area; however, they have many of the same characteristics in terms of the expression pattern of cell surface markers. Transplantation experiments of isolated mesenchymal cells into digit tip before digit-tip amputation revealed that both P2 and digit-tip mesenchymal cells contributed to blastema formation. At a later time point, digit-tip mesenchymal cells, but not P2 mesenchymal cells, were integrated in the regenerate into the bone cavity but did not differentiate into bone. At this time point, most of the grafted mesenchymal cells were found in the dermis and connective tissues around the regenerating bone. Conversely, mesenchymal cell transplantation into the P2 region before amputation showed that neither P2 nor digit-tip mesenchymal cells form a blastema. These results suggest that the difference of the ability of blastema formation and digit regeneration between the amputation at digit tip and P2 area is not due to the intrinsic difference in mesenchymal cells.

In addition to Sca1+ cells, Yu et al. (2012) found that all cultured cells isolated from the blastema after digit-tip amputation expressed chemokine (C-X-C motif) receptor 4 (CXCR4) and were attracted toward BMP2-soaked beads through SDF-1a/CXCR4 signaling. The CXCR4-positive cells were found in both P2 area and digit tip after amputation, and transplantation of SDF-1a-producing cells into the amputation site after amputation at the P2 level induce partial bone regeneration (Lee et al. 2013).

These results suggest that the mesenchymal progenitor cells with similar biochemical characteristics may emerge at amputation sites regardless of the level of amputations and that whether these cells contribute to the regeneration of digit-tip structure may depend on the environment or signals. On the other hand, the fate of blastemal differentiation may depend on the intrinsic difference between mesenchymal blastemal cells.

21.7 Concluding remarks

The study of mammalian digit/limb regeneration in the past decade greatly furthered our understanding of how the blastema forms during digit-tip regeneration in mammals at the cellular and molecular levels. Furthermore, these studies provided the concrete evidence to demonstrate the value of utilizing the knowledge of endogenous regeneration mechanisms. In addition, there has been remarkable progress toward understanding the biochemical nature of the blastema. An immediate next step in this research field includes understanding how molecular mechanisms identified thus far crosstalk with each other and how specific cell populations identified in the amputation sites communicate to achieve the coordinate regeneration of the tissues composing the digit tip. It is also important to thoroughly compare these mechanisms of mammalian digit-tip regeneration with those of salamander limb regeneration to understand how salamander limbs can regenerate without regional limitations.

References

Agrawal, V., S. A. Johnson, J. Reing, L. Zhang, S. Tottey, G. Wang, K. K. Hirschi, S. Braunhut, L. J. Gudas, and S. F. Badylak. 2010. Epimorphic regeneration approach to tissue replacement in adult mammals. *Proc Natl Acad Sci USA* 107 (8):3351–3355. doi:10.1073/pnas.0905851106.

Agrawal, V., B. F. Siu, H. Chao, K. K. Hirschi, E. Raborn, S. A. Johnson, S. Tottey, K. B. Hurley, C. J. Medberry, and S. F. Badylak. 2012. Partial characterization of the Sox2+ cell population in an adult murine model of digit amputation. *Tissue Eng Part A* 18 (13–14):1454–1463. doi:10.1089/ten.TEA.2011.0550.

Agrawal, V., S. Tottey, S. A. Johnson, J. M. Freund, B. F. Siu, and S. F. Badylak. 2011. Recruitment of progenitor cells by an extracellular matrix cryptic peptide in a mouse model of digit amputation. *Tissue Eng Part A* 17 (19–20):2435–2443. doi:10.1089/ten.TEA.2011.0036.

Badylak, S. F. 2007. The extracellular matrix as a biologic scaffold material. *Biomaterials* 28 (25):3587–3593. doi:10.1016/j.biomaterials.2007.04.043.

Becker, R. O. 1972. Stimulation of partial limb regeneration in rats. *Nature* 235 (5333):109–111.

Becker, R. O. and J. A. Spadaro. 1972. Electrical stimulation of partial limb regeneration in mammals. *Bull N Y Acad Med* 48 (4):627–641.

Biernaskie, J., M. Paris, O. Morozova, B. M. Fagan, M. Marra, L. Pevny, and F. D. Miller. 2009. SKPs derive from hair follicle precursors and exhibit properties of adult dermal stem cells. *Cell Stem Cell* 5 (6):610–623. doi:10.1016/j.stem.2009.10.019.

Borgens, R. B. 1982. Mice regrow the tips of their foretoes. *Science* 217 (4561):747–750.

Chen, Z. W. and Z. R. Chen. 1991. Reconstruction of the thumb and digit by toe to hand transplantation. *World J Surg* 15 (4):429–438.

Douglas, B. S. 1972. Conservative management of guillotine amputation of the finger in children. *Aust Paediatr J* 8 (2):86–89.

Fleming, M. W. and R. A. Tassava. 1981. Preamputation and postampuation histology of the neonatal opossum hindlimb: Implications for regeneration experiments. *J Exp Zool* 215 (2):143–149.

Fernando, W. A., E. Leininger, J. Simkin, N. Li, C. A. Malcom, S. Sathyamoorthi, M. Han. et al. 2011. Wound healing and blastema formation in regenerating digit tips of adult mice. *Dev Biol* 350 (2):301–310. doi:10.1016/j.ydbio.2010.11.035.

Han, M., X. Yang, J. E. Farrington, and K. Muneoka. 2003. Digit regeneration is regulated by Msx1 and BMP4 in fetal mice. *Development* 130 (21):5123–5132. doi:10.1242/dev.00710.

Han, M., X. Yang, J. Lee, C. H. Allan, and K. Muneoka. 2008. Development and regeneration of the neonatal digit tip in mice. *Dev Biol* 315 (1):125–135. doi:10.1016/j.ydbio.2007.12.025.

Han, M., X. Yang, G. Taylor, C. A. Burdsal, R. A. Anderson, and K. Muneoka. 2005. Limb regeneration in higher vertebrates: Developing a roadmap. *Anat Rec B New Anat* 287 (1):14–24. doi:10.1002/ar.b.20082.

Howard, D., L. D. Buttery, K. M. Shakesheff, and S. J. Roberts. 2008. Tissue engineering: Strategies, stem cells and scaffolds. *J Anat* 213 (1):66–72. doi:10.1111/j.1469-7580.2008.00878.x.

Illingworth, C. M. 1974. Trapped fingers and amputated finger tips in children. *J Pediatr Surg* 9 (6):853–858.

Johnston, A. P., S. A. Yuzwa, M. J. Carr, N. Mahmud, M. A. Storer, M. P. Krause, K. Jones, S. Paul, D. R. Kaplan, and F. D. Miller. 2016. Dedifferentiated Schwann cell precursors secreting paracrine factors are required for regeneration of the mammalian digit tip. *Cell Stem Cell* 19 (4):433–448. doi:10.1016/j.stem.2016.06.002.

Kragl, M., D. Knapp, E. Nacu, S. Khattak, M. Maden, H. H. Epperlein, and E. M. Tanaka. 2009. Cells keep a memory of their tissue origin during axolotl limb regeneration. *Nature* 460 (7251):60–65. doi:10.1038/nature08152.

Lee, J., L. Marrero, L. Yu, L. A. Dawson, K. Muneoka, and M. Han. 2013. SDF-1alpha/CXCR4 signaling mediates digit tip regeneration promoted by BMP-2. *Dev Biol* 382 (1):98–109. doi:10.1016/j.ydbio.2013.07.020.

Lee, L. P., P. Y. Lau, and C. W. Chan. 1995. A simple and efficient treatment for fingertip injuries. *J Hand Surg Br* 20 (1):63–71.

Lehoczky, J. A., B. Robert, and C. J. Tabin. 2011. Mouse digit tip regeneration is mediated by fate-restricted progenitor cells. *Proc Natl Acad Sci USA* 108 (51):20609–20614. doi:10.1073/pnas.1118017108.

Lehoczky, J. A. and C. J. Tabin. 2015. Lgr6 marks nail stem cells and is required for digit tip regeneration. *Proc Natl Acad Sci USA* 112 (43):13249–13254. doi:10.1073/pnas.1518874112.

Mizell, M. 1968. Limb regeneration: Induction in the newborn opossum. *Science* 161 (3838):283–286.

Mohammad, K. S., F. A. Day, and D. A. Neufeld. 1999. Bone growth is induced by nail transplantation in amputated proximal phalanges. *Calcif Tissue Int* 65 (5):408–410.

Mohammad, K. S. and D. A. Neufeld. 2000. Denervation retards but does not prevent toetip regeneration. *Wound Repair Regen* 8 (4):277–281.

Mu, X., I. Bellayr, H. Pan, Y. Choi, and Y. Li. 2013. Regeneration of soft tissues is promoted by MMP1 treatment after digit amputation in mice. *PLoS One* 8 (3):e59105. doi:10.1371/journal.pone.0059105.

Muller, T. L., V. Ngo-Muller, A. Reginelli, G. Taylor, R. Anderson, and K. Muneoka. 1999. Regeneration in higher vertebrates: Limb buds and digit tips. *Semin Cell Dev Biol* 10 (4):405–413. doi:10.1006/scdb.1999.0327.

Muneoka, K., C. H. Allan, X. Yang, J. Lee, and M. Han. 2008. Mammalian regeneration and regenerative medicine. *Birth Defects Res C Embryo Today* 84 (4):265–280. doi:10.1002/bdrc.20137.

Murai, M., H. K. Lau, B. P. Pereira, and R. W. Pho. 1997. A cadaver study on volume and surface area of the fingertip. *J Hand Surg Am* 22 (5):935–941. doi:10.1016/S0363-5023(97)80094-9.

Neufeld, D. A. 1980. Partial blastema formation after amputation in adult mice. *J Exp Zool* 212 (1):31–36. doi:10.1002/jez.1402120105.

Neufeld, D. A. and W. Zhao. 1993. Phalangeal regrowth in rodents: Postamputational bone regrowth depends upon the level of amputation. *Prog Clin Biol Res* 383A:243–252.

Neufeld, D. A. and W. Zhao. 1995. Bone regrowth after digit tip amputation in mice is equivalent in adults and neonates. *Wound Repair Regen* 3 (4):461–466. doi:10.1046/j.1524-475X.1995.30410.x.

Reginelli, A. D., Y. Q. Wang, D. Sassoon, and K. Muneoka. 1995. Digit tip regeneration correlates with regions of Msx1 (Hox 7) expression in fetal and newborn mice. *Development* 121 (4):1065–1076.

Rinkevich, Y., P. Lindau, H. Ueno, M. T. Longaker, and I. L. Weissman. 2011. Germ-layer and lineage-restricted stem/progenitors regenerate the mouse digit tip. *Nature* 476 (7361):409–413. doi:10.1038/nature10346.

Rinkevich, Y., D. T. Montoro, E. Muhonen, G. G. Walmsley, D. Lo, M. Hasegawa, M. Januszyk, A. J. Connolly, I. L. Weissman, and M. T. Longaker. 2014. Clonal analysis reveals nerve-dependent and independent roles on mammalian hind limb tissue maintenance and regeneration. *Proc Natl Acad Sci USA* 111 (27):9846–9851. doi:10.1073/pnas.1410097111.

Scharf, A. 1961. Experiments on regenerating rat digits. *Growth* 25:7–23.

Sebastin, S. J. and K. C. Chung. 2011. A systematic review of the outcomes of replantation of distal digital amputation. *Plast Reconstr Surg* 128 (3):723–737. doi:10.1097/PRS.0b013e318221dc83.

Selye, H. 1934. On the mechanism controlling the growth in length of the long bones. *J Anat* 68 (Pt 3):289–292.

Shieh, S. J. and T. C. Cheng. 2015. Regeneration and repair of human digits and limbs: Fact and fiction. *Regeneration* 2 (4):149–168. doi:10.1002/reg2.41.

Singer, M., E. C. Weckesser, J. Geraudie, C. E. Maier, and J. Singer. 1987. Open finger tip healing and replacement after distal amputation in rhesus monkey with comparison to limb regeneration in lower vertebrates. *Anat Embryol (Berlin)* 177 (1):29–36.

Spangrude, G. J., S. Heimfeld, and I. L. Weissman. 1988. Purification and characterization of mouse hematopoietic stem cells. *Science* 241 (4861):58–62.

Takeo, M., W. C. Chou, Q. Sun, W. Lee, P. Rabbani, C. Loomis, M. M. Taketo, and M. Ito. 2013. Wnt activation in nail epithelium couples nail growth to digit regeneration. *Nature* 499 (7457):228–232. doi:10.1038/nature12214.

Takeo, M., C. S. Hale, and M. Ito. 2016. Epithelium-Derived Wnt Ligands are essential for maintenance of underlying digit bone. *J Invest Dermatol* 136 (7):1355–1363. doi:10.1016/j.jid.2016.03.018.

Vidal, P. and M. G. Dickson. 1993. Regeneration of the distal phalanx. A case report. *J Hand Surg Br* 18 (2):230–233.

Waikakul, S., S. Sakkarnkosol, V. Vanadurongwan, and A. Un-nanuntana. 2000. Results of 1018 digital replantations in 552 patients. *Injury* 31 (1):33–40.

Wu, Y., K. Wang, A. Karapetyan, W. A. Fernando, J. Simkin, M. Han, E. L. Rugg. et al. 2013. Connective tissue fibroblast properties are position-dependent during mouse digit tip regeneration. *PLoS One* 8 (1):e54764. doi:10.1371/journal.pone.0054764.

Yu, L., M. Han, M. Yan, E. C. Lee, J. Lee, and K. Muneoka. 2010. BMP signaling induces digit regeneration in neonatal mice. *Development* 137 (4):551–559. doi:10.1242/dev.042424.

Yu, L., M. Han, M. Yan, J. Lee, and K. Muneoka. 2012. BMP2 induces segment-specific skeletal regeneration from digit and limb amputations by establishing a new endochondral ossification center. *Dev Biol* 372 (2):263–273. doi:10.1016/j.ydbio.2012.09.021.

Zhang, X., C. Xie, A. S. Lin, H. Ito, H. Awad, J. R. Lieberman, P. T. Rubery, E. M. Schwarz, R. J. O'Keefe, and R. E. Guldberg. 2005. Periosteal progenitor cell fate in segmental cortical bone graft transplantations: Implications for functional tissue engineering. *J Bone Miner Res* 20 (12):2124–2137. doi:10.1359/JBMR.050806.

Zhao, W. and D. A. Neufeld. 1995. Bone regrowth in young mice stimulated by nail organ. *J Exp Zool* 271 (2):155–159. doi:10.1002/jez.1402710212.

Zhao, W. and D. A. Neufeld. 1996. Bone regeneration after amputation stimulated by basic fibroblast growth factor in vitro. *In Vitro Cell Dev Biol Anim* 32 (2):63–65.

Chapter 22 Spinal cord repair and regeneration

Jennifer R. Morgan

Contents

Key concepts

a. After spinal cord injury (SCI), regeneration of the mammalian central nervous system (CNS) is extremely poor due to an inhospitable post-injury environment, which includes glial scarring, release of myelin inhibitors, inflammation, and neuronal death.

b. Unlike mammals, most other vertebrate species exhibit robust, spontaneous regeneration after traumatic SCI.

c. Knowledge of conserved cellular and molecular pathways that support CNS regeneration in vertebrates is beginning to emerge.

> d. Non-mammalian models could thus provide distinct advantages for understanding evolutionarily conserved mechanisms with the potential to promote greater CNS regeneration in mammals.
> e. Non-mammalian models could also provide higher-throughput means for screening new molecular therapies and bioengineered materials as treatments for SCI.

22.1 Introduction

Traumatic spinal cord injury (SCI) currently affects ~275,000 people in the United States, and 12,500 new cases are reported annually, with the most common causes being vehicle accidents and falls (https://www.nscisc.uab.edu). The sensory and motor damages caused by SCI are often permanent, because axon regeneration and plasticity within the human adult central nervous system (CNS) are very limited (Bradbury and McMahon, 2006; Blesch and Tuszynski, 2009). Thus, there is a great need for developing strategies to improve CNS regeneration. However, this has proven to be quite difficult, because CNS repair and regeneration are complex processes, and there are many dynamic changes that have to occur on the background of tissues that are relatively stable and mature. As a consequence, the field still needs to gain a deeper understanding of the post-injury environment and the cellular and molecular barriers to CNS regeneration in mammals. Although there has been some progress to move SCI treatments toward and into the clinic (Hug and Weidner, 2012), no single drug, molecular manipulation, or therapy has been successful in restoring normal function after SCI in preclinical animal models, emphasizing a need for combinatorial treatments that have greater efficacy (Benowitz and Yin, 2007; McCall et al., 2012; Zhao and Fawcett, 2013). However, it is not feasible to identify or test all possible combinations in mammalian SCI models, which are *a priori* so poor at regenerating, without first identifying more basic, fundamental mechanisms that support and promote CNS regeneration. In this respect, non-mammalian models such as fishes, amphibians, and reptiles, which undergo robust, spontaneous recovery from SCI, could provide distinct advantages for identifying key conserved pathways with the greatest potential for promoting CNS regeneration in mammals (Bloom, 2014; Morgan and Shifman, 2014; Rasmussen and Sagasti, 2016).

In this chapter, I discuss some of the differences in the injury and regeneration responses in mammalian versus non-mammalian SCI models and propose several ways in which highly regenerative models could be used to advance our goals. I focus primarily on the cellular processes underlying these events and discuss the molecular mechanisms only where essential. The detailed molecular mechanisms that affect CNS regeneration are reviewed elsewhere (Harel and Strittmatter, 2006; Giger et al., 2010; Bloom and Morgan, 2011; Gordon-Weeks and Fournier, 2014).

22.2 Post-injury environment in the mammalian spinal cord

22.2.1 The glial response after spinal cord injury and myelin inhibitors

In mammals, SCI leads to a complex cellular response involving invasion of resident and non-resident immune cells, inflammation, formation of a glial scar, release of inhibitory factors, and alterations in the extracellular matrix, which result in a post-injury environment that is inhospitable for neuronal regrowth, regeneration, and functional recovery (Figure 22.1a–c) (Yiu and He, 2006; Fawcett et al., 2012; Cregg et al., 2014; Silver et al., 2015). Within the first week, resident glial cells in the CNS, including microglia/macrophages and ependymal cells, migrate toward the injury site and begin to proliferate, filling the gap and providing wound-healing functions (Figure 22.1b). However, afterwards, a glial scar forms, in which astrocytes and their processes form an atypically dense network at the lesion periphery (Figure 22.1c) (Yiu and He, 2006; Cregg et al., 2014). Fibroblasts, which deposit extracellular matrix components, and oligodendrocyte precursors, which entrap regrowing axons, together form the lesion core (Cregg et al., 2014). This reactive gliosis thus creates a

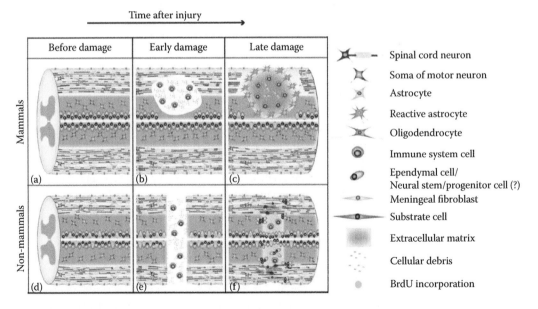

Figure 22.1 Different post-injury environments in the spinal cord in mammals versus non-mammals. (a–c) Spinal cord injury in mammals induces immune cells to infiltrate the lesion and form a glial scar, thus forming a major barrier to regeneration of spinal neurons. (d–f) In contrast, after SCI in non-mammals, the glial scar does not form and the lesioned area is permissive for regeneration. (Lee-Liu, D. et al.: Spinal cord regeneration: Lessons for mammals from non-mammalian vertebrates. *Genesis*. 2013. 51. 529–544. Copyright Wiley-VCH Verlag GmbH & Co. KGaA. Reproduced with permission.)

physical barrier that inhibits subsequent axon regeneration. Adding to the complexity of the problem, the microglia/macrophages that invade after SCI appear to have both pro-regenerative and anti-regenerative roles in the injured CNS (Bloom, 2014; Popovich, 2014; Gensel and Zhang, 2015).

The damage caused by SCI also induces myelinating oligodendrocytes to release several proteins that are well-established inhibitors of axon growth *in vitro* and *in vivo,* including Nogo-A, oligodendrocyte myelin glycoprotein (OMgp), and myelin-associated glycoprotein (MAG) (Yiu and He, 2006; Pernet and Schwab, 2012; Geoffroy and Zheng, 2014; Schwab and Strittmatter, 2014). A great deal of effort has been focused on developing ways to neutralize these myelin-associated inhibitors as a potential means for treating SCI (Domeniconi and Filbin, 2005). Supporting this approach, infusion of function-blocking anti-Nogo-A antibodies improved axon regeneration and functional recovery after SCI in both rodent and primate models (Liebscher et al., 2005; Freund et al., 2006; Pernet and Schwab, 2012). Positive effects on regeneration and recovery were also observed after Nogo-A deletion (Kim et al., 2003; Simonen et al., 2003), treatment with Nogo receptor antagonists (Li and Strittmatter, 2003; Li et al., 2005), or application of a function-blocking Nogo receptor ectodomain, which simultaneously inhibited Nogo, OMgp, and MAG (Li et al., 2004). In contrast, other studies showed little to no effect of Nogo manipulations on axon regeneration and functional recovery (Zheng et al., 2003; Zheng et al., 2005; Steward et al., 2008; Lee et al., 2010; Lee and Zheng, 2012). Such contradictory results have been attributed to genetic differences between the animal strains used in the experiments (Dimou et al., 2006). Despite these contradictory findings, anti-Nogo antibody infusion has been tested in phase I clinical trials as a potential therapeutic; however, results are not yet reported (https://clinicaltrials.gov/ct/show/NCT00406016). Thus, although myelin-associated inhibitors likely play a role in axon plasticity after SCI, their precise roles after injury require further investigation (Fawcett et al., 2012; Cregg et al., 2014; Geoffroy and Zheng, 2014). Whatever are their precise roles, reducing the glial scar and neutralizing myelin inhibitors remain among the primary targets for SCI.

22.2.2 The neuronal response after spinal cord injury

Spinal cord injury also damages neurons and often severs their axons. During the first few weeks post-injury, the proximal axons that are still connected to the cell body die back and form retraction bulbs, and the distal axons undergo Wällerian degeneration (Chang et al., 2016). While some CNS neurons survive the injury, others die by apoptosis (Hains et al., 2003; Lee et al., 2004; Shifman et al., 2008; Busch and Morgan, 2012). Within a month after SCI, some of the surviving CNS neurons mount a regeneration response and begin to regrow toward the lesion site. However, the vast majority of the axons are inhibited from regenerating by the post-injury environment in the lesioned area. Adding to the problem, there is also a diminished intrinsic capacity for axon regeneration in mature CNS neurons, compared with the greater growth potential of developing neurons (Afshari et al., 2009; Di Giovanni, 2009; Sun and He, 2010). Despite poor functional outcomes, both CNS and peripheral nervous system (PNS) neurons exhibit a range of capacities for post-injury regeneration and plasticity (Blesch and Tuszynski, 2009). Some mammalian spinal neurons exhibit a degree of spontaneous regeneration after SCI, including those in the reticulospinal (RS), rubrospinal, and raphe tracts, whereas the corticospinal tract that represents the major descending pathway in the mammalian spinal cord is rather poor at regenerating (Kwon et al., 2002; Blesch and Tuszynski, 2009). In contrast, PNS neurons, including sensory axons in the dorsal root ganglia (DRG) and neuromuscular junctions, are particularly good at regenerating (Son and Thompson, 1995; Smith et al., 2012; Kang and Lichtman, 2013). When a peripheral nerve bridge is present, spinal axons will regenerate into and beyond the bridge, indicating that mammalian CNS neurons are capable of regenerating when provided with a permissive environment (David and Aguayo, 1981). Thus, boosting the intrinsic growth capacity in adult CNS neurons is another possible means for improving outcomes after SCI (Di Giovanni, 2009; Sun and He, 2010; Fagoe et al., 2014).

The lack of successful CNS regeneration in mammals makes it difficult to identify the best strategies for improving functional recovery in the clinical setting. At present, bioengineering approaches are taking a lead role. Autologous peripheral nerve grafts are being used successfully to bridge spinal lesions and promote regeneration in mammalian SCI models (Houle et al., 2006; Cote et al., 2011; Khaing and Schmidt, 2012). Cellular transplantation therapies, including grafting of neural stem/progenitor cells to promote neuron replacement, olfactory ensheathing cells to promote reorganization of the glial scar and axon regrowth, and Schwann cells to promote remyelination, are being developed (Tetzlaff et al., 2011; Roet and Verhaagen, 2014). A number of molecular manipulations and delivery systems are also being considered, including those that increase the levels of the signaling molecule cyclic AMP (which improves CNS axon regeneration in many SCI models [Domeniconi and Filbin, 2005; Hannila and Filbin, 2008]), as are stimulation-based interventions (Roy et al., 2012). Despite recent progress, many questions remain about which treatment(s) will best support CNS repair and regeneration, because most single manipulations have modest effects, and therefore, no individual treatment has, as of now, been the key to a full recovery. Furthermore, reproducibility of experimental results has been problematic, in part, due to differences in the models and lack of clear documentation on the data collection methods, necessitating replication studies and broader re-evaluation (Steward et al., 2012; Biering-Sorensen et al., 2015). Thus, we are at a crossroad in the field of SCI and in need of higher throughput analyses to determine the best combinations of treatments that could promote recovery from SCI. Here, non-mammalian models could provide some advantages for delineating the basic, conserved mechanisms that support CNS repair and regeneration, which when complemented with mammalian studies, could contribute to solving some of the field's biggest problems.

22.3 Lessons on central nervous system regeneration from non-mammalian models

22.3.1 Nervous system regeneration is evolutionarily conserved across many vertebrate taxa

Unlike mammals, non-mammalian vertebrates, including lampreys, fishes, amphibians, and some reptiles, exhibit robust spontaneous regeneration after CNS injury, indicating that CNS repair is an evolutionarily conserved response (Sanchez Alvarado and Tsonis, 2006; Tanaka and Ferretti, 2009; Morgan and Shifman, 2014). For example, after a complete spinal cord transection, late larval lampreys spontaneously recover full locomotor capabilities (swimming) within 10–12 weeks post-injury (Figure 22.2a and b) (Rovainen, 1976; Selzer, 1978; Cohen et al., 1988; McClellan, 1990; Oliphint et al., 2010). Functional recovery in lampreys is supported by repair of the spinal lesion and regeneration of spinal axons several millimeters beyond the injury site (Figure 22.2c) (Yin and Selzer, 1983; Davis and McClellan, 1993; Oliphint et al., 2010). Adult zebrafish also recover swimming functions within 4–6 weeks after a complete spinal transection, accompanied by lesion repair and axon regeneration (Figure 22.2c) (Becker et al., 1997; Rasmussen and Sagasti, 2016). The recently developed larval zebrafish SCI model is particularly attractive because of the rapid time course of behavioral recovery, which occurs within 7–9 days post-injury, and the animal's transparency, which facilitates live imaging (Briona and Dorsky, 2014). Interestingly, despite the power of zebrafish genetics, no genetic screens have yet been published on zebrafish SCI models, indicating a future opportunity that could be undertaken. Another classical model for CNS regeneration is the adult goldfish, which also regenerates spinal axons and recovers locomotor behaviors, including the rapid startle response (Bernstein and Gelderd, 1970; Zottoli and Freemer, 2003). In frogs, there is a marked and rapid developmental switch in the capacity for spinal cord regeneration. After a complete spinal transection, pre-metamorphic tadpoles exhibit robust axon regeneration, and behavioral recovery occurs after they metamorphose into juvenile frogs (Forehand and Farel, 1982; Beattie et al., 1990; Gibbs and Szaro, 2006). However, during metamorphosis and throughout adulthood in frogs, axon regeneration and functional recovery do not occur after SCI (Beattie et al., 1990). This is due to a rise in levels of thyroid hormone, which triggers metamorphosis and causes large-scale

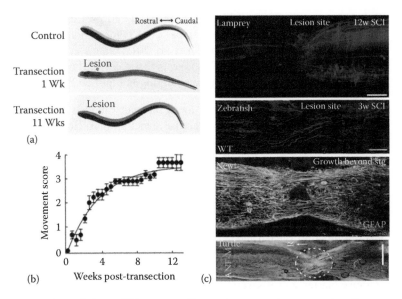

Figure 22.2 Robust CNS regeneration and recovery in non-mammalian vertebrates. (a–b) Lampreys recover near-normal swimming movements by 11 weeks post-injury after a complete spinal transection. (c) Extensive axon regeneration after SCI across non-mammalian taxa. Regenerating axons in lamprey (red), zebrafish(red), and newt (magenta) are labeled with fluorescent dyes. In turtle, regenerating axons are labeled with a neurofilament antibody (green). The image of the regenerating newt spinal cord also shows a close association between the regenerating axons and *glial bridges* (in green). Lesion site is indicated in center of each image. (Panel B: Oliphint, P.A. et al.: Regenerated synapses in lamprey spinal cord are sparse and small even after functional recovery from injury. *J Comp Neurol.* 2010. 518. 2854–2872. Copyright Wiley-VCH Verlag GmbH & Co. KGaA. Reproduced with permission. Panel C: Reprinted from *Exp Neurol*, 258, O. Bloom, Non-mammalian model systems for studying neuro-immune interactions after spinal cord injury, 130–140, Copyright (2014), with permission from Elsevier; Goldshmit, Y. et al., *J Neurosci*, 32, 7477–7492, 2012, with permission; Zukor, K.A. et al. *Neural Dev.* 6, 1, 2011, with permission; Rehermann, M.I. et al.: Neural reconnection in the transected spinal cord of the freshwater turtle *Trachemys dorbignyi. J Comp Neurol.* 2009. 515. 197–214. Copyright Wiley-VCH Verlag GmbH & Co. KGaA. Reproduced with permission.)

changes in gene expression that no longer support CNS regeneration (Gibbs et al., 2011). Thus, frogs could be a great model to further explore the mechanisms underlying age-related changes in regenerative capacity (Smith et al., 2011). In contrast, other amphibians such as newts and salamanders regenerate and recover as adults after a complete spinal transection (Figure 22.2c) (Davis et al., 1990; Zukor et al., 2011), including overground stepping behaviors that are more similar to mammalian locomotion (Chevallier et al., 2004; Cabelguen et al., 2013). Even reptiles such as turtles exhibit some regeneration and recovery after SCI (Figure 22.2c), albeit to a lesser degree than in other vertebrates

(Rehermann et al., 2009). Although changes in cell proliferation and gene expression are important for spinal cord regeneration in turtles (Rehermann et al., 2009; Rehermann et al., 2011; Garcia et al., 2012); their imperfect recovery after SCI suggests that some of the inhibitory factors observed in mammals maybe emerging in reptilian lineages.

The examples discussed previously focus on injury models where the spinal cord was damaged in an otherwise intact animal. However, the CNS and PNS in many adult vertebrates can also regenerate after a more severe injury such as tail or limb amputation, a topic that is discussed more fully in other chapters. As an example, after tail or limb amputation in the Mexican salamander, or the axolotl, these complex tissues, including the spinal cord or peripheral nerves and the surrounding bone, muscle, connective tissue, and skin, are rebuilt with high fidelity over the next few months (Tanaka and Ferretti, 2009; Voss et al., 2009; McCusker and Gardiner, 2011; Simon and Tanaka, 2013). During limb regeneration, the arm, hand, and digits are rebuilt from a developmentally labile, proliferative blastema over several months, successfully restoring the original anterior–posterior and proximal–distal organization using positional cues from the proximal stump (Maden and Turner, 1978; Muneoka and Bryant, 1982; Gardiner and Bryant, 1989; McCusker and Gardiner, 2013). Limb regeneration requires a contribution from the peripheral nerve stump, because regeneration no longer occurs when the proximal limb is denervated, and redirection of a peripheral nerve to a skin wound induces formation of an ectopic limb (Stocum, 2011). The role of the nerve is to maintain the blastema in a labile state, so that it can receive positional (e.g., anterior–posterior) information from the proximal limb (McCusker and Gardiner, 2013), and this likely occurs through release of fibroblast growth factor (FGF) and downstream signaling (Mullen et al., 1996). Similarly, tail regeneration in axolotls occurs with high fidelity. After regeneration, the original number of DRG and peripheral nerves returns, and they emanate from the regenerated spinal cord with the correct spacing, indicating tight control of tissue morphology and size (McHedlishvili et al., 2012). Although the molecular mechanisms of tail and limb regeneration are still under investigation (Ponomareva et al., 2015), a recent study identified a myristoylated alanine-rich C-kinase substrate (MARCKS)-like protein as possible regeneration initiator (Sugiura et al., 2016).

Beyond these examples, other CNS tissues, including the retina, optic nerve, and various brain regions, exhibit remarkable regeneration capacity and plasticity after injury in adult fishes, amphibians, and reptiles (Zupanc and Sirbulescu, 2011; Kizil et al., 2012; Gemberling et al., 2013; Gorsuch and Hyde, 2014). Even in avian and mammalian lineages, CNS regeneration occurs to some extent in young animals and diminishes greatly when they mature into adults (Tanaka and Ferretti, 2009). Together, these anatomical and behavioral studies illuminate the fact that spontaneous regeneration of CNS structures is a highly conserved response throughout vertebrate evolution (Tanaka and Ferretti, 2009; Ferretti, 2011).

22.3.2 Shared mechanisms that support successful central nervous system regeneration

We are only at the beginning stages of understanding the cellular and molecular mechanisms that support successful CNS regeneration in vertebrate species. However, several commonalities are emerging through studies in non-mammalian models, and these include a permissive glial response, lack of myelin inhibition, and high intrinsic growth capacity in CNS neurons. Unlike the situation in mammals where reactive gliosis and scarring occur, a permissive glial response has been documented in all other vertebrates that exhibit robust CNS regeneration, wherever examined thus far (Lurie et al., 1994; Goldshmit et al., 2008; Rehermann et al., 2009; Zukor et al., 2011; Rasmussen and Sagasti, 2016). In contrast to what occurs in mammals (Figure 22.1a–c), the glial cells in highly regenerative vertebrates appear to support and guide the regenerating spinal axons after injury (Figure 22.1d–f) (Lee-Liu et al., 2013). In lampreys, the glial cells nearest the lesioned area, which normally project transversely, reorient their processes along the longitudinal axis of the spinal cord within 4 weeks post-injury, and thereafter, the regenerating spinal axons can be seen in close proximity to the glial processes (Lurie et al., 1994). In addition, microglia/macrophages appear to be activated after SCI in lampreys (Shifman et al., 2009; Lau et al., 2013). However, the other glial cell types responding to injury in lampreys remain uncharacterized. In zebrafish, newts, and turtles, glial fibrillary acidic protein (GFAP)-positive glia also extend their processes across the spinal lesion, in close association with regenerating axons (see Figure 22.2c for an example in newts) (Goldshmit et al., 2008; Rehermann et al., 2009; Zukor et al., 2011; Goldshmit et al., 2012). In lampreys, application of db-cAMP (a cell-permeant analog of cyclic AMP) induced glial cells to acquire their longitudinal orientation at earlier post-injury times, and this correlated with a twofold increase in axon regeneration (Lau et al., 2013). Furthermore, loss of FGF function in zebrafish prevented the formation of the glial bridge and subsequently inhibited axon regeneration and swimming behaviors (Goldshmit et al., 2012). These data suggest that the *glial bridges* are required for CNS axon regeneration.

A second feature common to highly regenerative vertebrates is a lack of myelin inhibition. It is notable that with the exception of lampreys, whose nerves are unmyelinated (Bullock et al., 1984), all other highly regenerative vertebrates possess myelin in the CNS but are still able to regenerate. Interestingly, the myelin-associated inhibitors Nogo, MAG, and OMgp are robustly expressed during spinal cord regeneration in the axolotls, indicating that they are permissive for regeneration in this model (Hui et al., 2013; Welte et al., 2015). Furthermore, knockdown of the Nogo-A homolog in zebrafish inhibits the outgrowth of ganglion cell axons from retinal explants, indicating a positive role for Nogo-A in regeneration in this context (Welte et al., 2015). The differences between the functions of these proteins in mammals versus other vertebrates have been attributed to differences in their amino acid sequences and/or to downstream signaling events. Indeed, zebrafish and mammalian Nogo appear to have functionally diverged from each other, because mammalian Nogo inhibits growth of zebrafish axons, while the zebrafish homolog of Nogo does not inhibit the growth of fish or mammalian axons (Abdesselem et al., 2009; Hui et al., 2013; Rasmussen and Sagasti, 2016). Thus, across several non-mammalian taxa, myelin and its classical inhibitors do not impede CNS axon regeneration or functional recovery.

A third theme that is emerging is that adult CNS neurons of highly regenerative vertebrates have a greater intrinsic capacity for regrowth after injury than their mammalian counterparts. Here, gene expression analyses have been particularly illuminating. During spinal cord regeneration in zebrafish, a recent transcriptome study revealed that many genes previously identified as regeneration-associated genes in mammals are induced or upregulated (Hui et al., 2014). These include the transcription factors *Stat3, Atf3, Socs3,* and *Sox11*, which are all pro-regenerative in the mammalian PNS (Abe and Cavalli, 2008; Chandran et al., 2016). Conversely, in non-regenerative juvenile frogs, genes that regulate the axonal growth cone in developing neurons are significantly downregulated after SCI, suggesting a mechanism that actively limits CNS regeneration under these conditions (Lee-Liu et al., 2014). For understanding the intrinsic gene expression programs that promote or inhibit CNS neuron regeneration, the giant RS neurons in lamprey brains provide an excellent model. There are 30 giant RS neurons in lamprey brains, which are located in stereotypical positions and are identifiable due to their large size (100–200 μm in diameter) (Figure 22.3). After spinal transection, which axotomizes the RS neurons, a subset of the neurons predictably and reproducibly regenerates their axons across the spinal lesion (*good regenerators*), while another subset fails to regenerate and dies (*poor regenerators*) (Jacobs et al., 1997; Shifman et al., 2008; Busch and Morgan, 2012). Good regenerators recover higher levels of expression of the cytoskeletal element, neurofilament, after SCI, supporting greater axon regrowth (Jacobs et al., 1997; Zhang et al., 2015). In contrast, only the poor regenerators express measurable levels of neogenin, a receptor for repulsive guidance molecule, suggesting that chemorepulsive factors may play a role in inhibiting their regeneration (Shifman et al., 2009). Thus, different intrinsic molecular responses likely underlie the differences in regenerative capacities in these RS neuron subtypes. A growing body of data indicates that regenerating CNS neurons in non-mammalian vertebrates exhibit a gene expression profile that is similar to regenerating mammalian PNS neurons, suggesting the possibility of conserved gene programs that promote regeneration (Abe and Cavalli, 2008; Lerch et al., 2012; Hui et al., 2014; Chandran et al., 2016).

Given that so many species across diverse taxa are capable of robust CNS regeneration after injury suggests that successful regeneration is the evolutionary rule rather than the exception; however, it remains

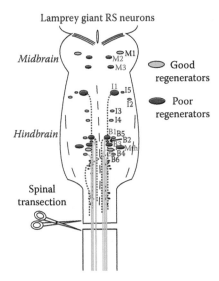

Figure 22.3 A model for investigating variable regeneration capacities among CNS neurons. There are 30 identified giant RS neurons in the lamprey midbrain and hindbrain, which exhibit known and reproducible capacities for regeneration after SCI. After spinal transection, the *good regenerators* typically regenerate 60%–90% of the time, while the *poor regenerators* regenerate less than 20% of the time. The RS neurons thus offer the possibility for studying the intrinsic responses, determining why some neurons regenerate and others do not.

unclear why avians and mammals exhibit reduced regenerative potential (Tanaka and Ferretti, 2009; Ferretti, 2011). This argues in favor of placing more emphasis on the highly regenerative models, so that we can take advantage of *nature's free lessons* to define the basic mechanisms underlying successful CNS and PNS regeneration and to identify relevant conserved molecular pathways.

22.4 How non-mammalian models can advance our understanding of central nervous system repair mechanisms

22.4.1 Comparative transcriptomics to identify conserved mechanisms of central nervous system regeneration

The injury response is reflected not only by changes in the cellular environment, as discussed earlier but also by acute and chronic changes in gene expression in and around the lesioned area (Nesic et al., 2002; Velardo et al., 2004; Chen et al., 2013). Unbiased approaches, such as next-generation transcriptome sequencing and microarray analyses, have already greatly advanced our understanding of pro-regenerative mechanisms in both mammalian and non-mammalian models. For example, in one RNA-Seq study, the resting, uninjured transcriptome of cultured mouse DRG neurons, which are highly regenerative, was compared with that of cerebellar granule neurons, which do not regenerate (Lerch et al., 2012). Consistent with a regenerative molecular profile, the DRG neurons exhibited significantly greater expression of genes related to growth factors, cytoskeleton, cell adhesion, and ion channels, compared with the poorly regenerative cerebellar neurons (Lerch et al., 2012). Similarly, after SCI in *Xenopus laevis,* large-scale gene programs predicted to support CNS regeneration, including those for neurogenesis, metabolism, and cell cycle, were differentially upregulated at early time points but only in the regenerative tadpoles (Lee-Liu et al., 2014). A microarray study in zebrafish revealed a common set of gene expression changes after SCI that are shared with other highly regenerative tissues, such as the PNS and optic nerve (Hui et al., 2014), which included an upregulation of genes involved in axogenesis and axon regrowth (Michaelevski et al., 2010; Fagoe et al., 2014; Chandran et al., 2016). In the axolotl, during the first week of spinal cord regeneration after tail amputation, genes related to immune function, cell cycle, cytoskeleton, and extracellular matrix were upregulated (Monaghan et al., 2007). A comparative tissue analysis performed in the newt on regenerating spinal cord, forelimb, hind limb, tail, and heart revealed a common set of activated genes, including those involved in immune function, cytoskeleton, and extracellular matrix; however, the temporal changes in expression between tissues differed (Mercer et al., 2012).

Taken together, these unbiased gene expression studies suggest the likelihood of conserved molecular mechanisms capable of promoting regeneration in the vertebrate PNS and CNS. However, it is still unknown which gene, combinations of genes, or pathways, when activated, could promote the greatest axon regeneration and recovery. Furthermore, except for studies that used pure populations of cultured neurons (Lerch et al., 2012), most transcriptomes on the regenerating CNS have been generated from complex tissues, with heterogeneous classes of neurons and glia whose molecular profiles can be extremely diverse (Cahoy et al., 2008; Usoskin et al., 2015; Zeisel et al., 2015). Likewise, since different types of CNS and PNS neurons exhibit distinct capacities for regeneration (Blesch and Tuszynski, 2009), there are also likely to be cell-specific injury and regeneration responses that differ substantially across neuronal and glial subtypes. Our lack of knowledge on these issues is a major barrier to progress.

One way to get at this problem would be to focus more efforts around single-cell transcriptome analyses on subclasses of identified, regenerating or non-regenerating, neurons or activated glial cells post-injury. For example, single-cell transcriptome analyses performed on the good and poor regenerators in lampreys could provide an excellent model for addressing the intrinsic mechanisms that control regenerative capacities among subclasses of CNS neurons (Smith et al., 2011). Where possible, this approach could be extended to other identified neuronal and glial subtypes in other injury models. Second, comparative transcriptomics between multiple vertebrate species—even at the tissue level—would allow us to determine common mechanisms that support CNS repair and regeneration. Despite its great potential, only a few such studies have been performed between highly regenerative vertebrates and mammals (Monaghan et al., 2007), and none has been performed where the injury type, location, and methods are uniform. Together, the outcomes of well-controlled comparative single-cell and tissue-level transcriptome studies would greatly expand our understanding of the basic, conserved molecular mechanisms that support CNS regeneration in vertebrates, and distinct, parallel mechanisms should they exist.

22.4.2 Non-mammalian models may offer more rapid means for testing new strategies for central nervous system repair

When considering a potential therapeutic for treating SCI, it is essential to perform extensive, well-controlled preclinical studies (Hug and Weidner, 2012). Mammalian SCI studies are by their nature relatively low-throughput studies because of the costliness of maintaining the animals and the lack of CNS regeneration. Here is where the rapid time course of recovery in non-mammalian SCI models, combined with the robustness of CNS repair and regeneration and knowledge of molecular pathways, could provide additional advantages by permitting higher-throughput testing of genetic manipulations, drugs, biomaterials, or perhaps even engineered devices. An elegant genetic screen on axon regeneration has been done in the worm *Caenorhabditis elegans*, which identified not only known, conserved pathways as regulators of axon regeneration (e.g., axon guidance and cytoskeletal genes) but also novel regeneration-associated genes, including those for several synaptic vesicle endocytosis proteins (Chen et al., 2011). Surprisingly, chemical and genetic screens have not been published in any vertebrate SCI model to date; however, this is currently being developed in the axolotl tail regeneration model (Ponomareva et al., 2015). Thus, extending the screening approaches into zebrafish or other vertebrate SCI models to identify novel regulators of regeneration would significantly advance the field. Here, the larval zebrafish SCI model, in which functional recovery occurs in just over a week, could provide an efficient means for performing genetic or chemical screens (Briona and Dorsky, 2014). Since genomes and transcriptomes for most of the highly regenerative vertebrates are already available, and gene editing with CRISPR/Cas9 or TALENs is feasible, validation of the targets from such screens in non-mammalian vertebrates will become increasingly easier (Ma and Liu, 2015; Tandon et al., 2017; Zu et al., 2016).

A second way in which non-mammalian models could contribute to advancing the field is as testers of new biomaterials for improving recovery after SCI. Any engineered material must be non-immunogenic and otherwise safe to apply to the injured CNS, without inducing unintended biological activities. It is for this reason, in part, that hydrogels based on natural materials have recently emerged as excellent candidates that can be used to provide scaffolding support to the injured PNS and CNS, deliver drugs, and graft cells (Khaing and Schmidt, 2012; Pakulska et al., 2012). Natural hydrogels typically comprise extracellular matrix components such as collagen, hyaluronic acid (HA), and/or fibrin, and synthetic hydrogels are also being developed. Demonstrating their versatility, collagen-based biomaterials can be formed into oriented filaments that permit directed re-growth of injured axons (Khaing and Schmidt, 2012). Fibrin- and HA-based hydrogels have been used to deliver neural precursor cells for cell replacement, Nogo or Nogo receptor antibodies, regeneration-promoting drugs, enzymes that degrade the extracellular matrix deposits in the lesion core, and growth factors (Khaing and Schmidt, 2012). Interestingly, HA hydrogels themselves also modify the post-injury environment to be more growth-permissive for CNS axons by significantly reducing microglial infiltration, astrocyte activation, and deposition of extracellular matrix (Khaing et al., 2011). Many studies now show that hydrogel-based therapies improved axon regeneration and functional recovery after PNS or CNS injury in both experimental and clinical settings (Khaing and Schmidt, 2012).

Some of the current challenges in engineering hydrogel-based therapies are to find the right combination of biomaterials that will integrate the existing nervous tissues with the proper tensile properties, fill larger gaps, and promote the greatest regeneration and functional recovery, without inducing deleterious responses from the endogenous environment. Here, non-mammalian models such as lampreys and fishes might offer good models for initial screening of biomaterials because the spinal transection surgeries are fast and reproducible enough to do large cohorts of animals that initiate substantial recovery within a few weeks post-SCI. Urodele amphibians would allow for further testing on recovery of locomotion patterns that include overground stepping in a higher vertebrate. Of course, these are just a few examples, but the same idea could be extended to testing new stem cell therapies, drugs and delivery systems, and other molecular manipulations as SCI therapeutics. Such a comparative approach could help speed up testing and narrow down the combinations of reagents that exhibit the greatest effects on regeneration and functional recovery, the best of which could then be tested more efficiently for efficacy in pre-clinical rodent SCI models.

22.5 Concluding remarks

The field of SCI is hard at work developing better therapeutics and biomaterials for improving the post-injury environment and regeneration in the injured mammalian CNS. Essential to this goal is a better understanding of the cellular and molecular pathways that promote successful CNS regeneration and plasticity, which could be accomplished in non-mammalian models. As transcriptome and other molecular studies are revealing, the genes and signaling pathways utilized by non-mammalian vertebrates for CNS repair and regeneration are generally highly conserved in mammals. Therefore, any shared mechanisms or strategies that further improve CNS regeneration among the non-mammalian models are likely to have relevance to improving regeneration in the mammalian CNS. Bridging this gap through comparative studies could therefore be the means to better treatments for SCI.

References

Abdesselem, H., A. Shypitsyna, G.P. Solis, V. Bodrikov, and C.A. Stuermer. 2009. No Nogo66- and NgR-mediated inhibition of regenerating axons in the zebrafish optic nerve. *J Neurosci.* 29:15489–15498.

Abe, N. and V. Cavalli. 2008. Nerve injury signaling. *Curr Opin Neurobiol.* 18:276–283.

Afshari, F.T., S. Kappagantula, and J.W. Fawcett. 2009. Extrinsic and intrinsic factors controlling axonal regeneration after spinal cord injury. *Expert Rev Mol Med.* 11:e37.

Beattie, M.S., J.C. Bresnahan, and G. Lopate. 1990. Metamorphosis alters the response to spinal cord transection in Xenopus laevis frogs. *J Neurobiol.* 21:1108–1122.

Becker, T., M.F. Wullimann, C.G. Becker, R.R. Bernhardt, and M. Schachner. 1997. Axonal regrowth after spinal cord transection in adult zebrafish. *J Comp Neurol.* 377:577–595.

Benowitz, L.I. and Y. Yin. 2007. Combinatorial treatments for promoting axon regeneration in the CNS: Strategies for overcoming inhibitory signals and activating neurons' intrinsic growth state. *Dev Neurobiol.* 67:1148–1165.

Bernstein, J.J. and J.B. Gelderd. 1970. Regeneration of the long spinal tracts in the goldfish. *Brain Res.* 20:33–38.

Biering-Sorensen, F., S. Alai, K. Anderson, S. Charlifue, Y. Chen, M. DeVivo, A.E. Flanders, L. Jones, N. Kleitman, A. Lans, V.K. Noonan, J. Odenkirchen, J. Steeves, K. Tansey, E. Widerstrom-Noga, and L.B. Jakeman. 2015. Common data elements for spinal cord injury clinical research: A National Institute for Neurological Disorders and Stroke project. *Spinal Cord.* 53:265–277.

Blesch, A. and M.H. Tuszynski. 2009. Spinal cord injury: Plasticity, regeneration and the challenge of translational drug development. *Trends Neurosci.* 32:41–47.

Bloom, O. 2014. Non-mammalian model systems for studying neuro-immune interactions after spinal cord injury. *Exp Neurol.* 258:130–140.

Bloom, O.E. and J.R. Morgan. 2011. Membrane trafficking events underlying axon repair, growth, and regeneration. *Mol Cell Neurosci.* 48:339–348.

Bradbury, E.J. and S.B. McMahon. 2006. Spinal cord repair strategies: Why do they work? *Nat Rev Neurosci.* 7:644–653.

Briona, L.K. and R.I. Dorsky. 2014. Radial glial progenitors repair the zebrafish spinal cord following transection. *Exp Neurol.* 256:81–92.

Bullock, T.H., J.K. Moore, and R.D. Fields. 1984. Evolution of myelin sheaths: Both lamprey and hagfish lack myelin. *Neurosci Lett.* 48:145–148.

Busch, D.J. and J.R. Morgan. 2012. Synuclein accumulation is associated with cell-specific neuronal death after spinal cord injury. *J Comp Neurol.* 520:1751–1771.

Cabelguen, J.M., S. Chevallier, I. Amontieva-Potapova, and C. Philippe. 2013. Anatomical and electrophysiological plasticity of locomotor networks following spinal transection in the salamander. *Neurosci Bull.* 29:467–476.

Cahoy, J.D., B. Emery, A. Kaushal, L.C. Foo, J.L. Zamanian, K.S. Christopherson, Y. Xing, J.L. Lubischer, P.A. Krieg, S.A. Krupenko, W.J. Thompson, and B.A. Barres. 2008. A transcriptome database for astrocytes, neurons, and oligodendrocytes: A new resource for understanding brain development and function. *J Neurosci.* 28:264–278.

Chandran, V., G. Coppola, H. Nawabi, T. Omura, R. Versano, E.A. Huebner, A. Zhang, M. Costigan, A. Yekkirala, L. Barrett, A. Blesch, I. Michaelevski, J. Davis-Turak, F. Gao, P. Langfelder, S. Horvath, Z. He, L. Benowitz, M. Fainzilber, M. Tuszynski, C.J. Woolf, and D.H. Geschwind. 2016. A systems-level analysis of the peripheral nerve intrinsic axonal growth program. *Neuron.* 89:956–970.

Chang, B., Q. Quan, S. Lu, Y. Wang, and J. Peng. 2016. Molecular mechanisms in the initiation phase of Wallerian degeneration. *Eur J Neurosci.* 44(4):2040–2048.

Chen, K., S. Deng, H. Lu, Y. Zheng, G. Yang, D. Kim, Q. Cao, and J.Q. Wu. 2013. RNA-seq characterization of spinal cord injury transcriptome in acute/subacute phases: A resource for understanding the pathology at the systems level. *PLoS One.* 8:e72567.

Chen, L., Z. Wang, A. Ghosh-Roy, T. Hubert, D. Yan, S. O'Rourke, B. Bowerman, Z. Wu, Y. Jin, and A.D. Chisholm. 2011. Axon regeneration pathways identified by systematic genetic screening in C. elegans. *Neuron.* 71:1043–1057.

Chevallier, S., M. Landry, F. Nagy, and J.M. Cabelguen. 2004. Recovery of bimodal locomotion in the spinal-transected salamander, Pleurodeles waltlii. *Eur J Neurosci.* 20:1995–2007.

Cohen, A.H., S.A. Mackler, and M.E. Selzer. 1988. Behavioral recovery following spinal transection: Functional regeneration in the lamprey CNS. *Trends Neurosci.* 11:227–231.

Cote, M.P., A.A. Amin, V.J. Tom, and J.D. Houle. 2011. Peripheral nerve grafts support regeneration after spinal cord injury. *Neurotherapeutics* 8:294–303.

Cregg, J.M., M.A. DePaul, A.R. Filous, B.T. Lang, A. Tran, and J. Silver. 2014. Functional regeneration beyond the glial scar. *Exp Neurol.* 253:197–207.

David, S. and A.J. Aguayo. 1981. Axonal elongation into peripheral nervous system "bridges" after central nervous system injury in adult rats. *Science.* 214:931–933.

Davis, B.M., J.L. Ayers, L. Koran, J. Carlson, M.C. Anderson, and S.B. Simpson, Jr. 1990. Time course of salamander spinal cord regeneration and recovery of swimming: HRP retrograde pathway tracing and kinematic analysis. *Exp Neurol.* 108:198–213.

Davis, G.R., Jr. and A.D. McClellan. 1993. Time course of anatomical regeneration of descending brainstem neurons and behavioral recovery in spinal-transected lamprey. *Brain Res.* 602:131–137.

Di Giovanni, S. 2009. Molecular targets for axon regeneration: Focus on the intrinsic pathways. *Expert Opin Ther Targets.* 13:1387–1398.

Dimou, L., L. Schnell, L. Montani, C. Duncan, M. Simonen, R. Schneider, T. Liebscher, M. Gullo, and M.E. Schwab. 2006. Nogo-A-deficient mice reveal strain-dependent differences in axonal regeneration. *J Neurosci.* 26:5591–5603.

Domeniconi, M. and M.T. Filbin. 2005. Overcoming inhibitors in myelin to promote axonal regeneration. *J Neurol Sci.* 233:43–47.

Fagoe, N.D., J. van Heest, and J. Verhaagen. 2014. Spinal cord injury and the neuron-intrinsic regeneration-associated gene program. *Neuromolecular Med.* 16:799–813.

Fawcett, J.W., M.E. Schwab, L. Montani, N. Brazda, and H.W. Muller. 2012. Defeating inhibition of regeneration by scar and myelin components. *Handb Clin Neurol.* 109:503–522.

Ferretti, P. 2011. Is there a relationship between adult neurogenesis and neuron generation following injury across evolution? *Eur J Neurosci.* 34:951–962.

Forehand, C.J. and P.B. Farel. 1982. Anatomical and behavioral recovery from the effects of spinal cord transection: Dependence on metamorphosis in anuran larvae. *J Neurosci.* 2:654–652.

Freund, P., E. Schmidlin, T. Wannier, J. Bloch, A. Mir, M.E. Schwab, and E.M. Rouiller. 2006. Nogo-A-specific antibody treatment enhances sprouting and functional recovery after cervical lesion in adult primates. *Nat Med.* 12:790–792.

Garcia, G., G. Libisch, O. Trujillo-Cenoz, C. Robello, and R.E. Russo. 2012. Modulation of gene expression during early stages of reconnection of the turtle spinal cord. *J Neurochem.* 121:996–1006.

Gardiner, D.M. and S.V. Bryant. 1989. Organization of positional information in the axolotl limb. *J Exp Zool.* 251:47–55.

Gemberling, M., T.J. Bailey, D.R. Hyde, and K.D. Poss. 2013. The zebrafish as a model for complex tissue regeneration. *Trends Genet.* 29:611–620.

Gensel, J.C. and B. Zhang. 2015. Macrophage activation and its role in repair and pathology after spinal cord injury. *Brain Res.* 1619:1–11.

Geoffroy, C.G. and B. Zheng. 2014. Myelin-associated inhibitors in axonal growth after CNS injury. *Curr Opin Neurobiol.* 27:31–38.

Gibbs, K.M., S.V. Chittur, and B.G. Szaro. 2011. Metamorphosis and the regenerative capacity of spinal cord axons in Xenopus laevis. *Eur J Neurosci.* 33:9–25.

Gibbs, K.M. and B.G. Szaro. 2006. Regeneration of descending projections in Xenopus laevis tadpole spinal cord demonstrated by retrograde double labeling. *Brain Res.* 1088:68–72.

Giger, R.J., E.R. Hollis, and M.H. Tuszynski. 2010. Guidance molecules in axon regeneration. *Cold Spring Harb Perspect Biol.* 2:a001867.

Goldshmit, Y., N. Lythgo, M.P. Galea, and A.M. Turnley. 2008. Treadmill training after spinal cord hemisection in mice promotes axonal sprouting and synapse formation and improves motor recovery. *J Neurotrauma.* 25:449–465.

Goldshmit, Y., T.E. Sztal, P.R. Jusuf, T.E. Hall, M. Nguyen-Chi, and P.D. Currie. 2012. Fgf-dependent glial cell bridges facilitate spinal cord regeneration in zebrafish. *J Neurosci.* 32:7477–7492.

Gordon-Weeks, P.R. and A.E. Fournier. 2014. Neuronal cytoskeleton in synaptic plasticity and regeneration. *J Neurochem.* 129:206–212.

Gorsuch, R.A. and D.R. Hyde. 2014. Regulation of Muller glial dependent neuronal regeneration in the damaged adult zebrafish retina. *Exp Eye Res.* 123:131–140.

Hains, B.C., J.A. Black, and S.G. Waxman. 2003. Primary cortical motor neurons undergo apoptosis after axotomizing spinal cord injury. *J Comp Neurol.* 462:328–341.

Hannila, S.S. and M.T. Filbin. 2008. The role of cyclic AMP signaling in promoting axonal regeneration after spinal cord injury. *Exp Neurol.* 209:321–332.

Harel, N.Y. and S.M. Strittmatter. 2006. Can regenerating axons recapitulate developmental guidance during recovery from spinal cord injury? *Nat Rev Neurosci.* 7:603–616.

Houle, J.D., V.J. Tom, D. Mayes, G. Wagoner, N. Phillips, and J. Silver. 2006. Combining an autologous peripheral nervous system "bridge" and matrix modification by chondroitinase allows robust, functional regeneration beyond a hemisection lesion of the adult rat spinal cord. *J Neurosci.* 26:7405–7415.

Hug, A. and N. Weidner. 2012. From bench to beside to cure spinal cord injury: Lost in translation? *Int Rev Neurobiol.* 106:173–196.

Hui, S.P., J.R. Monaghan, S.R. Voss, and S. Ghosh. 2013. Expression pattern of Nogo-A, MAG, and NgR in regenerating urodele spinal cord. *Dev Dyn.* 242:847–860.

Hui, S.P., D. Sengupta, S.G. Lee, T. Sen, S. Kundu, S. Mathavan, and S. Ghosh. 2014. Genome wide expression profiling during spinal cord regeneration identifies comprehensive cellular responses in zebrafish. *PLoS One.* 9:e84212.

Jacobs, A.J., G.P. Swain, J.A. Snedeker, D.S. Pijak, L.J. Gladstone, and M.E. Selzer. 1997. Recovery of neurofilament expression selectively in regenerating reticulospinal neurons. *J Neurosci.* 17:5206–5220.

Kang, H. and J.W. Lichtman. 2013. Motor axon regeneration and muscle reinnervation in young adult and aged animals. *J Neurosci.* 33:19480–19491.

Khaing, Z.Z., B.D. Milman, J.E. Vanscoy, S.K. Seidlits, R.J. Grill, and C.E. Schmidt. 2011. High molecular weight hyaluronic acid limits astrocyte activation and scar formation after spinal cord injury. *J Neural Eng.* 8:046033.

Khaing, Z.Z. and C.E. Schmidt. 2012. Advances in natural biomaterials for nerve tissue repair. *Neurosci Lett.* 519:103–114.

Kim, J.E., S. Li, T. GrandPre, D. Qiu, and S.M. Strittmatter. 2003. Axon regeneration in young adult mice lacking Nogo-A/B. *Neuron.* 38:187–199.

Kizil, C., J. Kaslin, V. Kroehne, and M. Brand. 2012. Adult neurogenesis and brain regeneration in zebrafish. *Dev Neurobiol.* 72:429–461.

Kwon, B.K., J. Liu, C. Messerer, N.R. Kobayashi, J. McGraw, L. Oschipok, and W. Tetzlaff. 2002. Survival and regeneration of rubrospinal neurons 1 year after spinal cord injury. *Proc Natl Acad Sci USA.* 99:3246–3251.

Lau, B.Y., S.M. Fogerson, R.B. Walsh, and J.R. Morgan. 2013. Cyclic AMP promotes axon regeneration, lesion repair and neuronal survival in lampreys after spinal cord injury. *Exp Neurol.* 250:31–42.

Lee, B.H., K.H. Lee, U.J. Kim, D.H. Yoon, J.H. Sohn, S.S. Choi, I.G. Yi, and Y.G. Park. 2004. Injury in the spinal cord may produce cell death in the brain. *Brain Res.* 1020:37–44.

Lee, J.K., C.G. Geoffroy, A.F. Chan, K.E. Tolentino, M.J. Crawford, M.A. Leal, B. Kang, and B. Zheng. 2010. Assessing spinal axon regeneration and sprouting in Nogo-, MAG-, and OMgp-deficient mice. *Neuron.* 66:663–670.

Lee, J.K. and B. Zheng. 2012. Role of myelin-associated inhibitors in axonal repair after spinal cord injury. *Exp Neurol.* 235:33–42.

Lee-Liu, D., G. Edwards-Faret, V.S. Tapia, and J. Larrain. 2013. Spinal cord regeneration: Lessons for mammals from non-mammalian vertebrates. *Genesis.* 51:529–544.

Lee-Liu, D., M. Moreno, L.I. Almonacid, V.S. Tapia, R. Munoz, J. von Marees, M. Gaete, F. Melo, and J. Larrain. 2014. Genome-wide expression profile of the response to spinal cord injury in Xenopus laevis reveals extensive differences between regenerative and non-regenerative stages. *Neural Dev.* 9:12.

Lerch, J.K., F. Kuo, D. Motti, R. Morris, J.L. Bixby, and V.P. Lemmon. 2012. Isoform diversity and regulation in peripheral and central neurons revealed through RNA-Seq. *PLoS One.* 7:e30417.

Li, S., J.E. Kim, S. Budel, T.G. Hampton, and S.M. Strittmatter. 2005. Transgenic inhibition of Nogo-66 receptor function allows axonal sprouting and improved locomotion after spinal injury. *Mol Cell Neurosci.* 29:26–39.

Li, S., B.P. Liu, S. Budel, M. Li, B. Ji, L. Walus, W. Li, A. Jirik, S. Rabacchi, E. Choi, D. Worley, D.W. Sah, B. Pepinsky, D. Lee, J. Relton, and S.M. Strittmatter. 2004. Blockade of Nogo-66, myelin-associated glycoprotein, and oligodendrocyte myelin glycoprotein by soluble Nogo-66 receptor promotes axonal sprouting and recovery after spinal injury. *J Neurosci.* 24:10511–10520.

Li, S. and S.M. Strittmatter. 2003. Delayed systemic Nogo-66 receptor antagonist promotes recovery from spinal cord injury. *J Neurosci.* 23:4219–4227.

Liebscher, T., L. Schnell, D. Schnell, J. Scholl, R. Schneider, M. Gullo, K. Fouad, A. Mir, M. Rausch, D. Kindler, F.P. Hamers, and M.E. Schwab. 2005. Nogo-A antibody improves regeneration and locomotion of spinal cord-injured rats. *Ann Neurol.* 58:706–719.

Lurie, D.I., D.S. Pijak, and M.E. Selzer. 1994. Structure of reticulospinal axon growth cones and their cellular environment during regeneration in the lamprey spinal cord. *J Comp Neurol.* 344:559–580.

Ma, D. and F. Liu. 2015. Genome editing and its applications in model organisms. *Genomics Proteomics Bioinformatics*. 13:336–344.

Maden, M. and R.N. Turner. 1978. Supernumerary limbs in the axolotl. *Nature*. 273:232–235.

McCall, J., N. Weidner, and A. Blesch. 2012. Neurotrophic factors in combinatorial approaches for spinal cord regeneration. *Cell Tissue Res*. 349:27–37.

McClellan, A.D. 1990. Locomotor recovery in spinal-transected lamprey: Role of functional regeneration of descending axons from brainstem locomotor command neurons. *Neuroscience*. 37:781–798.

McCusker, C. and D.M. Gardiner. 2011. The axolotl model for regeneration and aging research: A mini-review. *Gerontology*. 57:565–571.

McCusker, C.D. and D.M. Gardiner. 2013. Positional information is reprogrammed in blastema cells of the regenerating limb of the axolotl (Ambystoma mexicanum). *PLoS One*. 8:e77064.

McHedlishvili, L., V. Mazurov, K.S. Grassme, K. Goehler, B. Robl, A. Tazaki, K. Roensch, A. Duemmler, and E.M. Tanaka. 2012. Reconstitution of the central and peripheral nervous system during salamander tail regeneration. *Proc Natl Acad Sci USA*. 109:E2258–2266.

Mercer, S.E., C.H. Cheng, D.L. Atkinson, J. Krcmery, C.E. Guzman, D.T. Kent, K. Zukor, K.A. Marx, S.J. Odelberg, and H.G. Simon. 2012. Multi-tissue microarray analysis identifies a molecular signature of regeneration. *PLoS One*. 7:e52375.

Michaelevski, I., Y. Segal-Ruder, M. Rozenbaum, K.F. Medzihradszky, O. Shalem, G. Coppola, S. Horn-Saban, K. Ben-Yaakov, S.Y. Dagan, I. Rishal, D.H. Geschwind, Y. Pilpel, A.L. Burlingame, and M. Fainzilber. 2010. Signaling to transcription networks in the neuronal retrograde injury response. *Sci Signal*. 3:ra53.

Monaghan, J.R., J.A. Walker, R.B. Page, S. Putta, C.K. Beachy, and S.R. Voss. 2007. Early gene expression during natural spinal cord regeneration in the salamander Ambystoma mexicanum. *J Neurochem*. 101:27–40.

Morgan, J. and M.I. Shifman. 2014. *Non-mammalian Models of Nerve Regeneration*. Cambridge, UK: Cambridge University Press.

Mullen, L.M., S.V. Bryant, M.A. Torok, B. Blumberg, and D.M. Gardiner. 1996. Nerve dependency of regeneration: The role of Distal-less and FGF signaling in amphibian limb regeneration. *Development*. 122:3487–3497.

Muneoka, K. and S.V. Bryant. 1982. Evidence that patterning mechanisms in developing and regenerating limbs are the same. *Nature*. 298:369–371.

Nesic, O., N.M. Svrakic, G.Y. Xu, D. McAdoo, K.N. Westlund, C.E. Hulsebosch, Z. Ye, A. Galante, P. Soteropoulos, P. Tolias, W. Young, R.P. Hart, and J.R. Perez-Polo. 2002. DNA microarray analysis of the contused spinal cord: Effect of NMDA receptor inhibition. *J Neurosci Res*. 68:406–423.

Oliphint, P.A., N. Alieva, A.E. Foldes, E.D. Tytell, B.Y. Lau, J.S. Pariseau, A.H. Cohen, and J.R. Morgan. 2010. Regenerated synapses in lamprey spinal cord are sparse and small even after functional recovery from injury. *J Comp Neurol*. 518:2854–2872.

Pakulska, M.M., B.G. Ballios, and M.S. Shoichet. 2012. Injectable hydrogels for central nervous system therapy. *Biomed Mater*. 7:024101.

Pernet, V. and M.E. Schwab. 2012. The role of Nogo-A in axonal plasticity, regrowth and repair. *Cell Tissue Res*. 349:97–104.

Ponomareva, L.V., A. Athippozhy, J.S. Thorson, and S.R. Voss. 2015. Using Ambystoma mexicanum (Mexican axolotl) embryos, chemical genetics, and microarray analysis to identify signaling pathways associated with tissue regeneration. *Comp Biochem Physiol C Toxicol Pharmacol*. 178:128–135.

Popovich, P.G. 2014. Neuroimmunology of traumatic spinal cord injury: A brief history and overview. *Exp Neurol*. 258:1–4.

Rasmussen, J.P. and A. Sagasti. 2016. Learning to swim, again: Axon regeneration in fish. *Exp Neurol*. 287:318–330.

Rehermann, M.I., N. Marichal, R.E. Russo, and O. Trujillo-Cenoz. 2009. Neural reconnection in the transected spinal cord of the freshwater turtle Trachemys dorbignyi. *J Comp Neurol.* 515:197–214.

Rehermann, M.I., F.F. Santinaque, B. Lopez-Carro, R.E. Russo, and O. Trujillo-Cenoz. 2011. Cell proliferation and cytoarchitectural remodeling during spinal cord reconnection in the fresh-water turtle Trachemys dorbignyi. *Cell Tissue Res.* 344:415–433.

Roet, K.C. and J. Verhaagen. 2014. Understanding the neural repair-promoting properties of olfactory ensheathing cells. *Exp Neurol.* 261:594–609.

Rovainen, C.M. 1976. Regeneration of Muller and Mauthner axons after spinal transection in larval lampreys. *J Comp Neurol.* 168:545–554.

Roy, R.R., S.J. Harkema, and V.R. Edgerton. 2012. Basic concepts of activity-based interventions for improved recovery of motor function after spinal cord injury. *Arch Phys Med Rehabil.* 93:1487–1497.

Sanchez Alvarado, A. and P.A. Tsonis. 2006. Bridging the regeneration gap: Genetic insights from diverse animal models. *Nat Rev Genet.* 7:873–884.

Schwab, M.E. and S.M. Strittmatter. 2014. Nogo limits neural plasticity and recovery from injury. *Curr Opin Neurobiol.* 27:53–60.

Selzer, M.E. 1978. Mechanisms of functional recovery and regeneration after spinal cord transection in larval sea lamprey. *J Physiol.* 277:395–408.

Shifman, M.I., R.E. Yumul, C. Laramore, and M.E. Selzer. 2009. Expression of the repulsive guidance molecule RGM and its receptor neogenin after spinal cord injury in sea lamprey. *Exp Neurol.* 217:242–251.

Shifman, M.I., G. Zhang, and M.E. Selzer. 2008. Delayed death of identified reticulospinal neurons after spinal cord injury in lampreys. *J Comp Neurol.* 510:269–282.

Silver, J., M.E. Schwab, and P.G. Popovich. 2015. Central nervous system regenerative failure: Role of oligodendrocytes, astrocytes, and microglia. *Cold Spring Harb Perspect Biol.* 7:a020602.

Simon, A. and E.M. Tanaka. 2013. Limb regeneration. *Wiley Interdiscip Rev Dev Biol.* 2:291–300.

Simonen, M., V. Pedersen, O. Weinmann, L. Schnell, A. Buss, B. Ledermann, F. Christ, G. Sansig, H. van der Putten, and M.E. Schwab. 2003. Systemic deletion of the myelin-associated outgrowth inhibitor Nogo-A improves regenerative and plastic responses after spinal cord injury. *Neuron.* 38:201–211.

Smith, G.M., A.E. Falone, and E. Frank. 2012. Sensory axon regeneration: Rebuilding functional connections in the spinal cord. *Trends Neurosci.* 35:156–163.

Smith, J., J.R. Morgan, S.J. Zottoli, P.J. Smith, J.D. Buxbaum, and O.E. Bloom. 2011. Regeneration in the era of functional genomics and gene network analysis. *Biol Bull.* 221:18–34.

Son, Y.J. and W.J. Thompson. 1995. Schwann cell processes guide regeneration of peripheral axons. *Neuron.* 14:125–132.

Steward, O., P.G. Popovich, W.D. Dietrich, and N. Kleitman. 2012. Replication and reproducibility in spinal cord injury research. *Exp Neurol.* 233:597–605.

Steward, O., K. Sharp, K.M. Yee, and M. Hofstadter. 2008. A re-assessment of the effects of a Nogo-66 receptor antagonist on regenerative growth of axons and locomotor recovery after spinal cord injury in mice. *Exp Neurol.* 209:446–468.

Stocum, D.L. 2011. The role of peripheral nerves in urodele limb regeneration. *Eur J Neurosci.* 34:908–916.

Sugiura, T., H. Wang, R. Barsacchi, A. Simon, and E.M. Tanaka. 2016. MARCKS-like protein is an initiating molecule in axolotl appendage regeneration. *Nature.* 531:237–240.

Sun, F. and Z. He. 2010. Neuronal intrinsic barriers for axon regeneration in the adult CNS. *Curr Opin Neurobiol.* 20:510–518.

Tanaka, E.M. and P. Ferretti. 2009. Considering the evolution of regeneration in the central nervous system. *Nat Rev Neurosci.* 10:713–723.

Tandon, P., F. Conlon, J.D. Furlow, and M.E. Horb. 2016. Expanding the genetic toolkit in Xenopus: Approaches and opportunities for human disease modeling. *Dev Biol.* pii: S0012-1606(16)30063-X. doi:10.1016/j.ydbio.2016.04.009

Tetzlaff, W., E.B. Okon, S. Karimi-Abdolrezaee, C.E. Hill, J.S. Sparling, J.R. Plemel, W.T. Plunet, E.C. Tsai, D. Baptiste, L.J. Smithson, M.D. Kawaja, M.G. Fehlings, and B.K. Kwon. 2011. A systematic review of cellular transplantation therapies for spinal cord injury. *J Neurotrauma.* 28:1611–1682.

Usoskin, D., A. Furlan, S. Islam, H. Abdo, P. Lonnerberg, D. Lou, J. Hjerling-Leffler, J. Haeggstrom, O. Kharchenko, P.V. Kharchenko, S. Linnarsson, and P. Ernfors. 2015. Unbiased classification of sensory neuron types by large-scale single-cell RNA sequencing. *Nat Neurosci.* 18:145–153.

Velardo, M.J., C. Burger, P.R. Williams, H.V. Baker, M.C. Lopez, T.H. Mareci, T.E. White, N. Muzyczka, and P.J. Reier. 2004. Patterns of gene expression reveal a temporally orchestrated wound healing response in the injured spinal cord. *J Neurosci.* 24:8562–8576.

Voss, S.R., H.H. Epperlein, and E.M. Tanaka. 2009. Ambystoma mexicanum, the axolotl: A versatile amphibian model for regeneration, development, and evolution studies. *Cold Spring Harb Protoc.* 2009:pdb emo128.

Welte, C., S. Engel, and C.A. Stuermer. 2015. Upregulation of the zebrafish Nogo-A homologue, Rtn4b, in retinal ganglion cells is functionally involved in axon regeneration. *Neural Dev.* 10:6. doi:10.1186/s13064-015-0034-x.

Yin, H.S. and M.E. Selzer. 1983. Axonal regeneration in lamprey spinal cord. *J Neurosci.* 3:1135–1144.

Yiu, G. and Z. He. 2006. Glial inhibition of CNS axon regeneration. *Nat Rev Neurosci.* 7:617–627.

Zeisel, A., A.B. Munoz-Manchado, S. Codeluppi, P. Lonnerberg, G. La Manno, A. Jureus, S. Marques, H. Munguba, L. He, C. Betsholtz, C. Rolny, G. Castelo-Branco, J. Hjerling-Leffler, and S. Linnarsson. 2015. Brain structure. Cell types in the mouse cortex and hippocampus revealed by single-cell RNA-seq. *Science.* 347:1138–1142.

Zhang, G., L.Q. Jin, J. Hu, W. Rodemer, and M.E. Selzer. 2015. Antisense Morpholino Oligonucleotides Reduce Neurofilament Synthesis and Inhibit Axon Regeneration in Lamprey Reticulospinal Neurons. *PLoS One.* 10:e0137670.

Zhao, R.R. and J.W. Fawcett. 2013. Combination treatment with chondroitinase ABC in spinal cord injury—breaking the barrier. *Neurosci Bull.* 29:477–483.

Zheng, B., J. Atwal, C. Ho, L. Case, X.L. He, K.C. Garcia, O. Steward, and M. Tessier-Lavigne. 2005. Genetic deletion of the Nogo receptor does not reduce neurite inhibition in vitro or promote corticospinal tract regeneration in vivo. *Proc Natl Acad Sci USA.* 102:1205–1210.

Zheng, B., C. Ho, S. Li, H. Keirstead, O. Steward, and M. Tessier-Lavigne. 2003. Lack of enhanced spinal regeneration in Nogo-deficient mice. *Neuron.* 38:213–224.

Zottoli, S.J. and M.M. Freemer. 2003. Recovery of C-starts, equilibrium and targeted feeding after whole spinal cord crush in the adult goldfish Carassius auratus. *J Exp Biol.* 206:3015–3029.

Zu, Y., X. Zhang, J. Ren, X. Dong, Z. Zhu, L. Jia, Q. Zhang, and W. Li. 2016. Biallelic editing of a lamprey genome using the CRISPR/Cas9 system. *Sci Rep.* 6:23496.

Zukor, K.A., D.T. Kent, and S.J. Odelberg. 2011. Meningeal cells and glia establish a permissive environment for axon regeneration after spinal cord injury in newts. *Neural Dev.* 6:1.

Zupanc, G.K. and R.F. Sirbulescu. 2011. Adult neurogenesis and neuronal regeneration in the central nervous system of teleost fish. *Eur J Neurosci.* 34:917–929.

Chapter 23 Heart regeneration

Henrik Lauridsen

Contents

> **Key concepts**
>
> a. The human heart does not regenerate myocardium after injury; however, several vertebrates exist that are capable of this feat.
> b. The processes underlying heart development are strikingly similar in all vertebrates, and heart regeneration in regeneration-competent species share many similarities, inspiring hope that these species can ultimately teach us how to accomplish endogenous heart repair.
> c. Heart regeneration can also be approached by *de novo* techniques, preparing cardiac structures *in vitro* for later transplantation by repopulating decellularized matrix scaffolds or 3D bioprinting.
> d. Understanding the complex architecture of the myocardium and how this interacts with cardiac function is crucial to instruct regenerative therapies.
> e. Novel regenerative therapies for the heart need to be tested on both an anatomical level and a functional level to claim full cardiac regeneration.

23.1 The heart

Few organs have received as much attention and appreciation in lyrics and folklore as the heart. Once thought to be the home of the soul, this ever-beating organ, whose presence is easily realized by simple palpation or auscultation (in fact, the sound of mother's beating heart is likely to be the first sound that all of us hear some 19 weeks in gestation [Hepper and Shahidullah 1994]), must have inspired thoughts to its function ever since the first human being started wondering about the world. Although the symbolism of the heart as the source of emotions such as love, kindness, and bravery remains intact to this day, every school kid will also be well aware that the true function of the heart is not that of a spiritual kind, but instead, the heart is the main pump to ensure a sufficient supply/removal of gases and nutrients via the bloodstream. This by no means demeans the importance of the heart, and the steady beating of this organ lies at the foundation of all vertebrates' life. An average human being at the age of 70 years will host a heart that has contracted well more than two billion times and has pumped more than 60 Olympic-sized pools worth of blood through the vascular system. A monumental task viewed from an engineering perspective, considering the size of the pump and the fact that all maintenance needs to be done on the run, even a few minutes of pause is fatal. Therefore, it is no surprise that the anatomy, function, and possible malfunctions of the heart are some of the most thoroughly studied entities in medical science, and the potential to restore function of this crucial organ in case of failure is a long-thought dream of modern medicine.

23.1.1 Evolution of the heart

The evolutionary tale of the vertebrate heart is, like for any other organ, a complex story of evolving the machinery of need with the tools that are available. The pre-vertebrate heart possibly represented by present-day tunicates was a valveless, tubular structure that allowed for alternating flow directions. A similar arrangement was most likely found in those early vertebrates represented by present-day amphioxus (lancelets) in which a single contractile blood vessel performs peristaltic movements to perfuse blood through the vasculature at a very low pressure (Simões-Costa et al. 2005). The heart gradually became more sophisticated in the evolution of fish, with the development of valves and two chambers (atrium and ventricle), defining a unidirectional blood flow through, first, the gills and, later, the systemic circulation (Burggren et al. 2010). In addition, partial intracardiac separation arose in sarcopterygian fishes along with a pulmonary circulation, as is still found in present-day lungfishes. This was further evolved in amphibians, with two fully separated atria and a single non-separated ventricle (however, to some degree, functionally separated as separate oxygenated and deoxygenated blood flows leave the outflow tract, with little mixing) and even further evolved in reptiles, where partial division of the ventricle occurred (Jensen et al. 2010). Both birds and mammals independently evolved endothermy and likewise independently evolved a fully separated heart containing two atria and two fully separated ventricles, allowing for separate circulations with different pressures in the pulmonary system and the systemic system. Both classes have thick and compact left ventricular walls, and the force development of this chamber can sustain a higher blood pressure in the systemic circulation than that found in ectotherms (Jensen et al. 2013[*]). The sophistication of the cardiac anatomy in birds and mammals was accompanied by a several-fold increase in maximal heart rate. This, in combination with the separation of pulmonary and systemic blood flows, allows for the high cardiac output that is imperative for the high metabolism associated with an endothermic lifestyle (for an in-depth review of the evolution of the building plan of the heart, we refer the reader to Jensen et al. 2013).

It follows from this tale that the heart of a medium-sized mammal like humans is a pump that is working with quite a high output. This means that there is only little maneuver room for malfunctions in the heart and that diseases of this organ can be detrimental and very often fatal for the individual.

23.1.2 Heart disease—why we want to be able to regenerate heart structures

Besides a range of congenital heart defects that arise early in development, ischemic events are the most prevailing cause of failure of the adult human heart, and ischemic heart disease is a leading cause of fatalities globally

---

[*] An in-depth review of the evolution of the building plan of the vertebrate heart.

(Forouzanfar et al. 2012; Go et al. 2014). Ischemia-induced myocardial infarction results in extensive cell death of cardiomyocytes (Murry et al. 2006), and though rapid medical intervention can reduce the risk of immediate demise, the following fibrotic response that humans share with our traditional mammalian animal models in research (e.g., mouse, rat, rabbit, and pig) results in scar tissue formation that can ultimately lead to cardiac hypertrophy, arrhythmias, and heart failure (Laflamme and Murry 2011).

From a medicine perspective, there are logically three main strategies that can be applied to alleviate both the individual and society-related problems associated with ischemic heart disease.

23.1.2.1 Avoid that disease occurs   In essence, avoiding the occurrence of ischemic heart disease primarily means avoiding the buildup of arteriosclerosis. This is a highly prioritized research area in medical science but falls outside the scope of this book (for a recent review on arteriosclerosis prevention, we refer the reader to Shapiro and Fazio 2016[*]).

23.1.2.2 When disease occurs, try to minimize impact   To minimize the impact of an ischemic event in the heart, the source of the ischemia, for example, a ruptured plaque, should be removed as quickly as possible, and strategies to diminish the adverse effects of ischemia should be applied. The latter is the focus of the research area of remote ischemic conditioning. In this procedure, a short non-lethal period of ischemia in an extremity (e.g., the leg) protects against longer-lasting harmful periods of ischemia in another organ (e.g., the heart or the brain). Remote ischemic conditioning has been demonstrated to yield both early (minutes) and long-term (days) protection in several organs. Accordingly, the nomenclature pre-, per-, or post-remote ischemic conditioning describes this procedure performed before, during, or after the insult, respectively. The still-not-fully-unraveled mechanisms behind this procedure could prove valuable in the development of cardiac regenerative therapies, in which the patient may need to undergo a period of less-than-optimal cardiac function while the therapy is taking place (for further information about remote ischemic conditioning, we refer the reader to the original works by Bøtker et al. 2010[†] and Przyklenk et al. 1993[†] and the review by Kanoria et al. 2007[‡]).

23.1.2.3 Repair the injured heart to restore function   This strategy can be approached from several different angles, both mechanically (artificial heart) and biologically (tissue regeneration). The latter is the focus of the rest of this chapter (for an overview of regenerative strategies for the heart, see Figure 23.1).

---

[*] Review on arteriosclerosis prevention.
[†] Original work on remote ischemic conditioning.
[‡] Review on remote ischemic conditioning.

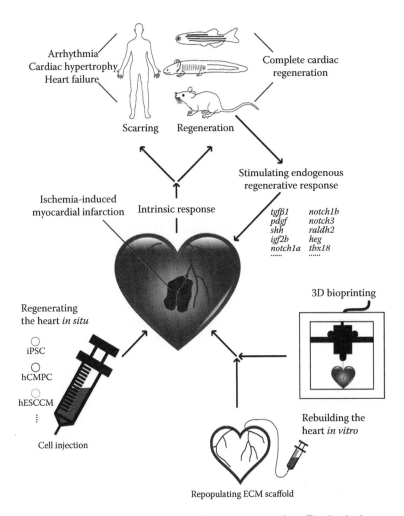

Figure 23.1 Overview of strategies of heart regeneration. The intrinsic response to ischemia-induced myocardial infarction in humans is scarring, with detrimental consequences such as arrhythmia, hypertrophy, and, ultimately, heart failure. Contrary to this, the zebrafish, the salamander, and the neonatal mouse are capable of complete cardiac regeneration. By investigating the molecular mechanisms of this capability, we may learn how to stimulate endogenous regeneration in human hearts. Alternatively, we may be able to develop *in situ* regenerative therapies by applying excessive amounts of stem cells (e.g., induced pluripotent stem cells [iPSCs], human cardiomyocyte progenitor cells [hCMPCs], and human embryonic stem cell–derived cardiomyocytes [hESCCM]). Finally, bioartificial heart construction maybe attempted *in vitro* by using repopulation of extracellular matrix (ECM) scaffolds or by using 3D bioprinting.

23.1.3 Innate regenerative potential of the heart

Before considering strategies to regenerate human hearts, it is instructive to consider the innate regenerative potential of our own heart and that of a few model animals in regenerative biology and medicine.

23.1.3.1 The regenerative potential of the human heart As it may have been expected, the human heart does not display extravagant regenerative potential. If this were the case, the detrimental consequences of ischemic heart disease on the personal level and the vast economic consequences to society that this disease is accountable for would be much less severe. However, unfortunately, neither the myocardium in the human heart recovers well after an ischemic infarction, nor does it regenerate new functioning myocardium. In fact, the turnover rate of cardiomyocytes in healthy human hearts is quite low. To apply numbers to this statement can be difficult due to the obvious ethical limitations in using human beings as test subjects in an investigation to measure turnover rates, which would normally be invasive and sacrificial. However, Bergmann et al. (2009[*]) elegantly overcame this obstacle by measuring ^{14}C concentration in flow-cytometry–purified cardiomyocytes from recently diseased people. After the nuclear bomb tests during the Cold War, high levels of the radioactive ^{14}C isotope were released into the atmosphere; later, this isotope incorporated into plants, along with the other carbon isotopes, and eventually incorporated into humans. By using mathematical modeling on the decay of ^{14}C, Bergman et al. demonstrated that human cardiomyocytes do, in fact, renew throughout life, with an annual turnover of ~1% at age 25 and a gradual decrease to 0.45% at 75 years of age (and thereby, Bergman et al. arguably put the nuclear bomb tests to the sole beneficial use for humankind so far). These results, in terms of annual cardiomyocyte turnover, have later been surpassed by other less ingenious methods deriving an annual turnover of cardiomyocytes of as much as 22% (Kajstura et al. 2010). Regardless of the exact level of annual turnover, the interesting lesson from these studies is that cardiomyocytes do proliferate in the adult human heart. Admittedly at a low level, but the fact that this process, which we will soon see is the underlying mechanism of myocardial regeneration in regeneration-competent animals, does take place suggests that pharmacological strategies to stimulate and upregulate cardiomyocyte proliferation maybe a valid alternative to strategies focused on transplanting differentiated cardiomyocytes or progenitor cells.

23.1.3.2 The regenerative potential of model species—the zebrafish The bony fish superclass Osteichthyes by far represents the most diverse and species-rich group of vertebrates on the planet. Still, only a few species have become successful research subjects in the biological and

---

[*] First evidence for cardiomyocyte renewal in humans with an ingenious technique.

biomedical laboratory. Of these, the zebrafish (*Danio rerio*) undoubtedly take the prize as the most widely applied piscine model species. This is partly due to the species being highly amenable to genetic and molecular approaches; a short generation time; near transparency of zebrafish embryos, facilitating optical imaging techniques; and, similarly to all other animals that we have brought into the laboratory, the fact that the species is easily adaptable to laboratory housing and breeding. In terms of cardiac regenerative research, the zebrafish may disputably be described as the most well-characterized model for successful heart regeneration. This is due to the efforts within the last two decades of several laboratories that have provided numerous thorough studies of both the heart and fin (evolutionary equivalent of the tetrapod limb) regeneration (for in depth review of zebrafish regeneration, we refer the reader to Gemberling et al. 2013[*]). The regenerative response of zebrafish to cardiac injury is robust, irrespective of the injury type—transection, cryoinjury, and inducible genetic ablation to the ventricular myocardium are all met with extensive tissue regeneration (Poss et al. 2002[†]; Chablais et al. 2011; Gonzalez-Rosa et al. 2011; Schnabel et al. 2011[‡]; Wang et al. 2011). The well-characterized events that follow after injury to myocardium of the zebrafish have taught us several aspects of vertebrate cardiac repair; however, two concepts stand out to be of most importance when we think about developing heart regeneration strategies with an engineering perspective. First, it has been demonstrated that most, if not all, newly formed cardiomyocytes in the zebrafish regenerative response arise from already-differentiated cardiomyocytes present in the uninjured portion of the heart, rather than local stem cells (Jopling et al. 2010[§]). This information, in combination with Bergmann et al.'s (2009) and Kajstura et al.'s (2010) findings that human cardiomyocytes renew throughout life, is important, as it implies that regenerative strategies in non-regenerative species, such as humans, may fare better when focusing on bringing already-differentiated cardiomyocytes into the cell cycle rather than attempting to activate a stem cell response. The second major concept of heart regeneration that the zebrafish has taught us is that there is a global, organ-wide response to a local injury, manifested as morphological changes and a shift in gene expression in myocardium, endocardium, and epicardium throughout the heart (Lepilina et al. 2006; Kikuchi et al. 2010; Kikuchi et al. 2011). This organ-wide response seems to underlie the capabilities of cardiac regeneration in regeneration-competent zebrafish, and similar but comparatively

---

[*] Review on zebrafish regeneration (multiple tissues).

[†] First demonstration of anatomical (but not physiological) regeneration in the zebrafish.

[‡] Together with Chablais et al (2011) and Gonzalez-Rosa et al. (2011), the first papers to establish the cryoinjury model in the zebrafish heart.

[§] First study describing that newly formed cardiomyocytes in regenerating zebrafish hearts originate from dedifferentiated and proliferating cardiomyocytes.

weak responses are also observed in the mammalian heart after injury (Porrello et al. 2011[*]; Aurora et al. 2014). Pinning down the mechanisms by which this process is initiated can be of great importance in the development of cardiac regenerative therapies.

23.1.3.3 The regenerative potential of model species—the salamander   As can be seen throughout this book, the caudate amphibians, the salamanders, are the vertebrate champions of tissue regeneration. Accordingly, the salamander heart shows a profound regenerative response to injury. Oberpriller and Oberpriller (1974[†]) laid the ground for several subsequent studies on salamander heart regeneration in the newt (*Notophthalmus viridescens*). Later, the axolotl (*Ambystoma mexicanum*) has also been used for this purpose (Cano-Martinez et al. 2010). Although molecular and genetic tools for investigating heart regeneration have seen tremendous development in the zebrafish in the last decade, this development has been slower in the salamander, with the result that the salamander is currently the preferred model in cardiac studies in only a few laboratories worldwide. This is unfortunate, since the salamander has several advantages over the zebrafish in terms of being the most appropriate model for intrinsic heart regeneration. Phylogenetically, the salamander represents the taxon closest to ourselves that express true heart regeneration without scarring (recently, cardiomyocyte proliferation at myocardial injury sites has been suggested in the alligator, *Alligator mississippiensis*; however, scar tissue formation was observed after several months, and these results remain unpublished). On the contrary, the zebrafish is a representative of a teleost fish, the most abundant, diverse, and species-rich infraclass of bony fish, but with a very different ancestry than the sarcopterygian background that we share with the salamander and all other tetrapods. Since heart developmental mechanisms are conserved to a very high degree in all vertebrates, this does not rule out the zebrafish and other teleosts as good models for heart regeneration, but it is important to realize that the evolution of the modern teleost heart has progressed to meet very different requirements than the tetrapod heart. In opposition to the zebrafish, most salamanders (and notably both the semi-terrestrial *N. viridescens* and the aquatic *A. mexicanum*) have lungs and the ability to breathe air. This results in the exposure to much higher concentrations of oxygen compared with what can be dissolved in water. The amphibian heart, containing fully separated atria and some flow separation of oxygenated and deoxygenated blood in the ventricle and outflow tract, also displays a marked increase in complexity compared with that of the zebrafish heart, with a single atrium and no flow separation. This situation better reflects the fully separated mammalian heart and could have important

---

[*] First description of the transient regenerative response of the neonatal mouse.
[†] First description of the regenerative response of the salamander heart.

implications when attempting to understand the physiological challenges of intrinsic heart regeneration. Finally, a practical advantage of salamanders, and in particular the axolotl, over zebrafish is the mere size difference. Adult and well-fed axolotls can be much larger than adult zebrafish, which allows for complicated surgical manipulations and the use of non-invasive imaging techniques, for example, magnetic resonance imaging (MRI) and echocardiography, for functional measurements that are not possible in the minute zebrafish. In summary, one can think of the zebrafish as currently the best model to understand genetic and molecular mechanisms in intrinsic heart regeneration and the salamander as the best model to understand the physiological and anatomical mechanisms involved in this feat (but with the potential for more in-depth genetic and molecular studies when these techniques mature in the salamander field).

23.1.3.4 The regenerative potential of model species—the mouse Describing the regenerative potential of the mouse (*Mus musculus*) heart would have been a repetition of the section regarding the human heart (and in fact all other mammalian hearts), had it not been the discovery by Porrello et al. (2011) that the neonatal mouse heart possesses a transient regenerative potential up to 7 days after birth. Interestingly, genetic fate mapping indicated that regenerated cardiomyocytes originated from pre-existing cardiomyocytes in the uninjured part of the heart, the same mechanisms as seen in zebrafish and salamanders. This early regenerative potential of neonatal mice maybe extended to other mammals and may pose an interesting, yet unresolved challenge of unraveling the mechanisms that ultimately terminate this seemingly advantageous capacity. The oxygen-rich environment imposed on neonatal mice (and placental mammals in general) after birth has been suggested as a reason for cardiomyocyte cell cycle arrest through DNA-damage response (Puente et al. 2014). This speculation has some support in the literature on zebrafish in an experiment where exposing zebrafish to an elevated level of oxygen impaired a cardiac regenerative response, whereas hypoxia seemed to induce cardiomyocyte proliferation (Jopling et al. 2012). However, as salamanders, which can be thought of as evolutionary intermediates between zebrafish and mouse, do not display this regenerative regression of cardiac tissue in the face of hyperoxia, the explanation regarding oxygen level for the termination of neonatal mouse regeneration maybe too simplistic.

Finally, it should be noted that cardiac regeneration in the mouse can be a controversial subject. At the brink of the millennium, the Murphy Roths Large (MRL) mouse strain was reported to display extraordinary heart regenerative capabilities in adult mice (Leferovich et al. 2001; Leferovich and Heber-Katz 2002). This inspired great promise of human regenerative therapies being just around the corner, and many resources were spent to elaborate on and extend these results. However, after considerable efforts in the next decade and more, it must be concluded that the MRL mouse does, in fact, not regenerate myocardial tissue after injury (neither ischemic nor cryoinjury) (Oh et al. 2004; Abdullah et al. 2005; Naseem et al. 2007; Cimini et al. 2008; Grisel et al. 2008; Robey and Murry 2008; Moseley et al. 2011; Smiley et al. 2014). A somewhat discouraging conclusion but an important reminder that great claims (e.g. claiming heart regeneration in a model) must be supported by great evidence should we not waste time on disproving them later.

23.2 Heart regenerative approaches

23.2.1 Setting the stage

An important realization when attempting cardiac regeneration is that the heart is a structure in which the importance of the organ is defined to a very high degree by its mechanical function and much less by the shape and architecture. This implies that any sort of pump that can match the cardiac output at need maybe sufficient to support life and is the rationale that allows for mechanical artificial hearts to bridge the gap between diseased heart malfunction and heart transplantation. Therefore, a tissue regenerative approach to rebuild cardiac structures should focus more on function than on anatomy, a realization that is very often neglected in studies on intrinsic regeneration in competent models, especially the zebrafish and salamander, where the sole focus is often on anatomical regeneration. The forgivingness of these low-metabolic animals to survive non-optimal heart function is, by ignorance or neglect, often assumed to indicate perfect regeneration, even though this has not been functionally tested (for examples of lacking or insufficient functional analysis to claim functional cardiac regeneration, see Poss 2002; Gonzalez-Rosa et al. 2011; Schnabel et al. 2011; Jopling et al. 2012; Mahmoud et al. 2015*, and examples of appropriate functional analysis see Cano-Martinez et al. 2010; Witman et al. 2011; Chablais et al. 2011; Porrello et al. 2011; Wang et al. 2011). The inconvenient truth of this neglect is that several of the instrumental studies on intrinsic regenerative phenomena in regeneration-competent species and, likewise, studies claiming rescue of regeneration after a regeneration dismissive manipulation are in fact inconclusive on functional regeneration of the heart, the one thing that really matters when it comes to heart regeneration. This is a fundamental different situation from the regeneration of, for example, a limb, where complete anatomical regeneration is often the goal and functional testing may not be as crucial.

However, with the recognition fresh in mind that function beats form when it comes to heart regeneration, one must also realize that the function of the blood pump that we call the heart is, in fact, very much defined by its architecture and microanatomy, as we shall see in Section 23.4.

---

* This study suggests nerve dependency in heart regeneration for the first time.

Therefore, heart regenerative approaches that solely focus on cardiomyo-cyte proliferation and not on subsequent organization of these cells are unlikely to generate fully functional hearts.

However, to build the house, we first need the bricks, and in the case of organs, the bricks are the cells that make up the organ. Therefore, the logical question arises: which cells can we use to build the heart?

23.2.2 Cell sources for cardiac regeneration

The cellular engines that run the pump (the heart) are the cardiac mus-cle cells, the cardiomyocytes. These contractive cells are connected in a functional syncytium via intercalated discs made up of three different types of cell junctions: the *fascia adherens*, the anchor points for actin filaments, aiding in the transmission of contractile forces; the *macula adherens* (desmosomes), adhesion sites between cardiomyocytes; and gap junctions that allow action potentials initially generated in the pacemaker region of the sinoatrial node and traveling through the cardiac conduc-tion system to spread between cardiomyocytes, causing depolarization and contraction of the heart in a controlled fashion. The inner surface of the myocardium is lined with endocardium, an epithelial layer of similar origin as the endothelium lining the inside of blood vessels. The endocar-dium provides a smooth and non-adherent surface for blood to pass by and provides a controlled barrier between the blood's extracellular fluid and the extracellular fluid that bathes the cardiomyocytes. The outer surface of the myocardium is lined by another epithelium, the epicardium, which is composed of connective tissue and protectively encompasses the heart. Both these epithelial layers play important roles during cardiac develop-ment and in the regulation of cardiac regeneration in species capable of such a feat (for detailed review, see Kikuchi and Poss 2012[*]). However, cardiomyocytes are the primary cells in the functioning heart and are what make the clock tick, so to speak. From an engineering perspective, the focus must first be on developing technologies to generate cardiomyo-cytes. This poses the question: from where do we get new cardiomyocytes in the case of major myocardial loss after an ischemic event?

Several approaches have been applied to compensate for the loss of car-diomyocytes after myocardial infarction (reviewed in Martin-Puig et al. 2012[†]). Overall, there are two methods that could yield cardiomyocytes after myocardial infarction: applying exogenous cell sources or stimu-lating endogenous regenerative capacity. Recently, Chong et al. (2014), applying the first strategy, reported successful integration of human embryonic stem cell–derived cardiomyocytes into infarcted macaque hearts and elegantly showed that these cardiomyocytes were electrically

[*] Review on i.a. the function of endocardium and epicardium during regeneration.
[†] Review on different approaches to generate cardiomyocytes.

coupled to the conduction system of the heart, although arrhythmias did occur. However, it was estimated that as many as 10^9 cells were sufficient for engraftment in each macaque heart, a number that requires a setup of a significant size of culture, and the promising prospects of this study should be viewed in the perspective that macaque hearts (37–52 g) are significantly smaller than human hearts (~300 g), therefore potentially requiring an order of magnitude more cells.

The second strategy for myocardial repair, stimulating endogenous regenerative capacity, is indicated in nature since all heart regeneration–competent species that we know of, for example, zebrafish and salamanders, apply this method. However, then the question becomes: how do we stimulate differentiated cardiomyocytes to re-enter the cell cycle, proliferate, organize, and ultimately rebuild the injured portion of the heart to restore cardiac performance?

23.2.3 Stimulating an endogenous regenerative response

Stimulating the potential endogenous regenerative response in hearts maybe seen as the ultimate way of rebuilding human cardiac tissue. This would rule out the risks and ethical aspects of administering stem cells and the complications of surgery involved in the use of bioartificial tissue. In addition, intrinsic regeneration is likely to restore the correct architecture of the myocardium, which can be of great importance to cardiac function. It may also be viewed as the simplest technology that would yield perfectly regenerated cardiac tissue, at least from the biomedical engineer's perspective, if assuming that once the correct stimulation has been applied to the injured area, regeneration will take its own path in reconstructing the damaged tissue in a similar fashion to embryonic development of the heart. Therefore, it is not surprising that a hallmark in heart regenerative research is to detect exactly the mechanisms that initiate an endogenous regenerative response in competent animal models. Unlike the case for limb regeneration in the salamander, where we, via the Accessory Limb Model, have a perfect model for testing which factors can replace the prerequisites for limb regeneration (wound surface, nerves, and fibroblasts with different positional identities), as described elsewhere in this book, no such model exists for cardiac regeneration in either of the animal models.

Several studies have addressed molecular events occurring shortly after injury to the heart in zebrafish. From these, we know that shortly after injury, retinoic acid and several developmental growth factor genes such as transforming growth factor β1 (*tgfβ1*), platelet-derived growth factor (*pdgf*), sonic hedgehog (*shh*), and insulin-like growth factor 2b (*igf2b*) are upregulated, promoting cardiomyocyte proliferation (Lien et al. 2006; Chablais and Jaźwińska 2012; Choi et al. 2013; Huang et al. 2013). Notch family genes *notch1a*, *notch1b*, and *notch3* are simultaneously expressed in local epicardial and endocardial cells, and an organ-wide endocardial expression of developmental marker genes *raldh2* and *heg* and embryonic epicardial markers *tbx18* and *raldh2* in epicardial cells can be observed in the entire heart (Raya et al. 2003; Lepilina et al. 2006; Kikuchi et al. 2011; Zhao et al. 2014). However, despite these efforts, we still do not know which factors are essential for the stimulation of the regenerative response. Recently, a highly interesting study reported that nerve function is required for zebrafish and mouse heart regeneration, thereby seemingly bridging the gap to what we know about limb regeneration (Mahmoud et al. 2015). However, when looking into detail, this study may raise more questions than it answers. Nerve dependency in limb regeneration was first demonstrated in a series of classic studies by Marcus Singer (Singer 1942, 1943, 1945, 1946a, 1946b, 1947a, 1947b). Singer clearly demonstrated that limb regeneration dependence on nerves is a threshold phenomenon; below a certain number of fibers (relative to wound surface area), regeneration will not take place; at the threshold, regeneration will occur in 50% of situations; and above the threshold, regeneration is likely to take place. However, once regeneration is initiated by the required number of nerves, it does take place, meaning that a full limb will be regenerated. In other words, the still-undefined neurotrophic factor is an initiator of regeneration. In opposition to this, Mahmoud et al. found that cardiac regeneration is gradually regulated by genetic ablation of cardiac innervation in the zebrafish, a radically different situation where a certain degree of ablation results in a corresponding, lower-than-normal, regenerative response. In addition, Mahmoud et al. (2015) reported that pharmacological inhibition of cholinergic nerve function with atropine reduces cardiomyocyte proliferation in injured zebrafish hearts. It is puzzling, to say the least, that the observation that atropine impairs heart regeneration has never been made before, given that this treatment is a very standard procedure in many physiological experiments on ectothermic (and regeneration-competent) animals and has been used extensively for research and teaching purposes in physiology laboratories around the world for decades. It will be interesting to see future studies replicating this observation in other models and to scrutinize nerve dependency for heart regeneration by using heterotopic heart transplantation models in which no cardiac innervation will be present.

23.2.4 Rebuilding the heart *in vitro*

The lesson from nature is that intrinsic in situ regeneration of myocardium is possible, and in combination with the realization that adult human cardiomyocytes do, in fact, replenish themselves to some degree, the strategy to stimulate innate myocardial regeneration in human patients suffering from ischemic heart disease is undoubtedly the most desirable strategy for future regenerative therapies of the heart. However, as the key to unlocking this intrinsic regenerative potential in adult mammals is

511

yet to be found, it is worth considering alternative strategies for rebuilding injured hearts.

A very different approach of constructing whole hearts or patches of myocardial tissue has arisen in the last decade: the construction of hearts *in vitro* that can, in theory, later be transplanted into patients. Since Dr. Barnard's successful heart transplantation on a December day in 1967, this procedure has become a well-established lifesaving treatment intended to extend and improve the quality of life for recipients. As of today, donor hearts come from recently deceased or clinically brain-dead organ donors, and the recipient's heart is either fully replaced with the donor heart (orthotopic procedure) or supported by an extra heart (heterotopic procedure). However, as suitable donor hearts are a more-than-scarce commodity, and there is an ever-increasing demand for hearts (and other organs), a novel approach receiving much attention in recent years is the construction of *de novo* bioartificial hearts, using either decellularization/repopulation of artificially constructed or extracellular matrix scaffolds or bottom-up constructions by using 3D bioprinting.

23.2.4.1 Repopulation of decellularized heart scaffolds The construction of artificial biocompatible scaffolds for cell seeding and transplantable tissue generation has long been an effort of bioengineering laboratories around the globe, and several successful procedures, especially in the field of orthopedics, have been developed (recently reviewed in Tatara and Micos 2016[*]). In myocardial regeneration, this approach is also applied on a research level; however, the complex cellular environment can be very hard to replicate in artificial scaffolds. To circumvent this limitation, natural extracellular scaffolds can be made by decellularization of excised hearts. This procedure of removing all living cells from an organ is a somewhat-complicated washing process that aims at leaving the supporting extracellular matrix as intact and non-modified as possible (for a detailed protocol on decellularization of whole organs, see Guyette et al. 2014[†]). The generated tissue scaffold can then be re-seeded with living cells, either by immersion in cell-containing medium or more efficiently by a slow perfusion procedure through the skeletonized vasculature of the decellularized organ. The cell source can either be differentiated cells that have previously been dissociated from their tissue state to a single-cell state or be stem cells that have been cultured before perfusion. The assumption of the decellularization/recellularization technique is that perfused cells will migrate through the extracellular matrix scaffold and for differentiated cells ultimately settle when they reach a suitable *milieu intérieur*, whereas undifferentiated cells may in fact start to differentiated in a manner defined by the

---

[*] Review on bioartificial tissues in orthopedics.
[†] Detailed protocol paper on decellularization of whole organs.

surrounding matrix components. After a maturation period, in which perfused cells proliferate and replenish the scaffold, the final result is a three-dimensional multi-cell-type culture that can be viewed as a breath of life into a once-dead-and-skeletonized organ. This technique would undoubtedly have been a source of great inspiration for late Mrs. Mary Shelley; however, the technique does not come without significant limitations. Ott et al. (2008[*]) applied the procedure to decellularize and recellularize rat hearts with neonatal cardiomyocytes, fibrocytes, endothelial cells, and smooth muscle cells and later heterotopically transplanted the constructed hearts into host rats. Although this study could remarkably report a contractile and drug-responsive heart after 8 days of maturation, the pump function of the generated construct was only 2% of adult rat heart function (equivalent to 25% of 16-week fetal human heart function). Likewise, Lu et al. (2013) succeeded in repopulating decellularized mouse hearts with induced pluripotent stem cell–derived cardiovascular progenitor cells from human, again yielding a heart construct that after a 20-day maturation period exhibited spontaneous contractions, generated mechanical force, and was responsive to drugs. However, like the attempt by Ott et al. (2008), this construct did only show minute function and was not able to propel blood through itself.

These and other attempts on rebuilding the heart and other organ systems (Song et al. 2013) are interesting because they suggest that constructing heart tissue by repopulation of decellularized hearts, a very engineering-minded and not biological-inspired way of regenerating hearts, does show great potential but is so far deficient in producing the requested amount of force that is necessary for the heart to provide its function. To overcome this obstacle, repopulation technologies may need to consider what we know about cardiac architecture (at the micro-, meso-, and macroscopic scale) and its influence on cardiac function, the focus of Section 23.4.

23.2.4.2 Three-dimensional bioprinted heart A completely different approach of constructing a complex organ such as the heart relative to that of stimulating endogenous regenerative capacity, stem cell therapies, and a cell repopulation of scaffolds is that of 3D bioprinting. The prospects of this method may have been overly hyped in recent years in TED Talks and other quasi-scientific fora, but the technology has proven very promising in terms of producing *de novo* tissue constructs.

Three-dimensional bioprinting has its origin in 3D printing (also known as rapid prototyping or more precisely as additive manufacturing) of prototypes and models in various materials such as plastic, wood, ceramics, and metals in industry. As the name suggests, this technology is defined by being additive in nature (starts from nothing and ends with

---

[*] Impressive example of decellularization/recellularization and transplantation of (somewhat) functioning rat hearts.

a structure after adding layers), as opposed to, for example, computer numerical control (CNC) drills that perform negative manufacturing by carving out a structure from an object. Additive manufacturing has existed for several decades (piloted in Kodama 1981[*]), but owing to patent issues, the technology has flourished only within the last decade, with 3D printers becoming better and more affordable. This technology has been applied to create organ models in medical and life sciences (see McGurk et al. 1997; Esses et al. 2011; McMenamin et al. 2014; Lauridsen et al. 2016) and even implantable constructs (Zopf et al. 2013). The American Society for Testing and Materials (ASTM) categorizes current 3D printing technologies into seven categories: binder jetting, directed energy deposition, material extrusion, material jetting, powder bed fusion, sheet lamination, and vat photopolymerization (ASTM Standard F2792-12a 2012). In addition to these technologies intended for non-biological materials, 3D printing has evolved into 3D bioprinting, in which cells and extracellular matrix are used as raw materials for printing (for a thorough review on 3D bioprinting, see Murphy and Atala 2014[†]). Drawing on two of the seven categories of non-biological 3D printing and one additional technology not used in industry, 3D bioprinting can currently be divided into three technologies. First, the inkjet method inspired by the material jetting method of 3D printers, and in principle the same inkjet technology applied in desktop paper printers, uses rapid electrical heating or piezoelectric/ultrasound-generated pressure pulses to propel cell-containing droplets from a nozzle to the build surface. Second, microextrusion printers use the same method as material extrusion 3D printers to pneumatically or mechanically dispense a continuous thread of cell-containing material. Finally, laser-assisted printers focus laser light on an absorbing substrate coated in cells and generate a pressure that propels cell-containing materials onto a collector substrate that becomes coated by cells in a pattern defined by the laser path. Subsequently, the layered construct of all three technologies is cured and hardened to support its own structure by the aid of well-designed polymers in the cell medium that respond to cooling, heating, chemical treatment, or, more commonly, light-activated polymerization in a similar fashion as vat photopolymerization in non-biological 3D printers.

All three 3D bioprinting technologies come with advantages and limitations. Inkjet bioprinting is the most simple and affordable technology (in fact, anybody with an engineering mindset could turn an inkjet desktop printer into a 2D bioprinter or, with a few more skills perhaps, even into a 3D bioprinter), and the printer speed is fast. In addition, cell viability is relatively high due to a relatively gentle printing process. However, spatial resolution of droplet size and cell density in the

---

[*] First report on 3D printing.
[†] A detailed review on 3D bioprintin.

construct are low. Microextrusion-based bioprinters can provide high cell densities but at the cost of viability (as low as 40%) due to sheer stress on cells during deposition, and printing is slow. Laser-assisted bioprinters work extremely fast, and cell viability is very high in these systems (>95%). In addition, spatial resolution is high, but these systems are highly expensive and challenging to maintain.

Three-dimensional bioprinting of cardiac tissue is still in its infancy. However, several attempts of printing different cardiac tissues have been made (for full review, see Duan 2016*). Most interestingly from a regenerative perspective, Gaetani et al. (2012) applied microextrusion to bioprint small alginate scaffolds with human fetal cardiomyocyte progenitor cells. These myocardial-like scaffolds showed high cell viability, and importantly, imbedded cardiomyocytes retained their commitment for the cardiac lineage and expressed both cardiac transcription factors and the sarcomere protein troponin T in the 3D culture. Gaebel et al. (2011) used the laser-assisted bioprinting technique to make a pattern of human umbilical vein endothelial cells and human mesenchymal stem cell on a polyester urethane urea cardiac patch. Some but not all vascular patterns were successful and resulted in the two cell types arranged into a capillary-like network. Patches with patterned cells were cultured and subsequently transplanted *in vivo* to infarcted areas of rat hearts. Interestingly, the study found an increased vessel formation and a significant functional improvement in infarcted hearts after transplantation.

So far, no successful attempts to 3D bioprint full organ constructs in a similar fashion as repopulated hearts have been reported, but efforts into using extracellular matrix as a building material have been made recently (Pati et al. 2014). However, if this technology matures to the state where the option of full-organ printing eventually becomes possible, the construct may suffer from the same deficiencies in terms of function and force production as described earlier for the repopulated heart constructs. To overcome these obstacles, some concerns about the architecture of the heart will have to be raised.

23.3 Concerns about architecture for biomedical engineering of heart tissue

From an overall engineering perspective, the heart is first and foremost a pump to propel blood through the vascular system. However, the cardiac pump stands out from a traditional engineered mechanical pump in the way that the engines running the cardiac pump, the cardiomyocytes, are also part of the overall architecture of the pump chambers and the

---

* Review on 3D bioprinting cardiac structures.

interconnectivity of these cardiomyocytes is crucial for optimal pump function. The implication of this is that in the quest of regenerating fully functional cardiac tissue or constructing bioartificial hearts *de novo*, one not only needs to be concerned about the overall structural architecture (e.g., the shape of cardiac chambers and surface properties of the interface between tissue and blood) but also need to think about the internal architecture, the wiring, of the heart.

23.3.1 The cardiac conduction system

The rapid heart rate, high blood pressure, and large cardiac output of the human heart (and endotherms in general) are possible to obtain because of specialized conductive tissue (the bundle of His and the Purkinje fibers) that ensures contraction of the heart in a controlled fashion. In endotherms, depolarization starts spontaneously in the sinoatrial node, the cardiac pacemaker region, and is propagated through the right atrium and through Bachmann's bundle to the left atrium, thus stimulating the contraction of the atria. In the atrioventricular node, conduction is delayed, facilitating a delay in atrial versus ventricular contraction before depolarization continues in the right and left branches of the bundle of His in the interventricular septum. The bundle braches taper out into numerous Purkinje fibers that finally stimulate the myocardial cells of the ventricles to contract in an apex-to-base fashion to facilitate maximal pump efficiency. In ectotherms, such as zebrafish and salamander, there is no anatomical evidence of a specific cardiac conduction system; however, they have a similar contraction pattern as described earlier but with the marked difference that the ventricular myocardium is stimulated in a base-to-apex fashion. Recently, it was suggested that the building blocks of the conduction system are similar between endothermic and ectothermic vertebrates and that the differentiated conduction system of endothermic heart, in fact, represents the primordial trabeculated myocardium of the ectothermic heart. Thus, the compact myocardial tissue of the endothermic hearts was secondarily derived to increase force production. In that sense, the base-to-apex stimulation of the bundle of His actually precisely matches that of the hearts of ectotherms (for in-depth presentation of this hypothesis, see Jensen et al. 2012, 2013).

This information on the conduction system of the endothermic human heart reveals two important concepts. First, since a differentiated conduction system is not present in ectotherms, we cannot rely solely on either the zebrafish model or the salamander model to study the endogenous regeneration of this structure. This makes the task of finding appropriate models tremendously more difficult, since what is needed is an endotherm with a cardiac conduction system that is also able to regenerate myocardial tissue. The solution to cut this Gordian knot maybe in the use of embryonic or neonatal endothermic models, since we know that at least the neonatal mouse does have a conduction system and can regenerate cardiac tissue. Second, when constructing cardiac structures either *de novo* or by seeding cells on scaffolds, the cardiac conduction system should be taken into account if cardiac function is to approach the function of a healthy heart. Recently, Takahashi et al. (2015) reported that brown adipose tissue–derived stem cells in mice can be differentiated into cardiomyocytes expressing the pacemaker markers of the cardiac conduction system, and after transplantation of these *in vivo* in atrioventricular blocked mice, some conduction properties of the heart could be restored. This approach and future techniques to construct the architecture of the conduction system may very well prove instrumental in the production of fully functional heart constructs.

23.3.2 Cardiomyocyte architecture

Myocardial tissue is not a random mesh of cardiomyocytes but rather an intricate structure of highly organized fibers. Currently, we do not have the full understanding of how the networks of fibers in the walls of the cardiac chambers are organized, and we lack the full structural picture of how end-to-end connections of myocytes are arranged. Several attempts have been made in recent years to obtain more information on this architecture by using high-resolution imaging technology such as micro-computed tomography (CT) of iodine-stained cardiac tissue, diffusion tensor MRI to create fiber tracks, and echocardiography (for in-depth review, see Stephenson et al. 2016[*]). For some time, the notion of a *helical ventricular myocardial band* (a compartmentalized ventricular band with selective regional innervation and well-defined deformation that the entire ventricles can theoretically be unwrapped into [Buckberg 2005]) existed. However, this view has been questioned recently (Stephenson et al. 2016). What remains is that the cardiomyocytes of compact endothermic hearts are organized in a highly ordered fashion, with circular and helical bands enclosing the heart, and that this architecture affects pump function. With the exception of decellularization/recellularization attempts, in which the cardiac matrix architecture is stored to some degree, biomedical engineers of cardiac tissues still remain to embrace this fact that, in a functioning heart, cardiomyocytes are not just seeded randomly in a matrix scaffold and that attempts to build bioartificial cardiac tissue is likely to succeed only if some control of cardiomyocyte orientation and interconnection is obtained.

23.4 Concluding remarks on heart regeneration

The human heart possesses very little intrinsic regenerative potential, but the lesson from nature is that endogenous cardiac repair may in fact be possible once we discover how to initiate this process like the

---

[*] In-depth review on the myocardial architecture of the heart.

zebrafish, the salamander, and the neonatal mouse. It maybe that *in vitro* approaches to rebuild the heart such as repopulation of decellularized matrix scaffolds, bioprinted hearts, and other bionic techniques outrun the development of endogenous therapies. In any case, the intricate architecture of the heart needs to be taken into account in efforts to rebuild injured heart tissue, and claims of regenerating cardiac structures on the anatomical level should always be followed up by functional analysis, for it is by virtue of its mechanical function and not its appearance that the heart is a vital organ.

References

ASTM Standard F2792-12a. 2012. Standard terminology for additive manufacturing technologies. ASTM International, West Conshohocken, PA, 2012, doi:10.1520/F2792-12A, www.astm.org.

Abdullah I, Lepore JJ, Epstein JA, Parmacek MS, Gruber PJ. 2005. MRL mice fail to heal the heart in response to ischemia-reperfusion injury. *Wound Repair Regen* 13: 205–208.

Aurora AB, Porrello ER, Tan W, Mahmoud AI, Hill JA, Bassel-Duby R, Sadek HA. et al. 2014. Macrophages are required for neonatal heart regeneration. *J Clin Invest* 124: 1382–1392.

Bergmann O, Bhardwaj RD, Bernard S, Zdunek S, Barnabé-Heider F, Walsh S, Zupicich J. et al. 2009. Evidence for cardiomyocyte renewal in humans. *Science* 324: 98–102.

Buckberg GD. 2005. Architecture must document functional evidence to explain the living rhythm. *Eur J Cardiothorac Surg* 27: 202–209.

Burggren W, Farrell A, Lillywhite H. 2010. Chapter 4: Vertebrate cardiovascular systems. In: *Comprehensive Physiology*. John Wiley & Sons, Inc.

Bøtker HE, Kharbanda R, Schmidt MR, Bøttcher M, Kaltoft AK, Terkelsen CJ, Munk K. et al. 2010. Remote ischaemic conditioning before hospital admission, as a complement to angioplasty, and effect on myocardial salvage in patients with acute myocardial infarction: A randomised trial. *Lancet* 375: 727–734.

Cano-Martínez A, Vargas-González A, Guarner-Lans V, Prado-Zayago E, León-Oleda M, Nieto-Lima B. 2010. Functional and structural regeneration in the axolotl heart (*Ambystoma mexicanum*) after partial ventricular amputation. *Arch Cardiol Mex* 80: 79–86.

Chablais F, Veit J, Rainer G, Jazwinska A. 2011. The zebrafish heart regenerates after cryoinjury induced myocardial infarction. *BMC Dev Biol* 11: 21.

    Together with Gonzalez-Rosa et al (2011) and Schnabel et al (2011) the first papers to establish the cryoinjury model in the zebrafish heart.

Chablais F, Jazwinska A. 2012. The regenerative capacity of the zebrafish heart is dependent on TGFβ signaling. *Development* 139: 1921–1930.

Choi W-Y, Gemberling M, Wang J, Holdway JE, Shen M-C, Karlstrom RO, Poss KD. 2013. *In vivo* monitoring of cardiomyocyte proliferation to identify chemical modifiers of heart regeneration. *Development* 140: 660–666.

Chong JJ, Yang X, Don CW, Minami E, Liu YW, Weyers JJ, Mahoney WM. et al. 2014. Human embryonic-stem-cell-derived cardiomyocytes regenerate non-human primate hearts. *Nature* 510: 273–277.

Cimini M, Fazel S, Fujii H, Zhou S, Tang G, Weisel RD, Li RK. 2008. The MRL mouse heart does not recover ventricular function after a myocardial infarction. *Cardiovasc Pathol* 17: 32–39.

Duan B. 2016. State-of-the-Art review of 3D bioprinting for cardiovascular tissue engineering. *Ann Biomed Eng* 45: 195–209.

Esses SJ, Berman P, Bloom AI, and Sosna J. 2011. Clinical applications of physical 3D models derived from MDCT data and created by rapid prototyping. *AJR Am J Roentgenol* 196: 683–688.

Forouzanfar MH, Moran AE, Flaxman AD, Roth G, Mensah GA, Ezzati M, Naghavi M. et al. 2012. Assessing the global burden of ischemic heart disease, part 2: analytic methods and estimates of the global epidemiology of ischemic heart disease in 2010. *Glob Heart* 7: 331–342.

Gaebel R, Ma N, Liu J, Guan J, Koch L, Klopsch C, Gruene M. et al. 2011. Patterning human stem cells and endothelial cells with laser printing for cardiac regeneration. *Biomaterials* 32: 9218–9230.

Gaetani R, Doevendans PA, Metz CH, Alblas J, Messina E, Giacomello A, Sluijter JP. 2012. Cardiac tissue engineering using tissue printing technology and human cardiac progenitor cells. *Biomaterials* 33: 1782–1790.

Gemberling M, Bailey TJ, Hyde DR, Poss KD. 2013. The zebrafish as a model for complex tissue regeneration. *Trends Genet* 29: 611–620.

Go AS, Mozaffarian D, Roger VL, Benjamin EJ, Berry JD, Blaha MJ, Dai S. et al. 2014. Executive summary: Heart disease and stroke statistics—2014 update: A report from the American Heart Association. *Circulation* 129: 399–410.

Gonzalez-Rosa JM, Martin V, Peralta M, Torres M, Mercader N. 2011. Extensive scar formation and regression during heart regeneration after cryoinjury in zebrafish. *Development* 138: 1663–1674.
Together with Chablais et al (2011) and Schnabel et al (2011) the first papers to establish the cryoinjury model in the zebrafish heart.

Grisel P, Meinhardt A, Lehr HA, Kappenberger L, Barrandon Y, Vassalli G. 2008. The MRL mouse repairs both cryogenic and ischemic myocardial infarcts with scar. *Cardiovasc Pathol* 17: 14–22.

Guyette JP, Gilpin SE, Charest JM, Tapias LF, Ren X, Ott HC. 2014. Perfusion decellularization of whole organs. *Nat Protoc* 9: 1451–1468.

Hepper PG, Shahidullah BS. 1994. Development of fetal hearing. *Arch Dis Child* 71: F81–F87.

Huang Y, Harrison MR, Osorio A, Kim J, Baugh A, Duan C, Sucov HM, Lien C-L. 2013. Igf signaling is required for cardiomyocyte proliferation during zebrafish heart development and regeneration. *PLoS One* 8(6): e67266

Jensen B, Nielsen JM, Axelsson M, Pedersen M, Lofman C, Wang T. 2010. How the python heart separates pulmonary and systemic blood pressures and blood flows. *J Exp Biol* 213: 1611–1617.

Jensen B, Boukens BJ, Postma AV, Gunst QD, van den Hoff MJ, Moorman AF, Wang TChristoffels VM. 2012. Identifying the evolutionary building blocks of the cardiac conduction system. *PLoS One* 7(9): e44231.

Jensen B, Wang T, Christoffels VM, Moorman AF. 2013. Evolution and development of the building plan of the vertebrate heart. *Biochim Biophys Acta* 1833: 783–794.

Jopling C, Sleep E, Raya M, Martí M, Raya A, Izpisúa Belmonte JC. 2010. Zebrafish heart regeneration occurs by cardiomyocyte dedifferentiation and proliferation. *Nature* 464: 606–609.

Jopling C, Suñé G, Faucherre A, Fabregat C, Izpisúa Belmonte JC. 2012. Hypoxia induces myocardial regeneration in zebrafish. *Circulation* 126: 3017–3027.

Kajstura J, Urbanek K, Perl S, Hosoda T, Zheng H, Ogórek B, Ferreira-Martins J. et al. 2010. Cardiomyogenesis in the adult human heart. *Circ Res* 107: 305–315.

Kanoria S, Jalan R, Seifalian AM, Williams R, Davidson BR. 2007. Protocols and mechanisms for remote ischemic preconditioning: A novel method for reducing ischemia reperfusion injury. *Transplantation* 84: 445–458.

Kikuchi K, Holdway JE, Werdich AA, Anderson RM, Fang Y, Egnaczyk GF, Evans T, Macrae CA, Stainier DY, Poss KD. 2010. Primary contribution to zebrafish heart regeneration by gata4+ cardiomyocytes. *Nature* 464: 601–605.

Kikuchi K, Holdway JE, Major RJ, Blum N, Dahn RD, Begemann G, Poss KD. 2011. Retinoic acid production by endocardium and epicardium is aninjuy esonse essential for zebrafish heart regeneration. *Dev Cell* 20: 397–404.

Kikuchi K, Poss KD. 2012. Cardiac regenerative capacity and mechanisms. *Annu Rev Cell Dev Biol* 28: 719–741.

Kodama H. 1981. Automatic method for fabricating a three-dimensional plastic model with photo hardening polymer. *Rev Sci Instrum* 52: 1770–1773.

Laflamme MA, Murry CE. 2011. Heart regeneration. *Nature* 473: 326–335.

Lauridsen H, Hansen K, Noergaard M, Wang T, Pedersen M. 2016. From tissue to silicon to plastic: 3D printing in comparative anatomy and physiology. *R Soc Open Sci* 3(3): 150643.

Leferovich JM, Bedelbaeva K, Samulewicz S, Zhang XM, Zwas D, Lankford EB, Heber-Katz E. 2001. Heart regeneration in adult MRL mice. *Proc Natl Acad Sci USA* 98: 9830–9835.

Leferovich JM, Heber-Katz E. 2002. The scarless heart. *Semin Cell Dev Biol* 13: 327–333.

Lepilina A, Coon AN, Kikuchi K, Holdway JE, Roberts RW, Burns CG, Poss KD. 2006. A dynamic epicardial injury response supports progenitor cell activity during zebrafish heart regeneration. *Cell* 127: 607–619.

Lien C-L, Schebesta M, Makino S, Weber GJ, Keating MT. 2006. Gene expression analysis of zebrafish heart regeneration. *PLoS Biol* 4(8): e260.

Lu TY, Lin B, Kim J, Sullivan M, Tobita K, Salama G, Yang L. 2013. Repopulation of decellularized mouse heart with human induced pluripotent stem cell-derived cardiovascular progenitor cells. *Nat Commun* 4: 2307.

Mahmoud AI, O'Meara CC, Gemberling M, Zhao L, Bryant DM, Zheng R, Gannon JB. et al. 2015. Nerves regulate cardiomyocyte proliferation and heart regeneration. *Dev Cell* 34: 387–399.

Martin-Puig S, Fuster V, Torres M. 2012. Heart repair: From natural mechanisms of cardiomyocyte production to the design of new cardiac therapies. *Ann N Y Acad Sci* 1254: 71–81.

McGurk M, Amis AA, Potamianos P, Goodger NM. 1997. Rapid prototyping techniques for anatomical modelling in medicine. *Ann R Coll Surg Engl* 79: 169–174.

McMenamin PG, Quayle MR, McHenry CR, Adams JW. 2014. The production of anatomical teaching resources using three-dimensional (3D) printing technology. *Anat Sci Educ* 7: 479–486.

Moseley FL, Faircloth ME, Lockwood W, Marber MS, Bicknell KA, Valasek P, Brooks G. 2011. Limitations of the MRL mouse as a model for cardiac regeneration. *J Pharm Pharmacol* 63: 648–656.

Murphy SV, Atala A. 2014. 3D bioprinting of tissues and organs. *Nat Biotechnol* 32: 773–785.

Murry CE, Reinecke H, Pabon LM. 2006. Regeneration gaps: Observations on stem cells and cardiac repair. *J Am Coll Cardiol* 47: 1777–1785.

Naseem RH, Meeson AP, Dimaio JM, White MD, Kallhoff J, Humphries C, Goetsch SC. et al. 2007. Reparative myocardial mechanisms in adult C57BL/6 and MRL mice following injury. *Physiol Genomics* 30: 44–52.

Oberpriller JO, Oberpriller JC. 1974. Response of the adult newt ventricle to injury. *J Exp Zool* 187: 249–260.

Oh YS, Thomson LE, Fishbein MC, Berman DS, Sharifi B, Chen PS. 2004. Scar formation after ischemic myocardial injury in MRL mice. *Cardiovasc Pathol* 13: 203–206.

Ott HC, Matthiesen TS, Goh SK, Black LD, Kren SM, Netoff TI, Taylor DA. 2008. Perfusion-decellularized matrix: Using nature's platform to engineer a bioartificial heart. *Nat Med* 14: 213–221.

Pati F, Jang J, Ha DH, Won Kim S, Rhie JW, Shim JH, Kim DH, Cho DW. 2014. Printing three-dimensional tissue analogues with decellularized extracellular matrix bioink. *Nat Commun* 5: 3935.

Porrello ER, Mahmoud AI, Simpson E, Hill JA, Richardson JA, Olson EN, Sadek HA. 2011. Transient regenerative potential of the neonatal mouse heart. *Science* 331: 1078–1080.

Poss KD, Wilson LG, Keating MT. 2002. Heart regeneration in zebrafish. *Science* 298: 2188–2190.

Przyklenk K, Bauer B, Ovize M, Kloner RA, Whittaker P. 1993. Regional ischemic 'preconditioning' protects remote virgin myocardium from subsequent sustained coronary occlusion. *Circulation* 87: 893–899.

Puente BN, Kimura W, Muralidhar SA, Moon J, Amatruda JF, Phelps KL, Grinsfelder D. et al. 2014. The oxygen-rich postnatal environment induces cardiomyocyte cell-cycle arrest through DNA damage response. *Cell* 157: 565–579.

Raya A, Koth CM, Büscher D, Kawakami Y, Itoh T, Raya RM, Sternik G. et al. 2003. Activation of Notch signaling pathway precedes heart regeneration in zebrafish. *Proc Natl Acad Sci USA* 100(Suppl 1): 11889–11895.

Robey TE, Murry CE. 2008. Absence of regeneration in the MRL/MpJ mouse heart following infarction or cryoinjury. *Cardiovasc Pathol* 17: 6–13.

Schnabel K, Wu CC, Kurth T, Weidinger G. 2011. Regeneration of cryoinjury induced necrotic heart lesions in zebrafish is associated with epicardial activation and cardiomyocyte proliferation. *PLoS One* 6(4): e18503.
the first papers to establish the cryoinjury model in the zebrafish heart.

Shapiro MD, Fazio S. 2016. From lipids to inflammation: New approaches to reducing atherosclerotic risk. *Circ Res* 118: 732–749.

Simões-Costa MS, Vasconcelos M, Sampaio AC, Cravo RM, Linhares VL, Hochgreb T, Yan CY, Davidson B, Xavier-Neto J. 2005. The evolutionary origin of cardiac chambers. *Dev Biol* 277: 1–15.

Singer M. 1942. The nervous system and regeneration of the forelimb of adult Triturus; the role of the sympathetics. *J Exp Zool* 90: 377–399.

Singer M. 1943. The nervous system and regeneration of the forelimb of adult Triturus; the role of the sensory supply. *J Exp Zool* 92: 297–315.

Singer M. 1945. The nervous system and regeneration of the forelimb of adult Triturus; the role of the motor supply. *J Exp Zool* 98: 1–21.

Singer M. 1946a. The nervous system and regeneration of the forelimb of adult Triturus; the stimulating action of a regenerated motor supply. *J Exp Zool* 101: 221–239.

Singer M. 1946b. The nervous system and regeneration of the forelimb of adult Triturus; the influence of number of nerve fibers, including a quantitative study of limb innervation. *J Exp Zool* 201: 299–337.

Singer M. 1947a. The nervous system and regeneration of the forelimb of adult Triturus; a further study of the importance of nerve number, including quantitative measurements of limb innervation. *J Exp Zool* 104: 223–250.

Singer M. 1947b. The nervous system and regeneration of the forelimb of adult Triturus; the relation between number of nerve fibers and surface area of amputation. *J Exp Zool* 104: 251–265.

Smiley D, Smith MA, Carreira V, Jiang M, Koch SE, Kelley M, Rubinstein J, Jones WK, Tranter M. 2014. Increased fibrosis and progression to heart failure in MRL mice following ischemia/reperfusion injury. *Cardiovasc Pathol* 23: 327–334.

Song JJ, Guyette JP, Gilpin SE, Gonzalez G, Vacanti JP, Ott HC. 2013. Regeneration and experimental orthotopic transplantation of a bioengineered kidney. *Nat Med* 19: 646–651.

Stephenson RS, Agger P, Lunkenheimer PP, Zhao J, Smerup M, Niederer P, Anderson RHJarvis JC. 2016. The functional architecture of skeletal compared to cardiac musculature: Myocyte orientation, lamellar unit morphology, and the helical ventricular myocardial band. *Clin Anat* 29: 316–332.

Takahashi T, Nagai T, Kanda M, Liu ML, Kondo N, Naito AT, Ogura T. et al. 2015. Regeneration of the cardiac conduction system by adipose tissue-derived stem cells. *Circ J* 79: 2703–2712.

Tatara AM, Mikos AG. 2016. Tissue engineering in orthopaedics. *J Bone Joint Surg Am* 98: 1132–1139.

Wang J, Panakova D, Kikuchi K, Holdway JE, Gemberling M, Burris JS, Singh SP. et al. 2011. The regenerative capacity of zebrafish reverses cardiac failure caused by genetic cardiomyocyte depletion. *Development* 138: 3421–3430.

Witman N, Murtuza B, Davis B, Arner A, Morrison JI. 2011. Recapitulation of developmental cardiogenesis governs the morphological and functional regeneration of adult newt hearts following injury. *Dev Biol* 354: 67–76.

Zhao L, Borikova AL, Ben-Yair R, Guner-Ataman B, MacRae CA, Lee RT, Burns CG. et al. 2014. Notch signaling regulates cardiomyocyte proliferation during zebrafish heart regeneration. *Proc Natl Acad Sci USA* 111: 1403–1408.

Zopf DA, Nelson ME, Ohye RG. 2013. Bioresorbable airway splint created with a three-dimensional printer. *N Engl J Med* 368: 2043–2045.

Chapter 24 Eye tissue regeneration and engineering

Konstantinos Sousounis,
Joelle Baddour, and
Panagiotis A. Tsonis*

Contents

Key concepts

  a. The eye relies on its layered and highly structured tissues to
     perform its function.

  b. Treatment of eye diseases heavily relies on new methods to
     increase comfort and quality of vision.

---

* Deceased

c. Regenerative engineering is rapidly growing in providing treatments for patients with eye diseases. New custom materials or matrices can be used in combination with advanced surgical procedures to enhance current methods used in the clinic. Alternatively, the innate regenerative ability of amphibians, fishes, and birds can inspire novel approaches for tissue replacement.

24.1 Introduction

The eye's primary function is vision, and sight is an essential function for carrying out our everyday activities. In the animal kingdom, numerous types of eyes can be found; be it simple or complex (Gehring 2012; Land and Nilsson 2002), the eye needs to fulfill three main tasks: (1) receive incoming light, which innately carries the information of the surrounding context; (2) focus the light to cells that can gather this information and translate it into electrical signals; and (3) transmit these signals to other neurons and eventually to the brain, where they can be sensed as an image.

The human eye is structured to deliver good-resolution images, a process that depends on numerous tissues specialized to perform all three main tasks: from capturing, transmitting, and focusing light to translating it into electrical signals that will reflect an image in the brain. From an anterior to a posterior direction, the three main tissues that this chapter will focus on are cornea, lens, and retina. These organs can fully function only when the integrity of their highly organized structure is preserved (Figure 24.1). However, when struck by disease or due to stress,

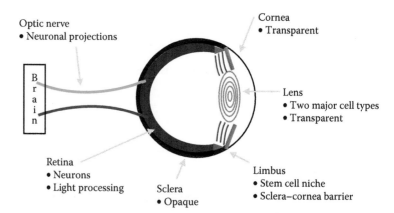

Figure 24.1 Basic eye anatomy and function. Major eye tissues discussed in this chapter are depicted with different colors and patterns. Cornea (thin black arc line), lens (concentric grey ovals), sclera (thick black arc line), limbus (area between cornea and sclera), retina (purple arc), and optic nerve (red/green lines) are shown along with their predominant roles.

Table 24.1 Common Disease Associated with the Cornea, Lens, or Retina

Tissue	Disease
Cornea	Allergic reactions
	Corneal infections: conjunctivitis, keratitis, herpes zoster, and ocular herpes
	Dry eye
	Corneal dystrophies: Fuchs' dystrophy, keratoconus, lattice dystrophy, map-dot-fingerprint dystrophy, and pterygium
	Iridocorneal endothelial syndrome and Stevens–Johnson syndrome
	Nearsightedness, farsightedness, and astigmatism
Lens	Cataracts
	Secondary cataracts
Retina	Age-related macular degeneration
	Glaucoma
	Retinoblastoma
	Retinal detachment
	Retinitis pigmentosa

the homeostasis and function of the organ collapse, leading to obscured vision (Table 24.1). Regenerative medicine and engineering have the mission to provide means that can restore normal tissue function. This chapter will primarily focus on reviewing existing methods and discussing potential future intervention needed to restore vision. These methods span from use of scaffolds, matrices, and micro-devices.

24.2 Cornea

The cornea covers the pupil and is in direct contact with the aqueous humor and the external environment. The cornea's main function is to allow the light to pass through, providing two-thirds of the total optical/focusing power of the eye (Meek et al. 2003). To accomplish this role, the cornea has five main layers composed of different cells and extracellular matrix (Figure 24.2). Bioengineering attempts have been mainly focused on repairing the damaged corneal endothelium and epithelium, along with their associative membranes (Navaratnam et al. 2015[*]). Positioned in the inner and outer-most areas of the cornea, these tissues are composed of metabolically and molecularly active cells that keep the transparency and homeostasis of the cornea. As in most of the eye-related diseases, disruption of the cornea's normal physiology leads to blurriness or even blindness. Treatment often relies on human donor tissues, thus limiting the availability and increasing the need for a method to overcome this problem.

---

[*] A review focusing on corneal endothelial cell expansion.

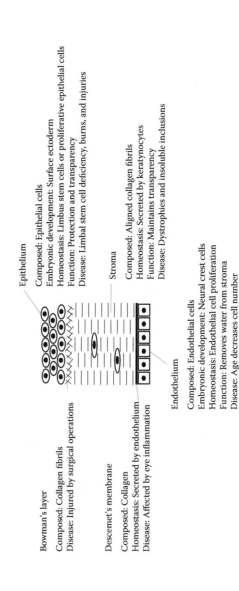

Epithelium

Composed: Epithelial cells
Embryonic development: Surface ectoderm
Homeostasis: Limbus stem cells or proliferative epithelial cells
Function: Protection and transparency
Disease: Limbal stem cell deficiency, burns, and injuries

Bowman's layer

Composed: Collagen fibrils
Disease: Injured by surgical operations

Stroma

Composed: Aligned collagen fibrils
Homeostasis: Secreted by keratynocytes
Function: Maintains transparency
Disease: Dystrophies and insoluble inclusions

Descemet's membrane

Composed: Collagen
Homeostasis: Secreted by endothelium
Disease: Affected by eye inflammation

Endothelium

Composed: Endothelial cells
Embryonic development: Neural crest cells
Homeostasis: Endothelial cell proliferation
Function: Removes water from stroma
Disease: Age decreases cell number

Figure 24.2 Cornea anatomy and function. Five basic corneal layers are shown. Information about their cellular and molecular compositions, embryonic origin, homeostasis, function, and how they are affected during disease is provided. Circles and squares depict cells with a black-filled circle nucleus in the middle. Lines depict extracellular matrix.

24.2.1 Corneal endothelium

Cells of the corneal endothelium draw water from the corneal stroma to the aqueous humor, an event required to keep the cornea transparent (Hodson 1971). The population of these cells declines with age but can also be affected by eye surgical procedures such as corneal transplantation and cataract treatment (Nucci et al. 1990) (Bourne et al. 1997). When the remaining cells fail to support the organ's function, corneal endothelial diseases occur, leading to reduced visual acuity and pain. Current standard of treatment requires transplantation of the corneal endothelium (endothelial keratoplasty) or of the whole cornea (penetrating keratoplasty) (Tan et al. 2012). The major caveat of these procedures is the dependence on human donors, who are not readily available for all patients.

More recent efforts have focused on using biocompatible materials to increase the proliferation of the residing corneal endothelial cells or coax induced pluripotent or embryonic stem cells to regenerate/increase the number of these cells (Table 24.2). Experimentation with human corneal endothelial cells also has a limitation of its own, as it is heavily dependent on the use of cadaver tissue or corneal tissue from patient surgeries. The former cannot be used in transplantation procedures, and the latter requires a donor to undergo a surgical intervention. In addition, the use of autologous induced pluripotent or embryonic stem cells in humans is still controversial for safety reasons. Thus, there is a need for a technology that combines safety and function for treating corneal endothelium diseases, as the only reliable treatment today is transplantation.

24.2.1.1 What to consider when designing your materials Descemet's membrane (Figure 24.2) is thought to play an important role in regulating the physiology of corneal endothelial cells. This membrane serves as a natural matrix where cells attach and grow. Materials and matrices with properties similar to the Descemet's membrane are thought to serve the purpose. Mimicking and usage of Descemet's membrane without having treated the underlying genetic, physiological, or environmental causes of the original problem may lead to recurrence of the disease. We thus believe that there is a need for matrices with properties that can better induce endothelial cell proliferation. Alternatively, materials that can induce de novo matrix secretion from corneal endothelial cells maybe more suitable in creating a physiologic niche for cell propagation.

Cell characterization and *in vivo* testing are important. Corneal endothelial cell proliferation *in vitro* can be enhanced by a variety of methods. However, cell division in an artificial environment having materials with which cells have not been in contact before may result in molecular and morphological cellular changes. Therefore, the meticulous validation of the end product is of utmost importance. Immunocytochemistry can be used to identify expression of proteins commonly found in endothelial cells (such as ZO-1) and can serve as an initial fast screening process. Transplantation

Table 24.2 Biological (Denuded) or Synthetic Materials Used to Propagate Corneal Endothelial Cells.

Category	Material	References
Denuded tissue	Amniotic membrane	Ishino et al. (2004)
	Corneal endothelium	Chen et al. (2001); Insler and Lopez (1986, 1991)
	Corneal stroma	Choi et al. (2010)
	Lens capsule	Kopsachilis et al. (2012); Yoeruek et al. (2009)
	Posterior lamellae of cornea	Bayyoud et al. (2012)
	Cornea	Yoeruek et al. (2012)
	Descemet's membrane	Mimura et al. (2004a)
In vitro culture	Culture media	Baum et al. (1979)
Collagen-based	Collagen vitrigel	Takezawa et al. (2007)
	Type I collagen sponges	Orwin and Hubel (2000)
	Type I collagen sheet	Mimura et al. (2004b)
	Type I or IV collagen-coated	Choi et al. (2013)
	Type IV collagen and temperature-responsive dishes	Sumide et al. (2006)
Gelatin-based	Gelatin	Nayak and Binder (1984)
	Temperature-responsive dishes and gelatin hydrogel	Lai et al. (2007)
Other combinations	Bovine extracellular matrix (ECM)-coated	Blake et al. (1997)
	Laminin and chondroitin sulfate	Engelmann et al. (1988)
	Temperature-responsive dishes	Ide et al. (2006)
	Type IV-coated and gelatin hydrogel	Watanabe et al. (2011)
	Laminin	Yamaguchi et al. (2011); Choi et al. (2013)
	Plastic compressed collagen	Levis et al. (2012)
	Corneal endothelium ECM	Amano (2003); Amano et al. (2005); Blake et al. (1997)
	Mesenchymal cell ECM	Numata et al. (2014)
	Fibronectin-coated	Blake et al. (1997); Choi et al. (2013)

in animal models of the disease should be used to test the efficiency, safety, and function of these experimentally generated cells. When the method is established, we believe that it is also critical to perform genomic and transcriptomic profile comparison of the transplanted cells with normal corneal endothelial cells to ensure correct cell potency months after the surgery.

24.2.2 Corneal epithelium and limbal stem cells

Cells in contact with the external environment are part of the corneal epithelium. The epithelium's main function is to keep the structural integrity of the cornea and protect it from particles and pathogens (Notara et al. 2010). Corneal epithelial cells exhibit some proliferation capabilities, but they are also dependent on their progenitors at the limbus to maintain homeostasis (Dua et al. 2009; Majo et al. 2008; Li et al. 2007). The limbus is the area where cornea and sclera connect. It has been shown that this progenitor cell population plays an important role in keeping non-transparent epithelial cells of the sclera from invading the cornea and causing blurriness. The same holds true for other tissues found in the sclera, such as blood vessels (Osei-Bempong et al. 2013). Diseases associated with the corneal epithelium are also directly related with the stem cells found at the limbus. After damage (e.g., burn injury), limbal stem cells might become reduced in number, allowing epithelial cells from the sclera to invade and cover the pupil in patients, leading to a progressive reduction in visual acuity (Osei-Bempong et al. 2013). Penetrating keratoplasty, a common transplantation method used to treat corneal endothelium–associated diseases, does not replace the number of limbal stem cells and, as such, is ineffective. Current methods to treat limbal stem cell deficiency depend on *ex vivo* expansion of an autologous limbal stem cell population followed by transplantation of these cells in the eye, a method called cultivated limbal epithelial transplantation (CLET) (Rama et al. 2010[*]; Pellegrini et al. 1997). When the limbus is severely damaged and biopsy cannot be performed, a method known as cultivated oral mucosa epithelium transplantation (COMET) has been successfully used to restore corneal epithelium homeostasis (Nishida et al. 2004).

Material engineers in this field are urged to investigate how procedures and manipulation during CLET or COMET that rely on xenogeneic materials can be substituted. This will decrease the chance of disease transmittance or adverse reactions to the healthy tissue. In addition, new methods that use novel materials should aim for lower failures rate than the currently adopted procedures (30% for CLET and 50% for COMET to be unsuccessful). Patients with repeated failed procedures with CLET or penetrating keratoplasty have the option to use artificial cornea (Polisetti et al. 2013). Boston Keratoprosthesis is commonly used in these cases (Ahmad et al. 2016).

24.2.2.1 How cultivated limbal epithelial transplantation works
Autologous cells are extracted from healthy limbus (or from oral mucosa in COMET) and cultured *ex vivo*. This expansion often uses cell carriers or feeder layers such as amniotic membranes, fibril gels, and fibroblasts. Next, cells are isolated and placed on a carrier substance, usually fibril-or collagen-based. After damaged epithelial tissue is removed, the transplant can be placed, which will eventually bring the healthy tissue into homeostasis (Shortt et al. 2010; Casaroli-Marano et al. 2015[†]).

---

[*] A comprehensive review on the uses of induced pluripotent stem cells in eye therapy.
[†] This paper provides important information on the use of limbal stem cells in the clinic.

24.2.2.1.1 What needs improvement?

1. The success rate: Products that increase the efficacy and quality of the transplants are likely to lower failure rates.
2. The usage of animal-derived materials: The nutrients/media from the *in vitro* culture, amniotic membranes, feeder layers, and even cell isolation and extraction methods contain xenogeneic materials that may negatively influence the quality of the process. They can also transmit diseases and induce inflammation or cancer.

These are the areas where new materials and matrices can play a vital role toward better treatments and increase patient comfort.

24.2.2.2 How Boston Keratoprosthesis works The device has two plates that hold a corneal donor graft and replace the cornea, as during a penetrating keratoplasty procedure.

24.2.2.2.1 Limitations and improvements Human corneal tissue is still needed for the procedure, which is still limited by donor availability. In addition, the lens is removed during the procedure because steroid treatment is required as a post-operative countermeasure, a treatment that may lead to cataract. Furthermore, rejection and inflammation are other common complications due to the non-biocompatibility of the materials used. Although these side effects are often severe, this procedure remains a patient's last resort after having lost vision and suffered multiple rounds of corneal transplantation failures. Materials that provide better efficiency are necessary to be developed to increase patient care (Chew et al. 2009).

24.2.3 What else to consider when designing your materials

1. Transparency: The end product should not obscure the light's path. If it cannot be avoided, then it needs to be designed to be easily removed from the eye after treatment is complete.
2. Safety: As in the case of treating corneal endothelium diseases, all protocols that propagate stem cells or progenitors *in vitro* need extensive validation of their cellular and molecular profiles and their cell potency. In addition, materials should be tested for potential inflammation responses *in vivo*.

24.3 Lens

The lens is the organ that focuses the light to the retina. It is composed of three main parts: the lens capsule that surrounds the whole organ, epithelial cells in the anterior side, and fibers in the posterior side and core (Figure 24.3). Epithelial cells proliferate and differentiate to lens fibers at the bow region throughout life. The most common lens disease is cataract, which is a generic word used to describe the obstruction of

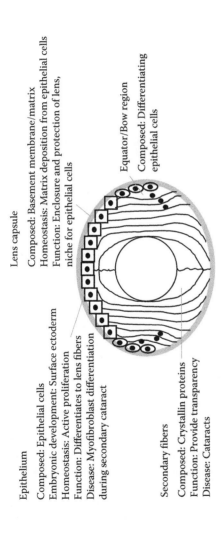

Epithelium

Composed: Epithelial cells
Embryonic development: Surface ectoderm
Homeostasis: Active proliferation
Function: Differentiates to lens fibers
Disease: Myofibroblast differentiation during secondary cataract

Lens capsule

Composed: Basement membrane/matrix
Homeostasis: Matrix deposition from epithelial cells
Function: Enclosure and protection of lens, niche for epithelial cells

Equator/Bow region

Composed: Differentiating epithelial cells

Secondary fibers

Composed: Crystallin proteins
Function: Provide transparency
Disease: Cataracts

Figure 24.3 Anatomy and function of the lens. Four lens structures or cell types are drawn. Information about their cellular and molecular compositions, embryonic origin, homeostasis, function, and how they are affected during disease is provided. Circles and squares depict cells with a black-filled circle nucleus in the middle. Lines depict fibers.

531

the light's path. Cataracts have many etiologies but mainly involve the deregulation of crystallins, the major proteins found in lenses (Francis et al. 1999; Sousounis and Tsonis 2012). Cataract surgery restores the light's path, and the implantation of an intraocular lens (IOL) inside the lens' capsule restores part of the optic power. The IOLs are the most permanent, effective, and convenient way to restore sight compared with the previously used bulky and thick glasses or troublesome contact lenses (Asbell et al. 2005). These implants are made from different materials, including polymethyl methacrylate (PMMA), silicone, hydrophobic or hydrophilic foldable acrylic, and Collamer. Implant coating is one area that can benefit from material engineers' input. Heparin-coated IOLs have been successfully used to reduce inflammation after the operation (Krall et al. 2014). Based on the pathophysiology of the disease in a patient-specific manner, coated IOLs may increase the success rate of the procedure and prevent unwanted side effects.

Posterior capsule opacification (PCO) may occur after IOL implantation. This condition involves proliferation of remaining epithelial cells, migration at the posterior side of the lens capsule behind the IOL, and differentiation to myofibroblasts that obscure the light's path. As a result, a second surgery is often required. This can be overcome by targeting the epithelial-to-mesenchymal transition of the lens cells during PCO (Spalton 1999; Nibourg et al. 2015; Wang et al. 2013). An IOL implant that can reduce or eliminate PCO will certainly be a step forward.

24.3.1 What to consider when designing your materials

1. Although there are controversial opinions surrounding the choice of the material that is best used for the IOL implants, PMMA has been proven to be effective for many years after the operation, and it is highly recommended for young patients.
2. Candidate materials may have different responses based on the type of the IOL used.
3. Recently, pharmacological treatments have been shown to reverse cataracts in animal models and seem to hold great promise for future treatments in humans (Makley et al. 2015; Zhao et al. 2015). That being said, it is necessary in this field to explore potential intraocular matrices or devices that are not dependent on IOLs but may serve to controllably deliver drugs in the eye, without risk of hyper- or hypo-dosing.
4. Surgical advances have been shown to allow lens regeneration from residing lens epithelial cells (Lin et al. 2016[*]).

---

[*] A recent paper describing clinical advances for cataract treatment.

Although this is a recently developed technique and long-term effects have not been determined, secondary cataracts are expected if the genetic basis of the initial cataract is not treated. Material engineers have the opportunity to develop novel biodegradable material that can be coupled with the cataract surgery and deliver molecular that can prolong the therapeutic outcome.

24.4 Retina

Located in the posterior side of the eye, the retina converts the light into electrical and chemical signals that will eventually be translated to an image in the brain. The retina has eight main layers, consisting of different neurons that play critical role in light capturing or electrical processing (Figure 24.4) (Masland 2012). Damage to the retina may lead to complete blindness. Treatment options are limited to stopping or decelerating disease progression. This is due to the fact that transplantation of the whole retina cannot be performed because the optic nerve and the neuronal axons from the ganglion cells (that transfer information to the brain) will not be regenerated and cannot connect to the transplanted retina. In the case where the retina's ability to process electrical signals is intact but the photoreceptors (light-capturing cells) are not functional, microchips have been developed and used to replace the nonfunctional cells. These subretinal microchips are activated with light to stimulate the remaining healthy retina and transmit information to the brain (Chuang et al. 2014).

24.4.1 What to consider when engineering retina replacements

1. The goal of this technology is to provide vision in patients that are completely blind. Since the glasses can bypass the whole eye structure, it can be used if blindness occurs from lens or cornea diseases. However, the vision clarity achieved by engineered retinas is far less than the current methods used to treat cataracts or corneal diseases.
2. In designing biocompatible materials for implantation in humans, there is always room for improvement. Inflammation reactions and rejection of the implants are the vast majority of the side effects of implantation procedures and material incompatibilities. Implants also need to be resistant to corrosion from exposure to bodily fluids, especially if they are expected to function for many years (Rizzo et al. 2014; Sabbah et al. 2014).

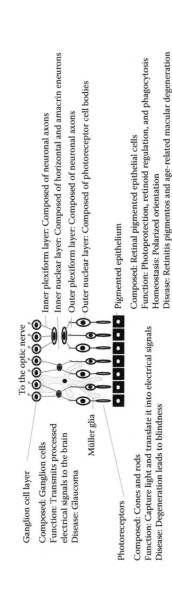

Figure 24.4 Retina anatomy and function. Seven retina layers are shown. Information about their cellular and molecular compositions, their homeostasis, function, and how they are affected during disease is provided. Circles and squares depict cells with a black (or white)-filled circle nucleus in the middle. Lines depict axons, and ovals with no nucleus depict rods or cones.

24.5 3D printing

In the last few years, 3D printers have emerged as a novel and promising tool in the medical field. Their application includes bioprinting of implants that are made from synthetic materials or a patient's own cells. In some fields, 3D printing has been used to engineer whole organs ready for human transplantation (Zopf et al. 2013). It has also been used to 3D print customized eyeglasses for patients suffering from vision problems. 3D printing devices have been reported to reduce surgical time for doctors performing corneal transplantations (Ruzza et al. 2015). In addition, radial electrospun scaffolds coupled with 3D printing of ganglion cells have been shown to improve the engineering of a functional retina (Kador et al. 2016). Despite the fact that stem cells and induced pluripotent stem cells have been explored for repairing eye tissues, their use in 3D printing on decellularized tissue (especially in lens and retina) has not yet been extensively investigated and is a field that holds great promise in developing treatments for human patients with vision problems. Providing eye tissue progenitor cells and matrix in specific layers using a 3D printer may provide alternative treatments for blind patients.

24.6 Tissue engineering of perfect organs: Lessons learned from vertebrates

Amphibians, fishes, and birds can regenerate many eye tissues after damage, and they have been extensively studied to decipher the mechanisms underlying their regenerative mechanisms. These studies include cornea regeneration in zebrafish and chick embryos (Spurlin and Lwigale 2013; Heur et al. 2013); lens regeneration in newts, axolotls, and frogs (Barbosa-Sabanero et al. 2012; Henry and Tsonis 2010[*]; Baddour et al. 2012); and retina regeneration in newts, zebrafish, and chick embryos (Goldman 2014; Barbosa-Sabanero et al. 2012).

Newts, for example, can regenerate a scar-free cornea, a whole lens, and even a retina after whole organ removal, and they have the ability to regenerate throughout their life (Sousounis et al. 2014; Baddour et al. 2012; Sousounis et al. 2015[†]). Studies in this field have aimed at understanding the reasons behind these animals' ability to regenerate and humans' inability to do so. It is now well understood that after injury, regeneration of the damaged tissue starts by a robust response that consists, in the remodeling, of the extracellular matrix or niche, trailed by cellular proliferation and differentiation (Sousounis et al. 2014).

---

[*] A review summarizing lens regeneration in frogs and newts.
[†] First paper suggesting that aging is not affected in a regenerating tissue.

These steps are translated into spatial- and temporal-specific expressions of proteins that perform crucial functions such as promoting cell division, migration, and changes in cell morphology. The process also requires epigenetic reprogramming (involving specific transcriptional factors) of adult cells to a more pluripotent-like state. Several decades of research have advanced our understanding of the genetic, molecular, and signaling events of regeneration and of how scientists can now use these findings to coax mammalian cell regeneration (Sousounis et al. 2013; Looso et al. 2013[*]). It is important to keep in mind that translation from amphibian to mammalian regeneration is not a straight path but requires extensive protocol validation using mammalian systems, a process that can be very time-consuming. However, a major question remains as to whether mammalian cells are susceptible to induction and whether they could respond by potentially inducing a regeneration cascade or whether this ability is intrinsic to certain specialized organisms (Looso et al. 2013). Reprogramming cells *in vivo*, the same or in a similar way that induced pluripotent stem cells are generated *in vitro*, is no easy process and it involves teratogenic side effects. Another area that can be further explored is the remodeling of the extracellular matrix that provides the niche where progenitors and stem cells can attach, grow, proliferate, and differentiate to regenerate the injured tissue. A synthetic or natural scaffold that can be engineered with properties mimicking those of the extracellular matrix can potentially be used to promote regeneration.

24.7 Concluding remarks

Regenerative medicine by means of engineering applications is a fast-growing field with much potential in treating eye diseases. In designing biocompatible materials, certain characteristics ought to be considered and improved, which include but are not limited to safety, compatibility, efficacy, and function. In addition, since the use of induced pluripotent stem cells is still controversial, scientists ought to make the decision between choosing to develop a scaffold that may potentially harbor such cells or to use native populations of cells and enhance their function. It is important to keep in mind that simple materials with clearly defined components, with or without a single autologous unmodified cell population, present a greater chance at receiving the U.S. Food and Drug Administration approval versus more complicated scaffolds that may contain matrices and *in vitro* manipulated cells. Taken together, this chapter has briefly presented the advantages of current eye disease treatments and areas where regenerative engineering can intervene and improve.

---

[*] First comprehensive newt transcriptome and proteome.

References

Ahmad, S., P. M. Mathews, K. Lindsley, M. Alkharashi, F. S. Hwang, S. M. Ng, A. J. Aldave, and E. K. Akpek. 2016. Boston Type 1 Keratoprosthesis versus repeat donor keratoplasty for corneal graft failure: A systematic review and meta-analysis. *Ophthalmology* 123 (1): 165–177. doi:10.1016/j.ophtha.2015.09.028.

Amano, S. 2003. Transplantation of cultured human corneal endothelial cells. *Cornea* 22 (7 Suppl): S66–S74. doi:10.1097/00003226-200310001-00010.

Amano, S., T. Mimura, S. Yamagami, Y. Osakabe, and K. Miyata. 2005. Properties of corneas reconstructed with cultured human corneal endothelial cells and human corneal stroma. *Japanese Journal of Ophthalmology* 49 (6): 448–452. doi:10.1007/s10384-005-0245-5.

Asbell, P. A., I. Dualan, J. Mindel, D. Brocks, M. Ahmad, and S. Epstein. 2005. Age-related cataract. *The Lancet* 365: 599–609. doi:10.1016/S0140-6736(05)17911-2.

Baddour, J. A., K. Sousounis, and P. A. Tsonis. 2012. Organ repair and regeneration: An overview. *Birth Defects Research Part C—Embryo Today: Reviews.*

Barbosa-Sabanero, K., A. Hoffmann, C. Judge, N. Lightcap, P. A. Tsonis, and K. Del Rio-Tsonis. 2012. Lens and retina regeneration: New perspectives from model organisms. *Biochemical Journal* 447 (3): 321–334. doi:10.1042/BJ20120813.

Baum, J. L, R. Niedra, C. Davis, and B. Y. Yue. 1979. Mass culture of human corneal endothelial cells. *Archives of Ophthalmology* 97 (6): 1136–1140. http://www.ncbi.nlm.nih.gov/pubmed/375894.

Bayyoud, T., S. Thaler, J. Hofmann, C. Maurus, M. S. Spitzer, K. U. Bartz-Schmidt, P. Szurman, and E. Yoeruek. 2012. Decellularized bovine corneal posterior lamellae as carrier matrix for cultivated human corneal endothelial cells. *Current Eye Research* 37 (3): 179–186. doi:10.3109/02713683.2011.644382.

Blake, D. A., H. Yu, D. L. Young, and D. R. Caldwell. 1997. Matrix stimulates the proliferation of human corneal endothelial cells in culture. *Investigative Ophthalmology & Visual Science* 38 (6): 1119–1129.

Bourne, W. M., L. I L Nelson, and D. O. Hodge. 1997. Central corneal endothelial cell changes over a ten-year period. *Investigative Ophthalmology and Visual Science* 38 (3): 779–782. doi:10.1016/S0002-9394(14)70810-4.

Casaroli-Marano, R., N. Nieto-Nicolau, E. Martínez-Conesa, M. Edel, and A. B. Álvarez-Palomo. 2015. Potential role of induced pluripotent stem cells (IPSCs) for cell-based therapy of the ocular surface. *Journal of Clinical Medicine* 4 (2): 318–342. doi:10.3390/jcm4020318.

Chen, K. H., D. Azar, and N. C. Joyce. 2001. Transplantation of adult human corneal endothelium ex vivo: A morphologic study. *Cornea* 20 (7): 731–737. doi:10.1097/00003226-200110000-00012.

Chew, H. F., B. D. Ayres, K. M. Hammersmith, C. J. Rapuano, P. R. Laibson, J. S. Myers, Y. P. Jin, and E. J. Cohen. 2009. Boston keratoprosthesis outcomes and complications. *Cornea* 28 (9): 989–996. doi:10.1097/ICO.0b013e3181a186dc.

Choi, J. S., E. Y. Kim, M. J. Kim, M. Giegengack, F. A. Khan, G. Khang, and S. Soker. 2013. In vitro evaluation of the interactions between human corneal endothelial cells and extracellular matrix proteins. *Biomedical Materials* 8 (1): 14108. doi:10.1088/1748-6041/8/1/014108.

Choi, J. S., J. K. Williams, M. Greven, K. A. Walter, P. W. Laber, G. Khang, and S. Soker. 2010. Bioengineering endothelialized neo-corneas using donor-derived corneal endothelial cells and decellularized corneal stroma. *Biomaterials* 31 (26): 6738–6745. doi:10.1016/j.biomaterials.2010.05.020.

Chuang, A. T, C. E Margo, and P. B. Greenberg. 2014. Retinal implants: A systematic review. *British Journal of Opthalmology* 98 (7): 852–856. doi:10.1136/bjophthalmol-2013-303708.

Dua, H. S., A. Miri, T. Alomar, A. M. Yeung, and D. G. Said. 2009. The role of limbal stem cells in corneal epithelial maintenance: Testing the dogma. *Ophthalmology* 116 (5): 856–863. doi:10.1016/j.ophtha.2008.12.017.

Engelmann, K., M. Bohnke, and P. Friedl. 1988. Isolation and long-term cultivation of human corneal endothelial cells. *Investigative Ophthalmology and Visual Science* 29 (11): 1656–1662.

Francis, P. J., V. Berry, A. T. Moore, and S. Bhattacharya. 1999. Lens biology: Development and human cataractogenesis. *Trends in Genetics* 15 (5): 191–196. doi:10.1016/S0168-9525(99)01738-2.

Gehring, W. J. 2012. The animal body plan, the prototypic body segment, and eye evolution. *Evolution and Development* 14 (1): 34–46. doi:10.1111/j.1525-142X.2011.00528.x.

Goldman, D. 2014. Müller glial cell reprogramming and retina regeneration. *Nature Reviews. Neuroscience* 15 (7): 431–442. doi:10.1038/nrn3723.

Henry, J. J. and P. A. Tsonis. 2010. Molecular and cellular aspects of amphibian lens regeneration. *Progress in Retinal and Eye Research* 29 (6): 543–555. doi:10.1016/j.preteyeres.2010.07.002.

Heur, M., S. Jiao, S. Schindler, and J. G. Crump. 2013. Regenerative potential of the zebrafish corneal endothelium. *Experimental Eye Research* 106: 1–4. doi:10.1016/j.exer.2012.10.009.

Hodson, S. 1971. Evidence for a bicarbonate-dependent sodium pump in corneal endothelium. *Experimental Eye Research* 11 (1): 20–29. doi:10.1016/S0014-4835(71)80060-X.

Ide, T., K. Nishida, M. Yamato, T. Sumide, M. Utsumi, T. Nozaki, A. Kikuchi, T. Okano, and Y. Tano. 2006. Structural characterization of bioengineered human corneal endothelial cell sheets fabricated on temperature-responsive culture dishes. *Biomaterials* 27 (4): 607–614. doi:10.1016/j.biomaterials.2005.06.005.

Insler, M. S. and J. G. Lopez. 1986. Transplantation of cultured human neonatal corneal endothelium. *Current Eye Research* 5 (12): 967–972.

Insler, M. S. and J. G. Lopez. 1991. Heterologous transplantation versus enhancement of human corneal endothelium. *Cornea* 10 (2): 136–148. http://www.ncbi.nlm.nih.gov/pubmed/2019124.

Ishino, Y., Y. Sano, T. Nakamura, C. J. Connon, H. Rigby, N. J. Fullwood, and S. Kinoshita. 2004. Amniotic membrane as a carrier for cultivated human corneal endothelial cell transplantation. *Investigative Ophthalmology & Visual Science* 45 (3): 800–806. doi:10.1167/iovs.03-0016.

Kador, K. E., S. P. Grogan, E. W. Dorthé, P. Venugopalan, M. F. Malek, J. L. Goldberg, and D. D'Lima. 2016. Control of retinal ganglion cell positioning and neurite growth: Combining 3D printing with radial electrospun scaffolds. *Tissue Engineering Part A* 22(3–4): 286–294. doi:10.1089/ten.TEA.2015.0373.

Kopsachilis, N., I. Tsinopoulos, T. Tourtas, F. E. Kruse, and U. W Luessen. 2012. Descemet's membrane substrate from human donor lens anterior capsule. *Clinical and Experimental Ophthalmology* 40 (2): 187–194. doi:10.1111/j.1442-9071.2011.02678.x.

Krall, E. M., E. M. Arlt, G. Jell, C. Strohmaier, A. Bachernegg, M. Emesz, G. Grabner, and A. K. Dexl. 2014. Intraindividual aqueous flare comparison after implantation of hydrophobic intraocular lenses with or without a heparin-coated surface. *Journal of Cataract and Refractive Surgery* 40 (8): 1363–1370. doi:10.1016/j.jcrs.2013.11.043.

Lai, J. Y., K. H. Chen, and G. H. Hsiue. 2007. Tissue-engineered human corneal endothelial cell sheet transplantation in a rabbit model using functional biomaterials. *Transplantation* 84 (10): 1222–1232. doi:10.1097/01.tp.0000287336.09848.39.

Land, M. F. and D. E. Nilsson. 2002. *Animal Eyes*. Oxford Animal Biology Series, 221. doi:10.1093/acprof:oso/9780199581139.001.0001.

Levis, H. J., G. S. L. Peh, K. P. Toh, R. Poh, A. J. Shortt, R. A. L. Drake, J. S. Mehta, and J. T. Daniels. 2012. Plastic compressed collagen as a novel carrier for expanded human corneal endothelial cells for transplantation. *PLoS ONE* 7 (11): e50993. doi:10.1371/journal.pone.0050993.

Li, W., Y. Hayashida, Y.T. Chen, S. C. G. Tseng. 2007. Niche regulation of corneal epithelial stem cells at the limbus. *Cell Research* 17 (17): 26–36. doi:10.1038/sj.cr.7310137.

Lin, H., H. Ouyang, J. Zhu, S. Huang, Z. Liu, S. Chen, G. Cao. et al. 2016. Lens regeneration using endogenous stem cells with gain of visual function. *Nature* 531 (7594): 323–328. doi:10.1038/nature17181.

Looso, M., J. Preussner, K. Sousounis, M. Bruckskotten, C. S. Michel, E. Lignelli, R. Reinhardt. et al. 2013. A de novo assembly of the newt transcriptome combined with proteomic validation identifies new protein families expressed during tissue regeneration. *Genome Biology* 14 (2). BioMed Central Ltd: R16. http://genomebiology.com/content/14/2/R16.

Majo, F., A. Rochat, M. Nicolas, G. A. Jaoudé, and Y. Barrandon. 2008. Oligopotent stem cells are distributed throughout the mammalian ocular surface. *Nature* 456 (November): 250–254. doi:10.1038/nature07406.

Makley, L. N., K. A. Mcmenimen, B. T. Devree, J. W. Goldman, B. N. Mcglasson, P. Rajagopal, B. M. Dunyak. et al. 2015. Pharmacological chaperone for a-crystallin partially restores transparency in cataract models. *Science* 350 (6261): 674–677. doi:10.1126/science.aac9145.

Masland, R. H. 2012. The neuronal organization of the retina. *Neuron.* doi:10.1016/j.neuron.2012.10.002.

Meek, K. M., S. Dennis, and S. Khan. 2003. Changes in the refractive index of the stroma and its extrafibrillar matrix when the cornea swells. *Biophysical Journal* 85 (4): 2205–2212. doi:10.1016/S0006-3495(03)74646-3.

Mimura, T., S. Amano, T. Usui, M. Araie, K. Ono, H. Akihiro, S. Yokoo, and S. Yamagami. 2004a. Transplantation of corneas reconstructed with cultured adult human corneal endothelial cells in nude rats. *Experimental Eye Research* 79 (2): 231–237. doi:10.1016/j.exer.2004.05.001.

Mimura, T., S. Yamagami, S. Yokoo, T. Usui, K. Tanaka, S. Hattori, S. Irie, K. Miyata, M. Araie, and S. Amano. 2004b. Cultured human corneal endothelial cell transplantation with a collagen sheet in a rabbit model. *Investigative Ophthalmology & Visual Science* 45 (9): 2992–2997. doi:10.1167/iovs.03-1174.

Navaratnam, J., T. P. Utheim, V. K. Rajasekhar, and A. Shahdadfar. 2015. Substrates for expansion of corneal endothelial cells towards bioengineering of human corneal endothelium. *Journal of Functional Biomaterials* 6 (3): 917–945. doi:10.3390/jfb6030917.

Nayak, S. K. and P. S. Binder. 1984. The growth of endothelium from human corneal rims in tissue culture. *Investigative Ophthalmology and Visual Science* 25 (10): 1213–1216.

Nibourg, L. M., E. Gelens, R. Kuijer, J. M. Hooymans, T. G. van Kooten, and S. A. Koopmans. 2015. Prevention of posterior capsular opacification. *Experimental Eye Research* 136:100–115. doi:10.1016/j.exer.2015.03.011.

Nishida, K., M. Yamato, Y. Hayashida, K. Watanabe, K. Yamamoto, E. Adachi, S. Nagai. et al. 2004. Corneal reconstruction with tissue-engineered cell sheets composed of autologous oral mucosal epithelium. *New England Journal of Medicine* 351 (12): 1187–1196. doi:10.1056/NEJMoa040455.

Notara, M., A. Alatza, J. Gilfillan, A. R. Harris, H. J. Levis, S. Schrader, A. Vernon, and J. T. Daniels. 2010. In sickness and in health: Corneal epithelial stem cell biology, pathology and therapy. *Experimental Eye Research* 90 (2): 188–195. doi:10.1016/j.exer.2009.09.023.

Nucci, P., R. Brancato, M. B. Mets, and S. K. Shevell. 1990. Normal endothelial cell density range in childhood. *Archives of Ophthalmology* 108 (2): 247–248. doi:10.1001/archopht.1990.01070040099039.

Numata, R., N. Okumura, M. Nakahara, M. Ueno, S. Kinoshita, D. Kanematsu, Y. Kanemura, Y. Sasai, and N. Koizumi. 2014. Cultivation of corneal endothelial cells on a pericellular matrix prepared from human decidua-derived mesenchymal cells. *PLoS ONE* 9 (2): e88169. doi:10.1371/journal.pone.0088169.

Orwin, E. J. and A.Hubel. 2000. In vitro culture characteristics of corneal epithelial, endothelial, and keratocyte cells in a native collagen matrix. *Tissue Engineering* 6 (4): 307–319. doi:10.1089/107632700418038.

Osei-Bempong, C., F. C. Figueiredo, and M. Lako. 2013. The limbal epithelium of the eye- A review of limbal stem cell biology, disease and treatment. *BioEssays: News and Reviews in Molecular, Cellular and Developmental Biology* 35 (3): 211–219. doi:10.1002/bies.201200086.

Pellegrini, G., C. E. Traverso, A. T. Franzi, M. Zingirian, R. Cancedda, and M. De Luca. 1997. Long-term restoration of damaged corneal surfaces with autologous cultivated corneal epithelium. *Lancet* 349 (9057): 990–993. doi:10.1016/S0140-6736(96)11188-0.

Polisetti, N., M. M. Islam, and M. Griffith. 2013. The artificial cornea. *Methods in Molecular Biology* 1014: 45–52. doi:10.1007/978-1-62703-432-6-2.

Rama, P., S. Matuska, G. Paganoni, A. Spinelli, M. De Luca, and G. Pellegrini. 2010. Limbal stem-cell therapy and long-term corneal regeneration. *The New England Journal of Medicine* 363 (2): 147–155. doi:10.1056/NEJMoa0905955.

Rizzo, S., C. Belting, L. Cinelli, L. Allegrini, F. Genovesi-Ebert, F. Barca, and E. Di Bartolo. 2014. The Argus II Retinal Prosthesis: 12-month outcomes from a single-study center. *American Journal of Ophthalmology* 157 (6): 1282–1290. doi:10.1016/j.ajo.2014.02.039.

Ruzza, A., M. Parekh, S. Ferrari, G. Salvalaio, Y. Nahum, C. Bovone, D. Ponzin, and M. Busin. 2015. Preloaded donor corneal lenticules in a new validated 3D printed smart storage glide for descemet stripping automated endothelial keratoplasty. *British Journal of Ophthalmology* 99 (10):1388–1395. doi:10.1136/bjophthalmol-2014-306510.

Sabbah, N., C. N. Authié, N. Sanda, S. Mohand-Said, J. A. Sahel, and A. B. Safran. 2014. Importance of eye position on spatial localization in blind subjects wearing an Argus II Retinal Prosthesis. *Investigative Ophthalmology and Visual Science* 55 (12): 8259–8266. doi:10.1167/iovs.14-15392.

Shortt, A. J., S. J. Tuft, and J. T. Daniels. 2010. Ex vivo cultured limbal epithelial transplantation. A clinical perspective. *The Ocular Surface* 8 (2): 80–90. doi:10.1016/S1542-0124(12)70072-1.

Sousounis, K., J. A. Baddour, and P. A. Tsonis. 2014. Aging and regeneration in vertebrates. *Current Topics in Developmental Biology* 108: 217–246.

Sousounis, K., R. Bhavsar, M. Looso, M. Krüger, J. Beebe, T. Braun, and P. A Tsonis. 2014. Molecular signatures that correlate with induction of lens regeneration in newts: Lessons from proteomic analysis. *Human Genomics* 8 (1): 22. doi:10.1186/s40246-014-0022-y.

Sousounis, K., M. Looso, N. Maki, C. J. Ivester, T. Braun, and P. A. Tsonis. 2013. Transcriptome analysis of newt lens regeneration reveals distinct gradients in gene expression patterns. *PLoS ONE* 8 (4): e61445.

Sousounis, K., F. Qi, M. C. Yadav, J. L. Millan, F. Toyama, C. Chiba, Y. Eguchi, G. Eguchi, and P. A. Tsonis. 2015. A robust transcriptional program in newts undergoing multiple events of lens regeneration throughout their lifespan. *Elife* 10 (November): 7554.

Sousounis, K., and P. A. Tsonis. 2012. Patterns of gene expression in microarrays and expressed sequence tags from normal and cataractous lenses. *Human Genomics* 6 (1): 14. http://www.pubmedcentral.nih.gov/articlerender.fcgi?artid=3563465&tool=pmcentrez&rendertype=abstract.

Spalton, D. J. 1999. Posterior capsular opacification after cataract surgery. *Eye (London, England)* 13 (Pt 3b: 489–492. doi:10.1038/eye.1999.127.

Spurlin, J. W. and P. Y. Lwigale. 2013. Wounded embryonic corneas exhibit nonfi-brotic regeneration and complete innervation. *Investigative Ophthalmology and Visual Science* 54 (9): 6334–6344. doi:10.1167/iovs.13-12504.

Sumide, T., K. Nishida, M. Yamato, T. Ide, Y. Hayashida, K. Watanabe, J. Yang. et al. 2006. Functional human corneal endothelial cell sheets harvested from temperature-responsive culture surfaces. *The FASEB Journal: Official Publication of the Federation of American Societies for Experimental Biology* 20 (2): 392–294. doi:10.1096/fj.04-3035fje.

Takezawa, T., A. Nitani, T. Shimo-Oka, and Y. Takayama. 2007. A protein-permeable scaffold of a collagen vitrigel membrane useful for reconstructing cross-talk models between two different cell types. *Cells Tissues Organs* 185: 237–241. doi:10.1159/000101325.

Tan, D. T. H., J. K. G. Dart, E. J. Holland, and S. Kinoshita. 2012. Corneal transplan-tation. *Lancet* 379 (9827): 1749–1761. doi:10.1016/S0140-6736(12)60437-1.

Wang, Y., W. Li, X. Zang, N. Chen, T. Liu, P. A. Tsonis, and Y. Huang. 2013. MicroRNA-204-5p regulates epithelial-to-mesenchymal transition during human posterior capsule opacification by targeting SMAD4. *Investigative Ophthalmology & Visual Science* 54 (1): 323–332. doi:10.1167/iovs.12-10904.

Watanabe, R., R. Hayashi, Y. Kimura, Y. Tanaka, T. Kageyama, S. Hara, Y. Tabata, and K. Nishida. 2011. A novel gelatin hydrogel carrier sheet for corneal endo-thelial transplantation. *Tissue Engineering. Part A* 17: 2213–2219. doi:10.1089/ten.tea.2010.0568.

Yamaguchi, M., N. Ebihara, N. Shima, M. Kimoto, T. Funaki, S. Yokoo, A. Murakami, and S. Yamagami. 2011. Adhesion, migration, and proliferation of cultured human corneal endothelial cells by Laminin-5. *Investigative Ophthalmology & Visual Science* 52: 679–684. doi:10.1167/iovs.10-5555.

Yoeruek, E., T. Bayyoud, C. Maurus, J. Hofmann, M. S. Spitzer, K. U. Bartz-Schmidt, and P. Szurman. 2012. Decellularization of porcine corneas and repopulation with human corneal cells for tissue-engineered xenografts. *Acta Ophthalmologica* 90 (2): e125–e131. doi:10.1111/j.1755 - 3768.2011.02261.x.

Yoeruek, E., O. Saygili, M. S. Spitzer, O. Tatar, K. U. Bartz-Schmidt, and P. Szurman. 2009. Human anterior lens capsule as carrier matrix for cultivated human corneal endothelial cells. *Cornea* 28 (4): 416–420. doi:10.1097/ICO.0b013e31818c2c36.

Zhao, L., X. J. Chen, J. Zhu, Y. B. Xi, X. Yang, L.D. Hu, H. Ouyang. et al. 2015. Lanosterol reverses protein aggregation in cataracts. *Nature* 523 (7562): 607–611. doi:10.1038/nature14650.

Zopf, D. A., S. J. Hollister, M. E. Nelson, R. G. Ohye, and G. E. Green. 2013. Bioresorbable airway splint created with a three-dimensional printer. *The New England Journal of Medicine* 368 (October 2015): 2043–2045. doi:10.1056/NEJMc1206319.

Chapter 25 Skin development and regeneration, and the control of fibrosis

Michael S. Hu,
H. Peter Lorenz, and
Michael T. Longaker

Contents

Key concepts

a. Skin is a complex structure consisting of epidermis, dermis, and epidermal glands that develop through intricate interactions, making regeneration particularly challenging.
b. Wound healing is a complicated process evolved to restore skin via a rapid fibroproliferative response at the cost of perfect regeneration.

> c. Wound repair *in utero* and in the oral mucosa occurs without the formation of a scar in a process resembling regeneration. Lessons learned from such physiologic processes can be applied for regenerative applications.
> d. Recent advances in stem cell and developmental biology offer a better understanding of scarring and fibrosis, giving hope for new ways to recapitulate skin regeneration.

25.1 Introduction

The skin is the largest organ in the human body. As such, there is an urgent need to regenerate skin to fill defects that may occur. Such defects can arise with traumatic injuries or burns or after surgeries to remove cancer. Moreover, with an aging population and a growing incidence of comorbidities, such as obesity, diabetes, and vascular disease, the healthcare system is increasingly burdened with chronic, nonhealing wounds. Advances in skin regeneration would help treat the millions of people who suffer from these conditions every year.

This chapter begins by briefly discussing skin development, with an emphasis on the role of stem cells in tissue maintenance. Next, we describe wound healing, as it occurs in both normal and pathologic states. Then, we elaborate on skin regeneration, in particular as it occurs naturally after *in utero* wounding and in the oral mucosa. Finally, we discuss recent advances in developmental biology that provide novel insight into scarring and fibrosis. This chapter concentrates heavily on mouse skin, as genetic engineering has accelerated the pace of scientific discovery in the mouse. An understanding of these topics will equip the reader with the background, goals, and challenges of skin regeneration.

25.2 Skin development

The skin is a fascinating organ that provides a barrier between humans and the environment. Continuously, it protects us from insults such as harmful ultraviolet radiation from the sun or deadly microbes. To accomplish this difficult feat, stem cells in the skin undergo continual self-renewal to repair damaged tissue and replace old cells. In adult skin, these stem cells reside in hair follicles, sebaceous glands, and the epidermis to maintain tissue homeostasis (Fuchs 2007). The formation of an organ with such capabilities is intricate and challenges regenerative efforts. However, understanding embryologic development and the role of stem cells in homeostasis provides a background to achieving skin regeneration.

25.2.1 Epidermal development

Mammalian skin consists of an epidermis and a dermis, separated by a basement membrane. It is a complex structure with an epidermis composed of several cell layers and epidermal appendages. The basal keratinocyte layer differentiates and stratifies to give rise to the spinous layer, granular layer, and acellular stratum corneum.

In embryogenesis, the skin epithelium arises after gastrulation. A single layer of neuroectoderm emerges and differentiates into the nervous system or skin based on Wnt signaling. Wnt inhibits ectoderm response to fibroblast growth factors (FGFs), resulting in the expression of bone morphogenetic proteins (BMPs), thereby fating cells to develop into epidermis. The resultant embryonic epidermis consists of a single layer of multipotent cells covered by a periderm. This protective layer of tightly connected squamous endodermis–like cells is transient and sheds as the epidermis differentiates and stratifies. In mice, a single epidermal layer is present from embryonic day (E)9.5 to E12.5, before the arrival of mesenchymal cells, which populate the dermis and instruct the formation of skin appendages (Blanpain and Fuchs 2009). Between E12.5 and E15.5, stratification of the epidermis begins, and it is completed by E17.5. The result is an inner layer of basal cells with proliferative potential and layers of terminally differentiating suprabasal cells. Epidermal appendages extend into the dermis and provide a home to skin-renewing stem cells.

25.2.2 Growth and homeostasis

In the adult epidermis, stem cells maintain homeostasis by replacing keratinocytes that are lost through normal turnover or cell death after injury. Current literature supports the presence of multiple lineage-restricted stem cells in distinct niches in the cutaneous epithelium (Ghazizadeh and Taichman 2001). These stem cells are found in the basal layer of the epidermis adjacent to the basement membrane, bulge region of the hair follicle, and base of the sebaceous gland and upper isthmus (Potten 1974; Cotsarelis et al. 1990; Nijhof et al. 2006).

Growth and homeostasis of the epidermis are currently explained by dividing the interfollicular epidermis into discrete proliferative units referred to as epidermal proliferating units (EPUs) (Potten and Morris 1988). It is thought that there is one self-renewing basal stem cell per EPU. All other basal cells in the EPU are transiently amplifying (TA) cells (Potten 1974). The TA cells have a limited life span and can divide only several times before leaving the basal layer and differentiating terminally. On the other hand, true epidermal stem cells are assumed to persist throughout the life of the organism.

Multipotent stem cells, able to give rise to all components of the epidermis, have been identified in the hair follicle in a specialized microenvironment of the outer root sheath known as the bulge (Taylor et al. 2000; Oshima et al. 2001). These stem cells do not contribute to the maintenance of the interfollicular epidermis during normal homeostasis. However, with injury, these cells rapidly migrate out of the bulge and give rise to a new epidermis (Levy et al. 2007). Despite a recent increase in the understanding of epidermal stem cells and their niche microenvironments, their role for skin regenerative purposes has not been defined.

25.3 Wound healing

Wound healing is a highly evolved process that is thought to quickly protect an organism from further injury and infection by rapid formation of a fibroproliferative scar. However, scar tissue is different from unwounded skin as it lacks appendages such as hair follicles and sebaceous glands and has a tensile strength of less than 70% of normal skin. Although much of the pathophysiology of wound healing has been elucidated, the exact molecular mechanisms underlying impaired repair are still poorly understood (Gurtner et al. 2008). Furthermore, the pathologic overhealing that is unique to humans is equally perplexing.

25.3.1 Stages of wound healing

Mammalian wound healing is traditionally depicted as occurring in three distinct yet overlapping stages: inflammation, proliferation, and remodeling (Epstein et al. 1999). Immediately after injury, inflammation occurs with the formation of a fibrin clot providing hemostasis and recruitment of multiple cell types to the wound (Baum and Arpey 2005). The first circulating immune cells to arrive are the neutrophils, peaking at 1 to 2 days after initial insult. Neutrophils prevent bacterial infection and activate keratinocytes, fibroblasts, and immune cells (Baum and Arpey 2005; Wilgus et al. 2013). As early as hours after wounding, keratinocytes will migrate along temporary fibrin extracellular matrix (ECM). About 2 to 3 days after injury, circulating monocytes arrive at the wound site and differentiate into macrophages. Macrophages function alongside neutrophils to prevent infection and debride necrotic tissue (Sindrilaru and Scharffetter-Kochanek 2013).

As the proliferation stage of wound healing begins, macrophages adopt a profibrotic, anti-inflammatory phenotype. In so doing, they secrete growth factors such as transforming growth factor (TGF)-beta, recruiting fibroblasts into the wound (Baum and Arpey 2005). In turn, endothelial cells are stimulated into the wound by factors such as vascular endothelial growth factor (VEGF) and undergo angiogenesis (Bao et al. 2009). Keratinocytes from the wound edges and hair follicle bulge regions continue to migrate, proliferate, and differentiate to complete re-epithelialization (Raja et al. 2007). Although stem cells are involved in the wound repair process, epidermal appendages are not regenerated (Baum and Arpey 2005). Extracellular matrix components such as fibronectin and cytokines such as TGF-β stimulate fibroblast activation into myofibroblasts (Wynn and Ramalingam 2012; Gabbiani 2003). These activated fibroblasts, named by their contractile ability due to the expression of alpha-smooth muscle actin (SMA), physically pull the edges of the wound together, decreasing the wound size (Gurtner et al. 2008). Although myofibroblasts play this desired role in wound healing, they are also responsible for the secretion of abnormal ECM, composed of increased levels of immature type III collagen, characteristic of scar tissue (Duffield et al. 2013).

About 3 weeks after initial insult, levels of wound collagen peak. This marks the remodeling stage that proceeds for years. There is a decrease in the cellular content of the wound due to apoptosis and migration. As the number of myofibroblasts decreases, so does collagen production, and the ECM slowly develops a more normal structure. Matrix metallo-proteinases (MMPs) and tissue inhibitors of metalloproteinases (TIMPs) work to reorganize collagen fibers into a stronger network as the ratio of type III to type I collagen decreases. The end product in normal physio-logic conditions is an imperfect fibrotic scar lacking the histologic char-acteristics and tensile strength of normal skin (Baum and Arpey 2005).

25.3.2 Pathologic wound healing

Although the fibroproliferative response to cutaneous injury is not equiv-alent to normal skin, it accomplishes the goal of quickly restoring the continuity of the integument. Pathologic wound healing relates to two ends of a spectrum: underhealing and overhealing, with normal physi-ologic wound healing in between. Chronic, nonhealing wounds in the form of diabetic, venous, arterial, and pressure ulcers are the examples of underhealing. Humans are uniquely burdened with pathologic overheal-ing, manifesting as keloids and hypertrophic scars. Although the patho-physiology underlying these aberrant forms of wound repair is not clearly understood, the consequent biomedical burden provides the impetus to advance skin regeneration efforts.

25.3.2.1 Chronic wounds   Chronic wounds are defined as nonheal-ing ulcers that persist for more than 6 weeks (Markova and Mostow 2012). They present in an aging population in the setting of multiple comorbidities, such as diabetes and atherosclerosis. In the United States alone, chronic ulcers account for about $10 billion in annual healthcare costs (Markova and Mostow 2012). The abnormal wound repair process is thought to occur by impaired neovascularization and alterations in the epidermal and dermal structures and proliferative capacity due to aging, diabetes, and vascular disease (Thangarajah et al. 2009). As such, chronic wounds are classified into three groups: diabetic, vascular, and pressure ulcers. Although each class of ulcer results from a markedly different comorbidity, all chronic wounds share the presence of common features. Chronic ulcers are characterized by increased inflammation, hypoxia, and, to some degree, microbial colonization and infection. For the latter reason, a large number of immune cells are recruited into the wound, resulting in increased levels of reactive oxygen species (ROS).

This results in impaired cellular migration and proliferation and damage to the ECM (Eming et al. 2007). Although there have been significant advancements in the understanding of these chronic wounds, effective treatments are lacking. However, the recent surge in novel stem cell–based skin substitutes hold promise for better therapies and advancement of regenerative approaches.

25.3.2.2 Fibroproliferative disorders On the other end of the wound-healing spectrum lie the fibroproliferative disorders, such as keloids and hypertrophic scars. Interestingly, humans are the only species burdened with the tendency to overheal wounds. Thus, there is no good animal model for the study of these aberrant scars, which has impeded research. Both keloids and hypertrophic scars present clinically as raised, firm, and thick scars that are often pruritic and painful (Halim et al. 2012). While hypertrophic scars stay within the borders of the original wound, keloids will extend beyond the borders and will even arise without an inciting event. Both forms of overhealing can be functionally debilitating due to contractures and when occurring across joints. In addition, they maybe cosmetically displeasing and result in psychosocial ramifications, particularly when they occur in the face. Dysfunctional inflammation and a prolonged inflammatory phase of wound healing have been implicated in hypertrophic scarring. On the other hand, keloids are predisposed by both genetic and environmental factors (Al-Attar et al. 2006). It is still unclear whether keloids and hypertrophic scars are varying degrees of the same disorder or two distinct disease entities. What is clear about both keloids and hypertrophic scars is that effective treatments are limited because of an incomplete understanding of the pathophysiology. Skin regeneration gives hope to the treatment of wound-healing pathology on both ends of the spectrum.

25.4 Skin regeneration

Since the early 1900s and the advent of cell culture, scientists have attempted to understand the intricate molecular and cellular processes responsible for scar formation. The goal of this research dates back to ancient times described in the Greek myth of Prometheus: to alter wound healing from an imperfect repair process to complete regeneration (Zielins et al. 2014). However, the idea of regeneration is not a myth, and it is seen not only among certain amphibian species but also in early-gestational mammalian fetal wounds and even in the oral mucosa of adult mammals.

25.4.1 Fetal wound healing

The final product of wound repair depends on the developmental stage and type of tissue. Cutaneous wounds created in the early- to mid-gestational fetus will heal in the absence of scar formation in a process resembling regeneration. Since this early discovery, many have tried to understand the mechanism behind fetal scarless wound regeneration in hopes to recapitulate this phenomenon.

Early- to mid-gestational fetal skin heals without formation of a scar in both human and animal models. Scarless fetal regeneration occurs rapidly and leads to complete regeneration of the dermis and epidermis (Adzick and Longaker 1991; Lorenz et al. 1992; Adzick and Longaker 1992). Gestational age dictates the transition from scarless repair to the postnatal scarring phenotype (Cass et al. 1997). In humans, this occurs at about 24 weeks of gestation. In mice, scarring begins at E18.5 (term = E20.5); however, studies have demonstrated that this varies based on the size of the defect (Colwell et al. 2006).

Although the exact mechanism of fetal regeneration is unknown as of now, research has provided a better understanding of this phenomenon. Fetal wound healing was once attributed to the tissue's environmental conditions. However, studies have demonstrated that the intrauterine environment is neither necessary nor sufficient for scarless healing (Longaker et al. 1994). Instead, the capacity for regeneration was proven intrinsic to the tissue. Scientists have identified distinct ECM components, differential expression of cytokines and growth factors, differences in the inflammatory response, and activation or inactivation of signaling cascades in fetal scarless versus adult scarring wound repair.

Lessons learned from the fetal skin have been difficult to translate to regenerative applications. However, one approach has been particularly promising. Exploiting a notable feature of fetal skin, decreased tension, researchers have elucidated focal adhesion kinase (FAK) as the pathway that ties mechanotransduction with fibrosis. Through extracellular-related kinase (ERK), FAK triggers the secretion of monocyte chemoattractant protein (MCP)-1, which is implicated in a number of human fibrotic diseases (Wynn 2008). By inhibiting or knocking out several components of the inflammatory FAK-ERK-MCP-1 pathway, scar formation was mitigated (Wong et al. 2011). This basic science work has already translated to the development of a tension-offloading device that has been shown in two randomized controlled clinical trials to reduce scarring (Longaker et al. 2014; Lim et al. 2014). This presents the first and only level I evidence for scar reduction. Further research may demonstrate the effectiveness of this new technology for keloids and hypertrophic scars.

25.4.2 Oral mucosal healing

Interestingly, wounds in the oral mucosa heal at an accelerated rate, with little to no scar formation in a process resembling fetal wound regeneration. Even fibroproliferative scars, such as keloids and hypertrophic scars, do not develop inside the oral cavity (Wong et al. 2009). In contrast to the

549

extensive literature on fetal scarless wound healing, relatively few studies have investigated the privileged repair that occurs in the oral mucosa. Mechanisms, both intrinsic and extrinsic to the oral mucosa, such as distinct fibroblast subpopulations, differential growth factors, diminished inflammatory response, reduced angiogenesis, increased capacity for ECM reorganization and remodeling, and saliva in the oral environment have been implicated in oral wound repair. A better understanding of the mechanism underlying oral mucosal wound healing may provide clues to skin regeneration.

25.5 Scarring and fibrosis

Scarring and fibrosis are enormous public health concerns, resulting in excessive morbidity and mortality and immeasurable financial loss in health care. Beyond cutaneous wound healing, fibrosis represents a tremendous biomedical burden. As in the skin, fibrosis occurs throughout the body when tissue is exposed to destructive stimuli. Fibrosis occurs after a myocardial infarction or spinal cord injury. It is also a pathological feature of many chronic inflammatory diseases where it occurs, both in specific organs, such as the kidney, liver, and lung, and outside of organs in the form of abdominal and pleural adhesions (LeBleu et al. 2013). Fortunately, lessons learned from fibroses have paved the way for advancements in skin regeneration.

Advances in stem cell and developmental biology, via lineage tracing and transplantation assays, have identified populations of cells that contribute to scarring and fibrosis. Transient expression of a disintegrin and metalloproteinase 12 (ADAM12) was utilized to distinguish a proinflammatory subset of perivascular cells that are activated in muscle and dermis on acute injury. Either ablating these cells or knocking down ADAM12 was sufficient to decrease scarring and fibrosis (Dulauroy et al. 2012).

Through the use of recent technology, a subpopulation of dermal fibroblasts, originating from Engrailed-1 (En1)-expressing progenitors, has been identified as responsible for the bulk of connective tissue deposition during embryonic development and cutaneous wound healing. CD26, also known as dipeptidyl peptidase-4 (DPP4), was found to enrich for this lineage of fibroblasts. Dipeptidyl peptidase-4 is a cell-surface serine exopeptidase that cleaves X-proline dipeptides from the N-terminus of polypeptides. Inhibition of the enzymatic activity of DPP4 with a small-molecule selective allosteric inhibitor of peptidase activity resulted in decreased scarring. Moreover, the subpopulation of fibroblasts labeled by expression of En1 was shown to be responsible for both acute and chronic progressive forms of fibrosis (Rinkevich et al. 2015). Identification and depletion of these cellular culprits of scarring and fibrosis are promising for regenerative efforts.

25.6 Future directions

Acute and chronic wounds continue to be a socioeconomic burden. As such, new technologies from science and engineering are being incorporated to develop innovative wound regeneration therapies. Although we cannot truly replicate lost or damaged skin, the rapidly evolving field of regenerative medicine promises advancement toward that goal.

Stem cell–based applications comprise the most promising avenue in regenerative efforts. Recent advances in stem cell biology provide scientists with cells that have already been shown to modulate and enhance the wound-healing response. Mesenchymal stem cells (MSCs) have been extensively studied, with mixed results, likely owing to poor survival in the hypoxic wound environment. Adipose-derived stem cells (ASCs) are promising as a stem cell source that can be harvested through liposuction, a less invasive means. Recently, the ability to form induced pluripotent stem cells (iPSCs) is especially exciting due to their increased capacity for differentiation into cells derived from all three germ layers.

Currently, cell-based applications are plagued by inefficiencies in delivery and risk of malignant transformation. However, novel biomimetic scaffolds and methods of stem cell harvest and derivation give promise to overcoming such challenges. As our knowledge of skin development, wound healing, and scarring and fibrosis continues to grow, new emerging technologies will allow for the development of safe, specialized methods for skin regeneration.

References

Adzick NS and MT Longaker. 1991. Animal models for the study of fetal tissue Repair. *The Journal of Surgical Research* 51 (3): 216–222.

Adzick NS and MT Longaker. 1992. Scarless fetal healing. Therapeutic implications. *Annals of Surgery* 215 (1): 3–7.

Al-Attar A, S Mess, JM Thomassen, CL Kauffman, and SP Davison. 2006. Keloid pathogenesis and treatment. *Plastic and Reconstructive Surgery* 117 (1): 286–300.

Bao P, A Kodra, M Tomic-Canic, MS Golinko, HP Ehrlich, and H Brem. 2009. The role of vascular endothelial growth factor in wound healing. *Journal of Surgical Research* 153 (2): 347–358. doi:10.1016/j.jss.2008.04.023.

Baum CL and CJ Arpey. 2005. Normal cutaneous wound healing: Clinical correlation with cellular and molecular events. *Dermatologic Surgery* 31 (6): 674–686; discussion 686.

Blanpain C and E Fuchs. 2009. Epidermal homeostasis: A balancing act of stem cells in the skin. *Nature Reviews Molecular Cell Biology* 10 (3): 207–217. doi:10.1038/nrm2636.

Cass DL, KM Bullard, KG Sylvester, EY Yang, MT Longaker, and NS Adzick. 1997. Wound size and gestational age modulate scar formation in fetal wound repair. *Journal of Pediatric Surgery* 32 (3): 411–415.

Colwell AS, TM Krummel, MT Longaker, and HP Lorenz. 2006. An in vivo mouse excisional wound model of scarless healing. *Plastic and Reconstructive Surgery* 117 (7): 2292–2296. doi:10.1097/01.prs.0000219340.47232.eb.

Cotsarelis G, TT Sun, and RM Lavker. 1990. Label-retaining cells reside in the bulge area of pilosebaceous unit: Implications for follicular stem cells, hair cycle, and skin carcinogenesis. *Cell* 61 (7): 1329–1337.

Duffield JS, M Lupher, VJ Thannickal, and TA Wynn. 2013. Host responses in tissue repair and fibrosis. *Annual Review of Pathology: Mechanisms of Disease* 8 (1): 241–276. doi:10.1146/annurev-pathol-020712-163930.

Dulauroy S, SE Di Carlo, F Langa, G Eberl, and L Peduto. 2012. Lineage tracing and genetic ablation of ADAM12(+) perivascular cells identify a major source of profibrotic cells during acute tissue injury. *Nature Medicine* 18 (8): 1262–1270. doi:10.1038/nm.2848.

Eming SA, T Krieg, and JM Davidson. 2007. Inflammation in wound repair: Molecular and cellular mechanisms. *The Journal of Investigative Dermatology* 127 (3): 514–525. doi:10.1038/sj.jid.5700701.

Epstein FH, AJ Singer, and RAF Clark. 1999. Cutaneous wound healing. *New England Journal of Medicine* 341 (10): 738–746. doi:10.1056/NEJM199909023411006.

Fuchs E. 2007. Scratching the surface of skin development. *Nature* 445 (7130): 834–842. doi:10.1038/nature05659.

Gabbiani G. 2003. The myofibroblast in wound healing and fibrocontractive diseases. *The Journal of Pathology* 200 (4): 500–503. doi:10.1002/path.1427.

Ghazizadeh S and LB Taichman. 2001. Multiple classes of stem cells in cutaneous epithelium: A lineage analysis of adult mouse skin. *The EMBO Journal* 20 (6): 1215–1222. doi:10.1093/emboj/20.6.1215.

Gurtner GC, S Werner, Y Barrandon, and MT Longaker. 2008. Wound repair and regeneration. *Nature* 453 (7193): 314–321. doi:10.1038/nature07039.

Halim AS, A Emami, I Salahshourifar, and TP Kannan. 2012. Keloid scarring: Understanding the genetic basis, advances, and prospects. *Archives of Plastic Surgery* 39 (3): 184–189. doi:10.5999/aps.2012.39.3.184.

LeBleu VS, G Taduri, J O'Connell, Y Teng, VG Cooke, C Woda, H Sugimoto, and R Kalluri. 2013. Origin and function of myofibroblasts in kidney fibrosis. *Nature Medicine* 19 (8): 1047–1053. doi:10.1038/nm.3218.

Levy V, C Lindon, Y Zheng, BD Harfe, and BA Morgan. 2007. Epidermal stem cells arise from the hair follicle after wounding. *The FASEB Journal* 21 (7): 1358–1366. doi:10.1096/fj.06-6926com.

Lim AF, J Weintraub, EN Kaplan, M Januszyk, C Cowley, P McLaughlin, B Beasley, GC Gurtner, and MT Longaker. 2014. The embrace device significantly decreases scarring following scar revision surgery in a randomized controlled trial. *Plastic and Reconstructive Surgery* 133 (2): 398–405. doi:10.1097/01.prs.0000436526.64046.d0.

Longaker MT, DJ Whitby, MW Ferguson, HP Lorenz, MR Harrison, and NS Adzick. 1994. Adult skin wounds in the fetal environment heal with scar formation. *Annals of Surgery* 219 (1): 65–72.

Longaker MT, RJ Rohrich, L Greenberg, H Furnas, R Wald, V Bansal, H Seify et al. 2014. A randomized controlled trial of the embrace advanced scar therapy device to reduce incisional scar formation. *Plastic and Reconstructive Surgery* 134 (3): 536–546. doi:10.1097/PRS.0000000000000417.

Lorenz HP, MT Longaker, LA Perkocha, RW Jennings, MR Harrison, and NS Adzick. 1992. Scarless wound repair: A human fetal skin model. *Development (Cambridge, England)* 114 (1): 253–259.

Markova A and EN Mostow. 2012. US skin disease assessment: Ulcer and wound care. *Dermatologic Clinics* 30 (1): 107–111. doi:10.1016/j.det.2011.08.005.

Nijhof JGW, KM Braun, A Giangreco, C van Pelt, H Kawamoto, RL Boyd, R Willemze et al. 2006. The cell-surface marker MTS24 identifies a novel population of follicular keratinocytes with characteristics of progenitor cells. *Development (Cambridge, England)* 133 (15): 3027–3037. doi:10.1242/dev.02443.

Oshima H, A Rochat, C Kedzia, K Kobayashi, and Y Barrandon. 2001. Morphogenesis and renewal of hair follicles from adult multipotent stem cells. *Cell* 104 (2): 233–245.

Potten CS. 1974. The epidermal proliferative unit: The possible role of the central basal cell. *Cell and Tissue Kinetics* 7 (1): 77–88.

Potten CS and RJ Morris. 1988. Epithelial stem cells in vivo. *Journal of Cell Science* (Suppl. 10): 45–62.

Raja KS, MS Garcia, and RR Isseroff. 2007. Wound re-epithelialization: Modulating keratinocyte migration in wound healing. *Frontiers in Bioscience: A Journal and Virtual Library* 12 (May): 2849–2868.

Rinkevich Y, GG Walmsley, MS Hu, ZN Maan, AM Newman, M Drukker, M Januszyk et al. 2015. Skin fibrosis. Identification and isolation of a dermal lineage with intrinsic fibrogenic potential. *Science (New York, N.Y.)* 348 (6232): aaa2151. doi:10.1126/science.aaa2151.

Sindrilaru A and K Scharffetter-Kochanek. 2013. Disclosure of the culprits: Macrophages-versatile regulators of wound healing. *Advances in Wound Care* 2 (7): 357–368. doi:10.1089/wound.2012.0407.

Taylor G, MS Lehrer, PJ Jensen, TT Sun, and RM Lavker. 2000. Involvement of follicular stem cells in forming not only the follicle but also the epidermis. *Cell* 102 (4): 451–461.

Thangarajah H, D Yao, EI Chang, Y Shi, L Jazayeri, IN Vial, RD Galiano et al. 2009. The molecular basis for impaired hypoxia-induced VEGF expression in diabetic tissues. *Proceedings of the National Academy of Sciences* 106 (32): 13505–13510. doi:10.1073/pnas.0906670106.

Wilgus TA, S Roy, and JC McDaniel. 2013. Neutrophils and wound repair: Positive actions and negative reactions. *Advances in Wound Care* 2 (7): 379–388. doi:10.1089/wound.2012.0383.

Wong JW, C Gallant-Behm, C Wiebe, K Mak, DA Hart, H Larjava, and L Häkkinen. 2009. Wound healing in oral mucosa results in reduced scar formation as compared with skin: Evidence from the red duroc pig model and humans. *Wound Repair and Regeneration* 17 (5): 717–729. doi:10.1111/j.1524-475X.2009.00531.x.

Wong VW, KC Rustad, S Akaishi, M Sorkin, JP Glotzbach, M Januszyk, ER Nelson et al. 2011. Focal adhesion kinase links mechanical force to skin fibrosis via inflammatory signaling. *Nature Medicine* 18 (1): 148–152. doi:10.1038/nm.2574.

Wynn TA. 2008. Cellular and molecular mechanisms of fibrosis. *The Journal of Pathology* 214 (2): 199–210. doi:10.1002/path.2277.

Wynn TA and TR Ramalingam. 2012. Mechanisms of fibrosis: Therapeutic translation for fibrotic disease. *Nature Medicine* 18 (7): 1028–1040. doi:10.1038/nm.2807.

Zielins ER, DA Atashroo, ZN Maan, D Duscher, GG Walmsley, M Hu, K Senarath-Yapa et al. 2014. Wound healing: An update. *Regenerative Medicine* 9 (6): 817–830. doi:10.2217/rme.14.54.

Chapter 26 Programming cells to build tissues with synthetic biology

A new pathway toward engineering development and regeneration

Leonardo Morsut

Contents

Key concepts

a. Synthetic biology tools and approaches allow for the control of complex cellular behaviors by engineering genetic programs in cells.
b. Tissue features are dictated by complex networks of cellular behaviors.
c. Cellular behaviors fall into two main categories that cells are executing concurrently: information processing-related and physical/material-related.
d. Tissues are built through development: a self-organized, timed orchestration of cellular behaviors that increase morphological and functional complexity.
e. Synthetic biology tools to control morphogenetic behaviors could be used to design user-defined tissue development and regeneration trajectories.
f. Increased control and complexity in tissue behaviors are needed for the translation from therapeutic promise to medical practice of tissue regeneration.

26.1 Introduction

Building tissues is one of the masterpieces of the building capacity of nature: starting from undifferentiated and amorphous (multi)-cellular structures, by self-organization, embryos grow tissues of sophisticated shapes and coherent functions. One of the aims of regenerative medicine can be framed as matching this tissue-building capacity for therapeutic purposes. This problem is being tackled with a multidisciplinary approach, with inputs from engineering, material science, computational biology, and developmental biology. More recently, another hybrid field, synthetic biology, is starting to contribute, by designing *smart* cells for enhanced control and complexity of tissue features. I will discuss here how the application of the constructive approach of synthetic biology to developmental systems can augment the current practices of tissue engineering and regeneration. I call this new approach *tissue development engineering*, as it consists of genetically programming mammalian cells to build tissues through designed developmental trajectory. By controlling the powerful building machine of development, it has the potential to generate complex tissues with novel structures and function, to interface with biomedical devices, and to direct regeneration instead of degeneration, among others. In addition, tissue development engineering has the fascinating potential to complement classical developmental biology inquiries with an *understanding-by-building* side, on the quest to understand tissue biology principles, one of the greatest challenges of modern biomedicine. The increased control and understanding of tissue generation principles are the keys to bridge the gap between therapeutic promise and medical practice in regenerative medicine.

This chapter is a combination of a discussion on molecular, cellular, and abstract concepts of tissue development and a discussion on ways in which synthetic biology tools can be used to control or replicate tissue behaviors. I will focus on some key areas and developments that will empower synthetic biology approaches for regenerative medicine in the coming years, especially focusing on what this approach can do for complex tissue regeneration. I first provide context comparing the tissue development engineering with other approaches for building tissues as a motivation for the discipline. After that, I introduce the synthetic biology framework, with selected examples. With this in mind, I then describe molecular, cellular, and multicellular behaviors underlying tissue development and homeostasis and a multiscale abstraction framework for how to connect them together. This chapter culminates with a discussion on principles of tissue development engineering and examples of specific strategies for controlling tissue formation from the outside with *therapeutic cells* and from the inside by engineering regeneration programs from the bottom-up.

26.2 Context and motivation for *tissue development engineering*

Understanding and controlling tissue biology is one of the great biomedical areas that twenty-first-century science hopes to address. The promises are to increase our scientific understanding and to produce therapeutic approaches to improve health (National Research Council [U.S.] Committee on a New Biology for the 21st Century, 2009; Reichert et al. 2010).

Several disciplines are contributing toward this endeavor. Developmental biology is studying tissues in model organism; stem-cell–based organoid and *in vitro* differentiation programs are producing tissues *in vitro* with stem cells by identifying chemical and mechanical protocols for differentiation; computational simulations are modeling these phenomena to understand them better and uncover the underlying principles; and tissue engineering is generating therapeutic units to increase regeneration in patients or provide organ substitutes. Each approach has trade-offs, and in an effort to combine their strengths, these practices have recently started to overlap and synergize (Matthys et al. 2016; Yin et al. 2016; Wu and Belmonte 2016). One example of common trade-off between control and complexity is: with tissue engineering, the control in the tissue structure can be very high, but complexity is not always comparable to endogenous tissues. In organoids and differentiation, cellular complexity can be very high, but control is quite limited. In the remainder of this paragraph, I discuss what synthetic biology can specifically bring to the table of tissue biology research.

One of the reasons why synthetic biology approaches are starting to converge (Davies and Cachat 2016) is to increase the control of cell behaviors. Synthetic biology tools, in fact, offer increased control of cellular behaviors (see later), thus providing an additional layer of control where it is currently lacking. In stem-cell–based organoids, for example, very complex tissue structures are formed, starting from stem cells by varying the chemical and mechanical environment of cells (Lancaster and Knoblich 2014). Having the possibility of enhancing the control of the cellular component of these *in vitro* tissues is poised to improve their maturity and homogeneity.

The specific area of convergence that I introduce in this chapter is between synthetic biology and developmental biology (Teague et al. 2016; Davies 2008). The idea is to use synthetic biology tools and approaches to genetically program cells to build tissues by design, encoding the developmental trajectory at the level of an engineered genetic program.

Tissue development engineering can help where increased complexity is needed, since it can tap into the endogenous complexity-building capacity of multicellular systems. Classical tissue engineering uses a combination of cells, scaffold materials, and soluble factors to generate tissue for transplantation and/or stimulate endogenous regeneration (Shea et al. 2014; Mao and Mooney 2015; Bajaj et al. 2014; Langer and Vacanti 2016; Guilak et al. 2014; Tibbitt et al. 2015). It has been tremendously successful for some applications and is poised to expand its scope in the coming decades (Oshima and Tsuji 2015; Peloso et al. 2015). With the synthetic biology tools, we could be able to provide cells with additional programs that will expand the scope of what is possible endogenously. Frontiers in the field, such as growing bigger structures *in vitro*, will require the ability to grow more sophisticated assemblies of cells, for example, with vascularization (Rouwkema and Khademhosseini 2016). The capacity of scaling growth is typical of development; hence, the tissue development engineering approach could allow building of larger and more complex tissue structures. In addition, with enhanced cellular computation, one could build smart tissues with user-defined functional properties. This approach has moved its early steps with a number of studies that use engineered cells to alleviate degenerative diseases (e.g., Ho et al. 2015; Jabbarzadeh et al. 2008; Glass et al. 2014; Diekman et al. 2015). The next step will be to engineer the cells to perform more complex tissue-level therapeutic actions, for example, recognize danger and actuate a self-repair program. Cells are especially well suited to perform these types of molecular sensing-and-response programs, as is seen in self-regenerating human tissues and in other animals. The convergence of technological advancements such as tissue 3D printing (Murphy and Atala 2014) will offer a possibility to integrate the best of the various approaches.

I am convinced that the tissue development engineering approach will also provide a complementary input to increase our knowledge of the principles that govern tissue development. Modern developmental biology is striving to put together molecular and cellular insights to comprehend emergence of tissue behaviors in specific systems (e.g., Tsiairis and Aulehla 2016; Durdu et al. 2014; Kuo and Krasnow 2015; Uygur et al. 2016). When programming cells to build tissues, we will necessarily be faced with problems regarding the principles of this type of cell programming as well. This *understanding-by-building* approach is one of the strengths of synthetic biology (Elowitz and Lim 2010).

I envision a research cycle for tissue development engineering that goes through iterations of the design/build/test/learn cycle (Figure 26.1). The cycle starts with the design of the genetic and cellular program for the developmental trajectory of the multicellular system. Several disciplines will converge in the phase of design: developmental biology, synthetic biology, tissue engineering, and computational biology. The design will be followed by implementation in cellular systems; these engineered cellular systems will be tested, and the result of these experiments will be used to learn more about the systems and the design principles. The cycle will generate two main outcomes: feedback on learning and products. Learning will generate more knowledge that can then be applied in the design phase. The products will be therapeutic units or interventions: the two that are most relevant for this discussion are smart tissues with enhanced functionalities and therapeutic cells for regeneration in case of disease or injury. In the next paragraph, I will share some examples and principles of synthetic biology practices and some lessons for applying them to tissue development engineering.

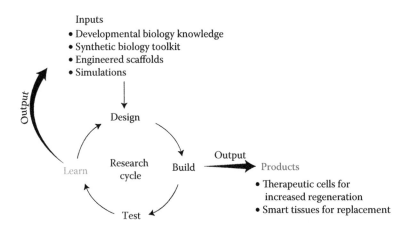

Figure 26.1 Overarching idea of tissue development engineering. The research is centered on a design/build/test/learn cycle. The inputs to the cycle come from developmental biology knowledge, synthetic biology toolkit, engineered scaffold, and computer simulations informing the design phase. There are two outputs (colored text) that justify the research cycles: products (as therapeutic cells for augmenting regeneration and smart tissues for replacement) and increased knowledge and tools that feed back to the inputs (in the form of papers, patents, and simulation environments).

26.3 Synthetic biology: Build with biology by designing molecular and cellular programs

The basic idea of synthetic biology is that we can modify and recombine genetic elements inside living cells to obtain useful behaviors. This approach has been used to make cells do things that are of interest to us, from producing drugs to biofuels to therapeutic cells. The novelty added to the classic idea of doing useful things with biological material is in the genetic engineering component; the molecular biology revolution gave us molecular tools to engineer the genetic makeup of cells. Thanks to this technical advancement, as soon as our understanding of a biological phenomenon is mature enough that we have an *operationable understanding* of it, synthetic biology approaches can be invoked. One important point is that since it involves engineering of the DNA of cells, especially for human health-related purposes, synthetic biology has always led to active discussions about ethics. The flexibility of synthetic biology approaches allows for creative thinking around ways of self-correcting, which can inform some of the practices, and discussions are always ongoing (Church et al. 2014; Boeke et al. 2016). For comprehensive reviews of the history and the current state of the synthetic biology field, the reader is referred to excellent works on the topic and references therein (Lienert et al. 2014; Purnick and Weiss 2009; Lim 2010; Cameron et al. 2014; Schwarz and Leonard 2016; Way et al. 2014; Khalil and Collins 2010). In the remaining of this first section, I will highlight a few examples and concepts of synthetic biology practice at different scales of organization that are relevant for the application to tissue development engineering.

At the molecular level, biology has evolved to catalyze chemical reactions through enzymes; synthetic biologists were able to tame this capacity and design enzymes that would make useful reactions instead of the original ones (Figure 26.2a). The understanding of the modularity of the functional structure of enzymes (in active site, catalytic domain, and substrate-specificity domain) on the one end and the availability of procedures of targeted or unbiased modification of protein sequences on the other end allowed this approach to be successful (Arnold 2015). Similar results are achieved with engineering of signaling enzymes to rewire information-processing networks inside cells and obtain new dynamical behaviors (Kiel et al. 2010; Lim 2010).

In general, protein engineering has shown the potential of using very sophisticated machines (proteins) as starting material to design new ones with novel functions; it is worth noting that this flexibility is what is thought to have underlain the acquisition of novelty in protein function during evolution (Lim and Pawson 2010; Di Roberto and Peisajovich 2013).

The synthetic biology framework of building with biology has been applied to more complex cellular behaviors as well (Figure 26.2b and c). Cells are very powerful machines themselves; they exert their functions

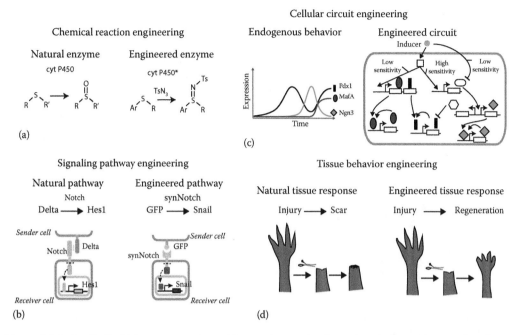

Figure 26.2 Synthetic biology approaches at increasing scales from chemical reaction, to signaling pathways, to cellular circuits, and eventually to tissue dynamics. (a) By altering the primary sequence of cytP450, instead of catalyzing the endogenous reaction on the left, the engineered version (cytP450*) catalyzes the reaction on the right. (Modified from Arnold, F. H., *Quarterly Reviews of Biophysics*, 48 (4), 404–410, 2015.) (b) Endogenous Notch signaling is a cell–cell contact signaling pathway that is activated by Delta cognate ligands on neighboring cells. An activated receiver cell induces Notch target genes such as *Hes1* (left). The synthetic Notch pathway can be activated by an orthogonal signal (in the example, an extracellular membrane-tethered green fluorescent protein [GFP]), and it results in activation of arbitrary target genes (in the example, the master transcription factor for epithelial–mesenchymal transition, Snail, right) (Modified from Morsut, L. et al., *Cell*, 164, 780–791, 2016.) (c) Left, complex temporal expression of master transcription factors is observed endogenously during differentiation of pancreatic precursor cells into glucose-sensitive insulin-secreting beta cells. The engineered circuit on the right was designed with a combination of orthogonal and endogenous parts to recapitulate the endogenous dynamic in a drug-controlled manner. (Based on Saxena, P. et al., *Nat. Commun.*, 7, 11247, 2016.) (d) Left, an endogenous tissue trajectory of scar formation after limb amputation; right, the endogenous tissue trajectory is re-routed to regeneration, thanks to tissue development engineering approaches.

by sensing their environment with receptors and activating responses intracellularly with their molecular machines. This process happens through cascades known as signaling pathways. Key to engineering cell behaviors then is the engineering of signaling pathways. To do that, the steps in the pathway are abstracted to extracellular recognition of a ligand, signal transduction, activation of transcription factors, and induction of target genes (Figures 26.3b). The recognition that there is modularity not only at the protein level but also at the pathway level allows one to start engineering the various parts separately and then combining them together (Gordley et al. 2016). One way to do this is

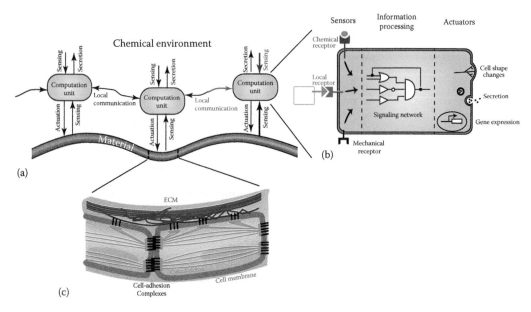

Figure 26.3 Tissues as robotic materials. (a) The tissue is made up of units of computation and material. Multiple computational units exist in a shared chemical environment and connect to form a continuous material. Computational units sense their neighbors (local communication), their mechanical environment, and their chemical environment (sensing) and actuate mechanically (actuation), chemically (secretion), or intracellulary, with changes in the information-processing circuits (Adapted from McEvoy, M. A. and N. Correll, *Science (New York, N.Y.)*, 347, 1261689, 2015.) (b) Expanded view of a single computation unit (cell). Each cell has sensors (receptors) that pass the extracellular information to the intracellular information–processing compartment (signaling networks), whose outputs can be mechanical, chemical, or genetic (Adapted from Lim, W. A., *Nat. Rev. Mol. Cell Biol.*, 11, 393–403, 2010.) (c) Cells also contribute to the material component of tissues. The distributed molecular skeleton (tissue–skeleton) is made of cytoskeleton + cell–cell and cell–matrix adhesions + membrane + ECM (Adapted from Jefferson, J. J. et al., *Nat. Rev. Mol. Cell Biol.*, 5, 542–553, 2004); see text for molecular details.

generating a pathway that is orthogonal, insulated from endogenous signaling. One simple yet powerful example is the implementation of a synthetic cell–cell contact-dependent pathway, the synNotch pathway (Figure 26.2b). It consists of a chimeric receptor and a DNA sequence containing target gene(s) downstream of a receptor-regulated promoter. Cells can be engineered with these two elements to respond to user-defined surface-bound ligands and activate user-defined genetic responses (Morsut et al. 2016). A few other examples of synthetic pathways insulated from endogenous signaling have been described (Payne and You 2013; Hennig et al. 2015; Schwarz and Leonard 2016). An alternative to engineer pathways that are orthogonal to endogenous signaling is to rewire endogenous pathways to new outputs or inputs (Schukur and Fussenegger 2016). The ultimate goal is to have a toolkit of modular pathway parts with different features (soluble vs. membrane or scaffold-bound input, different dynamics vs. number of genes in the output, and synthetic vs. rewired) that the researchers can choose from, according to the cellular behaviors that they want to achieve. The technologies of synthetic pathways are recent but have already been shown to be very powerful. Using engineered pathways, researchers are now able to implement complex cell behaviors such as homeostatic feedback and band-pass filters (Schwarz and Leonard 2016; MacDonald and Deans 2016) and perform therapeutic interventions. In one famous example, T-cells are being engineered to detect tumor antigens and kill the tumor cells (Miller and Sadelain 2015). In general, cells augmented with synthetic biology–based gene circuits are uniquely suited for the treatment of diseases with complex dynamics or with multiple steps and multiple players involved, since they can perform several actions, talk to the endogenous components (cells, matrix, and so on), and, for example, autonomously couple detection of disease biomarkers with production of therapeutic proteins (Schukur et al. 2015).

Complex endogenous cellular behaviors result from the crosstalk, convergence, and wiring of multiple intracellular pathways, to originate complex dynamics of gene expression. These complex dynamics are widespread during morphogenesis, for example, during cell differentiation. The endogenous dynamics are challenging to control, only tweaking the inputs of the pathways, with growth factors, for example. Synthetic biology approaches are being developed to replicate and control these dynamics. The aim is to precisely control the temporal dynamics of key cellular behaviors, for example, gene transcription. In one example (Figure 26.2c), the Fussenegger group has implemented a sophisticated synthetic genetic circuit to drive transcription factor expression dynamics that match the endogenous dynamics. Using the control of a drug inducer, they have been able to program human-induced pluripotent stem cells–derived pancreatic progenitor cells into glucose-sensitive insulin-secreting beta-like cells (Saxena et al. 2016).

One frontier of synthetic biology is moving toward multicellular systems and design of multicellular behaviors. The idea is that if by rewiring protein–protein interaction networks we can obtain new cell behaviors, then by rewiring cell–cell and cell–matrix interaction networks we can obtain new multicellular behaviors. Initial successes in the multicellular synthetic biology field came from the engineering of prokaryotic cells (Din et al. 2016; Teague et al. 2016). In these works, the researchers engineered not only circuits inside cells but also cell–cell communications, in order to obtain spatial and/or temporal self-organization. The difference with engineering single cells to detect and perform actions is that now we take advantage of multicellular phenomena, such as quorum sensing and long-distance signaling (Basu et al. 2005). Expanding this approach to mammalian cells is the natural extension of these works. By engineering cell circuits and cell–cell communications in multicellular mammalian systems, the promise is to control their multicellular behaviors. One facet of this is the new discipline of tissue development

563

engineering, the focus of this chapter. It consists of learning and implementing complex programs to design multicellular developmental trajectories, for example, rewiring a regeneration response instead of the endogenous degeneration response in degenerative diseases and injuries (Figure 26.2d).

In the effort of applying synthetic biology approaches to tissue biology, I think it is worth asking what we can learn from past synthetic biology successes. Specifically for tissue development engineering, the approach requires: (1) an operationable understanding of tissue formation and homeostasis; (2) computational tools to model and predict the complex system behavior; (3) molecular tools for controlling the input/out relationships of the constituents, i.e. the cells; (4) genome engineering tools to make all this implementable in cells; (5) ethical guidelines. I will talk about the first three points in the next sections; the last two points are beyond the scope of this chapter, and the reader is referred to recent discussions (Kim and Kim 2014; Boeke et al. 2016).

One of the key lessons from past synthetic biology practice is the importance of abstraction, or in other words, breaking down of biological systems into component parts and collections of parts into devices (Purnick and Weiss 2009). On the one hand, the abstraction helps formalizing the current knowledge and exposing gaps, and on the other hand, it offers a more tractable framework for engineering. Once there is a sufficiently developed abstraction of a system, synthetic biology approaches show that you can expand the possible behaviors and increase the control over them by exploiting the natural modularity of the system. In the next paragraph, I start by introducing an abstraction for tissue behaviors description rooted in cellular and molecular behaviors.

26.4 Tissue biology concepts

26.4.1 Tissues as robotic materials

Cells of metazoans are programmed to build tissues and complex organs through development, which means that there is the progressive acquisition of target morphology starting from a single cell. The program is written in the cells' DNA and is the outcome of evolutionary selection of trials and errors. We see the awe-inspiring execution of that program when we look at the coordinated dance of different cell types and multicellular structures during embryogenesis. We want to find an abstract and operationable answer to the question: how do cells make tissues? How does the program work? A lot of groundwork has been done to identify relevant molecular components and cellular behaviors during development and in mature tissues (Salazar-Ciudad et al. 2003). In parallel from computer science, it is as if we started to be familiar with the low-level machine code (the function of molecular and cellular components) and now we are on the quest to understand the higher-level programs (multicellular phenomena such as tissue formation), which are written in that code. It is starting to be possible to look for abstraction frameworks (Okabe and Medzhitov 2016; Vila et al. 2016). To be of help to the engineering efforts, these frameworks will need to be operationable, which means that they need to be based on principles that we can engineer and, at least in part, control.

I present here an abstraction of tissue function based on the realization that, in a very basic sense, tissues are made by cells that (1) communicate among themselves and with their extracellular environment and (2) execute mechanical and physiological functions. The magic of development (tissue morphogenesis) is, in fact, that the communication and the effectors are coupled together. A tissue can, in fact, be described in abstract as a robotic material (Mustard and Levin 2014; McEvoy and Correll 2015): a material that contains computation and effector units (Figure 26.3a). In this sense, one could say that tissues are the ultimate soft robots (Feinberg 2015).

In tissues, the computational units are the cells (Figure 26.3b). The cells process incoming information from the neighbors and their mechanical and their chemical environment through their receptor systems (sensing). The cells process this information with their intracellular molecular networks; eventually, the cells respond (actuation/secretion) in the form of (i) altered information exchange with neighbors, (ii) altered info processing inside, (iii) material actuation (e.g., change of cell shape and change in extracellular matrix [ECM] stiffness) or (iv) chemical actuation (e.g., secretion of enzymes and degradation of metabolites). The response at the level of the tissue can be more or less complex: for example, it can be the coordinated secretion of insulin in response to sensing of a certain glucose concentration or the massive coordinated changes in tissue architecture during metamorphosis in response to hormones.

The whole tissue is also a coherent material (Figure 26.3c) that is made up by the cells and the ECM that can be in the solid form of bone, the liquid form of blood, or the elastic form of muscle and epithelia. Single-cell shape is controlled intracellularly by the cellular skeleton (the cytoskeleton) and cell membrane and extracellularly by the ECM. Multiple cells are connected together by cell–cell adhesions, and cells are connected to the ECM to form a distributed structural unit that has features of a continuous material (see later for molecular details).

Cells in tissues then have a dual nature of information processors and material. The explicit distinction in the scheme of Figure 26.3 into a material component (C) and an information-processing component (B) is an abstraction that can be helpful in order to guide the engineering; however, *in vivo*, the distinction can be more blurred.

At any point in time, the computation units (cells) perform computation, see certain inputs, and make some responses as outputs. The outputs can change the structure and the function of the tissue itself. The output of the computation, as noted earlier, can be the change of the computation itself: either by changing the information processing by expressing different information-processing genes (e.g., genes of intracellular pathways) or by changing parameters that affect the information processing (e.g., neighbor disposition and material properties). Over time, for example, during a developmental trajectory to build a tissue, all these programs run concurrently, and the shape of the tissue is changed until the targeted morphology is generated.

It is evident that it is a very complex dynamic system. Rarely are all the components and the detailed programs known for a natural system (examples where there are efforts to produce comprehensive understanding include Wennekamp et al. 2013; Ben Tabou de Leon and Davidson 2009). This framework is an effort to focus on the common basic routines that cells run while building tissues.

The dynamics of the whole system can be described in terms of a complex physical system in the space of possible configurations, a morphospace defined by principal components of shape and complexity (Vila et al. 2016). Development would be the movement from point A to point B; regeneration would allow returning to point B after a perturbation, as in a stable equilibrium fixed point. There will be points of equilibrium that can work as attractors of the developmental dynamics, and each tissue would have a reachable space of configurations, depending on the program of the constituent cells.

The framework of tissues as robotic materials is introduced here to describe isolated transitions (e.g., early tissue development), when cells are uniform and there is not so much specialization, and when they have to self-organize to do morphogenesis. Later, when there are different parts of a tissue that cooperate with one another (e.g., vessel and nerves), the material is more composite and can be considered to be made of multiple units. With subcompartments connected via selected channels, not all cells necessarily perform all the actions, and there can be division of labor. Some compartments may do mainly material effector functions, while other compartments are specialized in information processing (Okabe and Medzhitov 2016). This concept relates to how the *external environment* is considered. Morphogenesis, in fact, can be templated or pre-patterned by external asymmetric inputs or can be autonomous and rely on self-organization to break the symmetry. Endogenously, pattern formation can be driven by autonomous force in same cases (e.g., Shaya and Sprinzak 2011), whereas other multicellular behaviors are definitively driven by an external stimuli (e.g., polyphenism in insects, which is the phenomenon whereby a single genome expresses different phenotypes in response to environmental cues; Corona et al. 2016). However, the boundary between these two situations is not always well defined, and an extensive discussion is beyond the scope of this chapter. The distinction is introduced here, since for engineering, these would be two distinct modes of implementation: one autonomous (all in one) and the other distributed with different cells having different roles (see later, division of labor in controlling morphogenesis).

In general, the engineering challenge is how to intervene and interface with this highly constrained system to achieve control of the behavior to build tissues by design. Pure tissue engineering intervenes at the level of the extracellular matrix with scaffolds inputs and by putting specific cells close by (e.g., with 3D printing techniques). Organoids and *in vitro* differentiation programs are based on changing chemical and mechanical environment and relying on endogenous cascades of differentiation and self-organization. What the synthetic biology approach can do here is engineering the computational unit behavior—the cell program.

Intuitively, it is quite hard to predict the behavior of such a complex system; that is why, we will need computational simulations to guide the intervention. What is the program that would generate the structure that I want? In the next section, I discuss the principles for implementing this model in simulation frameworks.

26.4.2 Simulations

The view of tissues as robotic materials and of tissue development as a trajectory of a robotic material in the morphospace should inspire the next generation of computer simulations. Specifically, simulation platforms that incorporate both the common cellular behavior of information processing and material actuation (Salazar-Ciudad et al. 2003) will be particularly useful. I will review here some features of the different models developed so far. Historically, the modeling paradigms can be distinguished according to whether they focus on the mechanical components, on the computational components, or on the crosstalk between the two. For extensive reviews of the various modeling paradigms, see Urdy (2012) and Morelli et al. (2012).

Treating cells as mechanical objects and coupling them in various ways allow one to study what happens when some subsets of cells start to change their shape or their physical properties (e.g., Fletcher et al. 2014; Okuda et al. 2012). For example, the consequence of loosening the strength of adhesion between cells or the tension of the cytoskeleton can be considered (Osterfield et al. 2013). These models do not necessarily ask which molecular network is producing the observed shape change and are abstract at the level of the mechanics. They are useful to map cellular mechanical actuation to tissue-level shape changes (or morphogenesis).

Another line of research in computational biology deals with the information-processing ability of collection of cells and how the communication between them can lead to stable or dynamic patterning (Rué and Garcia-Ojalvo 2013). Here too, often, the molecular details are abstracted, and the communication is not modeled with the chemical binding of the specific ligand with the receptor protein but by abstracting key feature of the signaling: the dose/response, the distance of the signal, and so on. For example, Simakov and Pismen (2013) explored the potential of contact-dependent signaling to pattern static hexagonal grids of epithelial cells. These models can explore the potential of pattern generation with various modes and networks of cell–cell and cell–ECM communication.

Some early models had already identified the relevance of the combination of cell signaling and morphogenetic movements (Cummings 1990), and recently, one quite comprehensive framework has been developed (Marin-Riera et al. 2016). In these works, both the mechanical properties and the communication properties are modeled, and the outcome of the computation inside cells can change the mechanical properties of the cells or the ECM. The next step would be to add dependence of cell behaviors on mechanical properties (e.g., signaling and differentiation [see later, cell–ECM communication]) to model all the behaviors influencing tissue dynamics (Figure 26.3).

These computational frameworks have been used to model endogenous tissue development. I suggest that these models, and in particular complete combination models, are needed to provide a more deterministic foundation of the tissue development engineering discipline. On one side, simulations will allow investigators to study the problem in theoretical terms, which is a fascinating self-organization problem of reachable space of the complex physical/computational system. The simulation platforms also can be used to see what the controllable space is and what kinds of control would endow the researcher with design capacity. In this sense, it is very similar to a pure robotic material problem. In fact, these models can come in at the level of design (Figure 26.1), where they can help screening a large set of programs to ask questions, such as: what is the combination of mechanical/functional changes and communication that makes an epithelial sheet fold into a sphere? With the simulations, one can search for this program and then use it as a starting point for the molecular design, contributing to the iterative cycle.

On the other side, the complete models will serve as the middle point between molecular details and tissue features. Once a program is found computationally, the next question is how to root it into molecular details for implementation in cells. For example, I may find that, in order to induce folding of an epithelium, I need a column of epithelial cells in a tissue to shrink one side of their membrane while maintaining contact with their neighbors. Then, I would turn to the molecular toolkit to implement this program in cells. These molecular tools can be both natural/endogenous, chimeric (patchwork of different endogenous proteins), or completely synthetic. In the example of epithelial folding, it is known that forced overexpression of a gene called *Shroom* in epithelial cells induces the type of membrane shrinking that is needed here (Haigo et al. 2003). One can use that output downstream of a synthetic program to achieve the localized folding. The need to control a functioning system implies that we will have the challenge of inferfacing with the existing systems, but it also means the opportunity that we can exploit or rewire some of the existing controls for user-defined goals.

To meet the need of the implementation, we will expand the molecular toolkit. It is deemed to expand, thanks to new discoveries and new protein-engineering works. It can also be expanded with specifically designed research to discover/invent the molecules that are needed, with screening for molecular or cellular behaviors. Now, I will revise some of the relevant molecular components that are found in endogenous systems and the current synthetic ones and present interesting directions to pursue for engineering novel ones.

26.4.3 Key molecular components

As I discussed, cells organize tissues as computational and mechanical units. Now, I will introduce a few examples of molecular components that contribute to these features. These are the components whose behavior needs to be mastered in order to achieve control over morphogenesis. They can be thought of as the molecular substrates that implement the tissue behaviors at the multicellular scale. For a broader, general introduction and discussion, the reader is referred to the corresponding chapters in Alberts et al. (2014).

26.4.3.1 Material molecular components

26.4.3.1.1 Cytoskeleton   Much of the cell shape and physical features is dictated by the cytoskeleton. It can be thought of as the intracellular skeleton of cells. It provides mechanical support, and its active components drive shape changes, and molecularly, it is composed of three families of protein filaments: actin filaments, microtubules, and intermediate filaments. Their behavior is modulated by a wealth of accessory proteins and by their dynamics of polymerization/depolymerization (Mandadapu et al. 2008; Luby-Phelps 2000; Lecuit et al. 2011). For example, the above-mentioned membrane deformation known as apical constriction is driven by cytoskeletal elements; actin–myosin network contraction generates force; and the attachment of actin networks to cell–cell junctions allows forces to be transmitted between cells. Apical constriction is a paradigmatic example of a single-cell shape change, rooted in the cytoskeleton, that is thought to underlie many epithelium-bending morphogenetic phenomena *in vivo* (Martin and Goldstein 2014).

In tissue biology, the cytoskeleton is studied for its relevance for morphogenesis and force sensing at the tissue level (Chanet and Martin 2014) via feedback with cell–cell adhesions and cell–matrix adhesions (see later, mechanotransduction). The status of the cytoskeleton is dictated by both: (1) the set of protein components that are expressed at a certain time, and (2) the post-translational modification of the components by other stimuli (e.g., signaling, the mechanical environment of the cells themselves, and focal adhesions).

The cytoskeleton presents a great challenge for synthetic biologists, since it is a very complex machine that elaborates responses to stimuli in a very sophisticated way. On the other end, it is a central player in multicellular morphogenesis, and an increased control would be very powerful. Developing a more elaborated picture of the cytoskeleton will help control its behavior. The starting point for synthetic biology will be the capacity to drive formation of cytoskeletal shapes in cells by controlled change in cell components. The goal could be to generate an atlas of molecular drivers of cell shape changes, akin to what can be done with overexpression of *Shroom* for apical constriction.

26.4.3.1.2 Cell–cell adhesion   Tissues are made of multiple cell types that adhere to each other to a greater or a lesser extent. From epithelia, where cells rarely exchange neighbors, to blood, where cell–cell adhesion is very sporadic, mechanical coupling between cells dictates many physical properties of tissues (Harris and Tepass 2010; Gumbiner 2005; Gomez et al. 2011). Cell–cell adhesion molecules also provide the driving force for boundary formation and cell sorting within tissues (Batlle and Wilkinson 2012), as they can be adhesive or repulsive (Halloran and Wolman 2006) and homotypic or heterotypic (Rikitake et al. 2012). Differential cell–cell adhesion is considered a strong morphogenetic force that drives reorganization at the tissue level, thanks to cell movements and minimization of surface energy (Steinberg 2007; Krens and Heisenberg 2011).

In tissue biology, cell–cell adhesions are studied for their relevance in development and morphogenesis (West and Harris 2016; Röper 2015). Where cells are adhering to each other to form tissues (as in epithelial sheets), mechanical coupling through cell–cell adhesion ensures mechanical coherence. It also allows the cytoskeleton to be connected to form a distributed multi-cytoskeleton, a structure that spans multiple cells connected across the membranes of two neighboring cells, thanks to cell–cell adhesion molecules that contact the cytoskeleton intracellularly (Röper 2015; Lecuit et al. 2011). The status of a single-cell adhesion machinery is dictated not only by the expressed proteins of the adhesion complexes but also by the tensional status of the cytoskeleton, endocytosis at the membrane, and the presence of extracellular binding partners (Collinet and Lecuit 2013; Lecuit and Sonnenberg 2011). The cell–cell adhesion complexes are also crucial sites of cell–cell communication: some signaling receptors live there, and signaling can happen at specific cell–cell junctions (e.g., polarity signaling [Shin et al. 2006]).

Adhesions offer a great potential for tissue engineering intervention. Scaffold-mediated ways of controlling cell–cell interactions (Kaji et al. 2011) and adhesions mediated through orthogonal degradable oligonucleotide *Velcro* are being developed (Todhunter et al. 2015). For synthetic biology, a foundational development would be to develop genetically encoded, synthetic cell–cell adhesion molecules orthogonal to the endogenous ones. Since the molecular description of endogenous adhesion complexes is quite mature, and their role in morphogenesis is well studied, controlling adhesion temporally and spatially with synthetic pathways seems a very promising way to control patterning of different cell populations in a tissue.

26.4.3.1.3 Extracellular matrix A variety of types of extracellular environments support tissues. From basal membrane to bones, the way in which cells produce and relate to their mechanical environment dictates many features of the shape, function, and development of the tissue. The ECM is composed of a complex and intricate network of macromolecules, locally secreted and assembled into an organized meshwork in close association with the surfaces of the cells that produce them (Hynes and Naba 2012). The status of an ECM is dictated by the activity of the cells that produced it and can be a very dynamic structure; for example, its composition can change or can be modified with extracellular membrane–bound or secreted enzymes produced by cells (Larsen et al. 2006; Lu et al. 2011). On the other side, the ECM itself is considered not only a mechanical support but also a signaling center (Yurchenco 2011).

Cells attach to the ECM with specialized receptors (e.g., integrins), thus forming a distributed skeleton made of cytoskeleton, cell–cell adhesion, and cell–matrix adhesion, which is a powerful machine for tissue morphogenesis (Lecuit et al. 2011). In tissue biology, the ECM is studied for its relevance in tissue development, differentiation, and wound repair (Daley et al. 2008; Phan et al. 2015). Multiple studies have shed light on what makes cells maintain healthy ECM as opposed to fibrotic or degenerating ECM (Lu et al. 2011; Goldsmith et al. 2014). As far as engineered controls, here is where the most is done in traditional tissue engineering focused on scaffolds (Cai and Heilshorn 2014; Lutolf and Hubbell 2005; Webber et al. 2016). The starting point for synthetic biology here would be to engineer cells to make different types of matrix structures and to remodel them in a programmed way.

26.4.3.1.4 Membrane Other determinants of cellular mechanics are the cell membrane (Escribá and Nicolson 2014) and the processes active at the cell surface. Many cell deformations relevant for migration, division, or morphogenesis are thought to be controlled by changes at the cell surface, by processes regulated by plasma membrane and the actin–myosin cortex cytoskeletal network (Clark et al. 2014).

The plasma membrane interacts with many other components discussed here, such as the ECM (Dong et al. 2014), cell–cell adhesion (see earlier), and cell–cell communication (Barbieri et al. 2016; McGough and

Vincent 2016). Tension developed by the cytoskeleton at the cortical membrane is studied as an important force of tissue patterning (Heller and Fuchs 2015). The starting point for synthetic biology control would be to be able to change cell shape, or cortical tension, with engineered proteins or by changing gene expression.

26.4.3.2 Molecular components for cellular communication

26.4.3.2.1 Cell–cell communication Cell–cell communication is mediated by several mechanisms. One is through signaling molecules produced by one cell (sender) and sensed by receptors on another cell (receiver). The signaling molecule could be membrane-tethered, secreted, ECM-tethered, or membrane-soluble. When the signal reaches the receiver, it is sensed by a receptor system that transduces the signal from the extracellular space to the inside of the cells in a process of signal transduction (Lim et al. 2014).

Cell–cell communication is thought to be at the basis of patterning and morphogenesis phenomena observed during tissue formation (Lewis 2008; Meinhardt and Gierer 2000; Rogers and Schier 2011; Lander 2013; Ochoa-Espinosa and Affolter 2012). Both the temporal patterning and the spatial patterning help a uniform mass of cells to install differential gene expression programs at different portions.

Another way in which cells in tissue communicate is through bioelectrical networks (Levin 2014). Here, the communication is enabled by gap junctions between cells, directed by ion channels on the membrane, and read by voltage-sensitive effectors (Mustard and Levin 2014). This seems to be a particularly important way of communication for starting cascades of (regenerative) morphogenesis.

Cell–cell communications are at the core of the toolkit for programming multicellular behaviors with the synthetic biology approach. A toolkit of synthetic cell–cell communications is being developed, and the next step will be to use them to pattern multicellular tissues with spatial (e.g., stripes and spots) or temporal (e.g., oscillations) structures. Similarly, for electrical potential–based communication, the starting point will be to develop ways to control membrane potential patterns in space and time. These synthetic communication pathways can be used to drive morphological effectors such as the ones described earlier (e.g., cytoskeleton and ECM), either concurrently or after the patterning step (see later, dynamics of morphogenesis).

26.4.3.2.2 Cell–matrix communication Cells make ECM, organize it, and degrade it. The ECM in turn influences cell behavior by engaging matrix receptors expressed by the cells. Integrins are the main receptors, and multi-protein complexes assemble around and with integrins. Intracellularly, they connect to the cytoskeleton (Wolfenson et al. 2013; Moser et al. 2009; Campbell and Humphries 2011).

For one mechanism of communication, mechanotransduction, the chemical nature of the binding is less important than the physical nature of the interaction (DuFort et al. 2011; Köster and Janshoff 2015; Hoffman et al. 2011). Much of this sensing is translated in cytoskeletal changes that then feed into changes in gene transcription that usually result in changes in the mechanical aspect of cells or of the matrix in a mechano-feedback mechanism studied in tissue biology and cancer (Wozniak and Chen 2009; Schedin and Keely 2011; Halder et al. 2012; Dupont 2016).

The cell–ECM communication is dictated by the ECM and also by the way in which cells adhere, for example, the type of complex that is assembled at the interface. The starting point for synthetic tissue engineering intervention would be to develop synthetic receptors for the mechanical environment and controllers of the mechano-feedback loop.

26.5 Controlling tissues morphogenesis

In the previous sections, we have seen how we can think of a tissue as a complex physical system made of cells that communicate and exert functions. In this section, we will try to think of ways in which we can gain control over this complex system to design tissue structure and function and control morphogenesis by programming cells at the genetic level. Can we learn how to genetically program cells to make developmental trajectories of our own design? The idea of the tissue development engineering approach is to influence developmental trajectories by engineering cells with custom-made programs, to either make completely new trajectories or to change the existing ones. For example, we want to be able to take stem cells and program them with genetic parts, so that they would autonomously build a tissue with four vessels of 25 μm of diameter interspersed in a matrix of 2 kPa of stiffness, a structure that they would not form in the absence of the engineered program. Alternatively, we want to make cells build a bone that would self-repair upon fracture or to make a degenerating tissue regenerate.

Considering the framework of tissue as robotic material (Figure 26.3), control on the trajectory of the tissue development can be exerted at several control points: (1) the program of the cells that constitute the tissue (computational units), or (2) the inputs that the cells are sensing in the form of the signaling molecules present in the extracellular environment, the material that the cells interact with, or additional cells introduced from the outside. As hinted before, various approaches are possible. Scaffold-based tissue engineering controls tissue structure by presenting structural and chemical signals in the environment of cells. Organoids and *in vitro* differentiation programs work by changing the chemical and mechanical environment and rely on endogenous cascades of differentiation and self-organization or overexpression of selected transcription factors. The tissue development engineering approach described here aims to target the computational unit program with synthetic biology tools. I anticipate that, ultimately, the combination of the different types of controls will provide the highest level of control.

I will present here some examples and related principles of tissue development engineering—how to design tissue development with engineered genetic programs. Common to all the examples is the unifying theme: engineered cells perform communication networks coupled with effector functions to generate novel multicellular behaviors. Each example underscores different aspects of the tissue development engineering approach.

26.5.1 Dynamics of morphogenesis

One very abstract way of thinking about designing a developmental trajectory is to start from a uniform population of cells, install a spatial pattern by cell–cell communication, and then have the different parts of the pattern actuate different programs. If, for example, the goal is to obtain a two-layered structure, starting from a uniform population of cells, one could design a developmental trajectory that first differentiates the population into two cell types A and B; one way to implement this could be a lateral-inhibition network (Matsuda et al. 2015). Once two stable cell fates are generated, the two cell types start to express adhesion molecules at different expression levels, so that cell type A will stick more strongly to other type A cells but only loosely to type B cells. This will then sort the two populations out, generating a two-layered structure, with A cells in the inside and B cells on the outside (Steinberg 2007).

In the framework of the tissue as a robotic material paradigm, the computational unit (the cells) executes a local communication network that installs a spatial pattern that is then coupled to a change in a material feature of the cells (stickiness of the cells) in an actuation step. In this example, the cell–cell communications install stable cell identities, starting from a uniform population of cells; then, this pattern is turned into functional differences by making different cells express effector genes and hence acquiring different mechanical properties. Patterning and effectors in this case are two sequential steps in the process. In general, patterning and effectors can happen sequentially or instead concurrently; in theoretical developmental/evolutionary biology, these two poles are defined as being morphostatic and morphodynamic, respectively (Salazar-Ciudad et al. 2003). The example of the two layers described earlier is morphostatic. A morphodynamic implementation would be obtained by starting the expression of adhesion molecules while the differentiation is being established, which would lead to a more complex dynamical evolution of the system. In morphodynamic systems, you can have different scenarios, depending on whether the patterning and the effectors favor the same morphology or not.

This distinction is useful in the analysis of both morphological development and evolution, and I think that it will be useful in phase of design.

What precisely these two modes of morphogenesis can do for synthetic tissue development will be best studied through complete simulations (see paragraph on simulations previously). Evolutionarily, it is thought that morphodynamic trajectories can give more variability with small changes, whereas morphostatic trajectories are more robust to variations (Salazar-Ciudad et al. 2003; Beloussov and Gordon 2006). Hence, morphodynamic development can be used to program trajectories that change more continuously and can generate more diversity with parameter variation. Morphostatic development on the other hand can be used to build more robust and stable morphologies.

As described, the difference of these two modes of morphogenesis lies in the dynamics of the various steps involved, and it will be important for synthetic biologists to develop tools to control dynamics of multicellular events. A large amount of work is done in synthetic biology to control single-cell dynamics (Inniss and Silver 2013; Purcell and Lu 2014; Gordley et al. 2016; Slusarczyk et al. 2012). It seems possible to extend these works toward multicellular dynamics, in particular working with recombinases and positive feedback circuits to install stable fates and working with negative feedback, protein-degradation domains, and transcriptional dampening to limit the duration of effector events.

26.5.2 Cascades

Programming morphogenesis by explicitly designing all the steps with simple and direct changes of mechanical effectors seems daunting, especially if you want to grow tissues as complex as a hand-out of a small group of amorphous cells. One of the design rules of endogenous tissue development that comes to our help is the presence of morphogenetic cascades or programs. Cells can do a lot on their own after just a kick-start; for example, stem cells can create organoids in appropriate cultures with little intervention, and tissue will undergo remodeling on injuries or other stimuli. This provides an opportunity for synthetic intervention to control these routines spatially or temporally, so that one does not have to implement all the steps explicitly. As an example, let us consider digit formation during limb development in mammals. Here, columns of cells in the primordium are selected and fated to form cartilage, and the differentiation of columns of cells is encoded by a self-organized patterning reminiscent of a Turing pattern (Cooper 2015). This pattern then dictates which cells will later generate the bones of the fingers. In fact, if you have engineered cells (or naturally occurring ones) that execute a different self-organization algorithm and form seven columns of cartilage precursors instead of five, the limb would end up with seven digits (Sheth et al. 2012). It is interesting to note that, again, this principle of master factors that trigger downstream cascade of morphogenesis is thought to have facilitated the evolution of forms by making development extremely evolvable (Gilbert et al. 1996; Kirschner and Gerhart 1998).

The phenomenon of cascades triggered by upstream kick-start occurs both at the level of single cells and at the level of multicellular systems. At the single-cell level, cell-fate reprogramming is one of such routines that is potentially very helpful for tissue design: overexpression of selected transcription factors leads to cells changing their differentiation state, for example, from fibroblasts to stem cells or from muscle to neurons (Takahashi and Yamanaka 2016). At the multicellular level, one spectacular example is given by metamorphosis, where a hormone trigger is able to start a complete reprogramming of organ structures, which is also a great example of the plasticity of tissue structure (Di Cara and King-Jones 2013).

In a sense then, the (multi)-cellular systems we work with come with pre-built routines. On the one end, this design principle constrains the possible developmental trajectories that a multicellular system can take, such that morphogenesis is a highly constrained phenomenon and an emergent property of entities (cells) that are themselves complex systems. The problem of how to deterministically control emergent properties in complex systems is studied across the board (Brueckner et al. 2005; Csete and Doyle 2002), and it is clear that the foundation for attacking this problem is to have control at the level of communication between entities. By instructing cells to communicate in a designed way with synthetic biology tools, we may be able to achieve control of collective behavior.

On the other hand, the availability of these cascades provides a fantastic opportunity for modular design: we already have the libraries of routines, so we only need a way to call them when we want them through a control system. In the examples of the digits, we could change morphogenesis without worrying about how to make a digit just by modifying a pre-pattern that activates a cascade of endogenous morphogenetic events. Hence, controlling the pre-pattern allows control of the final tissue form.

With a combination of changes in the control system and changes in the routines, we can start designing morphogenesis (Teague et al. 2016). That is why, one main focus in synthetic biology will be to seek synthetic ways to control such routines within cells. One strategy is to control how cells communicate with each other and with the environment through synthetic pathways, either completely orthogonal or by rewiring endogenous ones (see previously).

It will also be important to identify useful routines, for example, directed differentiation; invent new ones; and discover or invent minimal ways to call them.

26.5.3 Division of labor

Engineered cells can work as sensors and executors of morphological change while they are themselves developing. However, in general, for the problem of arbitrary tissue development engineering, there are two possible approaches at the extreme of a continuum: design the feature from the bottom-up (e.g., in the digits or in the two-layer experiment) or intervene from the outside to change an existing trajectory.

For example, we want to have a vascular tree that produces new vessels where the endothelial cells are experiencing stress. To achieve that goal, one could deliver a population of engineered *helper* cells to the vessels' environment. These helper cells would be engineered with receptors to detect stress signals in the endothelial cells and, in response, produce a timed cocktail of soluble growth factors (Clapp et al. 2009). These factors would induce the formation of new vessels by activating the endothelial cells to migrate and proliferate. The input signal could be directly on the endothelial cell or in the local microenvironment or the ECM. In this example, the change in morphology is carried out through the intervention of these helper cells that do not perform the morphogenesis themselves.

The two extremes (autonomous vs. division of labor) are two approaches that are used by endogenous biological systems. Completely autonomous developmental systems are more common in simpler organisms or isolated developmental transitions, whereas more complex systems tend to display higher degrees of division of labor, where helper cells carry out important morphogenetic or homeostatic effects for the tissue (Okabe and Medzhitov 2016). Endogenously, examples of helper cells are mesenchymal cells for epithelial morphogenesis. For example, in branching morphogenesis, it is thought that trachea and vasculature are mainly driven by external cells; lung and kidney are in the middle, since they are controlled by feedbacks from the epithelium to the mesenchyme and vice versa; and for salivary gland, it is mainly autonomous (Ochoa-Espinosa and Affolter 2012). Other types of endogenous helper cells are cells of the immune system that are emerging as key players for most of development and regeneration (Plaks et al. 2015; Godwin et al. 2016).

The synthetic biology approach here would be to design engineered helper cells that could interface with the endogenous tissue disease or normal trajectory of development to change it. The use of synthetic biology tools to build therapeutic cells is an exciting frontier of the field (Fischbach et al. 2013). It seems likely that one of the most powerful ways to communicate with tissues and tap onto endogenous morphogenesis program would be to send in cells that could talk in the language of the endogenous cells, so that they can *convince* them to do something they would not do normally, for example, regenerate (see later an example of tissue regeneration).

The problem of how to control multicellular dynamics with programmed helper cells can also be explored computationally: what is the program (sensing + effectors) that needs to be implemented to achieve control of the morphogenesis? Synthetic biology will provide sensors of specific timing of development or of disease states (e.g., through disease or tissue-specific antigens). It will also be important to identify which effectors, produced by the therapeutic cells, can change the features that are important for the morphogenesis, for example, voltage patterns, ECM structure, differentiation state, and cytoskeletal dynamics.

These are some of the design principles that I think will be important for engineering tissue development, and many other design principles will appear, thanks to the research cycle (Figure 26.1), telling us which application would benefit more from a focus on one or the other approach. In general, the questions will be: what are the network communication rules between cells and between cells and matrix, and which are the effectors that must be triggered to achieve the target morphogenesis? What are the boundary conditions? What is the reachable space, starting from a specific ensemble of cells? How can we extend the reachable space and reach a controlled space? How robustness against perturbation or unwanted behavior is obtained? In a sense, this problem is similar to cell-fate reprogramming, where you want to move deterministically cells in the Waddington landscape (Rajagopal and Stanger 2016). In the case of tissue reprogramming, the goal would be to move the tissue deterministically in the morphospace (Vila et al. 2016).

26.6 Engineering complex tissue formation

After having gone through some general principles of tissue development engineering, in the following section, I will focus on the application to complex tissue regeneration. Regeneration is the focus of this book, so I will just limit to say that with regeneration of complex tissues, we are trying to bring back a tissue to its original state after a perturbation due to injury or disease. I take into consideration two lines of possible synthetic biology intervention: build *in vitro* a self-regenerating organoid by engineering cells, and intervene to regenerate an amputated limb with externally provided scaffolds and engineered cells.

26.6.1 Engineering tissues that can self-regenerate

Tissue homeostasis is a feature of healthy adult tissues. It allows them to stay constant in the presence of normal use. On injury, or in case of disease, the homeostasis is lost and another tissue state is reached that is often non-healthy (Kotas and Medzhitov 2015). The goal is the generation of a smart tissue with enhanced homeostatic controls for transplantation, for example, one that could self-regenerate on injury or disease.

Engineered tissues could be endowed with improved homeostatic controls implemented in synthetic homeostatic programs. In this example, an organoid is formed with (some) engineered cells, and the cells are provided with programs that allow them to respond to injuries by regenerating the original shape, obtaining tissue structure homeostasis in response to injuries. In the morphospace dynamics formalism, it would mean to endow a tissue with a stable equilibrium control mechanism, which means that, in response to perturbations, it goes back to its initial state. This synthetic tissue structure homeostasis would build on earlier works that showed that it is possible to use synthetic controls to maintain homeostasis of selected populations of cells by controlling their differentiation based on local cues (Miller et al. 2012).

In general, to achieve homeostasis, a controller system and an effector system are needed. The controller checks the variables that must be kept constant, and the effector executes functions that restore the target value of the variables on deviation (Okabe and Medzhitov 2016). An example is the combination of thermostat (controller) and furnace (effector) to maintain temperature constant. The idea of how to implement this in the case of tissues is to trigger re-development (effector) upon sensing of disrupted tissue architecture by cells (controller) (Figure 26.4).

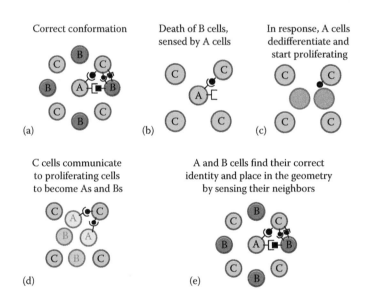

Figure 26.4 A self-regenerating multicellular system. Cells in the correct conformation constantly check on their neighbors with their receptor systems. (1) In this example, the correct conformation is: B cells should have two C neighbors and one A neighbor, C cells should have two B neighbors and one A neighbor, and A cell should have four B and four C neighbors. (2) Owing to injury or disease, all the B cells die. The remaining cells can sense the absence of the inputs and start a designed program of dedifferentiation and proliferation. (3) The remaining C cells communicate with the proliferating cells to execute differentiation in the two cells types A and B. (4) The differentiated cells A and B eventually come back to the positions that are encoded by the program to correspond to the correct conformations (5 and 1).

To achieve this control mechanism, the cells of the organoids are endowed with a set of synthetic receptors that checks on their neighborhood, specifically by binding the molecules that are expressed by their neighbors. If the receptors sense the presence of the normal neighbors, the output is to do nothing but continue the behavior that the cells are carrying out endogenously. When the inputs are not what are expected, then the cells are activated. The cells are programmed to activate only if a number of inputs are *wrong*, not if only one input is wrong. The synthetic program that is activated is dedifferentiation through pluripotency factors in a controlled and timed fashion. The resident cells would then start proliferating and re-establishing the target geometry and tissue structure after a redifferentiation program. The synthetic program of redifferentiation would depend on the tissue type and the cell types that are needed and will be influenced by the inputs present in the remaining cells. The program is active until the targeted structure is replicated, which is sensed again with the set of synthetic receptors. In this sense, it is a self-correcting mechanism that can also take care of inhomogeneity of the produced population. In this example, cells return to a stem-cell state to build back the original tissue architecture, modeling a regeneration type known as epimorphosis. It is observed in organisms that are able to regenerate complex organs through the passage through a stage of *stem tissue* known as a blastema (Brockes and Kumar 2002; Gardiner et al. 2002).

The underlying idea is to generate engineered cells that can be sentinel of tissue structure and are able to re-establish homeostasis when triggered by disruption of tissue structure. In general, sensing can be on the combinatorial discordant inputs from neighbors, ECM, and the microenvironment (e.g., lack of vascular closure, lack of terminal innervation, disruption of voltage gradients patterns, and lack of barrier function from the outside). It can be achieved with synthetic or with rewired endogenous receptors for the molecules produced in the context of the injury or specifically by dying cells. Then, the signals need to be integrated intracellularly, by processing information with artificial signaling cascades, to add up to a point of sufficient disruption. Synthetic logic circuits such as AND gates are potentially useful in this context (MacDonald and Deans 2016; Ausländer and Fussenegger 2016).

The responses are dedifferentiation and remaking of cell types and materials that are needed through a cascade of regeneration (tissue formation through patterning and differential behaviors) this is achieved by controlling morphogenesis (see previously). It is possible that not all the steps are to be implemented (see previously, Cascades). Moreover, we would have a built-in self-correction mechanism, such that when the goal structure is reformed, the input for the dedifferentiation is lost, and so, the system returns to its stable ground state. The system will try to make the correct structure, until it reaches the stable state, where the inputs are normal, at which point it stops.

As an alternative implementation, with the division-of-labor principle, dedicated sentinel cells could perform this program. In endogenous tissues, macrophages are thought to regulate different tissue homeostasis principles (Okabe and Medzhitov 2016). Engineered resident sentinel cells could be able to sense modifications in tissue structure and re-enact the developmental process to build the target structure by communicating with endogenous cells.

26.6.2 Synthetic blastema for limb regeneration

As another example of potential applications of the principles of engineering tissue development to complex tissue formation, I will sketch out the ideas for engineering synthetic epimorphic regeneration in the scenario of a mammalian limb amputation. Other approaches for achieving limb regeneration have recently been reviewed elsewhere (Laurencin and Nair 2016; Quijano et al. 2016).

After a limb amputation in adult humans, the endogenous response does not consist of regeneration but of scar formation. In other organisms, the endogenous program is regeneration (McCusker et al. 2015). Synthetic cells can be used to try to move the tissue response from scar formation to regeneration, by using the knowledge from the animals that endogenously regenerate. One obvious difference between regenerating animals and non-regenerating ones is the absence a blastema, a population of regeneration-competent limb progenitor cells. The idea is to provide non-regenerating tissues with a synthetic blastema, made of engineered cells, that is able to install a regeneration program (Figure 26.5).

The high-level steps that are necessary to synthesize a blastema are inhibition of fibrotic scar and engagement of endogenous precursors and mature cells to generate a synthetic blastema, an apical ectodermal cap (AEC), patterning, and direction of growth. The tenet is that we can find a way to make tissue re-enter the growth and development that it had during development, or throughout young age. This idea is shown to be valid in frogs (Lin et al. 2013), such that the transplantation of engineered tissue progenitor cells from a pre-metamorphic animal can be used to stimulate regeneration of an otherwise non-regenerating adult frog limb. Intriguingly, the regenerate is composed of both engineered cells and host cells. As discussed by the authors, this concept lends itself to a generalization in mammalian model systems, ideally using autologous induced pluripotent stem (iPS) cells.

26.6.2.1 Scarless repair   The first goal of the engineered cell blastema is to get past one of the first roadblocks to regeneration, fibrosis. This first step is aimed at generating an environment that is more conducive for the subsequent regeneration, allowing to tap onto endogenous regenerative growth programs that could be blocked by the early stages of fibrosis (Murawala et al. 2012). To this end, engineered cells secrete

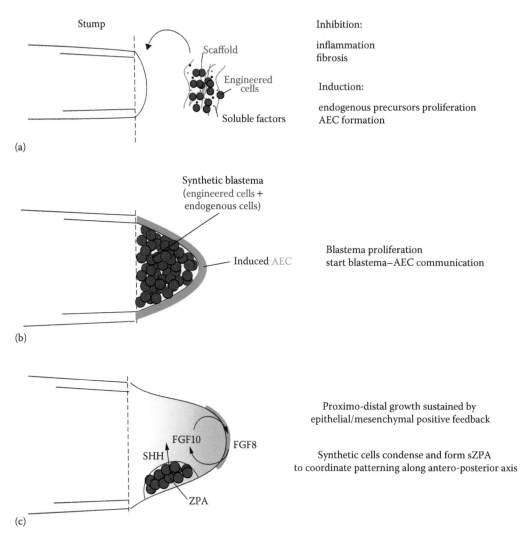

Stump

Scaffold

Engineered cells

Soluble factors

(a)

Inhibition:

inflammation
fibrosis

Induction:

endogenous precursors proliferation
AEC formation

Synthetic blastema
(engineered cells +
endogenous cells)

Induced AEC

(b)

Blastema proliferation
start blastema–AEC communication

FGF10   FGF8

SHH

ZPA

(c)

Proximo-distal growth sustained by
epithelial/mesenchymal positive feedback

Synthetic cells condense and form sZPA
to coordinate patterning along antero-posterior axis

Figure 26.5 Synthetic epimorphic regeneration. (a) The stump of the amputated limb is provided with a synthetic blastema inducer mix made of engineered cells, scaffold, and soluble factors. The cells are engineered to (1) produce specific factors in response to local environmental conditions to decrease inflammation and fibrosis, and (2) contact endogenous cells to tap onto endogenous regenerative programs. (b) Engineered cells remodel the environment and communicate with endogenous cells to induce proliferation and migration for the formation of an apical ectodermal cap (AEC) and a synthetic blastema made of engineered cells + activated endogenous precursors. (c) The AEC and the synthetic blastema start communicating with the synthetic blastema, secreting FGF10 to induce FGF8 from AEC in a positive-feedback loop. This self-sustaining loop between the AEC and underlying mesenchyme sustains the proximo-distal growth and patterning. Once the feedback is established, in response to FGF8 from the AEC, synthetic cells condense by increasing their adhesion and start secreting SHH, thus establishing a synthetic zone of polarizing activity (sZPA) that coordinates the patterning along the antero-posterior axis.

factors to minimize fibrosis (e.g., matrix metalloproteinase [MMP] and other enzymes to degrade the ECM and signaling molecules to influence local fibroblasts behavior). Specifically, in an attempt to try to make the adult wound healing more similar to the fetal one, cells would express factors such as epidermal growth factor (EGF), transforming growth factor beta 3 (TGF-β3), and interleukine 10 (IL-10) and inhibitors of basic fibroblast growth factor (bFGF), platelet-derived growth factor (PDGF), TGF-β1 and 2, vascular endothelial growth factor (VEGF), IL-6, and IL-8 (Ud-Din et al. 2014).

The engineered cells will come with controls to release these factors on demand. This temporal control can be implemented with engineered pathways that recognize injury signals and activate production of the remodeling factors. Alternatively, the cells could constitutively secrete the factors and then commit to apoptosis downstream of sensing that the first phase is terminated, or the production of the factors can be under drug control by using drug-gated promoters.

26.6.2.2 Talking to endogenous cells In addition to repressing fibrosis, the cells of the synthetic blastema will generate a regeneration-competent environment by calling endogenous cells to action. They do so by degrading the matrix, thus liberating the endogenous cells for proliferation, and through accessing endogenous cell–cell communication channels. In particular, engineered cells will secrete growth factors targeted at endogenous cells, connect via gap junctions with endogenous cells to establish voltage gradient patterns, connect with cell–cell junctions to establish new collective cytoskeletal patterns, and coordinate with cells of the immune system.

One important step to this end is to establish a positive feedback between the AEC and the underlying blastema/mesenchyme. This is established with blastema cells secreting FGF10 that is sensed by the AEC cells. In response to FGF10, the AEC cells are activated, and they produce FGF8, which is sensed by blastema cells that in turn respond by producing FGF10 in a positive-feedback manner. This can be achieved by engineering cells in the blastema to produce FGF10 constitutively at low levels and then boost its production through a positive-feedback loop triggered in response to FGF8.

26.6.2.3 Patterning: Control of growth The goal in the early steps in engineering a blastema is the recruitment of all the forces needed for the establishment of an engine of growth. Depending on how reliable the endogenous regeneration cascades that we activate are, we will ultimately need to control the patterning of the synthetic blastema–driven growth.

The engineered cells will perform patterning-installing routines to generate signaling centers for the localized production of growth factors. The aim is to replicate the 3D signaling centers that are guiding limb formation during development (Tabin and McMahon 2008) along the proximo-distal, antero-posterior, and dorso-ventral axes. This step of patterning has been problematic to achieve, even with a simple scaffold (Lin et al. 2013). Synthetic blastema cells have the potential of installing complex spatio-temporal dynamics in a self-organized fashion. For example, a cell–cell communication network of pole formation (e.g., positive feedback with lateral inhibition), a multicellular version of an intracellular polarization circuit (Chau et al. 2012), can be used to select and differentiate a group of cells to be the *synthetic zone of polarizing activity* (sZPA) to recapitulate the endogenous signaling center (ZPA) that patterns the developing limb along the antero-posterior axis. To do that, the polarization circuit is wired to effectors that increase homotypic cell–cell adhesion to condense the cells in a stable structure and to the secretion of sonic hedgehog (SHH), the main effector morphogen secreted by ZPA. Subsequently, non-sZPA cells will be executing a Turing pattern algorithm to generate columns of digit fields to specify the future sides of digit formation (Cooper 2015). The goal is to re-establish the signaling centers for limb patterning as they were used during development.

26.7 Outlook

The capacity of nature to build complex tissues with cellular programs written in their DNA is remarkable. We are on the verge of technological advancements that will allow us to start asking constructively with synthetic biology approaches if we understand the design principles of tissue morphogenesis. With the convergence of developmental biology understanding, tissue engineering technologies and practices, and stem cell biology and differentiation, it is tempting to speculate that we are at a hinge point toward the quest of controlling adult tissue regeneration for regenerative medicine applications.

Acknowledgments

I wish to thank Matt Thomson, Kole Roybal, and Wendel Lim for continuous exchange of creative ideas on the topic. I thank Satoshi Toda and Jonathan Brunger for reading and contributing to the final version of this chapter. Finally, I want to thank my wife Sabina and my son Gabriele for supporting me during these writing days, as they always do.

References

Alberts, B., A. Johnson, J. Lewis, D. Morgan, M. Raff, K. Roberts, and P. Walter. 2014. *Molecular Biology of the Cell*, 6th edition. New York: Garland Science.
Arnold, F. H. 2015. The nature of chemical innovation: New enzymes by evolution. *Quarterly Reviews of Biophysics* 48 (4): 404–410. doi:10.1017/S003358351500013X.

Ausländer, S. and M. Fussenegger. 2016. Engineering gene circuits for mammalian cell-based applications. *Cold Spring Harbor Perspectives in Biology* 8 (7): a023895. doi:10.1101/cshperspect.a023895.

Bajaj, P., R. M. Schweller, A. Khademhosseini, J. L. West, and R. Bashir. 2014. 3D Biofabrication strategies for tissue engineering and regenerative medicine. *Annual Review of Biomedical Engineering* 16 (1): 247–276. doi:10.1146/annurev-bioeng-071813-105155.

Barbieri, E., P. P. Di Fiore, and S. Sigismund. 2016. Science direct endocytic control of signaling at the plasma membrane. *Current Opinion in Cell Biology* 39: 21–27. doi:10.1016/j.ceb.2016.01.012.

Basu, S., Y. Gerchman, C. H. Collins, F. H. Arnold, and R. Weiss. 2005. A synthetic multicellular system for programmed pattern formation. *Nature* 434 (7037): 1130–1134. doi:10.1038/nature03461.

Batlle, E. and D. G. Wilkinson. 2012. Molecular mechanisms of cell segregation and boundary formation in development and tumorigenesis. *Cold Spring Harbor Perspectives in Biology* 4 (1): a008227. doi:10.1101/cshperspect.a008227.

Beloussov, L. V. and R. Gordon. 2006. Preface. Developmental Morphodynamics— Bridging the gap between the genome and embryo physics. *The International Journal of Developmental Biology* 50 (2–3): 79–80. doi:10.1387/ijdb.052137lb.

Ben Tabou de Leon, S. and E. H. Davidson. 2009. Experimentally based sea urchin gene regulatory network and the causal explanation of developmental phenomenology. *Wiley Interdisciplinary Reviews: Systems Biology and Medicine* 1 (2): 237–246. doi:10.1002/wsbm.24.

Boeke, J. D., G. Church, A. Hessel, N. J. Kelley, A. Arkin, Y. Cai, R. Carlson et al. 2016. Genome Engineering the genome project-write. *Science (New York, N.Y.)* 353 (6295): 126–27. doi:10.1126/science.aaf6850.

Brockes, J. P. and A. Kumar. 2002. Plasticity and reprogramming of differentiated cells in amphibian regeneration. *Nature Reviews Molecular Cell Biology* 3 (8): 566–574. doi:10.1038/nrm881.

Brueckner, S. A., G. D. M. Serugendo, A. Karageorgos, and R. Nagpal. 2005. Engineering Self-Organising Systems. In *Methodologies and Applications*, Lecture Notes in Computer Science. doi:10.1007/b136984.

Cai, L. and S. C. Heilshorn. 2014. Designing ECM-Mimetic materials using protein engineering. *Acta Biomaterialia* 10 (4): 1751–1760. doi:10.1016/j.actbio.2013.12.028.

Cameron, D. E., C. J. Bashor, and J. J. Collins. 2014. A brief history of synthetic biology. *Nature Reviews Microbiology* 12 (5): 381–390. doi:10.1038/nrmicro3239.

Campbell, I. D. and M. J. Humphries. 2011. Integrin structure, activation, and interactions. *Cold Spring Harbor Perspectives in Biology* 3 (3): a004994. doi:10.1101/cshperspect.a004994.

Chanet, S. and A. C. Martin. 2014. Mechanical Force Sensing in Tissues. *Progress in Molecular Biology and Translational Science* 126: 317. doi:10.1016/B978-0-12-394624-9.00013-0.

Chau, A. H., J. M. Walter, J. Gerardin, C. Tang, and W. A. Lim. 2012. Designing synthetic regulatory networks capable of self-organizing cell polarization. *Cell* 151 (2): 320–332. doi:10.1016/j.cell.2012.08.040.

Church, G. M., M. B. Elowitz, C. D. Smolke, C. A. Voigt, and R. Weiss. 2014. Realizing the potential of synthetic biology. *Nature Reviews Molecular Cell Biology* 15 (4): 289–294. doi:10.1038/nrm3767.

Clapp, C., S. Thebault, M. C. Jeziorski, and G. M. De La Escalera. 2009. Peptide hormone regulation of angiogenesis. *Physiological Reviews* 89 (4): 1177–1215. doi:10.1152/physrev.00024.2009.

Clark, A. G., O. Wartlick, G. Salbreux, and E. K. Paluch. 2014. Stresses at the cell surface during animal cell review morphogenesis. *Current Biology* 24 (10): R484–R494. doi:10.1016/j.cub.2014.03.059.

Collinet, C. and T. Lecuit. 2013. Chapter–2 Stability and dynamics of cell-cell junctions. *Progress in Molecular Biology and Translational Science* 116: 25–47. doi:10.1016/B978-0-12-394311-8.00002-9.

Cooper, K. L. 2015. Self-organization in the limb: A turing mechanism for digit development. *Current Opinion in Genetics & Development* 32: 92–97. doi:10.1016/j.gde.2015.02.001.

Corona, M., R. Libbrecht, and D. E. Wheeler. 2016. ScienceDirect molecular mechanisms of phenotypic plasticity in social insects. *Current Opinion in Insect Science* 13: 55–60. doi:10.1016/j.cois.2015.12.003.

Csete, M. E. and J. C. Doyle. 2002. Reverse engineering of biological complexity. *Science (New York, N.Y.)* 295 (5560): 1664–1669. doi:10.1126/science.1069981.

Cummings, F. W. 1990. A model of morphogenetic pattern formation. *Journal of Theoretical Biology* 144 (4): 547–566.

Daley, W. P., S. B. Peters, and M. Larsen. 2008. Extracellular matrix dynamics in development and regenerative medicine. *Journal of Cell Science* 121 (3): 255–264. doi:10.1242/jcs.006064.

Davies, J. A., and E. Cachat. 2016. Synthetic biology meets tissue engineering. *Biochemical Society Transactions* 44 (3): 696–701. doi:10.1042/BST20150289.

Davies, J. A. 2008. Synthetic morphology: Prospects for engineered, self-constructing anatomies. *Journal of Anatomy* 212 (6) 707–719. doi:10.1111/j.1469-7580.2008.00896.x.

Di Cara, F., and K. King-Jones. 2013. How clocks and hormones act in concert to control the timing of insect development. *Current Topics in Developmental Biology* 105: 1–36. doi:10.1016/B978-0-12-396968-2.00001-4.

Di Roberto, R. B. and S. G. Peisajovich. 2013. The role of domain shuffling in the evolution of signaling networks. *Journal of Experimental Zoology Part B: Molecular and Developmental Evolution* 322 (2): 65–72. doi:10.1002/jez.b.22551.

Diekman, B. O., P. I. Thakore, S. K. O'Connor, V. P. Willard, J. M. Brunger, N. Christoforou, K. W. Leong, C. A. Gersbach, and F. Guilak. 2015. Knockdown of the cell cycle inhibitor P21 enhances cartilage formation by induced pluripotent stem cells. *Tissue Engineering Part A* 21 (7–8): 1261–1274. doi:10.1089/ten.tea.2014.0240.

Din, M. O., T. Danino, A. Prindle, M. Skalak, J. Selimkhanov, K. Allen, E. Julio et al. 2016. Synchronized cycles of bacterial lysis for in vivo delivery. *Nature* 536 (7614): 81–85. doi:10.1038/nature18930.

Dong, B., E. Hannezo, and S. Hayashi. 2014. Balance between apical membrane growth and luminal matrix resistance determines epithelial tubule shape. *CellReports* 7 (4) 941–950. doi:10.1016/j.celrep.2014.03.066.

DuFort, C. C., M. J. Paszek, and V. M. Weaver. 2011. Balancing forces: Architectural control of mechanotransduction. *Nature Reviews Molecular Cell Biology* 12 (5): 308–319. doi:10.1038/nrm3112.

Dupont, S. 2016. Role of YAP/TAZ in cell-matrix adhesion-mediated signalling and mechanotransduction. *Experimental Cell Research* 343 (1): 42–53. doi:10.1016/j.yexcr.2015.10.034.

Durdu, S., M. Iskar, C. Revenu, N. Schieber, A. Kunze, P. Bork, Y. Schwab, and D. Gilmour. 2014. Luminal signalling links cell communication to tissue architecture during organogenesis. *Nature* 515 (7525): 120–124. doi:10.1038/nature13852.

Elowitz, M. and W. A Lim. 2010. Build life to understand it. *Nature* 468 (7326): 889–890. doi:10.1038/468889a.

Escribá, P. V. and G. L. Nicolson. 2014. Membrane structure and function: Relevance of lipid and protein structures in cellular physiology, pathology and therapy. *Biochimica et Biophysica Acta* 1838 (6): 1449–1450. doi:10.1016/j.bbamem.2014.03.008.

Feinberg, A. W. 2015. Biological soft robotics. *Annual Review of Biomedical Engineering* 17 (1): 243–265. doi:10.1146/annurev-bioeng-071114-040632.

Fischbach, M. A, J. A. Bluestone, and W. A. Lim. 2013. Cell-based therapeutics: The next pillar of medicine. *Science Translational Medicine* 5 (179): 179ps7. doi:10.1126/scitranslmed.3005568.

Fletcher, A. G., M. Osterfield, R. E. Baker, and S. Y. Shvartsman. 2014. Vertex models of epithelial morphogenesis. *Biophysical Journal* 106 (11): 2291–2304. doi:10.1016/j.bpj.2013.11.4498.

Gardiner, D. M., T. Endo, and S. V. Bryant. 2002. The molecular basis of amphibian limb regeneration: Integrating the old with the new. *Seminars in Cell and Developmental Biology* 13 (5): 345–352. doi:10.1016/S1084–9521(02)00090-3.

Gilbert, S. F, J. M. Opitz, and R. A Raff. 1996. Resynthesizing evolutionary and developmental biology. *Developmental Biology* 173 (2): 357–372. doi:10.1006/dbio.1996.0032.

Glass, K. A., J. M. Link, J. M. Brunger, F. T. Moutos, C. A. Gersbach, and F. Guilak. 2014. Tissue-engineered cartilage with inducible and tunable immunomodulatory properties. *Biomaterials* 35 (22): 5921–5931. doi:10.1016/j.biomaterials.2014.03.073.

Godwin, J. W., A. R. Pinto, and N. A. Rosenthal. 2016. Chasing the recipe for a pro-regenerative immune system. *Seminars in Cell and Developmental Biology* 1–9. doi:10.1016/j.semcdb.2016.08.008.

Goldsmith, E. C., A. D. Bradshaw, M. R. Zile, and F. G. Spinale. 2014. Myocardial fibroblast–matrix interactions and potential therapeutic targets. *Journal of Molecular and Cellular Cardiology* 70 (C): 92–99. doi:10.1016/j.yjmcc.2014.01.008.

Gomez, G. A., R. W. McLachlan, and A. S. Yap. 2011. Productive tension: Force-sensing and homeostasis of cell-cell junctions. *Trends in Cell Biology* 21 (9): 499–505. doi:10.1016/j.tcb.2011.05.006.

Gordley, R. M., L. J. Bugaj, and W. A. Lim. 2016. Modular engineering of cellular signaling proteins and networks. *Current Opinion in Structural Biology* 39 (July): 106–114. doi:10.1016/j.sbi.2016.06.012.

Guilak, F., D. L. Butler, S. A. Goldstein, and F. P. T. Baaijens. 2014. Biomechanics and mechanobiology in functional tissue engineering. *Journal of Biomechanics* 47 (9): 1933–1940. doi:10.1016/j.jbiomech.2014.04.019.

Gumbiner, B. M. 2005. Regulation of cadherin-mediated adhesion in morphogenesis. *Nature Reviews Molecular Cell Biology* 6 (8): 622–634. doi:10.1038/nrm1699.

Haigo, S. L., J. D. Hildebrand, R. M. Harland, and J. B. Wallingford. 2003. Shroom induces apical constriction and is required for hingepoint formation during neural tube closure. *Current Biology* 13 (24): 2125–2137. doi:10.1016/j.cub.2003.11.054.

Halder, G., S. Dupont, and S. Piccolo. 2012. Transduction of mechanical and cytoskeletal cues by YAP and TAZ. *Nature Reviews Molecular Cell Biology* 13 (9): 591–600. doi:10.1038/nrm3416.

Halloran, M. C. and M. A. Wolman. 2006. Repulsion or adhesion: Receptors make the call. *Current Opinion in Cell Biology* 18 (5): 533–540. doi:10.1016/j.ceb.2006.08.010.

Harris, T. J. C. and U. Tepass. 2010. Adherens junctions: From molecules to morphogenesis. *Nature Reviews Molecular Cell Biology* 11 (7): 502–514. doi:10.1038/nrm2927.

Heller, E. and E. Fuchs. 2015. Tissue patterning and cellular mechanics. *The Journal of Cell Biology* 211 (2): 219–231. doi:10.1083/jcb.201506106.

Hennig, S., G. Rödel, and K. Ostermann. 2015. Artificial cell-cell communication as an emerging tool in synthetic biology applications. *Journal of Biological Engineering* 1–12. doi:10.1186/s13036-015-0011-2.

Ho, CY., A. Sanghani, J. Hua, M. Coathup, P. Kalia, and G. Blunn. 2015. Mesenchymal stem cells with increased stromal cell-derived factor 1 expression enhanced fracture healing. *Tissue Engineering Part A* 21 (3–4): 594–602. doi:10.1089/ten.tea.2013.0762.

Hoffman, B. D., C. Grashoff, and M. A. Schwartz. 2011. Dynamic molecular processes mediate cellular mechanotransduction. *Nature* 475 (7356): 316–323. doi:10.1038/nature10316.

Hynes, R. O. and A. Naba. 2012. Overview of the matrisome—An inventory of extracellular matrix constituents and functions. *Cold Spring Harbor Perspectives in Biology* 4 (1): a004903. doi:10.1101/cshperspect.a004903.

Inniss, M. C. and P. A. Silver. 2013. Building synthetic memory. *Current Biology* 23 (17): R812–R816. doi:10.1016/j.cub.2013.06.047.

Jabbarzadeh, E., T. Starnes, Y. M. Khan, T. Jiang, A. J. Wirtel, M. Deng, Q. Lv, L. S. Nair, S. B. Doty, and C. T. Laurencin. 2008. Induction of angiogenesis in tissue-engineered scaffolds designed for bone repair: A combined gene therapy-cell transplantation approach. *Proceedings of the National Academy of Sciences* 105 (32): 11099–11104. doi:10.1073/pnas.0800069105.

Jefferson, J. J, C. L. Leung, and R. K. H. Liem. 2004. Plakins: Goliaths that link cell junctions and the cytoskeleton. *Nature Reviews Molecular Cell Biology* 5 (7): 542–553. doi:10.1038/nrm1425.

Kaji, H., G. Camci-Unal, R. Langer, and A. Khademhosseini. 2011. Engineering systems for the generation of patterned co-cultures for controlling cell–cell interactions. *Biochimica et Biophysica Acta—General Subjects* 1810 (3): 239–250. doi:10.1016/j.bbagen.2010.07.002.

Khalil, A. S. and J. J. Collins. 2010. Synthetic biology: Applications come of age. *Nature Reviews* 11 (5): 367–379. doi:10.1038/nrg2775.

Kiel, C., E. Yus, and L. Serrano. 2010. Engineering signal transduction pathways. *Cell* 140 (1): 33–47. doi:10.1016/j.cell.2009.12.028.

Kim, H. and J. S. Kim. 2014. A guide to genome engineering with programmable nucleases. *Nature Reviews Genetics* 15 (5): 321–334. doi:10.1038/nrg3686.

Kirschner, M. and J. Gerhart. 1998. "Evolvability. *Proceedings of the National Academy of Sciences* 95 (15): 8420–8427.

Kotas, M. E, and R. Medzhitov. 2015. Homeostasis, inflammation, and disease susceptibility. *Cell* 160 (5): 816–827. doi:10.1016/j.cell.2015.02.010.

Köster, S. and A. Janshoff. 2015. Editorial—Special issue on mechanobiology. *Biochimica et Biophysica Acta* 1853: 2975–2976. doi:10.1016/j.bbamcr.2015.08.002.

Krens, S. F. G. and C. P. Heisenberg. 2011. Cell sorting in development. *Current Topics in Developmental Biology* 95: 189–213. doi:10.1016/B978-0-12-385065-2.00006-2.

Kuo, C. S. and M. A. Krasnow. 2015. Formation of a neurosensory organ by epithelial cell slithering. *Cell* 163 (2): 394–405. doi:10.1016/j.cell.2015.09.021.

Lancaster, M. A. and J. A. Knoblich. 2014. Organogenesis in a dish: Modeling development and disease using organoid technologies. *Science (New York, N.Y.)* 345 (6194): 1247125. doi:10.1126/science.1247125.

Lander, A. D. 2013. How cells know where they are. *Science (New York, N.Y.)* 339 (6122): 923–927. doi:10.1126/science.1224186.

Langer, R. and J. Vacanti. 2016. Advances in tissue engineering. *Journal of Pediatric Surgery* 51 (1): 8–12. doi:10.1016/j.jpedsurg.2015.10.022.

Larsen, M., V. V Artym, J. A. Green, and K. M. Yamada. 2006. The matrix reorganized: Extracellular matrix remodeling and integrin signaling. *Current Opinion in Cell Biology* 18 (5): 463–471. doi:10.1016/j.ceb.2006.08.009.

Laurencin, C. T. and L. S. Nair. 2016. The quest toward limb regeneration: A regenerative engineering approach. *Regenerative Biomaterials* 3 (2): 123–125. doi:10.1093/rb/rbw002.

Lecuit, T. and A. Sonnenberg. 2011. Cell-to-cell contact and extracellular matrix editorial overview. *Current Opinion in Cell Biology* 23 (5): 505–507. doi:10.1016/j.ceb.2011.09.001.

Lecuit, T., P. F. Lenne, and E. Munro. 2011. Force generation, transmission, and integration during cell and tissue morphogenesis. *Annual Review of Cell and Developmental Biology* 27: 157–184. doi:10.1146/annurev-cellbio-100109-104027.

Levin, M. 2014. Endogenous bioelectrical networks store non-genetic patterning information during development and regeneration. *The Journal of Physiology* 592 (11): 2295–2305. doi:10.1113/jphysiol.2014.271940.

Lewis, J. 2008. From signals to patterns: Space, time, and mathematics in developmental biology. *Science (New York, N.Y.)* 322 (5900): 399–403. doi:10.1126/science.1166154.

Lienert, F., J. J. Lohmueller, A. Garg, and P. A. Silver. 2014. Synthetic biology in mammalian cells: Next generation research tools and therapeutics. *Nature Reviews Molecular Cell Biology* 15 (2): 95–107. doi:10.1038/nrm3738.

Lim, W. A. 2010. Designing customized cell signalling circuits. *Nature Reviews Molecular Cell Biology*, 1–11. doi:10.1038/nrm2904.

Lim, W. A. and T. Pawson. 2010. Phosphotyrosine signaling: Evolving a new cellular communication system. *Cell* 142 (5): 661–667. doi:10.1016/j.cell.2010.08.023.

Lim, W., B. Mayer, and T. Pawson. 2014. *Cell Signaling.* Taylor & Francis.

Lin, G., Y. Chen, and J. M. W. Slack. 2013. Imparting regenerative capacity to limbs by progenitor cell transplantation. *Developmental Cell* 24 (1): 41–51. doi:10.1016/j.devcel.2012.11.017.

Lu, P., K. Takai, V. M. Weaver, and Z. Werb. 2011. Extracellular matrix degradation and remodeling in development and disease. *Cold Spring Harbor Perspectives in Biology* 3 (12): a005058. doi:10.1101/cshperspect.a005058.

Luby-Phelps, K. 2000. Cytoarchitecture and physical properties of cytoplasm: Volume, viscosity, diffusion, intracellular surface area. *International Review of Cytology* 192: 189–221.

Lutolf, M. P. and J. A. Hubbell. 2005. Synthetic biomaterials as instructive extracellular microenvironments for morphogenesis in tissue engineering. *Nature Biotechnology* 23 (1): 47–55. doi:10.1038/nbt1055.

MacDonald, I. C. and T. L. Deans. 2016. Tools and applications in synthetic biology. *Advanced Drug Delivery Reviews* 105: 20–34. doi:10.1016/j.addr.2016.08.008.

Mandadapu, K. K., S. Govindjee, and M. R. K. Mofrad. 2008. On the cytoskeleton and soft glassy rheology. *Journal of Biomechanics* 41 (7): 1467–1478. doi:10.1016/j.jbiomech.2008.02.014.

Mao, A. S. and D. J. Mooney. 2015. Regenerative medicine: Current therapies and future directions. *Proceedings of the National Academy of Sciences of the United States of America* 112 (47): 14452–14459. doi:10.1073/pnas.1508520112.

Marin-Riera, M., M. Brun-Usan, R. Zimm, T. Välikangas, and I. Salazar-Ciudad. 2016. Computational modeling of development by epithelia, mesenchyme and their interactions: A unified model. *Bioinformatics (Oxford, England)* 32 (2): 219–225. doi:10.1093/bioinformatics/btv527.

Martin, A. C. and B. Goldstein. 2014. Apical constriction: Themes and variations on a cellular mechanism driving morphogenesis. *Development (Cambridge, England)* 141 (10): 1987–1998. doi:10.1242/dev.102228.

Matsuda, M., M. Koga, K. Woltjen, E. Nishida, and M. Ebisuya. 2015. Synthetic lateral inhibition governs cell-type bifurcation with robust ratios. *Nature Communications* 6: 6195. doi:10.1038/ncomms7195.

Matthys, O. B., T. A. Hookway, and T. C. McDevitt. 2016. Design principles for engineering of tissues from human pluripotent stem cells. *Current Stem Cell Reports* 2 (1): 43–51. doi:10.1007/s40778-016-0030-z.

McCusker, C., S. V. Bryant, and D. M. Gardiner. 2015. The axolotl limb blastema: Cellular and molecular mechanisms driving blastema formation and limb regeneration in tetrapods. *Regeneration* 2 (2): 54–71. doi:10.1002/reg2.32.

McEvoy, M. A. and N. Correll. 2015. Materials science. Materials that couple sensing, actuation, computation, and communication. *Science (New York, N.Y.)* 347 (6228): 1261689. doi:10.1126/science.1261689.

McGough, I. J. and J. P. Vincent. 2016. Exosomes in developmental signalling. *Development (Cambridge, England)* 143 (14): 2482–2493. doi:10.1242/dev.126516.

Meinhardt, H. and A. Gierer. 2000. Pattern formation by local self-activation and lateral inhibition. *BioEssays* 22 (8): 753–760. doi:10.1002/1521-1878(200008)22:8<753::AID-BIES9>3.0.CO;2-Z.

Miller, J. F. A. P. and M. Sadelain. 2015. The journey from discoveries in fundamental immunology to cancer immunotherapy. *Cancer Cell* 27 (4): 439–449. doi:10.1016/j.ccell.2015.03.007.

Miller, M., M. Hafner, E. Sontag, N. Davidsohn, S. Subramanian, P. E. M. Purnick, D. Lauffenburger, and R. Weiss. 2012. Modular design of artificial tissue homeostasis: Robust control through synthetic cellular heterogeneity. *PLoS Computational Biology* 8 (7): e1002579. doi:10.1371/journal.pcbi.1002579.

Morelli, L. G., K. Uriu, S. Ares, and A. C. Oates. 2012. Computational approaches to developmental patterning. *Science (New York, N.Y.)* 336 (6078): 187–191. doi:10.1126/science.1215478.

Morsut, L., K. T. Roybal, X. Xiong, R. M. Gordley, S. M. Coyle, M. Thomson, and W. A. Lim. 2016. Engineering customized cell sensing and response behaviors using synthetic notch receptors. *Cell* 164 (4): 780–791. doi:10.1016/j.cell.2016.01.012.

Moser, M., K. R. Legate, R. Zent, and R. Fässler. 2009. The tail of integrins, talin, and kindlins. *Science (New York, N.Y.)* 324 (5929): 895–99. doi:10.1126/science.1163865.

Murawala, P., E. M. Tanaka, and J. D. Currie. 2012. Regeneration: The ultimate example of wound healing. *Seminars in Cell and Developmental Biology* 23 (9): 954–962. doi:10.1016/j.semcdb.2012.09.013.

Murphy, S. V. and A. Atala. 2014. 3D bioprinting of tissues and organs. *Nature Biotechnology* 32 (8): 773–785. doi:10.1038/nbt.2958.

Mustard, J. and M. Levin. 2014. Bioelectrical mechanisms for programming growth and form: Taming physiological networks for soft body robotics. *Soft Robotics* 1 (3): 169–191. doi:10.1089/soro.2014.0011.

National Research Council (US) Committee on a New Biology for the 21st Century: Ensuring the United States Leads the Coming Biology Revolution. 2009. *A New Biology for the 21st Century: Ensuring the United States Leads the Coming Biology Revolution.* Washington (DC): National Academies Press (US). doi:10.17226/12764.

Ochoa-Espinosa, A. and M. Affolter. 2012. Branching morphogenesis: From cells to organs and back. *Cold Spring Harbor Perspectives in Biology* 4 (10): a008243. doi:10.1101/cshperspect.a008243.

Okabe, Y. and R. Medzhitov. 2016. Tissue biology perspective on macrophages. *Nature immunology* 17 (1): 9–17. doi:10.1038/ni.3320.

Okuda, S., Y. Inoue, M. Eiraku, Y. Sasai, and T. Adachi. 2012. Reversible network reconnection model for simulating large deformation in dynamic tissue morphogenesis. *Biomechanics and Modeling in Mechanobiology* 12 (4): 627–644. doi:10.1007/s10237-012-0430-7.

Oshima, M. and T. Tsuji. 2015. Whole tooth regeneration as a future dental treatment. In *Engineering Mineralized and Load Bearing Tissues*, edited by Luiz E Bertassoni and Paulo G Coelho, Vol. 881, pp. 255–269. Advances in Experimental Medicine and Biology. Cham, Switzerland: Springer International Publishing. doi:10.1007/978-3-319-22345-2_14.

Osterfield, M., X. X. Du, T. Schüpbach, E. Wieschaus, and S. Y. Shvartsman. 2013. Three-dimensional epithelial morphogenesis in the developing drosophila egg. *Developmental Cell* 24 (4): 400–410. doi:10.1016/j.devcel.2013.01.017.

Payne, S. and L. You. 2013. Engineered cell–cell communication and its applications. In *Productive Biofilms*, edited by Kai Muffler and Roland Ulber, Vol. 146, pp. 97–121. Advances in Biochemical Engineering/Biotechnology. Cham, Switzerland: Springer International Publishing. doi:10.1007/10_2013_249.

Peloso, A., A. Dhal, J. P. Zambon, P. Li, G. Orlando, A. Atala, and S. Soker. 2015. Current achievements and future perspectives in whole-organ bioengineering. *Stem Cell Research & Therapy* 6 (1): 107. doi:10.1186/s13287-015-0089-y.

Phan, A. Q., J. Lee, M. Oei, C. Flath, C. Hwe, R. Mariano, T. Vu et al. 2015. Positional information in axolotl and mouse limb extracellular matrix is mediated via heparan sulfate and fibroblast growth factor during limb regeneration in the axolotl (Ambystoma mexicanum). *Regeneration* 2 (4): 182–201. doi:10.1002/reg2.40.

Plaks, V., B. Boldajipour, J. R. Linnemann, N. H. Nguyen, K. Kersten, Y. Wolf, A. J. Casbon et al. 2015. Adaptive immune regulation of mammary postnatal organogenesis. *Developmental Cell* 34 (5): 493–504. doi:10.1016/j.devcel.2015.07.015.

Purcell, O. and T. K. Lu. 2014. Synthetic analog and digital circuits for cellular computation and memory. *Current Opinion in Biotechnology* 29: 146–155. doi:10.1016/j.copbio.2014.04.009.

Purnick, P. E. M. and R. Weiss. 2009. The second wave of synthetic biology: From modules to systems. *Nature Reviews Molecular Cell Biology* 10: 410–422. doi:10.1038/nrm2698.

Quijano, L. M., K. M. Lynch, C. H. Allan, S. F. Badylak, and T. Ahsan. 2016. Looking ahead to engineering epimorphic regeneration of a human digit or limb. *Tissue Engineering Part B: Reviews* 22 (3): 251–262. doi:10.1089/ten.teb.2015.0401.

Rajagopal, J. and B. Z. Stanger. 2016. Plasticity in the adult: How should the waddington diagram be applied to regenerating tissues? *Developmental Cell* 36 (2): 133–137. doi:10.1016/j.devcel.2015.12.021.

Reichert, W. M., B. D. Ratner, J. Anderson, A. Coury, A. S. Hoffman, C. T. Laurencin, and D. Tirrell. 2010. 2010 Panel on the biomaterials grand challenges. *Journal of Biomedical Materials Research Part A* 96A (2): 275–287. doi:10.1002/jbm.a.32969.

Rikitake, Y., K. Mandai, and Y. Takai. 2012. The role of nectins in different types of cell-cell adhesion. *Journal of Cell Science* 125 (Pt 16): 3713–3722. doi:10.1242/jcs.099572.

Rogers, K. W. and A. F. Schier. 2011. Morphogen gradients: From generation to interpretation. *Annual Review of Cell and Developmental Biology* 27: 377–407. doi:10.1146/annurev-cellbio-092910-154148.

Rouwkema, J. and A. Khademhosseini. 2016. Vascularization and angiogenesis in tissue engineering: Beyond creating static networks. *Trends in Biotechnology* 34 (9): 733–745. doi:10.1016/j.tibtech.2016.03.002.

Röper, K. 2015. Integration of cell-cell adhesion and contractile actomyosin activity during morphogenesis. *Current Topics in Developmental Biology* 112: 103–127. doi:10.1016/bs.ctdb.2014.11.017.

Rué, P. and J. Garcia-Ojalvo. 2013. Modeling gene expression in time and space. *Annual Review of Biophysics* 42 (1): 605–627. doi:10.1146/annurev-biophys-083012-130335.

Salazar-Ciudad, I., J. Jernvall, and S. A. Newman. 2003. Mechanisms of pattern formation in development and evolution. *Development (Cambridge, England)* 130 (10): 2027–2037. doi:10.1242/dev.00425.

Saxena, P., B. C. Heng, P. Bai, M. Folcher, H. Zulewski, and M. Fussenegger. 2016. A programmable synthetic lineage-control network that differentiates human IPSCs Into glucose-sensitive insulin-secreting beta-like cells. *Nature Communications* 7: 11247. doi:10.1038/ncomms11247.

591

Schedin, P. and P. J. Keely. 2011. Mammary gland ECM remodeling, stiffness, and mechanosignaling in normal development and tumor progression. *Cold Spring Harbor Perspectives in Biology* 3 (1): a003228. doi:10.1101/cshperspect.a003228.

Schukur, L. and M. Fussenegger. 2016. Engineering of synthetic gene circuits for (re-) balancing physiological processes in chronic diseases. *Wiley Interdisciplinary Reviews: Systems Biology and Medicine* 8 (5): 402–422. doi:10.1002/wsbm.1345.

Schukur, L., B. Geering, G. Charpin-El Hamri, and M. Fussenegger. 2015. Implantable synthetic cytokine converter cells with and-gate logic treat experimental psoriasis. *Science Translational Medicine* 7 (318): 318ra201. doi:10.1126/scitranslmed.aac4964.

Schwarz, K. A. and J. N. Leonard. 2016. Engineering cell-based therapies to interface robustly with host physiology. *Advanced Drug Delivery Reviews* 105: 55–65. doi:10.1016/j.addr.2016.05.019.

Shaya, O. and D. Sprinzak. 2011. From notch signaling to fine-grained patterning: Modeling meets experiments. *Current Opinion in Genetics & Development* 21 (6): 732–739. doi:10.1016/j.gde.2011.07.007.

Shea, L. D., T. K. Woodruff, and A. Shikanov. 2014. Bioengineering the ovarian follicle microenvironment. *Annual Review of Biomedical Engineering* 16 (1): 29–52. doi:10.1146/annurev-bioeng-071813-105131.

Sheth, R., L. Marcon, M. F. Bastida, M. Junco, L. Quintana, R. Dahn, M. Kmita, J. Sharpe, and M. A. Ros. 2012. Hox genes regulate digit patterning by controlling the wavelength of a turing-type mechanism. *Science* (*New York, N.Y.*) 338 (6113): 1476–1480. doi:10.1126/science.1226804.

Shin, K., V. C. Fogg, and B. Margolis. 2006. Tight junctions and cell polarity. *Annual Review of Cell and Developmental Biology* 22: 207–235. doi:10.1146/annurev.cellbio.22.010305.104219.

Simakov, D. S. A. and L. M. Pismen. 2013. Discrete model of periodic pattern formation through a combined autocrine-juxtacrine cell signaling. *Physical Biology* 10 (4): 046001. doi:10.1088/1478-3975/10/4/046001.

Slusarczyk, A. L., A. Lin, and R. Weiss. 2012. Foundations for the design and implementation of synthetic genetic circuits. *Nature Reviews Genetics* 13 (6): 406–420. doi:10.1038/nrg3227.

Steinberg, M. S. 2007. Differential adhesion in morphogenesis: A modern view. *Current Opinion in Genetics & Development* 17 (4): 281–286. doi:10.1016/j.gde.2007.05.002.

Tabin, C. J. and A. P. McMahon. 2008. Developmental biology: Grasping limb patterning. *Science* (*New York, N.Y.*) 321 (5887): 350–352. doi:10.1126/science.1162474.

Takahashi, K. and S. Yamanaka. 2016. A decade of transcription factor-mediated reprogramming to pluripotency. *Nature Reviews Molecular Cell Biology* 17 (3): 183–193. doi:10.1038/nrm.2016.8.

Teague, B. P., P. Guye, and R. Weiss. 2016. Synthetic morphogenesis. *Cold Spring Harbor Perspectives in Biology* a023929–16. doi:10.1101/cshperspect.a023929.

Tibbitt, M. W., C. B. Rodell, J. A. Burdick, and K. S. Anseth. 2015. Progress in material design for biomedical applications. *Proceedings of the National Academy of Sciences of the United States of America* 112 (47): 14444–14451. doi:10.1073/pnas.1516247112.

Todhunter, M. E., N. Y. Jee, A. J. Hughes, M. C. Coyle, A. Cerchiari, J. Farlow, J. C. Garbe, M. A. LaBarge, T. A. Desai, and Z. J. Gartner. 2015. Programmed synthesis of three-dimensional tissues. *Nature Methods* 12 (10): 975–981. doi:10.1038/nmeth.3553.

Tsiairis, C. D. and A. Aulehla. 2016. Self-organization of embryonic genetic oscillators into spatiotemporal wave patterns. *Cell* 164 (4): 656–667. doi:10.1016/j.cell.2016.01.028.

Ud-Din, S., S. W. Volk, and A. Bayat. 2014. Regenerative healing, scar-free healing and scar formation across the species: Current concepts and future perspectives. *Experimental Dermatology* 23 (9): 615–619. doi:10.1111/exd.12457.

Urdy, S. 2012. On the evolution of morphogenetic models: Mechano-chemical interactions and an integrated view of cell differentiation, growth, pattern formation and morphogenesis. *Biological Reviews* 87 (4): 786–803. doi:10.1111/j.1469-185X.2012.00221.x.

Uygur, A., J. Young, T. R. Huycke, M. Koska, J. Briscoe, and C. J. Tabin. 2016. Scaling pattern to variations in size during development of the vertebrate neural tube. *Developmental Cell* 37 (2): 127–135. doi:10.1016/j.devcel.2016.03.024.

Vila, A., S. Duran-Nebreda, N. Conde-Pueyo, R. Montanez, and Ricard Sole. 2016. A morphospace for synthetic organs and organoids: The possible and the actual. *Integrative Biology* 8: 485–503. doi:10.1039/c5ib00324e.

Way, J. C., J. J. Collins, J. D. Keasling, and P. A. Silver. 2014. Integrating biological redesign: Where synthetic biology came from and where it needs to go. *Cell* 157 (1): 151–161. doi:10.1016/j.cell.2014.02.039.

Webber, M. J., E. A. Appel, E. W. Meijer, and R. Langer. 2016. Supramolecular biomaterials. *Nature Materials* 15 (1): 13–26. doi:10.1038/nmat4474.

Wennekamp, S., S. Mesecke, F. Nédélec, and T. Hiiragi. 2013. A self-organization framework for symmetry breaking in the mammalian embryo. *Nature Reviews Molecular Cell Biology* 14 (7): 452–459. doi:10.1038/nrm3602.

West, J. J. and T. J. C. Harris. 2016. Cadherin trafficking for tissue morphogenesis: Control and consequences. *Traffic* 17(12): 1233–1243. doi:10.1111/tra.12407.

Wolfenson, H., I. Lavelin, and B. Geiger. 2013. Dynamic regulation of the structure and functions of integrin adhesions. *Developmental Cell* 24 (5): 447–458. doi:10.1016/j.devcel.2013.02.012.

Wozniak, M. A. and C. S. Chen. 2009. Mechanotransduction in development: A growing role for contractility. *Nature Reviews Molecular Cell Biology* 10 (1): 34–43. doi:10.1038/nrm2592.

Wu, J. and J. C. I. Belmonte. 2016. Stem cells: A renaissance in human biology research. *Cell* 165 (7): 1572–1585. doi:10.1016/j.cell.2016.05.043.

Yin, X., B. E. Mead, H. Safaee, R. Langer, J. M. Karp, and O. Levy. 2016. Engineering stem cell organoids. *Stem Cell* 18 (1): 25–38. doi:10.1016/j.stem.2015.12.005.

Yurchenco, P. D. 2011. Basement membranes: Cell scaffoldings and signaling platforms. *Cold Spring Harbor Perspectives in Biology* 3 (2): a004911. doi:10.1101/cshperspect.a004911.

Chapter 27 Recurrent concepts

David M. Gardiner

Contents

27.1 Introduction

The goal of this book is to begin the process of bringing developmental biology and regenerative engineering together in order to explore how to make regeneration work in humans. Our premise is that the answer lies somewhere in the middle between these two fields, and researchers from both sides need to participate. The challenges for the authors of this book were to examine what they know as developmental biologists and to communicate what they consider to be the developmental processes and principles that are most relevant to regenerative engineering.

As I began recruiting authors and asking them to think creatively about the most important concepts, I was repeatedly asked about what would happen if they wrote about some of the same topics as another author. At that point, I began to think of this book as an experiment: getting 44 scientists with expertise in embryonic and regenerative developmental biology to write about what they think is important, and if they are writing about different topics but still end up writing about some of the same ideas, then it is likely that those common ideas are important and worthy of note. In an attempt to capture these recurrent concepts, I decided to write a final chapter summarizing the concepts that were raised and discussed in multiple chapters.

27.2 The top-10 list of what is important to know for engineering regeneration

All authors indicated what they considered the most important points at the beginning of their respective chapters (key concepts), and thus, there are a large number of ideas to contemplate. My goal is to distill these down to a reasonably short list of the recurrent key concepts and elaborate on each briefly. For those readers who might like to start at the end, you can read this chapter first to get an idea of what is contained in the previous 26 chapters and then go back and read in more detail about what you are most interested in (I have included reference to some of the chapters that discuss these ideas). I apologize that I did not incorporate all the ideas from all the authors, and I am sure that I have missed points that others would consider to be equally or more important, so this is in reality my list of what I think is most important. To borrow from David Letterman, here is my top-10 list of what is important to know for engineering regeneration (in no particular order).

27.2.1 There are pattern-forming cells and pattern-following cells, and you need them both

(Chapters: 6, 7, 9, 12, 13, 14, and 19)

Cells need to have both the potential and the context to reactivate the programs that drive organ and appendage regeneration. These are two different requirements in as much as we need both the cells to rebuild the missing parts (the bits and pieces) and the cells that provide the information that coordinates the behavior of the bits and pieces in order to organize the final structure. Thus, regeneration requires both pattern-forming cells (with positional information [PI]) and pattern-following cells (the bits and pieces). The proper organ arrangement of tissues is critical for regenerating the proper function; otherwise, you end up with differentiated tissues that lack functional organization, as evidenced by the formation of teratomas. A lot is known about the biology of the pattern-following cells. These are the lineage-restricted cells that arise by the recruitment of differentiated cells (e.g., Schwann cells and vascular endothelial cells) or via activation of adult stem cells (e.g., satellite cells for muscle regeneration). Each is specialized to have a specific function that is associated with its differentiated phenotype in the regenerated structure.

In contrast, very little is known about the pattern-forming cells. Nevertheless, they also have a specialized function, which is to control pattern formation. They achieve this via short-range cell–cell interactions with other pattern-forming cells. Unlike the lineage-restricted pattern-following cells, they are multipotent cells that establish boundaries between tissue types and facilitate integration of the regenerated structure with the host. In this regard, it is not surprising that these cells are localized within the loose connective tissues and can give rise to other connective tissues in the regenerate (e.g., bone, cartilage, ligaments, tendons, and fascia). The pattern-forming cells obey the rules formalized in the Polar Coordinate Model (PCM) and are distributed throughout the connective tissues of the organ. Given this distribution, these cells are visualized as part of a PI grid, within which they extend cell processes that make contact with and communicate with other PI grid cells (e.g., via cytonemes, which are filopodial-like cell processes).

Even less is known about the nature of PI itself in the regenerating limb. Given that PI is essential for regeneration, this is a problem that needs more attention. It has been known for a long time from grafting experiments that there is PI and that it can be specifically and predictably altered by the application of precise concentrations of the developmental signaling molecule retinoic acid (RA). Recent evidence suggests that PI is encoded in the ECM and that spatial–temporal modifications of sulfated glycoproteins can regulate signaling by growth factors/morphogens that modify pattern formation.

27.2.2 Cells have a limited repertoire of behaviors and gene regulation functions to orchestrate those behaviors

(Chapters: 4, 6, 8, 9, 13, and 14)

Gene expression is critical to the success of regeneration, but the information for making regeneration to happen is not intrinsic to the genes but rather to how genes are expressed in time and space. Orchestrated gene expression allows each cell access to the information required to make (differentiate into) one of around 200 different cell types that make up the tissues, organs, and systems of our bodies, to enable them to perform their different functions, to replicate themselves when required, and to sacrifice themselves if it is appropriate to do so. It also appears that there are no special *regeneration genes*, but rather, regeneration is special because of the way highly conserved signals and pathways are regulated. Thus, the answer to controlling gene expression in order to induce regeneration lies in the regulatory elements of the genome (e.g., miRNAs and enhancers) rather than in the protein-coding regions.

Related to the idea that there is an orchestrated program of proliferation, migration, differentiation, and/or apoptosis that acts at the level of single cells, tissues and organs should be considered to be an emergent property of the cells, possessing a higher level of organization. Much like during embryonic development, there is a stepwise program, such that early events lead to outcomes that influence what occurs at later (yet to happen) events. The key to successful regenerative engineering is to direct cells along this pathway, to let the cells take over as early as possible, and to let the program play out. We can facilitate this process by providing necessary signals earlier rather than later in the process.

27.2.3 The key to regeneration is blastema formation, and the key to blastema formation is dedifferentiation

(Chapters: 5, 6, 8, 11, 15, 16, 17, 19, and 21)

At its core, the process of regeneration can be segregated into two important and distinct transformations: the transformation from an injury site composed of differentiated tissues into the blastema (a transient developmental structure) and the transformation of the blastema into the differentiated replacement structures. It is clear that understanding how the blastema is built and the microenvironment created by the blastema provides the best avenue for uncovering critical processes dictating whether or not a regenerative response is successful. In other words, all roads leading to regeneration must pass through the blastema, and because blastema characteristics are shared among vertebrates, they are likely to be highly relevant for human regeneration.

Blastemas are not homogenous in terms of their cell types of origin or their state of differentiation from one moment to the next. Early on, cells are predominantly recruited from the loose connective tissues, and among these are the pattern-forming cells. These early blastema cells also dedifferentiate, become developmentally plastic, and adopt the fate of cells at the host site when grafted. Subsequently, cells from other tissues are recruited, and these are the pattern-following cells. These later-arriving cells are lineage-restricted in terms of their developmental potential, for example, muscle from lineage-committed satellite cells and Schwann cells from Schwann cells. At later blastema stages, the cells in the apical region remain undifferentiated and developmentally plastic, comparable to the cells of the early-bud blastema. In contrast, the cells in the basal region have begun to differentiate in response to

PI cues, they become developmentally stabilized, and they form super-numerary structures corresponding to their site of origin when grafted to an ectopic host site. As a consequence of this spatial and temporal heterogeneity of blastemas, experiments that are based on the idea that blastemas are homogeneous (e.g., homogenize and analyze protein or gene expression of entire blastemas) have lost the critical positional and temporal information that makes a blastema function in the context of regeneration.

It is evident that you do not need pluripotent cells to make a blastema and regenerate, but you do need dedifferentiation. For many of the tissues of the limb that need to be regenerated, there are lineage-restricted progenitor cells. These cells exhibit limited plasticity rather than having a pluripotent nature and are sufficient for successful regeneration of most of the structures of the limb, and thus, regeneration competence is not the same as pluripotency. This is an important concept within the field of regenerative medicine, where the main focus has been on the establishment of pluripotent stem cell lines for therapeutic use. As natural systems do not make use of such cells, therapeutic strategies could, therefore, be simplified by taking advantage of cells with limited differentiation plasticity to facilitate successful regeneration. Although pluripotency is involved in the connective tissue lineage, the pluripotent state may not be particularly well suited overall for a regenerative response. Instead, a more effective approach might be to identify and expand a cell type or cell types that are better suited for a specific regenerative response.

Finally, there appears to be an inverse relationship between plasticity and differentiation, and by definition, you need plasticity of PI in order to get regeneration. The challenge is to start with something old (the tissues that remain after injury/amputation) and reprogram the progeny of those cells to make the information for something new (the missing parts). This cell plasticity has been well characterized during imaginal disc regeneration in *Drosophila*, which has established dedifferentiation and transdetermination as a paradigm of cell plasticity during regeneration. Conversely, differentiation is usually accompanied by a loss in cellular plasticity; therefore, as cells acquire a specific phenotype, they lose the possibility of acquiring other alternative phenotypes.

27.2.4 Regeneration and integration are not the same processes

(Chapters: 5, 6, 13, 18, 19, and 20)

The focus of most regeneration research is on how to make a new structure, and this is obviously required; however, unlike during embryogenesis, this new structure is developmentally out of sync with the old structures. In the embryo, all the tissues are progressing through the same stages of development, and the bits and pieces form and become associated with each other structurally and functionally as they form.

With regeneration, the old structures persist, and except for the cells that are within a short distance of the injury site, the old cells behave as if they do not know that the limb has been amputated. Meanwhile, the blastema forms and the missing parts are regenerated, but they still need to become integrated with the old parts. This problem is equivalent to the challenge of engineering prosthetic devices; our ability to engineer advance structures has progressed, but the ability to functionally integrate them with the injured host tissues has lagged.

Although tissue integration is a key step in the regeneration of a functional biological structure, and typically, it occurs in coordination with regeneration, it does not always happen, and regeneration is not dependent on integration. Ectopic limbs in the Accessory Limb Model (ALM) are formed with normal structures distally, but they do not integrate proximally into the host stump. Integration can be induced as a discrete step when the adjacent host tissues are injured, which presumably induces localized dedifferentiation of host cells. It is reasonable to assume that tissues involved in the regenerative response undergo some form of histolysis and dedifferentiation to establish an interface for integration between the mature tissues of the stump and the newly developing tissues of the regenerate. It will be important to better understand how histolysis is involved in establishing a functional interface between the mature stump tissues and the regenerated structure and potentially aiding in the release and/or activation of stem/progenitor cells from mature tissue, so they can participate in the response.

This process of integration appears to be a function of the state of developmental plasticity of PI. This is evident from experiments involved in blastema grafting to ectopic host sites. Early-bud blastemas and apical late-bud blastemas share the property of being composed of undifferentiated cells with plastic PI, and when grafted, they acquire the PI of the host site tissues. In doing so, they become integrated and differentiate in accordance with PI cues provided by the host cells. In contrast, once PI has become stabilized, as occurs in the proximal region of late-bud blastemas, the grafted blastema cells continue to develop and differentiate according to the PI from the donor location, resulting in supernumerary limb structures that are not integrated, comparable to accessory limbs in the ALM. Thus, regulating the plasticity of pattern-forming cells will be critical for solving the problem of how to integrate new structure (engineered to regenerate) into the host.

27.2.5 Nerves are important

(Chapters: 2, 4, 5, 19, and 22)

Regeneration is a stepwise process, and therefore, being successful at engineering a regenerative response will necessitate starting at the beginning and then progressing through the steps. Failure to progress will result in regenerative failure. It is abundantly evident that the early signals associated with nerves are critical to both the initiation of the regeneration cascade and the progression to subsequent steps. The good news is that the ALM has led to a number of insights about the signals and their targets, specifically the wound epithelium (WE)/apical epithelial cap (AEC), and we can replace the nerve with purified, recombinant human growth factor cocktails. Thus, at least with regard to the early signals that initiate blastema formation, the signals are not salamander-specific; humans have the right signals, and therefore, the challenge is to learn how to control those signals.

Nerve signals appear to have multiple critical functions, including controlling cell migration, proliferation, dedifferentiation, and positional plasticity of blastema cells. Little is known about the mechanisms involved in the latter function, but it is clear that the nerves maintain blastema cells in a labile state, so that they can receive PI from the proximal limb cells and thus make the new pattern to replace the missing parts. It maybe that these multiple functions are all different mechanistically, which would presumably involve different nerve-associated signals. Alternatively, if there is a single conserved primary function (e.g., chemoattraction), and the other cellular behaviors are secondary, indirect consequences, then the challenges for regenerative engineering are much simplified. Regardless, given the critical role of the nerve, it will be necessary to identify the pathways and the signals that activate those pathways.

27.2.6 The extracellular matrix is important
(Chapters: 3, 10, 12, 14, 17, 22, 24, and 25)

The fact that the extracellular matrix (ECM) is so important for regeneration is a very good news for regenerative engineering, since it is a physical structure that can be engineered. Historically, the ECM was assumed to play a role in all aspects of biology, including regeneration, but given the technical limitations associated with working with these large and complex molecules, little was known mechanistically. With advances in glycobiology and glycochemistry, it is now possible to tease the ECM apart, and the more that researchers do that, the more we come to appreciate how important the function of the ECM is in the regulation of regeneration. The ECM provides the niche where progenitors and stem cells can attach, grow, proliferate, and differentiate to regenerate the injured tissue. Progress has already been made in engineering scaffolds that can be engineered with properties mimicking those of the endogenous ECM that can be used to promote regeneration; for example, hydrogels based on natural materials can be used to provide scaffolding support to the injured peripheral nervous system (PNS) and central nervous system (CNS), deliver drugs, and graft cells.

The most important function of the ECM (aside from providing structure and stability) is in moderating the conversation between cells. Ultimately, the outcome of wound healing is determined by the balance between pro-regenerative extracellular matrix formation and the formation of a fibrotic scar. Fibroblasts are the critical cell types that shape the ECM environment during wounding and homeostasis. They are highly responsive to injury-associated signaling (e.g., inflammation and nerves) and continue to respond to later signaling associated with growth and pattern formation within the blastema. All of these signals are regulated by the ECM via specific chemical modifications (e.g., sulfation patterns) that can account for spatial and temporal differences in response to signals. By this view, the ECM is upstream of signaling and thus upstream of the multitude of proregenerative responses to those signals. It therefore may not be necessary to deal with the challenge of providing these signals, since they are already there in the extracellular milieu. The challenge will be to engineer the ECM to control the activity of the signals, which is the essential function of the PI grid, which in turn is made by cells (fibroblasts) in the connective tissue. Therefore, induced regeneration will likely necessitate engineering the PI grid, so that it can self-heal and guide pattern-following cells to replace the lost structures.

27.2.7 There are not a lot of signals involved; rather, the same, conserved signals are regulated in different ways

(Chapters: 2, 4, 5, 6, 17, 20, and 21)

As with embryonic development, there does not appear to be a multitude of signals and pathways involved in regenerative development. All of the most important regulators of development are used repeatedly, and the outcome is context-dependent. This aspect of context permeates the molecular biology of pattern regulation, where the same molecules (e.g., Wnts, bone morphogenetic protein [BMPs], and fibroblast growth factors [FGFs]) are used in many different structures for many different outcomes, revealing the need for informational context in how cells interpret chemical or physical stimuli. For example, *Hox* genes are involved in pattern formation along the primary body axis early in development, in patterning of the secondary axes of the appendages (limbs and genitalia) later in development, and then yet again in the regenerative development of limbs. Reaccess and reutilization of conserved developmental pathways appear to lie at the heart of regeneration. The ability to dedifferentiate into an earlier developmental stage and reaccess the same pathways that were used in the embryo to make the structure in the first place appears to distinguish animals that can regenerate from those that cannot. Another way to say this is that we as humans have the genes that are needed to regenerate an arm because we have the genes to make the arm in the first place as embryos (there are no magic regeneration genes). The challenge is to discover how to reaccess those genes and pathways, either endogenously or therapeutically.

To approach this challenge, it again is important to recognize that regeneration is a stepwise process. Certainly, the first step (initiation) is important, but all subsequent steps are equally important if you are going to achieve the desired outcome. Thus, it will be necessary to identify each of the steps, along with the signals and pathways that control each step. Herein lies a critical concept for thinking about regeneration: it is a stepwise series of interconnected and interdependent processes that must be firing on all cylinders to be successful. At this point, we have no idea how many of these steps there might be. Although it is obvious that at least one of these steps fails in humans, it is unlikely that we fail at all of them. The same pathways are used repeatedly in the regulation of multiple biological phenomena and cellular behaviors. These signals and their associated pathway components are all highly conserved, and thus, what is different is how they are regulated.

27.2.8 The goal is not to make humans regenerate like salamanders but to use those model organisms that can regenerate to identify the pathways that control regeneration

(Chapters: 5, 6, 8, 13, 14, 17, 22, 23, and 26)

Related to the above discussion (27.2.7) about conservation of the signals, the pathways involved also are conserved; however, there will be specific differences in how the pathways are regulated that will account for regenerative abilities, or lack thereof are regulated. The strategy then should be to study both regenerative and non-regenerative responses in the same animal (the same genetic background). For example, mammals can regenerate amputated digit tips but not amputations at the next most proximal level. Understanding why this occurs will lead to identifying the pathways that need to be targeted in order to turn a non-regenerative into a regenerative response. This is the basis for the approach in the axolotl, which can normally regenerate perfectly; however, in response to certain wounds, it does not regenerate (e.g., in the ALM, a skin wound does not regenerate a limb, but if provided with nerve signaling and a skin graft from the other side of the arm, it forms a blastema and makes an ectopic limb). By this approach, it is possible to identify the pathways that need to be activated to make a blastema and subsequently a limb. Identification of these pathways then provides the opportunity to intervene therapeutically in a non-regenerating wound (e.g., in a human). Since the pathways are conserved, it is likely that the outcome will be similar in terms of inducing a regenerative response from a non-regenerating wound. This argues in favor of placing more emphasis on the highly regenerative models, so that we can take advantage of *nature's free lessons* to define the basic mechanisms underlying successful regeneration and to identify relevant conserved molecular pathways.

The strategy of targeting conserved pathways raises the point that we are more likely to be successful by engineering how to moderate the existing processes rather than re-engineering the processes themselves. Rather than modifying existing gene regulatory networks in humans, the goal is to alter the outcome in response to injury. To be successful with the approach, it is important to recognize that although perfect regeneration (e.g., in an axolotl) would be great, it is not necessary. We should focus more on function than on anatomy, a realization that is very often neglected in studies on intrinsically regeneration-competent models, where the focus is on anatomical regeneration. For example, regeneration of mammalian digits tips is not perfect. Imperfections of the regenerated bone likely reflect the fact that the regenerative process is, in this case, not an example of redevelopment but is a response that successfully replaces physiological function at the expense of anatomical perfection. If there were multiple paths to achieve multiple functional regenerative outcomes, then it would be important to consider a strategy that allows cells to choose a pathway that might be more successful rather than forcing them along a path that should, but might not, work. This view is based on the hypothesis that cells and tissues are predisposed to regenerate and regenerative failure is a consequence of regeneration barriers. In regenerative engineering, we have to help cells overcome these barriers in order to achieve a desired outcome, and we have to let cells take over as early as possible. We can facilitate the process by providing necessary signals early rather than late in the process.

The approach of using animals that regenerate as an assay for proregenerative signaling to identify the target pathways is important, because it is not possible to do this in animal models that do not regenerate. It would be transformational to treat an amputated mouse limb with a single factor to make it regenerate but that is likely to happen only if regeneration is a simple process controlled by one pathway, which is almost certainly not the case. Given that multiple signals are involved, the identification of these multiple factors may require high throughput analyses to determine the best combinations of treatments that could promote regeneration. Here, non-mammalian models would provide the opportunity to delineate the basic, conserved mechanisms that support repair and regeneration, which when complemented with mammalian studies could contribute to solving some of the field's biggest problems. If there are some species-specific factors that are identified in these assays, advances in bioinformatics could identify targeted pathways that could be therapeutically manipulated to enhance mammalian regeneration.

27.2.9 There is a ying-and-yang duality that is intrinsic to regeneration

(Chapters: 3, 4, 5, 6, 11, 13, 17, 23, and 25)

The concept of ying and yang recognizes that seemingly opposite or contrary forces may actually be complementary and codependent, which appears to be the case for regeneration, cancer, and fibrosis. There is a tendency to see two sides as good or bad in terms of the outcome (e.g., regeneration is good and cancer is bad), but this view ignores the reality that these outcomes share common pathways. Regeneration is a dynamic process that has to happen during a period of time while the animal also needs to survive. In nature, and over the course of evolution, most animals cannot just stop and take time out to regenerate. Thus, function has to be sustained, whether it is locomotion to escape a predator or keeping your heart pumping blood. In modern times it will be possible to stop during the period of time required to regenerate (e.g. be a patient in a hospital and rehabilitation), but mechanisms to respond to injury have evolved to allow for immediate survival rather than long term regeneration. The response to injury involves hemostasis, inflammation, and re-epithelialization, and these processes are by necessity steps in regeneration. Connective tissue cells will resynthesize the ECM, which in some instances will lead to regeneration but in others, such as mammals, will lead to fibrosis and scarring. Regeneration-competent cells dedifferentiate, migrate, and proliferate, which will result in cancer if not done in a controlled manner. So each of the multiple, required injury responses will happen, and each of them can be regulated in multiple ways, leading to multiple outcomes, some of which are more desirable (regeneration) than others (cancer and fibrosis).

The ying-and-yang nature of regeneration and cancer has been noted for decades, since both blastemas and tumors are masses of undifferentiated, proliferative cells. It has long been assumed that cancer and regeneration share many commonalities; however, regenerating organs or tissues must have developed mechanisms to avoid overgrowth based on mechanisms such as PI to maintain the appropriate structure and function. Thus, a better understanding of the mechanisms controlling one will be highly relevant to the other. This is particularly the case for the role of epigenetics in regeneration. By definition, epigenetic modifications are required for regeneration, since cells with old information need to be reprogrammed to give rise to progeny with new information. Indeed, the fields of stem cell and cancer biology have much in common with regeneration biology, and epigenetic mechanisms uncovered in any of these three fields can provide important inroads to understanding the others.

Similarly, there appears to be a dichotomous relationship between the ECM and the behavior of fibroblasts in terms of either regeneration or fibrosis. In the regenerating animals, there is a transient deposition of ECM components that are later remodeled to allow scar-free regeneration. In contrast, ECM deposition continues in mammalian wounds, and though remodeled to some extent over time, the normal architecture and function of the ECM are not restored. It appears that the same cells (fibroblasts) control pattern formation during regeneration and fibrosis/scar formation during non-regenerative wound healing. If these responses are indicative of the possible range of fibroblast behaviors, then the goal is to learn how to talk to fibroblasts so as to influence their behaviors in response to injury.

There are two reasonable hypotheses that could account for the duality of fibroblast behaviors, as exemplified by the axolotl that can regenerate perfectly and humans who make scars rather than regenerate. One possibility is that these are two independently evolved states, and humans simply lack the ability to regenerate. However, regenerative failure is a negative result and therefore cannot be interpreted as evidence that the ability to regenerate is not present. It is just as likely that evolutionary selection pressure suppressed an intrinsic regenerative response to injury in favor of a fibrotic response that presumably led to increased survival. From an engineering perspective, uncovering the molecular underpinnings of regenerative potential associated with regenerative failure will prove to be as important as understanding the regeneration process itself, because it represents the chasm that needs to be bridged to achieve a functional outcome.

27.2.10 What we need to do next

Each of the authors either implicitly or explicitly identified the challenges we face moving forward and suggested what we need to do next to achieve the goal of engineering regeneration. I have highlighted many of those in this final section (in no particular order).

We need molecular markers for everything in model organisms that can regenerate. This is especially important for cells in the loose connective tissue, which is where the PI for regeneration is encoded. Presumably, there is a heterogeneous population of cells, and among those, there is a subpopulation of PI cells. In addition, as the blastema grows, it becomes increasingly heterogeneous in terms of both contribution and state of plasticity. The lack of validated markers prevents us from dealing with these heterogeneity issues and can lead to more confusion, rather than clarity. For example, most transcriptomes for regenerating tissues have been generated from complex tissues with heterogeneous cell types, whose molecular profiles can be extremely diverse. There almost certainly are cell-specific injury and regeneration responses that differ substantially between these cell subtypes. Our lack of knowledge on these issues is a major barrier to progress.

The role of the immune system has long been of interest, and intriguing correlations and observations have suggested its role in regeneration. Renewed interest and focus, along with new technologies and availability of molecular markers, are leading to advances in understanding the relationship between inflammation and regeneration. Moving forward, regenerative therapies will likely elicit acute or chronic inflammatory responses that will shape the success of the intervention and will require precise knowledge of how the host tissue and immune system will tolerate these interventions.

As new techniques become available and optimized for use in non-traditional model organisms, we can expect rapid advances in regenerative engineering. Combining established tools (e.g., microRNA [miRNA] and RA) with new genome editing technology (CRISPR-Cas9) will allow for functional studies in non-genetically tractable model organisms, such as the axolotl. Such studies will lead to the identification of regeneration-associated enhancers that will provide insights into how conserved pathways are differentially regulated between regenerative and non-regenerative species. Techniques for single-cell analyses such as single-cell transcriptome analysis will be possible for characterizing subclasses of identified regenerating or non-regenerating cells.

Advances in biomaterials will allow for quantitatively controllable delivery of reagents, including growth factors/morphogens, regulatory molecules, and miRNA inhibitors or mimics to the cells or tissues *in vivo*. It also will be possible to engineer smart materials that could regulate endogenous growth factor signaling, for example, sulfated glycoproteins. Since these molecules appear to encode PI, the strategy would be to engineer a PI grid to control regeneration. A first step to developing regenerative therapies with PI needs to be to map out the ECM composition across the entire body. In the way that the PCM assigned arbitrary numbers and letters to denote positional values, ECM composition–based coordinates can be assigned to each position (e.g., high 3-O sulfation, low N-sulfation = posterior forelimb mid-stylopod). The heparan sulfate composition can be cross-referenced with already described growth factor–binding sites (e.g., FGF2, 2-O, and N) and known areas of signaling (sonic hedgehog [SHH] in the Zone of Polarizing Activity [ZPA]).

It will be necessary to discover quantitative mechanisms of signaling to regulate proregenerative pathways, and this is not going to be possible with experiments with whole animals. As noted above, the heterogeneity and complexity of cell types and behaviors necessitate that we have appropriate markers. Thus *in vitro* models that maintain blastema cells in a regenerative state comparable to *in vivo*, and that allow specific cell-types to be studied are needed. There are enough data to indicate that once dissociated, blastema cells lose their characteristic *in vivo* behaviors, and thus, dissociated cell cultures are not the solution. Little has been done with 3D cultures; however, use of organotypic slice culture is promising. In the end, a detailed understanding of cellular behaviors within the regenerating limb blastema (particularly in the apical region) is necessary to target those behaviors via regenerative engineering strategies.

Nearly all research is focused on how to induce or jump start regeneration; whereas, there has no thought as to how to turn it off when finished. One likely pathway to target for this appears to be apical FGF signaling. Endogenously, the turning off of regeneration maybe a consequence of completion of restoring the lost pattern. The PCM predicts that in the absence of positional disparities at the end of regeneration, cell–cell interactions no long stimulate proliferation (intercalation ceases). This must result in AEC signaling being turned off. Regardless, both the turning on and the turning off of regeneration are related to PI and PI-mediated signaling, and much more needs to be discovered about this mechanism.

Finally, understanding PI and PI-mediated signaling will require a more detailed understanding of fibroblast biology, starting with having validated markers for the grid fibroblasts that control and induce pattern formation. This will allow for the next step, which is to determine if humans have PI cells and a functional PI grid. If so, then how can the PI grid be accessed, and if not, then how can it be made accessible? Understanding how to program and manipulate PI will follow from the use of RA or other molecules that are capable of reprogramming PI, and these will likely be important tools in the development of cell-based therapies. Given the key role of RA in PI, it is important to understand the mechanisms of RA release and transport between cells, which at present are completely unknown. In addition, more detailed analyses of the regulation of cell cycle kinetics at higher spatial and temporal resolution are needed given the body of evidence of a functional relationship between the control of growth and the control of pattern formation during regeneration.

Acknowledgments

Many people have put in tremendous effort to think and write about the problem of how to engineer regeneration. They have shared their views about how regeneration works in the context of their particular areas of expertise. They also have shared their best and most creative ideas. In the end, this volume has captured a view of where the field of regeneration biology is in 2017, and we will be able to look back and see how the field has moved forward and built upon these ideas. Most importantly, the goal of the book is to catalyze interactions between developmental biologists and regenerative engineers. On behalf of the many authors, I hope that we have been successful in communicating the view from the side of embryonic and regenerative developmental biology. I therefore want to thank all the authors for their efforts and their patience. Their contributions are very much appreciated.

I also want to thank Michael Slaughter, our production assistant at CRC Press for his patience and support from beginning to end, and Dr. Cato Laurence for his vision and efforts that led to the inception and completion of this project. Finally, I especially want to thank Dr. Susan Bryant for her never-ending support, encouragement, and patience as I worked through the process of bringing this book to completion.

Index

611

Printed and bound by CPI Group (UK) Ltd, Croydon, CR0 4YY

01/11/2024

01782601-0015